JOHN T. BROSNAN
Department of Biochemistry
Memorial University of Newfoundland
St. John's, Nfld. A1B 3X9

BIOCHEMICAL JOURNAL

REVIEWS 1996

BIOCHEMICAL JOURNAL
REVIEWS 1996

edited by A.E. Pegg

PORTLAND PRESS
London

The *Biochemical Journal* is published by Portland Press
on behalf of the Biochemical Society

Portland Press Ltd.
59 Portland Place
London W1N 3AJ
U.K.

ISBN 1 85578 126 3

British Library Cataloguing-in-Publication data
A catalogue record for this book
is available from the British Library

Cover illustration: Schematic ribbon drawing of bilin-binding protein. For
further details see the review by Flower in this Compendium. Illustration
reproduced by kind permission of Dr. D. R. Flower.

Printed in Great Britain by the University Press, Cambridge

CONTENTS

Biochem. J. (1996) **314**, 713–721 (Printed in Great Britain)

REVIEW ARTICLE
Myc oncogenes: the enigmatic family

Kevin M. RYAN* and George D. BIRNIE
The Beatson Institute for Cancer Research, Garscube Estate, Switchback Road, Bearsden, Glasgow G61 1BD, Scotland, U.K.

The *myc* family of proto-oncogenes is believed to be involved in the establishment of many types of human malignancy. The members of this family have been shown to function as transcription factors, and through a designated target sequence bring about continued cell-cycle progression, cellular immortalization and blockages to differentiation in many lineages. However, while much of the recent work focusing on the c-*myc* oncogene has provided some very important advances, it has also brought to light a large amount of conflicting data as to the mechanism of action of the gene product. In this regard, it has now been shown that c-*myc* is effective in transcriptional repression as well as transcriptional activation and, perhaps more paradoxically, that it has a role in programmed cell death (apoptosis) as well as in processes of cell-cycle progression. In addition, particular interest has surrounded the distinct roles of the two alternative translation products of the c-*myc* gene, c-Myc 1 and c-Myc 2. The intriguing observation that the ratio of c-Myc 1 to c-Myc 2 increases markedly upon cellular quiescence led to the discovery that the enforced expression of the two proteins individually showed that c-Myc 2 stimulates cell growth, whereas c-Myc 1 appears to be growth suppressing. Clearly, the disparities in the activities of c-Myc, together with the consistent occurrence of mutations of c-*myc* in human malignancies, means that, although reaching an understanding of the functions of the *myc* gene family might not be simple, it remains well worthy of pursuit.

INTRODUCTION

Since the discovery of c-*myc*, the *myc* gene family and the proteins they encode have probably formed part of the research programme of nearly all the cancer research institutes in the world. A recent survey of the Medline database, for the period 1990–1994, revealed 3343 references which include Myc in their title or abstract and 55 review articles specifically on these genes and proteins. However, in spite of this plethora of information, the role and mechanism of action of Myc proteins still remain enigmatic. In fact, although in recent years there have been considerable advances towards an understanding of the role played by these proteins, the data involved are becoming not only extremely complex, but in many cases highly contradictory. It is therefore the aim of this review to outline aspects of the current literature that are turning the Myc family of proteins into the subject of an intense and interesting debate.

HISTORY

The majority of work on the *myc* gene family has focused on three members which when activated have been shown to be important in the generation of various human malignancies [1]. The most widely studied of these, c-*myc*, was first discovered through its homology with the transforming gene (v-*myc*) of the avian myelocytomatosis virus MC29 [2]. The other two, N-*myc* and L-*myc*, were discovered later through their homology with v-*myc* in the amplified sequences of neuroblastoma cells [3] and a small cell lung tumour [4] respectively. In addition to these three proto-oncogenic family members, two other *myc* genes, S-*myc* and B-*myc*, have also been identified. Despite only having been partially characterized, these two genes appear highly interesting since they differ from c-*myc*, L-*myc* and N-*myc* in that the proteins they encode appear to suppress malignant transformation (Table 1).

In a variety of species the *myc* genes have been shown to be activated by gene amplification [5], chromosomal translocation [6], proviral insertion [7] and retroviral transduction [8] (Table 2). However, it is most frequently reported that the level of the c-Myc protein is elevated in many tumour types by a mechanism that is none of the above and is as yet not understood. Early research into the proteins indicated their expression to be associated with cell-cycle progression [9] and incompatible with terminal differentiation in a variety of cell lineages [10,11]. In addition, c-*myc* was shown to cause cellular immortalization [12] and was able to co-operate with an activated *ras* gene in the transformation of rat embryo fibroblast cells [13]. In 1985, Eisenman et al. [14] established that, in cells containing active c-Myc or v-Myc protein, the majority of the protein was associated with a nuclear fraction, termed the 'matrix lamin'. This observation was later substantiated by the identification of a domain within the c-Myc protein that was found to be effective as a nuclear localization signal [15]. As a result of these findings, two schools of thought were generated as to how the Myc proteins bring about phenotypic change. It was proposed that they either have a direct role in the DNA replication machinery [16] or are involved in the transcriptional control of genes involved in cellular replication [17].

Evidence for a role in the control of gene expression grew when the sequence of the c-Myc protein was shown to contain a series of motifs which were similar to ones previously described for known transcription factors. Leucine zipper motifs, like those found in the oncoproteins v-Fos and v-Jun, were the first to be identified and were shown to be located in the extreme C-terminus of the protein [18] (Figure 1). Subsequently, immediately upstream of the leucine zipper motif, similarity was then found to a second domain termed the 'helix-loop-helix' motif [19]. This domain had already been identified in a number of transcription factors, including the immunoglobulin enhancer binding proteins E12 and E47 [20]. However, although both of these motifs had

---

Abbreviations used: ODC, ornithine decarboxylase; Rb-1, retinoblastoma susceptibility protein; YY-1, Yin-Yang 1.
* To whom correspondence should be addressed.

Table 1   Summary of the mammalian *myc* genes described to date

myc gene	Ascribed functions	Amino acid sequence identity with c-*myc* (%)	Refs.
c-*myc*	Cellular transformation Apoptosis Continued cell-cycle progression	–	[1,12,19,89]
L-*myc*	Cellular transformation Continued cell-cycle progression	35	[1,12,19]
N-*myc*	As for L-*myc*	38	[1,12,19]
S-*myc*	Growth suppression Apoptosis	31	[109,110]
B-*myc*	Inhibition of neoplastic transformation	61	[111]
P-*myc*, L-*myc* Ψ	Pseudogenes	Derived from regions of L-*myc*	[112]

Table 2   Incidence, mechanism and effects of *myc* activation

Abstracted from Marcu et al. [74].

Mechanism of c-*myc* activation	Effects	Incidence
Amplification	Increased Myc protein abundance	Gastric adenocarcinoma; small cell lung carcinoma; glioblastoma; carcinoma of breast; carcinoma of colon; plasma cell leukaemia; promyelocytic leukaemia; granulocyte leukaemia
Proviral insertion	Deregulated expression of *myc* by viral long terminal repeat	Leukaemia resulting from infection by avian leucosis virus or Molony murine leukaemia virus
Retroviral transduction	Deregulated expression of viral Gag–Myc fusion protein	Feline leukaemia viruses, e.g. GT3 and FTT; avian leukaemia viruses, e.g. MC29 and MH2
Chromosomal translocations	Deregulated expression of full-length or truncated *myc*	Burkitt's lymphoma; mouse plasmacytoma

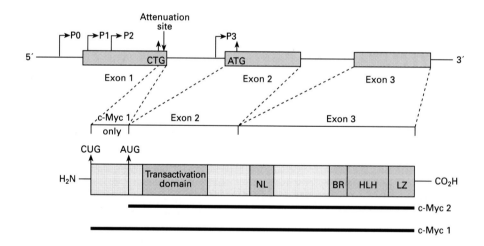

Figure 1   Schematic representation of the structure of the human c-*myc* gene and the two resultant protein products, c-Myc 1 and c-Myc 2

The sites of the gene's promoters are indicated as P0, P1, P2 and P3. The translation initiation sites of the two proteins are shown as the codons of the initial amino acid, i.e. CTG (CUG) and ATG (AUG). NL, nuclear localization; BR, basic region; HLH, helix-loop-helix; LZ, leucine zipper.

previously been shown to be involved in the formation of transcription factor complexes, attempts to detect complexes in which Myc was either homo- or hetero-oligomerized proved fruitless. In spite of this, further studies revealed that the Myc proteins also contained a tract of basic amino acids upstream of the helix-loop-helix motif (Figure 1). This 'basic region' motif had been previously identified in the myogenic transcription factor MyoD and was found to be the region involved in determining sequence-specific DNA binding [21]. As a final piece of indirect evidence for Myc family members being involved in the control of gene expression, it was shown that a region at the N-terminus of c-Myc (Figure 1) had the ability to act as a transcriptional transactivator. Fusion genes were constructed in which the c-*myc* N-terminal region was fused to the DNA binding domain of the gene for the yeast transcription factor GAL 4. This construct was then transiently co-transfected into cells, with a reporter construct containing the GAL 4 DNA binding site upstream of the chloramphenicol acetyltransferase gene. The results of these experiments indicated that the trans-activation potential of the fusion protein was both potent and highly specific for the N-terminal region of c-Myc [22].

A ROLE AS A TRANSCRIPTION FACTOR

Myc as an activator

As outlined above, many lines of indirect evidence accumulated to indicate that the Myc proteins functioned as transcription factors. However, problems arose in confirming this theory when it was found that Myc proteins could only form complexes with DNA at very high protein concentrations, implying that such interactions may not be physiologically significant [23]. It was therefore considered that Myc might require interaction with a second protein in order to achieve its role. With this in mind, Blackwood and Eisenman [24] screened a baboon expression library with a c-Myc protein labelled with ^{125}I. Subsequent sequencing of hybridizing colonies identified a small, novel, protein which they named Max (Figure 2). Analysis of this protein revealed that it was similar to Myc in that it also contained basic helix-loop-helix and leucine zipper motifs, and it was therefore considered to be a possible dimerization partner. Indeed, when assayed *in vitro* it was found that Max was able to form dimeric complexes with each of the Myc family members and at much lower protein concentrations than had previously been required to achieve Myc homodimerization. In addition, when a complex of c-Myc and Max was used to select preferred DNA sequences from a pool of partially randomized oligo-nucleotides, it was found that the dimer had specific DNA binding activity for the sequence CACGTG [25]. Further studies of this sequence using electrophoretic mobility shift assays showed that, although homodimers of Max were able to bind without the presence of Myc, appreciable binding of Myc was dependent on it forming a heterodimeric complex with Max [26].

Since the CACGTG motif had previously been identified as the binding site for other transcription factors containing basic region and helix-loop-helix motifs, it seemed encouraging that this might be the transcription target for Myc. This was analysed by a yeast reporter gene assay in which Myc and Max were co-expressed with a β-galactosidase gene under the control of a basal promoter linked to the CACGTG sequence. When Myc and Max were co-expressed in this system there was a considerable increase in the activity of the β-galactosidase gene [27].

The level of activity was dependent not only on the level of Myc but also on the presence of the domains which had previously implicated Myc as a transcription factor. Interestingly, although it was not a surprise that expression of Myc alone did not activate the reporter construct, it was intriguing to find that even though Max:Max homodimers bound to the CACGTG sequence they too did not activate the β-galactosidase gene [27] (Figure 3). Titration experiments revealed that, if the Max protein was expressed at high levels in conjunction with Myc, the β-galactosidase activity observed was less than when Myc and Max were expressed in roughly equivalent amounts. Although it was shown that Max preferentially formed a heterodimer with Myc as opposed to a homodimer [24,26] (Figure 3), this created a situation whereby the activity of genes regulated by Myc would be dependent not only on the levels of Myc but also on the levels of Max. The most likely explanation for the apparent lack of transcriptional activity by Max:Max homodimers has been given by Kato et al. [28]. In experiments designed to identify a transactivation domain in the Max protein, they constructed fusion proteins linking regions of Max with the DNA binding domain of GAL 4. When assayed for their ability to transactivate a reporter gene linked to the GAL 4 DNA binding site it was found that, unlike the Myc transactivation domain, no region of Max was effective in activating the gene's expression.

Although the CACGTG motif is present in the promoter regions of various genes, the search for Myc-activated genes has been somewhat without reward. In an attempt to address this problem Eilers et al. [9] generated a fusion protein between c-Myc and a portion of the oestrogen receptor. When this was introduced into mammalian cells, the activity of the exogenous Myc protein became dependent upon the presence of the steroid hormone β-oestradiol. cDNA libraries were then generated corresponding to mRNA species from control cells and from cells treated with β-oestradiol for 24 h. Comparison of these two libraries identified an mRNA that was induced by Myc as encoding α-prothymosin. Subsequent studies revealed that the gene could be activated by Myc in the absence of protein synthesis and that its activation was dependent upon the integrity of a CACGTG sequence found within the first intron of the gene [29]. As might be expected for a gene regulated by c-Myc, expression of α-prothymosin has been associated with pro-liferating cells [30,31] and it is expressed in nearly all tissues [32]. However, a role for α-prothymosin is yet to be found, and as a result its involvement in the phenotypes brought about by activated expression of Myc proteins remains a mystery.

A more appealing target for regulation by Myc is the enzyme ornithine decarboxylase (ODC). This enzyme catalyses the rate-limiting step in the production of polyamines [33]. Expression of

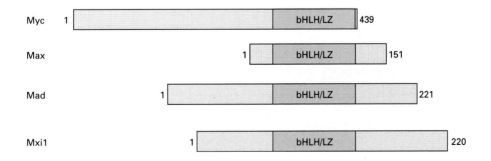

Figure 2 Alignment of Myc, Max, Mad and Mxi1

Domains that are common to each protein and are involved in heterodimerization are shown in colour. bHLH, basic helix-loop-helix; LZ, leucine zipper.

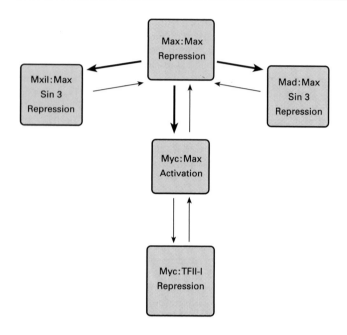

Figure 3   Transcriptional consequences of the formation of Myc-associated complexes

Arrow strength indicates preferential complex formation.

ODC is regulated in a cell-cycle-specific manner and is greatly elevated upon cellular transformation [34]. In addition, constitutive overexpression of ODC is able to bring about morphological transformation and its enhanced activity has been shown to be essential for transformation induced by the v-*src* oncogene [35]. The observation that the ODC gene contains two CACGTG sequences led to speculation that it too may be regulated by Myc. Studies using the Myc–oestrogen-receptor chimaeric protein revealed that ODC was indeed activated by Myc [36] and that its activation was dependent on the two Myc:Max binding sites within the first intron [37]. Further to these studies, it must be pointed out that while CACGTG is considered the primary target for the Myc:Max heterodimeric complex, Blackwell et al. [38] have also demonstrated DNA binding at a second canonical site (CATGTG) as well as to a series of non-canonical sequences. However, unlike the CACGTG sequence, the significance of these additional sites to the regulation of any genes by Myc:Max is yet to be established.

Myc as a repressor

The search for direct targets of c-Myc has perhaps been hindered by the relative difficulty of identifying genes which may be inactivated as opposed to activated by the protein. In this regard it is well established that the number of genes which have been shown to be repressed by Myc expression in *in vitro* reporter gene studies [17] or which are repressed in Myc-transformed cells is in excess of those which have been reported to be activated. Upon the discovery that Myc could interact with TFII-I [39] (Figure 4), a transcription initiation factor that activates core promoters through a sequence termed the initiator element [40], studies were undertaken to determine if Myc might participate in transcriptional control involving this protein. *In vitro* transcription assays using the adenovirus major late promoter revealed that, when introduced into a system involving activation by TFII-I, Myc protein was effective in bringing about transcriptional repression (Figure 3). Moreover, these effects were shown to be specific for TFII-I and were not apparent when Myc was introduced into systems being initiated by other factors, e.g. TFII-A. Subsequent analysis of the cyclin D1 gene, which had previously been shown to be repressed in Myc-transformed cells, indicated that the repression of this gene by Myc might also be mediated through TFII-I [41]. One might consider that the apparent repression of cyclin D1 by Myc seems paradoxical, bearing in mind the involvement that the cyclin is known to have in cell-cycle progression. In this regard, it must be pointed out that an independent study has implicated Myc in the activation of cyclin D1 transcription [42]. Further work is therefore required to establish the exact role, if any, that Myc is playing in the regulation of this gene.

Although a model for Myc action involving transcription repression is at first appealing, the probability of it being involved in the establishment of a transformed phenotype appears unlikely. Mutational studies of Myc and Max by Amati et al. [43] have indicated that cell-cycle progression and transformation by Myc proteins requires dimerization with Max. In contrast, inhibition of transcription through interaction with TFII-I occurs independently of Max and is refractory to forced expression of the Max protein [41]. Also, when mutants of the Myc protein were analysed for their ability to repress the promoter of the cyclin D1 gene, a deletion mutant involving amino acids 92–106 was ineffective in bringing about transcriptional repression [41], even though this region was previously shown to be dispensable during Myc-induced transformation of RAT1A cells [44]. However, these observations do not mean that transcriptional repression through TFII-I is artifactual. It would be naive to assume that the only transcriptional targets of c-Myc are those involved in transformation. For example, the ability of Myc

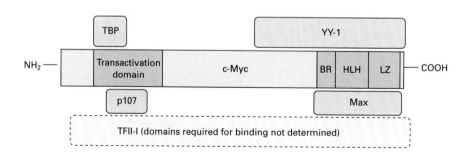

Figure 4   Representation of the sites of interaction of proteins known to bind c-Myc

BR, basic region; HLH, helix-loop-helix domain; LZ, leucine zipper domain; TBP, TATA binding protein component of the transcription initiator TFII-D.

proteins to inhibit processes of differentiation has not as yet been shown to be dependent on Max, and is therefore a possible Myc function in which TFII-I may be involved.

OTHER PROTEINS WHICH BIND TO MYC

An emerging theme arising from the study of transformed cells is that oncoproteins are frequently found to be associated with proteins having tumour suppressor or anti-proliferative function. Classically this observation comes from the analysis of the cellular proteins which co-precipitate with the transforming proteins of DNA tumour viruses. For example, the adenovirus E1A protein and the human papillomavirus type 16 and 18 E7 proteins have been shown to associate with the retinoblastoma protein (Rb-1) [45,46], while the respective E1B and E6 proteins of these viruses are believed to form complexes with p53 [47,48]. In addition, it is now becoming a frequent observation that, within the normal cellular environment, proteins having potential oncogenic activity are often found complexed with those with known tumour suppressor function, e.g. the association between the cyclin-dependent kinase 4 protein and the recently discovered tumour suppressor protein mts1/p16 [49]. It has now become apparent that the c-Myc protein forms such an interaction with the retinoblastoma family member p107 [50,51] (Figure 4). This protein had previously been shown to form complexes *in vivo* with the adenovirus E1A protein [52], the simian virus 40 large T antigen [53] and the human papillomavirus type 16 E7 protein [54]. Although it was known that this interaction formed an important part of the transforming ability of these viruses, its consequences for cellular homeostasis were not understood. Since the interaction between p107 and the N-terminus of c-Myc results in inactivation of the transactivation potential of the Myc protein, it seems feasible that binding of p107 by DNA tumour viruses might facilitate transformation by freeing the N-terminus of c-Myc and allowing the protein to interact with the TATA binding protein [55] (Figure 4). This would then allow Myc to transactivate its target genes. With respect to how this might be important to the establishment of tumours *in vivo*, the study included an analysis of mutated Myc proteins derived from Burkitt's lymphoma cell lines. Although these proteins retained an ability to form complexes with p107, they were unlike wild-type Myc in that their transactivation potential was unaffected by the interaction.

Perhaps more puzzling than the interaction with p107 is the report that c-Myc also interacts with the zinc finger protein Yin-Yang 1 (YY-1) [56] (Figure 4). Depending on the context, YY-1 has been shown to activate [57], repress [58] or initiate transcription [59]. However, when the YY-1 protein is bound to c-Myc both its ability to activate and its ability to repress transcription are reported to be inhibited [56]. Bearing in mind the work summarized above regarding the association between Myc and p107, it is interesting to note that among the targets for repression by YY-1 are the promoters of the adeno-associated virus P5 [59] and the human papillomavirus type 18 [60]. This, therefore, might be a possible reason for a negative regulator of Myc, i.e. p107, being a target for inhibition by human papillomavirus E7 [54]. Although this theory is highly speculative, it is somewhat reminiscent of the effect of the adenovirus E1A protein upon the interaction between Rb-1 and the transcription factor E2F [61]. Binding of Rb-1 by E1A results in the subsequent release of E2F such that it not only activates transcription of cellular genes involved in replication, but also causes activation of the adenovirus E2 promoter [62]. Since little is as yet known about the function and mechanism of action of YY-1, more work is required to establish the significance of its association with c-

Myc and the effect that this may have on the replication of DNA tumour viruses.

OTHER PROTEINS WHICH BIND TO MAX

It has been shown more recently that Max also forms complexes with two other proteins, Mad [63] and Mxi1 [64] (Figure 2). Whereas Mad:Max and Mxi1:Max complexes are similar to Myc:Max in that they have also been shown to bind the CACGTG motif, they are more like the Max:Max homodimer in that they are thought to act as transcription repressors (Figure 3). However, it has more recently been shown that, in contrast to the Max:Max homodimer, Mad:Max and Mxi1:Max are only effective in transcription repression as a ternary complex with homologues of the yeast transcription repressor Sin3 [65,66]. Although the role that these two new complexes may play within the cell is yet to be established, it is reasonable to assume that, since they are in competition with Myc for available Max protein, the levels or activities of Mad or Mxi1 may have a bearing on the ability of Myc to transactivate the promoters of its target genes. In fact, when these two genes were overexpressed in cells transformed by c-*myc* and *ras*, the number of cells scoring positive in an assay of transformation was greatly reduced [67]. In this regard, it is quite possible that loss or mutation of either *mad* or *mxi1* would result in release of Max protein, which would then be available to co-operate with Myc in situations where the abundance of Myc was elevated. However, since these genes have only recently been identified there has only been one report to date that would indicate that they might act as tumour suppressors in the genesis of human cancer. Eagle et al. [68] investigated the possibility that the *mxi1* gene, which maps to 10q24-q25, might be involved in a characteristic deletion of this area that occurs in a small number of cases of carcinoma of the prostate. The study reported that, in five tumours analysed, the *mxi1* locus was indeed reduced to hemizygosity. In addition, in four of the five cases the remaining allele involved was also mutated. This initial finding is highly provocative and it can now be considered an appropriate time to investigate whether *mxi1*, and *mad*, are targets for mutation in other tissues.

REGULATION OF THE ABUNDANCE OF THE MYC PROTEIN

Transcriptional control

It is reasonable to assume that the primary control of the abundance of a protein is largely determined by transcriptional initiation. Workers on Myc have not ignored this fact, and as a result a large amount of information has accumulated on regulation of the *myc* promoter. A complex story has evolved in which the gene is controlled not just by one promoter, but in fact by four. The two major promoters, P1 and P2 (Figure 1), contribute 75–90% and 10–25% of the cytoplasmic c-*myc* mRNAs respectively [69,70]. Approx. 1500 bp downstream of P1 and P2, and close to the translation start site, lies the third promoter P3. This promoter is thought to be less significant than P1 and P2, as it only contributes about 5% of c-*myc* mRNAs [71]. Of similar activity, but only present in the human c-*myc* gene, is a fourth promoter (P0) [72], which is located around 600 bp upstream of the major sites P1 and P2 (Figure 1). The significance of having the four promoters is as yet unknown, but their differential usage has been observed in many cell lineages and upon deregulated expression of the gene [73].

Analysis of this 5′ region of the c-*myc* gene for DNase I hypersensitive sites revealed multiple regions of potential protein interaction (reviewed by Marcu et al. [74]). One region of particular note is a binding site which lies −65 to −58 bp upstream of the P2 promoter. Deletion of this region has indicated that it is essential for the basal activity of P2 [75] and also for the activation of c-*myc* induced by the adenovirus E1A protein [76]. Studies of this region led to the discovery that it is bound by the cell-cycle-regulated transcription factor E2F [77]. In the light of this finding, it is encouraging that it has since been shown that activation of E2F is sufficient to direct cell-cycle progression [78], a characteristic that has also been attributed to the c-Myc protein. However, it is yet be established whether this function of the E2F protein is, at least in part, mediated through activation of c-*myc* transcription.

Whereas it is perhaps not very surprising that a region of the c-*myc* promoter is controlled by a cell-cycle-regulated transcription factor, studies of other regions have proved to be more intriguing. The analysis of a further DNase I hypersensitive site situated approx. 130 bp upstream of P2 revealed a region that is highly important for P1 activation as well as being modestly involved in the activation of P2 [79]. Partial fractionation of HeLa cell nuclear proteins by Postel et al. [80] revealed a DNA binding activity that was specific for this region and which they termed PuF. The surprising twist came when the binding site for PuF was used to screen a HeLa cell cDNA expression library. The subsequent sequence analysis of a hybridizing clone revealed that it was identical to nm23-H2 [81]. Surprisingly, the protein product of this gene had previously been implicated in suppressing the metastatic spread of certain tumours [82]. Admittedly, it is difficult to comprehend why a suppressor of metastatic disease should be involved in the activation of a gene that is frequently elevated in the late stages of human cancer. Further work is therefore required to establish the role that this protein is playing not only in the regulation of c-*myc* but also in the establishment of a metastatic state.

Other levels of control

Perhaps disheartening to the investigators involved in studying the transcriptional control of c-*myc* was the finding by Bentley and Groudine [83] that the initial control of c-*myc* mRNA levels during induced differentiation of myeloid leukaemic cells was by transcriptional attenuation. Using nuclear run-on analysis they found that the ratio of exon 1 transcription relative to that of exon 2 was increased from 3 to 15 following induction of the differentiation programme. Subsequent analysis of the sequences involved in this process identified a 95 bp segment, located at the 3' end of exon 1 (Figure 1), to be the site of the premature termination [84]. However, when this site was analysed for its ability to attenuate transcription directed by other promoters its effectiveness was highly variable [84]. In this regard, it has since been shown the sequences responsible for the premature termination are not located at the site of attenuation, but are in fact found proximal to the P2 promoter [85]. In this study it was postulated that at this site the RNA polymerase might undergo a pause, at which point it is modified in a way which determines how it will respond at the site of attenuation.

Although a large amount of data has accumulated as to the transcriptional regulation of the c-*myc* gene, it must be pointed out that the mRNA and protein are also subject to post-transcriptional control. Both the mRNA and protein have been shown to have extremely short half-lives, of 15 min [86] and 30 min [87] respectively. However, it is known that these values are by no means absolute. For example, Lacy et al. [88] demonstrated that the half-life of c-*myc* mRNA in an Epstein–Barr virus-negative lymphoma cell line was increased from < 36 to > 70 min following *in vitro* infection with Epstein–Barr virus.

However, the significance that these findings have to the regulation of c-*myc in vivo* has yet to be established.

PROMOTER OF CELL GROWTH OR OF CELL DEATH?

Undoubtedly, the most paradoxical revelation from work on Myc was the finding that under certain circumstances the protein is able to induce programmed cell death (apoptosis). Two separate studies, one in fibroblasts [89] and one in haemopoietic [90] cells, revealed that when Myc was constitutively expressed the cells required the presence of exogenous growth/survival factors in order to replicate effectively. If serum in the case of fibroblasts, or interleukin 3 in the case of haemopoietic cells, was withdrawn from the Myc-expressing cultures the cells continued to enter the cell cycle, but then underwent apoptosis.

Two possible explanations for how the cell might make a choice between life and death as a result of Myc expression are outlined in Figure 5. Initially it was thought possible that the absence of exogenous growth/survival factors was having an

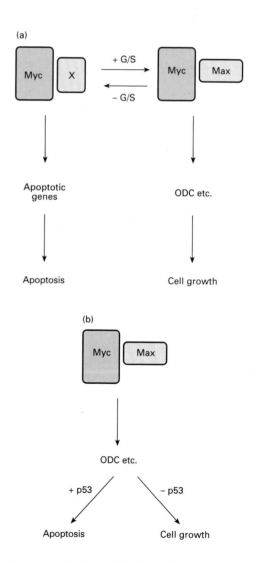

Figure 5   Possible mechanisms of Myc-induced apoptosis

(**a**) Pathway involving two sets of Myc-regulated genes, one set that regulates cell growth and one set that regulates apoptosis. (**b**) Pathway involving one set of Myc-regulated genes, in which the decision between cell growth and cell death is mediated by p53. G/S, growth/survival factors.

effect upon the transcriptional targets of c-Myc (Figure 5a). As described above, this could perhaps occur through a change in the relative abundances of transcriptional complexes involving the Myc protein, e.g. a change from a predominance of Myc:Max complexes to a predominance of Myc:TFII-I complexes. However, since it has now been shown that the apoptotic cell death induced by Myc requires dimerization with Max [91] and is dependent on an active ODC protein [92], it seems more likely that the same genes are transcribed in both the absence and presence of growth/survival factors and that the decision between life and death is made elsewhere (Figure 5b). The observation that the ability of Myc to induce apoptosis is abrogated in cells lacking a wild-type p53 protein [93] points to an involvement of p53 at some point during the apoptotic programme. In this regard it is interesting to note that apoptosis induced by the adenovirus E1A protein is also mediated by p53 [94]. However, whether the p53 protein acts via this mechanism to eradicate cells that contain an aberrant growth stimulus induced by other oncogenes is yet to be established.

TWO FORMS OF MYC PROTEIN: ONCOGENE VERSUS TUMOUR SUPPRESSOR?

A distinctive feature of the c-*myc* gene is that it encodes two alternatively translated protein products [95–97] (Figure 1). The proteins differ not only in size but also in the mechanism by which their translation is initiated [98]. The shorter and more predominant form of c-Myc, c-Myc 2, is initiated from a standard AUG site, whereas the longer protein, c-Myc 1, is initiated from a non-AUG site (CUG in humans) [98]. As it was noticed that c-Myc 1 was induced to levels comparable with those of c-Myc 2 as cells reached high densities during *in vitro* culture [99], it was considered that the two forms of the Myc protein may perform different roles. In order to investigate this, Hann et al. [100] generated clones of COS cells that expressed high levels of either c-Myc 1 or c-Myc 2. These cells were then transiently transfected with different regions of the Rous sarcoma virus long terminal repeat, to assess the possibility that the two Myc proteins possess distinct transcriptional activities [100]. The study revealed that one specific enhancer element, EFII, was activated by c-Myc 1 but not by c-Myc 2 (Figure 6). Surprisingly, when the sequence of EFII was analysed, it was observed that it did not contain the consensus Myc:Max binding site (CACGTG) and that activation by c-Myc 1 was dependent on a repeat sequence that contained a consensus binding site for a CCAAT/enhancer binding protein.

Subsequent electrophoretic mobility shift assays of this binding site showed, however, that it too was bound by Myc:Max heterodimers but not by Max:Max homodimers. This therefore opens up the possibility of a situation in which Myc:Max and the CCAAT/enhancer binding protein are in competition for the same binding site. However, the consequences that this might have *in vivo* are still unknown.

Although any cellular targets for transcriptional regulation by c-Myc 1 are yet to be discovered, the implications of this protein are extremely interesting, since COS cells in which it is over-expressed display a growth-inhibited phenotype [100]. It is also noteworthy that, in cases of Burkitt's lymphoma in which the c-*myc* gene is translocated to the immunoglobulin locus, the rearrangement frequently occurs such that only c-Myc 2 becomes activated [101,102]. Also, in cases in which the translocation results in both c-Myc 1 and c-Myc 2 activation, it has been reported that the c-Myc 1 protein is frequently mutated [103]. In addition, in all cases of Burkitt's lymphoma that display this reciprocal translocation, the non-translocated c-*myc* allele has been reported to be silenced [104]. In this regard, it is perhaps now worthwhile to re-investigate whether the c-*myc* gene is mutated in other tumours such that only c-Myc 2 becomes activated.

FUTURE PERSPECTIVES

Since the discovery of the Myc protein, much of the effort in cancer research has focused on the ability of oncogenes to induce continued cell-cycle progression. However, we are now aware that the establishment of disseminated disease is dependent not only on proliferative capacity but on a variety of contributing factors. As a result the oncogene field has come full circle, and groups are beginning to re-assess the roles that these genes are playing in processes such as apoptosis and metastasis. While it is now well established that Myc can also function as an inducer of programmed cell death, it would also be of interest to assess any involvement the protein might have in cellular invasion. The observations that the Myc proteins can induce morphological transformation and adhesion-independent growth can be considered as criteria of an invading cell. Coupled with the knowledge that the *myc* gene generally appears to be activated in the late stages of cancer [105], further investigation of a role for Myc in these areas might be particularly rewarding.

One of the major objectives for the further elucidation of the function of the Myc proteins must lie with the identification of

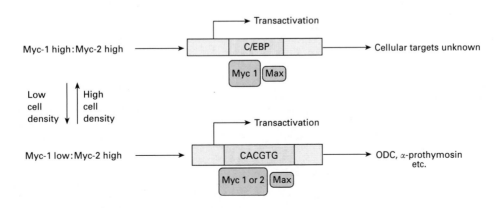

Figure 6 **Implications of the different properties of the two Myc proteins, c-Myc 1 and c-Myc 2**

C/EBP, CCAAT/enhancer binding protein recognition sequence.

additional target genes. If this is achieved, it will provide a means to establish how all of the phenotypic changes which are attributable to Myc actually occur. For example, at present the role of Myc in the regulation of cellular differentiation is still a contentious issue. It is widely reported that elevated expression of Myc protein can block differentiation in many lineages [10,11]. However, it is not known whether this is through direct interaction with the differentiation machinery or whether it is a result of the continued cell-cycle progression caused by Myc in cells which must undergo growth arrest for terminal differentiation. One report by La Rocca et al. [106] has indicated that the blockage of differentiation of myogenic cells by Myc can be dissociated from the ability of the protein to induce morphological transformation. However, the mechanistic differences that give rise to this observation are yet to be determined.

CONCLUDING REMARKS

In spite of the consistently conflicting data regarding the Myc proteins, the importance of their role in both the normal and the transformed cell cannot be overstated. Most strikingly, studies involving transgenic mice in which Myc was generally over-expressed showed the induction of widespread neoplasia [107], whereas when the gene was disrupted by homologous recombination the mice died before 10.5 days of gestation [108]. When this is considered alongside the frequent reports of *myc* mutation in malignant disease, it becomes obvious that, although the road to an understanding of Myc function may not always be smooth, it is certainly one worth following.

Note added in proof (received 5 February 1996)

Two further Mad-related proteins have since been discovered, Mad3 and Mad4 [113].

The Beatson Institute is supported by grants from the Cancer Research Campaign. K. R. is a recipient of an M.R.C. postgraduate studentship.

REFERENCES

1 Ingvarsson, S. (1990) Semin. Cancer Biol. **1**, 359–369
2 Vennstrom, B., Sheiness, D., Zabielski, J. and Bishop, J. M. (1982) J. Virol. **42**, 773–779
3 Schwab, M., Alitalo, K., Klempnauer, K-L., Varmus, H. E., Bishop, J. M., Gilbert, F., Brodeur, G., Goldstein, M. and Trent, J. (1983) Nature (London) **305**, 245–248
4 Nau, N. M., Brooks, B. J., Battey, J., Sausville, E., Gazdar, A. F., Kirsch, I. R., McBride, O. W., Bertness, V., Hollis, G. F. and Minna, J. D. (1985) Nature (London) **318**, 69–73
5 Dalla Favera, R., Wong-Staal, F. and Gallo, R. C. (1982) Nature (London) **299**, 61–63
6 Magrath, I. (1990) Adv. Cancer Res. **55**, 133–270
7 Payne, G. S., Bishop, J. M. and Varmus, H. E. (1982) Nature (London) **295**, 209–214
8 Neil, J. C., Forrest, D., Doggett, D. L. and Mullins, J. I. (1987) Cancer Surv. **6**, 117–137
9 Eilers, M., Schrim, S. and Bishop, J. M. (1991) EMBO J. **10**, 133–141
10 Coppola, J. A. and Cole, M. D. (1986) Nature (London) **320**, 760–763
11 Miner, J. H. and Wold, B. J. (1991) Mol. Cell. Biol. **11**, 2842–2851
12 Penn, L. J. Z., Laufer, E. M. and Land, H. (1990) Semin. Cancer Biol. **1**, 69–80
13 Land, H., Parada, L. and Weiberg, R. A. (1983) Nature (London) **304**, 596–602
14 Eisenman, R. N., Tachibana, C. Y., Abrams, H. and Hann, S. R. (1985) Mol. Cell. Biol. **5**, 114–126
15 Dang, C. V. and Lee, W. M. F. (1988) Mol. Cell. Biol. **8**, 4048–4054
16 Studzinski, G. P., Brelvi, Z. S., Feldman S. C. and Watt, R. A. (1986) Science **234**, 467–470
17 Kaddurah-Daouk, R., Greene, J. M., Baldwin, A. S. and Kingston, R. E. (1987) Genes Dev. **1**, 347–357
18 Landschulz, W. H., Johnson, P. F. and McKnight, S. L. (1988) Science **240**, 1759–1764
19 Luscher, B. and Eisenman, R. N. (1990) Genes Dev. **4**, 2025–2035

20 Murre, C., McCae, P. S., Vaessin, H., Caudy, M., Jan, Y. N., Cabrera, C. V., Buskin, J. N., Hauseka, S. D., Lassar, A. B., Weintraub, H. and Baltimore, D. (1989) Cell **58**, 537–544
21 Davis, R. L., Cheng, P.-F., Lassar, A. B. and Weintraub, H. (1990) Cell **60**, 733–746
22 Kato, G. J., Barrett, J., Villa-Garcia, M. and Dang, C. V. (1990) Mol. Cell. Biol. **10**, 5914–5920
23 Dang, C. V., McGuire, M., Buckmire, M. and Lee, W. M. F. (1989) Nature (London) **337**, 664–666
24 Blackwood, E. M. and Eisenman, R. N. (1991) Science **251**, 1211–1217
25 Solomon, D. L. C., Amati, B. and Land, H. (1993) Nucleic Acids Res. **21**, 5572–5576
26 Prendergast, G. C., Lawe, D. and Ziff, E. B. (1991) Cell **65**, 395–407
27 Amati, B., Dalton, S., Brooks, M. W., Littlewood, T. D., Evan, G. I. and Land, H. (1992) Nature (London) **359**, 423–426
28 Kato, G. J., Lee, W. M. F., Chen, L. and Dang, C. V. (1992) Genes Dev **6**, 81–92
29 Gaubatz, S., Meichle, A. and Eilers, M. (1994) Mol. Cell. Biol. **14**, 3853–3862
30 Eschenfeldt, W. H. and Berger, S. L. (1986) Proc. Natl. Acad. Sci. U.S.A. **83**, 9403–9407
31 Gomez-Marquez, J., Segade, F., Dosil, M., Bustelo, X. R. and Freire, M. (1989) J. Biol. Chem. **264**, 8451–8454
32 Clinton, M., Frangoulazaridis, M., Panneerselvam, C. and Horecker, B. L. (1989) Arch. Biochem. Biophys. **269**, 256–267
33 Tabor, C. W. and Tabor, H. (1984) Annu. Rev. Biochem. **53**, 749–790
34 Pegg, A. E. (1986) Biochem. J. **234**, 249–262
35 Auvinen, M., Paasinen, A., Andersson, L. C. and Holtta, E. (1992) Nature (London) **360**, 355–358
36 Wagner, A. J., Meyers, C., Laimins, L. A. and Hay, N. (1993) Cell Growth Differ. **4**, 879–883
37 Bello-Fernandez, C., Packham, G. and Cleveland, J. L. (1993) Proc. Natl. Acad. Sci. U.S.A. **90**, 7804–7808
38 Blackwell, T. K., Huang, J., Ma, A., Kretzner, L., Alt, F. W., Eisenman, R. N. and Weintraub, H. (1993) Mol. Cell. Biol. **13**, 5216–5224
39 Roy, A. L., Carruthers, C., Gutjahr, T. and Roeder, R. G. (1993) Nature (London) **365**, 356–361
40 Smale, S. T. and Baltimore, D. (1989) Cell **57**, 103–113
41 Phillip, A., Schneider, A., Vasrik, I., Finke, K., Xiong, Y., Beach, D., Alitalo, K. and Eilers, M. (1994) Mol. Cell. Biol. **14**, 4032–4043
42 Daksis, J. I., Lu, R. Y., Facchini, L. M., Markin, W. W. and Penn, L. J. Z. (1994) Oncogene **9**, 3635–3645
43 Amati, B., Brooks, M. W., Levy, N., Littlewood, T. D., Evan, G. I. and Land, H. (1993) Cell **72**, 233–245
44 Stone, J., de Lange, T., Ramsay, G., Jakobovits, E., Bishop, J. M., Varmus, H. and Lee, W. (1987) Mol. Cell. Biol. **7**, 1697–1709
45 Dyson, N., Howley, P. M., Munger, K. and Harlow, E. (1989) Science **243**, 934–937
46 Whyte, P., Buchkovich, K. J., Horwitz, J. M., Friend, S. H., Raybuck, M., Weinberg, R. A. and Harlow, E. (1988) Nature (London) **334**, 124–129
47 Werness, B. A., Levine, A. J. and Howley, P. M. (1990) Science **248**, 76–79
48 Sarnow, P., Ho, Y. S., Williams, J. and Levine, A. J. (1982) Cell **28**, 387–394
49 Serrano, M., Hannon, G. J. and Beach, D. (1993) Nature (London) **366**, 704–707
50 Beijersbergen, R. L., Hijmans, E. M., Zhu, L. and Bernards, R. (1994) EMBO J. **13**, 4080–4086
51 Gu, W., Bhatia, K., Magrath, I. T., Dang, C. V. and Dalla-Favera, R. (1994) Science **264**, 251–254
52 Whyte, P., Williamson, N. M. and Harlow, E. (1989) Cell **56**, 57–75
53 Dyson, N., Buchkovich, K., Whyte, P. and Harlow, E. (1989) Cell **58**, 249–255
54 Davies, R., Hicks, R., Crook, T., Morris, J. and Vousden, K. (1993) J. Virol. **67**, 2521–2528
55 Maheswaran, S., Lee, H. and Sonenshein, G. E. (1994) Mol. Cell. Biol. **14**, 1147–1152
56 Shrivastava, A., Saleque, S., Kalpana, G. V., Artandi, S., Goff, S. P. and Calame, K. (1993) Science **262**, 1889–1892
57 Riggs, K. J., Merrell, K. T., Wilson, G. and Calame, K. (1991) Mol. Cell. Biol. **11**, 1765–1769
58 Shi, Y., Seto, E., Chang, L. S. and Shenk, T. (1991) Cell **67**, 377–388
59 Seto, E., Shi, Y and Shenk, T. (1991) Nature (London) **354**, 241–245
60 Bauknecht, T., Angel, P., Royer, H.-D. and zur Hausen, H. (1992) EMBO J. **11**, 4607–4617
61 Shirodkar, S., Ewen, M., DeCaprio, J. A., Morgan, J., Livingston, D. M. and Chittenden, T. (1992) Cell **68**, 157–166
62 Kovesdi, I., Reichel, R. and Nevins, J. R. (1986) Cell **45**, 219–228
63 Ayer, D. E., Kretzner, L. and Eisenman, R. N. (1993) Cell **72**, 211–222
64 Zervos, A. S., Gyuris, J. and Brent, R. (1993) Cell **72**, 223–232
65 Ayer, D. E., Lawrence, Q. A. and Eisenman, R. N. (1995) Cell **80**, 767–776
66 Schreiber-Agus, N., Chin, L., Chen, K., Torres, R., Rao, G., Guida, P., Skoultchi, A. I. and DePinho, R. A. (1995) Cell **80**, 777–786

67 Lahoz, E.G, Xu, L., Schreiber-Agus, N. and DePinho, R. A. (1994) Proc. Natl. Acad. Sci. U.S.A. **91**, 5503–5507

68 Eagle, L. R., Yin, X., Brothman, A. R., Williams, B. J., Atkin, N. B. and Prochownik, E. V. (1995) Nature Genet. **9**, 249–255

69 Stewart, T. A., Bellve, A. R. and Leder, P. (1984) Science **226**, 707–710

70 Taub, R., Moulding, C., Battey, J., Latt, S., Lenoir, G. M. and Leder, P. (1984) Cell **36**, 339–348

71 Ray, D. and Robert-Lesenges, J. (1989) Oncogene Res. **5**, 73–78

72 Bentley, D. L. and Groudine, M. (1986) Mol. Cell. Biol. **6**, 3481–3489

73 Siebenlist, U., Hennighausen, L., Battey, J. and Leder, P. (1984) Cell **37**, 381–391

74 Marcu, K. B., Bossone, S. A. and Patel, A. J. (1992) Annu. Rev. Biochem. **61**, 809–860

75 Lipp, M., Schilling, R. and Bernhardt, G. (1989) Oncogene **4**, 535–541

76 Hiebert, S. W., Lipp, M. and Nevins, J. R. (1989) Proc. Natl. Acad. Sci. U.S.A. **86**, 3594–3598

77 Thalmeier, K., Synovik, H., Mertz, R., Winnacher, E.-L. and Lipp, M. (1989) Genes Dev. **3**, 527–536

78 Johnson, D. G., Schwarz, J. K., Cress, W. D. and Nevins, J. R. (1993) Nature (London) **365**, 349–352

79 Hay, N., Bishop, J. M. and Levens, D. (1987) Genes Dev **1**, 659–671

80 Postel, E. H., Mango, S. E. and Flint, S. J. (1989) Mol. Cell. Biol. **9**, 5123–5133

81 Postel, E. H., Berberich, S. J., Flint, S. J. and Ferrone, C. A. (1993) Science **261**, 478–480

82 Delarosa, A., Williams, R. L. and Steeg, P. S. (1995) BioEssays **17**, 53–62

83 Bentley, D. L. and Groudine, M. (1986) Nature (London) **321**, 702–706

84 Bentley, D. L. and Groudine, M. (1988) Cell **53**, 245–256

85 Krumm, A., Meulia, T., Brunvand, M. and Groudine, M. (1992) Genes Dev **6**, 2201–2213

86 Dani, C., Blanchard, J.-M., Piechaczyk, M., El Sabouty, S., Marty, L. and Jeanteur, P. (1984) Proc. Natl. Acad. Sci. U.S.A. **81**, 7046–7050

87 Rabbitts, P. H., Watson, J. V., Lamond, A., Forster, A., Stinson, M. A., Evan, G., Fischer, W., Atherton, E., Sheppard, R. and Rabbitts, T. H. (1985) EMBO J. **4**, 2009–2015

88 Lacy, J., Summers, W. P. and Summers, W. C. (1989) EMBO J. **8**, 1973–1980

89 Evan, G. I., Wyllie, A. H., Gilbert, C. S., Littlewood, T. D., Land, H., Brooks, M., Waters, C. M., Penn, L. Z. and Hancock, D. C. (1992) Cell **69**, 119–128

90 Askew, D. S., Ashmun, R. A., Simmons, B. C. and Cleveland, J. L. (1991) Oncogene **6**, 1915–1922

91 Amati, B., Littlewood, T. D., Evan, G. I. and Land, H. (1993) EMBO J. **12**, 5083–5087

92 Packham, G. and Cleveland, J. L. (1994) Mol. Cell. Biol. **14**, 5741–5747

93 Hermeking, H. and Eick, D. (1994) Science **265**, 2091–2093

94 Debbas, M. and White, E. (1993) Genes Dev. **7**, 546–554

95 Hann, S. R. and Eisenman, R. N. (1984) Mol. Cell. Biol. **4**, 2486–2497

96 Ramsay, G., Stanton, L., Schwab, M. and Bishop, J. M. (1986) Mol. Cell. Biol. **6**, 4450–4457

97 Dosaka-Akita, H., Rosenberg, J. D., Minna, J. D. and Birrer, M. J. (1991) Oncogene **6**, 371–378

98 Hann, S. R., King, M. W., Bentley, D. L., Anderson, C. W. and Eisenman, R. N. (1988) Cell **34**, 185–195

99 Hann, S. R., Sloan-Brown, K. and Spoots, G. D. (1992) Genes Dev. **6**, 1229–1240

100 Hann, S. R., Dixit, M., Sears, R. C. and Sealy, L. (1994) Genes Dev. **8**, 2441–2452

101 Dalla-Favera, R., Martinotti, S., Gallo, R. C., Erikson, J. and Croce, C. M. (1983) Science **219**, 963–967

102 Rabbitts, T. H., Hamlyn, P. H. and Baer, R. (1983) Nature (London) **306**, 760–765

103 Rabbitts, T. H., Forster, A., Hamlyn, P. H. and Baer, R. (1984) Nature (London) **309**, 592–597

104 Bernard, O., Cory, S., Gerondakis, S., Webb, E. and Adams, J. M. (1983) EMBO J. **2**, 2375–2383

105 Field, J. K. and Spandidos, D. A. (1990) Anticancer Res. **10**, 1–22

106 La Rocca, S. A., Crouch, D. H. and Gillespie, D. A. F. (1994) Oncogene **9**, 3499–3508

107 Leder, A., Pattengale, P. K., Kuo, A., Stewart, T. A. and Leder, P. (1986) Cell **45**, 485–495

108 Davis, A. C., Wims, M., Spotts, G. D., Hann, S. H. and Bradley, A. (1993) Genes Dev. **7**, 671–682

109 Sugiyama, A., Kume, A., Nemoto, K., Lee, S. Y., Asami, Y., Nemoto, F., Nishimura, S. and Kuchino, Y. (1989) Proc. Natl. Acad. Sci. U.S.A. **86**, 9144–9148

110 Asai, A., Miyagi, Y., Sugiyama, A., Nagashima, Y., Kanemitsu, H., Obinata, M., Mishima, K. and Kuchino, Y. (1994) Oncogene **9**, 2345–2352

111 Resar, L. M. S., Dolde, C., Barrett, J. F. and Dang, C. V. (1993) Mol. Cell. Biol. **13**, 1130–1136

112 DePinho, R. A., Schreiber-Agus, N. and Alt, F. W. (1991) Adv. Cancer Res. **57**, 1–46

113 Huslin, P. J., Quéva, C., Koskinen, P. J., Steingrimsson, E., Ayer, D. E., Copeland, N. G., Jenkins, N. A. and Eisenman, R. N. (1995) EMBO J. **14**, 5646–5659

Biochem. J. (1996) **316**, 361–371 (Printed in Great Britain)

REVIEW ARTICLE

Genomic mechanisms involved in the pleiotropic actions of 1,25-dihydroxyvitamin D₃

Sylvia CHRISTAKOS*, Mihali RAVAL-PANDYA, Roman P. WERNYJ and Wen YANG

Department of Biochemistry and Molecular Biology, University of Medicine and Dentistry of New Jersey, New Jersey Medical School, 185 South Orange Avenue, Newark, NJ 07103-2714, U.S.A.

The biologically active metabolite of vitamin D (cholecalciferol), i.e. 1,25-dihydroxyvitamin D_3 [$1,25(OH)_2D_3$], is a secosteroid hormone whose mode of action involves stereospecific interaction with an intracellular receptor protein (vitamin D receptor; VDR). $1,25(OH)_2D_3$ is known to be a principal regulator of calcium homeostasis, and it has numerous other physiological functions including inhibition of proliferation of cancer cells, effects on hormone secretion and suppression of T-cell proliferation and cytokine production. Although the exact mechanisms involved in mediating many of the different effects of $1,25(OH)_2D_3$ are not completely defined, genomic actions involving the VDR are clearly of major importance. Similar to other steroid receptors, the VDR is phosphorylated; however, the exact functional role of the phosphorylation of the VDR remains to be determined. The VDR has been reported to be regulated by $1,25(OH)_2D_3$ and also by activation of protein kinases A and C, suggesting co-operativity between signal transduction pathways and $1,25(OH)_2D_3$ action. The VDR binds to vitamin D-responsive elements (VDREs) in the 5′ flanking region of target genes. It has been suggested that VDR homodimerization can occur upon binding to certain VDREs but that the VDR/retinoid X receptor (RXR) heterodimer is the functional transactivating species. Other factors reported to be involved in VDR-mediated transcription include chicken ovalbumin upstream promoter (COUP) transcription factor, which is involved in active silencing of transcription, and transcription factor IIB, which has been suggested to play a major role following VDR/RXR heterodimerization. Newly identified vitamin D-dependent target genes include those for Ca^{2+}/Mg^{2+}-ATPase in the intestine and p21 in the myelomonocytic U937 cell line. Elucidation of the mechanisms involved in the multiple actions of $1,25(OH)_2D_3$ will be an active area of future research.

INTRODUCTION

It is known that the most active metabolite of vitamin D (cholecalciferol), i.e. 1,25-dihydroxyvitamin D_3 [$1,25(OH)_2D_3$], functions to regulate calcium homeostasis in intestine, bone and kidney [1,2]. $1,25(OH)_2D_3$ has also been found to have numerous other physiological functions, including effects on cell growth and differentiation [2,3], secretion of hormones [4–6], T-cell proliferation and cytokine production [7,8], sterol metabolism [9] and cardiovascular function [10] (see [1,2,8,10–14] for reviews). $1,25(OH)_2D_3$, similar to other steroid hormones, binds to a high-affinity, low-capacity receptor protein (the vitamin D receptor or VDR), resulting in concentration of the $1,25(OH)_2D_3$–VDR complex in the target cell nucleus. Exactly how $1,25(OH)_2D_3$ affects numerous different systems is a subject of continuing investigation; however, the VDR is clearly of major importance.

Although many gene products have been reported to be sensitive to $1,25(OH)_2D_3$ (see [15] for a list of 51 genes reported to be regulated by the active form of vitamin D), little is known about the molecular mechanisms involved in the regulation by $1,25(OH)_2D_3$ of most of these genes. Transcriptional regulation has been described for about 20 % of the genes reported to be regulated by $1,25(OH)_2D_3$, and only a small number of genes, including those encoding osteopontin, osteocalcin, 25-hydroxyvitamin D_3 24-hydroxylase (24-hydroxylase) and integrin $\alpha_v\beta_3$, have been reported to contain a vitamin D-responsive element (VDRE). It is likely that future research will result in the identification of a number of different mechanisms of regulation of these genes, including effects of $1,25(OH)_2D_3$ on stability and the possibility that induction by $1,25(OH)_2D_3$ may be secondary to a primary effect of $1,25(OH)_2D_3$ on other factors.

In this review, recent findings related to genomic mechanisms of $1,25(OH)_2D_3$ action will be discussed, including evidence for a role of transcription factor IIB (TFIIB) and other transcription factors in vitamin D-mediated transcription. An overview of newly identified vitamin D-responsive genes will also be presented. In addition, current controversies in the vitamin D field related to VDR regulation, the site of phosphorylation of the VDR and VDR homodimerization versus heterodimerization will be reviewed. It should be noted, however, that not all vitamin D-regulated biological responses may require nuclear-mediated mechanisms [16,17]; possible non-genomic mechanisms will not be discussed here (non-genomic effects of $1,25(OH)_2D_3$ are discussed as part of a recent excellent review by Bouillon et al. [18]).

THE VDR

General considerations

Although the VDR protein has been cloned from human [19], rat [20,21], mouse [22] and chicken [23] (for reviews concerning the VDR, see [24,25]), a complete analysis of the sequence and detailed structural and functional studies have not been published for a number of the VDRs. X-ray crystallographic data, which

---

Abbreviations used: $1,25(OH)_2D_3$, 1,25-dihydroxyvitamin D_3; (h)VDR, (human) vitamin D receptor; 24-hydroxylase, 25-hydroxyvitamin D_3 24-hydroxylase; VDRE, vitamin D-responsive element; TFIIB, transcription factor IIB; RAR, retinoic acid receptor; PMA, phorbol 12-myristate 13-acetate; CAT, chloramphenicol acetyltransferase; calbindin D_{28k}, 28 kDa form of calbindin D; NFAT, nuclear factor of activated T-cells; RXR, retinoid X receptor; tk, thymidine kinase; COUP-TF, chicken ovalbumin upstream promoter transcription factor.

* To whom correspondence should be addressed.

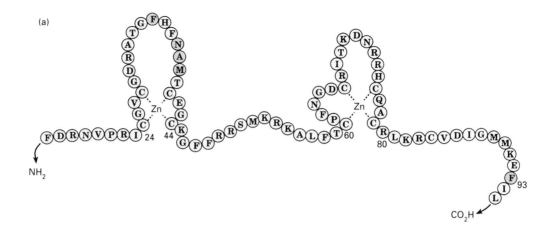

(a)

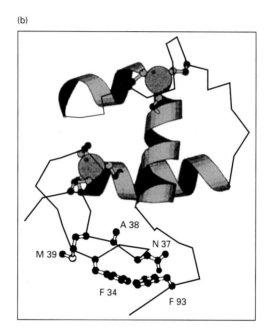

(b)

Figure 1   The VDR DNA-binding domain

(**a**) The zinc fingers of the DNA-binding domain of the hVDR (primary sequence and secondary structure [47]). Residues discussed in (**b**) are coloured. (**b**) A model for intramolecular interaction of the T-box of the VDR (approx. five residues beyond the C-terminus of the second zinc finger) with the tip of the first zinc finger. Helices (ribbons), the C^α backbone trace (line), side chains of selected residues of the T-box and the tip of the first zinc finger (ball and stick) and metal co-ordinating cysteines are shown. The two zinc ions are represented by spheres. The recognition α-helix of the first zinc finger, which lies in the major groove of the DNA, is oriented horizontally. The principal helix of the second zinc finger lies vertically. Phe-93 of the T-box may pack against Asn-37 and Phe-34, which is supported by Ala-38 and probably Met-39. Reproduced with permission from [114].

are not as yet available, should result in refined structural analysis, including tertiary structure and distance geometry, which should greatly facilitate structure/function studies. The molecular mass of the VDR has been calculated to be between 48 kDa {for the human VDR (hVDR) [26]} and 60 kDa (for the avian VDR [23]). Among species there is between 93 and 95 % sequence identity in the DNA- and hormone-binding domains. However, in the hinge region there was found to be significant variation in the sequences between species. Although detailed analysis of the structural organization of the VDR is not available as yet, 5′ and 3′ deletion of the hVDR were used to localize functional domains [26]. Data from these studies indicated that hormone binding was localized to a peptide fragment whose synthesis begins at residue 114 and ends at residue 373. The DNA-binding domain of the hVDR is the cysteine-rich region

located at the extreme N-terminus (residues 1–114), i.e. the conserved zinc finger region. There are two zinc fingers. The zinc atom is proposed to be in a tetrahedral arrangement with the cysteine residues, which are highly conserved (Figure 1). It has been proposed for the VDR, similar to the glucocorticoid receptor and the oestrogen receptor, that an α-helical region residing at the C-terminal base of the first zinc finger may be responsible for receptor specificity for binding a specific DNA sequence [27]. It has also been proposed that specificity can be conferred by spacing between two half-sites of a VDRE [28], although strict adherence to spacer regions of 3, 4 and 5 nucleotides for VDR, thyroid hormone receptor and retinoic acid receptor (RAR) recognition respectively has been questioned [25]. In addition, auxiliary factors may contribute to VDR specificity in specific target tissues.

A region in the second zinc finger (five amino acids between the first two co-ordinating systems), the D-box, which is essential for dimer binding to the glucocorticoid response element [29], has been described in the crystal structure of the glucocorticoid receptor [30]. Freedman and Towers [31] reported that when the glucocorticoid receptor D-box is replaced by residues in that region of the VDR DNA-binding domain, co-operative DNA binding usually observed with the glucocorticoid receptor is abolished, suggesting that the VDR and the glucocorticoid receptor use different strategies to bind half-sites (the VDR DNA-binding domain appears to bind half-sites non-cooperatively, without the free-energy contribution of dimerization that has been observed for the glucocorticoid receptor), and that the VDR DNA-binding domain binds DNA independently of a glucocorticoid receptor-type finger-mediated dimer.

Phosphorylation of the VDR

The VDR [32–34], similar to other steroid hormone receptors including those for progesterone [35], androgens [36], oestrogen [37,38] and glucocorticoids [39], is phosphorylated. Although it has been suggested that steroid receptor phosphorylation is involved in nuclear localization, hormone binding and transcriptional activation or repression, the exact functional role(s) of phosphorylation remains to be elucidated. Brown and DeLuca showed an increase in phosphorylation of the VDR within 15 min of $1,25(OH)_2D_3$ treatment, which continued for 1 h [33]. However, a low level of phosphorylated receptors was detectable even in the absence of $1,25(OH)_2D_3$. Thus ligand binding does not seem to be essential for phosphorylation, but does result in a substantial increase in phosphorylation. Several studies have shown that okadaic acid (an inhibitor of phosphatases 1 and 2A) can activate VDR-mediated transcription in the absence of ligand [40–42]. It is possible that activation of VDR-mediated transcription by okadaic acid is a result of direct phosphorylation of the VDR. However, it is also possible that okadaic acid may enhance the phosphorylation of an auxiliary transcription factor or of a member of the basal RNA polymerase II transcription machinery, which could further influence VDR-mediated transactivation. Although the exact nature of the effect of okadaic acid on VDR transactivation remains to be determined, these studies do suggest that ligand occupancy is not required for VDR-mediated transcription.

The subject of the site(s) of phosphorylation on the VDR has been somewhat controversial. The identification of the phosphoamino acid(s) of the hVDR was first reported by Hsieh et al. [43] and Jurutka et al. [44]. Hsieh et al. [43] reported that the hVDR is phosphorylated by protein kinase C-β on Ser⁵¹ in the zinc finger region, and Jurutka et al. [44] reported that the VDR

is phosphorylated in the ligand-binding domain by casein kinase II at Ser²⁰⁸ (Figure 2). The magnitude of the phosphorylation at Ser²⁰⁸ was not affected by the addition of $1,25(OH)_2D_3$. These results were obtained using site-directed mutagenesis. When Ser⁵¹ or Ser²⁰⁸ was mutated to glycine, phosphorylation of the hVDR was reduced. However, studies by Brown and DeLuca, using the rat VDR, were not in agreement with Ser⁵¹ as a phosphorylation site, since they found that phosphorylation was localized only in the hormone-binding domain [45]. In a later study, Hilliard et al. [46], using selective manual Edman degradation of phosphorylated peptides coupled with direct amino acid sequence analysis of the isolated peptides, reported that a phosphate was released only at Ser²⁰⁵ (note that residue 1 of the hVDR reported in that paper was equivalent to residue 4 of the hVDR sequence reported by Baker et al. [47] which was used in the papers of Hsieh et al. [43] and Jurutka et al. [44]). Unlike the results reported by Jurutka et al. [44], the phosphorylation of Ser²⁰⁵ was strongly dependent on $1,25(OH)_2D_3$ treatment. Hilliard et al. [46] also showed that when Ser²⁰⁵ was modified by site-directed mutagenesis to alanine, the mutated protein continued to be phosphorylated in a $1,25(OH)_2D_3$-dependent manner. Jurutka et al. [44] reported that, when Ser²⁰⁸ was replaced with glycine, at least one additional phosphoacceptor site existed and that this site (not Ser²⁰⁸) was increased in the presence of $1,25(OH)_2D_3$. Hilliard et al. [46] suggested that after site-directed mutagenesis an alternative site may become available due to structural alteration as a result of mutagenesis of the VDR. The authors therefore suggested that cautious interpretation should be made of data derived from mutagenesis experiments, particularly since the crystal structure of the VDR is unavailable at this time. Knowledge of this crystal structure will facilitate the resolution of the differences observed between the peptide mapping approach and mutagenesis experiments.

Regulation of the VDR

Up-regulation of VDRs by $1,25(OH)_2D_3$ has been demonstrated in several different systems, including rat intestine [48], pig kidney LLCPK-1 cells [49], 3T6 mouse fibroblasts [50], human HL60 leukaemia cells [51] and rat insulinoma cells [52]. However, whether the homologous up-regulation of the VDR involves induction of VDR mRNA is not clear. Mangelsdorf et al. [50] found an increase in VDR mRNA as well as an increase in VDR protein levels in mouse 3T6 cells in response to $1,25(OH)_2D_3$ treatment for 24–72 h. However, studies by Huang et al. [53] and Wiese et al. [54] indicated that administration of $1,25(OH)_2D_3$ to vitamin D-deficient rats did not alter VDR mRNA levels in intestine and kidney. Meyer et al. [55] reported that, when compared with control animals, chicks fed a low-calcium diet [which results in a marked increase in serum $1,25(OH)_2D_3$]

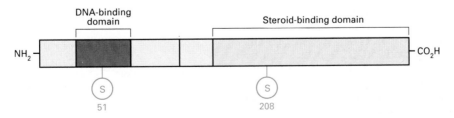

Figure 2   Schematic representation of the hVDR and proposed phosphorylation sites

Phosphorylation sites on the VDR are as reported by Hsieh et al. [43] and Jurutka et al. [44]. See the text for details.

exhibited a significant decrease in VDR mRNA. Reinhardt and Horst [56] reported that parathyroid hormone is a potent down-regulator of VDR mRNA. Thus it is possible that, in the low-calcium, vitamin D-deficient animals, up-regulation of VDR mRNA by $1,25(OH)_2D_3$ is not observed due, in part, to parathyroid hormone inhibition of VDR mRNA production. *In vitro* studies by Wiese et al. [54] indicated that VDR protein levels in 3T6 cells were induced 3-fold at 8 h after $1,25(OH)_2D_3$ treatment; however, VDR mRNA was not altered by $1,25(OH)_2D_3$ during this time. Similar findings indicating induction of VDR protein but not mRNA were noted in studies in rat epithelial cells [54], HL60 cells [51], rat insulinoma cells [52] and ROS 17/2.8 osteosarcoma cells [57], and for the VDR in transformed yeast cells [58]. It has been suggested that induction of $1,25(OH)_2D_3$ receptor protein following $1,25(OH)_2D_3$ administration is most likely due to altered stability of the occupied receptor [51,52,54,57,58].

Besides homologous up-regulation, the VDR has been reported to be regulated by a number of other factors, including glucocorticoids [53,59–63] and activation of protein kinase A [64–66] and protein kinase C [67]. Glucocorticoids have been reported to up-regulate as well as to down-regulate the VDR in a number of cells and tissues. However, it has been shown that up-regulation of rat intestinal VDR mRNA by glucocorticoids is not specific for VDR mRNA, but may reflect general effects of glucocorticoids on the intestine [53,63]. Concerning the effect of protein kinase C and protein kinase A, it has been suggested that the regulation of the VDR by signal transduction pathways may play an important role in modulating target cell responsiveness to $1,25(OH)_2D_3$. For example, Krishnan and Feldman [64] reported that in NIH3T3 mouse fibroblasts the elevation of intracellular cAMP by forskolin or dibutyryl-cAMP resulted in an 8–10-fold increase in VDR abundance. On the other hand, treatment of NIH3T3 cells with the phorbol ester phorbol 12-myristate 13-acetate (PMA), whose actions are mediated through the activation of protein kinase C, resulted in a time- and dose-dependent down-regulation of the VDR [67]. Up- and down-regulation of the VDR by cAMP and PMA respectively in NIH3T3 cells transfected with a plasmid containing the human osteocalcin VDRE fused to the reporter gene chloramphenicol acetyltransferase (CAT) resulted in a corresponding induction or attenuation in CAT activity. Thus the functional response corresponded to the changes in the VDR.

Treatment of mouse osteoblasts (MC3T3-E1 cells) and rat osteosarcoma (UMR-106-01) cells with forskolin or parathyroid hormone was also found to result in VDR up-regulation, and treatment with PMA was found to result in VDR down-regulation [66], similar to the results observed in NIH3T3 cells. Parathyroid hormone was found to enhance the $1,25(OH)_2D_3$-mediated induction of 24-hydroxylase mRNA in UMR cells. These findings have clinical implications, since in hyperparathyroid patients an augmented response to $1,25(OH)_2D_3$ may contribute to hypercalcaemia. It should be noted, however, that opposite findings have been reported by others. Reinhardt and Horst have shown parathyroid hormone down-regulates, and activation of protein kinase C up-regulates, the VDR in rat osteosarcoma cells (ROS 17/2.8) [56,68]. They also reported that PMA enhanced the $1,25(OH)_2D_3$ induction of the VDR in these cells [68]. Van Leeuwen et al. [69] reported that in UMR-106-01 osteosarcoma cells, treatment with PMA down-regulated the VDR at 2–6 h but resulted in up-regulation after 24 h. These findings suggest that proliferation state, cell type and stage of differentiation may affect the interaction between $1,25(OH)_2D_3$ and signal transduction pathways. Potentiation rather than inhibition of the effect of $1,25(OH)_2D_3$ by activation

of protein kinase C has also been observed using primary cultures of rat kidney cells [70] and intestinal epithelial cells [71]. In these cells, pharmacological doses of $1,25(OH)_2D_3$ resulted in induction of 24-hydroxylase mRNA. In the presence of PMA the dose–response curve shifted to the left, so that physiological concentrations of $1,25(OH)_2D_3$ were effective. The time course of the response was also shifted to the left. Thus PMA-induced second messenger pathways potentiated and accelerated the effect of $1,25(OH)_2D_3$ on 24-hydroxylase gene expression [70,71]. Potentiation of a response in the presence of both $1,25(OH)_2D_3$ and PMA was also observed by Chang and Prince [72], who reported that $1,25(OH)_2D_3$ is unable by itself to induce phosphorylation of osteopontin in JB6C141.5a epidermal cells or to induce tumorigenic transformation of these cells. However, $1,25(OH)_2D_3$ was found to enhance the synthesis and secretion of phosphorylated osteopontin induced by PMA and to enhance PMA-induced tumorigenic transformation of JB6C141.5a cells.

Collectively, these findings suggest co-operativity between signal transduction pathways and $1,25(OH)_2D_3$ action. The mechanism may involve an effect on the VDR, an effect on the promoter of the target gene and/or an effect on other transcription factors that may be interacting with the VDR to mediate $1,25(OH)_2D_3$-induced transcriptional activation. An interesting area of future research will be the study of the mechanisms and transcription factors involved in the interaction of these pathways.

VITAMIN D-MEDIATED TRANSCRIPTIONAL REGULATION

The genomic mechanism of $1,25(OH)_2D_3$ action involves the direct interaction of the VDR with DNA sequences. Although over 50 genes have been reported to be regulated by $1,25(OH)_2D_3$, only a small number have been reported to contain VDREs (Table 1). Although based on a very limited number of natural VDREs, in general the VDRE has been reported to consist of two direct imperfect repeats of the nucleotide sequence GGGTGA separated by three nucleotide pairs. The genes for osteocalcin, osteopontin and most recently 24-hydroxylase have provided the most information concerning transcriptional activation by $1,25(OH)_2D_3$. Although the exact function of osteocalcin is not known, its induction is positively correlated with new bone formation [73], whereas osteopontin promotes attachment of osteoclasts to the bone surface [74]. 24-Hydroxylase is thought to be the enzyme involved in the first step in the catabolism of $1,25(OH)_2D_3$ [75].

Both the human and rat osteocalcin VDREs have been well characterized [76–86]. Besides a VDRE, an AP-1 consensus sequence (TGACTCA) closely juxtaposed to the VDRE has been identified in the human osteocalcin promoter [78]. Ozono

Table 1 VDREs present in vitamin D-regulated genes

See the text for further details.

Gene product	VDRE sequence
Rat 24-hydroxylase	AGGTGA gtg AGGGCG (-151 to -137)
	CGCACC cgc TGAACC (-259 to -245)
Mouse osteopontin	GGTTCA cga GGTTCA (-757 to -743)
Human osteocalcin	GGGTGA acg GGGGCA (-499 to -485)
Rat osteocalcin	GGGTGA atg AGGACA (-455 to -441)
Mouse calbindin D_{28k}	GGGGGA tgtg AGGAGA (-198 to -183)
Rat calbindin D_{9k}	GGGTGT cgg AAGCCC (-489 to -475)
Avian integrin β_3	GAGGCA gaa GGGAGA (-770 to -756)

et al. [78] reported that the AP-1 site synergistically enhanced activation by 1,25(OH)$_2$D$_3$. However, expression of c-*jun* and c-*fos* in ROS 17/2.8 cells has been reported to reduce both the basal activity of the osteocalcin gene and the response to 1,25(OH)$_2$D$_3$ [86]. It has been proposed that the rat osteocalcin VDRE may also contain an AP-1 element within the VDRE [80] and that AP-1 proteins may function to induce or inhibit osteocalcin gene transcription, perhaps depending on the state of differentiation of the cell. Another well characterized VDRE, the mouse osteopontin VDRE, does not contain an AP-1 sequence [87]. Whether other vitamin D-responsive genes show similar cross-communication between the actions of jun and fos heterodimers remains an interesting area of investigation which may provide further insight concerning the inter-relationship between second messenger systems and steroid-receptor-mediated transcription.

Besides an AP-1 element, a glucocorticoid-responsive element has also been identified overlapping the TATA box in both the human and rat osteocalcin genes. It has been suggested that this element represses the inductive effect of 1,25(OH)$_2$D$_3$ [77,88].

Most recently, VDREs have been identified in the rat 24-hydroxylase gene. Initial studies seemed to give conflicting results. The studies of Zierold et al. [89] indicated a VDRE located between nucleotides -262 and -238. However, Ohyama et al. [90] and Hahn et al. [91] reported that a VDRE at positions -150 to -136 in the promoter of the rat 24-hydroxylase gene was important in mediating 1,25(OH)$_2$D$_3$-dependent transcription. This apparent conflict was resolved by recent findings by Jurutka et al. [92] and Ozono et al. [93], who reported that the 24-hydroxylase gene is actually the first vitamin D-stimulated gene to be controlled by two independent VDREs (at -264 to -238 and -150 to -136). Studies by Ozono et al. [93] suggested that the proximal VDRE has more of an effect on 1,25(OH)$_2$D$_3$ inducibility than the distal VDRE. Recently Chen and DeLuca [94] reported that two VDREs are also present in the promoter of the human 24-hydroxylase gene. In intestine and kidney, the most pronounced effect of 1,25(OH)$_2$D$_3$ is increased synthesis of 24-hydroxylase and of the vitamin D-dependent calcium-binding protein calbindin [95]. Unlike calbindin, which is only modestly transcriptionally responsive to 1,25(OH)$_2$D$_3$ [96,97] [suggesting that the large induction in calbindin mRNA by 1,25(OH)$_2$D$_3$ may be due primarily to post-transcriptional mechanisms], 24-hydroxylase is strongly responsive to 1,25(OH)$_2$D$_3$ at the level of transcription. Further studies using the 24-hydroxylase gene will be important in order to understand other factors that may be involved in regulating vitamin D metabolism.

The integrin $\alpha_v\beta_3$ has been reported to be expressed in the osteoclast plasma membrane. Since integrin $\alpha_v\beta_3$ binds to osteopontin and bone sialoprotein through the amino acid sequence RGD (arginine, glycine, aspartic acid), and since antibodies against $\alpha_v\beta_3$ block bone resorption, it has been suggested that this integrin has an important role in the bone resorptive process [98,99]. 1,25(OH)$_2$D$_3$ transcriptionally activates α_v and β_3 integrin genes, and recently a VDRE has been reported in the avian β_3 integrin gene (at positions -770 to -756). As for the genes encoding calbindin D$_{28k}$ (the 28 kDa form of calbindin D) [96] and calbindin D$_{9k}$ [97], the magnitude of the 1,25(OH)$_2$D$_3$-enhanced transcription is modest.

The first demonstration of a negative VDRE was the report by Demay et al. [100] of sequences in the human parathyroid hormone gene (positions -125 to -101) that mediate transcriptional repression in response to 1,25(OH)$_2$D$_3$. Unlike the other VDREs which are involved in induction of transcription by 1,25(OH)$_2$D$_3$, only a single-copy motif (AGGTTCA) was noted within this region. The sequences contained in the 25 bp oligo-nucleotide mediated transcriptional repression in GH4C1 cells but not in ROS 17/2.8 cells, suggesting the requirement of tissue-specific cellular factors in addition to the VDR for 1,25(OH)$_2$D$_3$-mediated transcription. In addition to the human parathyroid hormone gene, VDREs were also reported to be localized in the bovine parathyroid hormone gene (AGGTTA at -461 to -456 and AGTTCC at -449 to -444) [101]. However, these findings were obtained using South-western analysis and gel shift assays. Whether these sequences mediate transcriptional repression in response to 1,25(OH)$_2$D$_3$ remains to be determined.

In addition to regulating calcium homeostasis, one of the other important roles of 1,25(OH)$_2$D$_3$ is its effect on the immune system. Activation and proliferation of T-cells, resulting in the secretion of cytokines, are important steps in the initiation of the immune response. 1,25(OH)$_2$D$_3$ inhibits T-lymphocyte proliferation and the expression of interleukin-2 [102]. Recent studies by Alroy et al. [103] provided a mechanism for the repression by 1,25(OH)$_2$D$_3$ of interleukin-2. They reported that the VDR can bind to an important positive regulatory element of the interleukin-2 enhancer NFAT-1 (nuclear factor of activated T-cells), which is bound by the T-cell-specific transcription factor NFATp as well as by AP-1. Alroy et al. [103] showed that a VDR/retinoid X receptor (RXR) heterodimer can block NFATp–AP-1 complex-formation by inhibiting the interaction between NFATp and Jun/Fos and stably associating with the NFAT-1 element, providing for the first time a mechanism by which 1,25(OH)$_2$D$_3$ can act as an immunosuppressive agent.

VDR HOMODIMERIZATION VERSUS HETERODIMERIZATION

A number of early reports indicated that, for the VDR to bind to DNA, a nuclear accessory factor is required [104–108]. The RXR has been reported to be closely related and perhaps identical to this nuclear accessory factor. It was reported that highly purified VDR derived from baculovirus or yeast systems [105,109] or *in vitro* translated VDR [104] was unable to interact directly with VDREs, suggesting that the VDR is unable to form natural homodimers. However, more recently Carlberg et al. [110], using MCF-7 breast cancer cells and SL-3 *Drosophila* cells, reported that the mouse osteopontin VDRE (a perfect direct repeat of the motif GGTTCA spaced by three nucleotides) is preferentially activated by VDR/RXR heterodimers, whereas the human osteocalcin VDRE, an AP-1 site followed by inexact direct repeats (GGTGACTCAccGGGTGAacgGGGGCA), was found to confer vitamin D inducibility by the VDR alone, possibly as a VDR homodimer. In addition, using gel shift assays, Carlberg et al. [110] reported that, for the osteopontin VDRE, a prominent shifted complex is observed in the presence of both VDRs and RXRs. In contrast, for the osteocalcin VDRE, VDR translated *in vitro* showed a shifted complex which was not enhanced by the addition of RXR. Thus the authors concluded that, unlike the results observed with the osteopontin VDRE, the RXR does not increase the affinity of the VDR for the osteocalcin VDRE. These findings are controversial, since Nishikawa et al. [111,112] reported, on using gel shift assays, that purified VDRs could bind to the osteopontin VDRE in the absence of RXRs but that RXRs were required for binding to the osteocalcin VDRE. Freedman et al. [113] also reported that VDR monomers can bind to the mouse osteopontin VDRE. Further studies from Freedman's lab [114] indicated that the regions within the VDR DNA-binding domain that confer selectivity for the osteopontin VDRE are at the tip of the first zinc finger module and at the N- and C-termini of the second zinc finger module. Thus, unlike the work of Carlberg et al. [110], these studies suggest that the VDR alone can bind to the mouse osteopontin VDRE.

Additional controversy exists concerning the physiological relevance of VDR homodimers. In a recent study, Cheskis and Freedman [115] reported that the VDR exists as a monomer in solution and that homodimerization occurs upon binding to the osteopontin VDRE. $1,25(OH)_2D_3$ destabilizes homodimer formation, and VDR/RXR heterodimer formation is then favoured. 9-cis-Retinoic acid decreased heterodimer formation. It was suggested, as indicated previously by MacDonald et al. [116] using the osteocalcin VDRE, that 9-cis-retinoic acid drives the equilibrium from the VDR/RXR heterodimer to the RXR homodimer or the interaction of the RXR with other receptors. Thus the intracellular ligand concentration is important in regulating the response. The above results suggest that the heterodimer is the functional transactivating species. In recent studies by Nakajima et al. [117] a number of VDR mutants were constructed. None of the mutants lacking the capability to form a heterodimer with the RXR showed $1,25(OH)_2D_3$-dependent transcriptional activation, thus further suggesting the importance of heterodimerization for VDR transcriptional activation. The studies indicating the importance of heterodimerization were carried out using known VDREs. It is possible in future studies that VDR homodimers may be shown to have a functional role in VDR-mediated transcription of target genes yet to be identified.

Due to the realization of the importance of the interaction between the VDR and the RXR, attempts have been made through site-directed mutagenesis to map specific regions of the VDR that are involved in heterodimerization. hVDR residues between Cys-403 and Ser-427 were found to be required for ligand binding and transactivation, but not heterodimerization [117]. Regions in the C-terminus between Lys-382 and Arg-402 were suggested to have important roles in heterodimerization [117], as were regions between Phe-244 and Leu-262 [118,119]. Site-directed mutagenesis studies suggested that Phe-244, Leu-254, Gln-259 and Leu-262, but not Lys-246, of the VDR interact with an RXR isoform on DNA [118]. These important studies are suggestive, at this time, of contact points. Complete understanding of the interaction of the VDR with the RXR will be obtained after crystallographic analysis of the VDR, which will allow the visualization of the three-dimensional contacts.

In addition to the RXR, it has been suggested that the RAR and the thyroid hormone receptor can also heterodimerize with the VDR [120–122]. Using the region of the human osteocalcin VDRE (positions -510 to -492; GGTGACTCAccGGGT-GAac), Schrader et al. [121] reported binding of this element to a VDR/RAR heterodimer. However, MacDonald et al. [116], using the rat osteocalcin VDRE, showed that although RXRs can readily heterodimerize with VDRs, RARs do not. Concerning the possibility of VDR/thyroid hormone receptor heterodimerization, Schrader et al. [122] reported, using rat calbindin D_{9k} or mouse calbindin D_{28k} VDRE/thymidine kinase (tk) promoter/CAT reporter constructs transfected into human MCF-7 cells (which contain endogenous VDRs and thyroid hormone receptors), that either $1,25(OH)_2D_3$ or thyroid hormone could stimulate gene transcription. Stimulation by both hormones resulted in stimulation of CAT activity which was less than additive. In other studies, Schrader et al. [121] used the mouse osteopontin VDRE/tk/CAT construct or the -514 to -495 region of the human osteocalcin VDRE fused to tk/CAT co-transfected into *Drosophila* SL-3 cells transfected with VDRs and thyroid hormone receptors. They found that there was induction by thyroid hormone alone using both reporter constructs, and enhanced induction in the presence of both

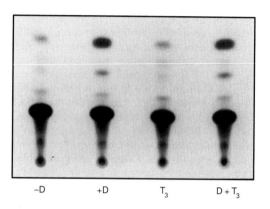

Figure 3   Activation of VDR-mediated transcription by $1,25(OH)_2D_3$ but not by 3,5,3′-tri-iodothyronine (T_3)

Human MCF-7 cells were transfected with a CAT reporter construct containing the rat osteocalcin VDRE [42] and treated with 10 nM $1,25(OH)_2D_3$ (D), 100 nM T_3 or a combination of $1,25(OH)_2D_3$ and T_3. CAT activity was assayed 16 h after stimulation with ligand(s). Three independent experiments showed similar results. Note that neither an activation with T_3 nor an enhanced response in the presence of both ligands was observed.

$1,25(OH)_2D_3$ and thyroid hormone using the -510 to -492 region of the human osteocalcin VDRE. These results suggested a VDR/thyroid hormone receptor signalling pathway. However, in MCF-7 cells transfected with natural calbindin D_{9k} promoter/ CAT constructs containing nucleotides -1009 to $+61$ or -590 to $+61$ (both containing the calbindin D_{9k} VDRE; -490 to -472) (M. Raval-Pandya and S. Christakos, unpublished work) or the rat osteocalcin VDRE (-457 to -430)/tk/CAT construct (Figure 3), neither activation by thyroid hormone nor an enhanced response in the presence of both ligands was observed. Thus further work is needed to examine whether the RAR and the thyroid hormone receptor are indeed nuclear accessory factors for the VDR which may participate in activation of certain VDREs.

OTHERS FACTORS INVOLVED IN VDR-MEDIATED TRANSACTIVATION

Chicken ovalbumin upstream promoter transcription factor (COUP-TF)

COUP-TF, initially characterized by its binding to the chicken ovalbumin upstream promoter, is a member of the steroid/ thyroid hormone nuclear receptor superfamily which is expressed ubiquitously [123]. COUP-TF shares sequence identity with other members of the steroid receptor family, including the presence of a double zinc finger DNA-binding domain. No ligand for COUP-TF has been identified, however. Thus it is classified as an orphan member of the family of steroid hormone receptors. Studies by Cooney et al. [124,125] have shown inhibition by COUP-TF of VDR-mediated transcription through the natural human osteocalcin VDRE. The mechanism of this repression was reported not to involve the formation of a functionally inactive VDR/COUP heterodimer, but rather to involve active silencing of transcription, direct competition with the VDR for binding to the VDRE and heterodimer formation with the RXR [125]. COUP-TF was also shown to repress thyroid hormone receptor- and RAR-mediated transcriptional activation [124,125]. The authors suggest that COUP-TF may play a master role in regulating transactivation by these steroid receptors. They propose that the levels of both COUP-TF and the steroid receptor will determine the overall magnitude of hormone-induced gene expression.

TFIIB

Transcription factors including steroid hormone receptors that bind to specific response elements need to interact with the basal transcription complex in order for transcription to be initiated. Recent studies have begun to address the mechanisms involved in VDR-mediated transcription following binding of the VDR/RXR heterodimer to DNA [126,127]. Initiation of basal transcription involves binding of TFIID [TATA box binding protein and factors associated with it known as TAFs] to the TATA element followed by the association of other factors (TFIIA, TFIIB, TFIIE, TFIIF, TFIIH and RNA polymerase II) with the complex [128]. Blanco et al. [126] reported that co-transfection of the VDR and TFIIB in P19 embryonal carcinoma cells activated 1,25(OH)$_2$D$_3$-dependent rat osteocalcin VDRE/CAT activity. In NIH3T3 cells, co-transfection with TFIIB resulted in repression of 1,25(OH)$_2$D$_3$-mediated transcription using the rat osteocalcin VDRE, suggesting that TFIIB and the VDR interact but this interaction is modulated physically and functionally by cell-type-specific factors. Additional studies by MacDonald et al. [127], using a yeast two-hybrid protein interaction assay, showed that the C-terminal ligand-binding domain of the VDR interacts with the N-terminal domain in TFIIB. TFIIB did not interact with similar regions of the RXR. These studies suggest that the VDR contacts the preinitiation complex through TFIIB and that TFIIB may play a major role in mediating transcription following VDR/RXR heterodimerization.

Calreticulin

Recent preliminary results have indicated that calreticulin, a calcium-binding protein in the endoplasmic reticulum and the nucleus, inhibits the formation of the VDR/RXR heterodimeric complex and blocks 1,25(OH)$_2$D$_3$-mediated transactivation of β_3 integrin [129] ($\alpha_v\beta_3$ integrin may mediate the attachment of osteoclasts to the bone matrix). These results suggested that calreticulin may be a modulator of 1,25(OH)$_2$D$_3$-mediated actions on osteoclast function. It is likely that in future studies new advances will be made in the identification of additional accessory factors as well as the mechanisms involved in transcriptional regulation beyond VDR/RXR binding to the VDRE.

NEWLY IDENTIFIED VITAMIN D-DEPENDENT GENES IN CLASSICAL TARGET TISSUES

Effects of 1,25(OH)$_2$D$_3$ in bone

Although 1,25(OH)$_2$D$_3$ is known to affect bone growth and mineralization and to increase intestinal calcium and phosphate absorption, the detailed mechanism of action of 1,25(OH)$_2$D$_3$ in these two classical target tissues has not been clearly understood. Evidence suggests that the anti-rachitic action of 1,25(OH)$_2$D$_3$ is indirect and is achieved by the effect of the hormone on increased intestinal absorption of calcium and phosphorus, thus resulting in increased availability of these minerals for incorporation into bone [130]. However, *in vitro* studies have indicated that 1,25(OH)$_2$D$_3$ can resorb bone [131]. Receptors for 1,25(OH)$_2$D$_3$ are not present in the bone-resorbing osteoclasts, however, but rather in osteoblasts [2]. It has been suggested that 1,25(OH)$_2$D$_3$ treatment may release a factor from osteoblasts that is responsible for stimulating osteoclast activity [132]. In addition, studies have indicated that immature progenitors can affect differentiation into osteoclasts in the presence of osteoblastic cells or bone

marrow-derived stromal cells [133]. Thus the increase in osteoclasts in the presence of 1,25(OH)$_2$D$_3$ may be an effect of the hormone on differentiation. Whether osteoblastic stromal cells are required for differentiation has been a matter of debate. Studies from Suda's group, which favours the involvement of accessory cells, have recently indicated that 1,25(OH)$_2$D$_3$ can up-regulate the third component of complement (C3) in murine osteoblastic cells and in bone marrow-derived stromal cells [134,135], as well as in bone *in vivo* [136]. Thus a new target gene in bone, in addition to the osteoblast proteins osteopontin and osteocalcin, whose regulation by 1,25(OH)$_2$D$_3$ has been studied in detail [76–86], has been identified. The up-regulation by 1,25(OH)$_2$D$_3$ of C3 in bone was found to be tissue-specific. *In vitro* studies indicated that the regulation of C3 was at the transcriptional level [135]. An antibody against C3 inhibited osteoclast formation *in vitro*, suggesting the involvement of C3, an important factor in the immune response, in 1,25(OH)$_2$D$_3$-mediated osteoclastic bone resorption.

In addition, coincident with the role of 1,25(OH)$_2$D$_3$ in the differentiation of cells towards functional osteoclasts, Billecocq et al. [137] found that 1,25(OH)$_2$D$_3$ increases the expression of carbonic anhydrase II protein and mRNA in bone marrow mononuclear cells. Carbonic anhydrase, which results in the formation of protons and bicarbonate from carbon dioxide and water, is expressed at high levels in the osteoclast and is involved in the process of bone resorption [138]. Billecocq et al. [137] also noted the need for accessory cells, since bone resorption in the presence of 1,25(OH)$_2$D$_3$ was only observed with cultures of total bone marrow including stromal cells. In order to begin to address the mechanisms involved in the induction of carbonic anhydrase by 1,25(OH)$_2$D$_3$, in preliminary studies binding sites on the chicken carbonic anhydrase II promoter were noted for c-Fos, JunD and EGR1 (early growth response gene). It was suggested that the up-regulation of carbonic anhydrase II transcription by 1,25(OH)$_2$D$_3$ may be secondary to a primary effect of 1,25(OH)$_2$D$_3$ on these transcription factors [139].

In addition to up-regulation by 1,25(OH)$_2$D$_3$, down-regulation of the Id gene in osteoblastic (ROS 17/2.8) cells has been reported [140]. Id is a member of the helix–loop–helix family of proteins that regulate differentiation and is classified as a ubiquitous suppressor [141]. As indicated by nuclear run-on assays, 1,25(OH)$_2$D$_3$ was found to down-regulate, by up to 80%, the transcription of the Id gene. Whether a VDRE is present in the Id gene or whether similar findings are observed in other osteoblastic cells has not as yet been determined. The authors suggest that Id could be a major determinant in osteoblastic differentiation.

The regulation of C3, carbonic anhydrase and Id by 1,25(OH)$_2$D$_3$ is of interest, since it suggests novel avenues of investigation related to the genomic effects of 1,25(OH)$_2$D$_3$ on bone. The exact roles of C3, carbonic anhydrase and Id, the significance of their regulation in mediating the effects of 1,25(OH)$_2$D$_3$ on bone and the mechanisms involved in 1,25(OH)$_2$D$_3$-dependent regulation remain to be determined.

Effects of 1,25(OH)$_2$D$_3$ in the intestine

One of the most extensively studied effects of 1,25(OH)$_2$D$_3$ is the stimulation of intestinal calcium absorption [142,143]. However, the exact mechanisms involved in this process have still not been definitively determined. It has been suggested that the intestinal calcium absorptive process is affected by 1,25(OH)$_2$D$_3$ in three phases. The first phase involves calcium transfer into the cell, and may not be dependent on genomic actions of 1,25(OH)$_2$D$_3$. The second phase responds more slowly to 1,25(OH)$_2$D$_3$ and involves the movement of calcium through the cell interior. It has been suggested that the vitamin D-induced calcium-binding protein, calbindin, reported to be a facilitator of calcium diffusion [144], has a role in this process [145,146]. The third phase involves calcium extrusion from the cell and involves calcium transport against a concentration gradient. The intestinal plasma membrane calcium pump protein and mRNA in vitamin D-deficient rats and chicks have been shown to be stimulated by $1,25(OH)_2D_3$, suggesting that the effect of $1,25(OH)_2D_3$ on intestinal calcium absorption may be mediated, at least in part, by a direct effect on calcium pump expression [147–150]. Although $1,25(OH)_2D_3$ has been reported to affect the transcription of the intestinal plasma membrane calcium pump gene [150], the mechanism involved remains to be determined. Studies concerning the inter-relationship between the vitamin D endocrine system and the calcium pump will be an important area of future research which will contribute to a better understanding of the process of $1,25(OH)_2D_3$-mediated intestinal calcium absorption.

EFFECTS OF $1,25(OH)_2D_3$ IN NON-CLASSICAL TARGET TISSUES

Besides bone and intestine, as indicated in the Introduction section $1,25(OH)_2D_3$ affects numerous other systems. Since $1,25(OH)_2D_3$ receptors have been identified in these non-classical target tissues, it has been suggested that the actions of $1,25(OH)_2D_3$ are mediated, at least in part, by genomic mechanisms. A well characterized action of $1,25(OH)_2D_3$ in a number of normal and malignant cells is its ability to inhibit proliferation and to stimulate differentiation [2,3]. The effect of $1,25(OH)_2D_3$ on the inhibition of proliferation and the stimulation of differentiation of keratinocytes is of interest, and has been related to the effective treatment of skin lesions found in psoriasis, a disease of abnormal growth of the epidermis, with $1,25(OH)_2D_3$ or its side-chain analogue MC903 [13,14,151,152]. However, the molecular basis of the effect of $1,25(OH)_2D_3$ on keratinocyte differentiation, which may involve an interaction with calcium [153,154], is not clearly understood.

In addition to its effects on the differentiation of keratinocytes, $1,25(OH)_2D_3$ has been found to inhibit proliferation and induce differentiation of leukaemia cells [3] and to inhibit the proliferation of a number of other malignant cells, including breast, prostate and colon cancer cells [155–158]. A very active area of current research is the development of analogues of $1,25(OH)_2D_3$ which inhibit growth and induce differentiation but do not affect serum calcium [18]. New insight concerning the mechanism involved in the effect of $1,25(OH)_2D_3$ on the differentiation of leukaemic cells into monocytes/macrophages was recently provided by Liu et al. [159]. They found that the cyclin-dependent kinase inhibitor p21, which is involved in blocking cell cycle progression, is transcriptionally induced in the myelomonocytic U937 cell line. A functional VDRE was identified in the human p21 promoter (AGGGAGattGGTTCA; −779 to −765). Transient overexpression of p21 in U937 cells in the absence of $1,25(OH)_2D_3$ resulted in the expression of monocyte/macrophage specific markers, suggesting a key role for p21 in the $1,25(OH)_2D_3$-mediated differentiation of leukaemic cells.

$1,25(OH)_2D_3$ has also been reported to affect the secretion of a number of hormones. It is known that $1,25(OH)_2D_3$ inhibits the secretion and synthesis of parathyroid hormone [1]. As indicated previously in this review, a negative VDRE has been identified in the promoter of the parathyroid hormone gene which may mediate the transcriptional repression in response to $1,25(OH)_2D_3$ [100]. In the pituitary, $1,25(OH)_2D_3$ has been

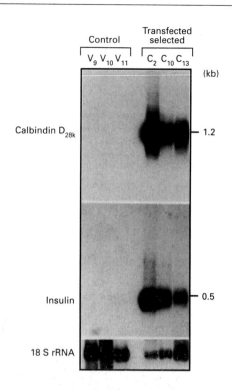

Figure 4   Northern analysis of islet β-cell calbindin D_{28k} and insulin mRNA in control and calbindin D_{28k}-transfected cells

Control cells were transfected with vector (V) alone. Results with positive clones C_2, C_{10} and C_{13} of calbindin D_{28k}-transfected cells are shown. Insulin mRNA is enhanced when calbindin D_{28k} is transfected and overexpressed in the islet β-cell line RIN 1046-38.

reported to enhance the agonist-induced secretion of both prolactin and thyroid-stimulating hormone [160,161]. In addition to affecting hormone secretion from the parathyroid and pituitary glands, $1,25(OH)_2D_3$ has also been reported to enhance pancreatic insulin secretion [162–164]. The pancreas was the first non-classical target tissue reported to possess receptors for $1,25(OH)_2D_3$ [165]. Although $1,25(OH)_2D_3$ has been reported to improve insulin secretion in vitamin D-deficient animals, it has been suggested that impairment of islet secretory function in such animals may be due to the hypocalcaemia, poor growth and inanition characteristic of these animals [163,166,167]. This is still a matter of debate, however. It is possible that $1,25(OH)_2D_3$ may act together with calcium to control β-cell function. A role for the vitamin D-dependent calcium-binding protein calbindin D_{28k} (which is present in islets in addition to kidney, intestine and brain) in insulin secretion and expression has been suggested [168]. When calbindin D_{28k} was transfected and overexpressed in the islet β-cell line RIN 1046-38, which contains receptors for $1,25(OH)_2D_3$, a marked induction (6–35-fold) of insulin secretion and mRNA expression was observed (Figure 4), suggesting a direct role for calbindin in insulin biosynthesis and secretion. The inter-relationship between $1,25(OH)_2D_3$, calbindin and insulin secretion remains to be resolved.

$1,25(OH)_2D_3$ has also been reported to have effects on the immune system. It suppresses T-cell proliferation and the expression of cytokines such as interleukin-2, as mentioned earlier, and interferon-γ [8,102,169]. This effect is in contrast with the effect of $1,25(OH)_2D_3$ to enhance macrophage phagocytic activity [169]. The effects of $1,25(OH)_2D_3$ on the immune system may be subtle, since vitamin D deficiency is not associated with major

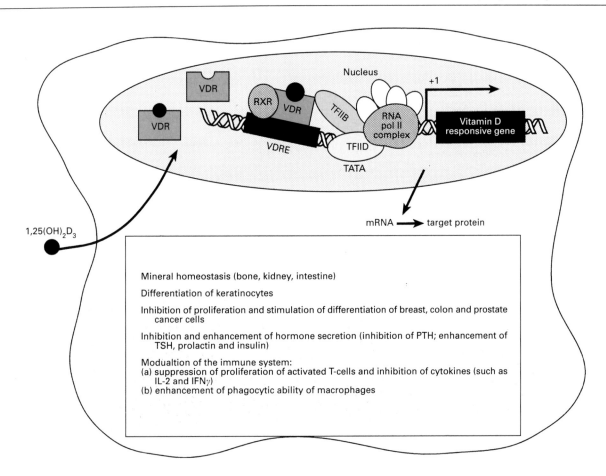

Figure 5   Genomic mechanism of action of 1,25(OH)₂D₃ in target cells

See the text for discussion of VDR/RXR heterodimerization and of the suggested interaction of TFIIB and the VDR. 1,25(OH)₂D₃ is known to affect mineral homeostasis, to promote differentiation of keratinocytes and leukaemia cells and to affect the immune system. Suggested roles for 1,25(OH)₂D₃ in the enhancement of hormone secretion and for differentiation of other malignant cells are also included in the model. Abbreviations: pol, polymerase; PTH, parathyroid hormone; TSH, thyroid-stimulating hormone; IL-2, interleukin-2; IFNγ, interferon γ.

deficiencies in the immune system. It has been suggested, however, that analogues of 1,25(OH)₂D₃ may have therapeutic potential when given in combination with classical immunosuppressive drugs.

NEW DIRECTIONS OF VITAMIN D RESEARCH

Exactly how 1,25(OH)₂D₃ acts to produce these effects on differentiation, proliferation, hormone secretion and the immune system is not known. The elucidation of the mechanisms involved will be an active area of further research. 1,25(OH)₂D₃ may mediate these effects, at least in part, by genomic mechanisms (Figure 5). New target genes and novel VDREs will undoubtedly be identified which should expand our understanding of the sequences involved in vitamin D-mediated genomic mechanisms. It is becoming increasingly evident that sequences divergent from the consensus response element may be physiologically important. New insight may also be obtained concerning different transcription factors which are involved in mediating these diverse biological responses. Although studies concerning vitamin D regulation have focused on transcriptional mechanisms, it is likely that post-transcriptional regulation by 1,25(OH)₂D₃ will be an important mechanism of control of a number of the newly identified target genes.

In addition to studies related to responsive elements and transcription factors, which have been a major focus of research in the vitamin D field in the past 5 years, studies related to the functional significance of target proteins in different systems are needed in the future. The use of transfected cells, knockout mice or transgenic mice that overexpress a particular vitamin D-dependent protein should result in a better understanding of the physiological significance of target proteins. It will be of interest, for example, to examine physiological abnormalities in VDR, calbindin, osteocalcin and 24-hydroxylase knockout mice. Such *in vivo* studies, combined with continuing studies concerning the basic molecular mechanism of 1,25(OH)₂D₃ action, will provide new insight into the role of 1,25(OH)₂D₃ in calciotropic and non-calciotropic target tissues.

We acknowledge the secretarial assistance of Ms. Connie Sheffield and Ms. Valerie Brooks, and the art work of Mr. Willy Kratel. Helpful discussions with Terri Towers and Dr. Leonard Freedman are also gratefully acknowledged.

REFERENCES

1   Darwish, H. and DeLuca, H. F. (1993) Crit. Rev. Eukaryotic Gene Expression **3**, 89–116

2   Suda, T., Shinki, T. and Takahashi, N. (1990) Ann. Rev. Nutr. **10**, 195–211

3   Suda, T. (1989) Proc. Soc. Exp. Biol. Med. **191**, 214–220

4   Kadowaki, S. and Norman, A. W. (1984) J. Clin. Invest. **73**, 759–766

5   D'Emden, M. C. and Wark, J. D. (1987) Endocrinology (Baltimore) **121**, 1192–1194

6   Wark, J. C. and Tashjian, Jr., A. H. (1982) Endocrinology (Baltimore) **111**, 1755–1757

7   Tsoukas, C. D., Provvedini, D. and Manolagas, S. C. (1984) Science **224**, 1438–1440

8   Manolagas, S. C., Yu, X. P., Girasole, G. and Bellido, T. (1994) Semin. Nephrol. **14**, 129–143

9   Henry, H. L. and Norman, A. W. (1974) J. Biol. Chem. **249**, 7529–7535

10 Weishaar, R. E. and Simpson, R. U. (1989) Endocr. Rev. **10**, 351–365

11 Walters, M. R. (1992) Endocr. Rev. **13**, 719–764

12 Walters, M. R. (1995) Endocr. Rev. Monogr. **4**, 47–55

13 Bikle, D. D. (1992) Endocr. Rev. **13**, 765–784

14 Bikle, D. D. (1995) Endocrin. Rev. Monogr. **4**, 77–83

15 Hannah, S. S. and Norman, A. W. (1994) Nutr. Rev. **52**, 376–382

16 Farach-Carson, M. C., Sergeev, I. and Norman, A. W. (1991) Endocrinology (Baltimore) **129**, 1876–1884

17 McLaughlin, J. A., Cantley, L. C. and Holick, M. F. (1990) J. Nutr. Biochem. **1**, 81–87

18 Bouillon, R., Okamura, W. H. and Norman, A. W. (1995) Endocr. Rev. **16**, 200–257

19 Baker, A. R., McDonnell, D. P., Hughes, M., Crisp, T. M., Mangelsdorf, D. J., Haussler, M. R., Pike, J.W, Shine, J. and O'Malley, B. W. (1988) Proc. Natl. Acad. Sci. U.S.A. **85**, 3294–3298

20 Burmester, J. K., Maeda, N. and DeLuca, H. F. (1988) Proc. Natl. Acad. Sci. U.S.A. **85**, 1005–1009

21 Pike, J. W., Kesterson, R. A., Scott, R. A., Kerner, S. A., McDonnell, D. P. and O'Malley, B. W. (1988) in Vitamin D: Molecular, Cellular and Clinical Endocrinology (Norman, A. W., Schaefer, K., Grigoleit, H. G. and Herrath, D. V. eds.), pp. 215–224, Walter de Gruyter, New York

22 Peleg, S., Pike, J. W. and O'Malley, B. W. (1988) in Vitamin D: Molecular, Cellular and Clinical Endocrinology (Norman, A. W., Schaefer, K., Grigoleit, H. G. and Herrath, D. V. eds.), pp. 242–243, Walter de Gruyter, New York

23 McDonnell, D. P., Mangelsdorf, D. J., Pike, J. W., Haussler, M. R. and O'Malley, B. W. (1987) Science **235**, 1214–1217

24 Pike, J. W. (1991) Annu. Rev. Nutr. **11**, 189–216

25 MacDonald, P. N., Dowd, D. R. and Haussler, M. R. (1994) Semin. Nephrol. **14**, 101–118

26 McDonnell, D. P., Scott, R. A., Kerner, S. A., O'Malley, B.W and Pike, J. W. (1989) Mol. Endocrinol. **3**, 635–644

27 Freedman, L. P. (1992) Endocr. Rev. **13**, 129–145

28 Umesono, K., Murakami, K. K., Thompson, C. C. and Evans, R. M. (1991) Cell **65**, 1255–1266

29 Dahlman-Wright, K., Wright, A., Gustafsson, D.-A. and Carlstedt-Duke, J. (1991) J. Biol. Chem. **266**, 3107–3112

30 Luisi, B. F., Xu, W. X., Otwinowski, S., Freedman, L. P., Yamamoto, K. R. and Sigler, P. B. (1991) Nature (London) **352**, 497–505

31 Freedman, L. P. and Towers, T. L. (1991) Mol. Endocrinol. **5**, 1815–1826

32 Pike, J. W. and Sleator, N. M. (1985) Biochem. Biophys. Res. Commun. **131**, 378–385

33 Brown, T. A. and DeLuca, H. F. (1990) J. Biol. Chem. **265**, 10025–10029

34 Jones, B. B., Jurutka, P. W., Haussler, C. A., Haussler, M. R. and Whitfield, G. K. (1991) Mol. Endocrinol. **5**, 1137–1146

35 Sheridan, P. L., Evans, R. M. and Horwitz, K. B. (1989) J. Biol. Chem. **264**, 6520–6528

36 van Laar, J. H., Berrevolts, C. A., Trapman, J., Zegers, N. D. and Brinkmann, A. O. (1991) J. Biol. Chem. **266**, 3734–3738

37 Washburn, T., Hocutt, A., Brautigan, D. L. and Korach, K. S. (1991) Mol. Endocrinol. **5**, 235–242

38 Migliaccio, A., Di Domenico, M., Green, S., de Falco, A., Kajtaniak, E. L., Blasi, F., Chambon, P. and Auricchio, F. (1989) Mol. Endocrinol. **3**, 1061–1069

39 Bodwell, J. E., Orti, E., Coull, J. M., Pappin, D. J. C., Smith, L. I. and Swift, F. (1991) J. Biol. Chem. **266**, 7549–7555

40 Power, R. F., Mani, S. K., Codina, J., Conneely, O. M. and O'Malley, B. W. (1991) Science **254**, 1636–1639

41 Darwish, H. M., Burmester, J. K., Moss, V. E. and DeLuca, H. F. (1993) Biochim. Biophys. Acta **1167**, 29–36

42 Matkovits, T. and Christakos, S. (1995) Mol. Endocrinol. **9**, 232–242

43 Hsieh, J. C., Jurutka, P. W., Galligan, M. A., Terpening, C. M., Haussler, C. A., Samuels, D. S., Shimizu, Y., Shimizu, N. and Haussler, M. R. (1991) Proc. Natl. Acad. Sci. U.S.A. **88**, 9315–9319

44 Jurutka, P. W., Hsieh, J. C., MacDonald, P. N., Terpening, C. M., Haussler, C. A., Haussler, M. R. and Whitfield, G. K. (1993) J. Biol. Chem. **268**, 6791–6799

45 Brown, T. A. and DeLuca, H. F. (1991) Arch. Biochem. Biophys. **286**, 466–472

46 Hilliard, G. M., Cook, R. G., Weigel, N. L. and Pike, J. W. (1994) Biochemistry **33**, 4300–4311

47 Baker, A. R., McDonnell, D. P., Hughes, M. R., Crisp, T. M., Mangelsdorf, D. J., Haussler, M. R., Pike, J. W., Shine, J. O. and O'Malley, B. W. (1988) Proc. Natl. Acad. Sci. U.S.A. **85**, 3294–3298

48 Strom, M., Sandgren, M. E., Brown, T. A. and DeLuca, H. F. (1989) Proc. Natl. Acad. Sci. U.S.A. **86**, 9770–9773

49 Costa, E. M., Hirst, M. A. and Feldman, D. (1985) Endocrinology (Baltimore) **117**, 2203–2210

50 Mangelsdorf, D. J., Pike, J. W. and Haussler, M. R. (1987) Proc. Natl. Acad. Sci. U.S.A. **84**, 354–358

51 Lee, Y., Inaba, M., DeLuca, H. F. and Mellon, W. S. (1989) J. Biol. Chem. **264**, 13701–13705

52 Lee, S., Clark, S. A., Gill, R. K. and Christakos, S. (1994) Endocrinology (Baltimore) **134**, 1602–1610

53 Huang, Y. C., Lee, S., Stolz, R., Gabrielides, C., Pansini-Porta, A., Bruns, M. E., Bruns, D. E., Mifflin, T. E., Pike, J. W. and Christakos, S. (1989) J. Biol. Chem. **264**, 17454–17461

54 Wiese, R. J., Uhland-Smith, A., Ross, T. K., Prahl, J. M. and DeLuca, H. F. (1992) J. Biol. Chem. **267**, 20082–20086

55 Meyer, J., Fullmer, C. S., Wasserman, R. H., Komm, B. S. and Haussler, M. R. (1992) J. Bone Miner. Res. **7**, 441–448

56 Reinhardt, T. A. and Horst, R. L. (1990) Endocrinology (Baltimore) **127**, 942–948

57 Arbour, N. C., Prahl, J. M. and DeLuca, H. F. (1993) Mol. Endocrinol. **7**, 1307–1312

58 Santiso-Mere, D., Sone, T., Hilliard, G. M., Pike, J. W. and McDonnell, D. P. (1993) Mol. Endocrinol. **7**, 833–839

59 Hirst, M. and Feldman, D. (1982) Endocrinology (Baltimore) **111**, 1400–1402

60 Chan, S. D., Chiu, D. K. and Atkins, D. (1984) J. Endocrinol. **103**, 295–300

61 Chen, T. L., Hauschka, P. V. and Feldman, D. (1986) Endocrinology (Baltimore) **118**, 1119–1126

62 Massaro, E. R., Simpson, R. U. and DeLuca, H. F. (1982) J. Biol. Chem. **257**, 13736–13739

63 Lee, S., Szlachetka, M. and Christakos, S. (1991) Endocrinology (Baltimore) **129**, 396–401

64 Krishnan, A. V. and Feldman, D. (1992) Mol. Endocrinol. **6**, 198–206

65 Van Leeuwen, J. P., Pols, H. A. P., Schilte, J. P., Visser, T. J. and Birkenhager, J. C. (1991) Calif. Tissue Int. **49**, 35–42

66 Krishnan, A. V., Cramer, S. D., Bringhurst, F. R. and Feldman, D. (1995) Endocrinology (Baltimore) **136**, 705–712

67 Krishnan, A. V. and Feldman, D. (1991) Mol. Endocrinol. **5**, 605–612

68 Reinhardt, T. A. and Horst, R. L. (1994) Mol. Cell. Endocrinol. **101**, 159–165

69 Van Leeuwen, J. P., Birkenhager, J. C., Buurman, C. J., Van den Bemd, G. J. C. M., Bos, M. P. and Pols, H. A. P. (1992) Endocrinology (Baltimore) **130**, 2259–2266

70 Chen, M. L., Boltz, M. A. and Armbrecht, H. J. (1993) Endocrinology (Baltimore) **132**, 1782–1788

71 Armbrecht, H. J., Hodam, T. L., Boltz, M. A. and Chen, M. L. (1993) FEBS Lett. **327**, 13–16

72 Chang, P. L. and Prince, C. W. (1993) Cancer Res. **53**, 2217–2220

73 Hauschka, P. V., Lian, J. B., Cole, D. E. and Gundberg, C. M. (1989) Physiol. Rev. **69**, 990–1047

74 Denhardt, D. T. and Guo, X. (1993) FASEB J. **7**, 1475–1482

75 Shinki, T., Jin, C. H., Nishimura, A., Nagai, Y., Ohyama, Y., Noshiro, M., Okuda, K. and Suda, T. (1992) J. Biol. Chem. **267**, 13757–13762

76 Kerner, S. A., Scott, R. A. and Pike, J. W. (1989) Proc. Natl. Acad. Sci. U.S.A. **86**, 4455–4459

77 Morrison, N. A., Shine, J., Fragonas, J. C., Verkest, V., McMenemy, M. L. and Eisman, J. A. (1989) Science **246**, 1158–1161

78 Ozono, K., Liao, J., Kerner, S. A., Scott, R. A. and Pike, J. W. (1990) J. Biol. Chem. **265**, 21881–21888

79 Demay, M. B., Gerardi, J. M., DeLuca, H. F. and Kronenberg, H. M. (1990) Proc. Natl. Acad. Sci. U.S.A. **87**, 369–373

80 Owen, T. A., Bortell, R., Yocum, S. A., Smock, S. L., Zhang, M., Abate, C., Shalhoub, V., Aronin, N., Wright, K. L., Van Wijnen, A. J., Stein, J. L., Curran, T., Lian, J. B. and Stein, G. S. (1990) Proc. Natl. Acad. Sci. U.S.A. **87**, 9990–9994

81 Markose, E. R., Stein, J. L., Spin, G. S. and Lian, J. B. (1990) Proc. Natl. Acad. Sci. U.S.A. **87**, 1701–1705

82 Terpening, C. M., Haussler, C. A., Jurutka, P. W., Galligan, M.A, Komm, B. S. and Haussler, M. R. (1991) Mol. Endocrinol. **5**, 373–385

83 Yoon, K. G., Rutledge, S. J., Buenagar, R. F. and Rodan, G. A. (1988) Biochemistry **27**, 8521–8526

84 Owen, T. A., Bortell, R., Shalhaub, V., Heinrichs, A., Stein, J. L., Stein, G. S. and Lian, J. B. (1993) Proc. Natl. Acad. Sci. U.S.A. **90**, 1503–1507

85 Demay, M. B., Kiernan, M. S., DeLuca, H. F. and Kronenberg, H. M. (1992) Mol. Endocrinol. **6**, 557–562

86 Schule, R., Umesono, K., Mangelsdorf, D. J., Bolado, J., Pike, J. W. and Evans, R. M. (1990) Cell **61**, 497–504

87   Noda, M., Vogel, R. L., Craig, A. M., Prahl, J., DeLuca, H. F. and Denhardt, D. (1990) Proc. Natl. Acad. Sci. U.S.A. **87**, 9995–9999

88   Stromstedt, P. E., Poellinger, L., Gustafsson, J. A. and Carlstedt-Duke, J. (1991) Mol. Cell. Biol. **11**, 3379–3383

89   Zierold, C., Darwish, H. M. and DeLuca, H. F. (1994) Proc. Natl. Acad. Sci. U.S.A. **91**, 900–902

90   Ohyama, Y., Ozono, K., Uchida, M., Shinki, T., Kato, S., Suda, T. Yamamoto, O., Noshiro, M. and Kato, Y. (1994) J. Biol. Chem. **269**, 10545–10550

91   Hahn, C. N., Kerry, D.M, Omdahl, J. L. and May, B. K. (1994) Nucleic Acids Res. **22**, 2410–2416

92   Jurutka, P. W., Hsieh, J. C., and Haussler, M. R. (1994) J. Bone Miner. Res. **9** (Suppl. 1), S160

93   Ozono, K., Ohyama, Y., Nakajima, S., Uchida, M., Yoshimura, M., Shinki, T., Suda, T. and Yamamoto, O. (1995) J. Bone Miner. Res. **10** (Suppl. 1), S288

94   Chen, K. S. and DeLuca, H. F. (1995) Biochim. Biophys. Acta **1263**, 1–9

95   Matkovits, T. and Christakos, S. (1995) Endocrinology (Baltimore) **136**, 3971–3982

96   Gill, R. K. and Christakos, S. (1993) Proc. Natl. Acad. Sci. U.S.A. **90**, 2984–2988

97   Darwish, H. M. and DeLuca, H. F. (1992) Proc. Natl. Acad. Sci. U.S.A. **89**, 603–607

98   Medhora, M. M., Teitelbaum, S., Chappel, J., Alvarez, J., Mimura, H., Ross, F. P. and Hruska, K. (1993) J. Biol. Chem. **268**, 1456–1461

99   Cao, X., Ross, F. P., Zhang, L., MacDonald, P. N., Chappel, J. and Teitelbaum, S. L. (1993) J. Biol. Chem. **268**, 27371–27380

100 Demay, M. B., Kiernan, M. S., DeLuca, H. F. and Kronenberg, H. M. (1992) Proc. Natl. Acad. Sci. U.S.A. **89**, 8097–8101

101 Hawa, N. S., O'Riordan, J. L. H. and Farrow, S. M. (1994) J. Endocrinol. **142**, 53–60

102 Bhalla, A. K., Amento, E. P. and Krane, S. M. (1986) Cell. Immunol. **98**, 311–322

103 Alroy, I., Towers, T. L. and Freedman, L. P. (1995) Mol. Cell. Biol. **15**, 5789–5799

104 Liao, J., Ozono, K., Sone, T., McDonnell, D. P. and Pike, J. W. (1990) Proc. Natl. Acad. Sci. U.S.A. **87**, 9751–9755

105 MacDonald, P. N., Haussler, C. A., Terpening, C. M., Galligan, M. A., Reeder, M. C., Whitfield, G. K. and Haussler, M. R. (1991) J. Biol. Chem. **266**, 18808–18813

106 Ross, T. K., Moss, V. E., Prahl, J. M. and DeLuca, H. F. (1992) Proc. Natl. Acad. Sci. U.S.A. **89**, 256–260

107 Sone, T., Ozone, K. and Pike, J. W. (1991) Mol. Endocrinol. **5**, 1578–1586

108 Sone, T., Kerner, S. and Pike, J. W. (1991) J. Biol. Chem. **266**, 23296–23305

109 Sone, T., McDonnell, D. P., O'Malley, B. W. and Pike, J. W. (1990) J. Biol. Chem. **265**, 21997–22003

110 Carlberg, C., Bendik, I., Wyss, A., Meier, E., Sturzenbecker, L. J., Grippo, J. F. and Hunziker, W. (1993) Nature (London) **361**, 657–660

111 Nishikawa, J., Matsumoto, M., Sakoda, K., Kitaura, M., Imagawa, M. and Nishihara, T. (1993) J. Biol. Chem. **268**, 19739–19743

112 Nishikawa, J., Kitaura, M., Matsumoto, M., Imagawa, M. and Nishihara, T. (1994) Nucleic Acids Res. **22**, 2902–2907

113 Freedman, L. P., Arce, V. and Fernandez, R. P. (1994) Mol. Endocrinol. **8**, 265–273

114 Towers, T. L., Luisi, B. F., Asianov, A. and Freedman, L. P. (1993) Proc. Natl. Acad. Sci. U.S.A. **90**, 6310–6314

115 Cheskis, B. and Freedman, L. P. (1994) Mol. Cell. Biol. **14**, 3329–3338

116 MacDonald, P. N., Dowd, D. R., Nakajima, S., Gallligan, M. A., Reeder, M. C., Haussler, C. A., Ozato, K. and Haussler, M. R. (1993) Mol. Cell. Biol. **13**, 5907–5917

117 Nakajima, S., Hsieh, J.-C., MacDonald, P. N., Galligan, M. A., Haussler, C. A., Whitfield, G. K. and Haussler, M. R. (1994) Mol. Endocrinol. **8**, 159–172

118 Whitfield, G. K., Hsieh, J.-C., Nakajima, S., MacDonald, P. N., Thompson, P. D. Jurutka, P. W., Haussler, C. A. and Haussler, M. R. (1995) Mol. Endocrinol. **9**, 1166–1179

119 Rosen, E. D., Beninghof, E. G. and Koenig, R. J. (1993) J. Biol. Chem. **268**, 11534–11541

120 Schrader, M., Bendik, I., Becker-Andre, M. and Carlberg, C. (1993) J. Biol. Chem. **268**, 17830–17836

121 Schrader, M., Muller, K. M. and Carlberg, C. (1994) J. Biol. Chem. **269**, 5501–5504

122 Schrader, M., Muller, K. M., Nayeri, S., Kahlen, J.-P. and Carlberg, C. (1994) Nature (London) **370**, 382–386

123 Wang, L.-H., Ing, N. H., Tsai, S. Y., O'Malley, B. W. and Tsai, M.-J. (1991) Gene Expression **1**, 207–216

124 Cooney, A. J., Tsai, S. Y., O'Malley, B. W. and Tsai, M.-J. (1992) Mol. Cell. Biol. **12**, 4153–4163

125 Cooney, A. J., Leng, X., Tsai, S. Y., O'Malley, B. W. and Tsai, M.-J. (1993) J. Biol. Chem. **268**, 4152–4160

126 Blanco, J. C. G., Wang, I. M., Tsai, S. Y., Tsai, M. J., O'Malley, B. W., Jurutka, P. W., Haussler, M. R. and Ozato, K. (1995) Proc. Natl. Acad. Sci. U.S.A. **92**, 1535–1539

127 MacDonald, P. N., Sherman, D. R., Dowd, D. R., Jefcoat, S. C. and DeLisle, R. K. (1995) J. Biol. Chem. **270**, 4748–4752

128 Buratowski, S. (1994) Cell **77**, 1–3

129 Cao, X., Teitelbaum, S., Ross, P., Onishi, T., Dedhar, S. and Hruska, K. (1995) J. Bone Miner. Res. **10** (Suppl. 1), S156

130 Underwood, J. L. and DeLuca, H. F. (1984) Am. J. Physiol. **246**, E493–E498

131 Raisz, L. G., Trummel, C. L., Holick, M. F. and DeLuca, H. F. (1972) Science **175**, 768–769

132 McSheehy, P. M. J. and Chambers, T. J. (1987) J. Clin. Invest. **80**, 425–429

133 Suda, T., Takahashi, N. and Martin, T. J. (1995) Endocr. Rev. Monogr. **4**, 251–265

134 Sato, T., Hong, M. H., Jin, C. H., Ishimi, Y., Udagawa, N., Shinki, T., Abe, E. and Suda, T. (1991) FEBS Lett. **285**, 21–24

135 Hong, M. H., Jin, C. H., Sato, T., Ishimi, Y., Abe, E. and Suda, T. (1991) Endocrinology (Baltimore) **129**, 2774–2779

136 Jin, C. H., Shinki, T., Hong, M. H., Sato, T., Yamaguchi, A., Ikeda, T., Yoshiki, S., Abe, E. and Suda, T. (1992) Endocrinology (Baltimore) **131**, 2468–2475

137 Billecocq, A., Emanuel, J. R., Levenson, R. and Baron, R. (1990) Proc. Natl. Acad. Sci. U.S.A. **87**, 6470–6474

138 Sly, W. S., Hewett-Emmett, D., Whyte, M. P., Yu, Y. S. and Tashian, R. E. (1983) Proc. Natl. Acad. Sci. U.S.A. **80**, 8752–8756

139 David, J. P., Rincon, M. and Baron, R. (1995) J. Bone Miner. Res. **10** (Suppl. 1), S393

140 Kawaguchi, N., DeLuca, H. F. and Noda, M. (1992) Proc. Natl. Acad. Sci. U.S.A. **89**, 4569–4572

141 Barinaga, M. (1991) Science **251**, 1176–1177

142 Wasserman, R. H. and Fullmer, C. S. (1983) Annu. Rev. Physiol. **45**, 375–390

143 DeLuca, H. F. (1985) Soc. Gen. Physiol. Ser. **39**, 159–176

144 Feher, J. J. (1983) Am. J. Physiol. **244**, C303–C307

145 Christakos, S., Gabrielides, C. and Rhoten, W. B. (1989) Endocr. Rev. **10**, 3–26

146 Christakos, S. (1995) Endocr. Rev. Monogr. **4**, 108–110

147 Zelinski, J. M., Sykes, D. E. and Weiser, M. M. (1991) Biochem. Biophys. Res. Commun. **179**, 749–755

148 Wasserman, R. H., Smith, C. A., Brindak, M. E., DeTalamoni, N., Fullmer, C. S., Penniston, J. T. and Kumar, R. (1992) Gastroenterology **102**, 886–894

149 Cai, Q., Chandler, J. S., Wasserman, R. H., Kumar, R. and Penniston, J. T. (1993) Proc. Natl. Acad. Sci. U.S.A. **90**, 1345–1349

150 Pannabecker, T. L., Chandler, J. S. and Wasserman, R. H. (1995) Biochem. Biophys. Res. Commun. **213**, 499–505

151 Kragballe, K. (1993) in Molecular Biology to Therapeutics (Bernard, B. A. and Schroot, B., eds.), pp. 174–181, Karger, Basel

152 Smith, E. I., Walworth, N. C. and Holick, M. F. (1986) J. Invest. Dermatol. **86**, 709–714

153 Su, M.-J., Bikle, D. D., Mancianti, M.-L. and Pillai, S. (1994) J. Biol. Chem. **269**, 14723–14729

154 Bikle, D. D. and Pillai, S. (1993) Endocr. Rev. **14**, 3–19

155 Frampton, R. J., Osmond, S. A. and Eisman, J. A. (1983) Cancer Res. **43**, 4443–4447

156 Peehl, D. M., Skowronski, R. J., Leung, G. K., Wong, S. T., Stamey, T. A. and Feldman, D. (1994) Cancer Res. **54**, 805–810

157 Shabahany, M., Buras, R. R., Davoodi, F., Schumaker, L. M., Nauta, R. J., Uskokovic, M. R., Brenner, R. V. and Evans, S. R. (1994) Cancer Res. **54**, 4057–4064

158 Zhou, J. Y., Norman, A. W., Lubbert, M., Collins, E. D., Uskokovic, M. R. and Koeffler, H. P. (1990) Proc. Natl. Acad. Sci. U.S.A. **87**, 3929–3932

159 Liu, M., Lee, M.-H., Cohen, M., Bommakanti, M. and Freedman, L. P. (1996) Genes Dev. **10**, 142–153

160 Murdoch, G. H. and Rosenfeld, M. G. (1981) J. Biol. Chem. **256**, 4050–4055

161 d'Emden, M. C. and Wark, J. D. (1989) J. Endocrinol. **121**, 441–450

162 Norman, A. W., Frankel, B. J., Heldt, A. M. and Grodsky, G. M. (1980) Science **209**, 823–825

163 Clark, S. A., Stumpf, W. E. and Sar, M. (1981) Diabetes **30**, 382–386

164 Chertow, B. S., Sivitz, W. I., Baranetsky, N. G., Clark, S. A., Waite, A. and DeLuca, H. F. (1983) Endocrinology (Baltimore) **113**, 1511–1518

165 Christakos, S. and Norman, A. W. (1979) Biochem. Biophys. Res. Commun. **89**, 56–63

166 Tanaka, Y., Seino, Y., Ishida, M., Yamaoka, K., Satomura, K., Yabuuchi, H., Seno, Y. and Imura, H. (1986) Endocrinology (Baltimore) **118**, 1971–1976

167 Beaulieu, C., Kestekian, R., Havrankova, J. and Gascon-Barre, M. (1993) Diabetes **42**, 35–43

168 Reddy, D., Pollock, A. S., Clark, S. and Christakos, S. (1994) J. Bone Miner. Res. **9** (Suppl. 1), S145

169 Rigby, W. F. C., Denome, S. and Fanger, M. W. (1987) J. Clin. Invest. **79**, 1659–1664

Biochem. J. (1996) **317**, 329–342 (Printed in Great Britain)

REVIEW ARTICLE
Multiple steps in the regulation of transcription-factor level and activity

Cor F. CALKHOVEN and Geert AB*
Department of Biochemistry, University of Groningen, Nijenborgh 4, 9747 AG Groningen, the Netherlands.

This review focuses on the regulation of transcription factors, many of which are DNA-binding proteins that recognize *cis*-regulatory elements of target genes and are the most direct regulators of gene transcription. Transcription factors serve as integration centres of the different signal-transduction pathways affecting a given gene. It is obvious that the regulation of these regulators themselves is of crucial importance for differential gene expression during development and in terminally differentiated cells. Transcription factors can be regulated at two, principally different, levels, namely concentration and activity, each of which can be modulated in a variety of ways. The concentrations of transcription factors, as of intracellular proteins in general, may be regulated at any of the steps leading from DNA to protein, including transcription, RNA processing, mRNA degradation and translation. The activity of a transcription factor is often regulated by (de)phosphorylation, which may affect different functions, e.g. nuclear localization, DNA binding and *trans*-activation. Ligand binding is another mode of transcription-factor activation. It is typical for the large super-family of nuclear hormone receptors. Heterodimerization between transcription factors adds another dimension to the regulatory diversity and signal integration. Finally, non-DNA-binding (accessory) factors may mediate a diverse range of functions, e.g. serving as a bridge between the transcription factor and the basal transcription machinery, stabilizing the DNA-binding complex or changing the specificity of the target sequence recognition. The present review presents an overview of different modes of transcription-factor regulation, each illustrated by typical examples.

INTRODUCTION

The conversion of abstract coded biological information stored in DNA into concrete physiologically active proteins, called gene expression, is tightly regulated. In a multicellular organism, all cell types with a few exceptions contain the same genetic information. Yet each cell type expresses only a unique subset of the total number of available genes. Differential gene expression is specified by unique epigenetic information which is present in the particular cell and determines its phenotype [1,2]. For many genes, control at the first step of expression, transcription, is paramount. The transcription profile is actually a convenient parameter for the identification of a particular cell type. Some genes are always turned on in all cells; they form the group of so-called 'housekeeping' genes, which encode structural proteins and enzymes catalysing the reactions of basic metabolism. Other genes are only transcribed in one or a few cell types, usually only during a particular stage of development or under the regime of particular extracellular and/or intracellular signals [2]. Differential gene expression is controlled by a complex regulatory network in which specialized transcription factors relay the signals to specific target genes. Many of these transcription factors are DNA-binding proteins that bind to regulatory DNA elements located *cis* to the target genes.

The levels of the DNA-binding transcription factors, or rather their activities, are decisive as to whether their target genes are transcribed and to what extent. This implies that these regulators

of gene expression in their turn must be tightly regulated. The question arises as to how this is accomplished without the need of an ever-increasing number of upstream regulatory genes. There are several ways by which cells extend the diversity of their regulatory repertoire. One way is to make use of the combinatorial action of a limited set of transcription factors. Another way is to modulate the activity of a transcription factor once it has been synthesized. The various ways by which transcription factor gene expression can be regulated are depicted in Figure 1. As for genes in general, transcription of regulatory genes is the prime level of control and provides the intermediates at which subsequent steps of control can be exerted. These include splicing, which may occur in alternative modes, transport to the cytoplasm and degradation of mRNA. Translation is another important level of control. Selection of alternative start sites may generate functionally distinct protein isoforms. Once the transcription factor has been synthesized, it has to be transported to the nucleus. Masking of the nuclear localization signal (NLS), e.g. by a sequestering protein or by phosphorylation, may hinder the factor from reaching the nuclear compartment. Finally, the functions that determine the transcription factor's activity, specifically its DNA-binding, dimerization and *trans*-activation functions, may be affected in a variety of ways, including post-translational modification (e.g. phosphorylation), ligand binding and interaction with other proteins. In the following sections of this review, examples of different levels of control will be discussed.

Abbreviations used: NLS, nuclear localization signal; NF-M, nuclear factor-myeloid. PPAR, peroxisome proliferator-activated receptor; HNF, hepatocyte nuclear factor; cAMP, cyclic AMP; CREM, cAMP-response element modulator; CRE, cAMP-responsive promoter element; ICER, inducible cAMP early repressor; 3'-UTR, 3'-untranslated region; ARE, AU-rich element; eIF, eukaryotic initiation factor; MAP kinase, mitogen-activated protein kinase; Met-tRNA$_i^{met}$, methionyl-initiator-tRNA; (u)ORF, (upstream) open reading frame; RARβ2, retinoic acid receptor β2; DBD, DNA-binding domain; HRE, hormone response element; LBD, ligind-binding domain; TR, thyroid receptor; VDR, vitamin D$_3$ receptor; RXR, retinoid X receptor; DR, direct repeat; 9-*cis*-RA, 9-*cis*-retinoic acid; bZIP, basic zipper; NF-AT, nuclear factor of activated T cells; GR, glucocorticoid receptor; CBP, CREB-binding protein; HTLV-I, human T-cell leukaemia virus type I; DCoH, dimerization cofactor of HNF-1; PCD, pterin-4a-carbinolamine dehydratase; C/EBP, CCAAT/enhancer-binding protein; GRE, glucocorticoid response element; Pit-1, pituitary specific factor 1; (b) HLH protein, (basic) helix–loop–helix protein.
* To whom correspondence should be addressed.

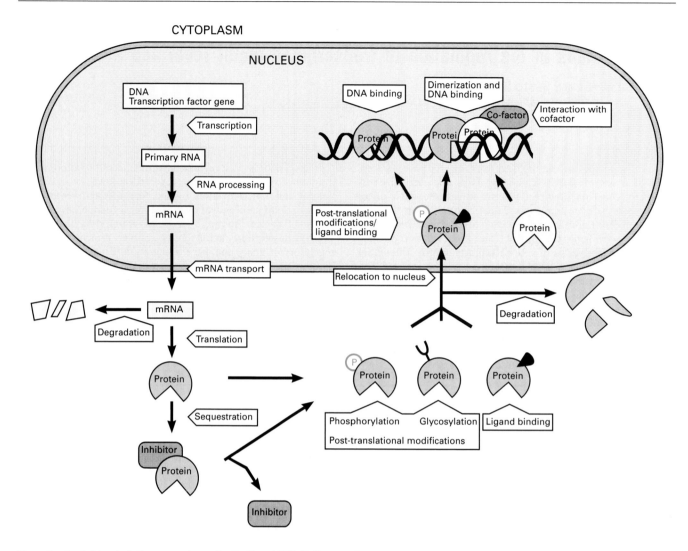

Figure 1   Control levels in the expression and activation of DNA-binding proteins

The scheme depicts crucial steps in the synthesis, activation and action of DNA-binding proteins. Along the pathway potential regulatory points are indicated (white labels). The concentration and activity of a particular DNA-binding protein may be regulated at several points down the pathway to its ultimate action in gene transcription as a transcription factor.

REGULATION OF TRANSCRIPTION-FACTOR LEVELS

Transcription and autoregulation

Synthesis of mRNA is the first level at which regulation can be exercised. Regulation of transcription-factor genes, as that of other genes, is accomplished through the combinatorial action of (other) transcription factors binding to promoter and enhancer sequences. Interestingly, many cell-type specific transcription factors behave as autoregulatory factors, being involved in the transcriptional activation of their own genes [2–12]. Genetic circuits with autoregulatory properties can be of a simple or complex type. A simple autoregulatory circuit consist of one single transcription factor that binds directly to the promoter of its own gene. A complex regulatory circuit comprises several transcription factors that bind to the promoters of one another's genes, thereby indirectly affecting the rate of transcription of their own genes [13]. Autoregulatory positive feedback provides a memory mechanism for maintaining the determined and/or differentiated state associated with a specific cell phenotype. Some examples of autoregulatory transcription factors are described below.

Pit-1

Pituitary specific factor 1 (Pit-1) is a tissue-specific transcription factor obligatory for the development of three cell types in the anterior pituitary gland: lactotrophes, somatotrophes and thyrotrophes. Studies with transgenic mice have revealed that the expression of Pit-1 is governed by a cell-specific enhancer located in the upstream *pit*-1-regulatory region [9]. The action of the enhancer depends on the concerted action of Pit-1-positive autoregulatory sites, a cell-specific element and morphogen response elements. Autoregulation is not required for the initial activation of *pit*-1 gene expression, but it is needed for the maintenance of *pit*-1 gene expression following birth [9].

Myogenic transcription factors

Differentiation of mammalian skeletal-muscle cells is regulated by members of the MyoD family of myogenic transcription factors. These include MyoD [14] and Myf5 [15], which are believed to be responsible for the determined myoblast state, myogenin [16,17], which has a unique role in the transition to the fully differentiated myotube, and MRF4 [18–20], which controls

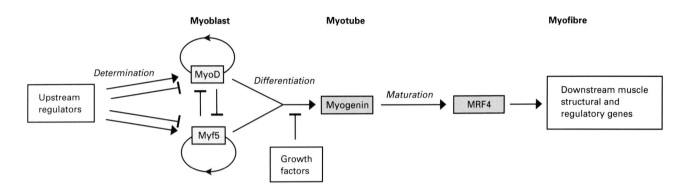

Figure 2   Regulatory circuit of myogenic transcription factors

Hypothetical regulatory circuit for murine myogenic genes comprising a core autoregulatory network of the MyoD transcription-factor family members MyoD, Myf5, myogenin and MRF4. The model implies that Myf5 and MyoD are primarily responsible for defining the myoblast state. Myf5 and MyoD autoregulate their expression, but may negatively regulate one another. Upon depletion of growth factors, MyoD and Myf5 activate myogenin, which induces the differentiated myotube state characterized by the expression myotube-specific genes. During myofibre maturation, MRF4 is up-regulated, which induces the myofibre-specific genes {adapted from Weintraub [21] and Olson and Klein [22]) and reproduced with the permission of the authors and the publishers (copyright Cell Press, 1993, and Cold Spring Harbor Laboraory Press, 1994, respectively)}.

the expression of myofibre-specific genes (reviewed in [21,22]). From experiments with transgenic mice it is inferred that MyoD and Myf5 act by turning on the myogenin gene [23–25]. In fibroblasts and various other cell types, the entire program of muscle differentiation can be turned on by transfection of MyoD or any one of the other members of the family [4,26–28]. Among the endogenous muscle-specific genes activated are the myogenic regulators themselves. These and other experiments have led to the postulation of an auto- and cross-regulatory network [21,22] (Figure 2). The positive-feedback loops tend to make expression of the genes self-sustaining. Maintenance of the determined myoblast in the undifferentiated state for a given length of time appears to be due to negative regulators that have MyoD and Myf5 as their primary targets. One of these inhibitors of differentiation, called Id, is, like MyoD, a helix–loop–helix (HLH) protein, but lacks the basic DNA-binding domain [29–31]. Dimerization with Id would prevent MyoD from binding to DNA. Triggering of the differentiation step may occur via inactivation of Id, allowing MyoD to heterodimerize with ubiquitous basic HLH (bHLH) proteins that potentiate its activity, known as E-proteins. This results in activation of the downstream myogenin gene [32–34]. The myogenin gene product, which has been shown to be essential for muscle development *in vivo*, activates muscle-specific genes, inducing the differentiated myotube state [22,24,35]. Finally, upregulation of MRF4 results in the induction of myofibre-specific genes causing the maturation of the myotube into the myofibre [22,36]. Within the myogenic bHLH regulatory network, non-bHLH protein muscle-specific enhancer factors of the MEF2 family [37,38] participate in the autoregulatory circuits and in activation of muscle-specific genes [39–42]. The core regulatory network in which a cell type (myocyte)-specific family of transcriptional regulators (myogenic bHLH proteins) synergize with unrelated factors (MEF2) in the induction of determination and differentiation is strikingly analogous to the regulatory network in adipocyte development, where CCAAT/enhancer-binding proteins (C/EBPs) act in synergy with the unrelated peroxisome proliferator-activated receptors (PPARs) (see below).

C/EBP

The family of C/EBP transcription factors consists of several proteins of which at least one, C/EBPα, appears to be subject to

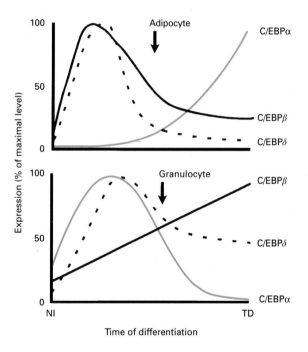

Figure 3   Temporal pattern of C/EBPα, C/EBPβ and C/EBPδ expression

Temporal patterns of C/EBPs expression during induced differentiation of 3T3-L1 preadipocytes (upper) [44] and 32D C13 myeloblasts (lower) [45]. The absolute levels of these proteins are unknown. Abbreviations: NI, non-induced; TD, terminally differentiated. Arrows indicate the approximate time when proliferation ceases in each pathway. This Figure is reproduced with the permission of A. D. Friedman [45] and the publishers (W. B. Saunders Co., Philadelphia, PA, U.S.A.).

autoregulation [8,11,43]. During cellular differentiation, each of the C/EBPs exhibits a temporal expression pattern which differs between gene types (α, β and δ) and cell type (adipocytes versus myelomonocytes). [44–46]. During hormone-induced differentiation of the 3T3-L1 adipoblast cell line into adipocytes, C/EBPδ and C/EBPβ have early catalytic roles leading to expression of C/EBPα concordant with the acquisition of the differentiated phenotype [44,46]. In differentiating myelomonocytic cells, expression of C/EBPα and C/EBPδ peaks during the proliferative state and the expression of C/EBPβ increases throughout until the terminally differentiated stage of polymorphonuclear leucocytes (neutrophils) is reached (Figure 3) [45]. The members of the C/EBP family recognize the same DNA targets. The different patterns of C/EBP expression may reflect partial functional redundancy of the related C/EBPα and -β transcription factors, as well as specialization in which the differentiated stage-specific factor in the adipocyte is C/EBPα and that in the myelomonocyte is C/EBPβ. Interestingly, each C/EBPα and C/EBPβ/NF-M (nuclear factor-myeloid) [47], together with the proto-oncogene Myb, can act as a combinatorial switch inducing myeloid-specific gene expression in heterologous cell types [48]. Conditional ectopic expression of C/EBPβ in NIH-3T3 cells initiates adipogenesis, converting these multipotent precursor cells into pre-adipocytes [49]. The concomitant induction of PPARδ by C/EBP, and its activation by ligand, induces the subsequent differentiation into adipocytes. C/EBPα, which together with PPARδ is expressed late in the differentiation process, is thought to be responsible for the establishment of the quiescent, terminally differentiated state [49–55]. C/EBPs and PPARs may be the core of an autoregulatory network responsible for the commitment of the adipocyte phenotype in a way similar to the autoregulatory network of myogenesis. The different levels of the C/EBPs in adipocytes and myelomonocytes may be considered as alternative steady states of an autoregulatory circuit [13] in which an additional synergistic partner is required for the establishment of the specific phenotype.

Lessons from *Drosophila*

Temporal transcriptional regulation of closely related transcription factors and loops of autoregulation appear to be a recurring theme in development and cell differentiation. Cascades of sequential transcriptional control are the basis for early *Drosophila* development. Numerous homeobox-containing genes in *Drosophila* control their own expression by positive autoregulation. It is beyond the scope of this review to describe the complex pattern of transcription-factor expression in this organism. For trancriptional regulation in *Drosophila*, we refer the reader to other publications [2,3,57–59].

RNA splicing

Before the RNA transcript of a eukaryotic gene is translocated to the cytosol to become translated, the non-coding sequences (introns) interrupting the coding sequences (exons) have to be excised from the primary transcript or pre-mRNA by the process of splicing. Alternative splicing may be constitutive, generating always the same isofoms, or regulated, resulting in different isoforms depending on the cell type and circumstances. Both negative-acting protein factors preventing the use of a particular splice site, and positive factors directing the splicing at an unconventional splice site, may play a part [2]. A peculiar mechanism involving inhibition by antisense mRNA of an alternative splice site has been proposed for the thyroid receptor TRα. One of the TRα splicing variants would be specifically suppressed by a partially overlapping antisense transcript [60]. Alternative splicing of transcription factor pre-mRNAs may generate multiple mRNAs that differ in their coding regions, yielding polypeptides with different, often opposing, activities, or in their untranslated regions, affecting mRNA stability, translation efficiency or intracellular localisation [61–64].

Hepatocyte nuclear factor (HNF)

In the case of the HNF homeodomain proteins, HNF1 (=HNF1α) and the closely related, by a different gene encoded variant vHNF1 (=HNF1β), differential use of polyadenylation sites and alternative splicing causes the formation of different isoforms [65]. All these HNF1/vHNF1 isoforms can mutually homo- and heterodimerize to attain the DNA-binding state. The *trans*-activating isoforms HNF1-A, HNF1-B, HNF1-C and vHNF1-A/B differ in the composition of their C-terminal domain, resulting in different *trans*-activating potentials: HNF1-B and HNF1-C > HNF1-A. The vHNF1-C isoform lacks most of the C-terminal amino acid sequences present in the vHNF1-A/B isoforms and behaves as a *trans*-dominant repressor when co-transfected with each of the *trans*-activating isoforms. The isoform mRNA levels vary between cell types and during organ development, suggesting regulation of both differential polyadenylation and splicing [65].

CREM

The CREM [cAMP (cyclic AMP)-response element modulator] gene encodes a family of activating and repressing isoforms binding to cAMP-responsive promoter elements (CREs) of genes involved in neuroendocrine processes and spermatogenesis. The CREM gene is an example of a gene which is very extensively regulated with respect to transcription, RNA processing and post-translational modifications. The gene has a modular structure containing two alternative promoters, two alternative DNA-binding domains and different *trans*-activation domains (Figure 4) [66–68]. Pre-mRNA transcribed from the non-inducible promoter P1 is differentially spliced in a tissue-specific way. This results in the synthesis of activators and repressors with alternative DNA-binding and *trans*-activation domain compositions (Figure 4) [66,69]. For example, during spermatogenesis, a developmental switch from the transcriptional repressors CREMα, CREMβ and CREMδ to the activator CREMτ takes place. CREMτ contains two additional glutamine-rich regions (Q1 and Q2) responsible for *trans*-activation [61,70,71]. The shorter transcript generated from the cAMP-inducible P2 promoter is processed into mRNAs encoding small repressors, so-called ICERs (inducible cAMP early repressors) [68,72]. Rhythmic adrenergic signals sent by the suprachiasmatic nucleus cyclically activate the P2 promoter in the pineal gland by stimulation of the cAMP signal-transduction pathway [72,73].

Other examples of differently acting transcription-factor isoforms generated by alternative splicing are the isoforms of *Bombyx mori* GATAβ [74], the Wilms'-tumour susceptible gene product WT1 [75], the acute-myeloid-leukaemia gene product AML1 [76], the lymphoid transcription factor LyF-1 [77], the upstream stimulatory factor ('USF') [78], the activating transcription factor-3 ('ATF3') [79], I kappa B gamma [80], the octamer motif-binding protein Oct-1 [81], Oct-2 [82], the *Drosophila* chorion transcription factor CF2 [83] and activating protein-2 ('AP-2') [84]. In many of these cases the alternative splicing is developmentally and/or spatially regulated.

Degradation of mRNA

mRNA turnover is an important aspect in the control of gene expression. Eukaryotic mRNAs have different turnover times that range from days to minutes, and often are influenced by environmental signals. Among the less stable mRNAs are those that encode proteins expressed transiently in response to extra-

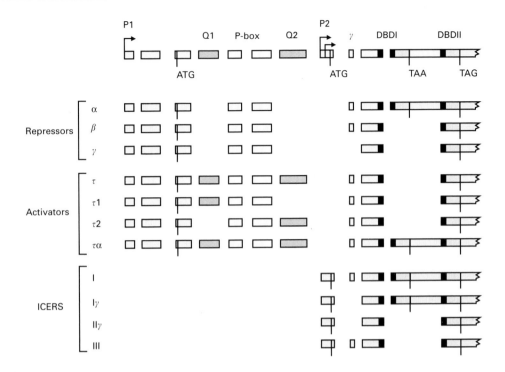

Figure 4   Activators and repressors from the CREM gene

Top: schematic representation of the CREM gene. Exons encoding the glutamine-rich domains (Q1 and Q2), the P-box, the δ-domain (δ) and the two alternative DNA-binding domains (DBDs) (DBDI and DBDII) are shown. Below the various activator and repressor isoforms which have been described to date are represented. The P1 promoter is GC-rich and directs a non-inducible pattern of expression. Also shown is the ICER family. All the ICER transcripts are derived from an internal start-site of transcription (P2) located between the Q2 and δ-exon. A family of four types of ICER transcript is generated by alternative splicing of the DBDs and δ-domain exons ICER-I, ICER-Iδ, ICER-II and ICER-IIδ. This Figure is reproduced with the permission of P. Sassone-Corsi [73] and the publishers (Oxford University Press, Oxford, U.K.).

cellular stimuli like growth factors, cytokines and transcription factors. Rapid mRNA degradation provides an efficient mechanism for transient protein expression because it links protein synthesis directly to the gene transcription rate. Unstable mRNAs may contain one or more specific sequences that stimulate their degradation [85,86].

Of the transcription factors, the proto-oncogene c-*fos* has been most extensively studied with respect to mRNA degradation. c-Fos (the c-*fos* gene product) is required during development and is induced rapidly and transiently by several extracellular stimuli [87,88]. Once synthesized, c-Fos mRNA is transported to the cytoplasm, where it is translated for only a brief period of time because of its rapid degradation [89]. c-Fos mRNA contains in its 3′-untranslated region (3′-UTR) a 75-nucleotide AU-rich element (ARE) containing the nonameric sequence UUAUUU-AUU, which has been shown to be critical for mRNA destabilization [90]. The presence of the destabilizing element triggers de-adenylation, which is an early step in mRNA decay. The consensus sequence UUAUUUA(U/A)(U/A) is present within the 3′-UTRs of many labile mRNAs and functions as a transplantable element; when inserted into a stable heterologous mRNA such as β-globin mRNA, it destabilizes the mRNA in a copy-dependent manner [90,91]. The mechanism by which the ARE-containing mRNAs are further degraded is still unknown, but studies in yeast suggest that deadenylation is followed by decapping of the 5′ terminus, making the mRNA accessible to a specific 5′ → 3′ exonuclease [85,92]. The short-lived mRNAs of two other transcription factors, c-Myc and c-Jun, which have AREs in their 3′-UTRs, are believed to be subject to the same degradation pathway [90,93]. Besides the determinant in the 3′-

UTR, there is an additional instability determinant in the coding region of the c-Fos mRNA. This 0.32 kb Coding Region Determinant of mRNA Instability ('CRDI') also works as an independent mRNA destabilization determinant when incorporated in a heterologous, stable mRNA. The region facilitates mRNA deadenylation and decay by a mechanism coupled to translation [94–96].

Translation

Translational regulation of transcription factors, as of proteins in general, usually occurs at the initiation level, specifically at two steps: (a) selection of the mRNA for translation by the ribosomal complex and (b) linear scanning of the mRNA from the 5′-end by the ribosomal complex to select the translation initiation codon. In both steps several eukaryotic initiation factors (eIFs) are involved [97,98], two of which, eIF-4E and eIF-2, are known to be important players in the regulation of translation.

Initiation factor eIF-4E is the cap-binding protein; it exerts its function as part of a complex termed eIF-4F, which, in addition, contains eIF-4δ of unknown function and eIF4A, an RNA helicase. Initiation factor eIF-4E is sequestered in an inactive complex by the 4E-binding proteins 4E-BP1 (homologous with PHAS-1 [99]) and 4E-BP2 [100], blocking cap-dependent translation. Release of eIF-4E is mediated by mitogen-activated protein (MAP)-kinase-catalysed phosphorylation of 4E-BP1/2, which occurs in response to insulin and/or growth factors [99,100]. An interesting hypothesis, which may have implications for transcription-factor regulation, has been proposed by Proud

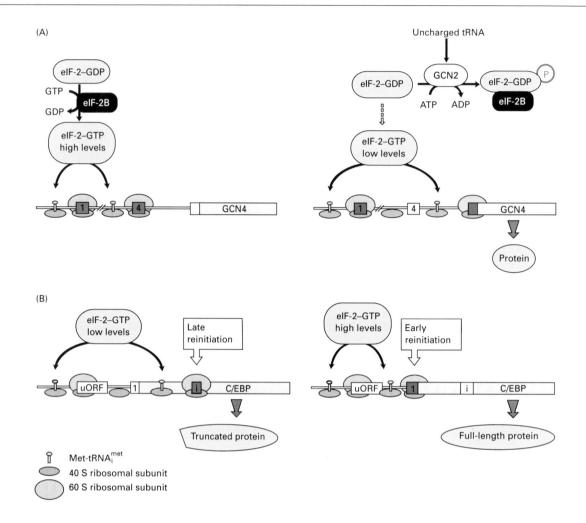

Figure 5   Translational regulation of GCN4 and C/EBP

(**A**) The Figure schematically depicts the action of eIF-2 on translation reinitiation at the GCN4 mRNA. Left: under non-starvation conditions eIF-2 is readily recycled by the guanine nucleotide exchange factor eIF-2B. The high level of eIF-2–GTP ensures frequent reinitiation at the upstream small ORFs with associated concomitant release of the small ribosomal subunit. Translation of the GCN4 ORF is thereby prevented. Right: under starvation conditions uncharged tRNAs accumulate, activating the eIF-2–GDP kinase GCN2. The phosphorylated eIF-2 sequesters the exchange factor eIF-2B in an inactive complex, resulting in low levels of active eIF-2–GTP. Ternary complex (eIF-2–GTP–Met-tRNA$_i^{met}$) assembly is delayed, causing by-passing of the upstream small ORFs and translation reinitiation at the GCN4 ORF {adapted from Hinnebusch [102]) and reproduced with the permission of the author and the publishers (Elsevier Trends Journals)}. (**B**) Hypothetical model for translational regulation of C/EBP mRNA. Analogous to the situation in yeast, high levels of eIF-2–GTP may cause efficient reinitiation after translation of the small upstream ORF, resulting in mainly full-length C/EBP protein (left). When eIF-2–GTP levels are low, delayed translation reinitiation may cause protein synthesis from an internal in-frame start codon (i), resulting in N-terminally truncated C/EBP protein (right).

[101], who stated that higher concentrations of eIF-4F would primarily facilitate the translation of mRNAs whose 5′-UTRs contain significant secondary structure. Interestingly, many mRNAs of regulatory proteins including transcription factors contain such highly structured CG-rich 5′-UTRs.

Modulation of translation initiation via eIF-2 in concert with a specific mRNA sequence is of particular interest because it controls the translation of at least one transcription factor, GCN4 [102], and possibly of the transcription factors C/EBPα and C/EBPβ as well. Initiation factor eIF-2 forms a ternary complex with GTP and methionyl-initiator-tRNA (Met-tRNA$_i^{met}$) and delivers the Met-tRNA$_i^{met}$ to the 40 S ribosomal subunit. Recycling of the eIF-2-GDP formed in the initiation step is inhibited by phosphorylation of eIF-2, sequestering the guanine nucleotide exchange factor GEF (eIF-2B) in an inactive complex with eIF–GDP [97,98,103–105]. Phosphorylation of eIF-2 occurs by specific kinases which are activated under various conditions, including the deprivation of growth factors, amino

acid starvation, viral infection, heat-shock, insulin stimulation and entry into the M-phase of the cell cycle [2,102,106–108].

GCN4

The *Saccharomyces cerevisiae* GCN4 gene encodes a transcriptional activator regulating a set of genes engaged in amino acid and purine biosynthesis. In response to starvation, the concentration of GCN4 is up-regulated against an overall reduction of protein synthesis. Regulation occurs at the translational level and acts via the combined action of eIF-2 and four *cis*-regulatory short upstream open reading frames (uORFs) in the 5′ leader sequence of the GCN4 mRNA. Under normal growth conditions the translation of the GCN4 coding region is restricted because the ribosomal subunits scanning the GCN4 mRNA are detached in the translation of the successive uORFs (Figure 5A, left). Under starvation conditions eIF-2 activity is reduced, causing the ribosomes that resume scanning after translation of the first uORF to ignore the subsequent small uORFs. This effectively enhances the chance of re-initiation at the downstream GCN4 ORF, because at the time the ribosome reaches the distant GCN4 start codon the ternary complex necessary for translation initiation has reassembled (Figure 5A, right) [102,109–112].

C/EBP

C/EBPα and C/EBPβ play a decisive role in the differentiation of a number of cell types, including adipocytes, hepatocytes, enterocytes and myelocytes [44,45,49,51,55,113–115]. The effects of C/EBPα and C/EBPβ proteins appear to be dual: firstly they induce the expression of tissue-specific genes in concert with other transcription factors, and secondly they can evoke growth arrest [51,116,117]. The magnitude of the transcriptional activation by C/EBPα and C/EBPβ transcription factors is modulated by the relative expression of two protein isoforms, a full-length and an N-terminally truncated isoform, both translated from the same mRNA [118–121]. The full-length isoform is a potent transcriptional activator in hepatocytes and adipocytes, and the truncated isoform acts as a repressor, or an activator with low activity, depending on the promoter context [118–122]. The generation of N-terminally truncated isoforms of C/EBPα from internal start codons depends on the presence of a small uORF, conserved in evolution, mediating delayed translation reinitiation [121]. The analogous organization of the C/EBPβ mRNA suggests that the generation of different translational isoforms from this particular mRNA may be governed by a similar mechanism [121].

We hypothesize that the C/EBP isoform ratio may be regulated by translation initiation factor eIF-2. Like the uORFs in GCN4, the C/EBP small uORF may actually sense the activity of eIF-2. Low eIF-2 activity would promote the ribosomes, which resume scanning after having read the uORF, to ignore the proximal C/EBP initiation codon and start at an internal AUG, yielding the truncated C/EBP isoform (Figure 5B, left). High eIF-2 activity would suppress 'leaky' scanning across the first AUG of the C/EBP coding region, as well as the associated internal initiation (Figure 5B, right). Such an eIF-2-sensing system would provide an interesting coupling between, on the one hand, growth and other factors (see above) that effect eIF-2 phosphorylation (and thus activity), and, on the other hand, C/EBP as a regulator of cell-type specific gene expression and growth control. Interestingly, the formation of the full-length C/EBP isoform from a transfected C/EBPα gene in COS cells is promoted by serum addition (C. F. Calkhoven and G. AB, unpublished work), a condition known to increase eIF-2 activity [106]. Moreover, insulin is known to stimulate eIF-2 activity [108] and causes a dramatic and rapid change in the C/EBPβ and a modest change in C/EBPα isoform ratio to the benefit of the full-length products in differentiating adipocytes [123]. C/EBP isoform ratio modulation may provide a mechanism for metabolic and hormone-imposed adaptation of C/EBP target genes. This is interesting in the light of the central roles that C/EBPα and C/EBPβ play in gluconeogenesis and lipogenesis, in liver and fat tissue [44,114,124–127]. Additionally, translational regulation via eIF-2 may be responsible for the temporary down-regulation of the full-length C/EBPα isoform in the M-phase of the cell cycle which is observed during liver regeneration [43].

c-myc

The proto-oncogene c-myc is a transcription factor having an important role in control of cell growth. The 5' mRNA leader sequence of murine and mice c-myc contains, besides an efficiently used AUG translation start codon, an upstream in-frame non-AUG start codon, CUG [128,129]. In cultured cells upon methionine depletion, the translation of c-myc shifts from the AUG codon to the upstream CUG codon. Although the physiological function of the two isoforms is not yet established, it could reflect a switch from cell growth promotion mediated by the smaller isoform to inhibition of cell growth mediated by the larger isoform [129]. Speculation on how the translational shift under methionine deprivation is brought about focus again on a role for eIF-2 in selecting the non-AUG codon [129]. An alternative explanation may be found in the CAP-binding eIF-4F complex, because increased levels of eIF-4F promote CAP-proximal AUG codon usage in cell-free extracts [130].

Retinoic acid receptor β2 (RARβ2)

The complex 5'-untranslated region containing five partially stacked uORFs of the mRNA is responsible for tissue-specific synthesis of its RARβ2 during mouse development. Transgenic mice containing an RARβ2–lacZ fusion construct including the 5'-UTR express no protein from the transcribed mRNA in heart and brain. By mutating part or all of the uORFs, the tissue-specific RARβ2 protein synthesis is lost, resulting in expression in heart and brain. The main conclusion drawn from this study is that uORFs in 5'-UTRs in combination with tissue-specific regulation of initiation factor level and/or activity play an important role in regulation of the tissue-specific expression of this regulatory protein [131].

Translation regulation based on start site selection has been implicated in the expression of other critical transcription factors involved in growth and differentiation control [132], e.g. isoforms of the rat hepatic leukaemia factor ('HLF') with different circadian levels, tissue distributions and target preferences [133], and the developmentally-regulated isoforms of the CREM isoforms, the CREMτ activator and the S-CREM repressor [134].

In by-passing the nucleus and relying on cis-regulatory elements in the mRNA proper, translation regulation enables fast and co-ordinated responses to external stimuli and provides an additional regulatory checkpoint [132,135–138]. Until now, the role of special structural mRNA features in translation regulation is far from clear. Future experiments should clarify their function and the mechanisms involved.

REGULATION OF TRANSCRIPTION-FACTOR ACTIVITY

Once a transcription factor has been synthesized, its activity can be controlled in a variety of ways, for instance by post-translational modification, for example phosphorylation or by binding of a ligand. These processes induce conformational changes in the transcription factor, exposing, masking or re-modelling a particular domain. Another means of activity modulation is by specific protein–protein interaction, either with a member of a related family of transcription factors to form a DNA-binding dimer, or with an unrelated factor.

Post-translational modification by phosphorylation

Transcription factors are important final targets of signal-transduction pathways in which transient signals generated by stimulation of cell-surface receptors are transmitted via phosphorylation cascades to the nucleus. An important facet of this type of modification is that it is reversible.

There are several ways by which phosphorylation can regulate a transcription factor. Firstly, sequestration of the transcription factor in an inactive complex, e.g. by an anchor protein, can be regulated. Phosphorylation or dephosphorylation of the transcription factor or of its anchor protein may result in dissociation of the complex, allowing translocation of the transcription factor to the nucleus. Secondly, phosphorylation may modulate the DNA-binding activity and/or the *trans*-activation potential of the transcription factor (reviewed in [139,140]).

Sequestration

The first example concerns the SWI5 protein of *Saccharomyces cerevisiae*, a transcription factor regulating cell-cycle-dependent expression of the specific HO endonuclease involved in mating-type switching. SWI5 localization is thought to be regulated by phosphorylation of three sites within or near the NLS through the cell-cycle regulatory protein kinase Cdc28 in conjunction with an activating cyclin subunit. In the G1-phase, SWI5 is located in the nucleus and activates HO gene transcription. During the other phases of the cell cycle, the SWI5 protein is sequestered in the cytoplasm as a consequence of the phosphoryl-ation of the NLS-proximal Cdc28 sites [141,142].

The second example concerns NF-κB of the Rel-related family of transcription factors [143–145]. NF-κB is ubiquitously expressed, but in most cells it is sequestered in the cytoplasm as an inactive complex with the inhibitory protein IκB, probably by masking of the NLS [146]. The IκB protein contains so-called ankyrin-like ('ANK') repeat motifs which are believed to be involved in protein–protein interactions [147,148] with the p65 subunit of NF-κB. In response to various signals such as mitogens, cytokines, viral double-stranded RNA and oxidative stress, IκB is inactivated by phosphorylation, triggering its degradation by the ubiquitin–proteinase pathway [149–153]. This liberates NF-κB, which then is translocated to the nucleus, where it can activate gene expression by binding to κB enhancer and promoter motifs.

DNA-binding and *trans*-activation

Both the DNA-binding and *trans*-activation functions of a transcription factor may be regulated by phosphorylation, either positively or negatively. An illustrative example is c-Jun, which is phosphorylated at five sites. Three of these are located just N-terminal to the DBD and are phosphorylated by a constitutively active kinase CKII, causing inhibition of DNA binding [154, 155]. Phorbol ester (phorbol 12-myristate 13-acetate), growth factors or expression of transforming oncogenes stimulate an unknown phosphatase which dephosphorylates c-Jun and in-creases its DNA-binding activity [139,154]. Phosphorylation-dependent DNA-binding activity is observed for many other transcription factors, for example Max [156], Oct1 [157] and SRF [158].

The two other phosphorylation sites of c-Jun, located in the N-terminal *trans*-activation domain, are phosphorylated in response to mitogenic stimulation and stress by distinct MAP kinases, c-Jun N-terminal kinases ('JNKs') and stress-activated protein kinases ('SAPKs'), resulting in elevated *trans*-activation potential [139,159–165].

Whereas c-Jun is activated by phosphorylation of the activation domain proper, C/EBPβ is activated in a different way. [166,167]. In chicken C/EBPβ (NF-M), the N-terminally located *trans*-activation domain is caught in an intramolecular in-teraction with inhibitory domains. Phosphorylation of these inhibitory domains liberates the *trans*-activation domain, in-ducing the *trans*-activation function [167]. One of the signal-transduction pathways leading to derepression of C/EBPβ includes MAP kinases that phosphorylate a conserved MAP-kinase site in the C/EBPβ proteins [166]. For murine C/EBPβ a

slightly different model was proposed in which both the exposure of the *trans*-activation domain and the DNA-binding domain is regulated by intra-molecular interactions with independent regu-latory domains. However, involvement of phosphorylation in this system has not been established [168]. Other conserved phosphorylation sites mapped are a serine in the leucine-zipper dimerization region of C/EBPβ that can be phosphorylated by Ca^{2+}-calmodulin-dependent protein kinase II ('CaMKII') con-ferring calcium-regulated stimulation of transcriptional activity [169], and a conserved serine in the DNA-contacting region of C/EBPs the phosphorylation of which by protein kinase C attenuates DNA-binding of C/EBPα *in vitro* [170].

It will be interesting to combine these biochemical studies with structural analysis of the differently phosphorylated transcription factors to see what the conformational consequences of these reversible phosphorylations are.

Ligand-dependent activation of nuclear hormone receptors

Lipophilic hormones, such as steroids, retinoids, thyroid hor-mones, vitamin D$_3$ and eicosanoids, are potent regulators of transcription. They exert their function within target cells by binding to specific intracellular receptors which function as ligand-activated transcription factors. Although the various ligands are chemically very different, the receptors exhibit a remarkable overall structural unity that permits their classifi-cation as one large family, namely the superfamily of nuclear receptors. The family also includes members with no apparent ligand, so-called 'orphan receptors'. Some of these may turn out to interact with novel ligands, while others may be constitutive factors. Excellent reviews on nuclear receptors have appeared recently [171–179]. Before we discuss the mechanism(s) of ligand-dependent activation, the common molecular design of nuclear receptors in terms of structural and functional domains will be described.

The N-terminal domain (A/B) exhibits little sequence simi-larity across the superfamily and is variable in length. The domain contains a ligand-independent *trans*-activation function with marked cell type and promoter specificity [180].

The most conserved central domain (C), which contains two zinc-co-ordinated modules ('zinc fingers') that fold together to a compact structure [181–185], functions as the DBD that targets the receptor to specific DNA sequences known as hormone response elements (HREs). The minimal target sequence recog-nized by the DBD consists of a six-base-pair sequence, the core recognition motif PuGGTCA (where Pu is a purine). Naturally occurring response elements frequently contain two copies (half-sites) of the core recognition motif, indicative of the fact that most nuclear receptors function as dimers, i.e. homo- or hetero-dimers. Receptor-specific differences in the contacts between DBDs permit the formation of a limited number of homodimeric or heterodimeric combinations. The precise sequence, orientation (direct versus inverted repeats) and spacing of the half-sites determine for which dimeric receptor combination an HRE is the target (see [186] and references cited therein). Dimerization between the DBDs results in a co-operative increase in the specificity and affinity of DNA binding. Recently, some orphan receptors recognizing their target DNA sequence in a monomeric mode have been identified [187–190]. In these cases, the affinity and specificity of DNA binding appears to be enhanced by a C-terminal extension of the zinc-finger domain contacting an extension of one to three base-pairs immediately 5′ of the core recognition motif [189,190].

C-terminal to the DBD, connected by a flexible hinge region (D), is the ligand-binding domain (LBD) formed by a large, hydrophobic, moderately conserved region (E). Together with the ligand-binding function, several other functions are integrated within this domain, including strong dimerization, *trans*-activation, nuclear localization and (in some cases) heat-shock-protein binding. The mechanisms by which ligands induce the receptors appear to be different, depending on the nuclear receptor class.

Steroid-hormone receptors

The steroid-hormone receptors, which function as ligand-induced homodimers (head-to-head configuration) and bind to inverted repeats of the core motif separated by three nucleotides, are associated with a large multiprotein complex of chaperones (Hsp90 and other heat-shock proteins) prior to hormone binding (reviewed in [191]). Binding of the hormone induces a conformational change of the LBD, causing dissociation of the multiprotein complex and exposure of the LBD/dimerization interface. The dimeric receptor can then bind to its palindromic HRE, consisting of AGGTCA or AGAACA half-sites for the oestrogen and the other steroid (glucocorticoids, mineralocorticoids, progesterone, androgen) receptors respectively [192–195]. The role of the chaperones is not simply sterical blocking of DNA binding; they are believed to keep the receptor in a poised, ligand-sensitive state, and help to fold the receptor in its transcriptionally active conformation. The transition exposes a ligand-dependent *trans*-activation function, AF2, which is located in the C-terminus of the LBD and appears to communicate with the transcription initiation complex via co-activators or bridging factors: RIP140 [196], RIP160/ERAP160 [197], TIF1 [198], Trip1 [199,200] and SRC-1 [201].

Non-steroid receptors

Unlike steroid receptors, the non-steroid-hormone receptors do not bind to heat-shock proteins and associate with DNA in the presence and absence of their respective ligands [191,202,203]. Although some of these receptors can also bind as homodimers, high-affinity binding of the RAR, the thyroid receptor (TR), the vitamin D_3 receptor (VDR) and the PPAR to their cognate HREs requires heterodimerization with the retinoid X receptor (RXR) [204–212]. The most potent of these HREs are direct repeats (DRs) of the AGGTCA half-site for which receptor specificity is determined by the so-called '1-to-5 rule' specifying the optimal half-site spacing of 1, 2, 3, 4 and 5 nucleotides for the PPAR, RAR, VDR, TR and RAR response elements respectively [176–205]. Structures of the unliganded human RXRα domain [213] and the liganded domains of the human RARδ [214] and TRα [215] receptors have recently been solved by X-ray crystallography. While the overall folds are similar, the unliganded and liganded domains show interesting differences that may not represent inherent differences between the receptor types but rather reflect ligand-induced conformational changes. Whereas the unliganded hRXRα DBD contains internal hydrophobic cavities, the ligand-bound LBD structures are more compact, with the hormone tightly packed within the core of the domain. In the unliganded receptor, the C-terminal amphipathic α-helix harbouring the transcriptional activation domain AF-2 protrudes from the LBD into the solvent. In the liganded domain, the particular α-helix is packed on to the body of the LBD, with the hydrophobic face contributing part of the hormone-binding cavity. The ligand-induced repositioning of AF-2 may facilitate the formation of transcriptionally active complexes with co-activators like those interacting with the steroid receptors.

The presence of receptor molecules with different ligand specificity within one and the same dimer raises the question of

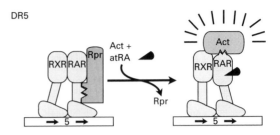

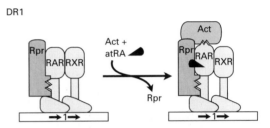

Figure 6   Allosteric control of nuclear receptor activity

The Figure shows a model for allosteric control of the nuclear receptor heterodimer RXR/RAR. Binding of the heterodimers to direct repeats follows strict polarity; if bound to a direct repeat spaced by five nucleotides (DR5), the polarity is 5′-RXR/RAR-3′; at DR1 the polarity is 5′-RAR/RXR-3′. In the absence of the ligand all-*trans* RA (atRA), the co-repressor (Rpr) is associated mediating repression on both DR5 and DR1. Ligand binding induces the recruitment of the co-activator (Act) and on the DR5 dissociation of the co-repressor (Rpr), leading to transcriptional activity. On the DR1, by contrast, the co-repressor is unable to dissociate, keeping the bound complex transcriptionally inactive {adapted from Perlmann and Vennström [225] and reprinted with permission from *Nature* [Copyright (1995) Macmillan Magazines Limited] and the authors}.

whether a heterodimeric receptor can be activated by both ligands and, if so, whether they act synergistically. Dual ligand-sensitivity of all the RXR-containing heterodimers would create a specificity problem, in the sense that the heterodimeric receptors would be activated by the RXR ligand 9-*cis*-retinoic acid (9-*cis*-RA) in addition to the cognate ligand of the dimerization partner. The actual situation appears to be complex and to be differently dependent on the particular receptor combination [216] (reviewed in [179]). Dual synergistic activation occurs in case of the PPAR/RXR heterodimer. In contrast, ligand-induced transcriptional activity of RXR by 9-*cis*-RA is suppressed when it is complexed with VDR, TR or RAR. The formation of the RXR/RAR heterodimer actually prevents the RXR subunit from binding to its ligand. These observations show that allosteric interactions among heterodimer partners create complexes with unique properties. Also interesting in this respect is the finding that the orphan receptor NGF1-B, which is capable of binding as a monomer to 9-*cis*-RA, can form a complex with RXR that is responsive to 9-*cis*-RA. Interestingly, DNA contact by RXR is not required for this effect [216].

In contrast with the steroid receptors, the unliganded TR and RAR bind as heterodimers with RXR to their cognate DNA sites and silence active promoters [217,218]. The suppressing action is mediated by co-repressors which bind to the unliganded receptors [219,220]. Recently two co-repressors, called NCoR and SMRT, have been cloned and further characterized [221–225]. Whether ligand binding is able to trigger the activation process depends on the half-site spacing of the binding site. On a DR5 site ligand binding results in activation, whereas on a DR1 site repression is maintained [226] (Figure 6). This difference finds its cause in the different polarities of the receptors in the heterodimer. On a DR5 site the RXR is in the upstream position, whereas on a DR1 site the orientation is reversed. In case of the heterodimer with the RXR/RAR (DR5) polarity, ligand binding results in the release of the co-repressor and binding of the co-activator, whereas with the heterodimer with the reversed polarity the co-repressor stays attached despite the fact that ligand and co-activator binding still takes place [224]. The co-repressor acts dominantly over the co-activator. These observations reveal allosteric interactions between the receptor subunits whose precise effect depends on the particular anisotropic configuration of the heterodimer imposed by the half-site spacing of the binding site. Interestingly the binding site for the co-repressor is located in the hinge region between the DBD and LBD of the RAR or TR; no interaction occurs with the RXR subunit. The hinge region must possess considerable flexibility to accommodate the head-to-tail orientation of the DBDs with the supposed head-to-head orientation of the LBDs on the DNA [213]. It could be envisaged that the binding-associated compaction of the LBD changes the LBD configuration in the dimer, since the DBDs are in a fixed configuration on the DNA. This would result in a conformational change of the hinge region and release of the co-repressor. Such a scheme would imply that the ligand-induced conformational change in the RAR or TR hinge region only occurs when these receptors are in the downstream position.

Protein–protein interactions

The activity of a transcription factor may be affected by the interaction with other proteins, which can be either DNA-binding proteins or non-DNA-binding accessory proteins. Both interactions with transcription factors of the same or a different family of DNA-binding proteins are possible. One example, the dimerization between related members of the nuclear-receptor family has already been discussed in the previous section. Alternatively, interaction takes place with non-DNA factors (accessory factors) that may mediate a divers range of functions, e.g. acting as a 'bridging factor' between the transcription factor and the basal transcription machinery, stabilizing the DNA-binding complex or changing the specificity of the target sequence recognition.

The basic-zipper (bZIP) DNA-binding proteins

The first well-characterized dimerization domain was the leucine zipper, an α-helix characterized by a heptad repeat of leucine residues [227–229]. When two such helices form a parallel coiled coil, the adjacent, positively charged, DNA-contacting regions are positioned in the proper orientation for DNA binding [230–232]. The subfamilies of C/EBP, Fos, Jun and ATF/CREB transcription factors all depend on a bZIP domain for DNA binding [233]. Within the C/EBP subfamily, the bZIP domain is highly conserved; all members can interact with each other, yielding a variety of dimers with very similar DNA-recognition characteristics. A special C/EBP protein called C/EBP-homologous protein CHOP (=C/EBPζ), which dimerizes avidly with other C/EBP proteins, has an unusual amino acid in its DNA-binding motif rendering it unable to bind to the classical C/EBP binding consensus [234]. Since heterodimers containing CHOP cannot bind to C/EBP-sites, CHOP acts as a dominant negative regulator of C/EBP DNA binding. CHOP is induced by a variety of cellular stresses caused by toxins, nutrient deprivation and metabolic inhibitors [234–237]. Its induction inhibits adipogenesis and attenuates C/EBPα and C/EBPβ gene expression, possibly by interference with the autoregulatory C/EBP cascade at an early stage [238].

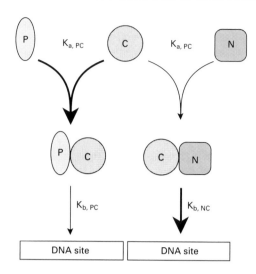

Figure 7   Theoretical model of dimerization-generated ultrasensitivity

The potency of the positive component P and the negative component N as a heterodimeric partner of the central component C for stimulating or repressing transcription is determined by their ability to associate with C and with the DNA target site. Ultrasensitivity can be obtained when, for example, the P–C association constant is higher than the N–C association constant ($K_{a,PC} > K_{a,NC}$) and the DNA–PC binding constant is lower than the DNA–NC binding constant ($K_{b,PC} < K_{b,NC}$). In this case, the P component may easily displace the N component from the C component. Nevertheless, as long as there is still repressor dimer (NC) present, the DNA target cannot be bound by the activator dimer (PC), which exhibits lower affinity for DNA. A near-maximal transcriptional response is induced abruptly, like a molecular switch, only if the majority of C molecules are bound by P in the PC dimer. For example, in the case of a 100-fold difference between $K_{a,NC}$ and $K_{a,PC}$, and between $K_{b,PC}$ and $K_{b,NC}$, a 2-fold increase in P concentration would give a shift from 10 to 90% of the maximum response [239]. (Reproduced with the permission of Swillens [239] and the publishers).

Heterodimerization of transcription factors can provide an explanation for the ultrasensitivity to fluctuations in the effector concentration. In a model outlined by Swillens and Pirson [239] ultrasensitivity can be generated by the reversible coupling of a central component (C) with low affinity to a positive acting (P) and high affinity to a negative acting (N) factor. Ultrasensitivity is obtained if the NC complex has a higher affinity for DNA than the PC complex (Figure 7).

A similar model may help us to understand the observed ultrasensitivity of C/EBP activity to the concentration of small inhibitory C/EBP isoforms [118,121]. Homodimers of small C/EBP isoforms, and probably heterodimers of small/large isoforms too, have higher DNA-binding affinities than homodimers of large isoforms [118,240]. This, with the assumption of a high dimerization constant for dimers containing the large isoform and a lower dimerization constant for the homodimer of the small isoforms, would theoretically explain the ultrasensitivity for small-isoform levels.

Transcription-factor supracomplexes

bZIP protein dimers are often cross-coupled to other transcription-factor dimers in so-called 'transcription-factor supracomplexes' [241]. Supracomplexes between bZIP and Rel transcription factors have physiological functions and DNA-binding characteristics distinct from those of the individual transcription-factor partner [241–243]. C/EBP-Rel supracomplexes do not bind to κB Rel-protein binding sites, but do bind to C/EBP binding sites with the Rel proteins not directly contacting the DNA [244]. The inability of C/EBP–Rel supracomplexes to bind κB sites is consistent with the observed repression of κB-dependent gene expression by C/EBP [243]. In another bZIP–Rel supracomplex of this kind, AP-1 proteins stabilize the interaction of nuclear factor of activated T cells (NF-AT) with weak NF-AT DNA sites [241,245,246]. Another example of transcription factor cross-coupling in which bZIP proteins play a part is the interaction between AP-1 transcription factors and the glucocorticoid receptor at a 'composite' glucocorticoid response element (GRE) [247]. Interaction between the GR or c-Jun homodimer strongly enhances promoter function, whereas interaction between the GR and the c-Jun–c-Fos heterodimer appears to repress promoter function [247]. Supracomplex formation between c-Jun and MyoD via their respective dimerization domains has been shown to inhibit the function of both proteins [248,249].

Co-activators and -repressors

Most transcription factors mediate diverse effects depending on cell type and the presence or absence of a particular stimulus. The effects depend on accessory proteins or co-factors that are expressed in a tissue-restricted manner and/or interact only with a particular state of the transcription factor. Clear examples of such co-factors are the co-activator and co-repressor proteins interacting with the nuclear receptor dimers as discussed in the previous section 'Ligand-dependent activation of nuclear hormone receptors'. Some other examples of what undoubtedly will become a growing group of transcription factor–cofactor interactions are discussed below.

CREB-binding protein

Signals that increase intracellular concentrations of cAMP activate genes that contain CREs. Target gene activation is mediated by the transcription factor CREB, which is activated by phosphorylation in response to the cAMP signal. The activation is not caused by a change in the intracellular localization, DNA binding or intrinsic *trans*-activation potential of CREB, but by recruitment of a co-activator CREB-binding protein (CBP). CBP only interacts with the phosphorylated CREB and functions as a bridging factor to the basal transcription factor TFIIB and is required for the recruitment of an active polymerase II complex [250–254].

Tax

Human T-cell leukaemia virus type I (HTLV-I) Tax proteins increase the DNA binding of many cellular transcription factors that contain a bZIP DNA-binding domain, including GCN4, ATF, AP-1, CREB and C/EBP. Tax interacts with the basic DNA-contacting region, increasing dimer stability and affinity for DNA. Tax also alters DNA-binding-site selectivity. Both effects are probably important for the ability of Tax to recruit the appropriate cellular bZIP proteins to the HTLV-I long terminal repeat during viral infection. The 'promiscuous' activation of cellular genes by the recruitment of bZIP proteins is probably the basis for Tax's oncogenic activity [255–257].

Dimerization cofactor of HNF-1 (DCoH)

The homeodomain protein HNF-1α regulates the expression of a large number of genes in the liver. For maximal transcriptional activation, the HNF-1α dimer requires the co-activator dimer DCoH. Although DCoH does not alter HNF-1α DNA affinity, it strongly enhances its transcriptional activity. How this is achieved is not known, but one of the effects is that it stabilizes the HNF-1α dimer in solution via interaction with the HNF-1α

dimerization domain. The intriguing aspect about DCoH is that it appears to be a bifunctional protein. DCoH was characterized as pterin-4a-carbinolamine dehydratase (PCD), a protein component of the phenylalanine hydroxylation system. PCD deficiency may be correlated with certain hyperphenylalaninaemias in children. Whether the dehydratase activity is also essential for the HNF-1α transcriptional enhancement is not known [258–261].

CONCLUSIONS AND PERSPECTIVES

In this review we have outlined control points in the expression of transcription factors. For these regulatory proteins collectively, any step in the complete sequence leading from the encoding gene to the active transcription factor may potentially be subject to control. For a given transcription factor, regulation is often exerted at more than one step. A clear example is C/EBP, which, during the course of cell differentiation, is primarily controlled at the transcriptional level. In addition, phosphorylation of the protein at specific sites provides links to signal-transduction pathways and translation regulation may be important in the response to metabolic signals. Finally, dimerization may affect the specificity and affinity of DNA binding and potentially integrates the regulatory pathways of both partners. Another example of multiple control is provided by the nuclear receptor family, where the restricted presence of the receptor in its target cell is determined at the transcriptional level, the actual activation occurring through the binding of a ligand, and further modulation of the activity may be achieved by the dimerization partner and phosphorylation [262,263]. There is probably a hierarchy in the importance of the control steps for a particular transcription factor, some of which may mainly serve for fine tuning by coupling to other regulatory pathways.

Gene duplication and mutation has generated a great deal of diversity within transcription-factor families, which may serve the purpose of redundancy as well as specialization. Regulatory diversity is further extended by processes like alternative splicing, translation-start-site multiplicity, phosphorylation and protein–protein interaction.

Insight into the action and regulation of transcription factors has been obtained by a variety of techniques. The recent results on transcription factors by X-ray crystallography and NMR methods has allowed us to project earlier, often fragmentary, data on to the three-dimensional structure. In the future more will be learned of how different regulatory pathways involving transcription factors are coupled. Our present knowledge of transcription-factor regulation is probably unbalanced, with a bias against translational regulation. We expect that future attention will be drawn more towards this level of regulation and how it regulates transcription factor activity.

Because of the width of the field we were forced to make a selection of the topics reviewed. Some interesting transcription factors and regulatory pathways are left undiscussed, e.g. proteins with unconventional properties like lactoferrin, which acts as a secretory transcription factor [264], the redox regulation of DNA-binding activity observed for many transcription factors ([265–267] and references therein), the activation of the 'STAT' (signal transducers and activators of transcription) family by cytokines [268] and transcription-factor gradients in embryonic development [59,269].

For a better understanding of the actual potential of a transcription factor in the cell, more detailed quantitative analyses will be necessary. Parameters like dimerization constants, DNA affinity constants, precise concentrations of transcription factors and threshold levels of transcription-factor concentrations may be used to generate mathematical models of gene regulation. These models will help to explain the dynamics and kinetics of transcription-factor action in gene expression and to generate new ideas.

We thank Onno Bakker for critical reading of the manuscript before its submission. We are grateful to A.D. Friedman, P. Sassone-Corsi and S. Swillens for their permission to use Figures from their publications in this review.

REFERENCES

1 Keller, A. D. (1994) J. Theor. Biol. **170**, 175–181
2 Alberts, B., Bray, D., Lewis, J., Raff, M., Roberts, K. and Watson, J. D. (1994) Molecular Biology of The Cell, 3rd edn., pp. 401–474, 1060–1063 and 1139–1193, Garland Publishing, New York
3 Serfling, E (1989) Trends Genet. **5**, 131–133
4 Thayer, M., Tapscott, S. J., Davis, R. L., Wright, W. E., Lasser, A. B. and Weintraub, H. (1989) Cell **58**, 241–248
5 Jiang, J., Hoey, T. and Levine, M. (1991) Genes Dev. **5**, 265–277
6 Regulski, M., Dessain, S., McGinnis, N. and McGinnis, W. (1991) Genes Dev. **5**, 278–286
7 Zhao, X.-Y. and Hung, M.-C. (1992) Mol. Cell. Biol. **12**, 2739–2748
8 Legraverend, C., Antonson, P., Flodby, P., Xanthopoulos, K. G. (1993) Nucleic Acids Res. **21**, 1735–1742
9 Rhodes, S. J., Chen, R., DiMattia, G. E., Scully, K. M., Kalla, K. A., Lin, S. C., Yu, V. C. and Rosenfeld, M. G. (1993) Genes Dev. **7**, 913–932
10 Shan, B., Chang, C.-Y., Jones, D. and Lee, W.-H. (1994) Mol. Cell. Biol. **14**, 229–309
11 Timchenko, N., Wilson, D. R., Taylor, L. R., Abdelsayed, S., Wilde, M., Sawadogo, M. and Darlington, G. J. (1995) Mol. Cell. Biol. **15**, 1192–1202
12 Walsh, M. J., Gongliang, S., Spidoni, K. and Kapoor, A. (1995) J. Biol. Chem. **270**, 5289–5298
13 Keller, A. D. (1995) J. Theor. Biol. **172**, 169–185
14 Davis, R. L., Weintraub, H. and Lassar, A. B. (1987) Cell **51**, 987–1000
15 Braun, T., Buschhausen, D. G., Bober, E., Tannich, E. and Arnold, H. H. (1989) EMBO J. **8**, 701–709
16 Edmondson, D. G. and Olson, E. N. (1989) Genes Dev. **3**, 628–640
17 Wright, W. E., Sassoon, D. A. and Lin, V. K. (1989) Cell **56**, 607–617
18 Rhodes, S. J. and Konieczny, S. F. (1989) Genes Dev. **3**, 2050–2061
19 Miner, J. H. and Wold, B. (1990) Proc. Natl. Acad. Sci. U.S.A. **87**, 1089–1093
20 Braun, T., Bober, E., Winter, B., Rosenthal, N. and Arnold, H. H. (1990) EMBO J. **9**, 821–831
21 Weintraub, H. (1993) Cell **75**, 1241–1244
22 Olson, E. N. and Klein, W. H. (1994) Genes Dev. **8**, 1–8
23 Rudnicki, M. A., Schnegelsberg, P. N., Stead, R. H., Braun, T., Arnold, H. H. and Jaenisch (1993) Cell **75**, 1351–1359
24 Hasty, P., Bradley, A., Morris, J. H., Edmondson, D. G., Venuti, J. M., Olson, E. N. and Klein, W. H. (1993) Nature (London) **364**, 501–506
25 Nabeshima, Y., Hanaoka, K., Hayasaka, M., Esumi, E., Li, S., Nonaka, I. and Nabeshima, Y. (1993) Nature (London) **364**, 532–535
26 Weintraub, H. (1991) Science **251**, 761–766
27 Buckingham, M. (1992) Trends Genet. **8**, 144–149
28 Zingg, J.-M., Pedraza-Alva, G. and Jost, J.-P. (1994) Nucleic Acids Res. **22**, 2234–2241
29 Benezra, R., Davis, R. L., Lockshon, D., Turner, D. L. and Weintraub, H. (1990) Cell **61**, 49–59
30 Kreider, B. L., Benezra, R., Rovera, G. and Kadesch, T. (1992) Science **255**, 1700–1702
31 Jen, Y., Weintraub, H. and Benezra, R. (1992) Genes Dev. **6**, 1466–1479
32 Hu, J. S., Olson, E. N. and Kingston, R. E. (1992) Mol. Cell. Biol. **12**, 1031–1042
33 Lassar, A. B., Davis, R. L., Wright, W. E., Kadesch, T., Murre, C., Voronova, A., Baltimore, D. and Weintraub, H. (1991) Cell **66**, 305–315
34 Zhuang, Y., Kim, G., Bartelmez, S., Cheng, P., Groudine, M. and Weintraub, H. (1992) Proc. Natl. Acad. Sci. U.S.A. **89**, 12132–12136
35 Cheng, T.-C., Tseng, B. S., Merlie, J. P., Klein, W. H. and Olson, E. N. (1995) Proc. Natl. Acad. Sci. U.S.A. **92**, 561–565
36 Mak, K.-L., To, R. Q., Kong, Y. and Konieczny, S. F. (1992) Mol. Cell. Biol. **12**, 4334–4346
37 Pollock, R. and Treisman, R. (1991) Genes Dev. **5**, 2327–2341
38 Yu, Y.-T., Breitbart, R. E., Smoot, L. B., Lee, Y., Mahdavi, V. and Nadal-Ginard, B. (1992) Genes Dev. **6**, 1783–1798
39 Edmondson, D. G., Cheng, T.-C., Cserjesi, P., Chakraborty, T. and Olson, E. N. (1992) Mol. Cell. Biol. **12**, 3665–3677
40 Yee, S.-P. and Rigby, P. W. J. (1993) Genes Dev. **7**, 1277–1289
41 Leibham, D., Wong, M., Cheng, T.-C., Schroeder, S., Weil, P. A., Olson, E. N. and Perry, M. (1994) Mol. Cell. Biol. **14**, 686–699
42 Breitbart, R., Liang, C., Smoot, L., Laheru, D., Mahdavi, V. and Nidal-Ginard, B. (1993) Development **118**, 1095–1106
43 Rana, B., Xie, Y., Mischoulon, D., Bucher, N. L. R. and Farmer, S. R. (1995) J. Biol. Chem. **270**, 18123–18132
44 Cao, Z., Umek, R. M. and McKnight, S. L. (1991) Genes Dev. **5**, 1538–1552
45 Scott, L. M., Civin, C. I., Rorth, P. and Friedman, A. D. (1992) Blood **80**, 1725–1735
46 Yeh, W.-C., Cao, Z., Classon, M. and McKnight, S. L. (1995) Genes Dev. **9**, 168–181
47 Katz, S., Kowentz-Leutz, E., Müller, C., Meese, K., Ness, S. A. and Leutz, A (1993) EMBO J. **12**, 1321–1332
48 Ness, S. A., Kowenz, L. E., Casini, T., Graf, T. and Leutz, A. (1993) Genes Dev. **7**, 749–759
49 Wu., Z., Xie., Y., Bucher, N. L. R. and Farmer, S. R. (1995) Genes Dev. **9**, 2350–2363
50 Samuelsson, L, Stromberg, K., Vikman, K., Bjursell, G. and Enerback, S. (1991) EMBO J. **10**, 3787–3793
51 Umek, R. M., Friedman, A. D. and McKnight, S. L. (1991) Science **251**, 288–292
52 Freytag, S. O. and Geddes, T. J. (1992) Science **256**, 379–382
53 Vasseur-Cognet, M. and Lane, M. D. (1993) Curr. Opin. Genet. Dev. **3**, 238–245
54 Lin, F. T. and Lane, M. D. (1994) Proc. Natl. Acad. Sci. U.S.A. **91**, 8757–8761
55 Freytag, S. O., Paielli, D. L. and Gilbert, J. D. (1994) Genes Dev. **8**, 1654–1663
56 Tontonoz, P., Hu, E. and Spiegelman, B. M. (1994) Cell **79**, 1147–1156
57 Ingham, P. W. (1988) Nature (London) **335**, 25–34
58 Driever, W. and Nüsslein-Volhard, C. (1989) Nature (London) **337**, 138–143
59 Jäckle, H. and Sauer, F. (1993) Curr. Opin. Cell Biol. **5**, 505–512
60 Lazar, M. A., Hodin, R. A., Cardona, G. and Chin, W. W. (1990) J. Biol. Chem. **265**, 12859–12863
61 Foulkes, N. S. and Sassone-Corsi, P (1992) Cell **68**, 411–414
62 Davis, I. and Ish-Horowicz, D. (1991) Cell **67**, 927–940
63 Kislauskis, H. and Singer, R. H. (1992) Curr. Opin. Cell Biol. **4**, 975–978
64 Sallés, F. J., Lieberfarb, M. E., Wreden, C., Gergen, J. P. and Strickland, S. (1994) Science **266**, 1996–1999
65 Bach, I and Yaniv, M (1993) EMBO J. **12**, 4229–4242
66 Foulkes, N. S., Borrelli, E. and Sassone-Corsi, P. (1991) Cell **64**, 739–749
67 Laoide, B. M., Foulkes, N. S., Schlotter, F. and Sassone-Corsi, P. (1993) EMBO J. **12**, 1179–1191
68 Molina, C. A. M., Foulkes, N. S., Lalli, E. and Sassone-Corsi, P. (1993) Cell **75**, 875–886
69 Mellström, B., Naranjo, J. R., Foulkes, N. S., Lafarga, M. and Sassone-Corsi, P. (1993) Neuron **10**, 655–665
70 Brindle, P., Linke, S. and Montminy, M. (1993) Nature (London) **364**, 821–824
71 Lalli, E and Sassone-Corsi, P (1994) J. Biol. Chem. **269**, 17359–17362
72 Stehle, J. H., Foulkes, N. S., Molina, C. A., Simonneaux, V. Pévet, P. and Sassone-Corsi, P. (1993) Nature (London) **365**, 314–320
73 Sassone-Corsi, P. (1994) EMBO J. **13**, 4717–4728
74 Drevet, J. R., Swevers, L. and Iatrou, K. (1995) J. Mol. Biol. **246**, 43–53
75 Wang, Z. Y., Qiu, Q., Huang, J., Gurrieri, M. and Deuel, T. F. (1995) Oncogene **10**, 415–422
76 Tanaka, T., Tanaka, K., Ogawa, S., Kurokawa, M., Mitani, K., Nishida, J., Shibata, Y., Teboul, M., Enmark, E., Li, Q., Wikstrom, A. C., Pelto-Huikko, M. and Gustafsson, J.-A. (1995) Proc. Natl. Acad. Sci. U.S.A. **92**, 2096–2100
77 Hahm, K., Ernst, P., Lo, K., Kim, G. S., Turck, C. and Smale, S. T. (1994) Mol. Cell. Biol. **14**, 7111–7123
78 Lin, Q., Luo, X. and Sawadogo, M. (1994) J. Biol. Chem. **269**, 23894–23903
79 Chen, B. P., Liang, G., Whelan, J. and Hai, T. (1994) J. Biol. Chem. **269**, 15819–15826
80 Grumont, R. J. and Gerondakis, S. (1994) Proc. Natl. Acad. Sci. U.S.A. **91**, 4367–4371
81 Das, G. and Herr, W. (1993) J. Biol. Chem. **268**, 25026–25032
82 Lillycrop, K. A. and Latchman, D. S. (1992) J. Biol. Chem. **267**, 24960–24965
83 Gogos, J. A., Hsu, T., Bolton, J. and Kafatos, F. C. (1992) Science **257**, 1951–1955
84 Buettner, R., Kannan, P., Imhof, A., Bauer, R., Yim, S. O., Glockshuber, R., van Dyke, M. W. and Tainsky, M. A. (1993) Mol. Cell. Biol. **13**, 4174–4185
85 Decker, C. J. and Parker, R. (1994) Trends Genet. **19**, 336–340
86 Beelman, C. A. and Parker, R (1995) Cell **81**, 179–183
87 Johnson, R. S., Spiegelman, B. M. and Papaioannou, V. (1992) Cell **71**, 577–586
88 Greenberg, M. E. and Ziff, E. B. (1984) Nature (London) **311**, 433–438
89 Muller, R., Bravo, R., Burckhardt, J. and Curran, T. (1984) Nature (London) **312**, 716–720
90 Zubiaga, A. M., Belasco, J. G. and Greenberg, M. E. (1995) Mol. Cell. Biol. **15**, 2219–2230
91 Winstall, E., Gamache, M. and Raymond, V. (1995) Mol. Cell. Biol. **15**, 3796–3804
92 Muhlrad, D., Decker, C. J. and Parker, R. (1994) Genes Dev. **8**, 855–866
93 Lagnado, C. A., Brown, C. Y. and Goodall, G. J. (1994) Mol. Cell. Biol. **14**, 7984–7995
94 Shyu, A.-B., Belaco, J. G. and Greenberg, M. E. (1991) Genes Dev. **5**, 221–234

95 Chen, C. Y., You, Y. and Shyu, A. B. (1992) Mol. Cell. Biol. **12**, 5748–5757

96 Schiavi, S. C., Wellington, C. L., Shyu, A. B., Chen, C. Y., Greenberg, M. E. and Belasco, J. G. (1994) J. Biol. Chem. **269**, 3441–3448

97 Hershey, J. W. B. (1991) Annu. Rev. Biochem. **60**, 717–755

98 Kozak, M. (1992) Annu. Rev. Cell Biol. **8**, 197–225

99 Lin, T.-A., Kong, X., Haystead, T. A. J., Pause, A., Belsham, G., Sonenberg, N. and Lawrence Jr., J. C. (1994) Science **266**, 653–656

100 Pause, A., Belsham, G. J., Gingras, A.-C., Donzé, O., Lin, T.-A., Lawrence, Jr., J. C. and Sonenberg, N. (1994) Nature (London) **371**, 762–767

101 Proud, C. G. (1994) Nature (London) **371**, 747–748

102 Hinnebusch, A. G. (1994) Trends Biochem. Sci. **19**, 409–414

103 Rowlands, A., Panniers, G. and Henshaw, E., C. (1988) J. Biol. Chem. **263**, 5526–5533

104 Rhoads, R. E. (1993) J. Biol. Chem. **268**, 3017–3020

105 Pantopoulos, K., Johansson, H. E. and Hentze, M. W. (1994) Prog. Nucleic Acids Res. Mol. Biol. **48**, 181–239

106 Montine, K. S. and Henshaw, E. C. (1989) Biochem. Biophys. Acta **1014**, 282–288

107 Sarre, T. F. (1989) Biosystems **22**, 311–325

108 Welsh, G. I. and Proud, C. G. (1993) Biochem. J. **294**, 625–629

109 Mueller, P. P. and Hinnebusch, A. G. (1986) Cell **45**, 201–207

110 Dever, T. E., Feng, L., Wek, R. C., Cigan, A. M., Donahue, T. F. and Hinnebusch, A. G. (1992) Cell **68**, 585–596

111 Dever, T. E., Chen, J.-J., Barber, G. N., Cigan, A. M., Feng, L., Donahue, T. F., London, I. M., Katze, M. G. and Hinnebusch, A. G. (1993) Proc. Natl. Acad. Sci. U.S.A. **90**, 4616–4620

112 Abastado, J.-P., Miller, P. F., Jackson, B. M. and Hinnebusch A. G. (1991) Mol. Cell. Biol. **11**, 486–496

113 Friedman, A. D., Landschulz, W. H. and McKnight, S. L. (1989) Genes Dev. **3**, 1314–1322

114 Descombes, P., Chojkier, M., Lichtsteiner, S., Falvey, E. and Schibler, U. (1990) Genes Dev. **4**, 1541–1551

115 Chandrasekaran, C. and Gordon, J. I. (1993) Proc. Natl. Acad. Sci. U.S.A. **90**, 8871–8875

116 Buck, M., Turler, H. and Chojkier, M. (1994) EMBO J. **13**, 851–860

117 Hendricks-Taylor, L. R. and Darlington, G. J. (1995) Nucleic Acids Res. **23**, 4726–4733

118 Descombes, P and Schibler, U (1991) Cell **67**, 569–579

119 Ossipow, V., Descombes, P. and Schibler, U. (1993) Proc. Natl. Acad. Sci. U.S.A. **90**, 8219–8223

120 Lin, F.-T., MacDougald, O. A., Diehl, A. M. and Lane, M. D. (1993) Proc. Natl. Acad. Sci. U.S.A. **90**, 9606–9610

121 Calkhoven, C. F., Bouwman, P. R. J., Snippe, L. and AB, G (1994) Nucleic Acids Res. **22**, 5540–5547

122 Nerlov, C. and Ziff, E. B. (1994) Genes Dev. **8**, 350–362

123 MacDougald, O. A., Cornelius, P., Liu, R. and Lane, M. D. (1995) J. Biol. Chem. **270**, 647–654

124 Birkenmeier, E. H., Gwynn, B., Howard, S., Jerry, J., Gordon, J. L., Landschulz, W. H. and McKnight, S. L. (1989) Genes Dev. **3**, 1146–1156

125 Manchado, C., Yubero, P., Vinas, O., Iglesias, R., Villarroya, F., Mampel, T. and Giralt, M. (1994) Biochem. J. **302**, 695–700

126 Williams, S. C., Cantwell, C. and Johnson, P. F. (1991) Genes Dev. **5**, 1553–1567

127 Wang, N. D., Finegold, M. J., Bradley, A., Ou, C. N., Abdelsayed, S. V., Wilde, M. D., Taylor, L. R., Wilson, D. R. and Darlington, G. J. (1995) Science **269**, 1108–1112

128 Hann, S. R., King, M. W., Bentley, D. L., Anderson, C. W. and Eisenman, N. (1988) Cell **52**, 185–195

129 Hann, S. R., Sloan-Brown, K. and Spotts, G. D. (1992) Genes Dev. **6**, 1229–1240

130 Tahara, S. M., Dietlin, T. A., Dever, T. E., Merrick, W. C. and Worrilow, L. M. (1991) J. Biol. Chem. **266**, 3594–3601

131 Zimmer, A., Zimmer, A. M. and Reynolds, K (1994) J. Cell Biol. **127**, 1111–1119

132 Hann, R. R. (1994) Biochimie **76**, 880–886

133 Falvey, E., Fleury-Olela, F. and Schibler, U. (1995) EMBO J. **14**, 4307–4317

134 Delmas, V., Laoide, B. M., Masquilier, D., Groot, R. P., Foulkes, N. S. and Sassone-Corsi, P. (1992) Proc. Natl. Acad. Sci. U.S.A. **89**, 4226–4230

135 Kozak, M. (1991) J. Cell. Biol. **115**, 887–903

136 Claret, F.-X., Chapel, S., Garcés, J., Tsai-Pflugfelder, M., Bertholet, C., Shapiro, D. J., Wittek, R. and Wahli, W. J. Biol. Chem. **269**, 14047–14055

137 Geballe, A. P. and Morris, D. R. (1994) Trends Biochem. Sci. **19**, 159–164

138 Hentze, M. W. (1995) Curr. Opin. Cell Biol. **7**, 393–398

139 Hunter, T and Karin, M. (1992) Cell **70**, 375–387

140 Hill, C. S. and Treisman, R. (1995) Cell **80**, 199–211

141 Moll, T., Tebb, G., Syrana, U., Robitsch, H. and Namyth, K. (1991) Cell **66**, 743–758

142 Jans, D. A., Moll., T., Nasmyth, K. and Jans, P. (1995) J. Biol. Chem. **270**, 17064–17067

143 Nolan, G. P. and Baltimore, D. (1992) Curr. Opin. Genet. Dev. **2**, 211–220

144 Blank, V., Kourilsky, P. and Israël, A. (1992) Trends Genet. **17**, 135–140

145 Beg, A. A. and Baldwin, A. S. (1993) Genes Dev. **7**, 2064–2070

146 Baeurle, P. A. and Baltimore, D. (1988) Science **242**, 540–546

147 Inoue, J.-I., Kerr, L. D., Rashid, D., Davis, N., Bose, Jr, H. R. and Verma, I. M. (1992) Proc. Natl. Acad. Sci. U.S.A. **89**, 4333–4337

148 Bork, P. (1993) Proteins **17**, 363–374

149 Ghosh, S. and Baltimore, D. (1990) Nature (London) **344**, 678–682

150 Henkel, T., Machleidt, T., Alkalay, I., Kronke, M., Ben-Neriah, Y. and Baeuerle, P. A. (1993) Nature (London) **365**, 182–185

151 Beg, A. A., Finco, T. S., Nantermet, P. V. and Baldwin, Jr., A. S. (1993) Mol. Cell. Biol. **13**, 3301–3310

152 Palombella, V. J., Rando, O. J., Goldberg, A. L. and Maniatis, T. (1994) Cell **78**, 773–785

153 Chen, Z., Hagler, J., Palombella, V. J., Melandri, F., Scherer, D., Ballard, D. and Maniatis, T (1995) Genes Dev. **9**, 1586–1597

154 Boyle, W. J., Smeal, T., Defize, L. H., Angel, P., Woodgett, J. R., Karin, M. and Hunter, T. (1991) Cell **64**, 573–584

155 Lin, A., Frost, J., Deng, T., Smeal, T., Al-Alawi, N., Kikkawa, U., Hunter, T., Brenner, D. and Karin, M. (1992) Cell **70**, 777–789

156 Berberich, S. J. and Cole, M. D. (1992) Genes Dev. **6**, 166–176

157 Kapiloff, M. S., Farkash, Y., Wegner, M. and Rosenfeld, M. G. (1991) Science **253**, 786–789

158 Janknecht, R., Hipskind, R. A., Houthaeve, T., Nordheim, A. and Stunnenberg, H. G. (1992) EMBO J. **11**, 1045–1054

159 Alvarez, E., Northwood, I. C., Gonzalez, F. A., Latour, D. A., Seth, A., Abate, C., Curran, T. and Davis, R. J. (1991) J. Biol. Chem. **266**, 15277–15285

160 Binétruy, B., Smeal, T. and Karin, M. (1991) Nature (London) **351**, 122–127

161 Pulverer, B. J., Kyriakis, J. M., Avruch, J., Nikolakaki, E. and Woodgett, J. R. (1991) Nature (London) **353**, 670–674

162 Smeal, T., Binétruy, B., Mercola, D., Birrer, M. and Karin, M. (1991) Nature (London) **354**, 494–496

163 Hibi, M., Lin, A., Smeal, T., Minden, A. and Karin, M. (1993) Genes Dev. **7**, 2135–2148

164 Minden, A., Lin, A., Smeal, T., Derijard, B., Cobb, M., Davis, R. and Karin, M. (1994) Mol. Cell Biol. **14**, 6683–6688

165 Kyriakis, J. M., Banerjee, P., Nikolakaki, E., Dai, T., Rubie, E. A., Ahmad, M. F., Avruch, J. and Woodgett, J. R. (1994) Nature (London) **369**, 156–160

166 Nakajima, T., Kinoshita, S., Sasagawa, T., Sasaki, K., Naruto, M., Kishimoto, T. and Akira, S. (1993) Proc. Natl. Acad. Sci. U.S.A. **90**, 2207–2211

167 Kowentz-Leutz, E., Twamley, G., Ansieau, S. and Leutz, A. (1994) Genes Dev. **8**, 2781–2791

168 Williams, S. C., Baer, M., Dillner, A. J. and Johnson, P. F. (1995) EMBO J. **14**, 3170–3183

169 Wegner, M., Cao, Z. and Rosenfeld, M. G. (1992) Science **256**, 370–373

170 Mahony, C. W., Shuman, J., McKnight, S. L., Chen, H.-C. and Huang, K.-P. (1992) J. Biol. Chem. **267**, 19396–19403

171 Wahli, W. and Martinez, E. (1991) FASEB J. **5**, 2243–2249

172 Glass, C. K. (1994) Endocr. Rev. **15**, 391–407

173 Beato. M., Herrlich, P. and Schütz, G. (1995) Cell **83**, 851–857

174 Kastner, P., Mark, M. and Chambon, P. (1995) Cell **83**, 859–869

175 Mangelsdorf, D. J., Thummel, C., Beato, M., Herrlich, P., Schütz, G., Umesono, K., Blumberg, B., Kastner, P., Mark, M., Chambon, P. and Evans, R. M. (1995) Cell **83**, 835–839

176 Mangelsdorf, D. J. and Evans, R. M. (1995) Cell **83**, 841–850

177 Thummel, C. S. (1995) Cell **83**, 871–877

178 Gronemeyer, G. and Miras, D. (1995) Nature (London) **375**, 190–191

179 Leblanc, B. P. and Stunnenberg, H. G. (1995) Genes Dev. **9**, 1811–1816

180 Tora, L., White, J., Bron, C., Tasset, D., Webster, N., Scheer, E. and Chambon, P. (1989) Cell **59**, 477–487

181 Hard, T., Kellenbach, E., Boelens, R., Maler, B. A., Dahlman, K., Freedman, L. P., Carlstedt-Duke, J., Yamamoto, K. R., Gustafsson, J. A. and Kaptein, R. (1990) Science **266**, 3107–3112

182 Schwabe, J. W., Neuhaus, D. and Rhodes, D. (1990 Nature (London) **348**, 458–461

183 Luisi, B. F., Xu, W. X., Otwinowski, Z., Freedman,L. P., Yamamoto, K. R. and Sigler, P. B. (1991) Nature (London) **352**, 497–505

184 Knegtel, R. M., Katahira, M., Schilthuis, J. G., Bovin, A. M., Boelens, R., Eib, D., van der Saag, P. T. and Kaptein, R. (1993) J. Biomol. NMR **3**, 1–17

185 Lee, M. S., Kliewer, S. A., Provencal, J., Wright, P. E. and Evans, R. M. (1993) Science **260**, 1117–1121

186 Rastinejad, F., Perlmann, T., Evans, R. M. and Sigler, P. B. (1995) Nature (London) **375**, 203–211

187 Harding, H. P. and Lazar, M. A. (1993) Mol. Cell. Biol. **13**, 3113–3121
188 Lala, D. S., Rice, D. A. and Parker, K. L. (1992) Mol Endocrinol. **6**, 1249–1258
189 Wilson, T. E., Fahrner, T. J. and Milbrandt, J. (1993) Mol. Cell. Biol. **13**, 5794–5804
190 McBroom, L. B. D., Flock, G. and Giguère, V. (1995) Mol. Cell. Biol. **15**, 796–808
191 Pratt, W. B. (1993) J. Biol. Chem. **268**, 21455–21458
192 Martinez, E., Givel, F. and Wahli, W. (1987) EMBO J. **6**, 3719–3727
193 Klock, G., Strahle, U. and Schutz, G. (1987) Nature (London) **329**, 734–736
194 Picard, D., Dhursheed, B., Garabedian, M. J., Fortin, M. G., Lindquist, S. and Yamamoto, K. R. (1990) Nature (London) **348**, 16–168
195 Bohen, S. P., Kralli, A. and Yamamoto, K. R. (1995) Science **268**, 1303–1304
196 Cavaillès, V., Cauvois, S., L'Horset, F., Lopez, G., Hoare, S., Kushner, P. J. and Parker, M. G. (1995) EMBO J. **14**, 3741–3751
197 Halachmi, S., Marden, E., Martin, G., MacKay, H., Abbondanza, C. and Brown, M. (1994) Science **264**, 1455–1458
198 Le Douarin, B., Zechel, C., Garnier, J. M., Lutz, Y., Tora, L., Pierrat, B., Heery, D., Gronemeyer, H., Chambon, P. and Losson, R. (1995) EMBO J. **14**, 2020–2033
199 Lee, J. W., Ryan, F., Swaffield, J. C., Johnston, S. A. and Moore, D. D. (1995) Nature (London) **374**, 91–94
200 vom Baur, E., Zechel, C., Heery, D., Heine, M., Garnier, J. M., Vivat, V., Le Douarin, B., Gronemeyer, H., Chambon, P. and Losson, R. (1995) EMBO J. **15**, 110–124
201 Oñate, S. A., Tsai, S. Y., Tsai, M.-J. and O'Malley, B. W. (1995) Science **270**, 1354–1357
202 Dalman, F. C., Koenig, R. J., Perdew, G. H., Massa, E. and Pratt, W. B. (1990) J. Biol. Chem. **265**, 3615–3618
203 Dalman, F. C., Sturzenbecker, L. J., Levin, A. A., Lucas, D. A., Perdew, G. H., Petkovitch, M., Chambon, P., Grippo, J. F. and Pratt, W. B. (1991) Biochemistry **30**, 5605–5608
204 Näär, A. M., Boutin, J. M., Lipkin, S. M., Yu, V. C., Holloway, J. M., Glass, C. K. and Rosenfeld, M. G. (1991) Cell **65**, 1267–1279
205 Umesono, K., Murakami, K. K., Thompson, C. C. and Evans, R. M. (1991) Cell **65**, 1255–1266
206 Yu, V. C., Delsert, C., Andersen, B., Holloway, J. M., Devary, O. M., Näär, A. M., Kim, S. Y., Boutin, J.-M., Glass, C. K. and Rosenfeld, M. G. (1991) Cell **67**, 1251–1266
207 Brugge, T. H., Pohl, J., Lonnoy, O. and Stunnenberg, H. G. (1992) EMBO J. **11**, 1409–1418
208 Kliewer, S. A., Umesono, K., Noonan, D. J., Heyman, R. A. and Evans, R. M. (1992) Nature (London) **358**, 771–774
209 Leid, M., Kastner, P., Lyons, R., Nakshatri, H., Saunders, M., Zacharewski, T., Chen, J.-T., Staub, A., Garnier, J.-M., Mader, S. and Chambon, P. (1992) Cell **68**, 377–395
210 Marks, M. S., Hallenbeck, P. L., Nagata, T., Segars, J. H., Apella, E., Nikodem, V. M. and Ozato, K. (1992) EMBO J. **11**, 1419–1435
211 Zhang, X. K., Hoffmann, B., Tran, P. B., Graupner, G. and Pfahl, M. (1992) Nature (London), **355**, 441–446
212 Isseman, I., Prince, R. A., Tugwood, J. D. and Green, S. (1993) J. Mol. Endocrinol. **11**, 37–47
213 Bourguet, W., Ruff, M., Chambon, P., Gronemeyer, H. and Miras, D. (1995) Nature (London) **375**, 377–382
214 Renaud, J.-P., Rochel, N., Ruff, M., Vivat, V., Chambon, P., Gronemeyer, H. and Moras, D. (1995) Nature (London) **378**, 681–689
215 Wagner, R. L., Apriletti, J. W., McGrath, M. E., West, B. L., Baxter, J. D. and Fletterick, R. J. (1995) Nature (London) **378**, 690–697
216 Forman, B. M., Umesono, K., Chen, J. and Evans, R. M. (1995) Cell **81**, 541–550
217 Damm, K., Thompson, C. C. and Evans, R. M. (1989) Nature (London) **339**, 593–597
218 Baniahmad, A., Köhne, A. C. and Renkawitz, R. (1992) EMBO J. **11**, 1015–1023
219 Baniahmad, A, Leng, X., Burris, T. P., Tsai, S. Y., Tsai, M. J. and O'Malley, B. W. (1995) Mol. Cell. Biol. **15**, 76–86
220 Casanova, J., Helmer, E., Selmi-Ruby, S., Qi, J. S., Au-Flieger, M., Desai-Yajnik, V., Koudinova, N., Yarm, F., Raaka, B. M. and Samuels, H. H. (1994) Mol. Cell. Biol. **14**, 5756–5765
221 Chen, J. D. and Evans, R. M. (1995) Nature (London) **377**, 454–457
222 Forman, B. M., Goode, E., Chen, J., Oro, A. E., Bradley, D. J., Perlmann, T., Noonan, D. J., Burka, L. T., McMorris, T., Lamph, W. W., Evans, R. M. and Weinberger, C. (1995) Cell **81**, 686–687

223 Hörlein, A. J., Näär, A. M., Heinzel, T., Torchia, J., Gloss, B., Kurokawa, R., Ryan, A., Kamei, Y., Söderström, M., Glass, C. K. and Rosenfeld, M. G. (1995) Nature (London) **377**, 397–404
224 Kurokawa, R., Söderström, M., Hörlein, A., Halachmi, S., Brown, M., Rosenfeld, M. G. and Glass, C. K. (1995) Nature (London) **377**, 451–454
225 Perlmann, T. and Vennström, B. (1995) Nature (London) **377**, 387–388
226 Kurokowa, R., DiRenzo, J., Boehm, M., Sugarman, J., Gloss, B., Rosenfeld, M. G., Heyman, R. A. and Glass, C. K. (1994) Nature (London) **371**, 528–531
227 Landschulz, W. H., Johnson, P. F. and McKnight, S. L. (1988) Science **240**, 1759–1763
228 Kouzarides, T. and Ziff, E. (1989) Nature (London) **340**, 568–571
229 Vinson., C. R., Sigler, P. B. and McKnight, S. L. (1989) Science **246**, 911–916
230 O'Shea, E. K., Klemm, J. D., Kim, P. S. and Alber, T. (1991) Science **254**, 539–544
231 Alber, T. (1992) Curr. Opin. Genet. Dev. **2**, 205–210
232 Ellenberger, T. E., Brandl, C. J., Struhl, K. and Harrison, S. C. (1992) Cell **71**, 1223–1237
233 Busch, S. L. and Sassone-Corsi, P. (1990) Trends Genet. **6**, 36–40
234 Ron, D. and Habener, J. F. (1992) Genes Dev. **6**, 439–453
235 Fornace, A. J., Neibert, D. W., Hollander, M. C., Luethy, J. D., Papathanasiou, M., Fragoli, J. and Holbrook, N. J. (1989) Mol. Cell. Biol. **9**, 4196–4203
236 Luethy, J. D. and Holbrook, N. J. (1992) Cancer Res. **52**, 5–10
237 Price, B. and Calderwood, S. (1992) Cancer Res. **52**, 3814–3817
238 Batchvarova, N., Wang, X.-Z. and Ron, D. (1995) EMBO J. **14**, 4654–4661
239 Swillens, S. and Pirson, I. (1994) Biochem. J. **301**, 9–12
240 Poli, V., Mancini, F. P. and Cortese, R. (1990) Cell **63**, 643–653
241 Nolan, G. P. (1994) Cell **77**, 795–798
242 LeClair, K. P., Blanar, M. A. and Sharp, P. A. (1992) Proc. Natl. Acad. Sci. U.S.A. **89**, 8145–8149
243 Stein, B., Cogswell, P. C. and Baldwin, Jr., A. S. (1993) Mol. Cell. Biol. **13**, 3964–3974
244 Diehl, J. A. and Hannink, M. (1994) Mol. Cell. Biol. **14**, 6635–6646
245 Jain, J., McCaffrey, P. G., Valge-Archer, V. E. and Rao, A. (1992) Nature (London) **356**, 801–804
246 Jain, J., Miner, Z. and Rao, A. (1993) J. Immunol. **151**, 837–848
247 Diamond, M. I., Miner, J. N., Yoshinaga, S. K. and Yamamoto, K. R. (1990) Science **249**, 1266–1272
248 Bengal, E., Ransone, L., Scharfmann, R., Dwarki, V. J., Tapscott, S. J., Weintraub, H. and Verma, I. M. (1992) Cell **68**, 507–519
249 Li, L., Chambard, J.-C., Karin, M. and Olson, E. N. (1992) Genes Dev. **6**, 676–689
250 Brindle, P. K. and Montminy, M. R. (1992) Curr. Opin. Genet. Dev. **2**, 199–204
251 Chrivia, J. C., Kwok, R. P., Lamb, N., Hagiwara, M., Montminy, M. R. and Goodman, R. H. (1993) Nature (London) **365**, 855–859
252 Arias, J., Alberts, A. S., Brindle, P., Claret, F. X., Smeal, T., Karin, M., Feramisco, F. and Montminy, M. (1994) Nature (London) **370**, 226–229
253 Kwok, R. P. S., Lundblad, J. R., Chrivia, J. C., Richards, J. P., Bächinger, H. P., Brennan, R. G., Roberts, S. G. E., Green, M. R. and Goodman, R. H. (1994) Nature (London) **370**, 223–226
254 Nordheim, A. (1994) Nature (London) **370**, 177–178
255 Wagner, S. and Green, M. R. (1993) Nature (London) **262**, 395–399
256 Baranger, A. M., Palmer, C. R., Hamm, M. K., Giebler, H. A., Brauweiler, A., Nyborg, J. K. and Schepartz, A. (1995) Nature (London) **376**, 606–608
257 Perini, G., Wagner, S. and Green, M. R. (1995) Nature (London) **376**, 602–605
258 Mendel, D. B., Khavari, P. A., Conley, P. B., Graves, M. K., Hansen, L. P., Admon, A. and Crabtree, G. R. (1991) Science **254**, 1762–1767
259 Citron, B. A., Davis, M. D., Milstien, S., Gutierrez, J., Mendel, D. B., Crabtree, G. R. and Kaufman, S. (1992) Proc. Natl. Acad. Sci. U.S.A. **89**, 11891–11894
260 Hansen, L. P. and Crabtree, G. R. (1993) Curr. Opin. Genet. Dev. **3**, 246–253
261 Ficner, R., Sauer, U. H., Stier, G. and Suck, D. (1995) EMBO J. **14**, 2034–2042
262 Tsai, M.-J. and O'Malley, B. W. (1994) Annu. Rev. Biochem. **63**, 451–486
263 Moudgil, V. K. (1990) Biochim. Biophys. Acta **1055**, 243–258
264 He, J. and Furmanski, P. (1995) Nature (London) **373**, 721–724
265 Abate, C., Patel, L., Rauscher, III, F. J. and Curran, T. (1990) Science **249**, 1157–1161
266 Bandyopadhyay, S. and Gronostajski, R. M. (1994) J. Biol. Chem. **269**, 29949–29955
267 Arnone, M. I., Zannini, M. and Di Lauro, R. (1995) J. Biol. Chem. **270**, 12048–12055
268 Ihle, J. N. and Kerr, I. M. (1995) Trends Genet. **11**, 69–74
269 Kerszberg, M. and Changeux, J.-P. (1994) Proc. Natl. Acad. Sci. U.S.A. **91**, 5823–5827

Biochem. J. (1996) **315**, 345–361 (Printed in Great Britain)

REVIEW ARTICLE
Peptide models for membrane channels

Derek MARSH

Max-Planck-Institut für biophysikalische Chemie, Abt. Spektroskopie, D-37077 Göttingen, Federal Republic of Germany

Peptides may be synthesized with sequences corresponding to putative transmembrane domains and/or pore-lining regions that are deduced from the primary structures of ion channel proteins. These can then be incorporated into lipid bilayer membranes for structural and functional studies. In addition to the ability to invoke ion channel activity, critical issues are the secondary structures adopted and the mode of assembly of these short transmembrane peptides in the reconstituted systems. The present review concentrates on results obtained with peptides from ligand-gated and voltage-gated ion channels, as well as proton-conducting channels. These are considered within the context of current molecular models and the limited data available on the structure of native ion channels and natural channel-forming peptides.

INTRODUCTION

Membrane channels provide one of the major avenues of communication across cell membranes. Of these, the ion channels that are responsible for nerve excitation and control of the transmembrane potential are the most studied and will be concentrated on here. The availability of amino acid sequences for many membrane channel proteins opens up the possibility of direct structural and functional studies on synthetic peptides that correspond to putative transmembrane and channel-lining sequences [1–3]. Such studies and their relevance to the intact channel assemblies are the subject of this review. This model membrane approach represents a considerable simplification of the system that allows detailed biophysical characterization, but carries with it attendant uncertainties in extrapolation to the natural state. Amongst the critical issues are the relative importance of different secondary structures and the mode of assembly of short transmembrane peptides in reconstituted systems.

The review focuses on the following topics, taken in order. First, methods for identification of putative transmembrane domains and pore-lining sequences are dealt with, before going on to a review of the limited amount of direct structural information that is available on intact channel proteins. Then the structures of natural channel-forming peptides are considered because currently it is for these that the most detailed information is available. The remainder of the review then concentrates on synthetic peptides that correspond to the putative transmembrane domains of intact channel proteins. Topics covered are secondary structure, channel-forming activity and mode of assembly in reconstituted systems. The different natural and synthetic peptides discussed are listed in Table 1.

DEDUCTIONS FROM PRIMARY SEQUENCE AND MUTAGENESIS

Hydropathy plots

The usual method for identifying putative transmembrane segments in a channel of known sequence is from hydropathy plots of the hydrophobicity of adjacent amino acid residues averaged over a moving window of suitable length [33–35]. The hydro-phobicity scale is defined from the transfer free energy of amino acids between organic solvents and water, and statistics on the distribution of residues in proteins of known structure. Typically, a stretch of 20–23 amino acids of high hydrophobicity is sufficient to define a transmembrane segment, if this is in an α-helical conformation. For a β-sheet conformation less than half this number would be required, if the β-strands are not tilted (e.g. see [36]). Putative transmembrane domains identified from hydropathy profiles are given in Figure 1 for the major families of ion channel proteins, both ligand-gated and voltage-gated. These schemes are consistent with the indicated segments being α-helical in conformation and they largely provide the current working models for the structure of these different families of channels. A notable exception is that of the nicotinic acetylcholine receptor for which recent structural data suggest that a large part of the transmembrane segments is not α-helical but possibly β-sheet [37]. For ionotropic glutamate receptors, recent evidence (reviewed in [38]; see also [39]) suggests that the C-terminus may be located on the opposite (i.e. cytoplasmic) side of the membrane from the N-terminus. This would imply an odd number of transmembrane segments, unlike the prototypic scheme indicated for ligand-gated channels in Figure 1 (see [39]).

Hydrophobic moment

The hydropathy profile is sufficient to identify putative transmembrane sequences if these are predominantly hydrophobic. The method has been verified for the bacterial reaction centre structure [40] and for bacteriorhodopsin [35]. In integral proteins that contain multiple transmembrane segments, however, this must not always be the case. For these more complex structures, all or part of certain transmembrane segments may contact other protein segments, rather than the hydrophobic lipid chains. In this case, depending on the internal protein contacts, the hydro-phobicity may be lower [41] and, for pore-lining segments, a partly hydrophilic face will be expected. The hydrophobic moment [42,43] may be used to identify amphipathic segments, such as those lining aqueous channels, and the direction of the hydrophobic moment can then give information on the orientation of the transmembrane segment within the membrane plane. For complex, polytopic, transmembrane proteins, how-

---

Abbreviations used: Aib, α-aminoisobutyric acid; FTIR, Fourier-transform IR; DCCD, dicyclohexylcarbodi-imide; GABA, γ-aminoisobutyric acid; CFTR, cystic fibrosis transmembrane conductance regulator; MDR, multidrug resistance; NMDA, N-methyl-D-aspartate.

Table 1   Natural channel-forming peptides of known structure and putative transmembrane segments of channel proteins

N, number of residues; $\langle H \rangle$, mean hydrophobicity; $\langle \mu_\alpha \rangle$ and $\langle \mu_\beta \rangle$, absolute values of mean hydrophobic moment of α-helical and β-sheet structures respectively, calculated with the normalized `consensus´ hydrophobicity scale [2,42]. Sequences are given using the standard single-letter code, with U=Aib, O=hydroxyproline and J=isovaline. Alignments for voltage-activated channels are given according to [32]. [a]Defininitions: gA, gramicidin A; alm, alamethicin; L-zrv, Leu-zervamicin; mel, melittin; δ-tx, δ-haemolysin; M1δAChR and M2δAChR, M1 and M2 segments of the δ-subunit of *Torpedo californica* nicotinic acetylcholine receptor; M2α4AChR, M2 segment of α_4 subunit of rat neuronal nicotinic acetylcholine receptor; MAβAChR, MA segment of the β-subunit of *Torpedo* nicotinic acetylcholine receptor (in the link between the M3 and M4 segments); M1GlyR, M2GlyR and M4GlyR, M1, M2 and M4 segments of the α_1-subunit of rat glycine receptor; M2NMDAR, M2 segment of NMDA receptor; M1CFTR–M6CFTR, M1–M6 segments of the CFTR Cl channel; CaIVS1–CaIVS6, S1–S6 segments from repeat IV of L-type voltage-gated Ca channel; CaIIS2, S2 segment from repeat II of dihydropyridine receptor Ca channel; KS2 and KS4, S2 and S4 segments of *Drosophila* Shaker voltage-activated K channel; NaIS3 and NaIS6, S3 and S6 segments from repeat I of rat brain voltage-activated Na channel; NaIS4 and NaIVS4, S4 segment of repeats I and IV of *Electrophorus electricus* voltage-activated Na channel; NaIS4(r), S4 segment of repeat I of rat brain voltage-activated Na channel; NaIVS4-S45, S4 segment with S4–S5 linker from repeat IV of *E. electricus* Na channel; CaIVH5, H5 segment from repeat IV of L-type voltage-gated Ca channel; NaIH5–NaIVH5, putative pore regions of repeats I–IV of *Electrophorus electricus* voltage-activated Na channel (designated PR-I to PR-IV); KH-5, KH-5-N, KH5 and KH5 pept., putative pore region of *Drosophila* Shaker A voltage-gated K channel; TM-minK and K26, apolar domain of rat IsK (minK) slow voltage-activated K channel; Pep-8a, central hydrophobic segment from subunit 8 of *Saccharomyces cerevisiae* proton F_1F_0-ATPase.

Peptide[a]	sequence	N	$\langle H \rangle$	$\langle \mu_\alpha \rangle$	$\langle \mu_\beta \rangle$	Ref
natural peptides:						
gA	f-VGALAVVVWLWLWLW-Etn	15	0.90	–	–	4
alm	Ac-UPUAUAQUVUGLUPVUUEQF-OH	20	0.54	0.23	–	5
L-zrv	Ac-LIQJITULUOQUOUPF-OH	16	0.56	0.34	–	6
mel	f-GIGAVLKVLTTGLPALISWIKRKRQQ-NH₂	26	0.10	0.35	–	7
δ-tx	f-MAQDIISTIGDLVKWIIDTVNKFTKK	26	0.12	0.60	–	8
transmembrane segments (ligand-gated):						
M1δAChR	LFYVINFITPCVLISFLASLAFY	23	0.71	0.09	0.12	9
M2δAChR	EKMSTAISVLLAQAVFLLLTSQR	23	0.24	0.25	0.03	9
M2α4AChR	EKVTLCISVLLSLTVFLLLITE	22	0.52	0.11	0.31	10
MAβAChR	GLTQPVTLPQDLKEAVEAIKYIAEQLK	27	0.06	0.34	0.21	11
M1GlyR	PYLIQMYIPSLLIVILSWISFWA	23	0.68	0.20	0.03	12
M2GlyR	PARVGLGITTVLTMTTQSSGSRA	23	0.10	0.21	0.20	12
	ARVGLGITTVLTMTTQSSGSRA	22	0.10	0.22	0.22	13
M4GlyR	RIGFPMAFLIFNMFYWIIYK	20	0.52	0.31	0.17	13
M2NMDAR	ALTLSSAMWFSWGVLLNSGIGE	22	0.48	0.12	0.17	3
transmembrane segments (phosphorylation regulated):						
M1CFTR	RFMFYGIFLYLGEVTKAVQPLLLG	24	0.42	0.28	0.18	14
M2CFTR	RSIAIYLGIGLCLLFIVRTLLL	22	0.55	0.20	0.19	14
M3CFTR	GLALAHFVWIAPLQVALLMGLI	22	0.74	0.08	0.10	14
M4CFTR	ASAFCGLGFLIVLALFQAGLG	21	0.71	0.10	0.01	14
M5CFTR	SAFFFSGFFVVFLSVLPYALI	21	0.79	0.09	0.09	14
M6CFTR	KGIILRKIFTTISFCIVLRMAV	22	0.33	0.04	0.34	14
transmembrane segments (voltage-gated):						
CaIVS1	TYFEYLMFVLILLNTICLAMQH	22	0.52	0.16	0.01	15
CaIIS2	Ac-DFANRVLLSLFTIEMLLKMYGL-NH₂	22	0.33	0.21	0.01	16
CaIVS2	IAMNILNMLFTGLFTVEMILK	21	0.52	0.33	0.16	15
KS2	ITDPFFLIETLCIIWFTFELTVRFLA	26	0.52	0.14	0.29	17
CaIVS3	DPWNVFDFLIVIGSIIDVILSE	22	0.52	0.23	0.12	15
NaIS3	DPWNWLDFTVITFAYVTEFVDL	22	0.39	0.15	0.13	18
NaIS4	RTFRVLRALKTITIFPGLKTIVRA	24	0.02	0.23	0.25	19
NaIS4(r)	ALRTFRVLRALKTISVIPGLK	21	0.06	0.20	0.16	20
NaIVS4	RVIRLARIARVLRLIRAAKGIR	22	-0.25	0.16	0.10	21
NaIVS4-S45	Ac-TLFRVIRLARIARVLRLIRAAKGIRTLLFAIMMS-NH₂	34	0.09	0.16	0.00	22
CaIVS4	NSRISITFFRLFRVMRLIKLLSR	23	-0.08	0.30	0.36	15
KS4	Me-ILRVIRLVRVFRIFKLSRHS-NH₂	20	-0.10	0.11	0.02	23
CaIVS5	YVALLIVMLFFIYAVIGMQMFGK	23	0.71	0.10	0.20	15
CaIVS6	VFYFISFYMLCAFLIINLFVAVI	23	0.85	0.20	0.00	15
NaIS6	IFFVLVIFLGSFTLINLILAVV	22	0.91	0.13	0.04	24
CaIVH5	FQTFPGAVLLLFRCATGEAWQ	21	0.32	0.28	0.10	15
NaIH5	DNFAWTFLCLFRLMLQDYWENLYQMT	26	0.18	0.31	0.27	25
NaIIH5	DFFHSFLIVFRALCGEWIETMWDCME	26	0.31	0.32	0.18	25
NaIIIH5	DNAGMGYLSLLQVSTFKGWMDIMYA	25	0.30	0.11	0.12	25
NaIVH5	KKQGGVDDIFNFETFGNSMICLFEITTSAGWDGLLL	36	0.25	0.08	0.03	25
KH-5	AFWWAVVTMTTVGYGDMTPVG	21	0.49	0.10	0.07	26
KH-5-N	SFFKSIPDAFWWAVVTMTTVGYG	23	0.44	0.29	0.05	26
KH5	f-DAFWWAVVTMTTVGYGDMT-NH₂	19	0.41	0.10	0.15	27
KH5 pept.	KSIPDAFWWAVVTMTTVGYGDMTPGK	26	0.26	0.21	0.04	28
TM-minK	SKLEALYILMVLGFFGFFTLGIMLSYIRSKKL	32	0.40	0.06	0.06	29
K26	K_EALYILMVLGFFGFFTLGIMLSYIR	26	0.54	0.19	0.03	30
transmembrane segment (proton ATPase):						
Pep-8a	Ac-GFLLMITLLILFSQFFLPMILR-NH₂	22	0.66	0.24	0.07	31

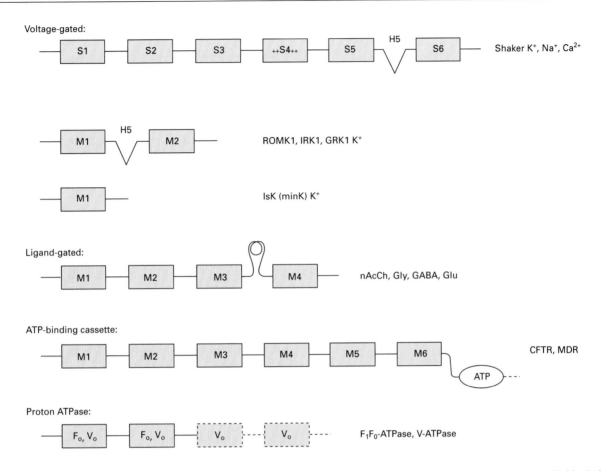

Figure 1   Arrangement of the putative transmembrane segments in the sequences of ion-channel proteins, or their subunits, as identified by hydropathy analysis

The N-terminus is on the left and all putative transmembrane sequences (boxes) consist of approx. 20 or more residues. For voltage-gated Na and Ca channels, the arrangement shown is for one of the four internal repeats in the sequence. For the voltage-gated channels, segment S4 does not have a high net hydrophobicity because every third residue is positively charged; the intervening residues are mainly hydrophobic and it is thought to be a transmembrane segment contributing to the voltage sensor. The region marked H5 is identified from mutational and blocking experiments as being associated with the pore of voltage-gated channels. For ligand-gated channels, the arrangement applies to all four (homologous) subunits. nAcCh, nicotinic acetylcholine. For ATP-binding cassette (ABC) channels, only the first of the two repeats is shown. For the CFTR the two repeats are linked by the intervening regulatory phosphorylation domain that is absent from the multidrug-resistance (MDR) P-glycoprotein. For F_1F_0-type ATPases, the proteolipid subunit c, which is thought to form the proton channel, corresponds to only half of that indicated for the 16 kDa polypeptide of the V-ATPases.

ever, not all highly amphiphilic stretches are necessarily transmembrane. A case in point is the MA segment of the nicotinic acetylcholine receptor [43], which is no longer thought to be a pore-lining sequence [11]. For the latter, the more hydrophobic M2 segment is accepted as constituting the lining of the pore (see below). The mean hydrophobicities and the absolute values of the mean hydrophobic moment (calculated for both α-helical and β-sheet structures) for various membrane peptides that have been studied are given in Table 1.

Mutagenesis and labelling

The prediction methods based on the sequence can be augmented by a variety of chemical labelling [44] and/or mutational experiments, in addition to proteolysis [45] and antibody labelling [46]. Cysteine-scanning mutagenesis can be used, in which label sites are introduced systematically at positions throughout the sequence. Examples of applications are the nicotinic acetylcholine receptor [47], voltage-gated K^+ channels [48,49] and the V-ATPase proton channel [50]. Some of these techniques, with

specific regard to channel structure, have been discussed recently [51].

In vitro mutagenesis is particularly helpful in establishing functionally important transmembrane segments. For instance, site-directed mutations and domain swapping, in conjunction with electrophysiological measurements and the use of channel blockers, have been employed to identify the pore region [52–54] and the voltage sensor [55–58] of voltage-gated ion channels. These two domains are marked by the H5 region and the S4 segment respectively in the schemes shown in Figure 1. Other details relating to voltage-gated channel structure obtained by mutagenesis are reviewed in refs. [59,60]. In the nicotinic acetylcholine receptor, and other ligand-gated channels, the M2 segment (see Figure 1) has been identified with the ion pore by similar mutagenesis and labelling experiments (reviewed in [37,61]).

On the one hand, the success of the above procedures is necessary for the identification of the correct sequences for use in the design of appropriate peptide models. On the other hand, the testing of these predictions with the isolated peptide domains in reconstituted membranes can further provide a valuable additional check on their reliability.

STRUCTURES OF INTACT CHANNELS

Nicotinic acetylcholine receptor

The structure of an intact ion channel in a membrane environment that currently has been obtained to highest resolution (9 Å; 0.9 nm) is that of the ligand-gated nicotinic acetylcholine receptor from a muscle-like synapse [37,62]. This structure has yielded some surprises. Whereas previously the intramembranous segments of the bacterial reaction centre [63] and of bacterio-rhodopsin [64] had been found to be entirely α-helical, which is also the case for the plant and bacterial light-harvesting complexes [65,66], it was found that part of the transmembrane segments of the nicotinic acetylcholine receptor are likely to be composed of β-sheet. The receptor channel structure is depicted in a sideways projection in Figure 2(a). The projection in the plane of the membrane (not shown) reveals electron density that corresponds to only one transmembrane α-helix from each of the four subunits that make up the $\alpha_2\beta\gamma\delta$ heteropentamer. These helices line the central pore in the channel structure and can be aligned with the M2 transmembrane segments in each of the subunits (cf. Figure 2b). They are enclosed on the lipid-facing side by a continuous rim of electron density that is insufficient to represent α-helices but could correspond to a β-sheet. It was suggested [37] that the overall transmembrane secondary structure resembles that of the B-subunit pentamers of the soluble enterotoxin [67] and verotoxin-1 [68] from *Escherichia coli*. Each

of these latter structures consists of a ring of five helices surrounded mainly by β-sheet, and the M1, M3 and M4 segments of the receptor subunits may form sets of three anti-parallel β-strands as are found in these proteins. Studies on the secondary structure of the nicotinic acetylcholine receptor by using IR spectroscopy have revealed that the proteinase-resistant transmembrane domain contains both α-helical and β structures [69], consistent with the proposal of a β-sheet scaffold surrounding pore-forming α-helices. Quantitatively, it was found that the intramembranous secondary structure was: α-helical, 50%; β structure and turns, 40%; random, 10%. This implies the existence of more than the four transmembrane stretches predicted by simple hydropathy plots.

The M2 helices constituting the pore are bent in the region of the highly conserved residue Leu-251 (dark pink in Figure 2b). Other (polar) residues that have been identified as lining the channel are the hydroxy-amino acids coloured light pink in Figure 2(b) and the two glutamates (E) at the lower and upper ends of M2. The conserved Leu-251, together with the leucine residues at homologous positions in the other subunits, form the gate of the closed channel, possibly as in a leucine zipper [37]. In the open channel configuration the kink in the M2 helix is preserved but has rotated away to the side, removing the gate that was formed by the hydrophobic leucine residues [62]. The narrowest constriction of the open pore is then formed by the lower face of M2, with the hydroxy-containing residues aligned almost parallel to the axis of the channel.

Bacterial toxins

The X-ray crystal structures of several bacterial toxins that are

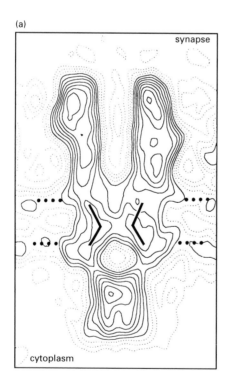

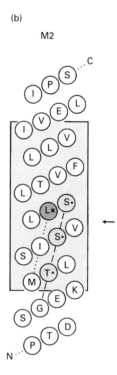

Figure 2   Channel structure of the nicotinic acetylcholine receptor from *Torpedo marmorata*

(a) Closed channel in profile obtained from the membrane structure at 9 Å resolution. The positions of the channel-lining transmembrane α-helices are marked by heavy lines and the estimated limits of the lipid bilayer by dotted lines 30 Å apart. (b) Helical net plot of the M2 segment of the α-subunit. The dashed line identifies a face composed entirely of small residues, flanked by bulky hydrophobic side chains (dotted line) used in aligning with electron densities in the three-dimensional map. The arrow shows the estimated position of the kink in the helix, and the bilayer-spanning segment is included in the shaded rectangle. Coloured residues are discussed in the text. Reproduced from [37], with the permission of Academic Press.

found to form membrane channels have been determined. One of the most significant of these is that for pro-aerolysin from *Aeromonas hydrophila*, because a direct connection with the pore structure was established in this case [70]. Proteolytic activation of the pro-toxin exposes a large hydrophobic patch on the major lobe of the protein. This lobe contains a high proportion of β-sheet structure composed in part of very long strands. This sheet adopts a helical twist such that the hydrophilic face undergoes a 180° rotation as it reaches the C-terminal domain. Fitting the X-ray structure to electron microscope images from two-dimensional membrane crystals suggests that the aerolysin channel is a heptameric structure. The C-terminal domains, which consist almost entirely of β-strands, traverse the bilayer, with the hydrophobic faces exposed to lipid and the hydrophilic faces lining the aqueous channel. The likely topology is that of a β-barrel, as in porins, resulting in a large transmembrane pore.

The membrane insertion of a variety of bacterial toxins is activated by large changes in conformation, possibly with molten globule structures as intermediates. The crystal structures of colicin, diphtheria toxin and δ-endotoxin have revealed their translocation domains to be characteristic of 'inside-out' membrane proteins (reviewed in [71]). The common structure of this domain consists of an α-helical bundle of between seven and ten helices, some of which are hydrophobic, are of sufficient length to span the membrane and are buried in the soluble toxins. Membrane insertion is suggested to be mediated by these helical hairpins of the hydrophobic core. Because the toxins can form pores in artificial membranes the structure of the inserted domain most probably also consists of a helical bundle rearranged such that the polar surfaces of the more amphiphilic helices face the channel lumen. The assembly and pore-forming ability of a putative membrane-spanning peptide (FLTTYAQAANTHLF-LLKDAQIYG) derived from the *Bacillus thuringiensis* δ-toxin has been studied [72].

Annexin V

The annexins are a family of cytosolic, calcium-dependent, membrane-binding proteins. The X-ray structure of annexin V, which is known to form voltage-gated Ca^{2+} channels in lipid bilayers, has been determined and found to be almost entirely α-helical [73]. Although this amphiphilic protein binds peripherally to negatively charged phospholipids at the membrane surface, without major structural rearrangement, the structure reveals a prominent central hydrophilic pore that has been associated with the calcium-sensitive channel [74]. The membrane permeability is proposed to arise from a destabilization of the lipid bilayer. Four repeats are arranged around the central protein pore, which consists of a four-helix barrel contributed by two helices each from repeats II and IV. The residues lining the pore are highly conserved and many are charged. Mutational analysis revealed that Glu-95 is essential for the cation selectivity filter, Glu-17 and Glu-78 outside the pore modulate the ion conductance, and Glu-112 is the main voltage sensor controlling the gating of the channel [74,75]. Like the B-subunits of enterotoxin and verotoxin-1 (see above), annexin V is, therefore, a further example of a soluble protein for which details of the configuration of a hydrophilic pore can be deduced from the high-resolution structure.

Porins

The porins, which form aqueous channels in the outer membrane of Gram-negative bacteria, constitute one of the few classes of integral membrane proteins for which the high-resolution structures have been solved by X-ray crystallography [76–78]. The secondary structure is predominantly β-sheet (approx. 60%), with reverse turns, and the transmembrane pore is formed by the lumen of an anti-parallel β-barrel that is constricted by an internal loop. The number of β-strands in the barrel is 16 or more, although OmpA, for which the structure has not yet been determined, is predicted to consist of only eight membrane-spanning strands [79]. The β-strands in porin from *Rhodobacter capsulatus* are composed of 7–18 residues, with a mean length of 11 residues, and are tilted at angles of 30–60° with respect to the membrane normal [76,77]. The porin molecules are present as stable trimers and therefore not all of the outward-facing residues are exposed to lipid, nor is the height of the barrel necessarily as great as the bilayer thickness over the whole of its perimeter. The membrane-facing outer surface is composed of hydrophobic or non-polar residues with a girdle of aromatic residues close to each bilayer surface, and the interior of the pore has a lining similar in composition to the outer surface of water-soluble proteins. Such structures would not be identified as membrane-spanning by conventional hydropathy analysis (cf. [79,80]).

In general, porins allow the transmembrane diffusion of small water-soluble molecules. The OmpF and PhoE porins, whose structure has been determined [78], are weakly cation-selective and weakly anion-selective respectively [81]. Both these porins (and others), when reconstituted in planar bilayers, show large distinct channel conductances that are voltage-gated [81–84]. The pore size is restricted by the internal loop within the barrel lumen, and is 7 Å × 11 Å for OmpF and 7 Å × 9 Å for PhoE, which is reduced to 3 Å × 6 Å by ordered water molecules [78]. At the level of the constriction, three arginine residues stacked at the barrel wall face a cluster of acidic residues situated between the internal loop and the wall. The mutation L125E at this region in PhoE reverses the ion selectivity to cationic [84]. In general, the selectivity seems to be determined by the net electric field at the mouth of the pore [85].

Channel models

In the absence of detailed structural information on any of the voltage-gated ion channels, current experimental strategies and channel design for these proteins are based partly on molecular modelling [32,86–88]. The fact that the homologous Na and Ca channels contain four internal repeats has led to the proposal of a four-fold tetrameric structure for these channels, and also for the A-type K channel subunit that contains only a single analogous repeat (cf. Figure 1). This supposition, which was supported by electrophysiological studies on the charybdotoxin sensitivity of hetero-oligomers [89], has been substantiated recently by low-resolution images of the purified Shaker K channel [90]. A three-dimensional model has been put forward for the K channel [86] that is consistent with mutagenesis and channel-blocking data, the most salient of which is the involvement of the H5 or P-region in the channel lining. In addition to the segments S1–S6, a putative amphipathic helix S45 in the linking region between S4 and S5 is postulated to span part of the membrane, and an amphipathic helix that contacts lipid chains is also postulated in the S2–S3 link. The structure is composed of three concentric cylinders, with the outer cylinder comprising α-helical segments S1–S3 and S5, all of which are exposed to lipid and span the bilayer. The intermediate cylinder is composed of α-helical segments S4, S45 and S6. The overall twisted superhelical structure is rather different from packing motifs involving helical bundles. Voltage gating is proposed to take place by the helical screw mechanism, in which the positively charged S4 segment moves a considerable way out of the membrane. The inner cylinder lines the pore and, at its most confined, is composed of the most conserved part of the H5 region in a short eight-stranded β-barrel which penetrates half-way through the membrane. The remaining, wider, section of the pore is composed of S45 and part of S6 from the middle cylinder. Subsequently it has been proposed that both ends of the P-segment are α-helical, forming the vestibule, and that the selectivity filter is formed by the linking sequence TTVGYGD (in Shaker) which contains β-turn and random coil conformations [91]. An anti-parallel β-hairpin model has been offered recently for the structure of the mouth of the Na channel [92]. This model, composed of decapeptides from the H5 regions of the four internal repeats, was aimed primarily at creating a structure for the toxin binding site. The hairpins do not form a barrel, but rather a cone-shaped outer vestibule. Questions of the unsatisfied hydrogen bonds in the backbone and the relationship with the rest of the protein structure were left unresolved.

Other models have concentrated on the pore region of the voltage-gated channels, most notably on long β-barrel structures. Secondary structure prediction and hydrophobic moment profiling have led to the proposal of a general β-turn–β-hairpin motif for the H5 segment of various voltage-gated and related K channels, independent of their numbers of putative transmembrane segments [88]. Not all predictions were compatible with a β-barrel of sufficient length to span the entire membrane. It is likely that there is variability amongst the different K channels. A long β-barrel model has been constructed for the pore of NGK2 (Kv3.1) K channels [87,93]. The 21-residue H5 pore segment (formerly called SS1-SS2) has the sequence PIGFWWAVVTMTTLGYGDMYP. The barrel has a right-handed twist with a 35° tilt (shear number +8) and is composed of eight anti-parallel strands contributed symmetrically by four β-hairpins, one from each subunit of the tetramer. The height of the barrel is 26 Å. The hairpin turn involves five residues centred on residue 10 of H5 (Thr-397 in NGK2; Thr-439 in Shaker) and has four unsatisfied hydrogen-bond pairs. Of the 12 possible hairpin configurations, on balance this was most in accord with the experimental data on pore-lining residues and channel blocking. It also has the most dense packing of side chains in the barrel lumen and therefore is most likely to display high ion selectivity. The channel is most constricted at the level of the aromatic residues at positions 4, 6 and 16, with the ring of hydroxy groups from the conserved tyrosine having a diameter of 3 Å. With the exception of one threonine, all polar side chains line the pore. The model provides an energetically feasible channel with plausible K^+ permeability [93]. This model has been described in some detail because it offers an alternative structure that may have a direct relevance to experiments on isolated peptides in lipid bilayers.

NATURAL PEPTIDE CHANNELS

In following the strategy of simplifying complex, polytopic and often multisubunit intrinsic channels down to single putative transmembrane segments, it is of interest to consider the known structures of small peptide antibiotics that form channels in membranes. The principal of these are gramicidin A and alamethicin, the electrical activity of which has been studied extensively in lipid bilayers. These two peptides are found to be prototypic of two quite different modes of membrane channel formation.

Gramicidin A

Gramicidin A from the bacterium *Bacillus brevis* is a linear 15-residue hydrophobic peptide, the amino acids of which have alternating L- and D-configurations. Both N- and C-terminal residues are blocked and both have the L-configuration; the amino acid sequence of the peptide is: HCO-L-Val-Gly-L-Ala-D-Leu-L-Ala-D-Val-L-Val-D-Val-L-Trp-D-Leu-L-Trp-D-Leu-L-Trp-D-Leu-L-Trp-NHCH$_2$CH$_2$OH.

Gramicidin A is selective for univalent cations and gives rise to well defined single-channel events in planar lipid bilayers, with a conductance that is comparable with that of natural channels. X-ray diffraction of single crystals grown from organic solvents has yielded structures that consist of left-handed, anti-parallel, double-stranded β-helices [94–96]. These structures are of sufficient length to span the hydrophobic thickness of a lipid bilayer, and have a central pore which is large enough to transport univalent cations. The double-helical structure is, however, incompatible with a variety of functional studies which indicate that the ion-conducting channel is formed by a head-to-head gramicidin dimer in lipid bilayers (reviewed in [4]). This plasticity in the structure of small peptides, a point that will be returned to later in connection with the α- and β-secondary structure of peptides composed solely of L-amino acids, has to be borne in mind.

The structure of gramicidin A in a phospholipid bilayer matrix has been determined from the orientational constraints provided by solid-state NMR measurements on isotopically labelled peptides in oriented samples, coupled with energy refinement [97]. The channel-forming structure consists of right-handed, single-stranded β-helices with six to seven residues per turn (Figure 3). The hydrogen bonding topology is that of a β-sheet in which the carbonyl oxygens of the peptide backbone are tipped towards the channel lumen and the hydrophobic side chains are all directed outwards towards the lipid. This amphipathic feature of the structure, which results from the alternating D- and L-amino acids in gramicidin, is not possible in membrane proteins, which are composed solely of L-amino acids. For the latter, the side chains of adjacent residues in a β-helix would alternate between channel-facing and lipid-facing orientations. More residues per turn would then be required to accommodate the bulk of the side chains that are situated within the channel lumen, and correspondingly more residues per peptide would be needed to span an equivalent width of the bilayer (cf. [98]). This second constraint could be alleviated in multiple-stranded β-helices such as are found in the crystals of gramicidin A. In the bilayer structure, the side chains of the tryptophan residues in the C-terminal parts of the head-to-head gramicidin A dimer are oriented with the indole -NH groups directed towards the hydrophilic bilayer surface (Figure 3). This is a feature that is found for tryptophan (and equivalently tyrosine) residues in the known high-resolution structures of integral membrane proteins [77,99] and is thought to assist in the vertical positioning of the protein/peptide in the membrane.

Alamethicin

Alamethicin is a 20-residue peptide from the fungus *Trichoderma viride* which contains a high proportion of the unusual amino acid α-aminoisobutyric acid (Aib), in addition to two proline residues, and is capable of forming voltage-gated channels of very high conductance in lipid membranes. The N-terminus of the peptide is blocked and the C-terminal residue is L-phenylalaninol (Phl). The amino acid sequence is: Ac-Aib-Pro-Aib-Ala-Aib-Ala-Gln-Aib-Val-Aib-Gly-Leu-Aib-Pro-Val-Aib-Aib-Glu-Glu-Phl.

Such peptaibol peptides which contain several Aib residues have a strong propensity to form helices, including 3$_{10}$-helices [100]. The X-ray crystal structure determined for alamethicin

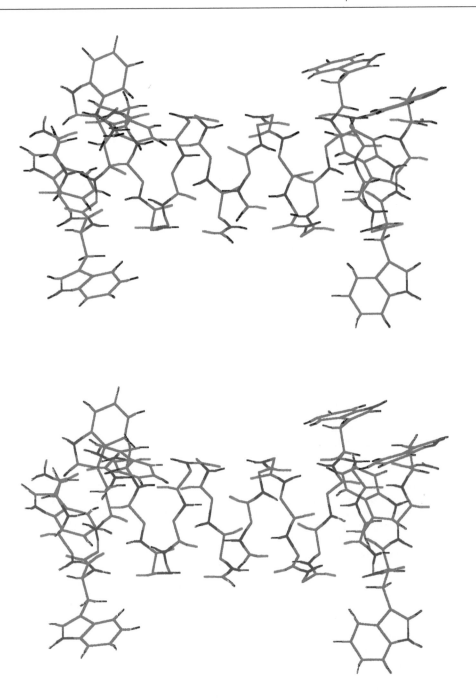

Figure 3   Channel structure of gramicidin A

Stereo view of the computationally refined structure in the lipid bilayer that is obtained from orientational restraints provided by solid-state NMR. The tryptophan indole-NH groups are clustered at the bilayer surface, and the N-terminus for each monomer is buried at the bilayer centre. Side chains of the central part of the helix are omitted for clarity. Reproduced courtesy of R. R. Ketchem and T. A. Cross. See also [97].

shows it to be predominantly α-helical with a bend, punctuated by a partial 3_{10}-helical turn, that is generated by the proline residue at position 14 [5] (Figure 4a). A somewhat similar structure was found by NMR for alamethicin in solution (methanol) [102]. The three polar residues, Gln-7, Glu-18 and Glu-19, have their side chains extended from the convex face of the helix, and the presence of Pro-14 causes the backbone carbonyls of Aib-10 and Gly-11 also to be exposed on this face. The latter increases the amphiphilic nature of the helix beyond that calculated from the amino acid sequence in Table 1 and contributes to its channel-forming properties.

The concentration dependence of the transmembrane current and the single-channel conductances supported by alamethicin indicate that the channel is composed of several alamethicin molecules, up to 10 or 11 (reviewed in [103]). Analysis of binding isotherms for the association of alamethicin with lipid vesicles has also indicated aggregation of the peptide within the membrane [104,105]. Based on the structural studies, it therefore seems likely that the channels are made up of helical barrels in which the polar faces of the alamethicin helices are oriented inwards to the central pore of the barrel, as proposed originally in the model of [5]. The most stable model channel structures are

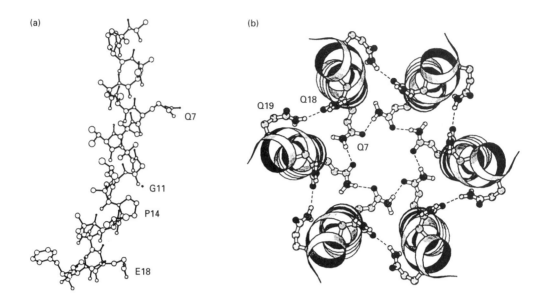

Figure 4   Structure of alamethicin and channel model

(**a**) Crystal structure of alamethicin [5]. There are three alamethicin molecules with closely related structures in the asymmetric unit for which the kink angle varies slightly (35° to 39°). The polar side chains Glu-18 and Gln-7, and the exposed carbonyl of Gly-11, as well as the proline residue at position 14, are indicated. Reprinted from Prog. Biophys. Mol. Biol. **55**, Sansom, M.S.P., pp. 139–235 [2], copyright (1991), with kind permission from Elsevier Science Ltd, The Boulevard, Langford Lane, Kidlington OX5 1GB, U.K. (**b**) Model of a hexameric helical barrel for the channel of the E18Q alamethicin variant based on the alamethicin crystal structure. The alamethicin molecules are oriented such that their N-terminal helices are parallel. Possible channel-stabilizing hydrogen bonds involving the side chains of residues Gln-7, Gln-18 and Gln-19 are indicated. Reproduced from [101], with permission, copyright Springer-Verlag (1993).

formed with the N-terminal helices aligned parallel [5,101]. This causes the pore to widen at the C-terminus as a result of the kink induced by the Pro-14 residue (Figure 4b). The side chains of adjacent N-terminal domains interdigitate in a zipper-like structure and the channel is stabilized by hydrogen bonds between the Gln-7 side chains of adjacent monomers and by Glu-18–Gln-19 H-bonds [5,101]. The (hydrated) annulus of the Gln-7 residue produces the greatest restriction in the channel diameter. Various models for the formation of helical bundles and barrels by alamethicin in the membrane and for voltage activation are discussed in [103].

Others: zervamicins and melittin

Other peptides with a similar structure and channel behaviour to that of alamethicin include the zervamicins from *Emericelloposis salmosynnemata* and melittin from bee venom. The zervamicins are also proline-containing peptaibols, the crystal structures of which have been determined and found similarly to be composed of bent helices, with the polar nature of the convex face enhanced by unsatisfied backbone hydrogen bonds [100]. Interestingly, embryonic water-associated channels have been identified in the crystal structure of Leu-zervamicin [6], a 16-residue peptaibol that also forms voltage-activated channels in lipid membranes [106]. The polar faces were found to associate in an anti-parallel fashion in the crystal, whereas the hydrophobic faces associate non-selectively in both parallel and anti-parallel modes [6]. Melittin is a 26-residue amphiphilic peptide that also contains a proline residue in the middle of the sequence but, unlike the peptaibols, does not contain Aib residues. The crystal structure of melittin [7] consists of two α-helical segments, without 3_{10}-helical elements, and like that of alamethicin has the shape of a bent rod. The solution structures of melittin in methanol [107] and bound to detergent micelles [108] obtained by NMR are similar to that in the crystal. Melittin also forms voltage-activated channels in planar lipid bilayers, but with a concentration

dependence that suggests that the channel is composed of four monomers [109].

The finding that ion channels are induced by strongly amphiphilic and surface-active peptides such as melittin, which have a pronounced lytic activity on membranes at higher concentrations, raises the question of the exact nature of the channel formation in these cases. Amphipathic helices are capable of associating with lipids with their axis parallel to the plane of the membrane, and this is the orientation found for melittin bound to micelles ([108]; see also [110]). It is possible, however, that the amphipathic peptide may change from a surface-aligned to a transmembrane orientation on application of a membrane potential (discussed in [3]). Another example of an extremely amphipathic cytolytic peptide is δ-haemolysin from *Staphylococcus aureus*. This 26-residue peptide does not contain proline residues, and the solution structure in methanol [8] and bound to lipid micelles [111] was found to have a central region that forms an extended amphipathic α-helix. This toxin also forms ion channels, but these are only weakly voltage dependent [2,112].

PEPTIDE SECONDARY STRUCTURE

Peptides corresponding to putative transmembrane channel segments that have been studied are listed in Table 1. Crystal structures are currently not available. Information on the secondary structure so far has come mostly from circular dichroism (CD), Fourier-transform IR (FTIR) and high-resolution NMR spectroscopies, often of the peptides in organic solvents. (Extrapolation of the latter to a membrane environment is uncertain and may account for the apparent variability in some results.) The application of these methods to channel peptides and proteins has been discussed recently [51]. Given the results summarized above for the structure of the acetylcholine receptor and the fact that current models for the pore region of voltage-gated ion channels propose this to involve a β-barrel type of structure [32,86–88], characterization of the secondary structure is a topic

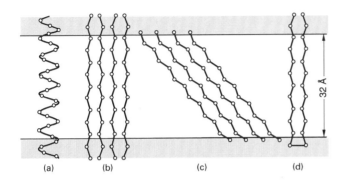

Figure 5 Possible secondary structures for transmembrane peptide segments

The thickness of the hydrophobic region of the membrane is taken as 32 Å. The number of residues in the hydrophobic span is: (**a**) 21 (α-helix), (**b**) 9 (untilted anti-parallel β-sheet), (**c**) 12 (35°-tilted anti-parallel β-sheet), (**d**) 20 (β-sheet hairpin).

of major importance. Possible allowed conformations and corresponding hydrophobic spans for transmembrane peptide secondary structures with fully satisfied backbone hydrogen bonding (as is required energetically in a hydrophobic environment) are indicated in Figure 5. Approx. 22 residues are required for a membrane-spanning α-helix, but only about nine residues are needed for the more extended β-sheet structure if the strands are oriented parallel to the membrane normal. However, for a β-strand tilt of 60°, or for an untilted β-hairpin structure, a comparable number of apolar residues to that for an α-helix is required. Tilts of 35° and 60° can be realized by staggering the hydrogen bonding between adjacent anti-parallel β-strands by one or two residues respectively relative to the untilted structure [36].

Ligand-gated channels

Of the peptides corresponding to putative transmembrane segments of the nicotinic acetylcholine receptor that have been studied in lipid bilayers, the CD of the M2δAChR peptide [113], and of the template-attached tetramers T$_4$M2δ and T$_4$M1δ [114], has been investigated in the helix-promoting solvent trifluoroethanol. For M2δAChR, corresponding to the M2 segment of the δ-subunit of the *Torpedo* receptor, the calculated secondary structural content is 57% α-helix and 43% random, and for the T$_4$M2δ tetramer it is 44% α-helix, 40% β-strand and 16% random. The latter includes the 9-residue template of the whole 101-residue structure. For mutant T$_4$M2δ with the single residue replacement S8A, 45% α-helix, 20% β-strand and 35% random structure was found. For the T$_4$M1δ tetramer, corresponding to the M1 segment of the δ-subunit, a higher helical content was obtained: 60% α-helix, 19% β-strand and 21% turns.

Of the glycine receptor peptides, CD spectra have been obtained from M2GlyR and M4GlyR in trifluoroethanol [13]. The M2GlyR peptide, corresponding to the M2 segment of the α_1-subunit of the rat receptor, and the variant in which the two arginine residues close to the termini were replaced by glutamic acid (R2E and R21E), had roughly 50% α-helical content, with random coil and some β-sheet conformations accounting for the majority of the residual secondary structure. The M4GlyR peptide, corresponding to the M4 segment of the α_1-subunit, contains too many aromatic residues for reliable analysis of the CD spectra, but ordered structure was indicated. In 10% aqueous

trifluoroethanol, random coil structures prevailed for all three peptides but the secondary structure of M4GlyR was found to be more stable against the conformational changes induced by addition of buffer.

High-resolution ^{1}H-NOESY NMR spectra of the M2δAChR, M2GlyR and M2NMDAR (glutamate receptor M2) peptides in dodecylphosphocholine micelles exhibit many inter-residue connectivities characteristic of an α-helix [3]. For M2GlyR, at least 15–20 of the 23 residues were deduced to be in an α-helical conformation. Solid-state ^{15}N-NMR of [A12-^{15}N]M2δAChR and of [A11-^{15}N]M2GlyR has been used to determine the orientation of these peptides in aligned phospholipid bilayers [3,115]. The chemical shift anisotropy demonstrated that the backbone N–H bonds of the labelled residues were parallel to the membrane normal. Assuming a helical conformation for the peptide, this would imply that the α-helices are incorporated spanning the membrane, with the helix axis perpendicular to the membrane plane. Interestingly, it was found by similar methods that magainin2, a highly amphiphilic channel-forming peptide, adopted an orientation suggesting that it lies parallel to the plane of the membrane [115].

Voltage-gated channels

Of the putative transmembrane segments of voltage-gated ion channels, the secondary structure of peptides corresponding to S2 and S3, to S4 with and without the linking region to S5, and to the H5 putative pore-lining segments, have been studied. In addition, the putative single transmembrane segment of the small minK (or IsK) K-channel protein has been investigated.

The CaIIS2 peptide, corresponding to the S2 segment of the second repeat from the dihydropyridine receptor Ca channel of skeletal muscle, has been studied by ^{1}H-NMR and CD spectroscopy [16]. Both indicated α-helical content for the peptide in the amphiphilic solvent methanol, although other structures also made some contribution to the CD spectra. The KS2 peptide, corresponding to the S2 segment of the Shaker K channel, has been studied by CD in trifluoroethanol and in detergent micelles [17]. In both environments the α-helical content was high, estimated to be 78% in trifluoroethanol/water (2:3, v/v) and 99% in dodecyl sulphate micelles. The NaIS3 peptide, corresponding to the S3 segment of the first repeat of the rat brain I Na channel, has been studied in trifluoroethanol by CD [113]. The calculated secondary structural content was 30% α-helix, 35% β-strand, 3% turns and 30% random coil. The CD of peptides corresponding to the two putative transmembrane segments, M1 and M2, of the ROMK1 delayed rectifier K channel has been examined in hydrophobic solvents [116,117]. Both peptides adopted a high degree of helical structure, estimated to be 88% for M1 and 73% for M2.

The secondary structure of the S4 voltage-sensor segment is of considerable interest because of the regular occurrence of positively charged residues at every third position. This motif would distribute the positive charges evenly around the face of an α-helix, as postulated in the helical screw model of voltage gating [86], but would align all the positive charges parallel to the helical axis in a 3$_{10}$-helix, creating a very strongly amphipathic structure. The structure of peptide NaIS4(r), the S4 segment from the first internal repeat of rat brain I and II Na channels, has been determined in trifluoroethanol/water (9:1, v/v) by ^{1}H-NMR and distance geometry/restrained molecular dynamics [20]. The peptide was found to be α-helical over the region from Leu-2 to Ile-14, with the C-terminus disordered and no regions of 3$_{10}$-helix.

The secondary structure of the blocked peptide KS4, based on the Shaker K-channel S4 sequence, has been investigated by FTIR spectroscopy in both aqueous and trifluoroethanol solutions, and reconstituted in phospholipid bilayers and micelles [23]. The peptide was found to be of random coil structure in water but α-helical in trifluoroethanol, the latter being consistent with the NMR results with the NaIS4(r) Na-channel peptide. The KS4 peptide was also predominantly α-helical in lysophosphatidylcholine micelles and in dimyristoyl phosphatidylcholine or phosphatidylglycerol bilayers, with relatively minor contributions from β-sheet or turn. The 34-residue NaIVS4-S45 Na-channel peptide, that contains the additional S45 section which forms the link with the S5 segment, has been examined by CD spectroscopy [22]. Calculated α-helical contents were approx. 50% in methanol, trifluoroethanol and vesicles containing anionic phospholipids, which was decreased slightly to approx. 40% in zwitterionic phospholipid vesicles. The remaining secondary structure was attributed to β-sheet and random coil, with the β-sheet component in lipid bilayers comprising 25% on average.

In several models for voltage-gated channels the putative pore-lining segment, H5, is assumed to have a secondary structure other than α-helical, often β-sheet (see above). Peptides NaIH5–NaIVH5, which correspond to the H5 or P-regions from all four internal repeats of *Electrophorus* Na channels, have been examined by CD spectroscopy [25]. The peptides from the first three repeats were estimated to have approx. 30% α-helical content both in trifluoroethanol/water (2:3, v/v) and in dodecyl sulphate micelles. For the longer NaIVH5 peptide this was decreased to approx. 15%, roughly corresponding to the same number of residues. Studies have been carried out on peptides of different lengths from the H5 region of Shaker K channels (see Table 1). The extended N-terminal section, KH-5-N, was estimated to have a fractional helicity of 0.38 in trifluoroethanol/water (2:3, v/v) [26]. The corresponding KH-5 peptide and its highly conserved region, consisting of the first 12 residues, were found to have low α-helical content, estimated at 20% and 15% respectively, in SDS micelles [17]. The H5 peptide from the ROMK1 delayed rectifier K channel was also found to have little α-helical content in a hydrophobic environment [116]. The full-length KH5 peptide was found from CD to be α-helical in 95% trifluoroethanol, but was interpreted to have a mixed secondary structure of low helical content in buffer and in the presence of liposomes [28]. In contrast, the shorter blocked peptide KH5 was found by FTIR spectroscopy to be predominantly α-helical both in lysophosphatidylcholine micelles and in phospholipid vesicles [27]. The latter method is more appropriate to vesicular systems if special corrections and precautions are not applied in CD spectroscopy [51]. It is also conceivable that blocked N- and C-termini may promote helix formation of peptide segments (cf. [31]).

The IsK (or minK) protein consists of only 130 residues and is encoded by a DNA clone that on expression induces slowly activating voltage-gated K channels [118]. It has been suggested that this may be a regulatory protein, because heterologous expression at high levels in oocytes also induces Cl channels, a property that is shared, however, with a variety of quite unrelated membrane proteins [119]. The available evidence is therefore in favour of IsK being a true channel protein. The 32-residue TM-minK peptide, which contains the single putative membrane-spanning segment of the IsK protein, has been studied by CD in methanol [29]. The α-helical content in this amphiphilic solvent was estimated to be 57%. A 63-residue 'truncated IsK' peptide containing nine additional residues at the N-terminus and 22 at the C-terminus, relative to TM-minK, was estimated to have 31% helicity that was attributed entirely to the putative trans-

membrane sequence [120]. Of peptides outside the putative transmembrane region, the 20-residue peptide from the N-terminus was found to have high helicity (estimated at 50%).

A shorter putative membrane-spanning peptide, K26, which contains the whole of the apolar sequence of IsK, has been reconstituted in zwitterionic phospholipid bilayers by dialysis from the helix-forming solvent 2-chloroethanol [30]. In bilayer membranes at high peptide concentrations, the K26 peptide was found by FTIR spectroscopy to be wholly in a β-sheet conformation. The latter is a further example of the possible plasticity in secondary structure of relatively short peptides that was mentioned above in connection with gramicidin A. It has long been known, for instance, that the M13 bacteriophage coat protein may be in either an α-helical or a β-sheet conformation, depending on the method of preparation [121]. This structural polymorphism has been characterized in detail for the M13 coat protein reconstituted in lipid bilayers [122]. Factors favouring the α-helical conformation were high lipid/peptide ratios, negatively charged lipids, high ionic strength and unsaturated lipid chains [123]. A concentration- and time-dependent conversion from α-helical monomers into β-sheet polymers was observed in lipid membranes. The equilibrium configuration at low lipid (high peptide) concentration can be in favour of the polymerized β-sheet structures. In general, all factors that can lead to aggregation of the peptide, not just high concentrations but also decreases in electrostatic repulsions, higher lipid chain packing densities and phase separation, will tend to favour the formation of β-sheet structures.

Proton channels

The structure of the whole 79-residue proteolipid c-subunit of the transmembrane section of the F_1F_o-ATPase from *E. coli* has been investigated in trifluoroethanol and in chloroform/methanol/water by two-dimensional NMR spectroscopy [124,125]. This proteolipid, which is identified with the dicyclohexylcarbodiimide (DCCD)-binding component of the proton channel, was found to contain two extended regions of predominantly α-helical character that were shown to be arranged in a hairpin configuration [125]. The 16 kDa DCCD-binding proteolipid from *Nephrops norvegicus* that corresponds to the V_o section of the vacuolar V-ATPases was estimated from FTIR spectroscopy to contain > 60% α-helix, with probably 20% extended-chain structures, 9% disordered structure and 4% turns [126]. A peptide corresponding to the single putative transmembrane segment of the small subunit 8 proteolipid (different from the c-subunit proteolipid analogue) of the F_1F_o-ATPase from *Saccharomyces cerevisiae* has been studied by FTIR and CD spectroscopy [31]. This proline-containing peptide, Pep-8a, was found from FTIR to be fully α-helical in the solid state (dried from chloroform or methanol), and from CD was estimated to contain approx. 60% α-helix when dissolved in hexafluoropropan-2-ol or its 1:1 (v/v) mixture with methanol.

Considered together, all these results suggest that measurements in organic solvents cannot necessarily be taken to indicate the secondary structure in a lipid membrane. Additionally, lipid composition and other parameters such as ionic strength and peptide concentration may be significant determinants of secondary structure.

PEPTIDE CHANNEL ACTIVITY

Single-channel electrical recordings have been made on planar bilayers, or bilayers formed on patch pipettes, with incorporated peptides and with their oligomers linked covalently to a multifunctional carrier template [24,127]. Ion selectivity was established by the reversal potentials obtained from current–voltage relationships that were determined in the presence of transmembrane salt gradients or in bi-ionic media (cf. [2]). The results for peptides from ligand-gated channels are considered first, followed by those for peptides from voltage-gated channels. Of the peptides studied, experimental data on secondary structure have not been reported in all cases (see previous section), although design principles involving prediction methods and molecular modelling have been used to identify putative helical sections and pore-lining sequences (cf. [15,113]).

Ligand-gated channels

Nicotinic acetylcholine receptor

The amphiphilic M2δAChR peptide, corresponding to the M2 segment of the *Torpedo* receptor δ-subunit that is believed to be one of the five homologous pore-lining segments, has been found to form cation-selective channels of heterogeneous conductance and lifetime [9,10]. Template-attached tetramers and pentamers ($T_4M2\delta$ and $T_5M2\delta$), on the other hand, formed cation-selective channels of homogeneous conductance which corresponded to the two primary channel populations that were observed with M2δAChR monomers [10,128]. It was concluded that the channels formed by the monomeric M2δAChR peptide corresponded to tetrameric and pentameric arrays. The pentameric $T_5M2\delta$ channels had the higher conductance (40 pS in 0.5 M KCl), which corresponded quite closely to that of the native reconstituted channels (45 pS) under similar conditions. Additionally, channels from the tethered peptides were blocked by the local anaesthetic analogue QX-222 and by chlorpromazine, as for the native channels. Replacing conserved residues expected to face the pore lumen (S8A and F16A) had similar effects on conductance (decrease and increase respectively) to those of site-directed mutagenesis of the native receptor. Peptides with the same amino acid composition as M2δAChR but with randomized sequence did not form discrete channels, although conductance was affected; neither did the hydrophobic M1δAChR peptide, corresponding to the highly conserved first putative transmembrane segment of the receptor, nor its tethered tetramer. Similar results were obtained with an M2α4AChR peptide that was based on the sequence of the M2 segment of the rat neuronal receptor α_4 subunit [10]. These findings are in rather good accord with current models, and with the structural data discussed above, in which the nicotinic acetylcholine receptor channel consists of a pentamer of bent α-helices contributed by the M2 segments from each subunit.

The peptide MAβAChR, which corresponds to the MA segment in the long sequence linking the M3 and M4 segments of the β-subunit of the *Torpedo californica* nicotinic acetylcholine receptor, has also been studied in lipid bilayers [11]. This MA-segment is not one of the putative transmembrane segments identified from hydropathy plots and is not found in the sequences of the γ-aminoisobutyric acid$_A$ (GABA$_A$) or glycine receptors, but was considered as a candidate for lining the channel because of its strongly amphipathic α-helical periodicity. Interestingly, the MAβAChR peptide monomer was found to give rise to long-lived channels with just two discrete conductance levels, which were, however, lower than that of native receptors. This behaviour was unlike that of the monomeric M2δAChR peptide, which gave a rather more continuous spectrum of conductances, and resembled more that of synthetic designed channels [129]. This MA-segment is currently not thought to be a candidate for the channel lining (see above).

Glycine receptor

Peptides corresponding to the amphiphilic M2 segment of the strychnine-binding α_1-subunit of the glycine receptor have been found to form randomly gated anion-selective channels [12,13]. In one case the channel fluctuations were not stationary, with conductances that were found to vary over a large range, and cation-selective channels were observed in addition to those that were anion-selective [13]. In the other case, the conductances of the most frequently formed anion-selective channels (25 pS and 49 pS in 0.5 M KCl) were comparable with those of the native inhibitory receptor Cl channel of the rat spinal cord [12]. Replacing the two arginine residues at the N- and C-termini of the M2 segment by glutamic acid (R2/3E and R21/22E) reversed the channel selectivity, giving rise solely to cation-selective channels. On the other hand, peptides corresponding to the more hydrophobic M1 [12] and M4 [13] segments were found not to form discrete gated channels, but M4GlyR increased the overall bilayer conductance. Both the M2 and M4 peptides, as well as M2 with the Glu substitutions, increased the permeability of lipid vesicles, prohibiting the establishment of a transmembrane potential or dissipating a pre-existing membrane potential [13].

A template-attached tetramer of the M2 peptide, $T_4M2GlyR$, formed homogeneous anion-selective channels with a single conductance level of 25 pS in 0.5 M KCl [12]. The primary monomer channels with lower conductance were therefore attributed to tetrameric assemblies and those of higher conductance to pentameric assemblies. The $T_4M2GlyR$ homotetramer channels were blocked by picrotoxin, which is effective with GABA receptors, and by the anion-channel blockers 9-anthracene carboxylic acid and niflumic acid, but not by the cation-channel blocker QX-222. On this basis, it was suggested that the native glycine receptor channel is composed of a pentameric assembly of α- and β-subunits with a central pore lined by the M2 segments [12]. This is in accord with current ideas on the receptor structure and its subunit stoichiometry [130].

Cystic fibrosis transmembrane conductance regulator (CFTR) channel

The channel-forming properties of peptides (M1CFTR–M6CFTR) corresponding to all six putative transmembrane domains of the first repeat of the CFTR phosphorylation-regulated Cl channel have been investigated [14]. The M1CFTR, M3CFTR, M4CFTR and M5CFTR hydrophobic peptides were found not to elicit discrete gated channels in lipid bilayers, but displayed irregular current fluctuations that indicated incorporation into the membrane. Both the M2CFTR and M6CFTR amphipathic peptides did, however, form anion-selective channels with two primary conductances and sporadic events of both higher and lower conductance. Peptides of scrambled sequence with the same amino acid composition as the M2 and M6 segments gave rise to erratic changes in conductance but did not elicit discrete channels.

Mixtures of the M2CFTR and M6CFTR peptides exhibited discrete channels with properties different from those of the single peptides, presumably indicating the formation of hetero-oligomers. These channels were of more uniform conductance and were long-lived (Figure 6). Both the conductance level and the strong anion selectivity, but not the gating properties, were similar to those of authentic CFTR channels under comparable conditions. It was therefore postulated that the M2 and M6 segments are constituents of the CFTR pore, which is possibly composed of the M2, M6, M10 and M12 segments, in agreement with mutagenesis studies. Secondary-structural characterization was not carried out, but the segments are of sufficient length to

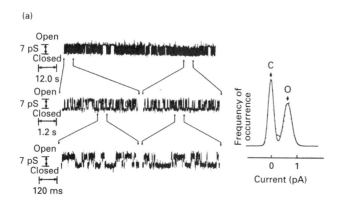
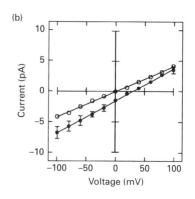

Figure 6   Channel formation and ionic selectivity in lipid bilayers containing an equimolar mixture of the M2CFTR and M6CFTR peptides corresponding to putative transmembrane segments of the CFTR Cl channel

(**a**) Single-channel recordings and current histogram for 100 mV membrane potential and bathing solutions of 0.15 M KCl, 1 mM CaCl$_2$ and 10 mM Hepes (pH 7.4). The open-channel conductance is 8 pS. (**b**) Current–voltage relationships for symmetrical 0.5 M KCl solutions (○) and under a 5-fold concentration gradient of KCl (●). For the latter, the reversal potential indicates that the channels are 95% anion-selective. From [14], with permission.

form a transmembrane helical bundle. Currently the CFTR channel is less well characterized than, for instance, the nicotinic acetylcholine receptor. These peptide studies therefore provide a valuable addition to the present knowledge that might also be relevant to other members of the superfamily of ABC transporters that contain conserved ATP-binding cassettes, for instance the multi-drug resistance P-glycoprotein.

Voltage-gated channels

Sodium channels

Early studies [18] on an amphipathic peptide, NaIS3, which corresponds to the S3 segment of the first repeat of the rat brain I sodium channel, showed that this was able to form cation-selective channels of heterogeneous conductance. The most frequent single-channel event had a conductance (25 pS in 0.5 M NaCl) similar to that of authentic sodium channels, but showed little discrimination between Na$^+$ and K$^+$ ions and was not sensitive to voltage. In contrast, a hydrophobic peptide (NaIS6) corresponding to the S6 segment of rat brain sodium channels did not form discrete channels [24].

The S4 channel segment has been identified with the voltage sensor and is characterized by a high density of positively charged residues, which occur at every third position in the sequence with the intervening residues being apolar. A peptide (NaIVS4) corresponding to the S4 segment in the fourth repeat of the sodium channel from *Electrophorus electricus* was found to give rise to conductances higher than those with the NaIS3 peptide [21]. A transmembrane potential induced a minimum conductance level of 300 pS with additional fluctuations of 200 pS, the relative population of which increased with increasing peptide concentration. The conductance was gated closed by prolonged negative potentials to a level of 70 pS (all in 0.5 M NaCl) and activated by positive potentials with an apparent gating charge of 3. Although this is a highly basic peptide, the conductance was found to be selective for cations, but with only slight selectivity for Na$^+$ over K$^+$ ions. As already noted, this is also the case for the NaIS3 peptide but, unlike the latter, the conductance of the NaIVS4 peptide was voltage-gated. The analogous NaIS4 peptide was found to be capable of dissipating a diffusion potential in vesicles containing anionic lipids (in which peptide aggregates were formed), but not in those com-

posed only of zwitterionic lipids (in which peptide aggregation did not take place) [19]. Similar results were obtained on leakage of calcein from vesicles.

An *E. electricus* NaIVS4 peptide that was increased in length to include the S45 region which constitutes the link with the S5 segment has also been studied [22]. The conductance elicited by this NaIVS4-S45 peptide was much reduced relative to that of NaIVS4 and was more comparable with that of reconstituted authentic Na channels. Voltage gating was retained by the NaIVS4-S45 peptide (gating charge of 3 or less) and, in contrast to the NaIVS4 peptide, the channels displayed a selectivity for Na$^+$ over K$^+$ ions, a property that is shared with intact Na channels. It was estimated that the channel was formed from 3–5 monomers. These results suggest a role for the S45 fragment in the channel lining and possibly in gating, as suggested in one of the proposed channel models [86]. It is possible that the far-reaching effects of the additional linker region could be a special feature of the unique composition of the S4 segment, which may require this additional stabilization. Other more hydrophobic putative transmembrane segments that have been studied may not prove to be so sensitive to the terminal extensions.

The finding that the NaIS3 and NaIVS4 peptides are capable of forming ion channels in lipid bilayers, although with somewhat different characteristics from those of the native Na channels, is rather at variance with the results of mutagenesis studies. The latter identify the H5 segment, and to some extent S6, as the main pore-lining elements of intact expressed channels. Structural simplification by concentrating on a single peptide segment may give rise to the formation of channels that are not realizable in the overall folding pattern of the whole polytopic protein. Conversely, peptide segments corresponding to sequences involved in the pore of native channels may lack some of the flanking elements essential to their intact three-dimensional configuration. On the other hand, direct evidence for the maintenance of native structure is lacking for the mutant proteins. Indeed, the structural information derived indirectly from mutagenesis studies often relies heavily on the perturbation of native function. Under these circumstances, the possibilities established by the peptide experiments need to be taken into account.

Potassium channels

Functional experiments on K-channel peptides have concentrated mostly on the putative pore region of the channel, the H5 sequence or P-region. This corresponds to the most highly conserved region of the voltage-gated K-channel family. The shorter KH5 peptide of Shaker channels in Table 1 was found to give rise to single-channel events with conductances for the monomer which ranged from 12 to 50 pS in 0.5 M KCl [27]. A tetramer that was obtained by tethering this sequence (with the additional terminal proline residues) to a branched lysine core elicited less frequent channel events with a conductance of.13 pS in 1.0 M KCl. These conductances are in the range of those found for native K channels but, as discussed above, the secondary structure of this peptide is at variance with current models for the pore lining. The longer KH5 peptide in Table 1, which contains additional residues flanking the P-region, was found to induce channels of greater conductance than that of the native Shaker channels [28]. These channels were preferentially conductive to anions ($P_K : P_{Na} : P_{Cl} = 1 : 0.4 : 3.2$) and were not blocked by either tetraethylammonium or the putative inactivating 'ball'-peptide. Ion selectivity was not determined for the shorter KH5 peptide. These results indicate that peptides corresponding to the H5 region are capable of producing ion channels in planar bilayers but, in one case at least, these have properties differing considerably from those of the native channels. In view of current interpretations of the mutagenesis data on this region of the channel protein, these results may imply that additional parts of the protein structure are necessary for correct assembly of the pore region and/or actually contribute to the pore lining. This is in line with at least one of the models put forward for the channel structure [86].

A peptide derived from an entirely different K channel, the small single-subunit IsK (or minK) protein, which is 130 residues in length and on expression induces a slowly activating, voltage-sensitive potassium current, has also been studied [29]. This protein has only one putative transmembrane domain and therefore is particularly suited for the study of derived peptides reconstituted in lipid bilayers. The peptide TM-minK containing the putative transmembrane sequence was found to elicit heterogeneous channels with major conductance levels of 12 and 29 pS in 0.5 M KCl, only with trans negative potentials. The latter corresponds to the rectification properties of the endogenous channels. No selectivity was found between K+ and Na+ ions, however. Similar results were obtained with a longer 'truncated-IsK' peptide of 63 residues in length which was extended by nine residues at the N-terminus and by 22 residues at the C-terminus, relative to the TM-minK sequence [120]. This truncated form had been expressed in oocytes and found to induce K channels that were characteristic of the mature form [131]. For this longer peptide, major single-channel conductances of 17, 38 and 50 pS in 0.5 M KCl were obtained in planar bilayers, but again with no discrimination between K+ and Na+, unlike the expressed truncated form. This latter result was attributed to the IsK protein being an activator rather than a channel-forming protein but, as discussed above [119], this explanation now seems less likely.

Calcium channels

A rather comprehensive survey has been conducted on peptides corresponding to the putative transmembrane segments of the dihydropyridine-sensitive, voltage-gated, L-type Ca channel [15,132]. Of the peptides derived from the fourth internal repeat (CaIVS1–CaIVS6), those with sequences of the amphipathic S2 and S3 segments were found to form discrete cation-sensitive channels in lipid bilayers, whereas those with sequences of the S1 segment, of the hydrophobic S5 and S6 segments, and also of the

S4 and H5 segments, did not. For the latter segments, incorporation into the bilayer was indicated, however, by the production of erratic variations in the membrane current. Similarly, bivalent-cation-selective channel activity has been found for a peptide (CaIIS2) that corresponds to the S2 segment of the second repeat from the dihydropyridine receptor Ca channel of the T-system of skeletal muscle [16]. These results are all in agreement with the more limited data from Na-channel peptides and, for the H5 segment, tend even more strongly in the direction of the conclusions drawn for the corresponding K-channel peptides.

Strikingly, a template-assembled tetramer of the CaIVS3 peptide was found to emulate many of the pharmacological and conductance properties of the native channels [132]. These cation-selective T_4CaIVS3 channels had saturable conductances for a range of bivalent ions that were similar to those of authentic channels, and exhibited blockade by nifedine, verapamil, QX-222 and Cd^{2+} at comparable concentrations to those for authentic channels. The T_4CaIVS3 channels also displayed conductance for Na+ and K+ ions that was blocked by low (micromolar) levels of Ca^{2+}, as for authentic channels. Most notably, the stereo-specific action of agonist and antagonist enantiomers of the dihydropyridine BayK 8644 was reproduced in the T_4CaIVS3 tetramer channels. As with the S5 monomer, the template-assembled T_4CaIVS5 tetramer did not elicit discrete channel activity. These extremely interesting results have to be viewed, however, in terms of the deterministic nature of the H5 segment in specifying the high-affinity Ca^{2+}-binding site [3]. The implications of somewhat similar discrepant results have been discussed already for Na-channel peptides, and this point is returned to again in the Conclusion section.

Proton ATPase

A peptide (Pep-8a) corresponding to the putative transmembrane segment of the subunit 8 proteolipid of the proton F_1F_o-ATPase from *S. cerevisiae* has been investigated in planar and patch-clamp bilayers [31]. Subunit 8 is not the DCCD-binding F_o-proteolipid (subunit 9) that is known to be involved in the proton channel [133], but is part of the F_o section of the ATPase, possibly involving the link with the catalytic F_1 section. Channels of very high conductance (in 1 M KCl) that were weakly voltage dependent were formed in planar bilayers. A single conductance level was observed in any given experiment but different conductance levels (most frequently 440 or 3000 pS) were observed in separate experiments. The ion selectivity, particularly that for protons, was not determined.

ASSEMBLY OF PEPTIDES IN BILAYERS

Several approaches have been used to study the assembly of putative transmembrane segments in lipid membranes (methods are reviewed in [51]). The first and perhaps most direct has been referred to above and consists of the synthesis of oligomers attached to designed templates [24]. The others involve the assembly of independent segments, partly in the spirit of the two-stage model of independent assembly proposed and established for certain α-helical integral membrane proteins, most notably bacteriorhodopsin [134,135]. One aspect of such studies, the concentration dependence of the membrane current, has been touched on already but has not been universally applied to the channel studies reviewed above. Further strategies are discussed below.

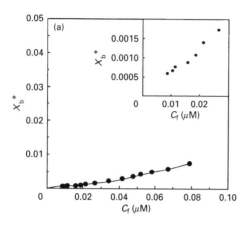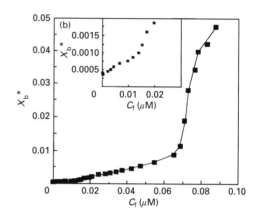

Figure 7   **Influence of lipid composition on the assembly of the NaIS4 peptide, corresponding to the S4 segment of the first repeat of *Electrophorus* Na channels, on binding to vesicles**

Shown are binding isotherms at 24 °C, where X_b^* is the molar ratio of bound peptide to total lipid and C_f is the equilibrium concentration of free peptide, obtained on binding of fluorophore-labelled NaIS4 to egg phosphatidylcholine vesicles (**a**), or to egg phosphatidylcholine/phosphatidylserine (1:1, mol/mol) vesicles (**b**). The sharp increase in binding to the latter, after an initial lag, is indicative of intramembranous aggregation of the peptide [104,105]. From [19], with permission, copyright American Chemical Society (1992).

Binding and resonance energy transfer

The assembly of peptides corresponding to putative transmembrane or pore-lining segments in lipid bilayers has been studied for a variety of channel proteins. This involves analysis of binding isotherms to reveal intramembranous aggregation processes (cf. [104,105]) and determining the resonance energy transfer between donor and acceptor fluorophores attached at the N- or C-terminus of the peptides. Fluorescence resonance energy transfer is a relatively long-range process (for 50% efficiency: $R_o \sim 50$ Å, depending on donor–acceptor pair), which ensures the reliable detection of peptide–peptide associations in oligomeric assemblies, but specific associations must be distinguished from random ones and from possible diffusional collisions between peptides [51,72].

The NaIS4 voltage sensor peptide derived from the first repeat of the *Electrophorus* Na channel has been studied by fluorescence labelling [19]. Shifts in the fluorescence maxima indicated the incorporation of the peptide in an apolar membrane environment. Both the shape of the binding isotherms and the fluorescence energy transfer measurements demonstrated that aggregation of the peptide monomers occurred readily in vesicles containing acidic lipids but not in those composed solely of zwitterionic lipids (Figure 7). The latter finding correlated with the ability of the peptide to induce permeation of the vesicles. Neither the insertion of the peptide nor its state of membrane aggregation were influenced, however, by the presence of a transmembrane diffusion potential. The S2-segment peptide, KS2, from the Shaker K channel has also been found to self-associate in lipid vesicles, but not to associate specifically with an unrelated membrane-bound amphipathic peptide [17]. The M2-segment peptide from the ROMK1 delayed rectifier K channel has also been reported to self-associate in lipid membranes, but the M1 peptide did not, nor did it associate with M2 [116,117].

The 'pore' peptides NaIH5–NaIVH5, corresponding to the H5 regions from all four internal repeats of the *Electrophorus* Na channel, have been studied by fluorescence spectroscopy [25]. From resonance energy transfer measurements, possible associations between the different pairs of H5 peptides were established. With the exception of the NaIH5/NaIIIH5 pair, all H5 peptides were found to co-assemble, but not with an unrelated

membrane-associated peptide. Fluorophore-labelled peptides corresponding to different sections of the H5 region from Shaker K channels were also studied [17,26]. It was found that the H5-derived peptides KH-5 and KH-5-N, and a peptide corresponding to the first 12 highly conserved residues of KH-5, can self-assemble in lipid bilayers, but do not co-assemble with an unrelated membrane-bound peptide. Interestingly, no association was found between the KH-5 peptides and the Na-channel peptide NaIS4. However, KH-5 and its conserved 12-mer were found to form heteroaggregates with the KS2 K-channel peptide. A peptide lacking residues 1–4 within the most conserved region of KH-5 did not bind to lipid membranes at all, and the single-residue substitution W4S reduced the partitioning of KH-5 into the membrane by a factor of five. The H5-segment peptide from the ROMK1 channel has also been found to self-associate, and additionally to co-associate with the M2 segment, in lipid membranes [116,117].

The 32-residue TM-minK peptide, which includes the single putative transmembrane segment from the IsK channel sequence, has also been shown from energy transfer studies to self-assemble in lipid bilayers [29]. The longer 63-residue peptide corresponding to the truncated form of IsK that can be functionally expressed [131] was also found to self-associate in lipid bilayer vesicles [120].

Taken together, the above results indicate that specific association is possible between putative transmembrane peptides, which is a prerequisite for them forming functional channel structures of the type thought to be present in the natural parent proteins. It should be noted, however, that most of the unrelated control peptides that did not co-assemble with the peptides of interest were highly amphiphilic in character.

Spin-label ESR

The study of the interactions between spin-labelled lipids and integral membrane proteins and of the rotational mobility of spin-labelled integral proteins by ESR spectroscopy is well established [136]. Lipid–protein interactions are characterized by the stoichiometry and by the selectivity, where the former is determined by the intramembranous perimeter of the protein assembly and the latter by the protein residues situated in the region of the lipid polar headgroups [36]. Rotational mobility is determined by the intramembranous cross-section of the protein and hence by its state of oligomerization. Both types of ESR experiment may therefore be used to investigate the assembly of transmembrane peptides in lipid bilayers [36,51].

The K26 peptide, corresponding to the putative transmembrane segment of the IsK K-channel protein, has been studied by ESR spectroscopy in phosphatidylcholine bilayers [30]. The peptide displays a pronounced selectivity for interaction with negatively charged lipids and a low stoichiometry of interaction of approx. 2.5 phospholipids per peptide monomer. The selectivity indicates close proximity of the positively charged residues at the peptide termini to the lipid headgroups. This and the low stoichiometry is consistent with an enclosed (e.g. barrel) structure in which the β-strands (cf. above) are either strongly tilted or have a reverse turn at their centre (see Figure 5). The oligomeric peptide exhibits rather slow rotational diffusion, which implies that the peptide units are aggregated in the lipid bilayer. Somewhat similar results have been obtained with the β-sheet form of the M13 coat protein [137]. In the latter case, marked differences were observed in the lipid–protein interactions with the α-helical and β-sheet forms of the coat protein [138].

The above ESR methods have also been used to determine the conditions under which removing the extramembranous domains of the Na,K-ATPase by extensive proteolysis preserves the structural integrity of the transmembrane domains [45]. This is an alternative but somewhat similar and complementary approach to simplifying the system to that of self-assembly of synthesized transmembrane peptides. As already mentioned, similar proteolytic shaving of the extramembranous portion has been used to determine the secondary structure of the membrane domains of the nicotinic acetylcholine receptor [69]. As expected, this was found to differ from that of the intact protein.

Lipid–protein interactions with the 16 kDa proteolipid from *Nephrops*, which corresponds to the V_o section of the V-ATPases, have been studied in membranous preparations by similar methods [139]. The lipid stoichiometry was found to be rather low (equivalent to 5.0–7.5 phospholipids per monomer), consistent with a hexameric arrangement deduced from low-resolution image reconstruction by electron microscopy. A selectivity was observed for negatively charged lipids, in agreement with the placement of positively charged residues in the vicinity of the lipid headgroups in model building. Additionally, the insertion of the channel-forming diphtheria toxin into anionic lipid bilayers has been studied with lipid spin labels [140]. The low-pH-triggered insertion was found to take place in two stages. At pH 6.2 the stoichiometry of lipid–protein interaction was 24 lipids per bound toxin, which was attributed to the insertion of only the translocation domain, whereas at pH 5.0 this was increased to 40 lipids/toxin, which was attributed to the insertion of the whole toxin. The change in folding of the translocation domain of diphtheria toxin on pH-induced membrane insertion has also been studied by ESR methods involving site-directed spin-labelling [141]. These results offer considerable promise for further studies on transmembrane channel peptides.

character. This extends even to the putative pore-lining faces of some channels. The philosophy of studying membrane-associated single peptide segments as channel models is further supported both by the observation that expression of homo-oligomers can result in the formation of functional channels and by the fact that independently folded domains of some integral proteins are capable of self-assembly to give the native form.

Functional studies on single-channel formation have met with some considerable success in reproducing the characteristics of the native channels. This has been spectacular even, in the case of the pharmacological properties of some template-assembled peptides. Characterization of the secondary structure in the membrane environment has not invariably gone hand-in-hand with the functional studies (and vice versa). This remains a critical issue in view of the proposals that some channels may depart from the α-helical paradigm on the one hand, and the fact that such peptides may be capable of displaying a plasticity in secondary structure on the other. Single-channel experiments are necessarily conducted at low peptide concentrations which will tend to favour retention of α-helical conformation, as opposed to assembly into polymeric β-sheet structures. Also, the assembly of peptide oligomers on predetermined templates allows some freedom of design in directing the type of secondary structure that will be adopted. Unfortunately, however, the low concentrations employed in single-channel measurements precludes direct application of most of the standard methods for structural determination.

The voltage-gated ion channels remain among those for which direct structural information is lacking and therefore they provoke intense interest. The finding that many of the properties of Ca channels may be reproduced by template-assembled tetramers of the S3 segment alone seems somewhat at variance with blocking, mutagenesis and inactivation experiments which implicate the H5 (or SS1–SS2) region and also part of S6 [59]. It has been suggested on the basis of peptide experiments that S3 may form the pore itself and that the other regions, specifically H5, may contribute to the mouth of the pore [3]. A similar discrepancy with regard to Na channels has already been discussed. The peptide experiments are able to identify a feasible pore configuration but do not necessarily establish this as the native pore structure. The final resolution of these points may prove to be a hybrid solution, guided partly by the results with reconstitution of transmembrane peptides.

Taking a more general view, experiments with isolated putative transmembrane segments indicate some of the ways in which such sequences of amino acids in integral proteins may assemble in membranes. They also reveal the features of such defined sequences that can contribute in detail to the properties of channel pores. Additionally, the contributions of specified peptide structures in defining the interactions of side chains at the intramembranous surface of integral proteins with the lipid milieu are available from such studies. It is likely that both solid-state NMR [142] and spin-label ESR [143] will be among the techniques that will contribute materially to such studies in the future.

CONCLUSION

Peptides corresponding to putative transmembrane channel segments are capable of self-assembly in lipid bilayers. The success in identifying such sequences by hydropathy analysis, and the at least partial success of the hydrophobic core of lipid membranes in mimicking the native environment, arise from the fact that many such sequences appear to have a markedly apolar

REFERENCES

1   Montal, M. (1990) FASEB J. **4**, 2623–2635
2   Sansom, M. S. P. (1991) Prog. Biophys. Mol. Biol. **55**, 139–235
3   Montal, M. (1995) Annu. Rev. Biophys. Biomol. Struct. **24**, 31–57
4   Woolley, G. A. and Wallace, B. A. (1994) in Membrane Protein Structure:
     Experimental Approaches (White, S. H., ed.), pp. 314–334, Oxford University Press,
     New York
5   Fox, R. O. and Richards, F. M. (1982) Nature (London) **300**, 325–330

6   Karle, I. L., Flippen-Anderson, J. L., Agarwalla, S. and Balaram, P. (1991) Proc. Natl. Acad. Sci. U.S.A. **88**, 5307–5311

7   Terwilliger, T. C. and Eisenberg, D. (1982) J. Biol. Chem. **257**, 6016–6022

8   Tappin, M. J., Pastore, A., Norton, R. S., Freer, J. H. and Campbell, I. D. (1988) Biochemistry **27**, 1643–1647

9   Oiki, S., Danho, W., Madison, V. and Montal, M. (1988) Proc. Natl. Acad. Sci. U.S.A. **85**, 8703–8707

10   Oblatt-Montal, M., Bühler, L. K., Iwamoto, T., Tomich, J. M. and Montal, M. (1993) J. Biol. Chem. **268**, 14601–14607

11   Ghosh, P. and Stroud, R. M. (1991) Biochemistry **30**, 3551–3557

12   Reddy, G. L., Iwamoto, T., Tomich, J. M. and Montal, M. (1993) J. Biol. Chem. **268**, 14608–14615

13   Langosch, D., Hartung, K., Grell, E., Bamberg, E. and Betz, H. (1991) Biochim. Biophys. Acta **1063**, 36–44

14   Oblatt-Montal, M., Reddy, G. L., Iwamoto, T., Tomich, J. M. and Montal, M. (1994) Proc. Natl. Acad. Sci. U.S.A. **91**, 1495–1499

15   Grove, A., Iwamoto, T., Tomich, J. M. and Montal, M. (1993) Protein Sci. **2**, 1918–1930

16   Reid, D. G., MacLachlan, L. K., Salter, C. J., Saunders, M. J., Jane, S. D., Lee, A. G., Tremeer, E. J. and Salisbury, S. A. (1992) Biochim. Biophys. Acta **1106**, 264–272

17   Peled, H. and Shai, Y. (1994) Biochemistry **33**, 7211–7219

18   Oiki, S., Danho, W. and Montal, M. (1988) Proc. Natl. Acad. Sci. U.S.A. **85**, 2393–2397

19   Rapoport, D., Danin, M., Gazit, E. and Shai, Y. (1992) Biochemistry **31**, 8868–8875

20   Mulvey, D., King, G. F., Cooke, R. M., Doak, D. G., Harvey, T. S. and Campbell, I. D. (1989) FEBS Lett. **257**, 113–117

21   Tosteson, M. T., Auld, D. S. and Tosteson, D. C. (1989) Proc. Natl. Acad. Sci. U.S.A. **86**, 707–710

22   Brullemans, M., Helluin, O., Dugast, J. Y., Molle, G. and Duclohier, H. (1993) Eur. Biophys. J. **23**, 39–49

23   Haris, P. A., Ramesh, B., Brazier, S. and Chapman, D. (1994) FEBS Lett. **349**, 371–374

24   Grove, A., Iwamoto, T., Montal, M. S., Tomich, J. M. and Montal, M. (1992) Methods Enzymol. **207**, 510–525

25   Pouny, Y. and Shai, Y. (1995) Biochemistry **34**, 7712–7721

26   Peled, H. and Shai, Y. (1993) Biochemistry **32**, 7879–7885

27   Haris, P. I., Ramesh, B., Sansom, M. S. P., Kerr, I. D., Srai, K. S. and Chapman, D. (1994) Protein Eng. **7**, 255–262

28   Shinozaki, K., Anzai, K., Kirino, Y., Lee, S. and Aoyagi, H. (1994) Biochem. Biophys. Res. Commun. **198**, 445–450

29   Ben-Efraim, I., Bach, D. and Shai, Y. (1993) Biochemistry **32**, 2371–2377

30   Horváth, L. I., Heimburg, T., Kovachev, P., Findlay, J. B. C., Hideg, K. and Marsh, D. (1995) Biochemistry **34**, 3893–3898

31   Molle, G., Dugast, J.-Y., Duclohier, H., Daumas, P., Heitz, F. and Spach, G. (1988) Biophys. J. **53**, 193–203

32   Guy, H. R. and Conti, F. (1990) Trends Neurosci. **13**, 201–206

33   Kyte, J. and Doolittle, R. F. (1982) J. Mol. Biol. **157**, 105–134

34   Argos, P., Rao, J. K. M. and Hargrave, P. A. (1982) Eur. J. Biochem. **128**, 565–575

35   Engelman, D. M., Steitz, T. A. and Goldberg, A. (1984) Annu. Rev. Biophys. Biophys. Chem. **15**, 321–353

36   Marsh, D. (1993) New Compr. Biochem. **25**, 41–66

37   Unwin, N. (1993) J. Mol. Biol. **229**, 1101–1124

38   Nakanishi, S. and Masu, M. (1994) Annu. Rev. Biophys. Biomol. Struct. **23**, 319–348

39   Wood, M. W., VanDongen, H. M. A. and Van Dongen, A. M. J. (1995) Proc. Natl. Acad. Sci. U.S.A. **92**, 4882–4886

40   Michel, H., Weyer, K. A., Gruenberg, H., Dunger, I., Oesterhelt, D. and Lottespeich, F. (1986) EMBO J. **5**, 1149–1158

41   Rees, D. C., DeAntonio, L. and Eisenberg, D. (1989) Science **245**, 510–513

42   Eisenberg, D., Schwarz, E., Komaromy, M. and Wall, R. (1984) J. Mol. Biol. **179**, 125–142

43   Finer-Moore, J. and Stroud, R. M. (1984) Proc. Natl. Acad. Sci. U.S.A. **81**, 155–159

44   Blanton, M. P. and Cohen, J. B. (1994) Biochemistry **33**, 2859–2872

45   Esmann, M., Karlish, J. D., Sottrup-Jensen, L. and Marsh, D. (1994) Biochemistry **33**, 8044–8050

46   Pedersen, S. E., Bridgman, P. C., Sharp, S. D. and Cohen, J. B. (1990) J. Biol. Chem. **265**, 569–581

47   Akabas, M., Kauffman, C., Archdeacon, P. and Karlin, A. (1994) Neuron **13**, 919–927

48   Kürz, L. L., Zühlke, R. D., Zhang, H.-J. and Joho, R. H. (1995) Biophys. J. **68**, 900–905

49   Lu, Q. and Miller, C. (1995) Biophys. J. **68**, A24

50   Jones, P. C., Harrison, M. A., Kim, Y.-I., Finbow, M. E. and Findlay, J. B. C. (1994) Biochem. Soc. Trans. **22**, 805–809

51   Findlay, J. B. C. and Marsh, D. (1995) in Ion Channels: A Practical Approach (Ashley, R., ed.), pp. 241–267, IRL Press, Oxford

52   Hartmann, H. A., Kirsch, G. E., Drewe, J. A., Taglialaterla, M., Joho, R. H. and Brown, A. M. (1991) Science **251**, 942–944

53   Yool, A. J. and Schwarz, T. L. (1991) Nature (London) **349**, 700–703

54   Heinemann, S. H., Terlau, H., Stühmer, W., Imoto, K. and Numa, S. (1992) Nature (London) **356**, 441–443

55   Stühmer, W., Conti, F., Suzuki, H., Wang, X., Noda, M., Yahagi, N., Kubo, H. and Numa, S. (1989) Nature (London) **339**, 597–603

56   Papazian, D. M., Timpe, L. C., Jan, Y. N. and Jan, L. Y. (1991) Nature (London) **349**, 305–310

57   Liman, E. R. and Hess, P. (1991) Nature (London) **353**, 752–756

58   Logothetis, D. E., Movahedi, S, Salter, C., Lindpaintner, K. and Nadal-Ginard, B. (1992) Neuron **8**, 531–540

59   Catterall, W. A. (1995) Annu. Rev. Biochem. **64**, 493–531

60   Pongs, O. (1993) J. Membr. Biol. **136**, 1–8

61   Changeux, J.-P., Galzi, J.-L., Devillers-Thiéry, A. and Bertrand, D. (1992) Q. Rev. Biophys. **25**, 395–432

62   Unwin, N. (1995) Nature (London) **373**, 37–43

63   Deisenhofer, J., Epp, O., Miki, K., Huber, R. and Michel, H. (1985) Nature (London) **318**, 618–624

64   Henderson, R., Baldwin, J. M., Ceska, T. A., Zemlin, F., Beckmann, E. and Downing, K. H. (1990) J. Mol. Biol. **213**, 899–929

65   Kühlbrandt, W., Wang, D. N., Fujiyoshi, Y. (1994) Nature (London) **367**, 614–621

66   McDermott, G., Prince, S. M., Freer, A. A., Hawthornthwaite-Lawless, A. M., Papiz, M. Z., Cogdell, R. J. and Isaacs, N. W. (1995) Nature (London) **374**, 517–521

67   Sixma, T. K., Pronk, S. E., Kalk, K. H., Wartna, E. S., Van Zanten, B. A. M., Witholt, B. and Hol, W. G. J. (1991) Nature (London) **351**, 371–377

68   Stein, P. E., Boodhoo, A., Tyrell, G. J., Brunton, J. L. and Read, R. J. (1991) Nature (London) **351**, 748–750

69   Görne-Tschelnokow, U., Streker, A., Kaduk, C., Naumann, D. and Hucho, F. (1994) EMBO J. **13**, 338–341

70   Parker, M. W., Buckley, J. T., Postma, J. P. M., Tucker, A. D., Leonard, K., Pattus, F. and Tsernoglou, D. (1994) Nature (London) **367**, 292–295

71   Parker, M. W. and Pattus, F. (1993) Trends Biochem. Sci. **18**, 391–395

72   Gazit, E. and Shai, Y. (1993) Biochemistry **32**, 12363–12371

73   Huber, R., Berendes, R., Burger, A., Schneider, M., Karshikov, A., Luecke, H., Römisch, J. and Paques, E. (1992) J. Mol. Biol. **223**, 683–704

74   Demange, P., Voges, D., Benz, J., Liemann, S., Göttig, P., Berendes, R., Burger, A. and Huber, R. (1994) Trends Biochem. Sci. **19**, 272–276

75   Burger, A., Voges, D., Demange, P., Ruiz Perez, C., Huber, R. and Berendes, R. (1994) J. Mol. Biol. **237**, 479–499

76   Weiss, M. S., Wacker, T., Weckesser, J., Welte, W. and Schulz, G. E. (1990) FEBS Lett. **267**, 268–272

77   Weiss, M. S. and Schulz, G. E. (1992) J. Mol. Biol. **227**, 493–509

78   Cowan, S. W., Schirmer, T., Rummel, R., Steiert, M., Ghosh, R., Pauptit, R. A., Jansonius, J. N. and Rosenbusch, J. P. (1992) Nature (London) **358**, 727–733

79   Jähnig, F. (1989) in Prediction of Protein Structure and Principles of Protein Conformation (Fasman, G. D., ed.), pp. 707–717, Plenum Press, New York

80   Cowan, S. W. (1993) Curr. Opin. Struct. Biol. **3**, 501–507

81   Benz, R., Schmid, A. and Hancock, R. E. W. (1985) J. Bacteriol. **162**, 722–727

82   Schindler, H. and Rosenbusch, J. P. (1978) Proc. Natl. Acad. Sci. U.S.A. **75**, 3751–3755

83   Lakey, J. H. and Pattus, F. (1989) Eur. J. Biochem. **186**, 303–308

84   Bauer, K., Struyvé, M., Bosch, D., Benz, R. and Tommassen, J. (1989) J. Biol. Chem. **264**, 16393–16398

85   Karshikov, A., Spassov, V., Cowan, S. W., Ladenstein, R. and Schirmer, T. (1994) J. Mol. Biol. **240**, 372–384

86   Durrell, S. R. and Guy, H. R. (1992) Biophys. J. **62**, 238–250

87   Bogusz, S., Boxer, A. and Busath, D. D. (1992) Protein Eng. **5**, 285–293

88   Soman, K. V., McCammon, J. A. and Brown, A. M. (1995) Protein Eng. **8**, 397–401

89   MacKinnon, R. (1991) Nature (London) **350**, 232–235

90   Li, M., Unwin, N., Stauffer, K. A., Jan, Y.-N. and Jan, L. Y. (1994) Curr. Biol. **4**, 110–115

91   Guy, H. R. and Durrell, S. R. (1994) Biophys. J. **66**, A248

92   Lipkind, G. M. and Fozzard, H. A. (1994) Biophys. J. **66**, 1–13

93   Bogusz, S. and Busath, D. (1992) Biophys. J. **62**, 19–21

94   Langs, D. A. (1988) Science **241**, 188–191

95   Langs, D. A., Smith, G. D., Courseille, C., Precigoux, G. and Hospital, M. (1991) Proc. Natl. Acad. Sci. U.S.A. **88**, 5345–5349

96   Wallace, B. A. and Ravikumar, K. (1988) Science **241**, 182–187

97   Ketchem, R. R., Hu, W. and Cross, T. A. (1993) Science **261**, 1457–1460

98   Kennedy, S. J. (1978) J. Membr. Biol. **42**, 265–279

99   Deisenhofer, J. and Michel, H. (1989) Science **245**, 1463–1473

100 Karle, I. L. (1994) in Membrane Protein Structure: Experimental Approaches (White, S. H., ed.), pp. 355–380, Oxford University Press, New York

101 Sansom, M. S. P. (1993) Eur. Biophys. J. **22**, 105–124

102 Esposito, G., Carver, J. A., Boyd, J. and Campbell, I. D. (1987) Biochemistry **26**, 1043–1050

103 Sansom, M. S. P. (1993) Q. Rev. Biophys. **26**, 365–421

104 Schwarz, G., Stankowski, S. and Rizzo, V. (1986) Biochim. Biophys. Acta **861**, 141–151

105 Rizzo, V., Stankowski, S. and Schwarz, G. (1987) Biochemistry **26**, 2751–2759

106 Balaram, P., Krishna, K., Sukumar, M., Mellor, I. R. and Sansom, M. S. P. (1992) Eur. Biophys. J. **21**, 117–128

107 Bazzo, R., Tappin, M. J., Pastore, A., Harvey, T. S., Carver, J. A. and Campbell, I. D. (1988) Eur. J. Biochem. **173**, 139–146

108 Inagaki, F., Shimada, I., Kawaguchi, K., Hirano, M., Terasawa, I., Ikura, T. and Go, N. (1989) Biochemistry **28**, 5985–5991

109 Tosteson, M. Y. and Tosteson, D. C. (1981) Biophys. J. **36**, 109–116

110 Altenbach, C., Froncisz, W., Hyde, J. S. and Hubbell, W. L. (1989) Biophys. J. **56**, 1183–1191

111 Lee, K. H., Fitton, J. E. and Wüthrich, K. (1987) Biochim. Biophys. Acta **911**, 144–153

112 Mellor, I. R., Thomas, D. H. and Sansom, M. S. P. (1988) Biochim. Biophys. Acta **942**, 280–294

113 Oiki, S., Madison, V. and Montal, M. (1990) Protein Struct. Funct. Genet. **8**, 226–236

114 Montal, M., Montal, M. S. and Tomich, J. M. (1990) Proc. Natl. Acad. Sci. U.S.A. **87**, 6929–6933

115 Bechinger, B., Kim, Y., Chirlian, L. E., Gesell, J., Neumann, J., Montal, M., Tomich, J., Zasloff, M. and Opella, S. J. (1991) J. Biomol. NMR **1**, 167–173

116 Shai, Y. and Ben-Efraim, I. (1994) Biophys. J. **66**, A427

117 Ben-Efraim, I. and Shai, Y. (1994) J. Gen. Physiol. **104**, 12a–13a

118 Takumi, T., Ohkubo, H. and Nakanishi, S. (1988) Science **242**, 1042–1045

119 Tzounopoulos, T., Maylie, J. and Adelman, J. (1995) Biophys. J. **69**, 904–908

120 Ben-Efraim, I., Strahilevitz, J., Bach, D. and Shai, Y. (1994) Biochemistry **33**, 6966–6973

121 Nozaki, Y., Reynolds, J. A. and Tanford, C. (1978) Biochemistry **7**, 1239–1246

122 Spruijt, R. B. and Hemminga, M. A. (1991) Biochemistry **30**, 11147–11154

123 Spruijt, R. B., Wolfs, C. J. A. M. and Hemminga, M. A. (1989) Biochemistry **28**, 9158–9165

124 Norwood, T. J., Crawford, D. A., Steventon, M. E., Driscoll, P. C. and Campbell, I. D. (1992) Biochemistry **31**, 6285–6290

125 Garvin, M. E. and Fillingame, R. H. (1994) Biochemistry **33**, 665–674

126 Holzenburg, A., Jones, P. C., Franklin, T., Páli, T., Heimburg, T., Marsh, D., Findlay, J. B. C. and Finbow, M. E. (1993) Eur. J. Biochem. **213**, 21–30

127 Grove, A., Mutter, M., Rivier, J. and Montal, M. (1993) J. Am. Chem. Soc. **115**, 5919–5924

128 Oblatt-Montal, M., Iwamoto, T., Tomich, J. M. and Montal, M. (1993) FEBS Lett. **320**, 261–266

129 Lear, J. D., Wasserman, Z. R. and DeGrado, W. F. (1988) Science **240**, 1177–1181

130 Betz, H. (1992) Q. Rev. Biophys. **25**, 381–394

131 Takumi, T., Moriyoshi, K., Aramori, I., Ishii, T., Oiki, S., Okada, Y., Ohkubo, H. and Nakanishi, S. (1991) J. Biol. Chem. **266**, 22192–22198

132 Grove, A., Tomich, J. M. and Montal, M. (1991) Proc. Natl. Acad. Sci. U.S.A. **88**, 6418–6422

133 Schindler, H. and Nelson, N. (1982) Biochemistry **21**, 5787–5794

134 Popot, J.-L. and Engelman, D. M. (1990) Biochemistry **29**, 4031–4037

135 Popot, J.-L., de Vitry, C. and Atteia, A. (1994) in Membrane Protein Structure: Experimental Approaches (White, S. H., ed.), pp. 41–96, Oxford University Press, New York

136 Knowles, P. F. and Marsh, D. (1991) Biochem. J. **274**, 625–641

137 Wolfs, C. J. A. M., Horváth, L. I., Marsh, D., Watts, A. and Hemminga, M. A. (1989) Biochemistry **28**, 9995–10001

138 Peelen, S. J. C. J., Sanders, J. C., Hemminga, M. A. and Marsh, D. (1992) Biochemistry **31**, 2670–2677

139 Páli, T., Finbow, M. E., Holzenburg, A., Findlay, J. B. C. and Marsh, D. (1995) Biochemistry **34**, 9211–9218

140 Montich, G. G., Montecucco, C., Papini, E. and Marsh, D. (1995) Biochemistry **34**, 11561–11567

141 Zhan, H., Oh, K. J., Shin, Y.-K., Hubbell, W. L. and Collier, R. J. (1995) Biochemistry **34**, 4856–4863

142 Opella, S. J. (1994) in Membrane Protein Structure: Experimental Approaches (White, S. H., ed.), pp. 249–267, Oxford University Press, New York

143 Hubbell, W. L. and Altenbach, C. (1994) in Membrane Protein Structure: Experimental Approaches (White, S. H., ed.), pp. 224–248, Oxford University Press, New York

Biochem. J. (1996) **313**, 353–368 (Printed in Great Britain)

REVIEW ARTICLE
Cellular consequences of thrombin-receptor activation

Roger J. A. GRAND,*‡ Andrew S. TURNELL* and Peter W. GRABHAM†
*CRC Institute for Cancer Studies, The Medical School, University of Birmingham, Edgbaston, Birmingham B15 2TJ, U.K., and †Department of Pharmacology,
Center for Neurobiology and Behavior, Columbia University College of Physicians and Surgeons, New York, NY 10032, U.S.A.

1. INTRODUCTION

To say that thrombin is a multi-functional protein is rather to understate the case. It is the key enzyme involved in haemostasis, playing important roles at all levels of complexity. First, in the coagulation cascade, thrombin converts fibrinogen into fibrin, which is readily cross-linked to form a clot (reviewed in [1,2]); secondly, thrombin activates blood platelets, causing aggregation and secretion [3,4]; and thirdly, thrombin can elicit mitogenic responses from vascular smooth-muscle cells [5,6]. This latter property is probably most significant in the renewal of damaged blood-vessel walls. Additionally, thrombin is able to elicit responses from cell types as diverse as macrophages [7], monocytes [8] and neutrophils [9]. Perhaps more surprisingly, it is able to regulate neurite outgrowth from cells of neuronal origin [10] and initiate resorption of bone cells [11].

All of these properties of thrombin appear to rely on its action as a serine proteinase, since modification (either chemically [12–14] or mutationally [15–17]) which destroys its proteolytic activity leads to a loss of biological activity. These crucial findings have been explained to a large extent by the recent elegant characterization of a novel, widely expressed thrombin receptor which is activated by proteolytic cleavage rather than by ligand (protein) binding [18,19]. At the same time as elucidation of the mechanism of thrombin signalling there has been a dramatic increase in our understanding of how thrombin signals are mediated within the cell. This therefore seems an appropriate time to assess our current knowledge and perhaps to try to predict areas where the greatest advances will be made in the immediate future. In this review, because of limitations in length, emphasis will be placed on recent advances, in particular on the characterization of the thrombin receptor and in the mechanisms of intracellular signalling. Whilst most studies of thrombin have concentrated on its action on cells involved in blood clotting and wound healing, it is now becoming apparent that it can modulate the growth and differentiation status of cells of neuronal origin. A consideration of recent advances in this area of study will form the third major theme of this review.

2. STRUCTURE OF THROMBIN

Thrombin is generated in circulating plasma by the cleavage of prothrombin, when it forms part of the prothrombinase complex. The other components of this complex, which are essential for the proteolytic reaction, are activated Factors X and V, Ca^{2+} and membrane phospholipid (reviewed, for example, in [1]). Pro-thrombin itself is synthesized in the liver and represents a unique class of vitamin K-dependent zymogens in that it has two 'kringle domains' of approximately eight amino acids each and an N-terminal 'Gla domain' containing ten γ-carboxyglutamic acid (Gla) residues [1,20] (Figure 1a). Conversion of prothrombin into thrombin involves the cleavage of peptide bonds C-terminal to Arg^{271} and Arg^{320}. The N-terminal region of prothrombin containing the Gla and kringle domains is inactive and appears to have no further biological role [21]. Thus thrombin is about half the size of prothrombin (39 kDa compared with 71.6 kDa) and comprises an A (light) chain (49 residues) and a B (heavy or catalytic) chain (259 residues) joined by a disulphide link between Cys^{22} (A chain) and Cys^{119} (B chain) (throughout this review the N-terminal amino acid of either prothrombin or the A and B chains of thrombin is designated residue 1 of the respective protein, as in [20] and Figure 1a). Further cleavage of the thrombin A chain can occur, removing a further 13 amino acids. Three carbohydrate moieties are present on each prothrombin molecule, with one being present in thrombin (linked to Asn^{53} on the B chain).

The amino acid sequence similarities between thrombin and other serine proteinases such as trypsin encouraged a number of structural predictions to be made. More recently, a high-resolution X-ray-crystallographic study of thrombin bound to various small molecules has allowed the structure of human and bovine thrombin to be determined at high resolution (see, for example, [22–24]). These studies have been reviewed in considerable detail by Stubbs and Bode [25], and readers in need of a detailed account of thrombin structure are referred to that excellent article. It is, however, worth briefly describing here a few of the major structural features of the molecule so that its mode of action can be better understood. Thrombin is a highly structured globular protein and is ellipsoid in shape. A deep narrow groove containing the active site runs across the molecule with the three catalytic amino acids, His^{43}, Asp^{99} and Ser^{205}, lying at its base (Figure 1b). The short A chain is held in place on the side of the protein opposite to that of the active site. Until the cloning of the thrombin receptor, the primary natural substrates for thrombin had been considered to be the A and B chains of fibrinogen, although hydrolysis of peptide bonds in a number of other proteins occurs *in vivo*. In all cases of mammalian proteins thrombin cleaves a peptide bond C-terminal to an Arg residue. A number of amino acids can occupy the P1′ position (C-terminal to the Arg), but they are generally small and hydrophilic, such as glycine in the fibrinogen A and B β chains or serine in the thrombin receptor, Factor V and Factor VIII. The P2 position (N-terminal to the Arg) is usually occupied by a Pro, but not invariably so. Restrictions can also be seen in the amino acids allowed for P3 (usually hydrophilic and/or small) and P4 (usually large and hydrophobic) in thrombin substrates. The much greater substrate specificity shown by thrombin over other serine proteinases is probably attributable to the depth and narrowness of the active-site cleft, making the active-site amino acids much less accessible to protein substrates in general (Figure 1b).

A second important structural feature involved in the interaction of thrombin with its substrates is the anion-binding exosite (basic patch), which is centred around the loop between Lys^{65} and Lys^{77}. Thus a series of basic amino acids (Arg^{68}, Arg^{70} and Arg^{73}) are in close proximity to Arg^{20}, Lys^{21} and Lys^{154} in the

‡ To whom correspondence should be sent.

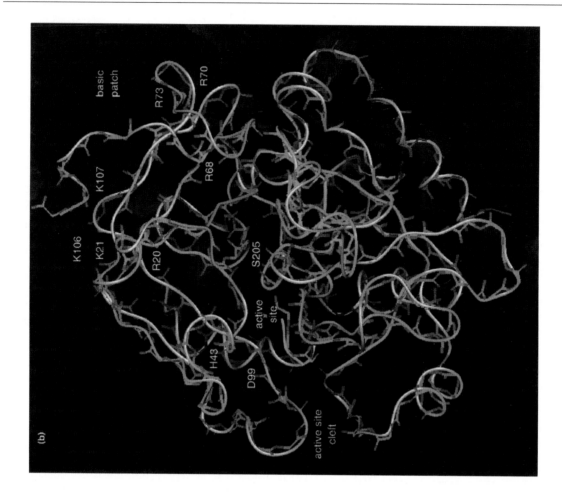

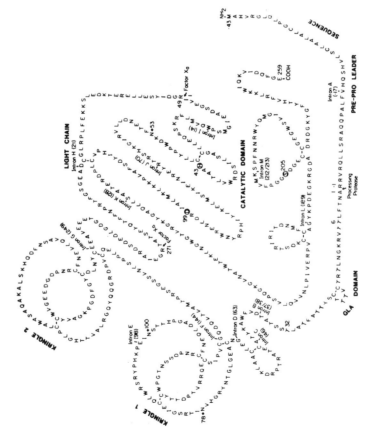

Figure 1   The structure of thrombin

(a) The amino acid sequence of preprothrombin is shown. The peptide bond at the C-terminus of the preproleader peptide is cleaved by a signal peptidase, and the bonds required for conversion of prothrombin to thrombin are cleaved by Factor Xa. The positions of 13 introns (A–M) in the prothrombin gene sequence are indicated, as are the Gla and kringle domains. The active-site amino acids His[43], Asp[99] and Ser[205] are sites of glycosylation (◆). Amino acid numbering is as follows: preproleader sequence −43 to −1; prothrombin fragment 1, 1–271; thrombin light (A) chain, 1–49; thrombin heavy (B) chain, 1–259. Reproduced from Friezner, Degen and Davie [20] with the permission of the publishers of *Biochemistry*. (b) The three-dimensional structure of the thrombin heavy chain. The ribbon structure of the heavy chain of human thrombin is shown (from the P1DWE PDB entry). The positions of the active-site amino acids (His[43], Asp[99] and Ser[205]) and the basic patch amino acids (Arg[20], Lys[21], Arg[68], Arg[70], Arg[73], Lys[106] and Lys[107]) are shown, as is the position of the active-site cleft.

folded protein, with Lys[77], Lys[106] and Lys[107] lying somewhat further away (Figure 1b). These residues give rise to a positively charged patch which forms a site of interaction with acidic areas on fibrinogen, fibrin, thrombomodulin, hirudin and the thrombin receptor. The interaction of thrombin with its receptor is discussed in more detail in section 3 below.

3. STRUCTURE AND MODE OF ACTION OF THE THROMBIN RECEPTOR

(i) Structure of the mammalian thrombin receptor

The ability of thrombin to elicit cellular responses has been well documented for many years. During that time a number of thrombin-binding proteins were identified [26], but none would initiate a second-messenger response and was therefore unlikely to represent a receptor molecule. Many of these inconsistencies and anomalies were resolved when thrombin receptors from human platelets [18,19], hamster lung fibroblasts [27], rat aortic smooth-muscle cells [28] and, later, *Xenopus laevis* [29], were cloned and sequenced. Largely on the basis of the derived amino acid sequences of these proteins a model was proposed for the mechanism of receptor activation which has stood up well to experimental analysis.

The human thrombin receptor comprises 425 amino acids and has a number of structural features in common with the classical G-protein-linked receptors (Figure 2), being most similar to receptors for neuropeptides and glycoprotein hormones [18]. Seven helical hydrophobic transmembrane regions have been proposed [18] giving rise to three intra- and three extra-cellular loops and a C-terminal intracellular tail and a long N-terminal extracellular domain (Figure 2). Within this latter region are a number of structural determinants which are essential for the correct functioning of the receptor. A thrombin cleavage site [LDPR↓SFLL (one-letter amino acid code)] similar to that present in Protein C [18] is present between residues 41 and 42 (in the human protein). Two thrombin interaction sites have been proposed on the receptor: one immediately N-terminal to Arg[41] is involved in recognition of the cleavage site, whilst the second, between residues 53 and 64, shows sequence similarity to a C-terminal region of hirudin and is involved in interaction with the thrombin anion-binding exosite ([23] and Figure 1b). Both of these sites are highly conserved between the cloned mammalian thrombin receptors, but rather less so for the *Xenopus* protein [29] (Figure 3).

Three-dimensional structural studies of the thrombin receptor have been limited by the lack of sufficient protein, and therefore use has been made of the peptide-mimetic approach. Thus an examination of the structure of peptides identical with parts of the N-terminal extracellular domain has allowed the identification of overlapping turns in the region between residues 47 and 51 stabilized by a hydrogen bond 48_{CO}–51_{NH} [30] as well as a 3_{10} helix covering the seven residues between Pro[40] and Arg[46]. This latter structural motif is stabilized by a charged hydrogen bond between the side chains of Asp[39] and Arg[41] [30]. Obviously, an element of doubt exists as to whether such structures, based on short synthetic peptides, exist in the intact protein, although computer modelling has indicated that it is possible to fit the proposed peptide structure into the substrate-binding cleft of thrombin (see below).

NMR spectroscopic and crystallographic studies of synthetic receptor peptides in the presence of thrombin have allowed the sites of interaction to be closely defined. In the bound state a receptor peptide comprising amino acids Leu[38]–Glu[60] adopts an S-shaped conformation with three anti-parallel strands (Leu[38]–Pro[40], Arg[41]–Leu[45] and Asp[50]–Pro[54]) about 0.7 nm (7 Å)

apart [31]. These data are reasonably compatible with the helical structures between Pro[40] and Arg[46] and turns around Asn[47]–Lys[51] reported by Smith et al. [30]. The three strands of the receptor peptide appear to correspond to the thrombin cleavage site, the agonist peptide and the hirudin-like domain. It should be noted, however, that in the crystallographic studies reported by Mathews et al. [31], each receptor peptide binds to two adjacent thrombin molecules, one contributing to the catalytic binding site and the second to the anion-binding exosite. So far there is no evidence for such a structure existing *in vivo*. Interaction of the receptor with the thrombin catalytic (active centre) site primarily involves the few residues on the receptor immediately N-terminal to the cleavage site (Leu[38]–Arg[41]) [31,32]. In thrombin-receptor peptide crystals the side chain of Leu[38] in the receptor occupies a hydrophobic site formed by Ile[179] and Trp[227] of thrombin, with a hydrogen bond being formed between the Leu backbone and Tyr[37] [31,32]. Perhaps more importantly, a salt bridge is formed between the receptor Asp[39] and His[43] in the thrombin catalytic site (Figure 1b). A hydrogen bond has also been demonstrated between Asp[39] or Pro[40] and thrombin Gly[230] [31,32], orienting Asp[39] towards Arg[233] of thrombin and reducing the mobility of the receptor cleavage site. Interaction of receptor peptides at the anion-binding exosite of thrombin appears to involve primarily the D[50]KYEPF[55] motif in a manner analogous to that seen for the hirudin homologues hirugen and hirulog 1 in complexes with thrombin [33]. Additional evidence for the receptor KYEPF sequence being the primary area of interaction with the anion-binding exosite is provided by the observations that substitutions for Tyr[52], Glu[53] and Phe[55] in the receptor result in loss of receptor response to thrombin [19] and that appropriate synthetic receptor peptides can displace hirudin from the exosite [34].

Thus it seems probable that, in the interaction of thrombin with its receptor, an initial contact is made between the anion-binding exosite and the receptor sequence KYEPF. In the second step the orientated receptor peptide binds to the active site with receptor Leu and Pro interacting with the apolar pocket. It has been proposed that, before cleavage, the helical structure around the cleavage site must be unwound, disrupting the hydrogen bond between Asp and Arg, thus freeing the Arg residue [30] and allowing it to move into the specificity pocket. Cleavage of the receptor then occurs, generating a new N-terminal sequence which is free to trigger the intracellular response. It is believed that this occurs by interaction of the first five or six amino acids with a binding site possibly located on extracellular loop 2 (amino acids 244–268) and/or the N-terminal exodomain (amino acids 76–93) [29,35] (Figure 2).

(ii) Thrombin-receptor-activating peptides (TRAPs)

One of the most telling pieces of evidence in favour of the proposed model for receptor activation has been the observation that short peptides identical with the sequence C-terminal to the receptor cleavage site (Arg[41]↓Ser[42]) can duplicate the actions of thrombin. Thus TRAPs can cause platelet aggregation [18,36–42], release of intracellular Ca[2+] stores [38,42–44], 5-hydroxytryptamine release [36], adenylate cyclase inhibition [38,45,46], stimulation of DNA synthesis and mitogenesis [46,47], neurite retraction [48], activation of MAP (mitogen-activated protein) kinase [49,50] and a number of other effects associated with thrombin stimulation. On the basis of these studies, it is possible to define quite precisely the structural determinants within the peptides (and therefore in the N-terminal region of the receptor) which are essential for receptor activation. Thus substitutions or 'deletions' outside the amino acid sequence SFLLRN (residues

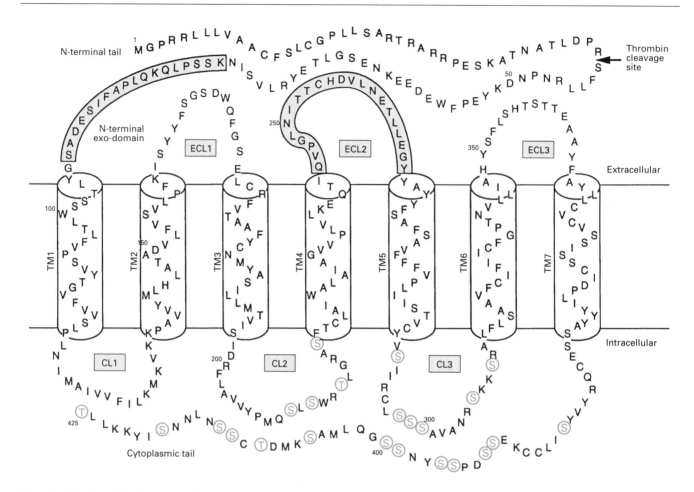

Figure 2   Structure of the human thrombin receptor

The amino acid sequence is arranged through the membrane, as suggested by Vu et al. [18] and Gerszten et al. [29]. The thrombin cleavage site (←), extracellular domain, extracellular loops (ECL) and cytoplasmic loops (CL) are indicated. Potential intracellular phosphorylation sites (encircled red letters) and amino acid residue numbers are marked. The extracellular domains considered to be the sites of interaction with the N-terminal receptor peptide are indicated by boxed-in stretches of amino acids shaded pink.

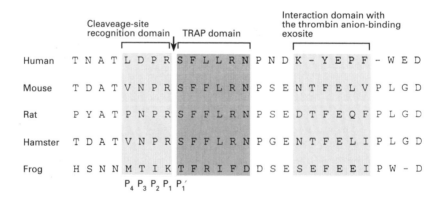

Figure 3   Comparison of the amino acid sequences of part of the extracellular N-terminal domain of vertebrate thrombin receptors

Sequences have been aligned for maximum identity. The thrombin cleavage site (↓), the cleavage recognition domain, the hirudin-like domain which interacts with the thrombin anion-binding exosite and the receptor N-terminal domain (TRAP) are all identified. − indicates a gap in the sequence introduced to allow for better alignment. References for the sequences are given in the text.

1–6 of the cleaved receptor) appear to make very little difference to the biological activity of the peptide [36–39] and even deletion of amino acid 6 (Asn) reduces the potency of the peptide only marginally [36,37,39]. Within the region Ser–Arg, however, only very conservative replacements are allowed. A free N-terminal amino group is essential for activity, N-acetyl derivatives being unable to elicit a response from platelets [36,37,40]. There is a requirement for a small, neutral hydrophilic amino acid at the N-terminus, such that Gly, Ala, Thr, Cys and Ile can substitute for Ser with only limited loss of activity [36–38,40,46]. There appears to be an absolute requirement for an aromatic residue at position 2, with even the conservative Tyr-for-Phe substitution causing a reduction in activity [41,46]. Replacement of the Leu residues at positions 3 and 4 with Ala also results in a loss of the peptide's ability to cause platelet aggregation and 5-hydroxytryptamine release [37,41], although substitution at position 3 appears to have a somewhat more deleterious effect on the activity [36,38]. Similarly, considerable constraints exist on the amino acid present at position 5 in the peptide, there being a requirement for a basic residue for the retention of activity [36,41,46].

NMR-spectroscopic studies carried out on synthetic peptides covering the N-terminus of the receptor exposed after thrombin cleavage indicate little secondary structure attributable to the first five amino acids [30]. Thus it seems likely that recognition of the tethered ligand by other region(s) of the receptor probably involves 'docking' of the relatively unstructured and possibly extended peptide into a pocket where side-chain recognition is all important.

The amino acid sequences of mammalian thrombin receptors are conserved in the proposed thrombin-binding sites (both the region which binds to the thrombin active centre and that which is recognized by the anion-binding exosite) and in the sequence of the tethered ligand exposed after cleavage. The only difference in this latter area is the substitution of Phe for Leu at position 3 in all the rodent sequences (Figure 3). This substitution does not appear to impart a species specificity, as the weight of evidence suggests that there is no difference in the ability of the rodent or human peptides to elicit a response (in either human or rodent cells) [46,51]. It should be noted, however, that, in at least one recent study [52], rabbit and rat platelets did not respond fully to the human TRAP. The amino acid sequence of the *Xenopus* receptor is quite different to the mammalian protein and is unresponsive to human peptides [29]. This is probably a result of substitution in the as-yet-unidentified ligand-binding site. The two major changes in the frog tethered ligand are Arg for Leu at position 3 and Phe at position 5 (Figure 3). As expected, this peptide is unable to activate the human receptor [29].

(iii) Different responses to thrombin and TRAPS

Not surprisingly, differences have been observed in the concentrations of thrombin and receptor peptides required to elicit a similar level of response (see, for example, [46]). Naïvely, this may be explained on the basis of either very high local concentrations of tethered ligand present after thrombin stimulation, or a more appropriate orientation adopted by the tethered peptide compared with one free in solution. Perhaps more interestingly, different cellular responses to receptor activation by thrombin and by peptides have been observed. For example, whilst thrombin is mitogenic for hamster lung fibroblasts, the synthetic N-terminal peptide is not. Treatment with growth factors (fibroblast growth factor) in addition to peptide is required for re-entry into the cell cycle [49]. This observation has been explained by the ability of thrombin to cause sustained activation of MAP kinase, whilst stimulation with the peptide only gives a transient increase in activity [49]. Additionally, differences in intracellular Ca^{2+} signals in osteosarcoma-like cells have been reported in response to thrombin and a thrombin-receptor-agonist peptide (the former inducing a transient increase in $[Ca^{2+}]$, whilst the latter produced a biphasic response [17]). Even in the human platelet system, different responses to thrombin and receptors peptides have been seen. Thus the extent of aggregation, activation and association of the functional

integrin glycoprotein IIb-IIIa and Src with the cytoskeleton, phosphatidate production, arachidonic acid release and phosphorylation of phospholipase A_2 (PLA_2) (see section 4) is always greater following thrombin stimulation [53]. These sorts of data have been variously interpreted as evidence for the existence of one or more thrombin receptors in addition to the cloned molecule or for the inability of peptides to mimic completely the action of thrombin due to signals which may be generated by interaction of the intact protein with the receptor further to those produced by proteolysis [49,52,54,55].

It has long been acknowledged that thrombin can produce graded responses in a number of different cell types [56,57]. How this might be achieved through a receptor activated only by a single proteolytic step has proved a matter of some debate, since it might be expected that all receptor extracellular domains would eventually be cleaved, even by very small amounts of thrombin. However, it has recently been demonstrated that total phosphoinositide hydrolysis over a particular time interval correlated well with the total extent of receptor cleavage [58]. It has therefore been suggested that stimulation of a receptor molecule generates a single release ('quantum') of second messenger before becoming inactive. Thus the cellular response to thrombin is regulated by the number of quanta released (i.e. the concentration) and the subsequent rate of second-messenger breakdown (removal), which presumably stays constant [58].

Support for the view that the receptor is capable of considerable subtlety has been provided by an examination of the effects of prothrombin on neuronal cells, where it has been shown that even though responses are only triggered after conversion into thrombin (and then presumably stimulation of the thrombin receptor), prothrombin will not duplicate all of the biochemical events observed after thrombin treatment [59]. It has been concluded that this is due to the fact that the concentration of prothrombin-generated thrombin at the cell surface never reaches that found after addition of the purified protein and is therefore not able to generate the full range of intracellular signals [59].

(iv) Thrombin-receptor inactivation and cellular processing

It has been recognized for some time that desensitization of the thrombin receptor requires the activity of an intracellular kinase [60], but it has only recently been directly demonstrated that rapid phosphorylation of serine or threonine residues in the cytoplasmic tail occurs very soon after thrombin or peptide stimulation and that this correlates well with receptor inactivation [61] (Figure 2). Phosphorylation is probably attributable to a G-protein-coupled receptor kinase [61] and not protein kinase C (PKC) [61,62]. Differences in susceptibility of the receptor to inactivating phosphorylation could explain the differences between thrombin and peptide activation. If, for example, the cleavage step, rather than ligand binding, is more important in inducing a conformational change allowing receptor phosphorylation and desensitization, a more prolonged response might be expected after peptide stimulation [17]. This would not, of course, explain the more transitory effects observed in some systems following peptide activation [49]. Although a specific thrombin-receptor kinase has not yet been identified, the observation that the β-adrenergic receptor kinase 2 (BARK2) is much more active than BARK1 or the rhodospin kinase (all kinases involved in the down-regulation of G-protein-linked receptors [62]) lends weight to the notion that such a specific kinase could be present in the cytoplasm of thrombin-sensitive cells. The well-characterized G-protein-coupled receptor kinases appear to interact with the activated receptors, either through the cytoplasmic loops or the C-terminal tail, although it should be

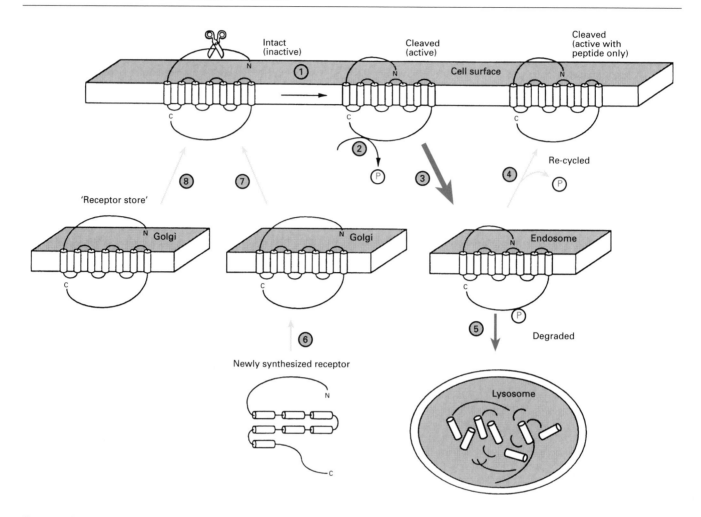

Figure 4   Thrombin-receptor trafficking

Surface thrombin receptors are normally activated by proteolytic cleavage by thrombin (1), and after an appropriate time are inactivated by phosphorylation by a receptor kinase (2). Internalization (heavy red arrows) occurs via coated pits and endosomes (3). Receptor can then be either recycled (4) to the surface, possibly following dephosphorylation (where it can only be activated by TRAP), or degraded in the lysosomes (5). Newly synthesized receptor (6) is processed in the Golgi and transported to the surface (7), but additional molecules are kept in a 'receptor store' ready for rapid activation (8), thus restoring the cells' responsiveness to thrombin after an initial stimulation. This scheme is based on models proposed by Coughlin and co-workers and Brass and co-workers [63–65].

noted that the sites of phosphorylation probably reside in this latter part of the molecule (reviewed in [62]) (Figure 2). These observations are consistent with the reported properties of the thrombin-receptor kinases [61].

A second mechanism for down-regulation of the thrombin receptor following activation has been identified. Soon after (within 1 min) thrombin stimulation of HEL and CHRF cells, most of the receptor molecules appear to be internalized in coated pits in the plasma membrane [63] (Figure 4). Over the course of about 30 min receptors can be detected in, sequentially, endosomes, prelysosomes and mature lysosomes [63,64]. In some cell types (e.g. HELs) considerable time needs to elapse between receptor activation (and internalization) and the regeneration of thrombin-sensitivity [63,65]. This correlates well with the *de novo* synthesis of receptors and is blocked by protein-synthesis inhibitors [63,64]. However, in cells of endothelial origin and fibroblasts, a population of thrombin receptors is present within the cell (but not on the surface), possibly located in the Golgi [64] (Figure 4). Thus activation of the surface thrombin receptor initiates transport of these intracellular molecules to the cell surface, where they replace cleaved molecules and are then ready to allow the cell to respond to a second round of thrombin

stimulation. It seems likely that this ability distinguishes endothelial cells, which may need to respond to thrombin repeatedly, from platelets, which generally are required to make a single response (in the form of granule release and aggregation) [64].

After internalization, most the thrombin-receptor molecules are degraded in lysosomes. However, in CHRF and HEL cells a proportion are recycled to the cell surface over a period of 1–3 h [16]. If the cell has been stimulated with receptor peptide, the recycled receptors are sensitive to either thrombin or additional peptide (presumably because an inactivating phosphorylation step has not taken place). If the cells have been stimulated with thrombin, however, they are only responsive to receptor peptide. At a later time, full responsiveness to thrombin stimulation returns as new receptor molecules are synthesized (Figure 4).

It is important to remember that considerable differences in thrombin-receptor internalization and trafficking have been observed in different cell types. As noted above, the data obtained from a study of endothelial cells does not duplicate that from the HEL and CHRF288 cell lines [16,63–65]. Additionally, recent evidence suggests that receptor internalization in platelets occurs to only a very limited extent [65,66].

(v) The contribution of growth factors to thrombin activity

Whilst stimulation of the thrombin receptor can, in many cell types, induce mitogenesis, in some circumstances additional ligands are required for cell growth. For example, there is a synergistic effect produced by thrombin and various growth factors [epidermal growth factor, platelet-derived growth factor and basic fibroblast growth factor (bFGF)] on the proliferation of hamster fibroblasts [67], human vascular endothelial cells [68,69] and vascular smooth-muscle cells [70]. It has been suggested that these growth factors may activate tyrosine kinases essential for mitogenesis [71], thus complementing the action of thrombin. Additionally, it is possible that thrombin may elicit an autocrine response from cells, namely triggering release of growth factors which can then activate surface receptors [70]. The weight of evidence, however, supports the view that, for most cell types, there is no requirement for ligands in addition to thrombin to develop the full range of responses.

(vi) Other thrombin receptors

Whilst the cloned receptor described here is now accepted as the major, if perhaps not the only, receptor, a body of evidence exists for additional molecules responsive to thrombin on the surface of cells. These so called 'high-affinity sites' are activated by thrombin binding rather than by proteolysis (reviewed, for example, in [26,72]). It has been suggested that peptides equivalent to regions of thrombin can bind to the cell surface and initiate cellular responses such as mitogenesis with no requirement for proteolytic activity [73].

4. INTRACELLULAR RESPONSES TO THROMBIN-RECEPTOR ACTIVATION

The thrombin receptor stimulates a number of phospholipid-directed enzymes, generating both lipid-soluble and water-soluble second messengers. Although originally identified as a member of the G-protein-coupled receptor family, studies with specific inhibitors of tyrosine kinases and monomeric, low-molecular-mass G-proteins suggest that these molecules also participate in thrombin-stimulated phospholipid metabolism. Similarly, thrombin-stimulated MAP kinase activation is dependent on heterotrimeric G-proteins, tyrosine kinases and low-molecular-mass G-proteins (summarized in Figure 5). However, the precise relationship between G-protein activation, tyrosine kinase activation and Ras activation remains ill-defined.

Since gross morphological changes accompany thrombin-mediated mitogenesis and differentiation, thrombin must also regulate actin cytoskeleton assembly and disassembly. The ability of the thrombin receptor to initiate changes in cytoskeletal reorganization defines more subtle roles for thrombin in controlling physiological processes in terminally differentiated cells, for example thrombin-stimulated aggregation and granular release in platelets [74], thrombin-stimulated cell motility [75] and thrombin-stimulated neurite retraction [10]. The roles of polyphosphoinositides and Ca^{2+}, heterotrimeric G-proteins, tyrosine kinases and monomeric G-proteins in the control of cytoskeletal dynamics have recently been comprehensively covered in two outstanding reviews [76,77].

(i) Thrombin stimulation of phosphoinositidase C (PIC) activity

(a) Role of heterotrimeric G-proteins

Many receptors possessing seven transmembrane domains, including the thrombin receptor, couple through heterotrimeric G-proteins to stimulate a PICβ activity, directed towards PtdIns(4,5)P_2, generating the second-messenger molecules Ins(1,4,5)P_3 and diacylglycerol (DAG) (reviewed in [78]). PICβ1 and PICβ4 are regulated predominantly by GTP-ligated Gα subunits of the G$_q$ subfamily, whereas PICβ2 and PICβ3 are more responsive to G$\beta\gamma$ subunits derived predominantly from G$_i$ (reviewed in [79]). Agonists which couple to G$_i$ are sensitive to inhibition by pertussis toxin, which inhibits G$_i$ function specifically by ADP-ribosylating Gα_i at a Cys residue four amino acids from its C-terminus [80]. G$_q$, G$_{12}$ and G$_s$ lack this critical Cys residue and are therefore unaffected by the reagent [81–83].

Thrombin stimulation of PIC activity in platelets [84], HEL cells [85], Chinese-hamster CCL39 cells [86], chick embryonic heart cells [87] and vascular smooth-muscle cells [88] is sensitive to pretreatment with pertussis toxin, suggesting a role for G$\beta\gamma$ subunits derived from G$_i$ in stimulating PIC activity in these cells. In contrast, thrombin-stimulated PIC activity in IIC9 Chinese-hamster embryonic fibroblasts [89], 3T3-fibroblasts [90], VMR 106-H5 osteosarcoma cells [91] and human umbilical-vein endothelial cells [92] is pertussis-toxin-insensitive, potentially implicating members of the G$_q$ subfamily in stimulating PIC activity. The role of Gα subunits in thrombin-responsive cells has been studied by microinjecting inhibitory anti-α_q, anti-α_o and anti-α_i monoclonal antibodies into CCL39 cells and analysing their effects on TRAP-induced Ca^{2+} mobilization and DNA synthesis [93]. Both anti-α_q and anti-α_o monoclonal antibodies significantly inhibit intracellular Ca^{2+} release and DNA synthesis, whilst anti-α_i monoclonal antibodies have no effect, implicating α_q and α_o in thrombin-stimulated mitogenesis.

PICβ1 is involved in thrombin-stimulated PtdIns(4,5)P_2 hydrolysis in CCL39 cells. A CCL39 derivative that expresses constitutively low levels of PICβ1 compared with the parental cell line has impaired coupling to effectors in response to thrombin [94]. Thus thrombin-stimulated inositol phosphate production and cytosolic Ca^{2+} mobilization are reduced, particularly when the external Ca^{2+} concentration is low. Moreover, thrombin activation of effectors that lie downstream of PICβ1 activation in these cells, notably phospholipase D (PLD) and cytosolic PLA$_2$ [see section 4(v)], are also reduced in activity. The overexpression of Gα_q, PICβ1 and PICβ2 in *Xenopus* oocytes significantly enhances thrombin-stimulated Ca^{2+} release, further implicating these molecules in thrombin signalling [95]. Similarly, overexpression of PICδ1 in Chinese-hamster ovary (CHO) cells greatly enhances thrombin-stimulated PtdIns(4,5)P_2 hydrolysis [96]. This response is potentiated by ionomycin in intact cells and guanosine 5'-[γ-thio]triphosphate (GTP[γS]) in permeabilized cells, indicating roles for both G-proteins and Ca^{2+} in coupling thrombin receptor activation to the regulation of PICδ1.

(b) Role of non-receptor protein tyrosine kinases

The thrombin receptor can activate non-receptor tyrosine kinases. Thus thrombin stimulation induces the tyrosine-specific phosphorylation of a large number of proteins [97,98] as judged by Western blotting using tyrosine phosphate-specific antibodies. A few of these polypeptides have been identified, although the identity of the majority remains unknown at present. The best characterized of the substrates are Src and FAK, but recent evidence has indicated that JAK2, implicated in transcriptional regulation, and Syk, found predominantly in haemopoietic cells, are phosphorylated in response to thrombin-receptor activation [97–102].

Thrombin stimulates a rapid and transient increase in the specific activity of Src, followed by the translocation of the activated protein to a cytoskeleton-rich fraction [103]. In the

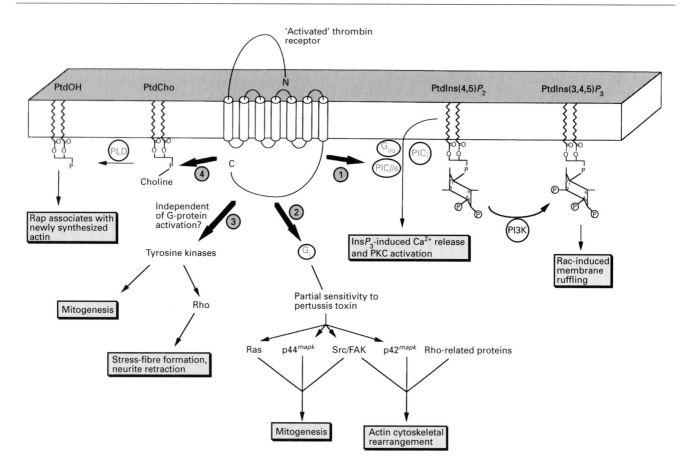

Figure 5 Pictorial depiction of the roles of G-proteins, non-receptor tyrosine kinases and Ras-related proteins in signal transducing events initiated by thrombin receptor activation

(1) PIC and PI3K utilize PtdIns(4,5)P_2 to generate the second messengers InsP_3 and PtdIns(3,4,5)P_3 respectively. (2) Role of G$_i$ in activating other effector molecules, in particular Ras, MAP kinases and tyrosine kinases. (3) Role of tyrosine kinases in thrombin-mediated responses. (4) PLD-catalysed degradation of PtdCho to phosphatidic acid (PtdOH). See the text for a more detailed explanation of the potential physiological roles of the effector molecules that are stimulated by thrombin.

initial 15 s of thrombin stimulation, Src is dephosphorylated on Tyr[527] and is subsequently phosphorylated at Tyr[416], both residues which are thought to be critical in controlling activity [104,105]. Other studies indicate that thrombin-stimulated PKC might be important in stimulating Src activity by directly phosphorylating Ser[12] [106]. Src activation by thrombin is partially inhibited by pertussis toxin, suggesting that molecules other than G$_i$ may participate in its activation [107].

The tyrosine kinase inhibitor tyrphostin AG-213 has been used to define the role of PICγ in thrombin-stimulated platelets [108]. Thrombin-stimulated inositol phosphate formation, platelet aggregation and 5-hydroxytryptamine secretion were substantially inhibited by tyrphostin AG-213. Thrombin also stimulated the appearance of PICγ in anti-phosphotyrosine immunoprecipitates. This work has recently been extended to demonstrate the tyrosine kinase-dependent recruitment of PICγ1 to the cytoskeleton in thrombin-stimulated platelets [109]. PICγ2 is also a substrate for tyrosine phosphorylation in human platelets in response to thrombin [110]. Src can couple to PICγ to stimulate the tyrosine kinase-dependent hydrolysis of PtdIns(4,5)P_2 [111], though the precise role of Src in thrombin-stimulated PtdIns(4,5)P_2 hydrolysis is not fully understood.

(c) Role of Ras-related proteins

The Ras-related protein Rho, which is integral to mediation of

stress-fibre formation [112], is implicated in thrombin-induced responses [see section 4(iii)]. Moreover, it has been suggested that Rho may contribute indirectly to the regulation of thrombin-stimulated PtdIns(4,5)P_2 hydrolysis by stimulating a PtdIns4P 5-kinase activity, thus serving to maintain the levels of agonist-responsive PtdIns(4,5)P_2 [113]. The evidence that Rho, in its GTP-bound form, stimulates PtdIns4P 5-kinase activity in lysates obtained from Swiss-3T3 cells is based on the observation that botulinum C3 exozyme (which specifically ADP-ribosylates Rho in its effector domain, inhibiting its function), inhibits GTP[S]-stimulated PtdIns4P 5-kinase activity. Moreover, lova-statin, which inhibits post-translational acylation and thus membrane targeting of Ras-related proteins, substantially reduces thrombin-stimulated Ca^{2+} mobilization in Swiss 3T3 cells. The botulinum C3 exozyme has a similar effect on thrombin-stimulated Ca^{2+} mobilization when microinjected into C3H10T$\frac{1}{2}$ cells. The reduction in thrombin-stimulated Ca^{2+} mobilization in these circumstances has been suggested to be due to reduced PtdIns4P 5-kinase activity [113].

(ii) Thrombin stimulation of phosphoinositide 3-kinase (PI3K)

(a) Synthesis of 3-phosphorylated inositol lipids

PI3K specifically phosphorylates PtdIns(4,5)P_2, following agonist stimulation [114,115]. Thus far, two agonist-stimulated PI3K isoforms, both with a preferred substrate specificity towards

$PtdIns(4,5)P_2$, have been identified in cells. The PI3K harnessed predominantly by tyrosine kinases is a heterodimer, consisting of a p85 regulatory subunit and a p110 catalytic subunit [116]. A 215 kDa PI3K activity has also been purified from human neutrophils and is specifically activated by the $\beta\gamma$ subunits of G-proteins [117].

Thrombin stimulates the synthesis of $PtdIns(3,4,5)P_3$ in platelets [118] by activating a PI3K directed towards $PtdIns(4,5)P_2$ [119]. Thrombin-stimulated $PtdIns(3,4)P_2$ formation is due to either $PtdIns(3,4,5)P_3$ dephosphorylation or $PtdIns4P$ 3-kinase activation [119]. The possibility of a thrombin-activatable $PtdIns4P$ 3-kinase cannot be ignored, since it appears that the major 3-phosphorylated inositol lipid of unstimulated cells, $PtdIns3P$, is synthesized by a PI3K directed towards $PtdIns$ [120]. Indeed, some reports suggest that thrombin can stimulate $PtdIns(3,4)P_2$ formation independently of $PtdIns(3,4,5)P_3$ synthesis [121,122], implying that a $PtdIns(4,5)P_2$-directed PI3K might not be essential for $PtdIns(3,4)P_2$ synthesis. Thrombin-stimulated $PtdIns(3,4)P_2$ synthesis correlates well with tyrosine kinase activation [122].

(b) Regulation of thrombin-stimulated PI3K

Some efforts have been made to determine the regulatory mechanisms underlying thrombin stimulation of PI3K activity in platelets. The Ca^{2+}-ionophore A23187 and phorbol dibutyrate, a potent activator of the Ca^{2+}/DAG-sensitive PKC isoenzymes, were used in an attempt to duplicate thrombin activation of PI3K. However, elevation of intracellular Ca^{2+} or activation of PKC were insufficient to activate PI3K fully [118]. A later report from the same laboratory suggested that protein kinases and G-proteins contributed to the thrombin activation of PI3K; thus staurosporine inhibited the production of the 3-phosphorylated inositol lipids, whilst pertussis toxin inhibited $PtdIns(3,4,5)P_3$ and $PtdIns(3,4)P_2$ formation by about 25% [123].

Upon thrombin stimulation of platelets, increased PI3K activity can be detected in anti-phosphotyrosine, anti-Src and anti-Fyn immunoprecipitates [124], implicating tyrosine kinases and p85/p110 PI3K in the thrombin-stimulated synthesis of 3-phosphorylated inositol lipids. Indeed, thrombin can stimulate the translocation of both Src [section 4(i)(b)] and p85/p110 PI3K to the cytoskeleton [125,126]. Tyrosine kinases are further implicated in this process, since the tyrosine kinase inhibitor tyrphostin AG-213 inhibits thrombin-stimulated $PtdIns(3,4)P_2$ production in platelets by half [108]. Increases in p85 levels have also been detected in anti-phosphotyrosine immunoprecipitates, suggesting that thrombin stimulates either the recruitment of p85 to phosphotyrosine complexes or tyrosine phosphorylation of p85.

The Ras-related protein Rho might also be involved in mediating thrombin stimulation of PI3K. Activation of PI3K by GTP[γS] in platelet lysates was blocked by specific ADP-ribosylation of endogenous Rho by botulinium C3 exozyme. These effects were overcome by addition of exogenous recombinant Rho, but not by recombinant Rac [127]. Rho is also implicated in protein tyrosine phosphorylation and PI3K activation in Swiss 3T3 cells [128]. Rho-mediated stress-fibre formation in Swiss 3T3 cells is inhibited by genistein, suggesting that tyrosine kinase(s) could be downstream effectors of Rho [129].

$p125^{FAK}$ (a **F**ocal **A**dhesion **K**inase) lies downstream of Rho activation [129]. Thrombin provokes FAK phosphorylation and activation in human platelets, in an integrin-dependent manner [99]. Previously identified as a cytosolic tyrosine kinase that associates with focal adhesion plaques upon stimulation by integrins and Src, this protein is implicated in the control of cytoskeletal reorganization processes. FAK is also implicated in PI3K activation [130]. It binds specifically to the p85 regulatory subunit of PI3K, and autophosphorylation of FAK increases the affinity of the association. Src and FAK appear to be functionally related, since FAK is a major tyrosine phosphorylated protein in Src-transformed cells [131], and autophosphorylation of FAK on Tyr^{397} (upon receptor binding) creates a high-affinity binding site for Src, so that FAK is recruited to an SH2 region of Src to initiate downstream signalling [132]. Ras binds specifically to p110, the catalytic subunit of PI3K, through its effector domain to stimulate PI3K activity [133,134]. The role of Ras, however, in the regulation of thrombin-stimulated PI3K remains uncertain.

A recent report has attempted to dissect the relative contributions of heterotrimeric G-proteins and monomeric G-proteins in regulating PI3Ks in thrombin-stimulated platelets [135]. A Rho-activatable p85/p110 PI3K, and a novel $G\beta\gamma$-sensitive $p110\gamma$ PI3K that does not bind to p85 and thus functions independently of it, are activated in platelets in response to thrombin. The relationship, however, between this novel $G\beta\gamma$-sensitive $p110\gamma$ PI3K and the $G\beta\gamma$-sensitive p85/p110 PI3K, also identified in platelets [136], remains to be elucidated.

(iii) Thrombin stimulation of Ras and Ras-related proteins

Thrombin stimulates Ras rapidly and transiently in a pertussis-toxin-sensitive manner in CCL39 cells, suggesting a role for the $G\beta\gamma$ subunits derived from G_i as upstream regulators of Ras [137]. Tyrosine kinases are also implicated in G_i activation of Ras, since thrombin stimulation of Ras is inhibited by the tyrosine kinase inhibitor genistein. Ras activation has been shown to be critical in thrombin-stimulated mitogenesis in 1321N1 astrocytoma cells [138], since microinjection of either a dominant negative Asn^{17} Ha-Ras mutant or an inhibitory antibody to Ras inhibits thrombin-stimulated DNA synthesis. It is noteworthy, however, that $PtdIns(4,5)P_2$ hydrolysis and intracellular Ca^{2+} mobilization were not dependent on endogenous Ras function.

Upon thrombin stimulation of human platelets the Ras-related protein Rap1b translocates to the cytoskeleton [139]. In the presence of extracellular Ca^{2+}, conditions that favour platelet aggregation, the thrombin-stimulated incorporation of Rap1b in the platelet cytoskeleton is biphasic and characterized by an initial rapid incorporation of approx. 20% of the protein, followed by a sustained, but slower, second phase of incorporation lasting approx. 5 min [140]. In the absence of extracellular Ca^{2+}, conditions which inhibit aggregation, only the initial phase of Rap1b incorporation is observed [140]. The precise role of Rap1b in cytoskeletal dynamics is not known, although functionally Rap can compete with Ras for RasGAP and therefore might attenuate Ras signalling [141]. In this instance, however, p120RasGAP association with the cytoskeleton was not observed during thrombin stimulation. Rap2b also becomes associated with the platelet cytoskeleton upon thrombin stimulation [142]. Agonist-induced actin polymerization is required for the recruitment of Rap2b to the cytoskeleton, suggesting that Rap2b associates with newly synthesized actin filaments [143]. This association is dependent on glycoprotein IIb-IIIa, since inhibitory monoclonal antibodies directed towards this receptor inhibit Rap2b recruitment to the cytoskeleton [143].

Rho is important in regulating integrin-dependent formation of stress fibres at focal adhesion complexes [112] and has been implicated in thrombin-stimulated neurite retraction in N1E-115 and NG108-15 neuronal cells [144] and in thrombin-induced platelet aggregation mediated by glycoprotein IIb-IIIa [145]. The molecules linking thrombin-receptor activation to Rho activation are not yet known, although it is apparent that Rho can regulate multiple effectors with tyrosine kinases, PI3K, PLD and PtdIns4P 5-kinase all lying downstream of Rho activation. The Rho-related protein Rac, integral in lamellipodia formation and membrane ruffling [146], has also been implicated as a downstream effector of PI3K [147,148], although it is not known whether the thrombin receptor can thus couple to Rac activation.

(iv) Thrombin stimulation of MAP kinases

MAP kinases, which possess serine, threonine and tyrosine kinase activity, are important in mediating responses to both tyrosine kinase receptors and G-protein-coupled receptors. They are activated by kinase cascades that are initiated following Raf activation by Ras (reviewed in [149]). Raf is a specific MAP kinase kinase kinase (MAPkkk), which phosphorylates (on serine and threonine residues) and activates MAP kinase kinase (MAPkk) [150], which in turn phosphorylates and activates p42^{mapk} and p44^{mapk} [151].

Thrombin has been shown to stimulate both p42^{mapk} and p44^{mapk} activity, although it differentially stimulates p44^{mapk} in G$_0$-phase-arrested CCL39 cells [49]. p44^{mapk} stimulation is defined by an initial transient activation that is maximal after 5 min, followed by a second period of activation that lasts up to 4 h. The prolonged phase of MAP kinase activation correlates well with thrombin-stimulated DNA synthesis. Interestingly TRAP transiently stimulates p44^{mapk}, but fails to induce the prolonged phase of activation and fails to stimulate DNA synthesis [see section 3(iii)]. Conversely, in human platelets, thrombin selectively stimulates p42^{mapk}, but not p44^{mapk} [152], an activation that parallels p90^{rsk} activation. These responses were shown to be independent of integrin-mediated platelet aggregation [152]. This suggests that p42^{mapk} and p90^{rsk} might be important in regulating cytoskeletal changes that accompany platelet activation, whereas p44^{mapk} might be more directly involved in mitogenesis.

(v) Thrombin stimulation of phosphatidylcholine (PtdCho) breakdown

Thrombin stimulation of human platelets causes activation of PLA$_2$. This enzyme is responsible for the liberation of arachidonic acid, predominantly from PtdCho, which is required for the synthesis of thromboxane A$_2$ and leukotrienes. PLA$_2$ is found in two forms: sPLA$_2$, a secreted protein that is activated by millimolar Ca^{2+}, and cPLA$_2$, a cytosolic species that is activated by micromolar Ca^{2+} and is characterized by a Ca^{2+}-dependent phospholipid-binding motif in its N-terminal domain. An sPLA$_2$ that is secreted within minutes of thrombin stimulation of human platelets has been cloned [153], although since thromboxane A$_2$ production in human platelets is maximal at low external Ca^{2+}, this form is perhaps not critical in early arachidonic acid generation. An 85 kDa cPLA$_2$ species present in human platelets is also stimulated by thrombin. p42^{mapk} has been shown to stimulate cPLA$_2$ activity by directly phosphorylating cPLA$_2$ on serine residues. Ca^{2+} is also required for activity, which serves to localize cPLA$_2$ to membranes [154,155]. Activation of cPLA$_2$ positively correlates with thrombin-stimulated phosphorylation of cPLA$_2$ [156]. G-proteins are also implicated in cPLA$_2$ activation. Thrombin-stimulated cPLA$_2$ in CHO cells has been shown to be dependent on functional Gα_{i2} proteins, since a Gα_{i2} mutant, G203T, substantially inhibits cPLA$_2$ activation [157].

PLD stimulates the generation of phosphatidate primarily from PtdCho and exists as both membrane-associated and cytosolic species [158]. Two distinct membrane-associated PLDs have been identified [159]. One form is oleate-dependent [159,160], whereas the other is stimulated by the low-molecular-mass G-proteins ARF and Rho [159,161] and requires PtdIns(4,5)P_2 as an obligatory cofactor [159,162]; this serves to accelerate GDP dissociation from ARF [163]. Phosphatidate is involved in thrombin-stimulated stress-fibre formation in IIC9 fibroblasts [164], characterized by an increase in actin polymerization. Tyrosine kinases are implicated in thrombin stimulation of PLD in platelets, since the tyrosine kinase inhibitors genistein and the tyrphostins A25 and A47 inhibit the activity of the thrombin-stimulated enzyme. Moreover, in saponin-permeabilized platelets, tyrosine kinase inhibitors markedly reduce GTP[γS]-stimulated PLD activity, suggesting that tyrosine kinase activation of PLD lies downstream of a regulatory G-protein [165].

Thrombin also stimulates DAG accumulation in IIC9 fibroblasts, characterized by the transient translocation of PKCα and the biphasic translocation of PKCϵ to a membrane fraction. DAG from PtdIns(4,5)P_2 hydrolysis accounts for the initial phase of PKCα and PKCϵ activation, with DAG derived from PtdCho responsible for prolonged phase of PKCϵ activation [166]. The differences in selective PKC isoenzyme activation do not reside in DAG species differences [166], but can be explained by the requirement for Ca^{2+} for PKCα activation but not for PKCϵ activation [166]. The thrombin-stimulated accumulation of nuclear DAGs in IIC9 fibroblasts correlates positively with the translocation of PKCα to the nucleus [167].

(vi) Thrombin inhibition of adenylate cyclase

Agonists such as prostaglandins, which couple to adenylate cyclase to elevate intracellular cyclic AMP levels, inhibit platelet activation [168]. Thrombin inhibits agonist-stimulated adenylate cyclase activity in both membrane preparations and intact human platelets, thus reducing intracellular cyclic AMP levels [85]. Moreover, thrombin markedly reduces forskolin-stimulated adenylate cyclase activity in CCL39 fibroblasts, whilst pertussis toxin pretreatment abolishes thrombin-mediated inhibition, implicating G$_i$ in the control of adenylate cyclase [169]. Gα_z, found predominantly in neuronal tissue and platelets, shows 65% sequence similarity at the amino acid level with Gα_i members, but is insensitive to pertussis-toxin-catalysed ADP-ribosylation [170,171]. Thrombin and phorbol 12-myristate 13-acetate stimulate Gα_z phosphorylation in platelets, and purified recombinant Gα_z is a substrate for PKC *in vitro*, whereas Gα_i members are not [172]. Gα_z has recently been shown to inhibit adenylate cyclase activity [173].

Thrombin-mediated inhibition of adenylate cyclase effectively reduces intracellular cyclic AMP levels, presumably reducing cyclic AMP-dependent protein kinase A activity. Elevation of intracellular cyclic AMP also functions to inhibit Raf activity [174] such that thrombin might mediate Raf activation not only through Ras, but also by relieving cyclic AMP-mediated inhibition. Thrombin can also potentiate prostaglandin-stimulated adenylate cyclase activity in HEL cells by a mechanism necessitating PKC activation [175].

5.   CELLULAR RESPONSES TO THROMBIN

As we have mentioned above, thrombin has the remarkable ability to interact with a large variety of cell types. Furthermore, depending on the cellular target, it can rapidly elicit a range of physiological responses in addition to the events leading to cell division [176–183]. Many of these actions of thrombin are precisely those that would be required during inflammation, tissue remodelling and eventual wound repair. For example, thrombin is involved in the inflammatory response by causing chemotaxis and adhesion of inflammatory cells [177,184], it invokes contraction and tissue remodelling by inducing morphological changes in endothelial cells [185] and fibroblasts [186], and it contributes to wound healing by stimulating mitogenesis, either directly [187] and/or via its ability to induce the secretion of other growth factors [183]. Thus thrombin plays a central role in a second 'cascade' of events following injury, a cascade of post-clotting cellular effects whereby the enzyme acts, in concert with other molecules, as a hormone and growth factor. Limitations of space preclude a detailed discussion of all of the cell types which respond to thrombin, and so we have concentrated in this section on a consideration of the role of thrombin in the regulation of growth of cells of neuronal origin, as this is an area which seems to have received rather less attention than it merits.

(i) Thrombin in the nervous system

Although the blood/brain barrier segregates macromolecules between the brain and the vascular system, evidence has accumulated for an important function for thrombin and other serine proteinases in the brain (reviewed in [188]). Most graphic, perhaps, is the effect of thrombin on the morphology of various cultured cell types derived from the nervous system. The response is typified by a rapid (visible within minutes) retraction of neurites or processes and can be induced by picomolar concentrations of thrombin or nanomolar concentrations of its precursor prothrombin. Furthermore, it is specific to thrombin, since other serine proteinases are either ineffective or result in general proteolysis and cell detachment. Interestingly, another serum component, namely lysophosphatidic acid (LPA), which also couples to a G-protein-coupled receptor, triggers rapid neuronal shape changes that are indistinguishable from those induced by thrombin [189]. To date, neurite retraction has been demonstrated in a variety of neuronal cells, including mouse neuroblastoma cells [10], the SK-N-SH human neuroblastoma cell line, human transformed fetal retinoblasts (HER 10), primary cultures of human fetal cerebellar, hippocampal and midbrain neurons [190], primary cultures of neonatal rat brain (Figure 6), hippocampal pyramidal cell cultures [191] and primary cultures of fetal dopaminergic neurons [192]. A comparable response has also been demonstrated in cells of glial lineage. Astrocytes and glioma cells rapidly lose their stellate morphology and acquire a flat epithelial shape upon addition of picomolar concentrations of thrombin [193–195]. Astrocytes participate in several processes crucial to brain function (reviewed in [196]). They may provide

Control   Thrombin

FSLLRN   SFLLRN

Figure 6   Thrombin- and TRAP-induced neurite retraction in primary cultures of rat neonatal brain

Cells were differentiated by treatment with 30 μM forskolin in the absence of serum for 4 h. Thrombin (28 pM), TRAP-SFLLRNP (10 μM) or control peptide FSLLRNP (10 μM) were added to cultures and observed 2 h later (as shown). The bar represents 50 μm (P. W. Grabham and D. D. Cunningham, unpublished work).

trophic support for neuronal growth and differentiation *in vivo* by secreting neurotrophins such as nerve growth factor (NGF) [197] and FGF [198], and it has been suggested that astrocytes regulate synaptic remodelling of neurosecretory neurons [199,200]. Moreover, astrocytes are believed to contribute to the blood/brain barrier [201]. In addition to the reversal of stellation, thrombin is known to induce other responses in astrocytes. For example, it causes secretion of the potent vasoconstrictor endothelin-1 [202], it has recently been demonstrated to enhance secretion and synthesis of NGF [203] and it can act as a growth factor on astrocytes [193–204]. Thus many of the post-clotting cellular effects of thrombin seen in the vasculature may be reproduced within the brain. Furthermore, morphological changes in neurons and astrocytes and mitosis in astrocytes have been shown to be mediated by the tethered ligand receptor [48,205–207], indicating that these responses in brain cells are not due to general proteolytic effects, but involve the receptor-signalling mechanisms outlined above.

The precise second-messenger pathway(s) by which activation of the cloned receptor leads to morphological change in neuronal cells remains to be elucidated. Pharmacolgical studies have shown the involvement of serine/threonine kinase(s), whilst experiments using tyrosine kinases and phosphatase inhibitors indicate that a receptor-mediated pathway associated with these enzymes might transmit the response [48]. The reversal of astrocyte stellation, however, is not sensitive to the tyrosine kinase inhibitor herbimycin A, suggesting that certain tyrosine kinases are probably not involved in this process [208]. Additionally, in adenovirus-transformed HER10 cells, thrombin-induced neurite retraction is accompanied by PIC and PI3K activation [59], suggesting that at least some of the responses to thrombin in neuronal cells are the same as those seen in platelets and fibroblasts. In any event it is likely that regulation of the actin-based cytoskeleton is crucial to thrombin and LPA-induced morphological changes [189], and recent studies on N1E-115 and NG108-15 neuronal cells indicated that Rho mediates this process [144] [section 4 (iii)].

Evidence for the importance of thrombin effects on neuronal cells *in vivo* comes from studies of proteinase nexin-1 (PN-1), also known as glia-derived nexin (GDN) (reviewed in [208]), a serine-proteinase inhibitor that has highest affinity for thrombin and is found primarily in the brain [209,210]. PN-1/GDN inactivates thrombin by forming a tight complex which subsequently binds back to the cells, where it is rapidly internalized and degraded [211]. Thus PN-1/GDN counteracts the cellular effects of thrombin, inhibiting mitogenic activity and inducing neurite extension in neurons and stellation in astrocytes. Indeed, the neurite-promoting action of PN-1/GDN [212,213], which corresponds directly with its inhibition of the enzyme activity of thrombin [214], led to the hypothesis that neuronal morphology is governed by a delicate interplay of proteinase and inhibitor [215].

(ii) Injury-related actions of thrombin in the brain

A compromise of the blood/brain barrier during disease or injury would, in theory, expose cells in the brain to relatively large amounts of thrombin and other plasma proteins, with obvious deleterious effects [216–218]. Recent studies have demonstrated expression of thrombin-receptor mRNA in rat brain, strongly suggesting that the extensive morphological alterations caused by thrombin on neurons and astrocytes in culture also occur *in vivo*. Such activity could disrupt critical interactions between neurons (synapses) and between astrocytes and either neurons or capillaries. Furthermore, the mitogenic activity of thrombin on astrocytes in culture might, *in vivo*,

contribute to the gliosis commonly observed after central-nervous-system (CNS) injury. Indeed, it has recently been shown that infusion of thrombin into rat brain results in histological changes that resemble the inflammation, scar formation and reactive gliosis seen in the CNS following injury [219].

In the brain PN-1/GDN is secreted mainly by astrocytes [220] and is found at sites that indicate a role in maintaining the integrity of the blood/brain barrier under normal conditions by excluding thrombin from the brain. For example, it is concentrated around cerebral blood vessels and associated with the endfeet of astroglia that form tight junctions with endothelial cells around capillaries [221]. Since the concentration of pro-thrombin in plasma is in the micromolar range and the concentration of PN-1/GDN in the brain is relatively low [221], extravasation of plasma through injury or disease would be expected to result in an imbalance in favour of thrombin and lead to the deleterious effects described above. There is, however, evidence that injury and disease-related events trigger PN-1/GDN production to levels which ultimately negate the effects of thrombin. First, a marked increase in PN-1/GDN mRNA has been found in biopsies of human glioblastoma and astrocytoma compared with normal brain [222]. Secondly, it has been shown that forebrain ischaemia in gerbils leads to alterations in the blood/brain barrier, selective degeneration of hippocampal CA1 pyramidal neurons and increased immunoreactivity to PN-1/GDN [223]. Thirdly, astroglial cells in the substantia nigra synthesize PN-1/GDN *de novo* following cytotoxic insult [224]. Finally, glial cells in culture have been shown to upregulate their secretion of PN-1/GDN in response to injury-related factors such as interleukin-1 and tumour necrosis factor-β [225].

While excess, unregulated concentrations of thrombin are undoubtedly damaging, the half-maximal concentration of thrombin required for proliferation in astrocytes (500 pM) is 100-fold higher than that required for morphological changes in astrocytes (2 pM) and neurons (2–10 pM) [10,190,193]. These observations suggest that thrombin might have a differential effect *in vivo*, depending on its concentration. It has therefore been postulated that a low concentration of thrombin may, in fact, contribute to the repair process by increasing plasticity and aiding the reconstruction of damaged synapses and other intercellular connections. Furthermore, PN-1/GDN might participate in this process by regulating the levels of active thrombin in the immediate environment of both neurons and astrocytes [208].

(iii) A function for thrombin in the normal brain

A major unanswered question concerning the action of thrombin in the brain is whether it has a function under normal physiological conditions. Although there is convincing evidence for a type of wound response and ultimately reactive gliosis, as described in the previous section, the evidence for a function under normal circumstances is scant and conjectural. Perhaps most persuasive are observations that prothrombin mRNA is present in brain and in cell lines of neuronal and glial origin [226], indicating that, in addition to injury-related extravasated thrombin, there is a potential endogenous source in the brain. It should, however, be noted that this has not yet been demonstrated directly.

The distribution of prothrombin mRNA in brain provides clues to the biology of a possible normal function of thrombin in the CNS. Expression has been observed in a number of embryonic and early postnatal rat brain structures, including the olfactory bulb, cortex, hippocampus, striatum, colliculi, thalamus and substantia nigra [217,226]. A comparison of the pattern of prothrombin and thrombin-receptor mRNA has revealed a distribution that is distinct, but overlapping, in select brain regions. For example, in cerebellum, thrombin-receptor mRNA is localized to the Purkinje-cell layer, whereas prothrombin mRNA is localized to both Purkinje- and granule-cell layers [217]. In most cases, thrombin mRNA and receptor mRNA expression are in close proximity, indicating that locally synthesized thrombin could act in an autocrine or paracrine manner to regulate specific processes. Further clues to a normal function for thrombin in the brain can, once again, be found in studies on PN-1/GDN. In addition to the protective role it might play in pathological situations, it is likely that the inhibitor also modulates the activity of locally synthesized thrombin during normal processes such as synaptic remodelling. Expression of PN-1/GDN mRNA and protein have distinct spatial and temporal patterns that support this theory [210,227,228]. For example, PN-1/GDN levels are high in the olfactory bulb, where glial cells and neurons proliferate postnatally and there is continual ingrowth of sensory axons and formation of new synapses into adulthood [229]. Furthermore, levels of prothrombin mRNA and thrombin-receptor mRNA are also high in this region [217].

Perhaps the most convincing role for thrombin is during development. This is a period when cell division among both neuronal and glial cells is maximal, remodelling and migration occur widely, and the blood/brain barrier is not yet intact. Indeed, formation of the blood/brain barrier in the rat occurs gradually during embryogenesis, and complete closure to macromolecules is not achieved until the third postnatal week [230]. Studies on the expression of rat thrombin-receptor mRNA [217,218], prothrombin mRNA [240] and PN-1/GDN [209,228] all indicate a correlation with age, where levels are highest during ontogeny and thence decrease to a basal level in adult life.

The precise nature of a physiological action for thrombin at the cellular level remains unknown. Current knowledge, however, suggests that thrombin-induced morphological changes represent a response that occurs under normal conditions. Although PN-1 is abundant around vessels, it is present to a lesser extent in the brain [221], indicating that very low concentrations of thrombin are normally present (this is probably why prothrombin has not, to date, been detected either *in vivo* or in neuronal cell lines in culture). Tissue-culture experiments show that much lower (picomolar) concentrations of thrombin are required for neurite retraction and the reversal of stellation than for mitogenesis ([10,59,190,193]; D. P. Brant and R. J. A. Grand, unpublished work). Moreover, the notion that neurite retraction is separate from other responses is supported by the observation that stimulation of distinct second-messenger pathways occurs in these responses [48,59,205,207]. Thus it is possible that small quantities of prothrombin are produced locally and converted into thrombin, which then acts locally to modify the morphology of neurons and astrocytes. Such activity, in exquisite co-ordination with PN-1/GDN, might contribute to cytoarchitectural remodelling and cellular migration in both the developing and adult brain.

(iv) Biological activity of prothrombin in the brain

Prothrombin appears to have no inherent proteolytic activity and has generally been considered to be biologically inactive prior to cleavage. However, it has been known for some time that it can induce neurite retraction [190,231] and the reversal of astrocyte stellation (P. W. Grabham and D. Cunningham, unpublished work) in the absence of other serum factors. This raises the possibility that the proenzyme can act directly on cells

in the brain to elicit a biological effect. However, the weight of evidence so far accumulated suggests that prothrombin must be converted into thrombin for it to evoke a response. Thus GDN and hirudin inhibit prothrombin-mediated neurite retraction even though they are incapable of binding to the proenzyme and only interact with thrombin [59,232]. Additionally, prothrombin elicits many of the intracellular responses seen with thrombin. For example, stimulation of cells with either protein rapidly activates a similar set of protein tyrosine kinases and PI3K. However, certain differences in intracellular responses to the two proteins have been noted. In neurite-retraction assays, appreciably higher concentrations of prothrombin than thrombin are required to produce a similar response [59,190]. Stimulation of cells with prothrombin does not cause detectable activation of PIC nor rapid synchronized Ca^{2+} mobilization [59]. On the basis of these data it could be suggested that cellular metabolic responses to prothrombin duplicate those elicited by very low thrombin concentrations. This is consistent with small amounts of thrombin being formed by cleavage of prothrombin at the cell surface [59,190] and, indeed, recent evidence has indicated the Ca^{2+}-dependent conversion of proenzyme into enzyme on the surface of human neuronal cells [59] and feline kidney fibroblasts [233]. It is not clear whether a specific enzyme, analogous to Factor Xa, is present on the surface of responsive cells or whether a more general proteinase is involved in activation.

6. FUTURE PROSPECTS

One of the most engaging features of thrombin research is the realization that an enzyme which is usually encountered as simply a member of the coagulation cascade can markedly influence the behaviour of a wide range of mammalian cell types which ostensibly have little to do with the clotting process. The observation that cells as diverse as platelets and astrocytes will respond to thrombin has confirmed that the protein has a hormone-like activity, allowing it to regulate the growth patterns in an array of targets. The elegant characterization of the thrombin receptor by groups led by S. R. Coughlin, J. Pouyssegur and L. F. Brass over the past 5 years has gone a long way towards explaining how the proteinase is able to produce this range of cellular responses. Additionally this work has provided a lucid account of the events which occur at the cell surface as a result of thrombin proteinase activity.

It now seems that the major areas of uncertainty in our understanding of thrombin signalling lie downstream of receptor activation. In particular, in the immediate future it will be important to understand whether separate pathways are involved in mitogenesis (in which transcriptional activation occurs) and those responses which only result in modification of the cytoskeletal architecture (i.e. cellular shape changes through reorganization of actin filaments). Indeed, preliminary evidence has already been presented suggesting that p44^{mapk} and p42^{mapk} may have distinct roles in separate thrombin-stimulated signalling pathways, only some of which are involved in proliferation [49,153]. Additionally, we need to question the significance of the roles played by Ras-related proteins in thrombin responses. It may be that some components (e.g. Ras itself) are involved in transduction of signals to the nucleus, whilst others may be central to 'shape changes', for example the role of Rho in neurite retraction [144].

A further point of considerable interest is the relative importance of G-protein-coupled and tyrosine kinase-mediated responses triggered by the thrombin receptor. Do these biochemical events form part of the same signalling pathway or are they quite distinct? In the event of the latter possibility does this mean that proteinase-activated receptors can (uniquely?) activate tyrosine kinases without recruiting G-proteins?

Whilst most interest has obviously focussed on the action of thrombin on platelets, in this review we have tried to emphasize the ubiquitous nature of the thrombin response. Although there have been demonstrations of the regulation by thrombin of growth and differentiation in cells of neuronal origin in tissue-culture systems, these experiments have not yet been extended to show unequivocally that the proteinase can determine cell morphology *in vivo*. This remains a priority. In this context it is a matter of some urgency to understand how prothrombin might be processed on the surface of cells, since it is likely that, under certain circumstances *in vivo* (for example in the brain), this might serve as the only source of thrombin.

A number of cell-surface proteinases have been reported, however (see, for example [234,235]). Although their biochemical functions are not clear, they play roles in cell growth, cell invasion and tissue rearrangement. Specific protein substrates for these proteinases remain unknown, but it might be supposed that one or more might be able to convert prothrombin into thrombin. Hepsin [237,238], for example, has been shown to be essential for cell growth of human hepatoma cells [238], and it is possible that it is required for the generation of thrombin from the added bovine serum in culture.

With the recent cloning of a trypsin-sensitive receptor homologous with the thrombin receptor [239], it now seems possible that a whole family of proteinase-activated cell-surface receptors may be present in mammalian cells [240]. An understanding of their mode of action and the relationship between them and the thrombin receptor opens up a large new area of exciting research possibilities.

Note added in proof (received 18 October 1995)

Hartwig et al. [241] have recently proposed a role for the D3 and D4 polyphosphoinositides in mediating thrombin-stimulated actin polymerisation in human platelets. Specifically, PtdIns4P, PtdIns(4,5)P_2, PtdIns(3,4)P_2 and PtdIns(3,4,5)P_3 stimulate the 'uncapping' of barbed (fast-growing) actin filaments, allowing for the further extension of actin polymers. The low-molecular-mass G-protein Rac plays a crucial role in this process by stimulating PtdIns(4,5)P_2 synthesis, thus confirming its role in thrombin-mediated morphological changes.

We thank Professor Dennis Cunningham and Dr. Patrick Vaughan (University of California) for their encouragement and help. We are also most grateful to Dr. Julian Parkhill and Dr. David Molloy, Institute for Cancer Studies, for assistance with the computer graphics, Dr. Matt Hodgkin for critical reading of parts of the manuscript before its submission, and Nicola Waldron and Deborah Williams for excellent secretarial assistance. We thank the Cancer Research Campaign for financial support (to R. J. A. G. and A. S. T.).

REFERENCES

1 Mann, K. G., Nesheim, M. E., Church, P., Haley, P. and Krishnaswamy, S. (1990) Blood **76**, 1–16

2 Davie, E. W., Fujikawa, K. and Kisiel, W. (1991) Biochemistry **30**, 10363–10369

3 Davey, M. G. and Luscher, E. F. (1967) Nature (London) **216**, 857–858

4 Shuman, M. A. and Levine, S. P. (1978) J. Clin. Invest. **61**, 1102–1106

5 McNamara, C. A., Sarembok, I. J. and Gimple, L. W. (1992) J. Clin. Invest. **91**, 94–98

6 Chen, L. B. and Buchanan, J. M. (1975) Proc. Natl. Acad. Sci. U.S.A. **72**, 131–135

7 Bar-Shavit, R., Kahn, A., Mann, K. G. and Wilner, G. D. (1986) Cell. Biochem. **32**, 261–272

8 Bar-Shavit, R., Kahn, A., Wilner, G. D. and Fenton, J. W., II (1983) Science **220**, 728–731

9 Bizios, R., Lai, L., Fenton, J. W., II and Malik, A. B. (1986) J. Cell. Physiol. **128**, 485–490

10 Gurwitz, D. and Cunningham, D. D. (1988) Proc. Natl. Acad. Sci. U.S.A. **85**, 3440–3444

11 Gustafson, G. T. and Lerner, U. (1983) Biosci. Rep. **3**, 255–261

12 Miller, K. D. and Van Vunakis, H. (1956) J. Biol. Chem. **223**, 227–237

13 Lundblad, R. L. (1971) Biochemistry **10**, 2501–2506

14 Kettner, C. and Shaw, E. (1979) Thromb. Res. **14**, 969–973

15 Wu, Q., Sheehan, J. P., Tsiang, M., Lentz, S. R., Birktoft, J. J. and Sadler, J. E. (1991) Proc. Natl. Acad. Sci. U.S.A. **88**, 6775–6779

16 Brass, L. F., Pizarro, S., Ahuja, M. et al. (1994) J. Biol. Chem. **269**, 2943–2952

17 Jenkins, A. L., Bootman, M. D., Berridge, M. J. and Stone S. R. (1994) J. Biol. Chem. **269**, 17104–17110

18 Vu, T.-K. H., Hung, D. T., Wheaton, V. I. and Coughlin, S. R. (1991) Cell **64**, 1057–1068

19 Vu, T.-K.H, Wheaton, V. I., Hung, D. T., Charo, I. and Coughlin, S. R. (1991) Nature London) **353**, 674–677

20 Friezner Degen, D. J. and Davie, E. W. (1987) Biochemistry **26**, 6165–6177

21 Aronson, D. L., Ball, A. P., Franza, R. B., Hugli, T. E. and Fenton, J. W., II (1980) Thromb. Res. **20**, 239–253

22 Bode, W., Mayr, I., Baumann, U., Huber, R., Stone, S. R. and Hofsteenge, J. (1989) EMBO J. **8**, 3467–3475

23 Rydel, T. J., Ravichandran, K. G., Tulinsky, A. et al. (1990) Science **249**, 277–280

24 Brandstetter, H., Turk, D., Hoeffken, W. et al. (1992) J. Mol. Biol. **226**, 1085–1099

25 Stubbs, M. T. and Bode, W. (1993) Thromb. Res. **69**, 1–58

26 Greco, N. J. and Jamison, G. A. (1991) Proc. Soc. Exp. Biol. Med. **198**, 792–799

27 Rasmussen, U. B., Vouret-Craviari, V., Jalla, S. et al. (1991) FEBS Lett. **288**, 123–128

28 Zhong, C., Hayzer, D. J., Corson, M. A. and Runge, M. S. (1992) J. Biol. Chem. **267**, 16975–16979

29 Gerszten, R. E., Chen, J., Ishii, M. et al. (1994) Nature (London) **368**, 648–651

30 Smith, K. J., Trayer, I. P. and Grand, R. J. A. (1994) Biochemistry **33**, 6063–6073

31 Mathews, I. I., Padmanabhan, K. P., Ganesh, V., Tulinsky, A., Westbrook, M. and Maraganore, J. (1991) Biochemistry **33**, 3266–3279

32 Ni, F., Ripoll, D. R., Martin, P. D. and Edwards, B. F. P. (1992) Biochemistry **31**, 11551–11557

33 Skrzypczak-Jankun, E., Carperos, V., Ramachandran, K. G., Tulinsky, A., Westbrook, M. and Maraganore, J. (1991) J. Mol. Biol. **221**, 1379–1393

34 Liu, L. W., Vu, T.-K. H., Esmon, C. T. and Coughlin, S. R. (1991) J. Biol. Chem. **266**, 16977–16980

35 Bahou, W. F., Kutok, J. L., Wong, A., Potter, C. L. and Coller, B. S. (1994) Blood **84**, 4195–4202

36 Sabo, T., Gurwitz, D., Motola, L., Brodt, P., Barak, R. and Elhanaty, E. (1992) Biochem. Biophys. Res. Commun. **188**, 604–610

37 Scarborough, R. M., Naughton, M. A., Teng, W. et al. (1992) J. Biol Chem. **267**, 13146–13149

38 Vassallo, R. R., Kieber-Emmons, T., Cichowski, K. and Brass, L. F. (1992) J. Biol. Chem. **267**, 6081–6085

39 Hui, K. Y., Jakubowski, J. A., Wyss, V. L. and Angleton, E. L. (1992) Biochem. Biophys. Res. Commun. **184**, 790–796

40 Coller, B. S., Ward, P., Ceruso, M. et al. (1992) J. Biol. Chem. **268**, 11713–11720

41 Chao, B. H., Kalkante, S., Maraganore, J. M. and Stone, S. R. (1992) Biochemistry **31**, 6175–6178

42 Rasmussen, U. B., Gachet, C., Schlessinger, Y. et al. (1993) J. Biol. Chem. **268**, 14322–14328

43 Tiruppathi, C., Lum, H., Andersen, T. T., Fenton, J. W., II and Malik, A. B. (1992) Am. J. Physiol. **263**, L595–L601

44 Jenkins, A. L., Bootman, M. D., Taylor, C. W., Mackie, E. J. and Stone, S. R. (1993) J. Biol. Chem. **268**, 21432–21437

45 Seiler, S. M., Michel, I. M. and Fenton, J. W., II. (1992) Biochem. Biophys. Res. Commun. **182**, 1296–1302

46 van Obberghen-Schilling, E., Rasmussen, U. B., Vouret-Craviari, V. et al. (1993) Biochem. J. **292**, 667–671

47 Reilly, C. F., Connolly, T. M., Feng, D. M., Nutt, R. F. and Mayer, E. J. (1993) Biochem. Biophys. Res. Commun. **190**, 1001–1008

48 Suidan, H. S., Stone, S. R., Hemmings, B. A. and Monard, D. (1992) Neuron **8**, 363–375

49 Vouret-Craviari, V., van Obberghen-Schilling, E., Scimeca, J. C., van Obberghen, E. and Pouyssegur, J. (1993) Biochem J. **289**, 209–214

50 Pumiglia, K. M. and Feinstein, M. B. (1993) Biochem. **294**, 253–260

51 Shimohigashi, Y., Nose, T., Okazaki, M. et al. (1994) Biochem. Biophys. Res. Commun. **203**, 366–372

52 Kinlough-Rathbone, R. L., Rand, M. L. and Packham, M. A. (1993) Blood **82**, 103–106

53 Lau, L.-F., Pumiglia, K., Cote, Y. P. and Feinstein, M. B. (1994) Biochem. J. **303**, 391–400

54 Seiler, S. M., Goldenberg, H. J., Michel, I. M., Hunt, J. T. and Zavoico, G. B. (1991) Biochem. Biophys. Res. Commun. **181**, 636–643

55 Muramatsu, I., Lantyonu, A., Moore, G. J. and Hollenberg, M. D. (1992) Can. J. Physiol. Pharmacol. **70**, 996–1003

56 Detwiler, T. C. and Feinman, R. D. (1973) Biochemistry **12**, 282–289

57 Martin, B. M., Feinman, R. D. and Detwiler, T. C. (1975) Biochemistry **14**, 1308–1314

58 Ishii, K., Hein, L., Kobilka, B. and Coughlin, S. R. (1993) J. Biol. Chem. **268**, 9780–9786

59 Turnell, A. S., Brant, D. P., Brown, G. R. et al. (1995) Biochem. J. **308**, 965–973

60 Brass, L. F., Vassallo, R. R., Belmonte, E., Ahuja, M., Cichowski, K. and Hoxie, J. A. (1992) J. Biol. Chem. **267**, 13795–13798

61 Ishii, K., Ishii, M., Koch, W. J., Freedman, N. J., Lefkowitz, R. J. and Coughlin, S. R. (1994) J. Biol. Chem. **269**, 1125–1130

62 Palczewski, K. and Benovic, J. L. (1991) Trends Biochem. Sci. **16**, 387–391

63 Hoxie, J. A., Ahuja, M., Belmonte, E. et al. (1993) J. Biol. Chem. **268**, 13756–13763

64 Hein, L., Ishii, K., Coughlin, S. R. and Kobilka, B. K. (1994) J. Biol. Chem. **269**, 27719–27729

65 Brass, L. F., Ahuja, M., Belmonte, E. et al. (1994) Semin. Hematol. **31**, 251–260

66 Norton, K. J., Scarborough, R. M., Kutok, J. L., Escobedo, M.-A, Nannizzi, L. and Coller, B. S. (1993) Blood **82**, 2125–2136

67 Cherington, P. V. and Pardee, A. B. (1980) J. Cell. Physiol. **105**, 25–32

68 Gospodarowitz, D., Brown, K. D., Birdwell, C. R. and Zetter, B. R. (1978) J. Cell Biol. **77**, 774–788

69 Zetter, B. R. and Antoniades, H. N. (1979) J. Supramol. Struct. **11**, 361–370

70 Weiss, R. H. and Nuccitelli, R. (1992) J. Biol. Chem. **267**, 5608–5613

71 Weiss, R. H. and Maduri, M. (1993) J. Biol. Chem. **268**, 5724–5727

72 Carney, D. H., Herbosa, G. J., Stiernberg, J. et al. (1986) Semin. Thromb. Hemostasis **12**, 231–240

73 Glenn, K. C., Frost, G. H., Bergmann, J. S. and Carney, D. H. (1988) Peptide Res. **1**, 65–73

74 Seiss, W. (1989) Physiol. Rev. **69**, 58–178

75 Joesph, S. and MacDermot, J. (1992) Biochem. J. **286**, 945–950

76 Janmey, P. A. (1994) Annu. Rev. Physiol. **56**, 169–191

77 Hall, A. (1994) Annu. Rev. Cell Biol. **10**, 31–54

78 Berridge, M. J. (1993) Nature (London) **361**, 315–325

79 Exton, J. H. (1994) Annu. Rev. Cell. Physiol. **56**, 349–369

80 Lochrie, M. A. and Simon, M. I. (1988) Biochemistry **27**, 4957–4965

81 Strathmann, M. and Simon, M. I. (1990) Proc. Natl. Acad. Sci. U.S.A. **87**, 9113–9117

82 Wilkie, T. M., Scherle, P. A., Strathmann, M. P., Slepak, V. Z. and Simon, M. I. (1991) Proc. Natl. Acad. Sci. U.S.A. **88**, 10049–10053

83 Casey, P. J., Fong, H. K. W., Simon, M. I. and Gilman, A. G. (1990) J. Biol. Chem. **265**, 2383–2390

84 Crouch, M. and Lapetina, E. G. (1988) J. Biol. Chem. **263**, 3363–3371

85 Brass, L. F., Manning, D. R., Williams, A. G., Woolkalis, M. J. and Poncz, M. (1991) J. Biol. Chem. **266**, 958–965

86 Paris, S. and Pouyssegur, J. (1986) EMBO J. **5**, 55–60

87 Chien, W. W., Mohabar, R. and Clusin, W. T. (1990) J. Clin. Invest. **85**, 1436–1443

88 Huang, C.-L. and Ives, H. E. (1989) J. Biol. Chem. **264**, 4391–4397

89 Raben, D. M., Yasuda, K. M. and Cunningham, D. D. (1987) J. Cell. Physiol. **130**, 466–473

90 Murayama, T. and Ui, M. (1985) J. Biol. Chem. **260**, 7226–7233

91 Babich, M., King, K. L. and Nissenson, R. A. (1990) Endocrinology (Baltimore) **126**, 948–954

92 Brock, T. A. and Capasso, E. L. (1989) Am. Rev. Respir. Dis. **140**, 1121–1125

93 Baffy, G., Yang, L., Raj, S., Manning, D. R. and Williamson, J. R. (1994) J. Biol. Chem. **269**, 8483–8487

94 Fee, J. A., Monsey, J. D., Handler, R. J. et al. (1994) J. Biol. Chem. **269**, 21699–21708

95 Chen, J., Ishii, M., Wang, L., Ishii, K. and Coughlin, S. R. (1994) J. Biol. Chem. **269**, 16041–16045

96 Banno, Y., Okano, Y. and Nozawa, Y. N. (1994) J. Biol. Chem. **269**, 15846–15852

97 Ferrell, J. E. and Martin, G. S. (1988) Mol. Cell. Biol. **8**, 3603–3610

98 Golden, A. J. and Brugge, J. S. (1989) Proc. Natl. Acad. Sci. U.S.A. **86**, 901–905

99 Lipfert, L., Haimovich, B., Schaller, M. D., Cobb, B. S., Parsons, J. T. and Brugge, J. S. (1992) J. Cell Biol. **119**, 905–912

100 Sada, K., Yanagi, S. and Yamamura, H. (1994) Biochem. Biophys. Res. Commun. **200**, 1–7

101 Rodriguez-Linares, B. and Watson, S. P. (1994) FEBS Lett. **352**, 335–338

102 Schuia, K. (1994) Curr. Opin. Cell Biol. **6**, 253–259

103 Clarke, E. A. and Brugge, J. S. (1993) J. Mol. Cell. Biol. **6**, 1863–1871

104 Roussel, R. R., Brodeur, S. R., Shalloway, D. and Laudano, A. P. (1991) Proc. Natl. Acad. Sci. U.S.A. **88**, 10696–10700

105 Kmiecik, T. E., Johnson, P. J. and Shalloway, D. (1988) Mol. Cell. Biol. **8**, 4541–4546

106 Liebenhoff, U., Brockmeier, D. and Presk, P. (1993) Biochem. J. **295**, 41–48

107 Chen, Y.-H., Pouyssegur, J., Courtneidge, S. A. and Van Obberghen-Schilling, E. (1994) J. Biol. Chem. **269**, 27372–27377

108 Guinebault, T. C., Payrastre, B., Sultan, C. et al. (1993) Biochem. J. **292**, 851–856

109 Guinebault, T. C., Payrastre, B., Mauco, G., Breton, M., Plantavid, M. and Chap, H. (1994) Cell. Mol. Biol. **40**, 687–693

110 Tate, B. F. and Rittenhouse, S. E. (1993) Biochim. Biophys. Acta **1178**, 281–285

111 Nakanishi, O., Shibasaki, F., Hidaka, M., Homma, Y. and Takenawa, T. (1993) J. Biol. Chem. **268**, 10754–10759

112 Ridley, A. J. and Hall, A. (1992) Cell **70**, 389–399

113 Chong, L. D., Traynor-Kaplan, A., Bokoch, G. M. and Shwartz, M. A. (1994) Cell **79**, 507–513

114 Stephens, L. R., Hughes, K. T. and Irvine, R. F. (1991) Nature (London) **351**, 33–39

115 Hawkins, P. T., Jackson, T. R. and Stephens, L. R. (1992) Nature (London) **358**, 157–159

116 Carpenter, C. L., Duckworth, B. L., Auger, K. R., Cohe, B., Schaffhausen, B. S. and Cantley, L. C. (1990) J. Biol. Chem. **265**, 19704–19711

117 Stephens, L. R., Smrcka, A., Cooke, F. T., Jackson, T. R., Sternweiss, P. C. and Hawkins, P. T. (1994) Cell **72**, 83–93

118 Kucera, G. L. and Rittenhouse, S. E. (1990) J. Biol. Chem. **265**, 5345–5348

119 Carter, A. N., Huang, R. S., Sorisky, A., Downes, C. P. and Rittenhouse, S. E. (1994) Biochem. J. **301**, 415–420

120 Stephens, L. R., Cooke, F. T., Walters, R. et al. (1994) Curr. Biol. **4**, 203–214

121 Yatomi, Y., Ozaki, Y. and Kume, S. (1994) Biochem. Biophys. Res. Commun. **186**, 1480–1486

122 Yatomi, Y., Ozaki, Y., Satoh, H. K. and Kume, S. (1994) Biochim. Biophys. Acta **1212**, 337–344

123 King, W. G., Kucera, G. L., Sorisky, A., Zhang, J. and Rittehouse, S. E. (1991) Biochem. J. **278**, 475–480

124 Gutkind, J. S., Lacal, P. M. and Robbins, K. C. (1990) Mol. Cell. Biol. **10**, 3806–3809

125 Grondin, P., Plantavid, M., Sultan, C., Breton, M., Mauco, G. and Chap, H. (1991) J. Biol. Chem. **266**, 15705–15709

126 Zhang, J., Fry, M. J., Waterfield, M. D., Jaken, S., Liao, L., Fox, J. E. B. and Rittenhouse, S. E. (1992) J. Biol. Chem. **267**, 4686–4692

127 Zhang, J., King, W. G., Dillon, S., Hall, A., Feig, L. and Rittenhouse, S. E. (1993) J. Biol. Chem. **268**, 22251–22254

128 Kumagi, N., Morri, N., Fujisawa, K., Nemoto, Y. and Narumiga, S. (1993) J. Biol. Chem. **268**, 24535–24538

129 Ridley, A. J. and Hall, A. (1994) EMBO J. **13**, 2600–2610

130 Chen, H. L. and Guan, J. L. (1994) Proc. Natl. Acad. Sci. U.S.A. **91**, 10148–10152

131 Guan, J.-L. and Shalloway, D. (1992) Nature (London) **358**, 690–692

132 Schaller, M. D., Hildebrand, J. D., Shannon, J. D., Fox, J. W., Vines, R. R. and Parsons, J. T. (1994) Mol. Cell. Biol. **14**, 1680–1688

133 Rodriguez-Viciani, P., Warne, P. H., Dhand, R. et al. (1994) Nature (London) **370**, 527–532

134 Kodaki, T., Woscholski, R., Hallberg, B., Rodriguez-Viciani, P. R., Downward, J. and Parker, P. J. (1994) Curr. Biol. **4**, 798–806

135 Zhang, J., Zhang, J., Benovic, J. C. et al. (1995) J. Biol. Chem. **270**, 6589–6594

136 Thomason, P. A., James, S. R., Casey, B. J. and Downes, C. P. (1994) J. Biol. Chem. **269**, 16525–16528

137 van Corven, E. J., Hordijk, P. L., Medema, R. H., Bos, J. L. and Moolenaar, W. H. (1993) Proc. Natl. Acad. Sci. U.S.A. **90**, 1257–1261

138 LaMorte, V. I., Kennedy, E. D., Collins, L. R. et al. (1993) J. Biol. Chem. **268**, 19411–19415

139 Fischer, T. H., Gatling, M. N., Lacal, J.-C. and White, G. C. (1990) J. Biol. Chem. **265**, 19405–19408

140 Fischer, T. H., Gatling, M. N., McCormick, F., Duffy, C. M. and White, G. C. (1994) J. Biol. Chem. **269**, 17257–17261

141 Zhang, K., Noda, M., Vass, W. C., Papgeorge, A. G. and Lowry, R. (1990) Science **249**, 162–165

142 Torti, M., Ramaschi, G., Sinigaglia, F., Lapetina, E. G. and Balduini, C. (1993) Proc. Natl. Acad. Sci. U.S.A. **90**, 7553–7557

143 Torti, M., Ramaschi, G., Sinigaglia, F., Lapetina, E. G. and Balduini, C. (1994) Proc. Natl. Acad. Sci. U.S.A. **91**, 4239–4243

144 Jalink, K., van Corven, E. J., Hengeveld, T., Morii, N., Narumiya, S. and Moolenaar, W. H. (1994) J. Cell Biol. **126**, 801–810

145 Morii, N., Teruuchi, T., Tominaga, T. et al. (1992) J. Biol. Chem. **267**, 20921–20926

146 Ridley, A. J., Paterson, H. F., Johnston, C. L., Dickmann, D. and Hall, A. (1992) Cell **70**, 401–410

147 Wennstrom, S., Hawkins, P., Cooke, F. et al. (1994) Curr. Biol. **4**, 385–393

148 Kotani, K., Yonezawa, K., Hara, K. et al. (1994) EMBO J. **13**, 2313–2321

149 Blenis, J. (1993) Proc. Natl. Acad. Sci. U.S.A. **90**, 5889–5892

150 Kyriakis, J. M., App, H., Zhang, X.-F. et al. (1992) Nature (London) **358**, 417–420

151 Kosako, H., Nishida, E. and Goth, Y. (1993) EMBO J. **12**, 787–794
152 Papkoff, J., Chen, R.-H., Blenis, J. and Forsman, J. (1994) Mol. Cell. Biol. **14**, 463–472
153 Kramer, R. M., Hession, C., Johansen, B. et al. (1989) J. Biol. Chem. **264**, 5768–5775
154 Nemenhoff, R. A., Winitz, S., Qian, N.-X, van Putten, V., Johnson, G. L. and Heasley, L. E. (1993) J. Biol. Chem. **268**, 1960–1964
155 Lin, L.-L, Wartmann, M., Lin, A. Y., Knopf, J. L., Seth, A. and Davies, R. J. (1993) Cell **72**, 269–278
156 Kramer, R. M., Roberts, E. F., Manetta, J. V., Hyslop, P. A. and Jakobowski, J. A. (1993) J. Biol. Chem **268**, 26976–26804
157 Winitz, S., Gupta, S. K., Qian, N.-X., Heasley, L. E., Nemenhoff, R. A. and Johnson, G. L. (1994) J. Biol. Chem. **269**, 1889–1891
158 Exton, J. H. (1994) Biochim. Biophys. Acta **1212**, 26–42
159 Massenburg, D., Han, J.-S., Liyange, M. et al. (1994) Proc. Natl. Acad. Sci. U.S.A. **91**, 11718–11722
160 Okamura, S.-I. and Yamashita, S. (1994) J. Biol. Chem. **269**, 31207–31213
161 Bowmann, E. P., Uhlinger, D. J. and Lambeth, J. D. (1993) J. Biol. Chem. **268**, 21509–21512
162 Brown, H. A., Gutowski, S., Moomaw, C. R., Slaughter, C. and Sternweiss, P. C. (1993) Cell **75**, 1137–1144
163 Terui, T., Kahn, R. A. and Randazzo, P. A. (1994) J. Biol. Chem. **269**, 28130–28135
164 Ha, K.-S. and Exton, J. H. (1993) J. Cell Biol. **123**, 1789–1796
165 Martinson, E. A., Scleible, S. and Presek, P. (1994) Cell. Mol. Biol. **40**, 627–634
166 Ha, K.-S. and Exton, J. H. (1993) J. Biol. Chem. **268**, 10534–10539
167 Leach, K. L., Ruff, V. A., Jarpe, M. B., Adams, L. D., Fabbro, D. and Raben, D. M. (1992) J. Biol. Chem. **267**, 21816–21822
168 Feinstein, M. B., Zavoico, G. B. and Halenda, S. P. (1985) in The Platelets: Physiology and Pharmacology (Longnecker, G. L., ed.), pp. 237–269, Academic Press, New York
169 Hung, D. T., Wong, Y.-T., Vu, T.-K. H. and Coughlin, S. R. (1992) J. Biol. Chem. **267**, 20831–20834
170 Carlson, L. L., Weaver, D. R. and Reppert, S. M. (1989) Endocrinology (Baltimore) **125**, 2670–2676
171 Gagnon, A. W., Manning, D. R., Catani, I., Gerwitz, A., Poncz, M. and Brass, L. F. (1991) Blood **78**, 1247–1253
172 Lounsbury, K. M., Sclegel, B., Poncz, M., Brass, L. F. and Manning, D. R. (1993) J. Biol. Chem. **268**, 3494–3498
173 Kozasa, T. and Gilman, A. G. (1995) J. Biol. Chem. **270**, 1734–1741
174 Cook, S. J. and McCormick, F. (1993) Science **262**, 1069–1072
175 Turner, J. T., Camden, J. M., Kansra, S., Shelton James, D., Wu, H. and Halenda, S. P. (1992) J. Pharmacol Exp. Ther. **263**, 708–716
176 Fenton, J. W. II (1988) Semin. Thromb. Hemostasis **14**, 234–240
177 Bar-Shavit, R., Benezra, M., Eldor, A., Hy-Am, E., Fenton, J. W., II., Wilner, G. D. and Vlodavsky, I. (1990) Cell. Regul. **1**, 453–463
178 Carney, D. H. (1992) in Thrombin (Berliner, L. J., ed.), pp. 351–396. Plenum. New York
179 Berk, B. C., Taubman, M. B., Cragoe, E. J., Fenton, J. W., II. and Greindling, K. K. (1990) J. Biol. Chem. **265**, 17334–17340
180 Bar-Shavit, R. (1992) in Thrombin (Berliner, L. J., ed.), pp. 315–350, Plenum, New York
181 Michel, M. C., Brass, L. F. Williams, A., Bokoch, G., LaMorte, V. J. and Motulsky, H. J. (1989) J. Biol. Chem. **264**, 4986–4991
182 He, C. J., Rondeau, E., Medcalf, R. L., Lecave, R., Schleuning, W. D., and Sraer, J. D. (1991) J. Cell. Physiol. **146**, 131–140
183 Garcia, J. G. N., Aschner, J. L. and Malik, A. B. (1992) in Thrombin (Berliner, L. J., ed.), pp. 397–437, Plenum, New York
184 Hattori, R., Hamilton, K. K., Fugate, R. D., McEver, R. P. and Sims, P. J. (1989) J. Biol. Chem. **264**, 7768–7771
185 Sago, H. and Iinuma, K. (1992) Thromb. Haemostasis **67**, 331–334
186 Kolondey, M. S. and Wysolmerski, R. B. (1992) J. Cell Biol. **117**, 73–82
187 Glenn, K., Carney, D., Fenton, J. W., II. and Cunningham, D. (1980) J. Biol. Chem. **255**, 6609–6616
188 Smirnova, I. V., Ho, G. J., Fenton, J. W., II. and Festoff, B. W. (1994) Semin. Thromb. Hemostasis **20**, 426–432
189 Jalink, E., Eichholtz, T., Postma, F. R., van Corven, E. J. and Moolenaar, W. H. (1993) Cell Growth Differ. **4**, 247–255
190 Grand, R. J. A., Grabham, P. W., Gallimore, M. J. and Gallimore, P. H. (1989) EMBO J. **8**, 2209–2215
191 Farmer, L., Sommer, J. and Monard, D. (1990) Dev. Neurosci. **12**, 73–80
192 Detta, A., Grabham, P. and Hitchcock, E. (1992) Restor. Neurol. Neurosci. **4**, 41–46
193 Cavanaugh, K. P., Gurwitz, D., Cunningham, D. D. and Bradshaw, R. A. (1990) J. Neurochem. **54**, 1735–1743
194 Nelson, R. B. and Siman, R. (1990) Dev. Brain Res. **54**, 93–104
195 Tas, P. W. L. and Koschel, K. (1990) Exp. Cell Res. **189**, 22–27

196 Vernadakis, A. (1988) Int. Rev. Neurobiol. **30**, 149–224
197 Furukawa, S., Furukawa, Y., Samyoshi, E. and Hayashi, K. (1986) Biochem. Biophys. Res. Commun. **136**, 57–63
198 Hatten, M. E., Lynch, M., Rydel, R. E. et al. (1988) Dev. Biol. **125**, 280–289
199 Perlmutter, L. S., Tweedle, C. D. and Hatton, G. I. (1984) Neuroscience **13**, 768–779
200 Tweedle, C. D. and Hatton, G. I. (1984) Brain Res. **309**, 373–376
201 Janzer, R. C. and Raff, M. C. (1987) Nature (London) **325**, 253–257
202 Ehrenreich, H., Costa, T. and Clouse, K. A. (1993) Brain Res. **600**, 201–207
203 Neveu, I., Jehan, F., Jandrot-Perrus, M., Wion, D. and Brachet, P. (1993) Neurochemistry **60**, 858–867
204 Perraud, F., Besnard, F., Sensenbrenner, M. and Labourdette, G. (1987) Int. J. Invest. Neurosci. **5**, 181–188
205 Jalink, K. and Moolenaar, W. H. (1992) J. Cell. Biol. **118**, 411–419
206 Beecher, K. L., Anderson, T. T., Fenton, J. W., II. and Festoff, B. W. (1994) J. Neurosci. Res. **37**, 108–115
207 Grabham, P. W. and Cunningham, D. D. (1995) J. Neurochem. **64**, 583–591
208 Cunningham, D. D., Pulliam, L. and Vaughan, P. J. (1993) Thromb. Hemostasis **70**, 168–171
209 Wagner, S. L., Van Nostrand, W. E., Lau, A. L. et al. (1993) Brain Res. **626**, 90–98
210 Reinhard, E., Suidan, H., Pavlik, A. and Monard, D. (1994) J. Neurosci. Res. **37**, 256–270
211 Low, D. A., Baker, J. B., Koonce, W. C. and Cunningham, D. D. (1981) Proc. Natl. Acad. Sci. U.S.A. **78**, 2340–2344
212 Monard, D., Niday, E., Limat, A. and Solomon, F. (1983) Prog. Brain Res. **58**, 359–364
213 Gloor, S., Odink, K., Guenther, J., Nick, H. and Monard, D. (1986) Cell **47**, 687–693
214 Gurwitz, D. and Cunningham, D. D. (1990) J. Cell. Physiol. **142**, 155–162
215 Monard, D. (1988) Trends Neurosci. **11**, 541–544
216 Soifer, S. J., Peters, K. J., O'Keefe, J. and Coughlin, S. R. (1994) Am. J. Pathol. **144**, 60–69
217 Weinstein, J., Gold, S. J., Cunningham, D. D. and Gall, C. M. (1995) J. Neurosci. **15**, 2906–2919
218 Niclou, S., Suidan, H., Brown-Luedi, M. and Monard, D. (1994) Cell. Mol. Biol. **40**, 421–428
219 Nishino, A., Suzuki, M., Ohtani, H. et al. (1993) J. Neurotrauma **10**, 167–179
220 Rosenblatt, D. E., Cotman, C. W., Nietro-Sampedro, M., Rowe, J. W. and Knauer, D. J. (1987) Brain Res. **45**, 40–48
221 Choi, B. H., Suzuki, M., Taisong, K., Wagner, S. L. and Cunningham, D. D. (1990) Am. J. Pathol. **137**, 741–747
222 Rao, J. S., Suzuki, R. and Festoff, B. W. (1990) in Serine Proteases and Their Serpin Inhibitors in the Nervous System (Festoff, B. W., ed.), pp. 301–311, Plenum, New York
223 Hoffman, M.-C, Nitsch, C., Scotti, A., Reinhard, E. and Monard, D. (1992) Neuroscience **49**, 397–408
224 Scotti, A. L., Monard, D. and Nitsch, C. (1994) Neurosci. Res. **37**, 155–168
225 Vaughan, P. J. and Cunningham, D. D. (1993) J. Biol. Chem. **268**, 3720–3727
226 Dihanich, M., Kaser, M., Reinhard, E., Cunningham, D. D. and Monard, D. (1991) Neuron **6**, 575–581
227 Reinhard, E., Meier, R., Halfter, W., Rovelli, G. and Monard, D. (1988) Neuron **1**, 387–394
228 Mansuy, I. M., van der Putten, H., Schmid, P., Meins, M., Botteri, F. M. and Monard, D. (1993) Development **119**, 1119–1134
229 Brunjes, P. C. and Frazier, L. L. (1986) Brain Res. **396**, 1–45
230 Risau, W. and Wolberg, H. (1990) Neuroscience **37**, 155–168
231 Grabham, P. W., Grand, R. J. A. and Gallimore, P. H. (1989) Cell. Signalling **1**, 269–281
232 Grabham, P. W., Monard, D., Gallimore, P. H. and Grand, R. J. A. (1991) Eur. J. Neurosci. **3**, 663–668
233 Sekiya, F., Usui, H., Inoue, K., Fukodome, K. and Morita, T. (1994) J. Biol. Chem. **269**, 32441–32445
234 Tanaka, K., Nakamura, T. and Ichhara, A. (1986) J. Biol. Chem. **261**, 2610–2615
235 Aoyama, A. and Chen, W. T. (1990) Proc. Natl. Acad. Sci. U.S.A. **87**, 8296–8300
236 Leytus, S. P., Loeb, K. R., Hagen, F. S., Kurachi, K. and Davie, E. W. (1988) Biochemistry **27**, 1067–1074
237 Tsuji, A., Torres-Rosado, A., Arai, T., Le Beau, M. M., Lemons, R. S., Chou, S.-H. and Kurachi, K. (1991) J. Biol. Chem. **266**, 16948–16953
238 Torres-Rosado, A., O'Shea, K. S., Tsuiji, A., Chou, S.-H. and Kurachi, K. (1993) Proc. Natl. Acad. Sci. U.S.A. **90**, 7181–7185
239 Nystedt, S., Emilsson, K., Wahlestedt, C. and Sundelin, J. (1994) Proc. Natl. Acad. Sci. U.S.A. **91**, 9208–9212
240 Coughlin, S. R. (1994) Proc. Natl. Acad. Sci. U.S.A. **91**, 9200–9202
241 Hartwig, J. H., Bokoch, G. M., Carpenter, C. L., Janmey, P. A., Taylor, L. A, Toker, A. and Stossel, T. P. (1995) Cell **82** 643–653

Biochem. J. (1996) **318**, 361–377 (Printed in Great Britain)

REVIEW ARTICLE
CD28: a signalling perspective

Stephen G. WARD

Department of Pharmacology, School of Pharmacy and Pharmacology, University of Bath, Bath BA2 7AY, U.K.

CD28 and the related molecule cytotoxic T lymphocyte-associated molecule-4 (CTLA-4), together with their natural ligands B7.1 and B7.2, have been implicated in the differential regulation of several immune responses. CD28 provides signals during T cell activation which are required for the production of interleukin 2 and other cytokines and chemokines, and it has also been implicated in the regulation of T cell anergy and programmed T cell death. The biochemical signals provided by CD28 are cyclosporin A-resistant and complement those provided by the T cell antigen receptor to allow full activation of T cells. Multiple signalling cascades which may be independent of, or dependent on, protein tyrosine kinase activation have been demonstrated to be activated by CD28, including activation of phospholipase C, p21^{ras}, phosphoinositide 3-kinase, sphingo-myelinase/ceramide and 5-lipoxygenase. The relative contributions of these cascades to overall CD28 signalling are still unknown, but probably depend on the state of activation of the T cell and the level of CD28 activation. The importance of these signalling cascades (in particular the phosphoinositide 3-kinase-mediated cascade) to functional indications of CD28 activation, such as interleukin 2 gene regulation, has been investigated using pharmacological and genetic manipulations. These approaches have demonstrated that CD28-activated signalling cascades regulate several transcription factors involved in interleukin 2 transcriptional activation. This review describes in detail the structure and expression of the CD28 and B7 families, the functional outcomes of CD28 ligation and the signalling events that are thought to mediate these functions.

INTRODUCTION

The term 'T cell activation' refers to events triggered by antigen stimulation of T cells that culminate in interleukin 2 (IL-2) production and IL-2 receptor expression (G_0–G_1 cell cycle transition) [1]. A two-signal model of effective activation of resting T lymphocytes has been proposed [2]: Signal 1 is provided by engagement of the T cell antigen receptor (TCR)/CD3 complex with foreign antigen associated with MHC molecules; Signal 2 is provided by a co-stimulatory molecule(s) present on antigen presenting cells such as dendritic cells, monocytes and activated B cells [1,2]. Considerable attention has focused on the CD28 molecule as providing a potential 'second signal' required for T cell activation, since ligation of the CD28 T cell antigen by anti-CD28 monoclonal antibodies or the natural ligand B7.1 (also termed CD80) or B7.2 (also termed CD86) can lead to cyclosporin-resistant IL-2 production and cellular proliferation in the presence of an additional signal provided by the phorbol ester phorbol 12-myristate 13-acetate (PMA) [3]. Subsequently, the identification of additional family members has given the potential for at least four interactions in T cell co-stimulation which involve the proteins B7.1 [4,5], B7.2 [6,7], CD28 [3,8] and cytotoxic T lymphocyte-associated molecule-4 (CTLA-4) [6,7,9]. The signalling events and functional outcomes that are elicited following B7–CD28 interaction are the best understood, and this review will describe the structure and expression of the CD28 and B7 families, the functional effects of CD28 ligation and the signalling events which mediate these functions.

STRUCTURE AND EXPRESSION OF CD28 AND B7 RECEPTOR FAMILIES

CD28 and CTLA-4

The known members of this family of receptors have now been cloned and analysed in some detail at both the DNA and protein level [3,10,11]. The genes for CD28 and CTLA-4 are closely linked on human chromosome 2 (2q33-34) and share the same genomic organization, suggesting a common evolutionary origin [10]. However, the overall amino acid conservation between the genes is as little as 30%. CD28 is a homodimeric cell surface glycoprotein expressed on 95% of CD4$^+$ T cells and approx. 50% of CD8$^+$ T cells [3]. Although the extracellular domains have limited sequence identity, mapping of the ligand binding sites of CD28 and CTLA-4 has localized the site of interaction with B7.1 and B7.2 to a conserved sequence Met-Tyr-Pro-Pro-Pro-Tyr in the complementary determining region 3-like region (CDR3)-like region of both CD28 and CTLA-4 2 [3].

CD28 is composed of two glycosylated 44 kDa chains which are members of the immunoglobulin superfamily, each containing a single disulphide-linked extracellular Ig variable-like (V) domain. The mature CD28 polypeptide contains 202 amino acids giving a molecular mass of 23 kDa, which is then glycosylated at five sites to give the molecular mass of the mature protein. The extracellular domain is linked via a single transmembrane region to a 41-amino-acid cytoplasmic domain which is presumed to be responsible for initiating co-stimulatory signals [3]. Sequence

Abbreviations used: [Ca^{2+}]$_i$, cytosolic free calcium concentration; CTLA-4, cytotoxic T lymphocyte-associated molecule-4; ERK, extracellular-signal-regulated kinase; FRAP, FK506 binding protein/rapamycin-associated protein; GAP, GTPase-activating protein; GM-CSF, granulocyte/macrophage colony-stimulating factor; Grb, growth-factor-receptor binding protein; ICAM, intracellular adhesion molecule; IκB, inhibitor of NFκB; Ig V and Ig C domains, Ig variable-like and constant-like domains respectively; IL-2, interleukin 2; JNK, c-Jun N-terminal kinase; mAb, monoclonal antibody; MAP kinase, mitogen-activated protein kinase; NFAT, nuclear factor of activated T cells; NFκB, nuclear transcription factor κB; PI 3-kinase, phosphoinositide 3-hydroxy kinase; PKC, protein kinase C; PLC, phospholipase C; PMA, phorbol 12-myristate 13-acetate; p70^{s6k}, p70 ribosomal protein-S6 kinase; PTK, protein tyrosine kinase; RANTES, regulated on activation, normal T cell expressed and secreted; SAP kinase, stress-activated protein kinase; SH domain, *src*-homology domain; Sos, mammalian Son of sevenless; TCR, T cell antigen receptor; TOR, target of rapamycin.

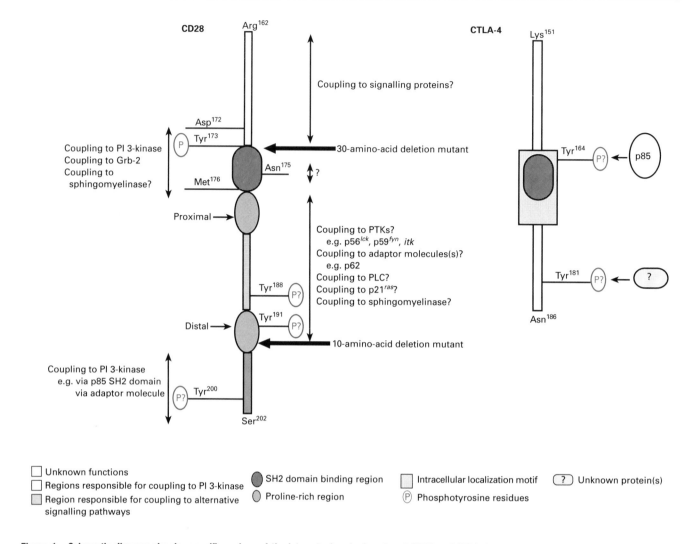

Figure 1   Schematic diagram showing specific regions of the intracytoplasmic domains of CD28 and CTLA-4

comparison between human, rat, mouse and chicken CD28 cytoplasmic domains demonstrates high interspecies conservation, suggesting a crucial role for this domain in signal transduction. The cytoplasmic domain of CD28 lacks any direct enzymic activity and is therefore presumed to signal via the recruitment of cellular enzymes. Of specific interest, there is a consensus sequence motif within the cytoplasmic domain of CD28 [(p)Tyr^{173}-Met-Asn-Met] which forms a potential binding site for interaction with signalling proteins, including the *src*-homology 2 (SH2) domains of the p85 subunit of phosphoinositide 3-hydroxy kinase (PI 3-kinase) and growth-factor-receptor binding protein (Grb-2), a ubiquitous adaptor protein [3,12] (Figure 1). Moreover, CD28 also contains two proline-rich motifs (Pro^{178}-Arg-Arg-Pro and Pro^{190}-Tyr-Ala-Pro) which conform to the Pro-Xaa-Xaa-Pro SH3 binding consensus sequence [13], and these regions of the CD28 tail may mediate interactions with signalling proteins.

CTLA-4 is also a disulphide-linked homodimer of predicted molecular mass 20 kDa, with only a single glycosylation site. CD28 and CTLA-4 are expressed on the cell surface as either monomeric or homodimeric forms [3,8]. Both CD28 and CTLA-4 can bind to the same physiological ligands, namely the 60 kDa B7.1 and the 70 kDa B7.2 [3,9], but CD28 is a low-avidity receptor for both B7.1 and B7.2, whereas CTLA-4 is a high-avidity receptor [3]. CTLA-4 was originally identified in a screen for genes involved in T-cell-mediated cytolysis [14], but CTLA-

4 transcripts have been detected in both CD4$^+$ and CD8$^+$ T cell clones [15]. CTLA-4 contains a single disulphide-linked extracellular Ig V domain, a transmembrane region and a cytoplasmic domain of 36 amino acids. Interestingly, there is 100 % conservation between the cytoplasmic domains of human and mouse CTLA-4, suggesting a strongly conserved function [3]. In contrast, there is only limited sequence conservation (approx. 30 % amino acid similarity) between CD28 and CTLA-4 in the cytoplasmic domain, perhaps suggesting an ability to generate different signals. Although the biochemical signals generated by CTLA-4 are poorly defined, CTLA-4 does contain a consensus binding site (Tyr^{164}-Val-Lys-Met) for the SH2 domains of the p85 subunit of PI 3-kinase. Indeed, anti-CTLA-4 monoclonal antibodies (mAbs) trigger binding of the p85 subunit of PI 3-kinase to this site in the CTLA-4 cytoplasmic domain [16]. To date, tyrosine phosphorylation of CTLA-4 has not been demonstrated, but phosphopeptide binding analysis has confirmed that CTLA-4 can interact with PI 3-kinase via the Tyr^{164}-Val-Lys-Met motif [16]. There is evidence that these sites may not be functionally equivalent in CD28 and CTLA-4, since the CTLA-4 motif (but not the CD28 motif) is located within an intracellular localization motif (Thr^{160}-Thr-Gly-Val-Tyr-Val-Lys-Met-Pro-Pro-Thr; Figure 1) that restricts CTLA-4 expression to intracellular membranes such as the perinuclear Golgi or post-Golgi compartment [17].

B7.1 and B7.2

B7.1 and B7.2 are also members of the immunoglobulin super-family with two Ig-like domains [Ig V domain and Ig constant-like (C) domain], and share approx. 25% amino acid conservation [3]. Unlike their counter-receptors CD28 and CTLA-4, B7.1 and B7.2 are not disulphide-linked and are not expressed as homodimers. There is also evidence that an additional B7 molecule may exist which has been provisionally termed B7.3 [18]. B7.1 is a 60 kDa glycoprotein (30 kDa polypeptide) which consists of two extracellular Ig-like domains with a trans-membrane region and a short 19-amino-acid cytoplasmic domain. In contrast, B7.2 is a 70 kDa glycoprotein (34 kDa polypeptide) and also consists of two Ig-like domains, but its major feature is an extended cytoplasmic domain which contains phosphorylation sites for protein kinase C (PKC), possibly indicating a signalling function in antigen presenting cells. B70 is an alternate splice variant of B7.2 that is identical to B7.2, except for six additional amino acids at the N-terminus of the B7.2 sequence [7]. B7.1 and B7.2 bind CTLA-4 with 20–100-fold higher affinity than they bind CD28 [3–8], with B7.2 being at most 2–3-fold less active than B7.1 [19]. Site-directed mutagenesis of the B7.1 and B7.2 Ig V and Ig C domains has revealed that full CD28 and CTLA-4 binding to B7 molecules is dependent upon residues in the GFCC'C' β-sheet face of the Ig V domain and the ABED β sheet face of the Ig C domain [20]. Another study reported two hydrophobic amino acid residues, Tyr^{87} and Trp^{84}, in the Ig V domain which are critical for binding to CD28 and CTLA-4. These residues lie in the sequence Trp-Pro-Xaa-Tyr-Lys-Asn-Arg-Thr, which is conserved across species [21].

FUNCTIONAL OUTCOME OF B7–CD28 INTERACTIONS

Up-regulation of lymphokines and chemokines

Ligation of CD28 alone has little effect on resting T cell proliferation, but CD28 ligation in the presence of limited concentrations of anti-CD3 or antigen promotes T cell proliferation and IL-2 production (Figure 2) in both CD4[+] and CD8[+] T cell subsets, by regulating IL-2 mRNA at the level of both transcription and translation [4,5]. Co-stimulation via CD28 also mediates strong up-regulation of IL-4, IL-5, IL-13, γ-interferon, tumour necrosis factor α and granulocyte/macrophage colony-stimulating factor (GM-CSF) [22–25], up-regulation of the IL-2 receptor α [26] and β [27] chains, and up-regulation of CD40 ligand expression [24,28]. CD28 also induces up-regulation of the chemokine IL-8 [29] and the regulated on activation, normal T cell expressed and secreted (RANTES) chemokine (L.Turner, S. Ward and J. Westwick, unpublished work). The gene promoters of IL-2, GM-CSF, IL-3, γ-interferon [30], IL-8 [29] and RANTES [31] have conserved sequences which are responsive to CD28 signal transduction.

CD28 can prevent anergy

Studies using soluble antagonists such as CTLA-4–Ig fusion proteins or antibodies to block CD28–B7.1 or –B7.2 interactions *in vitro* have shown that T cells which are deprived of CD28-mediated co-stimulatory signals enter into a state of proliferative hyporesponsiveness [32–35]. This long-term antigen-specific unresponsiveness is termed anergy *in vitro* and tolerance *in vivo*.

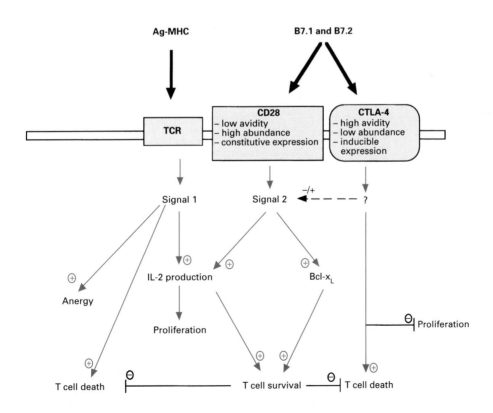

Figure 2   Model of potential outcomes of CD28/CTLA-4 ligation

+ denotes activation; − denotes inhibition; ? denotes unknown intermediates. The broken line represents a putative interaction.

Anergic T cells could be rescued from this non-functional state by the addition of exogenous IL-2 or by the delivery of a co-stimulatory signal via CD28, indicating that the anergized cells were functionally intact and that unresponsiveness was limited to a specific antigen [32]. The B7-family-mediated co-stimulation appears to be unique in its ability to prevent T cells from entering the anergic state. Co-stimulation induced by intracellular adhesion molecule (ICAM), which produced equivalent levels of proliferation to those induced by B7, was sensitive to cyclosporin A and did not result in IL-2 production; furthermore, upon subsequent antigen-specific rechallenge, T cells were anergic when ICAM had delivered the initial co-stimulatory signal [32]. These data, along with *in vivo* CD28$^{-/-}$ experiments [36], provide ample evidence of a critical role for CD28 signals in controlling T cell responsiveness. Moreover, blockade of CD28 signalling can result in suppression of humoral responses, delay allograft and block xenograft rejection, and inhibit the progression of autoimmune diseases [37–39]. In addition, a number of *in vivo* experimental systems have suggested that B7–CD28 interactions generate anti-tumour immune responses (reviewed in [8]).

CD28 can signal for T cell survival

Although much attention has focused on the role of CD28 in the prevention of anergy, it may also play a role in the modulation of programmed cell death (also termed apoptosis). During T cell development in the thymus, activation of immature thymocytes results in apoptosis, which may be a mechanism for negative selection of immature T cells [40–42]. In contrast, mature naive T cells usually produce cytokines and proliferate upon activation by the TCR and co-stimulatory molecules such as CD28. The

level of CD28 expression is regulated during T cell development as well as during T cell activation and proliferation, suggesting that CD28 may play a role in determining the effect of TCR activation during T cell ontogeny. Indeed, anti-CD28 mAbs prevent apoptosis in immature thymocytes and peripheral blood T cells induced by TCR mAbs [43,44]. There are, however, conflicting studies regarding the effect of CD28 on activation-induced apoptosis in mature and immature T cells ranging from no effect [45] to induction of apoptosis [46]. Nevertheless, CD28 co-stimulation of human T cells has been demonstrated to augment the expression of the intrinsic cell survival factor Bcl-x$_L$ and IL-2 (Figure 2), which correlates with enhanced resistance to apoptosis induced by γ-irradiation, antibodies to Fas or CD3, or IL-2 withdrawal [47]. The expression of Bcl-x$_L$ induced in T cells co-stimulated through CD28 could not be mimicked by treatment of cells with lymphokines such as IL-2, suggesting that CD28-induced expression of Bcl-x$_L$ may be through a direct signalling mechanism and not via the enhanced production of lymphokines [47].

DIFFERENTIAL REGULATION OF IMMUNE RESPONSES

There is considerable evidence that the interactions involving CD28 and B7 family members function to differentially regulate the immune response (Figure 2). For instance, CTLA-4, unlike CD28, is not expressed constitutively and is only expressed maximally 2–3 days after T cell activation by TCR/CD3 and CD28 ligation [48]. The biological consequences of CTLA-4–B7.1/B7.2 interactions are only now being clearly defined, but CTLA-4 cannot replace CD28 in providing co-stimulation in

Table 1   Biochemical signals elicited by anti-CD28 mAbs, B7.1 and B7.2 in T cells

ND, not determined; +, activated; —, no detectable effect; * denotes that cross-linking is required.

Response	mAb	B7.1	B7.2	Remarks	References
1. Early signals					
Tyrosine phosphorylation	+	+	+	Phosphorylation of unidentified substrates in activated T cells and cell lines	[3,82,83,87,166]
PLCγ phosphorylation	+	+	ND	Activated T cells, Jurkat and murine hybridomas	[3,87]
[Ca^{2+}]$_i$ elevation	+*	+	—	Cross-linking is a prerequisite in resting T cells, activated T cells and murine hybridomas but not necessarily Jurkat	[3,46,83–89]
Ins(1,4,5)P_3 elevation	+*	ND	ND	Jurkat or activated T cells; B7.1 elicits no detectable PtdIns(4,5)P_2 degradation	[3,85,86,91]
PI 3-kinase association	+	+	+	CD28 associates with the p85α subunit	[70,92,114–117]
PI 3-kinase activation	+	+	+	Occurs in resting T cells, activated T cells, Jurkat and murine hybridomas	[70,91,92,114]
Association with Grb-2	+*	ND	ND	Resting T cells, Jurkat and HPB-ALL	[173]
p21^{ras} activation	+	—	—	Jurkat	[96]
Sphingomyelinase activation	+*	+	+	Acidic isoform activated in resting T cells, activated T cells and Jurkat	[125,126]†
5-Lipoxygenase	+	ND	ND	Resting T cells	[128]
2. Intermediate events					
Vav	+	+	+	Tyrosine phosphorylated in Jurkat	[96]
p36	+	—	ND	Tyrosine phosphorylated in Jurkat	[96]
p62 (p120^{GAP}-associated)	+	+	+	Tyrosine phosphorylated in Jurkat	[96]
c-*cbl*	+	+	+	Tyrosine phosphorylated in Jurkat	[166]
Raf-1 activation	+	—	ND	Jurkat	[96]
ERK activation	+	—	—	Jurkat	[96,97]
JNK activation	+	+	+	Synergizes with anti-CD3 mAb in Jurkat cells; cyclosporin A-sensitive Synergizes with PMA/ionomycin; inactive alone	[215] [97]
p70^{s6k} activation	+*	ND	ND	Resting T cells; rapamycin-sensitive and cyclosporin A-insensitive	[197]
3. Nuclear events					
IκBα down-regulation	+	ND	ND	Jurkat and resting T cells; requires PMA; rapamycin-sensitive and cyclosporin A-insensitive	[193,194]
AP-1 regulation	+	+	+	Requires TCR signal in murine T cells; down-regulated during anergy	[211,212]
4. Co-stimulation (e.g. IL-2 secretion)	+	+	+	Resting T cells and Jurkat	[70,92,97,114]

† Also D. Sansom, personal communication.

CD28 mutant mice, ruling out functional redundancy [36]. Antibodies to CTLA-4 can enhance T cell proliferation, suggesting that CTLA-4 may synergize with CD28 and sustain co-stimulation [49]. However, cross-linked antibodies to CTLA-4 inhibit proliferation of naive T cells [50,51]. These data suggest that blockade of CTLA-4 removes, whereas aggregation of CTLA-4 provides, inhibitory signals that down-regulate T cell responses [49–51]. Recent evidence from CTLA-4 knock-out mice further suggests that CTLA-4 mediates immune response suppression [52,53]. In addition, anti-CTLA-4 mAbs mediate antigen-specific apoptosis [54] and can allow for, and potentiate, effective immune responses against tumour cells [55]. Furthermore, it is interesting to note that CD28 is transiently down-regulated upon binding to its ligand, and may become hyporesponsive to subsequent stimulation that occurs at approximately the same time that CTLA-4 is normally expressed [56].

A number of differences between B7.1 and B7.2 have also been reported that suggest differential immune regulation by these molecules. (i) Whereas both B7.1 and B7.2 are expressed on activated monocytes, activated B and T cells and activated natural killer cells, only B7.2 is constitutively expressed on dendritic cells and resting monocytes [3,6–8,18,57,58]. (ii) B7.2 is rapidly expressed on B cells following activation, whereas maximum B7.1 expression appears significantly later [6,7,18,57,59]. (iii) In contrast to anti-B7.1 mAbs, anti-B7.2 mAbs are potent inhibitors of T cell proliferation and cytokine production *in vitro* [58,60,61], and *in vivo* B7.2 is the major CD28 ligand active in clonal expansion systems [62], development of diabetes in non-obese diabetic mice [63] and in the stimulation of T-cell-dependent antibody responses [59]. (iv) B7.1 is superior to B7.2 co-stimulation in the induction and maintenance of T-cell-mediated anti-leukaemia immunity [64]. (v) The B7.2–CTLA-4–Ig complex dissociates much faster than the B7.1–CTLA-4–Ig complex [19]. Thus the fast on–off nature of B7.2–CTLA-4 interactions may function to limit an immune response by allowing more efficient disengagement of B7.2 from the high-avidity CTLA-4 receptor. In support of the hypothesis that B7.1 and B7.2 can regulate different functions, several groups have presented *in vivo* evidence to show that B7.1 and B7.2 may play distinct roles in the differentiation of CD4$^+$ Th cells. CD4$^+$ Th0 cells can differentiate into either Th1 cells (which produce IL-2, γ-interferon and tumour necrosis factor α, and mediate cell-mediated immune responses) or Th2 cells (which produce IL-4, IL-5 and IL-10, and which regulate humoral immune responses). It has been proposed that Th2 immune responses are dependent on high doses of initial antigen and CD28 co-stimulation, whereas Th1 immune responses are less dependent on CD28 co-stimulation (reviewed in [65]). Hence recent reports have shown that B7.1 preferentially acts as a co-stimulator of Th1 cells, whereas B7.2 co-stimulates and induces Th2 cells [66,67]. The molecular basis for the distinct regulation of differentiation remains obscure, since there are no significant intrinsic differences between B7.1 and B7.2 (Table 1) in their ability to activate second messengers and to co-stimulate proliferation, cytokine secretion *in vitro* from CD4$^+$ T cells and cell-mediated cytotoxicity [68,69].

CD28-ACTIVATED SIGNALLING CASCADES

The two-signal concept

Individually, neither the TCR nor CD28 can induce levels of IL-2 production that are comparable with those resulting from the combination of TCR and CD28 receptor ligation, unless second messenger agonists such as phorbol esters are also added [70]. A hallmark of CD28-mediated signal transduction is the production of IL-2 that is resistant to cyclosporin A and FK506 [3]. Accordingly, CD28 has been postulated to provide the co-stimulatory Signal 2 to T cells, which is thought to consist, at least in part, of a unique biochemical signal that is distinct from that provided by the TCR-induced cyclosporin A-sensitive Signal 1 [2,3]. Thus, in order to determine the signals generated by CD28, it is first necessary to understand the signals generated by the TCR.

Signal transduction pathways elicited by the TCR have been reviewed extensively elsewhere [71,72] and will only be described in brief detail here. Triggering of the TCR/CD3 complex initiates a cascade of biochemical events, the earliest of which is the activation of intracellular protein tyrosine kinases (PTKs) [71,72]. This activation of PTKs appears to be absolutely required for all subsequent T cell responses [73] and couples the TCR to an array of signal transduction and adaptor molecules [72]. The TCR does not possess intrinsic enzymic activity, and a specific motif, Glu-Xaa$_2$-Tyr-Xaa$_2$-Leu/Ile-Xaa$_7$-Tyr-Xaa$_2$-Leu/Ile, termed an immunoglobulin receptor family tyrosine-based activation motif, is crucial for TCR coupling to PTKs and hence subsequent TCR signalling processes [72]. This motif, after phosphorylation by *src* kinases, is responsible for coupling the TCR to non-receptor tyrosine kinases such as ZAP-70 [71]. Several cellular substrates have been identified as substrates for TCR-activated PTKs [72], including components of at least two signalling pathways believed to be important pivotal links in the control of cell activation. The first of these is phospholipase Cγ (PLC) [74], which regulates the hydrolysis of phosphatidylinositol lipids, releasing diacylglycerols which activate the serine/threonine kinase PKC, and Ins(1,4,5)P_3 which regulates intracellular calcium release [75]. Secondly, TCR activation of the guanine nucleotide binding protein p21^{ras} is PTK-dependent and is essential for the coupling of the TCR/CD3 complex to mitogen-activated protein kinase (MAP kinase) [also known as extracellular-signal-regulated kinase 1/2 (ERK1/2)] [76]. The calcium signalling system co-operates, via the calcium phosphatase calcineurin, with p21^{ras} for induction of the transcription factor nuclear factor of activated T cells (NFAT) which, along with AP-1, nuclear transcription factor κB (NFκB) and Oct-1, is involved in IL-2 gene expression [76]. A third signalling pathway, which is putatively mediated by PI 3-kinase, has also been implicated in cell growth and activation [77]. This may also be controlled by TCR-activated PTKs, since the p85 subunit of PI 3-kinase binds to the TCR ζ chain [78], although this has not been consistently observed to correlate with PI 3-kinase activation [79].

The use of distinct signal transduction pathways by CD28 would lead to a complementation between the TCR and CD28 to allow full activation of T cells, which would fit well with the idea of co-stimulation. However, it is possible that CD28 may enhance the amplitude or duration of a TCR-triggered signal, thereby crossing a threshold to activate downstream signalling cascades, or it may provide signals similar to those provided by the TCR, but at different time points, in different cellular compartments. Although the CD28 co-signal appears to be insensitive to inhibition by cyclosporin A and thus independent of Ca^{2+} and calcineurin [80,81], CD28 signal transduction does appear to be regulated by PTKs [82,83]. The tyrosine phosphorylation pattern induced by CD28 ligation is similar to that induced by TCR ligation, but unique CD28-induced tyrosine phosphorylation of 62 kDa, 75 kDa and 100 kDa proteins is also observed [82,83], suggesting that unique pathways may be coupled to CD28. Until recently, studies addressing CD28-mediated signal transduction events were based on the use of specific anti-CD28 mAbs as stimulators, but the advent of specific ligands (B7.1 and B7.2) has provided interesting results and has highlighted the fact that the ability of CD28 to activate certain signalling pathways depends on whether the activating stimulus is provided by an anti-CD28 mAb or by ligand stimulation (summarized in Table 1). Several signalling cascades which may be dependent on PTK activation have been demonstrated to be activated by CD28, and these will be reviewed here.

The PLC pathway

It has been described that CD28 regulates the tyrosine phosphorylation of PLCγ, PtdIns hydrolysis and the cytosolic free calcium concentration ($[Ca^{2+}]_i$) [84–87]. There appears to be a difference, however, depending on the type of T cell model used, in the relative ability of CD28 to elicit these responses (summarized in Table 1). For instance, in Jurkat cells, PtdIns hydrolysis, increases in $[Ca^{2+}]_i$ and diacylglycerol generation can be elicited in the absence of cross-linking [88]. However, it has been noted that the ability of anti-CD28 mAbs to elicit these signals is dependent on the epitope recognized by the mAb [88]. Moreover, other groups have reported that CD28 cross-linking is a prerequisite for eliciting PLC activity in purified resting T cells, activated T cells and human CD28-transfected murine T cell hybridoma [46,84–86]. Interestingly, ligation of CD28 by B7.1, but not B7.2, has been shown to result in elevation of $[Ca^{2+}]_i$ in activated T cells [89]. Human CD28 intracytoplasmic tail deletion mutants (Figure 1), which have been stably transfected into a murine hybridoma cell line, have been used to assess the regions of the CD28 cytoplasmic domain required for coupling to PLC. Thus deletion of the last 10 amino acid residues disrupts the distal proline-rich region and Tyr^{200}, but has no effect on the anti-CD28-mAb-induced elevation of $[Ca^{2+}]_i$ (D. Olive, personal communication). In contrast, deletion of the last 30 amino acids additionally disrupts the SH2 binding motif, the proximal proline-rich region, Tyr^{188} and Tyr^{191}, and prevents the CD28-induced elevation of $[Ca^{2+}]_i$. This implies that the proximal proline-rich region and/or the Tyr^{188} and Tyr^{191} residues are involved in regulating the coupling of CD28 to PLC (Figure 1). It is possible that the SH2 binding motif is also involved in coupling CD28 to PLC, although this is unlikely since CD28 does not form complexes with SH2 domains derived from PLCγ [90].

In marked contrast to anti-CD28 mAbs, B7.1 failed to elicit any detectable activation of PLC as assessed by PtdIns(4,5)P_2 degradation and phosphatidic acid generation [91] (Table 1), although B7.1 has been reported to induce PLCγ phosphorylation [87] and IL-2 production [70,92]. This suggests that PLC activation following mAb activation of CD28 may be functionally redundant. There is evidence to suggest, however, that there are both cyclosporin A-sensitive and -resistant components of the CD28 signalling responses, depending on the activation state of the T cell, such that the cyclosporin A-sensitive pathway is dominant in activated T cells and tumour cells [3,46], whereas the cyclosporin A-resistant pathway is dominant in resting T cells [80]. Recent studies have shown that the CD28-induced calcium signal was mainly limited to T cells of the CD4$^+$ subset [93], indicating that the co-stimulatory signal delivered by CD28 may have different biochemical properties in CD4$^+$ and CD8$^+$ T cell subsets and therefore that the role of CD28 may differ in these cells. The role of PKC activation in CD28 signalling is also unclear, since CD28 activation of Jurkat cells is not associated with PKC translocation [85], although levels of diacylglycerol are elevated following CD28 ligation by mAbs [88]. Moreover, PKC inhibitors have been reported to either inhibit [88] or have no effect on [94] CD28-dependent IL-2 production.

The p21^{ras} pathway

The intracellular proteins linking receptor-associated PTKs to the regulation of p21^{ras} have been defined: the adaptor molecule Grb-2, which is composed of two SH3 domains that bind to mammalian Son of sevenless (Sos), a guanine nucleotide exchange protein that activates p21^{ras}, and an SH2 domain that interacts with tyrosine phosphoproteins including Shc and a membrane-located tyrosine phosphoprotein p36 [76,95]. Two PTK substrates, Vav (a guanine nucleotide exchange factor) and a 36 kDa membrane protein, are potential mediators of the TCR/p21^{ras} link [76,95]. Interestingly, at least one 100 kDa substrate for CD28-induced tyrosine phosphorylation has been identified as Vav [96], suggesting that CD28 may also activate p21^{ras}. Indeed, an anti-CD28 mAb (CD28.2) can activate the p21^{ras} pathway, since it induced an increase in Ras–GTP complexes and in the phosphorylation of Vav and of the Grb-2/Sos-associated protein p36 [96]. Moreover, the MAP kinase cascade is also activated by CD28.2, resulting in hyperphosphorylation of Raf-1 and activation of ERK2 [96]. In marked contrast (Table 1), B7.1 stimulation of CD28 is unable to induce tyrosine phosphorylation of p36, although it can tyrosine phosphorylate Vav to a greater extent than CD28.2 or CD3 ligation [96]. Another anti-CD28 mAb (CD28.5), which binds CD28 but is unable to induce Ca^{2+} mobilization or IL-2 production, is devoid of an effect on p21^{ras} [96]. Thus CD28 does have the capacity to couple to the p21^{ras} pathway, as demonstrated by the effect of the CD28.2 mAb, although CD28 does not detectably activate p21^{ras} following ligation by B7.1 [96]. Accordingly, the physiological relevance of the anti-CD28-mAb-induced activation of ERK2 is not resolved, but it is probably not essential for CD28 co-stimulation of IL-2 production, since B7.1 and B7.2 can function equally in this context [97].

PI 3-kinase

PI 3-kinase controls an alternative inositol phospholipid metabolic pathway which phosphorylates the membrane phosphatidylinositol lipids at the 3-position on the inositol ring, resulting in the generation of PtdIns3P, PtdIns(3,4)P_2 and PtdIns(3,4,5)P_3 [77]. At present, at least three biochemically distinct PI 3-kinases have been described in mammalian cells: (i) a p85/p110 heterodimer, which associates with tyrosine-phosphorylated proteins through SH2 domains in the p85 subunit [77]; (ii) PI 3-kinase γ, which is activated by G-protein βγ subunits [98]; and (iii) a mammalian PtdIns-specific 3-kinase [99]. Use of the term PI 3-kinase should be taken to mean the PTK/SH2-coupled form, unless otherwise stated. This form is the best characterized and comprises a heterodimer of a 85 kDa regulatory subunit (α, β and γ isoforms) containing two SH2 domains, one SH3 domain and two proline-rich sequences that mediate interactions with other signalling molecules [100]. The p85 subunit is tightly associated with a 110 kDa catalytic subunit (α and β isoforms) [101], which is the first reported incidence of a dual-specificity kinase that possesses both protein serine and lipid kinase activities [102].

The D-3 phosphoinositide lipids represent only a minor fraction of cellular phosphatidylinositol lipids and are not substrates for PLC [77,103]. However, considerable importance has been attached to this pathway, particularly concerning the putative role of these lipids as regulatory molecules [77,103]. Kinetic evidence has suggested that PtdIns(3,4,5)P_3 formed by the PtdIns(4,5)P_2-specific 3-kinase is the probable physiological intracellular mediator of downstream signalling cascades, and this is subsequently degraded and inactivated by sequential dephosphorylation to PtdIns(3,4)P_2 and PtdIns3P [103]. The downstream effector molecules of PI 3-kinase-activated signalling cascades in T cells remain relatively obscure, although several proteins such as PKC ϵ, δ, η, λ and ζ isoforms [104–107], protein kinase B/Akt [108,109], p70 ribosomal protein S6 kinase (p70^{s6k}) [110,111] and Rac [112] and SH2 domains [113] have been demonstrated to lie downstream of PI 3-kinase in other systems. These proteins may be direct targets for PI 3-kinase or its lipid metabolites, or there may be intermediate target proteins involved in the activation of these proteins. PtdIns3P may have another specialized function, since it is the major product of mammalian PtdIns-specific 3-kinase, which shares extensive sequence conservation and substrate specificity with the yeast lipid kinase Vps34p [77,99]. Vps34p shares sequence conservation with mammalian p110 [101], and mutagenesis experiments have demonstrated that the PtdIns 3-kinase activity of Vps34p is required for protein sorting and vesicular trafficking [77,103]. Accordingly, it has been postulated that PtdIns3P may have a distinct function related to the trafficking of cell surface receptors [77,103], rather than acting as a signalling molecule. This function of PtdIns3P may be relevant to the mechanisms which regulate CTLA-4 expression, since the CTLA-4 p85 SH2 binding motif [16] is located in an intracellular localization motif within the CTLA-4 cytoplasmic domain [17].

An interaction between PI 3-kinase and CD28 was suggested by the presence of the p85 SH2 domain binding motif Tyr^{173}-Xaa-Xaa-Met around Tyr^{173} [3] (Figure 1). The first indication that CD28 could indeed couple to PI 3-kinase was the demonstration that ligation of CD28 by B7.1 induced the accumulation of D-3 phosphoinositide lipids, the products of PI 3-kinase activation, in the leukaemic T cell line Jurkat [91]. Later studies revealed that B7.2 or mAb ligation of CD28 also induced formation of D-3 phosphoinositide lipids in Jurkat cells [70] (Table 1) and that the association of PI 3-kinase with CD28 occurred in CD4$^+$ but not CD8$^+$ cells following CD28 ligation [114]. The accumulation of PtdIns(3,4,5)P_3 following CD28 ligation occurs with delayed kinetics, greater magnitude and prolonged duration when compared with the accumulation in response to ligation of the TCR alone [70]. The p85α subunit of PI 3-kinase can associate with the cytosolic domain of CD28 via the (p)Tyr^{173}-Met-Asn-Met motif [90,115,116], as predicted. The interaction of the (p)Tyr^{173}-Met-Asn-Met motif with the SH2 domains of the p85α subunit depends on phosphorylation of the tyrosine residue within the motif, and CD28 is indeed phosphorylated on tyrosine following ligation [90,117]. Phosphopeptide binding studies have confirmed that PI 3-kinase binds directly to CD28 via the (p)Tyr^{173}-Met-Asn-Met motif [115,116]. Interestingly, CD28 binds to the C-terminal SH2 domain of p85α with 10-fold greater affinity than to the N-terminal domain [116], but the relevence of this observation is not presently known.

The sphingomyelinase/ceramide signalling pathway

The sphingomyelinase pathway is a ubiquitous signalling pathway coupled to receptors for tumour necrosis factor α, γ-interferon, IL-1, nerve growth factor and Fas [118,119], and is initiated by hydrolysis of the phospholipid sphingomyelin by the specific PLC sphingomyelinase, resulting in ceramide production. Ceramide serves as a second messenger, and may activate downstream targets such as PKCζ[120], serine/threonine-specific protein kinases [121] and phosphatases [122]. Sphingomyelinase exists in two forms: a Mg^{2+}-dependent membrane-bound enzyme with a neutral pH optimum, and a lysosomal acidic form [118]. The acidic form has been shown to be activated by Fas [123] and the tumour necrosis factor receptor [124]. Since ligation of Fas and the tumour necrosis factor receptor generally results in cell

death, ceramide has been suggested to mediate death signals. Recent reports have shown that anti-CD28 mAbs induce the activation of acidic sphingomyelinase [125,126]; this was prevented by agents shown to be inhibitory to the activation of acidic sphingomyelinase, such as chloroquine. Moreover, this pathway appears to be required for co-stimulation, since the cell-permeable ceramide analogue C-6-ceramide mimicked the CD28 signal by inducing T cell proliferation and IL-2 gene transcription [126]. The natural ligands B7.1 and B7.2 also appear to activate sphingomyelinase (D. Sansom, personal communication). The ability of CD28 to potently activate both the PI 3-kinase and sphingomyelinase pathways (Table 1) may be a key distinguishing feature between the signals provided by CD28 and the so-called 'death' receptors such as Fas and the tumour necrosis factor receptor [127], which lack any known PI 3-kinase consensus binding motif but activate the acidic sphingomyelinase signalling pathway. Thus there may be key synergy between the CD28-activated PI 3-kinase and ceramide-mediated signalling pathways in protecting the cell from apoptosis and promoting cell survival and/or IL-2 production.

5-Lipoxygenase

It has been reported recently that NFκB activation and IL-2 gene expression following addition of anti-CD28 mAbs requires leukotriene B4 production by 5-lipoxygenase [128]. In addition, it was demonstrated that leukotriene B4, when added in combination with anti-CD3 mAbs, could mimic anti-CD28 effects [128]. However, these observations have not been confirmed and the effects of the natural ligands B7.1 and B7.2 on this pathway (Table 1) have yet to be determined.

CD28 interactions with PTKs

The primary structure of CD28 indicates that the cytoplasmic tail has no obvious enzymic activity (Figure 1), but CD28 clearly couples to PTKs, as indicated by the tyrosine phosphorylation of a number of substrates [3,8,82,83]. In addition, CD28 must also act as a substrate for PTKs, since it is tyrosine phosphorylated [90,117] and this tyrosine phosphorylation appears critical for its interaction with PI 3-kinase [90]. From analogy with the TCR, CD4 and CD8 T cell surface molecules, which couple to *src*-family non-receptor tyrosine kinases such as p59^{fyn} and p56^{lck} [129], it was postulated that CD28 also couples to similar non-receptor PTK(s). Deletion of the last 30 amino acids (but not the last 10 amino acids) of the CD28 cytoplasmic domain disrupts the tyrosine phosphorylation of substrates by CD28 ligation (D. Olive, personal communication). This implies that the proximal proline-rich region and/or the Tyr^{188} and Tyr^{191} residues are involved in regulating the coupling of CD28 to PTK(s) as well as coupling to PLC, as described earlier (Figure 1). It is unlikely that the SH2 binding motif is involved in coupling CD28 to PTKs, since CD28-induced tyrosine phosphorylation of substrates is still observed in mutants where the motif has been disrupted by site-specific mutagenesis of Tyr^{173} (D. Olive, personal communication).

The involvement of PTK(s) in CD28 signalling events has been investigated using the tyrosine kinase inhibitor herbimycin A, which prevents tyrosine phosphorylation of substrates [82,83], elevation of [Ca^{2+}]$_i$ [83,89], PI 3-kinase activation (S. Ward and C. June, unpublished work), leukotriene B4 release [128] and IL-2 production [82,83,128] induced by anti-CD28 mAbs. However, recent evidence indicates that herbimycin A induces non-specific covalent modification of thiol groups [130], and results using herbimycin should therefore be interpreted with extreme caution. Nevertheless, there is other evidence that suggests the involvement of PTK(s) in CD28 signalling. For instance, $p56^{lck}$ has been demonstrated to be activated following cross-linking of anti-CD28 mAbs, but not by anti-CD28 mAbs alone [131], and anti-CD28 mAbs have been demonstrated to induce the association of CD28 with $p56^{lck}$ and $p59^{fyn}$ in Jurkat cells as detected by *in vitro* kinase assays [132]. Anti-CD28 mAbs also induce the association with, and immediate tyrosine phosphorylation and activation of, the Tec family kinase $p72^{ITK/EMT}$ in the Jurkat T cell line [133]. Accordingly, it was postulated that $p72^{ITK/EMT}$ activation by CD28 may result in tyrosine phosphorylation of CD28 and the recruitment of signalling proteins such as PI 3-kinase. A novel member of the Tec/itk/Btk family of PTKs has been reported to exist in resting T cells and has appropriately been termed resting lymphocyte kinase [134]. This newly identified protein kinase may prove to play an important role in T cell co-stimulation, although this has yet to be verified. Recent evidence from a transfected insect cell model (*Spodoptera frugiperda*) has clearly demonstrated that $p56^{lck}$ and $p59^{fyn}$ (but not ZAP-70 or ITK/EMT) both phosphorylate CD28 at Tyr^{173} and induce PI 3-kinase, Grb-2 and ITK/EMT binding to CD28 [135]. In the human Jurkat leukaemic cell line, however, although functional $p56^{lck}$ is required for optimal CD28-mediated activation of ITK/EMT [136], the association of CD28 with PI 3-kinase and the subsequent activation of PI 3-kinase can still occur in $p56^{lck}$-deficient cells [137,138]. In contrast, anti-CD28 mAbs are unable to elicit Ca^{2+} mobilization and tyrosine phosphorylation of certain substrates in these $p56^{lck}$-deficient cell lines [137]. Taken together, the data from human T cell models implies that there are at least two PTK pathways coupled to CD28: $p56^{lck}$-dependent event(s) (e.g. ITK/EMT phosphorylation and activation, as well as Ca^{2+} mobilization) and $p56^{lck}$-independent event(s) (e.g. PI 3-kinase association and activation) [137]. However, given the specificity problems associated with the use of herbimycin A [130], it is unwise to speculate that the herbimycin A-sensitive and a herbimycin A-insensitive pathway correlate with the $p56^{lck}$-dependent and a $p56^{lck}$-independent pathways respectively.

ROLE OF PI 3-KINASE IN CD28-MEDIATED CO-STIMULATION

In contrast to the sphingomyelinase/ceramide pathway, the PI 3-kinase pathway has been studied extensively with a view to defining its role in CD28-mediated IL-2 production, using a combination of pharmacological and genetic manipulations, which have yielded controversial results and conclusions.

Effect of inhibition of PI 3-kinase on IL-2 production

The fungal metabolite wortmannin at concentrations up to 100 nM irreversibly inhibits the lipid and serine kinase activity of PI 3-kinase, by covalent interaction with the p110 catalytic site [139]. Wortmannin is cell permeable and commercially available, and has been widely used to assess the biological role of PI 3-kinase (reviewed in [140]). The highly electrophilic C-20 position of the furan ring of wortmannin is known to be sensitive to amines and is thought to be the alkylating part of the molecule [141]. Since site-directed mutagenesis of $p110\alpha$ revealed that a Lys^{802}-to-Arg mutation abolished wortmannin binding to p110, it seems that nucleophilic attack of Lys^{802} results in the formation of an enamine on the furan ring (at C-20) of wortmannin, leading to alkylation and irreversible inhibition of PI 3-kinase [142]. Equivalent residues to Lys^{802} are present in all lipid kinases, such as $p110\beta$, PI 3-kinase γ, PtdIns 3-kinase, Vps34 and human and yeast PtdIns 4-kinases, as well as in protein kinases such as myosin light chain kinase, protein kinase A, DNA-dependent protein kinase catalytic subunit and the target of rapamycin

(TOR)-related proteins [142]. Wortmannin can be predicted, therefore, to react with other known or not yet identified protein and lipid kinases if they show identity within the PI 3-kinase Lys^{802} region. The concentration of wortmannin required to inhibit these enzymes varies (reviewed in [143]), and may depend on the level and efficiency of initial non-covalent interactions with the substrate binding sites of individual kinases and/or the presence of amino acid residues such as Glu^{821} in p110, which may increase the nucleophilicity of Lys^{802} [142]. Although PI 3-kinase is the only high-affinity target for wortmannin in mammalian cells (K_i 1–10 nM), with the possible exception of a soluble PtdIns 4-kinase [144] and phospholipase A_2 [145], non-specific interactions are to be expected at pH values above 8, when the nucleophilicity of lysine is increased by deprotonation. Indeed, wortmannin at concentrations above 1 μM has been reported to inhibit other enzymes such as PKC and phospholipase D (reviewed in [140,143]). Hence information gained from the use of wortmannin to assess the biological implications of PI 3-kinase may have limited value, at least under some circumstances.

Wortmannin has been used to assess the functional relevance of PI 3-kinase activation with respect to CD28-mediated co-stimulation of IL-2 production, but it has yielded controversial results. Hence in the leukaemic T cell line Jurkat, wortmannin inhibits the CD28-stimulated accumulation of *D*-3 phospho-inositide lipids with an IC_{50} of less than 10 nM [70,92]. However, similar concentrations of wortmannin (10–100 nM) actually potentiate the CD28-mediated co-stimulation of IL-2 production by these cells [70]. In addition, nanomolar concentrations of wortmannin can potentiate NFAT induction (D. Cantrell, personal communication) and induce low amounts of IL-2 production from Jurkat cells [70] in the absence of any other stimulus, implying that PI 3-kinase may be constitutively active in Jurkat cells. In marked contrast, other groups were unable to detect any effect of wortmannin (at concentrations up to 10 μM) on CD28-induced IL-2 production by Jurkat cells [146,147]. The apparent resistance of CD28-mediated co-stimulation to wortmannin under certain conditions is not unique to Jurkat cells, since similar observations were made using freshly isolated murine $CD4^+$ cells [148]. These findings have led to much confusion and considerable speculation that PI 3-kinase has no functional role in the induction of IL-2 production by CD28 [146–148]. The use of Jurkat cells to study the CD28-mediated co-stimulation of IL-2 production could be misleading, however, since transformed cell lines may be poor models of Signal 2 because they have been selected to be independent of co-stimulatory signals. Of major importance, therefore, is the observation that nanomolar concentrations of wortmannin (< 100 nM) inhibit the CD28-mediated co-stimulation of IL-2 production in purified resting T cells and human T lymphoblasts [70,92,114]. In these cells, optimal IL-2 production is usually much more dependent on the provision of Signal 2, thus making them much more relevant models in which to study the co-stimulatory signalling cascades regulating IL-2 production.

The quercitin derivative LY294002 has been reported to competitively inhibit PI 3-kinase at micromolar concentrations, and is without effect on PtdIns 4-kinase and several other protein and lipid kinases at concentrations up to 50 μM [149]. This compound is structurally unrelated to wortmannin and has no site for nucleophilic attack. Thus LY294002 is unlikely to have the same non-specific actions as wortmannin and provides an alternative to wortmannin in the pharmacological assessment of the cellular role of PI 3-kinase. Interestingly, micromolar concentrations of LY294002 (< 3 μM) also potentiate the CD28-mediated co-stimulation of IL-2 production in Jurkat cells, mimicking the effect of wortmannin [70]. The effects of wortmannin on resting T cells and T cell lymphoblasts argue strongly in favour of PI 3-kinase having an important role in regulating CD28-induced IL-2 production. It is possible, however, that non-specific actions of wortmannin may contribute to its observed effects on IL-2 production. Moreover, the effect of LY294002 on the CD28-mediated co-stimulation of IL-2 production in purified resting T cells or T lymphoblasts has yet to be reported. Despite the obvious limitations of using wortmannin as a PI 3-kinase inhibitor, the effects of this agent on IL-2 production in normal T cells [70,92,114], as well as in Jurkat cells under certain conditions [70], imply that PI 3-kinase may deliver either positive (normal T cells) or negative/suppressive (Jurkat) signals for IL-2 production. The apparent ability of PI 3-kinase to transmit either a negative or a positive signal may also be important in understanding (i) the ability of CTLA-4 to co-operate with CD28 in the induction of production of IL-2 [49] but to generate negative signals under other circumstances [50,51,55], and (ii) the contrasting effects of CD28 on apoptosis [43–47].

Effect of mutagenesis of the SH2 domain binding motif on IL-2 production

Other controversial observations concerning the role of PI 3-kinase in CD28-mediated co-stimulation have been obtained from site-specific mutagenesis of the CD28 cytoplasmic tail. For instance, mutation of the human CD28 cytoplasmic tail residue Tyr^{173} to Phe^{173} and subsequent expression in a murine T cell hybridoma prevents CD28–PI 3-kinase association, activation of PI 3-kinase [90,150] and production of IL-2 in response to CD28 ligation. It should be noted that the murine T cell hybridoma is unusual in that CD28 ligation alone is sufficient to elicit IL-2 production in these cells. PI 3-kinase and Grb-2/Sos bind the same CD28 cytoplasmic (p)Tyr^{173}-Met-Asn-Met motif, and mutation of Tyr^{173} disrupts both PI 3-kinase binding and Grb-2 binding [150]. Thus this mutation does not distinguish the roles played by PI 3-kinase and Grb-2/Sos in CD28 signalling. In contrast, mutation of Met^{176} to Cys^{176} disrupts only PI 3-kinase binding, and this mutant is also unable to support CD28-driven IL-2 production [150].

The use of Jurkat cells to study the effect of site-specific mutagenesis of murine CD28 on IL-2 production has produced contrasting data. Thus mutation of Tyr^{170} of murine CD28 (which corresponds to Tyr^{173} of human CD28) to Phe^{170} prevented the ligation-induced association of CD28 with PI 3-kinase, but CD28 ligation still resulted in IL-2 production in these models [148,151]. These studies also identified different regions which were required for CD28 co-stimulatory activity. (i) The minimal elements required for co-stimulatory activity were suggested to be the Asp^{169} and Asn^{172} residues (corresponding to Asp^{172} and Asn^{175} respectively of human CD28) which reside in or near the consensus p85 binding motif. The Asp^{169} mutant failed to bind PI 3-kinase and did not co-stimulate IL-2 production, whereas the Asn^{172} mutant did bind PI 3-kinase but failed to co-stimulate IL-2 production [151]. (ii) CD28 co-stimulatory activity requires the integrity of one or more of the other three C-terminal tyrosine residues at Tyr^{185}, Tyr^{188} and Tyr^{197} (which correspond to Tyr^{188}, Tyr^{191} and Tyr^{200} respectively of human CD28; Figure 1), since mutation of these residues (while preserving Tyr^{170}) was sufficient to inhibit CD28-induced IL-2 production in Jurkat cells, but had no effect on PI 3-kinase activation by CD28 [148]. The effect of mutation of the three C-terminal tyrosine residues and Asp^{172} or Asn^{175} on other signalling pathways activated by CD28 (e.g. $p21^{ras}$ activation, PLC activation, sphingomyelinase activation and PTK activation) has

yet to be elucidated. However, at least one of these C-terminal tyrosine residues is involved in modulating the coupling of PI 3-kinase to human CD28 in the murine hybridoma model, since: (i) site-specific mutation of human CD28 Tyr^{200} (while preserving Tyr^{173}) is sufficient to induce an approx. 90% decrease in CD28-induced PI 3-kinase activation and also reduced CD28-mediated IL-2 production [152]; and (ii) phosphopeptides corresponding to residues 186–202 of CD28 with Tyr^{200} phosphorylated interact directly with the SH2 and SH3 domains of p85 *in vitro* [152]. Hence, in some models at least, other tyrosine residues within the CD28 cytoplasmic tail may be involved in the recruitment and/or activation of PI 3-kinase [152] (Figure 1).

Although IL-2 gene expression has traditionally been used as a functional indication of CD28-mediated signal transduction, it has been suggested that CD28 function may also be assessed by changes in β_1-integrin-mediated adhesion, which is up-regulated following CD28 ligation in the absence of any signal from the TCR [153]. Similarly, ligation of a chimaeric construct containing the CD28 cytoplasmic domain fused to the extracellular and transmembrane domains of CD2 results in increased β_1-integrin-dependent adhesion of HL60 transfectants to fibronectin, which was completely inhibited by wortmannin [154]. Furthermore, cross-linking of a CD2/CD28 chimaera containing a Tyr to Phe substitution corresponding to position 173 of the CD28 cytoplasmic tail prevented both ligation-induced association of the chimaera with PI 3-kinase and up-regulation of β_1-integrin function, implying a critical role for PI 3-kinase in the regulation of β_1-integrin activity by CD28 [154]. It is interesting to note that wortmannin does not inhibit the CD28-mediated up-regulation of integrin adhesiveness on Jurkat cells [154]. These data further emphasize the differences that have been observed in the role of PI 3-kinase in CD28-mediated stimulation in Jurkat cells, and indicates that PI 3-kinase may mediate multiple functions of CD28.

Is PI 3-kinase a specific CD28 signal?

The accumulation of D-3 phosphoinositide lipids has been reported to occur in response to ligation of several other T cell surface molecules, such as (i) the TCR [155], CTLA-4 [16], CD2 [155] and CD4 [156,157], which modulate the early events (G_0 to G_1 transition) in T cell activation; (ii) CD7 [158], which has a relatively undefined role; and (iii) in response to interactions between IL-2 and the IL-2 receptor [159–161], which control later events (e.g. the G_1 to S transition) in T cell growth. The D-3 phosphoinositide lipids generated by molecules such as the TCR [155] may target the same proteins as the D-3 phosphoinositides generated by CD28, but may be generated at insufficient levels or at inappropriate times to elicit full-scale activation events independently of the CD28 signal. Interestingly, TCR ligation alone or in combination with phorbol esters can induce small but significant levels of wortmannin-sensitive IL-2 production [70,114]. The significance of PI 3-kinase activation in the events leading to T cell growth mediated by the IL-2 receptor [159–161] is also unclear, since wortmannin has no effect on the IL-2-induced proliferation of normal T lymphoblasts [92] but does inhibit the IL-2-induced proliferation of tumour cells [161]

Several notable features of the enzyme may allow the participation of PI 3-kinase in diverse and specific cellular responses. These include the following. (i) Several biochemically distinct PI 3-kinases exist that exhibit different substrate specificities and wortmannin sensitivities [98,99,140,143]. (ii) The different lipid products PtdIns3P, PtdIns(3,4)P_2 and PtdIns(3,4,5)P_3 may mediate different functions, such as trafficking of cell-surface receptors (PtdIns3P) and cell signalling [PtdIns(3,4)P_2 and

PtdIns(3,4,5)P_3] [77,103]. (iii) Heterogeneity of each distinct PI 3-kinase, similar to the heterogeneity observed for p85 and p110 subunits [77,100,101], would allow differential and specific regulation of this enzyme by different receptors. At least two forms of p85 (α and β) have been shown to be expressed in T cells and to be differentially regulated by the TCR [162]. (iv) The *D*-3 phosphoinositides, particularly PtdIns(3,4,5)P_3, are generally restricted to the membrane compartments in which they are made [163], giving rise to a high degree of compartmentalization and possibly different functional consequences. (v) The additional wortmannin-sensitive protein kinase activity of PI 3-kinase [102,164] also allows for diversity in PI 3-kinase function and regulation. The role of the protein serine kinase function of PI 3-kinase, as well as the involvement of PtdIns 3-kinase and PI 3-kinase γ in T cell activation, have yet to investigated.

REGULATION OF ADAPTOR PROTEINS BY CD28

It is becoming clear that a growing number of proteins that do not exhibit catalytic functions can act as specialized adaptor molecules which facilitate the coupling of receptors to downstream signalling cascades. Generally, these proteins feature domains that are important for protein–protein interactions such as SH2/SH3 domains and/or proline-rich regions. Recent evidence suggests that the multi-functional SH2 and SH3 domain binding protein p62 [165], complexed with p120-Ras-GTPase-activating protein (GAP), is a substrate for CD28- but not TCR-activated PTK(s) [166]. In addition, tyrosine phosphorylation of p62 is also triggered by the accessory receptor CD2 [167]. This raises the possibility that, in T cells at least, p62 is involved selectively in accessory receptor signal transduction mechanisms. Although the role of p62 in T cells has yet to be fully determined, another tyrosine phosphorylated protein, p190, is present in the p62-Ras-GAP complexes, which contains a domain with GTPase activity for the GTP binding protein Rho and a second domain that has identity to Rho-GAP [168]. Accordingly, p62 should be considered to have the potential to link receptor-activated PTKs

to the Rho family of GTP binding proteins. The role of Rho in T cells is not known, but Rho and the related protein Rac are essential for the control of the actin cytoskeleton in fibroblasts [169], and it is recognized that Rho may be equally as important as p21^{ras} and Rac in the regulation of protein kinase cascades that culminate in transcription factor phosphorylation [170]. Moreover, Rho can regulate the activity of PtdIns 5-kinase [171] and can activate PI 3-kinase [172]. Thus Rho may be involved in modulating the established effect of CD28 on PI 3-kinase and/or possible effects of CD28 ligation on PtdIns 5-kinase.

The PI 3-kinase binding motif (p)Tyr^{173}-Met-Asn-Met present in the CD28 cytoplasmic tail also corresponds to the consensus binding motif [(p)Tyr-Xaa-Asn-Xaa] for Grb-2 [12]. Indeed, a recent report demonstrated that CD28 binds to the Grb-2/Sos complex by means of the cytoplasmic (p)Tyr^{173}-Met-Asn-Met motif, although the Grb-2 binding has much lower affinity (up to 100-fold lower) than CD28/PI 3-kinase binding [173]. It should also be noted that the CD28/Grb-2 association was observed using 1.5×10^8 cells (resting T cells, Jurkat or HPB-ALL) in the presence of extensively cross-linked CD28 mAbs [173], and has not so far been reported following ligation of CD28 by either of the natural ligands. The relevence of this association has not yet been determined, but it may be responsible for mediating CD28 coupling to the p21^{ras} pathway and/or PLCγ [76] in response to anti-CD28 mAbs. The demonstration that CD28 ligation induces tyrosine phosphorylation of p62 provides an alternative mechanism by which CD28 may be coupled to p21^{ras}, since p62 co-associates with p120-Ras-GAP [166]. This, therefore, implicates p62 in the regulation of p21^{ras}, although there is no obvious correlation between p62 tyrosine phosphorylation and p21^{ras} activation. The functional importance of p21^{ras} activation by anti-CD28 mAbs has yet to be demonstrated, although it is interesting to note that certain anti-CD28 mAbs which can induce p21^{ras} activation can also induce small but significant levels of IL-2 production [96,174] in the absence of any detectable association with PI 3-kinase; this is resistant to wortmannin [114]. Thus the p21^{ras} signalling pathway may be a relatively minor, but

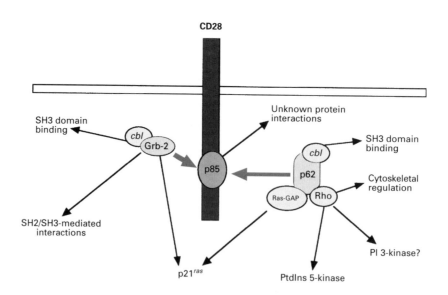

Figure 3   Putative interactions of CD28 with adaptor molecules

The p85 subunit of PI 3-kinase is bound via the (p)Tyr^{173}-Met-Asn-Met motif to the CD28 cytoplasmic tail upon ligation. Grb-2, p62 and/or unknown proteins, together with their associated proteins (e.g. *cbl*, Ras-GAP, Rho), are recruited to p85 via SH2/SH3 interactions (denoted by coloured arrows). These p85-bound proteins facilitate coupling of CD28 with other proteins and signalling cascades as shown. The possible direct interaction of Grb-2 with CD28 via the (p)Tyr^{173}-Met-Asn-Met motif is not shown. Refer to the text for further details.

nonetheless significant, contributing pathway in the CD28-derived signals required for IL-2 production, at least in some T cell models. Another putative role for p21^{ras} activation by CD28 may be to enhance or facilitate the activation of PI 3-kinase induced by CD28, in the light of recent reports that show that p21^{ras} can stimulate PI 3-kinase [175–177]. Such p21^{ras} stimulation of PI 3-kinase may also be an important function, at least in part, of TCR-induced p21^{ras} activation, mediating a kind of 'cross-talk' between the TCR and CD28. Conversely, CD28 may also influence TCR activation of p21^{ras} in a reciprocal manner, since some cellular responses to constitutively active PI 3-kinase are Ras-dependent [178], and PI 3-kinase has been demonstrated to inhibit GAP activation in adipocytes, allowing insulin to fully activate p21^{ras} [179].

It is becoming clear that Grb-2 SH3 domains can bind molecules other than Sos and may therefore regulate multiple effector pathways. In particular, recent studies have identified at least three other potential Grb-2 SH3 binding effector proteins in T cells: a proline-rich and SH2-domain-containing 75 kDa protein [180], a 116 kDa protein [181], PI 3-kinase [182] and the proto-oncogene c-cbl [183]. The functions of p75, p116 and c-cbl are not known, but p75 and p116 are both substrates for TCR-activated PTKs [180,181], while c-cbl is a substrate for CD28- and TCR-activated PTKs [166,183]. TCR ligation has recently been observed to modulate the interaction of p85 with the putative SH3 binding molecule cbl, as well as with Grb-2 [182]. The interaction of PI 3-kinase with Grb-2 has been observed in other systems [184] and may provide an additional mechanism for coupling PI 3-kinase to receptors, such as the TCR, which lack the ability to bind PI 3-kinase directly via p85 SH2 binding motifs.

The SH2, SH3 and proline-rich regions of p85 can putatively mediate a plethora of protein–protein interactions [77], such that the p85 subunit may play an important adaptor-like role in other signal transduction cascades quite unrelated to the lipid or serine kinase activities of p110; these adaptor functions would thus be expected to be resistant to inhibition by wortmannin (Figure 3). In this respect, p85 has been reported to function as an adaptor molecule linking the activated insulin receptor to the multifunctional SH2 and SH3 domain binding protein p62, and with Ras-GAP [185]. PI 3-kinase, by virtue of its adaptor molecule potential, may therefore participate in coupling CD28 to other signalling cascades such as the activation of p21^{ras}.

DOWNSTREAM EFFECTOR TARGETS FOR CD28-ACTIVATED SIGNALLING CASCADES

The 5' upstream region of the IL-2 promoter between position −325 and the transcriptional start site contains a number of sites for the binding of transcription factor complexes including AP-1, NFκB, NFAT and Oct for transcriptional activation [30]. Since IL-2 gene expression cannot be induced by a single signalling pathway, studies of transcriptional regulation utilize a combination of potent signals such as those induced by cross-linked mAbs, PMA with ionomycin or mitogens to mimic activation signals. Moreover, much of the information on IL-2 transcription has been obtained from studies on tumour cell lines such as Jurkat and EL-4, since it is difficult to measure the low transcriptional rate of the IL-2 gene in normal T cells by conventional assays, and normal T cells are generally resistant to transfection by reporter genes. However, a recent study has succeeded in optimizing conditions for transient transfection of normal human peripheral blood T cells with mutagenized promoters [186]. This approach has revealed the interesting observation that the important IL-2 cis-regulatory elements in normal T cells are the proximal AP-1 site and the NFκB site [186], while the NFAT sites are much less important in normal T cells than in Jurkat cells. Thus IL-2 transcriptional regulation differs in tumour cell lines compared with normal T cells [186], further highlighting the problems of using transformed cell lines as models of T cell activation. To date, few data are available relating to the effects of natural CD28 ligands on transcriptional regulation, and there is little direct evidence concerning the effects of CD28 on the known effector targets of the PI 3-kinase and sphingomyelinase/ceramide pathways. However, it is possible to gauge the involvement of these putative downstream targets on transcriptional regulation by studying the IL-2 transcriptional events modulated by CD28 in the presence of pharmacological and/or genetic manipulations of PI 3 kinase or sphingomyelinase/ceramide, or by studying the putative effector molecules themselves.

The CD28 response complex and NFκB

Anti-CD28 mAbs induce the formation of a CD28-responsive complex that binds to a site on the IL-2 gene distinct from previously described binding sites [30,187]. This binding site has been termed the CD28 response element, and is located between positions −160 and −152 relative to the transcription start site. The IL-8 [29], RANTES [31], IL-3 and GM-CSF [30,188] promoters also contain distinct elements of similar sequence which specifically bind a CD28-induced nuclear complex. Mutation of the CD28 response element results in the loss of CD28-induced activity without affecting TCR-induced IL-2 promoter activity [187]. The CD28 response element is distinct from, but related to, the NFκB element [30]. The NFκB/Rel family of transcription factors consists of five family members: NFκB1 (p50), NFκB2 (p52), c-Rel, p65 RelA and RelB [189]. The Rel family proteins p50, p65 and c-Rel are among the components of the CD28 response complex [190], and the recruitment of these proteins was shown to be resistant to cyclosporin A [190]. Evidence for the importance of p50 in the immune response can be seen from knock-out mice which show multiple B cell and T cell defects, including defective CD28 responses [191]. Activation of NFκB can be brought about by a variety of agents, including PMA and ionomycin, which result in the rapid and transient degradation of the cytoplasmic inhibitor of NFκB (IκBα) and dissociation of active NFκB from the inhibitory NFκB–IκBα complex [189,192]. Interestingly, anti-CD28 mAbs, in the presence of PMA, can induce sustained, cyclosporin A-resistant, down-regulation of IκBα which is associated with an increase in the nuclear translocation of c-Rel [193,194].

Regulation of NFκB by CD28-activated signalling cascades

The signalling mechanisms by which CD28 regulates NFκB have still to be fully elucidated. However, there is evidence that the signalling cascades mediated by both the sphingomyelinase and PI 3-kinase pathways are involved, at least to some degree. For instance, ceramide has been shown to modulate the activity of several intracellular proteins including NFκB [195], which may be mediated via ceramide-activated PKCζ [119,120], ceramide-activated protein kinase [121] or a ceramide-activated protein phosphatase [122]. It is particularly interesting to note that overexpression of sphingomyelinase substitutes for CD28 with regard to NFκB activation [125].

The signalling cascade(s) regulated by PI 3-kinase may also be involved in regulating NFκB activation. PI 3-kinase may exert regulation of NFκB via the activation of the serine/threonine protein kinase p70^{s6k} [196]. There is considerable evidence that PI 3-kinase and its D-3 phosphoinositide lipid products may mediate p70[s6k] activation, since the PI 3-kinase inhibitors wortmannin and LY294002 block platelet-derived growth factor- and insulin-mediated p70[s6k] activation [110]. Moreover, PI 3-kinase mediates activation of p70[s6k] *in situ* through site-specific threonine phosphorylation [111]. Given that CD28 elicits marked activation of PI 3-kinase [70,91,92,114], several groups have investigated whether p70[s6k] is a target for CD28-activated signalling cascades. Accordingly, cross-linked CD28 mAbs (but not soluble CD28 mAbs) have been found to induce activation of p70[s6k] in resting purified T lymphocytes [197]. Similar to the CD28-induced elevation of *D*-3 phosphoinositides [70], the activation of p70[s6k] occurs with slower kinetics than the kinetics observed for anti-CD3-mAb-induced activation of p70[s6k] [197]. In addition, both TCR- and CD28-induced effects are inhibited by the macrocyclic compound rapamycin (but not by FK506 or cyclosporin A) which inhibits activation of p70[s6k] [197]. Since the activation of p70[s6k] by the TCR and CD28 appears to be additive [197], this dual stimulation of a common pathway may be sufficient to cross critical thresholds that are not crossed upon ligation of individual molecules, and this pathway may be a necessary component of growth regulation via either cell surface receptor.

While not sufficient to drive proliferation, the induction of p70[s6k] activity may be necessary for the initiation of cell cycle progression in resting T cells. Indeed, a requirement for p70[s6k] activity has already been demonstrated for the G1 to S progression triggered by IL-2 [198]. Rapamycin inhibits anti-CD28-mAb-induced IκBα down-regulation and c-Rel translocation [194], but it has no effect on CD28-induced PI 3-kinase activation [199]. In contrast, wortmannin has been demonstrated to inhibit the activation of p70[s6k] induced by anti-CD28 mAbs [199]. Thus there may be upstream regulation of IκBα and/or c-Rel by a signalling cascade involving the lipid products of PI 3-kinase and p70[s6k]. However, experiments assessing whether PI 3-kinase inhibitors prevent CD28-induced activation of NFκB and/or c-Rel translocation have not yet been reported. Blocking of p70[s6k] either by injection of antibodies or by the use of rapamycin has suggested a role for p70[s6k] in G1–S phase transition [200,201]. Given the importance of CD28-derived signals in regulating the proliferation of resting T cells, a role for CD28-induced signals in p70[s6k] activation therefore seems appropriate.

It should be noted that the activation of p70[s6k] by CD28 has only been observed in response to anti-CD28 mAbs [197] and, to date, there are no reports concerning the effect of ligand activation of CD28 on p70[s6k] activity. Moreover, there is some doubt as to whether CD28-activated p70[s6k] alone is sufficient to mediate IκBα down-regulation, since the signals generated by PMA alone (which include p70[s6k] activation [110]) induced rapamycin-resistant, transient, down-regulation of IκB (as opposed to rapamycin-sensitive, sustained, down-regulation in the presence of anti-CD28 mAbs [193,194]). This discrepancy could be due, in part, to the activation by CD28 of a phorbol ester-insensitive PKC isoform such as PKCζ, which is activated *in vitro* by *D*-3 phosphoinositides [105]. PKCζ has also been suggested as a direct target for ceramide and thus may play a critical role in sphingomyelinase signalling [119,120]. Hence PKCζ could be a potential target for integration of the CD28-induced PI 3-kinase signals and sphingomyelinase signals. It should be noted, however, that active mutants of PKCζ have no effect on p70[s6k] activity when transfected into COS cells [202].

The PKC responsive element in the IL-2 gene enhancer contains sites for the transcription factors NFκB, AP-1 and NFAT [76]. The role of specific PKC isoenzymes is poorly understood, although PKCε and PKCθ can regulate the transcription factors AP-1 and NFAT in leukaemic T cells [203,204], while PKCζ modulates NFκB induction in fibroblasts [120]. In addition, PKC has been implicated in the control of many signalling pathways, including p21[ras] and the MAP kinases such as ERK1/2, and c-Jun N-terminal kinase (JNK-1/2) [76,97,205]. JNK1/2, along with p38, belongs to the stress-activated protein (SAP) kinase subfamily of MAP kinases, which are subject to distinct upstream regulatory kinases, such as SAP kinase/JNK(s) rather than MAP kinase kinase(s) [205,206]. Whereas the MAP kinase cascade can be activated by anti-CD28 mAbs in the absence of PMA [96], CD28 ligation (by anti-CD28 mAbs or natural ligands) only activates the SAP kinase/JNK cascade in the presence of PMA [97]. While both of these cascades appear to be important for the activation of κB-dependent promoters [202], active mutants of PKCζ have been demonstrated to activate the MAP kinase cascade but not the SAP kinase/JNK cascade [202]. Thus the effects of CD28 on NFκB probably require the integration of PKCζ-dependent and -independent signalling cascades. The PKCζ-independent cascades may be regulated by other phorbol ester-insensitive PKC isoforms or by undefined targets of CD28-activated PI 3-kinase and/or sphingomyelinase.

The intermediate signals that mediate the effect of CD28 on NFκB may also involve the mammalian equivalent of the TOR protein, which is the site of action of rapamycin in yeast [207]. One such protein, FK506 binding protein/rapamycin-associated protein (FRAP), is a member of the PI 3-kinase family of proteins that is expressed in T cells [208]. FRAP, like PI 3-kinase, exhibits serine kinase activity that results in autophosphorylation. This kinase activity is rapamycin-sensitive and is required for regulation of p70[s6k] [209]. Although the kinase activity of FRAP is insensitive to wortmannin [209], the possibility remains that another wortmannin-sensitive TOR-like protein(s) lies upstream of p70[s6k]. Thus it will be interesting to determine the effect of CD28 ligation on TOR-like protein kinases such as FRAP.

Regulation of AP-1 and JNK by CD28-activated signalling kinases

In addition to binding a functionally important AP-1 site in the IL-2 promoter, AP-1 proteins participate in the formation of NFAT and NF–IL-2 complexes [210]. Interestingly, AP-1 transcriptional activity requires both TCR- and CD28-mediated signals [211], and a lack of AP-1 induction has been suggested as one of the abnormalities found in anergic murine T cells [212]. Given that CD28 is the proposed co-stimulus required to prevent anergy, this suggestion appears to be somewhat at odds with a recent observation that CD28 stimulation also causes down-regulation of AP-1 [213]. AP-1 activity is regulated both at the level of *jun* and *fos* gene transcription and by post-translational modification of their products [210]. A key event in c-*fos* induction is phosphorylation of the transcription factor Elk-1 by the MAP kinases ERK1 and ERK2 [210]. Accordingly, the role of p21[ras] in coupling the TCR to ERK2 and the ability of ERK2 to translocate to the nucleus, where it can directly regulate transcriptional factors such as Elk-1, could explain the role of p21[ras] in TCR signal transduction [76]. In contrast, induction and regulation of c-Jun appears to be mediated by the SAP kinases JNK-1/2 and p38 [214].

JNK, unlike ERK, is not activated by phorbol ester alone, but rather requires phorbol ester in combination with calcium ionophore [97,215]. Triggering of CD3 and CD28 also results in cyclosporin A-sensitive activation of JNK, whereas each stimulus alone results in little or no activation [215]. This observation prompted speculation that the integration of signals that lead to T cell activation and IL-2 induction occurs at the level of JNK, although subsequent studies have failed to detect synergistic activation of JNK after CD3 and CD28 ligation ([97]; S. Ward,

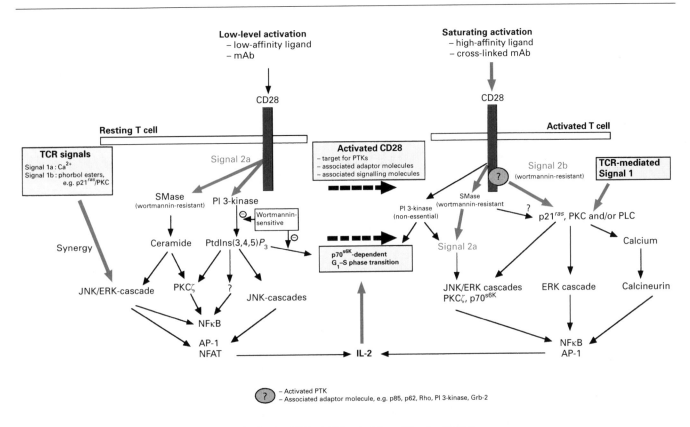

Figure 4   Schematic representation of signal transduction by CD28 at different states of T cell activation

Coloured arrows represent the proposed importance of the major signalling cascades in resting and activated T cells. Broken arrows represent the transition from the resting state to the activated state of the T cell. The major target of PI 3-kinase in resting T cells is $p70^{s6k}$. This wortmannin-sensitive activation of $p70^{s6k}$ by CD28 (and subsequently by IL-2, produced in response to the synergistic action of Signal 1 and wortmannin-resistant components of Signal 2a such as sphingomyelinase) facilitates cell cycle progression. In activated T cells, the wortmannin-sensitive activation of $p70^{s6k}$ by CD28 is non-essential. Instead, $p70^{s6k}$ activity is maintained by IL-2, produced in response to Signal 1, Signal 2b and possibly wortmannin-resistant components of Signal 2a. These wortmannin-resistant components may comprise sphingomyelinase-dependent signals and/or adaptor functions of p85 which facilitate coupling of CD28 to $p21^{ras}$ via Grb-2 and/or p62-associated Ras-GAP. Other signalling pathways (e.g. PLC) may also be coupled to CD28 via Grb-2 or unknown adaptor-protein–protein interactions. Abbreviation: SMase, sphingomyelinase. Refer to the text for further details.

unpublished work). Rather, the combination of CD28 ligation (following B7.1, B7.2 or anti-CD28 mAb stimulation) plus PMA and ionomycin was found to markedly increase JNK activity above the levels observed in response to PMA and ionomycin alone [97]. Taken together, these observations suggest that CD28-activated signals co-operate with calcium ionophores and phorbol esters to regulate JNK. Calcium signals mediated by calcineurin are important for JNK activation in T cells [215]. The TCR controls calcium homoeostasis during T cell activation, and this is one mechanism whereby the TCR may contribute to JNK regulation [215]. However, calcium signals and CD28 signals are not sufficient for JNK activation in T cells, implying that there is a minimal requirement for the convergence of three signals for JNK activation. The third signal can be provided by phorbol esters, and may thus be mediated by $p21^{ras}$ and/or PKC.

The TCR- and CD28-activated signals that contribute to JNK activation have not so far been fully elucidated. There are at least two distinct MAP kinase cascades in mammalian cells which control ERK and JNK activation [205,206], and the point of convergence of calcium/TCR/CD28 signals in the JNK regulatory cascade is at present unknown. It is interesting to note, however, that the PI 3-kinase and sphingomyelinase signalling pathways, which are potently activated by CD28, have been demonstrated to modulate both the ERK and JNK cascades in other systems and may therefore be involved in the regulation of c-*fos* or c-*jun*. For instance, a constitutively active p110 mutant

activates *fos* transcription in a Ras-dependent manner [178], implying that PI 3-kinase can directly modulate the $p21^{ras}$ pathway and thus the eventual modulation of ERKs that are thought to be involved in the transmission of signals from $p21^{ras}$ to the nucleus. Moreover, the Ras-related small GTP binding protein Rac, which has been implicated as an upstream regulator of the kinase cascades involved in the activation of JNK [214,216], appears to be a downstream effector for PI 3-kinase [112]. Furthermore, the sphingomyelinase/ceramide signalling pathway is also known to contribute to the regulation of ERK and JNK, since ceramide has been demonstrated to activate components of the ERK cascade such as Raf-1 [217], MAP kinase [218] and JNK [219–221].

CONCLUDING REMARKS

An updated view of Signal 2

It has previously been suggested that CD28-activated Signal 2 consists of two distinct signals: Signal 2a is cyclosporin A-resistant and occurs in naive T cells after low-level CD28 activation (subsaturating bivalent anti-CD28 mAbs or low-affinity natural ligands), whereas Signal 2b is cyclosporin A-sensitive, is transmitted in activated T cells and requires saturating CD28 activation (cross-linked anti-CD28 mAbs and high-affinity natural ligands) [3]. It is clear that both Signal 2a and Signal 2b may actually consist of several components. For instance, the components of Signal 2a defined to date probably consist of the signals mediated by PI 3-kinase and sphingomyelinase (Figure 4), given that the TCR triggers accumulation of D-3 phosphoinositide lipids with different stoichiometries and kinetics in comparison with CD28 and does not activate sphingomyelinase. The PI 3-kinase- and sphingomyelinase-mediated signals may act synergistically, although the level at which this putative synergy occcurs has yet to be elucidated. Hence cyclosporin A-resistant Signal 2a is, at least in part, wortmannin-sensitive. In contrast, Signal 2b is likely to consist predominantly of the p21^{ras} and PLC signalling pathways, and these appear to be shared signals that can be activated by both the TCR and CD28 (Figure 4). Thus Signal 2b is, at least in part, cyclosporin A-sensitive and wortmannin-resistant. One of the major targets of components (e.g. PI 3-kinase) of Signal 2a may be p70^{s6k}, which has been implicated in regulating G$_1$–S phase transition [200,201]. Therefore it is likely that, in activated T cells, p70^{s6k} is a non-essential downstream target of PI 3-kinase, since the activity of p70^{s6k} is sufficiently maintained by IL-2, produced as a result of T cell activation. In the activated state only, Signal 1 and Signal 2b may be identical, additive and sufficient to drive further IL-2 production. Alternatively, the wortmannin-resistant components of Signal 2a (e.g. sphingomyelinase and adaptor functions mediated by the p85 subunit of PI 3-kinase) may still be required for activation of other downstream targets in synergy with Signal 2b and/or Signal 1. For instance, p85 may be required for recruitment of Signal 1 or Signal 2b signalling proteins such as Grb-2 [182,184] and p120-Ras-GAP associated with p62 [166]. In addition, sphingomyelinase has been shown to modulate the ERK cascade [217,218] as well as the JNK cascade [219–221].

Recent evidence has demonstrated that the JNK signalling pathway contributes to apoptosis, whereas the ERK signalling pathway promotes cell survival in neuronal cells [222]. As described above, the relative contributions of PI 3-kinase and sphingomyelinase to CD28 signalling probably depend on the state of activation of the T cell and the level of CD28 activation. This will determine the effect of CD28 ligation on the overall balance of the ERK and JNK signalling pathways, which in turn will direct the cell to apoptosis or cell survival. This may explain why CD28 has been reported to protect from [47], have no effect on [45] or induce [46] apoptosis. Since several signalling cascades, including those regulated by PI 3-kinase [178], p21^{ras} [76] and sphingomyelinase [217–221], contribute to the regulation of ERK and/or JNK, the occurrence of cell survival or apoptosis probably requires complex integration of multiple signals from several receptors and cascades.

Future prospects

The known downstream effector targets of the PI 3-kinase- and sphingomyelinase-mediated signalling cascades have yet to be fully evaluated in terms of their regulation by CD28 and their functional importance with respect to cytokine gene regulation. Thus it will be important in the future to determine the effect of CD28 ligation on the downstream PI 3-kinase targets (protein kinase B, PKC isoforms and Rac) and ceramide-activated targets (PKCζ, phosphatases and kinases), and the effects of mutation of these targets on CD28-mediated co-stimulation. Other outstanding questions concern the role of p62 and p85 as specialized adaptor molecules in CD28 signalling cascades and the possible role of Rho in CD28-activated protein kinase cascades. Given the large number of cytokines/chemokines and cell surface molecules reportedly influenced by CD28, it is perhaps not surprising that CD28 couples to multiple signalling cascades, as this multiple signal capacity provides CD28 with the mechanistic

potential to regulate diverse events. It will therefore be particularly interesting to determine how multiple CD28-activated signalling cascades are integrated and whether distinct signalling cascades are responsible for regulating distinct transcriptional regulation of cytokine and chemokine genes.

S.W. is the recipient of a Wellcome Trust Career Development Award. I thank Carl June, Doreen Cantrell, Daniel Olive and David Sansom for sharing unpublished data and/or for enlightening discussions.

REFERENCES

1 Mueller, D. L., Jenkins, M. K. and Schwartz, R. H. (1989) Annu. Rev. Immunol. **7**, 445–475

2 Bretscher, P. (1982) Immunol. Today **13**, 74–76

3 June, C. H., Bluestone, J. A., Nadler, L. M. and Thompson, C. B. (1994) Immunol. Today **15**, 321–331

4 Gimmi, C. D., Freeman, G. J., Gribben, J. G., Sugita, K., Freedman, A. S., Morimoto, C. and Nadler, L. M. (1991) Proc Natl. Acad. Sci. U.S.A. **88**, 6575–6579

5 Linsley, P. S., Brady, W., Grosmaire, L., Aruffo, A., Damle, N. K. and Ledbetter, J. A. (1991) J. Exp. Med. **173**, 721–730

6 Freeman, G. J., Gribben, J. G., Boussiotis, V. A., Ng, J. W., Restivo, V. A., Lombard, L. A., Gray, G. S. and Nadler, L. M. (1993) Science **262**, 909–911

7 Azuma, M., Phillips, J. H., Lanier, L. and Somoza, C. (1993) Nature (London) **366**, 76–78

8 Guinan, E. C., Gribben, J. G., Boussiotis, V. A., Freeman, G. J. and Nadler, L. M. (1994) Blood **84**, 3261–3282

9 Linsley, P. S., Brady, W., Urnes, M., Grosmaire, L., Damle, N. K. and Ledbetter, J. A. (1991) J. Exp. Med. **174**, 561–569

10 Harper, K., Balzano, C., Rouvier, E., Mattei, M., Luciani, M. and Golstein, P. (1991) J. Immunol. **147**, 1037–1044

11 Aruffo, A. and Seed, B. (1987) Proc. Natl. Acad. Sci. U.S.A. **84**, 8573–8577

12 Songyang, Z., Shoelson, S.E, Chaudhuri, M, Gish, G., Pawson, T., Haser, W. G., King, F., Roberts, T., Ratnofsky, S., Lechleider, R. J., et al. (1993) Cell **72**, 767–778

13 Ren, R., Mayer, B. and Baltimore, D. (1992) Science **259**, 1157–1159

14 Brunet, J. F., Denizot, F., Luciani, M. F., Roux-Dosseto, M., Suzan, M., Mattei, M. G. and Golstein, P. (1987) Nature (London) **328**, 267–270

15 Freeman, G. J., Lombard, D. A., Gimmi, C. D., Brod, S. A., Lee, K., Laning, J. C., Hafler, D. A., Dorf, M. E., Gray, G. S. and Reiser, H. (1992) J. Immunol. **149**, 3795–3801

16 Schneider, H., Prasad, K. V., Shoelson, S. and Rudd, C. (1994) J. Exp. Med. **181**, 351–355

17 Leung, H. T., Bradshaw, J., Cleaveland, J. S. and Linsley, P. S. (1995) J. Biol. Chem. **270**, 25107–24114

18 Boussiotis, V. A., Freeman, G. J., Gribben, J. G., Daley, J., Gray, G. S. and Nadler, L. M. (1993) Proc. Natl. Acad. Sci. U.S.A. **90**, 11059–11063

19 Linsley, P., Greene, J. L., Brady, W., Bajorath, J., Ledbetter, J. A. and Peach, R. (1994) Immunity **1**, 793–801

20 Peach, R., Bajorath, J., Naemura, J., Leytze, G., Greene, J., Aruffo, A. and Linsley, P. S. (1995) J. Biol. Chem. **270**, 21181–21187

21 Fargeas, C. A., Truneh, A., Reddy, M., Hurle, M., Sweet, R. and Sekaly, R. P. (1995) J. Exp. Med. **182**, 667–675

22 Thompson, C. B., Lindsten, T., Ledbetter, J. A., Kunkel, S. L., Young, H. A., Emerson, S. G., Leiden, J. M. and June, C. H. (1989) Proc. Natl. Acad. Sci. U.S.A. **86**, 1333–1337

23 Minty, A., Chalon, P., Derocq, J. M., Dumont, X., Guillemot, J. C., Keghad, M., Labit, C., Leplatois, P., Liauzun, P., Miloux, B., et al. (1993) Nature (London) **362**, 248–250

24 De-Boer, M., Kasran, A., Kwekkeboom, J., Walter, H., Vandenberghe, P. and Cueppens, J. L. (1993) Eur. J. Immunol. **23**, 3120–3125

25 Seder, R. A., Germain, R. N., Linsley, P. S. and Paul, W. E. (1994) J. Exp. Med. **179**, 299–304

26 Cerdan, C., Martin, Y., Courcoul, M., Brailly, H., Mawas, C., Birg, F. and Olive, D. (1992) J. Immunol. **149**, 2255–2261

27 Cerdan, C., Martin, Y., Courcoul, M., Mawas, C., Birg, F. and Olive, D. (1995) J. Immunol. **154**, 1007–1013

28 Ding, J., Green, J. M., Thompson, C. B. and Shevach, E. M. (1995) J. Immunol. **155**, 5124–5132

29 Weschler, A. S., Gordon, M. C., Dendorfer, U. and LeClair, K. P. (1994) J. Immunol. **153**, 2515–2523

30 Fraser, J. D., Straus, D. and Weiss, A. (1993) Immunol. Today **14**, 357–362

31 Nelson, P. J., Kim, H. T., Manning, W. C., Goralski, T. J. and Krensky, A. M. (1993) J. Immunol. **151**, 2601–2612

32 Boussiotis, V. A., Freeman, G. J., Gray, G. S., Gribben, J. G. and Nadler, L. M. (1993) J. Exp. Med. **178**, 1753–1763

33 Harding, F., McArthur, J. G., Gross, J. A., Raulet, D. H. and Allison, J. P. (1992) Nature (London) **356**, 607–609

34 Gimmi, C. D., Freeman, G. J., Gribben, J. G., Gray, G. and Nadler, L. M. (1993) Proc. Natl. Acad. Sci. U.S.A. **90**, 6586–6590

35 Tan, P., Anasetti, C., Hansen, J. A., Melrose, J., Brunvand, M., Bradshaw, J., Ledbetter, J. A. and Linsley, P. S. (1993) J. Exp. Med. **177**, 165–173

36 Shahinian, A., Pfeffer, K., Lee, K. P., Kundig, T. M., Kishihara, K., Wakeham, A., Kawai, K., Ohashi, P. S., Thompson, C. B. and Mak, T. (1993) Science **261**, 609–611

37 Lenschow, D. J., Zeng, Y., Thistlethwaite, J. R., Montag, A., Brady, W., Gibson, M.G., Linsley, P. S. and Bluestone, J. A. (1992) Science **257**, 789–792

38 Lin, H., Bolling, S. F., Linsley, P. S., Wei, R. Q., Gordon, D., Thompson, C. B. and Turka, L. A. (1993) J. Exp. Med. **178**, 1801–1806

39 Turka, L. A., Linsley, P. S., Lin, H., Brady, W., Leiden, J. M., Wei, R. Q., Gibson, M. L., Zheng, X. G., Myrdal, S., Gordon, D., et al. (1992) Proc. Natl. Acad. Sci. U.S.A. **89**, 11102–11105

40 Smith, C. A., Williams, G. T., Kingston, R., Jenkinson, E. J. and Owen, J. J. T. (1989) Nature (London) **337**, 181–184

41 Shi, Y., Sahai, B. and Green, D. R. (1989) Nature (London) **339**, 625–628

42 Murphy, K. M., Heimberger, A. B. and Loh, D. Y. (1991) Science **250**, 1720–1723

43 Groux, H., Torpier, G., Monte, D., Mouton, Y., Capron, A. and Amieson, J. C. (1992) J. Exp. Med. **175**, 331–340

44 Shi, Y., Radvanyi, L. G., Sharma, A., Shaw, P., Green, D. R., Miller, R. G. and Mills, G. B. (1995) J. Immunol. **155**, 1829–1837

45 Tan, R., Teh, S. H., Ledbetter, J. A., Linsley, P. S. and Teh, H. S. (1992) J. Immunol. **149**, 3217–3224

46 Couez, D., Pages, F., Ragueneau, M., Nunes, J., Klasen, S., Mawas, C., Truneh, A. and Olive, D. (1994) Mol. Immunol. **31**, 47–56

47 Boise, L. H., Minn, A. J., Noel, P. J., June, C. H., Accavitti, M. A., Lindsten, T. and Thompson, C. B. (1995) Immunity **3**, 87–98

48 Lindsten, T., Lee, K. P., Harris, E. S., Petryniak, N., Craighead, N., Reynolds, P. J., Lombard, D. B., Freeman, G. J., Nadler, L. M., Gray, G. S., Thompson, C. B. and June, C. H. (1993) J. Immunol. **151**, 3489–3499

49 Linsley, P. S., Greene, J. L., Tan, P., Bradshaw, J., Ledbettter, J. A., Anasetti, C. and Damle, N. (1992) J. Exp. Med. **176**, 1595–1604

50 Walunus, T. L., Lenschow, D. J., Bakker, C. Y., Linsley, P. S., Freeman, J. M., Thompson, C. B. and Bluestone, J. A. (1994) Immunity **1**, 405–413

51 Krummel, M. F. and Allison, J. P. (1995) J. Exp. Med. **182**, 459–465

52 Waterhouse, P., Penninger, J. M., Timms, E., Wakeham, A., Shahinian, A., Lee, K. P., Thompson, C. B. and Mak, T. W. (1995) Science **270**, 985–988

53 Tivol, E. A., Borriello, F., Schweitzer, A. N., Lynch, W. P. and Bluestone, J. A. (1995) Immunity **3**, 541–547

54 Gribben, J., Freeman, G., Boussiotis, V., Rennert, P., Jellis, E., Greenfield, M., Barber, M., Restivo, V. A., Ke, G. S., Gray, G. and Nadler, L. (1995) Proc. Natl. Acad. Sci. U.S.A. **92**, 811–815

55 Leach, D. R., Krummel, M. F. and Allison, J. P. (1996) Science **271**, 1734–1736

56 Linsley, P. S., Bradshaw, J., Urnes, M., Grosmaire, L. and Ledbetter, J. A. (1993) J. Immunol. **150**, 3161–3169

57 Lenschow, D. J., Sperling, A. L., Cooke, M. P., Freeman, G., Rhee, L., Decker, D. C. Gray, G., Nadler, L. M., Goodnow, C. C. and Bluestone, J. A. (1994) J. Immunol. **153**, 1990–1997

58 Lenschow, D. J., Su, G. H., Zuckerman, L. A., Nabavi, N., Jellis, C. L., Gray, G. S., Miller, J. and Bluestone, J. A. (1993) Proc. Natl. Acad. Sci. U.S.A. **90**, 11054–11058

59 Hathcock, K. S., Laszlo, G., Pucillo, C., Linsley, P. and Hodes, R. J. (1994) J. Exp. Med. **180**, 631–640

60 Chen, L., Faherty, D. A., Gault, A., Connaughton, S. E., Powers, G. D., Godfrey, D. I. and Nabavi, N. (1994) J. Immunol. **152**, 2105–2114

61 Wu, Y., Guo, Y. and Liu, Y. (1993) J. Exp. Med. **178**, 1789–1793

62 Kearney, E. R., Pape, K. A., Loh, D. Y. and Jenkins, M. (1994) Immunity **1**, 327–339

63 Lenschow, D. J., Ho, S. C., Sattar, H., Rhee, L., Gray, G., Nabavi, N., Herold, K. C. and Bluestone, J. A. (1995) J. Exp. Med. **181**, 1145–1155

64 Matulonis, U., Dosiou, C., Freeman, G., Lamont, C., Mauch, P., Nadler, L. M. and Griffin, J. D. (1996) J. Immunol. **156**, 1126–1131

65 Thompson, C. B. (1995) Cell **81**, 979–982

66 Freeman, G. J., Boussiotis, V. A., Anamanthan, A., Bernstein, G. M., Rennert, P. D., Gray, G. S., Gribben, J. G. and Nadler, L. M. (1995) Immunity **2**, 1–20

67 Kuchroo, V. K., Das, M. P., Brown, J. A., Ranger, A. M., Zamvil, S. S., Sobel, R. A., Weiner, H. L., Nabavi, N. and Glimcher, L. H. (1995) Cell **80**, 707–718

68 Levine, B. L., Ueda, Y., Craighead, N., Huang, M. L. and June, C. H. (1995) Int. Immunol. **7**, 891–904

69 Lanier, L. L., O'Fallon, S., Somoza, C., Phillips, J. H., Linsley, P. S., Okumura, K., Ito, D. and Azuma, M. (1995) J. Immunol. **154**, 97–105

70 Ueda, Y., Levine, B., Freeman, G. J., Nadler, L. M., June, C. H. and Ward, S. G. (1995) Int. Immunol. **7**, 957–966

71 Weiss, A. and Littman, D. R. (1994) Cell **76**, 263–274

72 Cantrell, D. A. (1996) Annu. Rev. Immunol. **14**, 259–274

73 Samelson, L. E. and Klausner, R. D. (1992) J. Biol. Chem. **267**, 24913–24916

74 Park, D., Rho, H. and Rhee, S. G. (1991) Proc. Natl. Acad. Sci U.S.A. **99**, 5453–5456

75 Berridge, M. J. (1993) Nature (London) **361**, 315–325

76 Izquierdo, M., Reif, K. and Cantrell, D. A. (1995) Immunol. Today **16**, 159–164

77 Kapeller, R and Cantley, L. C. (1994) BioEssays **16**, 565–576

78 Carrera, A., Rodriguez-Borlado, L., Martinez-Alonso, C. and Merida, I. (1994) J. Biol. Chem. **269**, 19435–19440

79 Ward, S. G., Reif, K., Ley, S., Fry, M. J., Waterfield, M. D. and Cantrell, D. A. (1992) J. Biol. Chem. **267**, 23862–23869

80 June, C. H. Ledbetter, J. A., Gillespie, M. M., Lindstenm, T. and Thompson, C. B. (1987) Mol. Cell. Biol. **7**, 4472–4481

81 Sigal, N. and Dumont, F. J. (1992) Annu. Rev. Immunol. **10**, 509–560

82 Vandenberghe, P., Freeman, G. J., Nadler, L. M., Fletcher, M. C., Kamoun, M., Turka, L. A., Ledbetter, J. A., Thompson, C. B. and June, C. H. (1992) J. Exp. Med. **175**, 951–960

83 Lu, Y., Granelli-Piperno, A., Bjorndahl, J. M., Phillips, C. A. and Trevillyan, J. M. (1992) J. Immunol. **149**, 24–29

84 Ledbetter, J. A., Parson, M., Martin, P. J., Hansen, J. A., Rabinovitch, P. S. and June, C. H. (1986) J. Immunol. **137**, 3299–3305

85 Weiss, A., Manger, B. and Imboden, J. (1986) J. Immunol. **137**, 819–825

86 Ledbetter, J. A., Imboden, J. B., Schieven, G. L., Grosmaire, L. S., Rabinovitch, P. S., Lindsten, T., Thompson, C. B. and June, C. H. (1990) Blood **75**, 1531–1539

87 Ledbetter, J. A. and Linsley, P. S. (1992) Adv. Exp. Med. Biol. **323**, 23–27

88 Nunes, J., Klasen, S., Franco, M. D., Lipcey, C., Mawas, C., Bagnasco, M. and Olive, D. (1993) Biochem. J. **293**, 835–842

89 Ueda, Y., Freeman, G., Levine, B., Ward, S. G., Huang, M. L., Abe, R., Nadler, L. M. and June, C. H. (1994) Clin. Res. **42**, 309

90 Pages, F., Ragueneau, M., Rottapel, R., Truneh, A., Nunes, J., Imbert, J. and Olive, D. (1994) Nature (London) **369**, 327–329

91 Ward, S. G., Westwick, J., Hall, N. and Sansom, D. M. (1993) Eur. J. Immunol. **23**, 2572–2577

92 Ward, S. G., Wilson, A., Turner, L., Westwick, J. and Sansom, D. M. (1995) Eur. J. Immunol. **25**, 526–532

93 Abe, R., Vandenberge, P., Craighead, N., Smoot, D., Lee, K. P. and June, C. H. (1995) J. Immunol. **154**, 985–997

94 Van Lier, R. A. W., Brouer, M., De Groot, E., Kramer, I., Aarden, L. A. and Verhoeven, A. J. (1991) Eur. J. Immunol. **21**, 1775–1778

95 McCormick, F. (1993) Nature (London) **363**, 15–16

96 Nunes, J., Collette, Y., Truneh, A., Olive, D. and Cantrell, D. A. (1994) J. Exp. Med. **180**, 1067–1076

97 Nunes, J., Battifora, M., Woodgett, J. R., Truneh, A., Olive, D. and Cantrell, D. A. (1996) Mol. Immunol. **33**, 63–70

98 Stoyanov, B., Volinia, S., Hanck, T., Rubio, I., Loubtchenkov, M., Malek, D., Stoyanova, S., Vanhaesebroeck, B., Dhand, R., et al. (1995) Science **269**, 690–692

99 Volinia, S., Dhand, R., Vanhaesebroeck, B., MacDougall, L. K., Stein, R., Zvelebil, M. J., Domin, J., Panaretou, C. and Waterfield, M. D. (1995) EMBO J. **14**, 3339–3348

100 Otsu, M., Hiles, I., Gout, I., Fry, M. J., Ruiz-Larrea, F., Panayatou, G., Thompson, A., Dhand, R., Hsuan, J., Totty, N., et al. (1991) Cell **65**, 91–104

101 Hiles, I. D., Otsu, M., Volinia, S., Fry, M. J., Gout, I., Dhand, R., Panayatou, G., Ruiz-Larrea, F., Thompson, A., Totty, N., et al. (1992) Cell **70**, 419–429

102 Dhand, R., Hiles, I., Panayotou, G., Roche, S., Fry, M. J., Gout, I., Totty, N. F., Truong, O., Vicendo, P., Yonezawa, K., et al. (1994) EMBO J. **13**, 522–533

103 Stephens, L., Jackson, T. and Hawkins, P. (1993) Biochim. Biophys. Acta **1179**, 27–75

104 Toker, A., Meyer, M., Reddy, K., Falck, J. R., Aneja, R., Aneja, S., Parra, A., Burns, D., Ballas, L. M. and Cantley, L. C. (1994) J. Biol. Chem. **269**, 32358–32367

105 Nakanishi, H., Brewer, K. A. and Exton, J. H. (1993) J. Biol. Chem. **268**, 13–16

106 Akimoto, K., Takahashi, R., Moriya, S., Nishioka, N., Takayanagi, J., Kimura, K., Fukui, Y., Osada, S., Mizuno, K., Hirai, S., Kazlauskas, A. and Ohno, S. (1996) EMBO J. **15**, 788–798

107 Moriya, S., Kazlauskas, A., Akimoto, K., Hirai, S., Mizuno, K., Takenawa, T., Fukui, Y., Watanabe, Y., Ozaki, S. and Ohno, S. (1996) Proc. Natl. Acad. Sci. U.S.A. **93**, 788–798

108 Franke, T. F., Yang, S. I., Chan, T. O., Datta, K., Kazlauskas, A., Morrison, D. K., Kaplan, D. R. and Tsichlis, P. N. (1995) Cell **81**, 727–736

109 Burgering, B. T. and Coffer, P. J. (1995) Nature (London) **376**, 599–602

110 Chung, J., Grammer, T., Lemon, C., Kazlauskas, A. and Blenis, J. (1994) Nature (London) **370**, 71–73

111 Weng, Q. P., Andrabi, K., Klippel, A., Kozlowski, M. T., Williams, L. T. and Avruch, J. (1995) Proc. Natl. Acad. Sci. U.S.A. **92**, 5744–5748

112 Hawkins, P. T., Eguinoa, A., Qui, R. G., Stokoe, D., Cooke, F. T., Walters, R., Wennstrom, S., Claesson-Welsch, L., Evans, T., Symons, M. and Stephens, L. (1995) Curr. Biol. **5**, 393–403

113 Rameh, L. E., Chen, C. S. and Cantley, L. C. (1995) Cell **83**, 821–830

114 Ghioti-Ragueneau, M., Battifora, M., Truneh, A., Waterfield, M. D. and Olive, D. (1996) Eur. J. Immunol. **26**, 34–41

115 Truitt, K., Hicks, C. M. and Imboden, J. B. (1994) J. Exp. Med. **179**, 1071–1076

116 Prasad, K. V., Cai, Y., Raab, M., Duckworth, B., Cantley, L., Shoelson, S. and Rudd, C. R. (1994) Proc. Natl. Acad. Sci. U.S.A. **91**, 2834–2838

117 August, A., and Dupont, B. (1994) Int. Immunol. **6**, 769–774

118 Kolesnik, R. N. and Fuks, Z. (1995) J. Exp. Med. **181**, 1949–1952

119 Hannun, Y. A. and Obeid, L. M. (1995) Trends Biochem. Sci. **20**, 73–77

120 Lozano, J., Berra, K., Municio, M. M., Diaz-Meco, M. T., Dominguez, I., Sanz, L. and Moscat, J. (1994) J. Biol. Chem. **269**, 19200–19202

121 Mathias, S., Dressler, K. A. and Kolesnick, R. N. (1992) Proc. Natl. Acad. Sci. U.S.A. **88**, 10009–10013

122 Dobrowski, R. T. and Hannun, Y. A. (1992) J. Biol. Chem. **267**, 5048–5051

123 Cifone, M. C., De Maria, R., Roncailoi, R., Rippo, M. R., Azuma, M., Lewis, L. L., Santoni, A. and Testi, R. (1994) J. Exp. Med. **177**, 1547–1552

124 Schutze, S., Potthof, K., Macleidt, T., Berkovic, D., Wiegmann, K. and Kronke, M. (1992) Cell **71**, 765–776

125 Boucher, L. M., Wiegman, K., Futterer, A., Pfeffer, K., Mak, T. W. and Kronke, M. (1995) J. Exp. Med. **181**, 2059–2068

126 Chan, G., and Ochi, A. (1995) Eur. J. Immunol. **25**, 1999–2004

127 Cleveland, J. L. and Ihle, J. N. (1995) Cell **81**, 479–482

128 Los, M., Schenk, H., Hexel, K., Baeuerle, P. A., Dronge, W. and Schulz-Osthoff, K. (1995) EMBO J. **14**, 3731–3740

129 Rudd, C. E., Janssen, O., Cai, Y. C., da Silva, A., Raab, M. and Prasad, K. V. S. (1994) Immunol. Today **15**, 225–234

130 Mahon, T. and O'Neill, L. A.J (1995) J. Biol. Chem. **270**, 28557–28564

131 August, A. and Dupont, B. (1994) Biochem. Biophys. Res. Commun. **199**, 1466–1473

132 Hutchcroft, J. and Bierer, B. (1994) Proc. Natl. Acad. Sci. U.S.A. **91**, 3260–3264

133 August, A., Gibson, S., Kawakami, Y., Kawakami, T., Mills, G. and Dupont, B. (1994) Proc. Natl. Acad. Sci. U.S.A. **91**, 9347–9351

134 Hu, Q., Davidson, D., Schwartzberg, P. L., Macchiarini, F., Lenardo, M., Bluestone, G. and Matis, L. (1995) J. Biol. Chem. **270**, 1928–1934

135 Raab, M., Cai, Y. C., Bunnell, S. C., Heyeck, S., Berg, L. J. and Rudd, C. E. (1995) Proc. Natl. Acad. Sci. U.S.A. **92**, 8891–8895

136 Gibson, S., August, A., Branch, D., Dupont, D. and Mills, G. B. (1996) J. Biol. Chem. **271**, 7079–7083

137 Lu, Y., Phillips, C. A., Bjorndahl, J. M. and Trevillyan, J. M. (1994) Eur. J. Immunol. **24**, 2732–2739

138 Stein, P. H., Fraser, J. D. and Weiss, A. (1994) Mol. Cell. Biol. **14**, 3392–3402

139 Yano, H., Nakanishi, S., Kimura, K., Hanai, N., Saitoh, Y., Fukui, Y., Nonomura, Y. and Matsuda, Y. (1993) J. Biol. Chem. **268**, 25846–25856

140 Ui, M., Okada, T., Hazeki, K. and Hazeki, O. (1995) Trends Biochem. Sci. **20**, 303–307

141 Norman, B. H., Paschal, J. and Vlahos, C. J. (1995) Bioorg. Med. Chem. Lett. **5**, 1183–1186

142 Wymann, M. P., Bugaelli-Leva, G., Zvelebil, M. J., Pirola, L., Vanhaesebroeck, B., Waterfield, M. D. and Panayotou, G. (1996) Mol. Cell. Biol. **16**, 1722–1733

143 Ward, S. G., June, C. H. and Olive, D. (1996) Immunol. Today **17**, 187–197

144 Nakanishi, S., Catt, K. and Balla, T. (1995) Proc. Natl. Acad. Sci. U.S.A. **92**, 5317–5321

145 Cross, M. J., Stewart, A., Hodgkin, M. N., Kerr, D. J. and Wakelam, M. J. O. (1995) J. Biol. Chem. **270**, 25352–25355

146 Lu, Y., Phillips, C. A. and Trevillyan, J. M. (1995) Eur. J. Immunol. **25**, 533–537

147 Hutchcroft, J. E., Franklin, D. P., Tsai, B., Harrison-Findik, D., Varticovski, L. and Bierer, B. E. (1995) Proc. Natl. Acad. Sci. U.S.A. **92**, 8808–8812

148 Truitt, K. E., Shi, J., Gibson, S., Segal, L. G., Mills, G. B. and Imboden, J. B. (1995) J. Immunol. **155**, 4702–4710

149 Vlahos, C. J., Matter, W. F., Hui, K. Y. and Brown, R. F. (1994) J. Biol. Chem. **269**, 5241–5248

150 Cai, Y. C., Cefai, D., Schneider, H., Raab, H., Nabavi, N. and Rudd, C. E. (1995) Immunity **3**, 417–426

151 Casey-Crooks, M., Littman, D. R., Carter, R. H., Fearon, D. T., Weiss, A. and Stein, P. H. (1995) Mol. Cell. Biol. **15**, 6820–6828

152 Pages, F., Ragueneau, M., Klasen, S., Battifora, M., Couez, D., Sweet, R., Truneh, A., Ward, S. G. and Olive, D. (1996) J. Biol. Chem. **271**, 9403–9409

153 Shimizu, Y., van Seventer, G. A., Ennis E., Newman, W., Horgan, K. J. and Shaw, S. (1992) J. Exp. Med. **175**, 577–582

154 Zell, T., Hunt, S. W., Mobley, J. L., Finkelstein, L. D. and Shimizu, Y. (1996) J. Immunol. **156**, 883–886

155 Ward, S. G., Ley, S., MacPhee, C. and Cantrell, D. A. (1992) Eur. J. Immunol. **22**, 45–49

156 Prasad, K. V., Kapeller, R., Janssen, O., Repke, H., Duke-Cohan, J., Cantley, L. and Rudd, C. (1993) Mol. Cell. Biol. **13**, 7708–7717

157 Thompson, P. A., Gutkind, J. S., Robbins, K. C., Ledbetter, J. A. and Bolen, J. B. (1992) Oncogene **7**, 719–725

158 Ward, S. G., Parry, R., Lefeuvre, C., Sansom, D. M., Westwick, J. and Lazarovits, A. (1995) Eur. J. Immunol. **25**, 502–507

159 Remillard, B., Petrillo, R., Maslinski, W., Tsudo, M., Strom, T. B., Cantley, L. and Varticovski, L. (1991) J. Biol. Chem. **266**, 14167–14170

160 Augustine, J. A., Sutor, S. L. and Abraham, R. T. (1991) Mol. Cell. Biol. **11**, 4431–4440

161 Karnitz, L. M., Burns, L. A., Sutor, S. L., Blenis, J. and Abraham, R. T. (1995) Mol. Cell. Biol. **15**, 3049–3057

162 Reif, K., Gout, I., Waterfield, M. D. and Cantrell, D. A. (1993) J. Biol. Chem. **268**, 10780–10788

163 Parker, P. (1995) Curr. Biol. **5**, 577–579

164 Lam, K., Carpenter, C. L., Ruderman, N. B., Friel, J. C. and Kelly, K. L. (1994) J. Biol. Chem. **269**, 20648–20652

165 Richard, S., Yu, D., Blumer, K. J., Hausladen, D., Olszowy, M. W., Connelly, P. A. and Shaw, A. S. (1995) Mol. Cell. Biol. **15**, 186–187

166 Nunes, J., Truneh, A., Olive, D. and Cantrell, D. A. (1996) J. Biol. Chem. **271**, 1591–1598

167 Hubert, P., Debre, P., Boumsell, L. and Bismuth, G. (1993) J. Exp. Med. **178**, 1587–1596

168 Settleman, J., Narasimhan, V., Foster, L. C. and Weinberg, R. A. (1992) Cell **69**, 539–549

169 Takai, Y., Sasaki, T., Tanaka, K. and Nakanishi, H. (1995) Trends Biochem. Sci. **20**, 227–231

170 Hill, C. S., Wynne, J. and Treisman, R. (1995) Cell **81**, 1159–1170

171 Chong, L. D., Traynor-Kaplan, A., Bokoch, G. M. and Schwartz, M. A. (1994) Cell **79**, 507–513

172 Zhang, J., King, W. G., Dillon, S., Hall, A., Feig, L. and Rittenhouse, S. E. (1993) J. Biol. Chem. **268**, 22251–22254

173 Schneider, H., Cai, Y. C., Prasad, K. V. S., Shoelson, S. E. and Rudd, C. E. (1995) Eur. J. Immunol. **25**, 1044–1050

174 Nunes, J., Klasen, S., Ragueneau, M., Pavon, C., Couez, D., Mawas, C., Bagnasco, M. and Olive, D. (1993) Int. Immunol. **5**, 311–315

175 Rodriguez-Viciana, P., Warne, P. H., Dhand, R., Van Haesebroeck, B., Gout, I., Fry, M. J., Waterfield, M. D. and Downward, J. (1994) Nature (London) **370**, 494–499

176 Kodaki, T., Woscholski, R., Hallberg, B., Rodriguez-Viciana, P., Downward, J. and Parker, P. (1994) Curr. Biol. **4**, 798–806

177 Rodriguez-Viciana, P., Warne, P. H., Van Haesebroeck, B., Waterfield, M. D. and Downward, J. (1996) EMBO J. **15**, 2442–2451

178 Hu, Q., Klippel, A., Muslin, A., Fantl, W. and Williams, L. T. (1995) Science **268**, 100–102

179 DePaolo, D., Reusch, J. E. B., Carel, K., Bhuripanyo, P., Leitner, J. W. and Draznin, B. (1996) Mol. Cell. Biol. **16**, 1450–1457

180 Jackman, J. K., Motto, D. G., Sun, Q., Tanemoto, M., Turck, C. W., Peltz, G. A., Koretzky, G. A. and Findell, P. R. (1995) J. Biol. Chem. **270**, 7029–7032

181 Motto, D. G., Ross, S. E., Jackman, J. K., Sun, Q., Olson, A. L., Findell, P. R. and Koretzky, G. A. (1994) J. Biol. Chem. **269**, 21608–21613

182 Meisner, H., Conway, B. R., Hartley, D. and Czech, M. P. (1995) Mol. Cell. Biol. **15**, 3571–3578

183 Donovan, J. A., Wange, R. L., Langdon, W. A. and Samelson, L. E. (1994) J. Biol. Chem. **269**, 22921–22924

184 Wang, J., Auger, K. R., Jarvis, L., Shi, Y. and Roberts, T. M. (1995) J. Biol. Chem. **270**, 12774–12780

185 Sung, C. K., Sanchez-Margalet, V. and Goldfine, I. D. (1994) J. Biol. Chem. **269**, 12503–12507

186 Hughes, C. W. and Pober, J. S. (1996) J. Biol. Chem. **271**, 5369–5377

187 Fraser, J. D., Irving, B. A., Crabtree, G. R. and Weiss, A. (1991) Science **251**, 311–316

188 Fraser, J. D. and Weiss, A. (1992) Mol. Cell. Biol. **12**, 4357–4363

189 Liou, H. and Baltimore, D. (1993) Curr. Opin. Cell Biol. **5**, 477–487

190 Ghosh, P., Tan, T., Rice, N. R., Sica, A. and Young, H. A. (1993) Proc. Natl. Acad. Sci. U.S.A. **90**, 1696–1700

191 Sha, W. C., Liou, H., Tuomanen, E. I. and Baltimore, D. (1995) Cell **80**, 321–330

192 Henkel, T., Machleidt, T., Alkalay, I., Kronke, M., Ben-Neriah, Y. and Baeuerle, P. A. (1993) Nature (London) **365**, 182–185

193 Bryan, R. G., Li, Y., Lai, J. H., Van, M., Rice, N., Rich, R. and Tan, T. H. (1994) Mol. Cell. Biol. **14**, 7933–7942

194 Lai, J. H. and Tan, T. H. (1994) J. Biol. Chem. **269**, 30077–30080

195 Hannun, Y. A. (1994) J. Biol. Chem. **269**, 3125–3128

196 Downward, J. (1994) Nature (London) **371**, 378–379

197 Pai, S., Calvo, V., Wood, M. and Bierer, B. (1994) Eur. J. Immunol. **24**, 2364–2368

198 Kuo, C. J., Chung, J., Fiorentino, D. F., Flanagan, W. M., Blenis, J. and Crabtree, G. R. (1992) Nature (London) **358**, 70–73

199 Parry, R. V. and Ward, S. G. (1996) Biochem. Soc. Trans. **24**, 88S

200 Lane, H. A., Fernandez, A., Lamb, N. J. C. and Thomas, G. (1993) Nature (London) **370**, 71–75

201 Reinhard, C., Fernandez, A., Lamb, N. J. C. and Thomas, G. (1994) EMBO J. **13**, 1557–1565

202 Berra, E., Diaz-Meco, M. T., Lozano, J., Frutos, S., Municio, M. M., Sanchez, P., Sanz, L. and Moscat, J. (1995) EMBO J. **14**, 6157–6163

203 Genot, E., Parker, P. and Cantrell, D. A. (1995) J. Biol. Chem. **270**, 9833–9839

204 Baier-Bitterlich, G., Uberall, F., Bauer, B., Fresser, F., Wachter, H., Grunicke, H., Uterman, G., Altman, A. and Baier, G. (1996) Mol. Cell. Biol. **16**, 1842–1850

205 Cano, E. and Mahadevan, L. C. (1995) Trends Biochem. Sci. **20**, 117–122

206 Lin., A., Minden, A., Martinetto, H., Claret, F. X., Lange-Carter, C., Mercurio, F., Johnson, G. L. and Karin, M. (1995) Science **268**, 286–288

207 Kunz, J., Henriquez, R., Schneider, U., Deuter-Reinhard, M., Movva, N. R. and Hall, M. N. (1993) Cell **73**, 585–596

208 Brown, E., Albers, M., Shin, T. B., Ichikawa, K., Keith, C. T., Lane, W. S. and Schreiber, S. L. (1994) Nature (London) **369**, 756–758

209 Brown, E. J., Beal, P. A., Keith, C. T., Chen, J., Shin, T. B. and Schreiber, S. L. (1995) Nature (London) **377**, 441–446

210 Angel, P. and Karin, M. (1991) Biochim. Biophys. Acta **1072**, 129–157

211 Rincon, M. and Flavell, R. A. (1994) EMBO J. **13**, 4370–4381

212 Kang, S., Beverly, B., Tran, A., Brorson, K., Schwartz, R. H. and Lenardo, M. J. (1992) Science **257**, 1134–1138

213 Los, M., Dronge, W. amd Schulze-Osthoff, K. (1994) Biochem. J. **302**, 119–123

214 Minden, A., Lin, A., Claret, F. X., Abo, A. and Karin, M. (1995) Cell **81**, 1147–1157

215 Su, B., Jacinto, E., Hibi, M., Kailunki, T., Karin, M. and Ben-Neriah, Y. (1994) Cell **77**, 727–736

216 Coso, O. A., Chiariello, M., Yu, J. C. Teramoto, H., Crespo, P., Xu, N., Miki, T. and Gutkind, J. S. (1995) Cell **81**, 1137–1146

217 Raines, M. A., Kolesnick, R. N. and Golde, D. W. (1993) J. Biol. Chem. **268**, 14572–14575

218 Westwick, J. K., Bielawska, A. E., Dbaibo, G., Hannun, Y. A. and Brenner, D. A. (1995) J. Biol. Chem. **270**, 22689–26692

219 Westwick, J. K., Bielawska, A. E., Dbaibo, G., Hannun, Y. A. and Brenner, D. A. (1995) J. Biol. Chem. **270**, 22689–26692

220 Kyriakis, J., Banerjee, P., Nikolakaki, E., Dal, T., Rubie, E., Ahmad, A., Avruch, J. and Woodgett, J. R. (1994) Nature (London) **369**, 156–159

221 Verheij, M., Bose, R., Lin, X. H., Yao, B., Jarvis, W. D., Grant, S., Birrer, M. J., Szabo, E., Zon, L. I., Kyriakis, J. M., et al. (1996) Nature (London) **380**, 75–79

222 Xia, Z., Dickens, M., Raingeaud. J., Davis, R. J. and Greenberg, M. E. (1995) Science **270**, 1326–1331

Biochem. J. (1996) **318**, 729–747 (Printed in Great Britain)

REVIEW ARTICLE

Signal-transducing protein phosphorylation cascades mediated by Ras/Rho proteins in the mammalian cell: the potential for multiplex signalling

David T. DENHARDT

Department of Biological Sciences, P.O. Box 1059, Rutgers University, Piscataway, NJ 08855, U.S.A.

The features of three distinct protein phosphorylation cascades in mammalian cells are becoming clear. These signalling pathways link receptor-mediated events at the cell surface or intracellular perturbations such as DNA damage to changes in cytoskeletal structure, vesicle transport and altered transcription factor activity. The best known pathway, the Ras → Raf → MEK → ERK cascade [where ERK is extracellular-signal-regulated kinase and MEK is mitogen-activated protein (MAP) kinase/ERK kinase], is typically stimulated strongly by mitogens and growth factors. The other two pathways, stimulated primarily by assorted cytokines, hormones and various forms of stress, predominantly utilize p21 proteins of the Rho family (Rho, Rac and CDC42), although Ras can also participate. Diagnostic of each pathway is the MAP kinase component, which is phosphorylated by a unique dual-specificity kinase on both tyrosine and threonine in one of three motifs (Thr-Glu-Tyr, Thr-Pro-Tyr or Thr-Gly-Tyr), depending upon the pathway. In addition to activating one or more protein phosphorylation cascades, the initiating stimulus may also mobilize a variety of other signalling molecules (e.g. protein kinase C isoforms, phospholipid kinases, G-protein α and $\beta\gamma$ subunits, phospholipases, intracellular Ca^{2+}). These various signals impact to a greater or lesser extent on multiple downstream effectors. Important concepts are that signal transmission often entails the targeted relocation of specific proteins in the cell, and the reversible formation of protein complexes by means of regulated protein phosphorylation. The signalling circuits may be completed by the phosphorylation of upstream effectors by downstream kinases, resulting in a modulation of the signal. Signalling is terminated and the components returned to the ground state largely by dephosphorylation. There is an indeterminant amount of cross-talk among the pathways, and many of the proteins in the pathways belong to families of closely related proteins. The potential for more than one signal to be conveyed down a pathway simultaneously (multiplex signalling) is discussed. The net effect of a given stimulus on the cell is the result of a complex intracellular integration of the intensity and duration of activation of the individual pathways. The specific outcome depends on the particular signalling molecules expressed by the target cells and on the dynamic balance among the pathways.

INTRODUCTION

The 21 kDa mammalian Ras proteins (p21) consist of 188 or 189 amino acids. They are integral to signal transduction pathways connecting events at many cell surface receptors to intracellular processes. Mammalian cells contain three very similar *ras* genes (H-*ras*, N-*ras* and K-*ras*, which has two splice variants, A and B, that differ in their C-terminal sequences). In this review, Ras is used to refer collectively to the products of these three genes, even though in different cell types one or the other may predominate. At the protein level Ras is present in all cells, with the highest levels in proliferating cells. At the mRNA level H-*ras* is highest in skin and muscle, K-*ras* in gut and thymus, and N-*ras* in thymus and testes. For signalling purposes, there are no known differences among these Ras proteins, though cell-specific differences in the transforming potential of the codon 12 Gly → Asp mutants have been reported [1]. Recent reviews of the Ras-controlled signalling pathways contain references to background material that is assumed or summarized here [2–4].

These three *ras* genes are part of a large superfamily of genes encoding small GTP-binding proteins (G-proteins) [5]. Closely related to Ras are several species of Rap proteins found in granules in the Golgi and endoplasmic reticulum. Rap1A can antagonize the transforming ability of Ras, a function that led to its isolation as a suppressor, called Krev-1, of the K-*ras* oncogene. Also closely related to Ras are the Ral-A and Ral-B proteins, which appear to regulate the activity of exocytic and endocytic vesicles. This Ras/Rap/Ral group is important in growth and development, and certain members are implicated in exocytosis, anabolic processes and/or regulation of the oxidative burst. Less closely related to Ras are several other families in the Ras superfamily [5–10]. The Rho family, which includes Rho-A, -B and -C, Rac-1 and -2, CDC42, Rho-G and TC10, embraces small G-proteins that play dynamic roles in the regulation of the actin cytoskeleton and focal contacts, mediating formation of filopodia and lamellipodia. Rac also controls NADPH oxidase activity in phagocytes. Ran proteins are involved in the transport of RNA and proteins across the nuclear membrane. ARF/SAR proteins are important for vesicle formation and budding. Members of the large and extensively studied Rab/YPT family are involved both in regulating intracellular vesicle trafficking between donor and acceptor membrane-enclosed compartments and in controlling the exocytosis and endocytosis of different types of vesicles.

Abbreviations used: see Table 1 for the names and abbreviations of most of the signalling intermediates mentioned in this review; CREB, cAMP response element-binding protein; GAP, GTPase-activating protein; GEF, guanine nucleotide exchange factor; GSK-3, glycogen synthase kinase-3; Hsp90, heat-shock protein of 90 kDa; IL-1, interleukin-1; JAK, Janus kinase; LPS, lipopolysaccharide; MAPK, mitogen-activated protein kinase; NFκB, nuclear factor κB; PH domain, pleckstrin homology domain; PI3-K, phosphoinositide 3-kinase; PKA, protein kinase A; PKC, protein kinase C; PLC and PLD, phospholipases C and D respectively; PTB domain, phosphotyrosine-binding domain; SH domain, Src homology domain; SRF, serum response factor; STAT, signal transducer and activator of transcription; TNFα, tumour necrosis factor α.

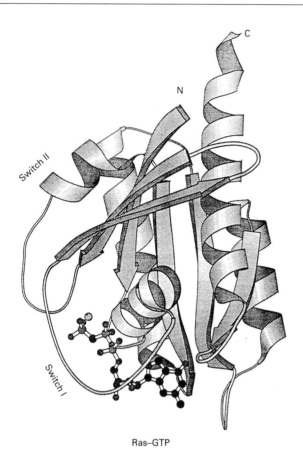

Ras–GTP

Figure 1   Structure of p21 Ras–GTP

See [11] and the text for details. Figure generously provided by Dr. Sung-Hou Kim.

Figure 1 illustrates the structure of the GTP–Ras protein [11–13]. Presumably all members of the Ras superfamily have a very similar structure. Notable features include the highly conserved catalytic domain made up of amino acids 1–164, the GTP-binding site (constituted from amino acids 12–18, 57–63, 116–119 and 144–147) and the two 'switch' regions (amino acids Asp-30–Asp-38 and Gly-60–Glu-76). The switch regions of the Ras protein are close to the γ-phosphate group of the activating GTP and exhibit different conformations depending upon whether GDP or GTP is bound. This N-terminal proximate region is the effector binding site (involving amino acids Tyr-32–Tyr-40) responsible in part for interactions with the GTPase-activating protein (GAP) and with downstream effectors of Ras action such as Raf, which competes with GAP for binding to Ras. The two Ras-GAPs, p120Ras[GAP] and neurofibromin, appear to interact somewhat differently with Ras in that substitution of the amino acids Lys-Arg-Val with Leu-Ile-Arg at positions 101–103 had no effect on the intrinsic hydrolytic activity of Ras or its sensitivity to neurofibromin, but did reduce p120Ras[GAP]-stimulated hydrolysis 2–3-fold [14]. The interaction of the guanine nucleotide exchange factor (GEF) with Ras is, in part, with the switch 2 region [15].

The Ras proteins are bound to the inner surface of the plasma membrane by several lipophilic interactions involving post-translational modifications of the C-terminus [5,12]. A cysteine residue very near to the C-terminus, in a CAAX (where A = aliphatic amino acid) motif, typically has a polyisoprenyl moiety attached to it, either C_{15} (farnesyl) for Ras or C_{20} (geranylgeranyl)

for some Rap proteins and members of the Rho superfamily. (These modifications are determined by the specific C-terminal amino acid sequences and they may determine the specific cell membrane targeted by the protein.) Subsequently the three C-terminal amino acids are removed and the newly generated farnesylated Ras C-terminal cysteine is carboxymethylated. H-Ras, N-Ras and K-Ras(A) have in addition a palmitoyl group added to a cysteine residue in the hypervariable region (amino acids 165–185) near the C-terminus; K-Ras(B) does not have a cysteine residue that can be palmitoylated, but does have a polybasic domain that fulfils a similar function, presumably by forming an amphipathic helix.

Activation of Ras induces proliferation in many types of cells, and in its mutant oncogenic form Ras transforms many immortal cell lines, conferring on them a malignant phenotype [16]. Oncogenic Ras can also co-operate with an 'immortalizing oncogene' to transform primary cells. In some cell types, e.g. pheochromocytoma PC12 cells, Ras activation induces differentiation and neurite outgrowth. Oncogenic Ras bearing a mutation that alters the Ras structure so as to prolong the lifetime of the active GTP-bound form (in amino acids 12, 13, 59, 61 and 63 in the phosphate-binding region) generates a constitutive signal that is a major factor in many human cancers. Mutations that increase the guanine nucleotide exchange rate (amino acids 116, 117, 119 and 146) are in the base-binding region and also produce a transforming Ras. Ras is also responsible for transmitting the proliferative signal generated by a number of oncogenes, including the *src*, *fms*, and *fes* oncogenes.

ACTIVATION OF RAS- AND RHO-GTPase CASCADES

Ras and its relatives are activated in response to an extracellular or intracellular signal that generates the GTP-bound form and energizes the signal transducing ability. Typically the Ras–GTP level will increase 2–3-fold and remain high for at least 30 min after receptor activation. Hydrolysis of the bound GTP by an intrinsic GTPase activity relaxes the conformation and terminates the signal. Figure 2 illustrates the cycling of Ras between these two forms. Table 1 lists many of the proteins discussed in this review.

GAPs enhance the GTPase activity of normal Ras (but typically not oncogenic Ras) and thus shorten the lifetime of the signalling form. Presumably Ras–GTP continues to signal as long as the GTP remains intact, although association with GAP may alter the signal. The mammalian GAPs include the well studied p120Ras[GAP] and the quite distinct and larger (~ 290 kDa) neurofibromin, NF1, whose expression is largely restricted to neural tissues (including neural-crest-derived tissues such as Schwann cells) and gonadal tissues [17]. There are two members of the GAP1 family, p100GAP1[m] and GAP1[IP4BP], which via a PH (pleckstrin homology) domain bind to and are stimulated by inositol 1,3,4,5-tetrakisphosphate [18,19]. GAPs active on Rho family members include Ral-BP1 (CDC42-GAP), p190Rho[GAP] and BCR, which are respectively preferentially active on CDC42, Rho and Rac [10,20] (see Table 1 and Figure 3).

Studies with homozygous null mice ('knock-outs') have revealed the importance of some of these proteins. Animals unable to make p120Ras[GAP] are embryonic lethals that exhibit increased Ras–GTP signalling and possess major defects in endothelial cell organization [21]. Mice unable to produce NF1 also die as embryos. It appears that one function of NF1 is to promote the death of certain neurons when an appropriate signal from certain nerve growth factors (neurotrophins) is not received [22].

GEFs [also known as guanine nucleotide releasing proteins, or guanine nucleotide dissociation stimulators (GDSs)], catalyse

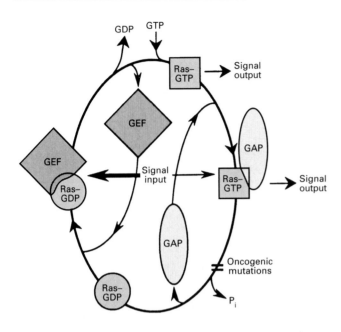

Figure 2 The Ras cycle

Inactive, 'nucleotide-free', Ras (top of cycle), which may exist only as a transition state during which GTP is replacing the GDP, associates with GTP and is activated as a consequence. Ras–GTP continues to interact with and activate effector proteins (as discussed in the text) until GAP enforces hydrolysis of the GTP by the low-level intrinsic Ras-GTPase activity. The inactive Ras–GDP complex associates with a GEF, which when activated catalyses the exchange of GTP for GDP, again activating Ras. The signal from activated Ras is transmitted by protein–protein interactions, which may differ depending upon whether GAP is bound or not. Input signals leading to Ras activation are thought to be transmitted more often via GEF and less often via GAP. See the text for references and further details.

the release of GDP from inactive p21 proteins complexed with GDP; they are major players in signal activation [10,15,22]. Release of GDP allows GTP, present at much higher levels than GDP in the healthy cell, to bind and activate the p21 protein once again. SOS1 and SOS2, CDC25 and C3G are the four known mammalian GEFs with activity towards Ras–GDP; DBL, OST and smg-GDS are known GEFs acting to activate various Rho family members. DBL, isolated from a diffuse B-cell lymphoma, stimulates exchange activity for Rho-A, Rac-1 and CDC42 [15]. OST, isolated from an osteosarcoma, is an effector for Rac-1, acting as a GEF for Rho-A and CDC42, but not for Rho-B or Rho-C [23]. A domain found in DBL, known as the dibble homology domain, along with a PH domain, appears to be characteristic of the Rho-GEFs. To what extent different GEFs act differentially on members of the Ras and Rho (including Rac and CDC42) families *in vivo*, perhaps co-ordinately, remains to be established. Their importance is underscored by the fact that several known or putative Rho-GEFs are proto-oncoproteins (e.g. DBL, VAV, OST) that can be activated by N-terminal truncations. VAV is a 95 kDa protein expressed in haematopoietic cells that is tyrosine phosphorylated in response to the activation of a variety of receptors (e.g. the T-cell antigen receptor, the IgM receptor or c-Kit). It has a dibble domain, suggesting it may be a GEF, and it is capable of binding both to tubulin and to the nuclear protein Ku-70, the DNA-binding element of the DNA-dependent protein kinase [24,25].

Rho-GDI (guanine nucleotide dissociation inhibitor) binds equally well to both the GDP- and GTP-bound forms of CDC42, inhibiting both GDP dissociation and GTP hydrolysis [25a]. It inhibits the interaction of the p21Rac, Rho and CDC42 proteins

with GAP and stimulates their release from cell membranes, perhaps to allow them to relocate to another membrane site [10].

The interactions among Ras, Ral, CDC42, Rac and Rho and both the downstream effectors of their actions and their diverse upstream activators, particularly physical and chemical stress, are poorly understood [9,10,15]. Via their respective and sometimes shared GEFs and GAPs, they have the potential to socialize as suggested in Figure 3, although which interactions are important is likely to depend on the abundance and location of relevant proteins in the cell in question. An important consequence of these shared interactions may be to place a particular GAP or GEF in a location where it can act more efficiently on its target p21 protein. p120Ras[GAP] interacts with p190Rho[GAP], which can deactivate GTP-bound Rho, Rac and CDC42. The precise consequences of the interaction between these two GAP proteins is not known. BCR, which contains a serine/threonine kinase domain, also contains both a dibble domain that may function as a GEF for certain Rho family members (CDC42 > Rho-A > Rac in terms of relative activity) and a GAP domain targeted to CDC42 and Rac, but not to Rho-A [25b]. N-chimerin and B-chimerin are Rho-GAPs specific for Rac. Ral is a downstream effector of Ras, as shown by the ability of oncogenic Ha-Ras to activate the guanine nucleotide dissociation stimulator Ral-GDS, which (as a GEF) activates Ral-A and Ral-B; this appears to enhance the transforming activity of Ras (and Raf). Ral interacts with Ral-binding protein 1 [Ral-BP1; also known as Ral interacting protein 1 (RIP1)], which is a GAP protein for CDC42 [26–28]. Elegant microinjection experiments by Nobes and Hall [29] and Kozma et al. [30] have revealed that CDC42 can activate Rac, which in turn activates Rho. Each of these members of the Rho family can generate specific changes in cytoskeletal elements in response to specific extracellular factors (bradykinin for CDC42, PDGF or insulin for Rac and lyso-phosphatidic acid for Rho) [29,30]. In this so-called 'GTPase cascade', activation of CDC42 promotes the formation of shaft-like filopodia and microspikes, activated Rac promotes membrane ruffling and the formation of curtain-like lamellipodia, and activated Rho promotes the formation of stress fibres and focal adhesions. The formation of these different cytoskeletal structures presumably reflects the specific cytoskeletal components recruited in each case. The different ways in which these signalling elements can be arranged, determined by which proteins are actually present, generate specific cell surface structures and choreograph complex activities such as cell migration and neurite outgrowth [9].

Khosravi-Far et al. [31] have presented compelling evidence that oncogenic Ras, possibly acting via CDC42, requires both Rac-1 and Rho-A in order to establish a fully transformed state. They showed that dominant negative mutants of Rac-1 and Rho-A blocked Ras transformation, and that constitutively activated but weakly transforming Rac-1 and Rho-A could synergize with a weakly transforming Raf-1 to transform cells efficiently. Rho-A appeared to be downstream of Rac-1 on a pathway that had its major impact on the actin cytoskeleton, as contrasted with the action of the Raf-1 → MEK → ERK pathway on gene expression.

Signalling initiated by receptor tyrosine kinases

In one-pass receptor-initiated signal transduction pathways, activation of the cell surface receptor tyrosine kinase by its ligand generally stimulates receptor homo- or hetero-dimerization and/or a conformational change that typically results in autophosphorylation of multiple tyrosine residues in the cytoplasmic portion of the receptor [3,32]. Receptor dimerization can be accomplished

Table 1   Names, abbreviations and terminology

Receptors and growth factors
TrkA: 'tropomyosin receptor kinase'; the receptor for NGF (nerve growth factor)
TrkB and TrkC bind NGF-related neurotrophins
EGF: epidermal growth factor
PDGF: platelet-derived growth factor
FGF: fibroblast growth factor
IGF: insulin-like growth factor
CSF: colony-stimulating factor

Small-molecular-mass p21 GTPase proteins
Ras: rat sarcoma virus
Rac: Ras-related C_3-botulinum toxin substrate
Rho: Ras homologous
Ral: Ras-related protein
cdc42: homologous to yeast cell division cycle gene 42
ARF: ADP-ribosylation factor
SAR: secretion-associated and ras-superfamily-related gene
YPT: yeast protein two

GAPs (GTPase-activating proteins)
p120Ras^{GAP}: 120 kDa Ras-GAP
NF1: neurofibromin; also a Ras-GAP
p100GAP1^m: a rat Ras-GAP1
GAP1^{IP4BP}: a human Ras-GAP1
CDC42-GAP (Rho-GAP)
p190Rho^{GAP}: 190 kDa protein; binds p120Ras^{GAP}
BCR: product of breakpoint cluster region gene
ABR: active BCR-related

GDS (guanine nucleotide dissociation stimulator)

GEFs (guanine nucleotide exchange factors)
SOS: son of sevenless (a *Drosophila* gene product); SOS1 and 2 are mammalian homologues
CDC25: mammalian homologue of yeast cdc25
C3G: Crk SH3-binding GEF
DBL: diffuse B cell lymphoma
VAV: named from the sixth letter of the Hebrew alphabet
OST: oncogene from rat osteosarcoma cells
smgGDS: small G-protein GDP dissociation stimulator; possible GEF for K-Ras
Ral-GDS: GDS protein active on Ral

Adaptor proteins
GRB2: growth-factor-receptor-bound protein 2
SHC: SH2-domain-containing α2-collagen-related
NCK: a novel cytoplasmic protein

MAPKKKs (mitogen*-activated protein kinase kinase kinases)
Raf-1, A-Raf, B-Raf
PAK: p21-activated kinase
MEKK: MEK kinase
TAK: transforming-growth-factor-β-activated kinase

MAPKKs (mitogen*-activated protein kinase kinases)
MEK1 (MKK1) and MEK2 (MKK2): MEK = MAPK/ERK kinase; MKK = MAP kinase kinase
JNKK: JNK kinase
SEK1: SAPK/ERK kinase 1; vastly prefers SAPK as substrate
MKK3 and MKK 4: MAP kinase kinases 3 and 4
RKK: RK kinase

MAPKs (mitogen*-activated protein kinases)
ERK: extracellular-signal-regulated kinase; ERK1 = p44; ERK2 = p42
FRK: Fos-related kinase
JNK: Jun N-terminal kinase, also known as:
SAPK: stress-activated protein kinase; several splice variants are known, producing proteins of about 46 and 54 kDa; thus there are p46 and p54 versions of SAPKα1, SAPKα2, SAPKβ and SAPKγ (p46SAPKα1 = JNK2)
p38: RK (reactivating kinase)
CSBP: cytokine-suppressive anti-inflammatory drug binding protein

TCFs (ternary-complex factors)
Elk-1: Ets-like transcription factor
SAP-1: SRF accessory protein 1 or stress-activated protein 1

MAPKAPK (mitogen-activated protein kinase activated protein kinase)
MAPKAPK-1: RSK, the ribosomal S6 kinase p90^{rsk}
MAPKAPK-2: 50 kDa protein kinase that phosphorylates Hsp25/Hsp27

Phosphatases
SHP-1: also known as SHP, PTP1C, SHPTP-1 and HCP
SHP-2: also known as SYP, PTP1D, SHPTP-2, SHPTP-3 and PTP2C
SH-PTP: SH-domain-containing protein tyrosine phosphatase
MKP-1: MAP kinase phosphatase-1 (also known as 3CH134 and CL100)
PTP: protein tyrosine phosphatase
PAC: phosphatase of activated cells
PP1G: protein phosphatase-1 associated with hepatic glycogen

* Only some of these kinases are strongly activated by mitogens.

---

either by dimeric ligands such as PDGF or by monomeric ligands such as EGF, presumably because there are two ligand-binding sites on the monomer. One of the first tyrosines to be autophosphorylated in the PDGF receptor is in the kinase activation domain of the receptor; this enhances the activity of the kinase. Different growth factor receptors (e.g. the PDGF, EGF, FGF and NGF receptors) present phosphorylated tyrosines in different constellations of amino acids. Thus each receptor interacts with its own characteristic set of proteins, endowing each receptor type with a unique composite signal. Cytokine receptors present a variation on this theme in that the receptor is not a tyrosine kinase itself, but rather upon activation stimulates an associated protein, a Janus kinase (JAK), to phosphorylate tyrosine residues on both itself and the receptor, thereby providing docking sites for various proteins including STATs (signal transducer and activator of transcription) and other signalling molecules such as SHC, SHP-2, phospholipase Cγ (PLCγ) and p85 [33].

Some of the proteins known to associate with particular tyrosine phosphates in the cytoplasmic domain of the dimerized and activated PDGF receptor are depicted in Figure 4 [34]. The proteins interact with the tyrosine phosphates via the SH2 (Src homology 2) domain or the more recently described PTB (phosphotyrosine-binding) sequence [35–37]. Structural studies have revealed a similarity between the PTB and PH domains that suggests a possible role for the PTB domain in membrane localization, a conclusion that is supported by its ability to bind acidic phospholipids [38]. Only a subset of those proteins potentially able to associate with the fully phosphorylated receptor are actually able to bind with any one particular activated receptor molecule. In some cases (e.g. binding of p85 to the PDGF receptor) more than one interaction may be involved (e.g. two SH2–phosphotyrosine interactions, or one SH2 and one PTB interaction), presumably reinforcing what is otherwise a fairly weak single interaction ($K_d = 0.3$–$4.0\ \mu M$) [39]. In other cases two different proteins may compete for the same phosphotyrosine. For example, NCK can bind to pTyr-751; SHC, which has more relaxed binding site characteristics, competes for pTyr-579, -740, -751 and -771 [34]. Src, with only one SH2 site, appears able to bind to either Tyr-579 or, less strongly, Tyr-581. All these proteins contribute to the consequences of PDGF activation. Some of them have the potential to interact with each other, and other proteins, via SH3 domains, which recognize a left-handed polyproline type II helix domain in their partner [40]. Via such an interaction, both NCK and SHC (non-enzymic adaptor proteins) form a complex with SOS (SHC can also associate with SOS via GRB2) [41,42]. van der Geer et al. [3] have reviewed in detail the

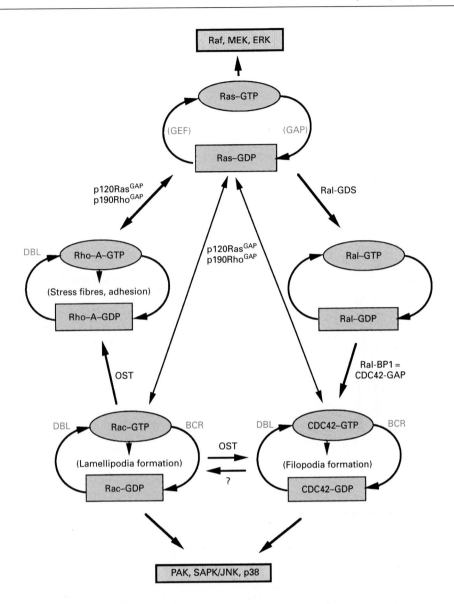

Figure 3   Potential interactions, or 'GTPase cascades', among the p21 GTP-binding proteins of the Ras/Rho/Ral family

Which interactions, indicated by the arrows, actually occur will depend on which proteins are present. Depending upon the circumstances, the GTP-bound and activated form of one p21 protein may promote the activation or deactivation of a second p21 protein via an interaction with the appropriate GEF or GAP. Known or suspected interactions involve the indicated GAP and GEF proteins, which interact respectively with the GTP-bound and GDP-bound forms of the enzymes. p190Rho^{GAP} acts more efficiently on Rho than on CDC42 [10]. BCR (and ABR) have GAP activity towards Rac and CDC42, while their DBL homology domains are the GEFs CDC42, Rho-A and Rac [25b]. DBL is a GEF for CDC42 and Rho [10]. OST interacts specifically with the GTP-bound form of Rac-1 and is a GEF for Rho-A and CDC42 [23]. Output signals from Ras–GTP and CDC42/Rac–GTP are described in Figure 6.

properties of the various proteins that can bind to tyrosine-phosphorylated receptors, and they suggest that activation of several distinct species of receptor-binding proteins is essential for a full response to any given stimulus.

The EGF receptor possesses five tyrosines located in a cluster spanning residues 992–1173 near the C-terminus that are strongly autophosphorylated after receptor activation and are thus assumed to be important in signal transmission. However, a mutant EGF receptor in which these tyrosines have been deleted or mutated to phenylalanine can still deliver mitogenic and transforming signals, albeit not as vigorously [43,44]. Although protein tyrosine phosphorylation in general was substantially reduced, SHC appeared to be fully phosphorylated after EGF stimulation of the mutant receptor, probably accounting for the activation of Ras and the ERKs. The process leading to SHC phosphorylation

remains obscure, but could be due to interaction of the mutant EGF receptor, which still contained an active kinase domain, with susceptible target proteins (e.g. cErbB2). Alternatively, Tyr-845 may become phosphorylated by Src in an EGF-dependent manner, thus enabling the mutant protein to transmit a signal [45]. When Ras action is blocked by a dominant negative Ras mutant (p21$^{rasAsn-17}$), EGF-induced phosphorylation of ERK2 can still occur, possibly as a result either of PKC activation or of an elevation of the free intracellular Ca^{2+}concentration, depending upon the cell type [46]. In contrast, ERK2 phosphorylation induced by insulin or PDGF was abrogated by expression of the dominant negative Ras mutant. The ERK proteins, reviewed below, are mitogen-activated protein kinases (MAPKs) that are responsible for the phosphorylation of many target proteins in response to mitogen stimulation.

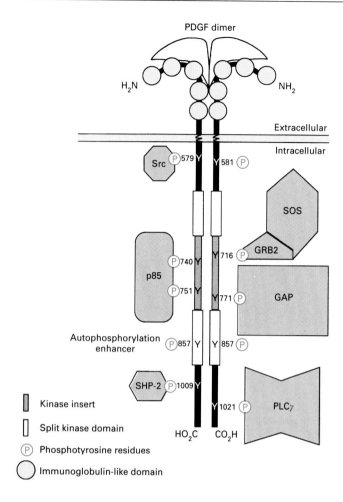

PDGF dimer

H₂N

NH₂

Extracellular

Intracellular

Src (P)579Y   Y581 (P)

SOS

GRB2

(P)740Y   Y716 (P)

p85   (P)751Y   Y771 (P)   GAP

Autophosphorylation enhancer   (P)857Y   Y857 (P)

SHP-2 (P)1009Y

Y1021 (P)   PLCγ

HO₂C   CO₂H

▮ Kinase insert

▯ Split kinase domain

(P) Phosphotyrosine residues

◯ Immunoglobulin-like domain

Figure 4   Cartoon of the activated human PDGFβ receptor dimer

A PDGF dimer brings two PDGF receptor monomers into close proximity, allowing cross-phosphorylation to occur at multiple tyrosine residues. Various proteins are able to bind to the phosphotyrosines with a specificity determined by the neighbouring amino acids. As discussed in the text, the proteins binding to the activated receptor may also be subject to tyrosine phosphorylation by the receptor. They also have the potential to interact with each other. Since the stoichiometry of receptor autophosphorylation is not high, it is likely that on average any one receptor molecule will only be phosphorylated at a few sites and be able to bind only a subset of all the proteins with the potential to bind.

Proteins that interact with protein tyrosine phosphates

Proteins with SH2-binding sites, such as Src, p85, SHP-2 (= syp), PLCγ, GAP, GRB2, SHC and NCK, bind via these domains only to those tyrosine phosphates that are found in the appropriate amino acid sequence context. This context is usually determined by the 3–6 amino acids on the C-terminal side of the phosphotyrosine. In contrast, the ability to bind the PTB sequence, which has been identified in nine proteins [37], appears to be determined largely by the 3–5 amino acids on the N-terminal side of the phosphotyrosine. Certain of the amino acids in the SH2 domains that are important in determining which phosphotyrosines will be most strongly bound have been identified and used to classify SH2 domains into groups which distinguish the various contexts in which the phosphotyrosine in found [47].

Like the phosphotyrosine-binding proteins, the protein tyrosine kinases (both receptor and non-receptor kinases) target different tyrosines depending upon the local amino acid context.

Cytosolic protein tyrosine kinases prefer to phosphorylate tyrosines in binding motifs recognized by group I SH2 domains (e.g. Src, GRB2 and GAP), whereas receptor protein tyrosine kinases phosphorylate tyrosines in motifs recognized by group III SH2 domains (e.g. SHC, p85 and PLCγ) [48]. (These groups are distinguished on the basis of the βD5 residue in the SH2 domain.) Proteins containing phosphotyrosine-binding sites (i.e. SH2 or PTB domains) compete with protein phosphotyrosine phosphatases for the tyrosine phosphate, and by doing so prolong the lifetime of those phosphotyrosine residues that are particularly strongly bound. Some of these SH2-containing proteins are enzymes [e.g. p120Ras^GAP, SHP-2, phosphoinositide 3-kinase (PI3-K), Src and PLCγ] whose enzymic activity is enhanced, whereas others are adaptor proteins (GRB2, SHC and NCK) that bind a specific target protein and deliver it to the activated receptor.

Enzymes regulated by phosphotyrosine interactions

Association of p120Ras^GAP with tyrosine-phosphorylated receptors is mediated by its SH2 domains. By virtue of its ability to activate the Ras GTPase it is a negative regulator of Ras function [49]. p120Ras^GAP can also acquire a signalling function, mediated by the SH2 or SH3 domains in its N-terminal region, as a consequence of the interaction of its C-terminal region with Ras. This signalling may involve two phosphorylated proteins: p62, a membrane-associated RNA-binding protein, and p190Rho^GAP, a protein that possesses GTPase-activating activity towards proteins in the Rho/Rac family. As discussed above (see Figure 3), p120Ras^GAP may serve to co-ordinate the activity of the Ras and Rho signalling pathways [9]. The N-terminal domain of p120Ras^GAP can modify cellular cytoskeletal structure (disruption of the actin cytoskeleton and focal contacts), possibly as a result of its interaction with p190Rho^GAP [50]. NF1 may also have a signalling function separate from its GTPase-activating ability, since it can inhibit transformation by v-ras, even though the GTPase activity of v-Ras cannot be stimulated by NF1 [51]. Deficits in NF1-GAP activity have been detected in a number of human tumours, including von Recklinghausen's neurofibromatosis, consistent with an important signalling role.

SHP-2 (the recently proposed name for the independently isolated but identical SYP, PTP1D, SH-PTP2, and PTP2C protein tyrosine phosphatases [52]) is a protein tyrosine phosphatase that contains two N-terminal SH2 domains through which it can associate with tyrosine-phosphorylated receptors, e.g. the PDGF receptor, and become activated. Phosphorylation of a C-terminal tyrosine in SHP-2 produces a docking site for GRB2, thus facilitating activation of SOS [53–55]. SHP-2 can activate Src by dephosphorylating Tyr-527 in inactive Src. When phosphorylated, Tyr-527 inhibits Src activity by forming an intramolecular complex with the Src SH2 domain [55]; freed, the Src SH2 domain can, for example, interact with an appropriately phosphorylated PDGF receptor (see Figure 4). (Alternatively, interaction of the Src SH2 domain with the phosphorylated receptor, should it be able to compete, would expose the Tyr-527 phosphate to the action of SHP-2 [3].) GRB2 has been reported to bind tyrosine-phosphorylated focal adhesion kinase, thus providing a potential link between integrin-mediated signalling and Ras-mediated signalling. This may explain the activation of the p42 and p44 ERKs when integrins are engaged by their ligands [56,57].

PI3-K is a dimeric enzyme (subunits of 110 and 85 kDa) capable of phosphorylating suitable phosphoinositides in the 3-position; it is also a serine/threonine protein kinase able to phosphorylate various proteins, including its own regulatory

85 kDa subunit and the insulin receptor substrate-1 [58,59]. The relationship between the lipid- and peptide-binding sites is not known. p85 possesses one SH3 group, two SH2 groups and a domain that may have GAP activity towards an unidentified target. It probably activates the kinase holoenzyme by an allosteric effect and by fostering relocation of the p110 catalytic subunit to the membrane. Tyrosine phosphorylation of the p85 subunit down-regulates PI3-K activity. p110 can also interact directly with activated, membrane-bound, Ras and thereby itself be activated. One function of the 3-phosphorylated phosphatidylinositol phosphates is to participate with Rac in inducing membrane ruffling. PtdIns$(3,4,5)P_3$ interacts with the SH2 domains of some proteins, including p85 and pp60$^{c\text{-}src}$, competing directly with receptor protein tyrosine phosphates for binding to these proteins [60]. PtdIns$(3,4,5)P_3$ also directly activates protein kinase (PKC) ζ, thereby stimulating phosphorylation of serine/threonine residues in proteins targeted by this kinase in a calcium- and diacylglycerol-independent process [61,62].

Src is a protein tyrosine kinase that is attached to the plasma membrane by its N-terminal myristate modification and is essential for the mitogenic action of certain receptors, e.g. that for PDGF. When activated, it phosphorylates a number of cytoskeleton-associated proteins (focal adhesion protein tyrosine kinase, paxillin) and probably contributes to the activation of both Raf-1 and PI3-K [63]. v-Src has been found to activate phospholipase D (PLD) by inducing Ras, possibly by phosphorylation of SHC with subsequent activation of GRB2/SOS, to activate a Ral-GDS activity, the exchange factor responsible for the activation of Ral [64] (see Figure 3). PLD is found in a complex with Ral-A, and when it is activated as a consequence of the Ras–Ral interaction it generates several lipid second messengers, e.g. phosphatidic acid derived from the action of PLD on phosphatidylcholine. v-Src augments both the Ras/MEKK1/JNK and Ras/Raf-1/ERK pathways, which among other actions contribute to the activation of a cAMP response element in the mouse prostaglandin synthase-2 promoter to enhance transcription of the gene [65].

Adaptor proteins can mediate phosphotyrosine interactions

SOS is a cytoplasmic Ras-GEF that is typically constitutively associated with the adaptor protein GRB2 and can be stimulated to activate Ras in various ways [66]. When GRB2 interacts with a tyrosine-phosphorylated membrane receptor, it positions SOS at the plasma membrane where it can promote activation of Ras. Emphasizing the importance of the cellular location of an enzyme is the observation that SOS derivatives engineered to localize at the membrane are able to activate Ras in a receptor-independent manner and to transform cells. Cell transformation by SOS (achieved by targeting SOS to the cell membrane by providing it with a farnesylation or myristoylation signal) is accentuated by deletion of the C-terminal region, suggesting that this region can fold to inhibit the interaction of SOS with Ras [67]. Alternatively, instead of GRB2 binding directly to a tyrosine-phosphorylated receptor such as the activated EGF receptor (Figure 5a), the SOS–GRB2 complex may interact with a phosphotyrosine on SHC, which in turn can associate with a receptor phosphotyrosine, as illustrated for the NGF receptor TrkA (Figure 5b). The importance of the SHC–GRB2–SOS pathway in *ErbB*2 (p185, or *HER-2/neu*) signalling was demonstrated by the finding that GRB2 in which the N-terminal SH3 domain was deleted was able to block substantially the oncogenic signal from activated *ErbB*2; although still able to bind SHC, the ΔN-GRB2 could no longer bind SOS [68].

The SHC gene encodes several variant proteins that become tyrosine phosphorylated when various receptors are activated. SHC can itself transform fibroblasts and promote PC12 differentiation in a Ras-dependent fashion. It appears that SHC first binds to a tyrosine-phosphorylated receptor, which phosphorylates a tyrosine in SHC that can then serve as bait for GRB2 docking and consequent SOS activation. The FGF receptor-1 stimulates the GRB2–SOS complex not only via their mutual interactions with SHC but also via p89, another adaptor protein that is in addition membrane-associated [69]. Although the SH2 domain of SHC has been considered as its primary receptor interaction domain, a recently identified PTB domain that recognizes NPXY(P) may mediate this interaction instead [70–72]. By computer analysis, Bork and Margolis [36] detected this domain in a number of other proteins and designated it PID (for phosphotyrosine interaction domain).

Signalling via the heterotrimeric G-proteins

Activation of heterotrimeric G-proteins (consisting of α, β and γ subunits), typically by stimulated serpentine (7-pass) membrane receptors [74], is effected by the replacement of GDP with GTP on the Gα component and the consequent dissociation of Gα from the membrane-bound G$\beta\gamma$ heterodimer. Like Ras, Gα is active only when bound to GTP; it is inactivated by its intrinsic GTPase. Research during the past few years has revealed that signals from the heterotrimeric G-proteins, particularly those stimulated by pertussis toxin-inhibitible receptors (G$_i$), can impact on the Ras/MAPK signal transduction cascade [75–77]. The G$\beta\gamma$ subunit, by virtue of its ability to interact with certain PH domains, may influence the activity of either SOS or Ras-GAP, both of which have PH domains, perhaps by attracting them to the plasma membrane. In COS-7 cells, agonists acting on G$_s$-, G$_q$- or G$_i$-coupled receptors stimulated p44^{MAPK} (ERK1) expressed as an epitope-tagged molecule in transiently transfected cells; both the Gα and G$\beta\gamma$ subunits were implicated in the signalling process [78].

Several groups have transfected COS-7 cells with various expression vectors in order to elucidate the mechanisms of G-protein-coupled receptor action. Crespo et al. [79] found that G$\beta\gamma$ generated by stimulation of a muscarinic m2 receptor activated ERK2 via a Ras-dependent pathway, possibly involving SHC. Touhara et al. [80] used anti-SHC antibodies to demonstrate directly that the pertussis toxin-sensitive G$_i$-coupled receptor (the α2-C10 adrenergic receptor) stimulated tyrosine phosphorylation of the $\sim$ 50 kDa SHC protein via an action of G$\beta\gamma$. These and other results established that the G$\beta\gamma$ heterodimer, or at least certain species thereof [77], can stimulate phosphorylation of SHC, facilitating the association of the latter with GRB2–SOS, activating Ras and resulting in Raf-dependent ERK activation [81]. This same group also reported that, in Rat 1 fibroblasts, activation of the IGF receptor by IGF-1 also required input from G$\beta\gamma$ subunits (e.g. derived from the G$_i$-coupled receptor for lysophosphatidic acid) in order for full ERK activation to occur [82]. Signalling from the β-adrenergic receptor, assessed by ERK activation, involves both a positive signal conveyed by G$\beta\gamma$ acting on a Ras-dependent pathway and a negative signal resulting from activation of adenylate cyclase by Gα_s. The increase in cAMP activates protein kinase A (PKA), which suppresses activation of the Ras pathway by an inhibitory phosphorylation of Raf-1 [77,83].

The PH domain (named for its presence in the abundant platelet protein pleckstrin) is an approx. 120-amino-acid module with a characteristic structure; it is found in many signalling proteins and mediates protein–protein interactions and membrane attachment [84]. The membrane-bound G$\beta\gamma$ complex is

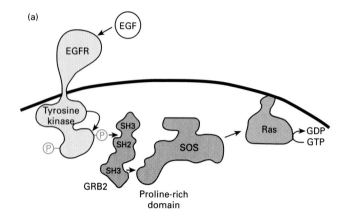

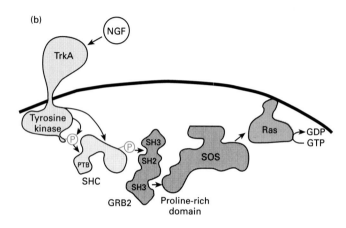

Figure 5   Model for the regulation of Ras by activated receptors

(**a**) The EGF receptor (EGFR) transmits its signal directly through GRB2 to SOS and on to Ras. (**b**) The TrkA receptor, unable to bind GRB2 directly, interacts with SHC via a PTB domain. SHC then provides a mechanism for GRB2 and SOS to be activated. In this representation the locations of tyrosine phosphates (P), PTB, SH2, SH3 and the proline-rich SH3-binding domain are for illustrative purposes only. (Modified from FEBS Lett., **338** Downward, J., 113–117, ©1994, with kind permission of Elsevier Science – NL, Sara Burgerhartstraat 25, 1055 KV Amsterdam, The Netherlands.)

believed to interact with PH domains in target proteins, causing them to relocate to the cell membrane [76]. The PH domain has also been implicated in the binding of both $PtdIns(4,5)P_2$ [85–87] and PKC (both calcium-dependent and calcium-independent isoforms) [88]. The PH domain in $PLC\delta_1$, but not in certain other proteins, can bind to $Ins(1,4,5)P_3$ with stereospecific high affinity [89]. Lee and Rhee [90] have reviewed the significance of $PtdIns(4,5)P_2$ hydrolysis by PtdIns-specific PLC isoenzymes with regard to the role of $PtdIns(4,5)P_2$ in regulating PLC isoenzymes, activating PLD and controlling actin polymerization. There are four $PLC\beta$, two $PLC\gamma$ and four $PLC\delta$ enzymes, all possessing a PH domain near the N-terminus. The $PLC\gamma$ isoenzymes are activated by receptor (and non-receptor) tyrosine kinases, which phosphorylate three tyrosines, one of which (Tyr-783) is essential for activating the phospholipase activity and also for association with the cytoskeleton. The $PLC\beta$ isoenzymes are activated by GTP-bound α_q and $\beta\gamma$ subunits of receptor-activated $G\alpha\beta\gamma$. Activated PLC not only generates diacylglycerol and $Ins(1,4,5)P_3$ but also reduces the level of $PtdIns(4,5)P_2$ and consequently the number of membrane attachment sites for proteins with PH domains.

THE p21 SIGNALLING PATHWAYS

Ras is normally found attached to the inner face of the plasma membrane by the lipid interactions discussed above. Mobilized Ras–GTP is capable of interacting with and activating a number of proteins; known direct targets include p120Ras^{GAP}, the p110 subunit of PI3-K, Ral-GDS, PKCζ, Raf-1, A-Raf, B-Raf, and possibly certain MEKKs [91,92]. Each of these proteins is likely to be activated by Ras–GTP via a unique interaction with a particular site on the p21 protein, and each contributes to the overall consequence of Ras activation. With regard to interaction sites, Joneson et al. [93] reported recently that the mutation Tyr-40 → Cys abrogated ERK activation but not membrane ruffling, whereas the Thr-35 → Ser mutation impaired membrane ruffling but not ERK activation when quiescent fibroblasts were transfected with the indicated mutant of Ha-Ras. This is consistent with the idea that distinct downstream consequences of Ras activation may be caused by different Ras effectors generating different signals.

Figure 6 shows a generic schematic of the Ras → Raf → MEK → ERK pathway, along with what is known about two related pathways that for the most part operate independently of Ras. These pathways are not equally activated by any single stimulus. Growth factors and mitogens preferentially activate ERK1 and ERK2, whereas inflammatory cytokines and various forms of stress preferentially activate SAPK/JNK and/or p38/RK (see Table 1 for details of nomenclature). Different laboratories using different cells and different activating strategies have identified various different components in these cascades and have often given them different names. The 'take-home message' is that there are several parallel tracks down which a signal can be transmitted. These three phosphorylation cascades are activated to different extents depending upon the particular stimulus and the various other signalling components that are activated (e.g. PI3-K, PKC isotypes, non-receptor Src-family tyrosine kinases and phospholipases). The degree of activation of a particular signalling pathway depends upon the number of activated receptors and the specific upstream effectors that are activated. Phosphorylation of specific residues in some proteins may be inhibitory, thus allowing modulation of the signal intensity. As discussed later (see Figure 9), if different phosphorylated sites in a protein convey separate signals, for example to phosphorylate different kinases, then one has the potential for a pathway to transmit more than one signal simultaneously (multiplex signalling).

The categorizing of various 'levels' in the different pathways in Figure 6 more likely reflects a human desire than nature's intent to arrange things this systematically. A comparison of these pathways with the five or so known pathways in yeast suggests that additional signalling cascades remain to be defined in mammalian cells [94,95]. Consistent with this thought is the existence of several enzymes that might be parts of as-yet-undefined mammalian pathways (e.g. ERK3 [96] and MEK5 [97]). Although 'mitogen-activated protein kinase' has been used as a generic term to describe collectively all of the MAP kinases that are activated by dual phosphorylation in Figure 6, only a subset of them (the ERKs) are typically strongly activated by a proliferative stimulus.

The determination of which proteins are true *in vivo* substrates for the various kinases is not trivial. Conclusions drawn from experiments with purified enzymes and substrates can be extrapolated to the intracellular situation only with certain caveats, particularly concerning protein concentrations and macromolecular associations. Since the concentrations of a particular kinase and its substrate, and their mutual affinity, determine

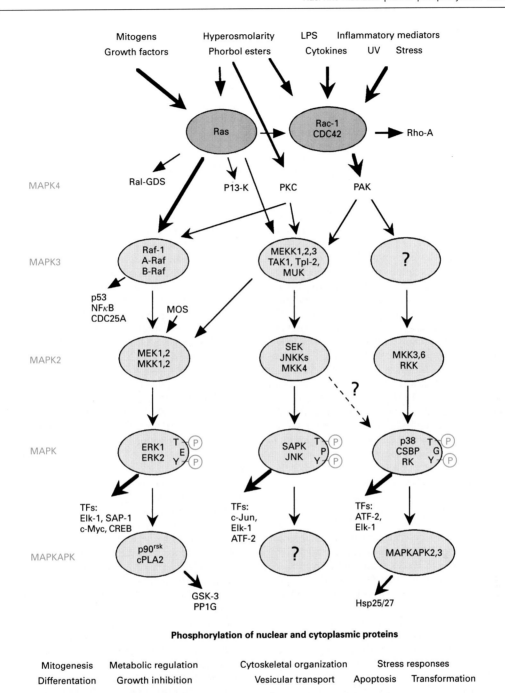

Figure 6   Multiple signal transduction pathways initiated by various stimuli

Considerable evidence suggests the existence of three reasonably discrete protein phosphorylation cascades in mammalian cells, although with an unknown amount of cross-talk and variable activation relative to each other depending upon the stimulus. Arrows indicate some of the major routes of signal transmission; many interactions are not shown. As discussed in the text, localization of the components in the pathway and their recruitment into macromolecular complexes by factors that nucleate specific interactions [116] may endow the pathways with a specificity that is not apparent when efforts are made to dissect the pathways *in vitro*. Phosphorylation of upstream proteins by downstream effector kinases (not illustrated) completes the signalling circuit. Abbreviations: TF, transcription factor; MOS, cellular homologue of the oncogene of Moloney murine sarcoma virus; cPLA2, cytosolic phospholipase A2.

whether that substrate is phosphorylated, *in vitro* experiments should in principle be performed using *in vivo* concentrations, which of course cannot easily be known if the proteins are not homogeneously distributed in the cell. There is also the potential that consecutive elements in a phosphorylation cascade exist in a specific complex, thus providing exquisite specificity with regard to transmission of the signal. Furthermore, if the signalling

molecules are normally organized in a complex (a 'signalsome') with a particular scaffold protein present in limiting quantities, then results obtained using transient transfection protocols may not mimic the real situation *in vivo*, thus further clouding the interpretation of various experiments [93]. Signals transmitted within a complex of proteins will not be amplified. Amplification does occur at the level of the downstream effectors, particularly the MAPKs, and also as the result of positive synergistic interactions of individual signalling pathways.

The 'MAPK4' level

This 'level' encompasses everything between the receptor and MAPK3, i.e. mitogen-activated protein kinase kinase kinase. In this group, which is not as well defined a level as those that follow, are the GAPs and GEFs, the PKC isoforms, PI3-K and p65[pak]. One could argue that PKA and assorted phospholipases should be included also.

The serine/threonine PKC family is divided into three sub-groups depending on whether both Ca^{2+} and diacylglycerol are required, whether only diacylglycerol is required, or whether some other lipid (e.g. ceramide) is required [59,62,98]. Second messengers generated from the breakdown of various membrane lipids, especially phospholipids, enhance PKC activity, for example by recruiting the cytosolic protein to the plasma membrane or stimulating proteolysis of the inhibitory pseudosubstrate. The tumour-promoting phorbol esters, analogues of the natural activator diacylglycerol, are potent activators of most PKC isoforms. PKC is regulated by phosphorylation, which is necessary for activity, and by lipids such as phosphoserine and diacylglycerol, the product along with $Ins(1,4,5)P_3$ of the action of PtdIns-specific PLC on $PtdIns(4,5)P_2$. This PLCγ is activated by many receptor tyrosine kinases via its SH2-mediated association with the appropriate phosphotyrosine (Figure 4), and it may make a definite contribution to the mitogenic response. *In vitro* the PKC isoforms are rather promiscuous with regard to which substrates, typically Arg-rich proteins, they will phosphorylate. Although the critical targets of the various PKC isoforms *in vivo* remain for the most part to be discovered, it is clear that PKC is an important constituent of many signalling pathways.

PAK (p65[pak]; p21-activated kinase) was discovered as a brain serine/threonine kinase that is a potential downstream target for Rac-1 and CDC42 signalling. PAK binds to the activated GTP-bound p21 proteins CDC42 and Rac-1, inhibiting their intrinsic GTPase activity and promoting its own phosphorylation [99]. Because PAK-I is active in *Xenopus* oocytes and in quiescent and serum-starved cells, and because (when microinjected) it can inhibit cleavage in blastomers, Jakobi et al. [100] have suggested that PAK-I may be involved in the regulation of cytostasis and the response of the cell to stress. PAK is activated in phagocytes as a result of its interaction with Rac–GTP, and it can phosphorylate the p47[phox] NADPH oxidase in a Rac–GTP-dependent manner. Rac–GTP is itself generated via an interaction with a Gα protein activated, for example, by a chemoattractant receptor [101]. Coso et al. [102] have suggested that PAK might be an intermediate linking Rac-1 and CDC42 to MEKK1 on the basis of their observation that mutationally activated CDC42 or Rac-1 stimulated JNK/SAPK, but not ERK, activity. *In vivo* experiments with constitutively activated CDC42 and PAK3 revealed that PAK3 was activated by CDC42, and that JNK1 and p38 were in turn activated more efficiently than ERK2 [103]. Studies by Minden et al. [104] and Qiu et al. [105] using constitutively active or dominant interfering alleles of the various signalling intermediates placed Rac-1, possibly followed by PAK, down-stream of Ras and upstream of MEKK1 in a signalling cascade resulting in JNKK and p38 activation; attempts to locate CDC42 in the pathway were inconclusive. Studies of interleukin-1 (IL-1) signalling suggested a pathway (Rac/CDC42) → PAK → (un-identified intermediates) leading to p38 and JNK activation [106]. The observation that a dominant negative CDC42 mutant inhibited IL-1 signalling was consistent with such a pathway [103]. Mutationally activated Rac-1 and CDC42 GTPases, and also the GEFs for these Rho family proteins, activate JNK/SAPK without affecting the ERKs [102] (see Figure 3). In addition to PAK, CDC42 may activate activated CDC42-associated kinase (ACK), a non-receptor tyrosine kinase that inhibits the GTPase activity of CDC42, and PI3-K [9,10].

The 'MAPK3' level

Although Ras can interact with a number of effector proteins in a GTP-dependent manner, it is likely that there will be competition among effectors, with certain interactions being preferred over other interactions. Each of these Ras effectors also has its own differing target protein affinities. Thus A-Raf can activate MEK1 but not MEK2 in EGF-stimulated HeLa cells [107]. Ras and Raf-1 form a signalling complex with MEK1 but not MEK2 [108], whereas the Ras–MEKK1 and –MEKK2 interactions preferentially activate JNKK [109]. MEKK1, 2 and 3 are all able to stimulate the ERK and SAPK/JNK pathways, although MEKK1 and 2 preferentially stimulate SAPK/JNK, whereas MEKK3 preferentially activates MEK/ERK [92]. MEKK1 is activated in resting PC12 cells by EGF, and to a lesser extent by NGF, phorbol 12-myristate 13-acetate and oncogenic Ras [110]. EGF-induced activation of MEKK1 and B-Raf can be inhibited by a dominant negative Ras and also by PKA activation (which inhibits Raf). Both the Raf-1 and MEKK1 pathways are stimulated in rat 3Y1 fibroblasts by hyperosmolar (0.5 M NaCl) shock, with consequent activation of ERK1 and ERK2 (p43/44[MAPK] and p41/42[MAPK]) and the SAPKs (p54 and p46) [111]. How osmotic shock activates these pathways, including that leading to p38 phosphorylation, and the extent to which Ras is involved, is not known.

Downstream elements in Ras-controlled signalling pathways may be activated by various forms of stress-initiated signals that do not function by activating Ras itself. However, Ras does appear to be an important sensor of oxidative stress. Lander et al. [112] observed in PC12 and Jurkat cells that inhibitors of Ras function (a dominant negative mutant p21 and an inhibitor of farnesyltransferase) blocked nuclear factor κB (NFκB) activation by various oxidative agents that modulate cellular redox stress (peroxide, haemin, Hg^{2+}). The authors concluded from these and other experiments that free radicals can activate p21[ras] and generate a nuclear signal. Other studies have found that the MEK/ERK pathway is activated in neutrophils exposed to an oxidant (diamide or peroxide with a catalase inhibitor) [113] and in HeLa cells exposed to lysophosphatidic acid, which is believed to elevate reactive oxygen species via lysophosphatidic acid-induced arachidonic acid release [114]. However, in these latter two studies the mechanism of MEK activation was not determined.

The interaction of MEKK1 with Ras, at least in the test tube, involves the kinase domain of MEKK1 and the effector domain of Ras. When MEKK1 is complexed with activated Ras it is capable of phosphorylating MEK1 [115]. Xu et al. [116] have shown that a recombinant fragment of MEKK1 containing the catalytic domain could phosphorylate MEK1 and MEK2, and that MEK1 and MEK2 could be activated when co-transfected with MEKK1. However, ERK2 was not activated to the full extent expected, possibly, the authors suggest, because of an inhibitory effect of the up-regulated MEKK1 activity upon ERK2 activity. The relevance of these *in vitro* observations to the *in vivo* situation remains unclear, since it is possible that the highly specific interactions of MEK1/2 with Raf-1 and with

ERK1/2 are the consequence of the formation of macromolecular complexes. It remains unclear to what extent Ras can activate MEKK1 directly, or whether it normally proceeds indirectly via Rac and PAK, or even by a paracrine process involving a secreted extracellular mediator.

In proliferating cells, transforming growth factor-β is usually a negative regulator of proliferation. However, this factor can deliver a mitogenic signal to quiescent (confluent, serum-starved) 3T3 cells in the absence, surprisingly, of detectable tyrosine phosphorylation of the 41 or 43 kDa ERKs (ERK activation typically is associated with a proliferative stimulus) [117]. A kinase called TAK1 has recently been identified as a possible mediator of transforming growth factor-β and bone morphogenetic protein signalling; consistent with this possibility was its inability to activate the MEK → ERK pathway and its ability (*in vitro* at least) to activate SEK1 [118]. Another recently identified MEKK, a proto-oncogene called Tpl-2, is of interest in that it has the potential to activate, independently of Ras and Raf, both MEK1 and SEK1 (but not p38/RK); how Tpl-2 is activated is not known [119]. The so-called mixed-lineage kinases have sequence motifs that are associated with both serine/threonine kinases and tyrosine kinases, and at least one of them (called MUK) has the properties of a MEKK that is a selective activator of the SAPK/JNK pathway [120,121].

Reuter et al. [122] have recently characterized several MEK activators, including the ∼ 95 kDa B-Raf and an unidentified 40–50 kDa MEK activator, and shown that they, rather than Raf-1, are the major serum-stimulated MEK activators in fibroblasts. (Lysophosphatidic acid, which stimulates G-protein-coupled receptors, may be the active serum component [122a].) Raf-1 does respond well to both PDGF and phorbol esters. In PC12 cells B-Raf was responsible for the NGF-stimulated, p21^{ras}-mediated, activation of MEK [123]. Pritchard et al. [124] expressed the protein kinase domains of Raf-1, A-Raf and B-Raf coupled to the oestrogen receptor and found that after oestradiol activation the B-Raf construct was more effective than the Raf-1 or A-Raf constructs in activating MEK and the ERKs. All three constructs could (after oestradiol activation) morphologically transform 3T3 cells, but only the A-Raf construct could cause quiescent (G$_o$) 3T3 cells to enter the cell cycle.

The interaction of Ras with Raf-1, illustrated in Figure 7, has been extensively studied [125]. Raf-1 is normally located in the cytosol in association with Hsp90 (heat-shock protein of 90 kDa) and Hsp50, which appear to stabilize the protein and facilitate a proper interaction with Ras [126]. Activation of the 74 kDa Raf-1 protein by Ras involves relocation to the cell membrane via direct association of two regions in the N-terminal conserved region 1 of Raf-1, a previously recognized Ras-binding domain and a more recently recognized cysteine-rich region, with the switch 1 and switch 2 domains respectively of Ras [127,128]. Full activation of Raf-1 requires both an interaction with the plasma membrane and phosphorylation of key amino acids [129–131]. Phosporylation events include tyrosine phosphorylation by membrane-bound Src [132] and serine/threonine phosphorylation by a PKC isoform [133] or by KSR (kinase suppressor of Ras) [134]. The latter is a protein kinase with some similarity to Raf-1 that has been implicated in Ras signalling either upstream of or in parallel to Raf-1. Unlike Raf-1, which requires Ras-induced membrane localization and tyrosine phosphorylation, B-Raf contains aspartic acids in place of the phosphorylatable tyrosines (at 340 and 341) and is not localized to the plasma membrane upon Ras activation [135].

Sites of serine phosphorylation on Raf-1 include Ser-43, Ser-259 and Ser-621; the latter two occur in a consensus sequence RSXSXP, which is conserved among all Raf family members

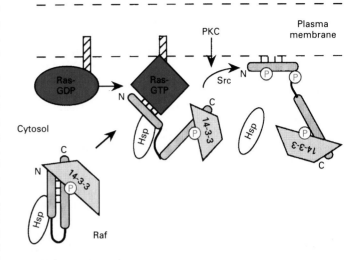

Figure 7   Proposed mechanism for the activation of Raf by Ras and protein phosphoylation

Activated Ras binds to the conserved region near the N-terminus of Raf-1 and attracts it to the plasma membrane, where it undergoes further conformational changes and phosphorylation by PKC and a protein tyrosine kinase, enhancing its ability to phosphorylate MEK. The involvement of 14-3-3 is discussed in the text. The location of phosphate groups, known to include both Ser and Tyr residues, is for illustrative purposes only.

[136]. Phosphorylation of at least some of the serine/threonine residues [those just mentioned and also Ser-499 and Thr-268 (the site of *in vitro* autophosphorylation)], and Tyr-340 and Tyr-341 is essential for complete Raf-1 activation [137]. Ser-43 and Ser-621 are both constitutively phosphorylated, whereas Ser-259 is phosphorylated in response to PDGF. The conformational changes induced in Raf-1 activate its serine/threonine kinase activity, which is specific for Ser-218 and Ser-222 in human MEK1 and for comparable conserved residues in other MEK family kinases [138]. The specificity appears to be provided by a proline-rich sequence in MEK1 and MEK2 that is missing from the other MAPKKs (JNKK/SEK1/MKK4 and RKK/MKK3, described below) and appears essential for the interaction of the MEKs with Raf-1 and downstream signalling proteins [139]. Other possible *in vivo* Raf-1 substrates include the NFκB inhibitor IκB [140], p53 [141] and the dual-specificity phosphatase CDC25A [142]. Thus Raf-1 activation can directly stimulate the expression of genes under the control of NFκB and p53, and can initiate cell cycle progression by dephosphorylation and activation of the cyclin kinases. Two of the three CDC25 isoforms associate with Raf-1 and are phosphorylated when Ras is activated, perhaps accounting for the synergism between CDC25A or CDC25B and Ras in transforming primary cells [143]. CDC25 is a dual-specificity phosphatase that regulates CDC2 kinase, activating CDC2 by dephosphorylating adjacent pThr-14 and pTyr-15 residues.

Hafner et al. [144] showed in studies with purified proteins that phosphorylation of Raf-1 by PKA was inhibitory both to Ras binding and to the functioning of the kinase domain. However, *in vivo* the situation becomes more complex. For example, Raf-1 and the downstream signalling pathway are inhibited by cAMP in fibroblasts, whereas in PC12 cells cAMP synergizes, in the presence of serum, with NGF to enhance neurite outgrowth induced by B-Raf. This difference in sensitivity is the result of the presence only in Raf-1 of Ser-43 in a consensus site for phosphorylation by PKA [145]. Because Raf-1 is rapidly and negatively regulated by PKA phosphorylation *in vivo*, Burgering and

Bos [77] have suggested that Raf-1 activation is a reiterative process and that Ras action is repeatedly required to maintain Raf-1 activity. Raf-1 is also subject to an activating phosphorylation on Thr-269 by a membrane-bound and proline-directed kinase that is activated by ceramide, the second messenger of the sphingomyelin pathway [146]. Both tumour necrosis factor α (TNFα) and IL-1β signal through this pathway, enhancing the activity of Raf-1 towards MEK; presumably Ras–GTP is required to localize Raf-1 to the cell membrane, although this was not investigated. If the different phosphorylations on Raf-1 send different signals downstream, differentially activating MEK1 or MEK2 for example, then Raf-1 would be an example of a multiplex signalling element.

Proteins known as 14-3-3 proteins are typically acidic, dimeric proteins of about 60 kDa that are found in a broad range of organisms. Several isoforms are known, some of which are simply phosphorylation variants. They are found to associate with, and often to modify the activity of, a number of other proteins (e.g. tyrosine hydroxylase, PKC), not unlike chaperones [147]. The catalytic p110 subunit of PI3-K binds to and is inhibited by 14-3-3 [148]. Various isoforms of 14-3-3 have different effects on the different species of PKC; both activation and inhibition of PKC isoforms by preparations of 14-3-3 have been reported [149]. Binding of 14-3-3 to target proteins is via a phosphoserine residue in the motif RSXSXP [150], identical to the conserved consensus sequence for serine phosphorylation in Raf-1 noted above [136]. 14-3-3 associates efficiently with Raf-1 (via the constitutively phosphorylated Ser-621), but does not activate its kinase activity in immunoprecipitates [151]. The association of 14-3-3 (β and ζ isoforms) with inactive Raf-1 may stabilize that conformation, and upon Raf-1 activation it may facilitate activation, perhaps via a specific interaction with the kinase domain [152,153]. As indicated in Figure 7, there are two sites of interaction of 14-3-3ζ with Raf-1, one of which can be competed for by activated Ras [154]. Dent et al. [137] showed that both 14-3-3ζ and Hsp90 could inhibit the inactivation of purified Raf-1 by either serine/threonine or tyrosine protein phosphatases. Because of the dimer structure of 14-3-3, the protein has the potential to foster interactions between proteins to which it can bind, and Braselmann and McCormick [154a] have demonstrated that 14-3-3 can indeed perform such an adaptor function, generating a complex with BCR and Raf, particularly the membrane-bound form. This may be important in facilitating the action of BCR on target proteins (see Figure 3).

Buscher et al. [155] investigated in the macrophage-like line BAC-1.2F5 the signal transduction pathways used by CSF-1 and lipopolysaccharide (LPS) to induce phosphorylation of the two ternary-complex factors Elk and SAP. Although CSF-1, a mitogen, stimulated Ras–GTP formation, Raf-1 activation curiously did not seem to depend on Ras activation, as judged by its insensitivity to a dominant negative Ras. LPS, a stimulator of macrophage differentiation, clearly did not utilize Ras at all, and instead activated Raf-1, MEK and ERK via a PKC and phosphatidylcholine-specific PLC pathway. Elk, but not SAP, activation was dependent upon ERK activation. The CSF-1 receptor appeared to activate Src, which in turn activated Raf-1, probably by tyrosine phosphorylation; full mobilization of the downstream ERKs and Elk was accomplished synergistically with other signals generated simultaneously by the CSF receptor. The LPS receptor CD14 was also able to induce activation of Src family kinases.

The 'MAPK2' level

The MAPKKs are typically activated by serine/threonine phos-

phorylation catalysed by one of the MAPK3s (Raf, MEKK). The MAPKKs (MEKs, MKKs, JNKKs, SEKs, RKK; see Figure 6 and Table 1) are a subset of protein kinases that possess unique dual (or mixed-lineage?) kinase specificities. When activated (e.g. by phosphorylation of Ser-222 in MKK1a [156]) the MAPKKs phosphorylate only very specific native MAPKs on both Thr and Tyr residues in a Thr-Xaa-Tyr motif, where Xaa is Glu, Pro or Gly.

One subgroup of the MAPKKs includes MEK1a, MEK1b and MEK2, which phosphorylate different ERKs [157] but nevertheless appear to elicit similar transcriptional and morphological responses [158]. Using dominant negative and constitutively active forms of MEK1a (MKK1a), Seger et al. [156] demonstrated that this signalling element controlled NIH-3T3 cell proliferation and morphology. Mansour et al. [159] have shown that NIH-3T3 cells expressing constitutively active forms of MEK were transformed, as judged from the formation of both foci in culture and tumours in nude mice. Cowley et al. [160] have shown that, in the appropriate cell context, activation of MEK is necessary and sufficient for growth-factor-induced proliferation (NIH-3T3 cells) or differentiation (PC12 cells).

A second subgroup of these dual-specificity kinases includes SEK/JNKK/MKK4. These are preferentially activated by MEKK1 and 2, but not by the Raf proteins, and their target proteins are the SAPKs/JNKs [161,162]. JNK was identified on the basis of its ability to phosphorylate the N-terminal region of c-Jun on Ser-63 and Ser-73; its activity is stimulated by inflammatory mediators (e.g. LPS, interferon, TNFα, IL-1) and UV light, and can be further augmented by oncogenic Ras [163]. The SAPKs were identified independently as protein kinases that are activated by stress (e.g. heat shock, inflammatory cytokines, inhibitors of protein synthesis, and DNA-damaging agents such as UV light, ionizing radiation, arabinosylcytosine, alkylating agents and topoisomerase inhibitors). For some of these the activating signal is conveyed by the c-Abl non-receptor tyrosine kinase, a nuclear and cytoplasmic protein that shares structural features with Src and activates SAPK via SEK1 [164]. The presence of both SAPK and SEK1 in the nucleus suggests that this cascade may be activated independently of strictly cytosolic or plasma-membrane-bound kinases [164]. Activated (e.g. by AraC) SAPK binds to the SH3 domain of GRB2, which in turn can bind to the (tyrosine phosphorylated) p85 subunit of PI3-K via its SH2 group, forming a SAPK–GRB2–PI3-K complex that effectively inhibits the lipid kinase and protein serine kinase activities of PI3-K [165]. The significance of this may be to favour the SEK $\rightarrow$ SAPK pathway over the MEK $\rightarrow$ ERK pathway.

The JNKs and SAPKs are the same set of proteins, and they are activated by SEK1, the same enzyme as JNKK [166]. Because of the greater efficiency with which MEKK1 activates JNKK/SEK/MKK4 as compared with MEK, it appears that (despite its name) these may be the normal substrates for MEKK1 in the cell, while the 'true' activators of MEK are the Rafs and possibly MEKK3. Since MEKK1 can at high concentrations activate MEK, it is not excluded that there are circumstances whereby MEK is a substrate for MEKK1 in vivo. Ras and the Ras-coupled agonist Raf-1 are inefficient activators of SEK and SAPK. Thus there appear to be two pathways under the influence of Ras (see Figure 6): (1) the Raf $\rightarrow$ MEK/MKK $\rightarrow$ ERK pathway activated by growth factors, mitogens and tumour promoters that is highly dependent upon Ras activation; and (2) the MEKK $\rightarrow$ SEK/JNKK/MKK4 $\rightarrow$ SAPK/JNK pathway that can be stimulated through Ras but is more strongly activated by various stressors and cytokines in a largely Ras-, Raf- and MEK-independent manner.

MKK3, along with MKK4, was first cloned from human brain tissue [167]. It, and the closely related MKK6, appear to be related to a yeast gene involved in responding to hyperosmotic conditions (pBS2), and are responsible for phosphorylating p38, which was cloned on the basis of its enhanced phosphorylation in a mouse pre-B cell line by LPS [168,169]. The sequence of p38 most closely resembles that of the yeast kinase HOG-1, which is involved in adaptation to osmotic stress and is the substrate for pBS2. IL-1 and hyperosmolar conditions also induce p38 phosphorylation. MKK3/6 preferentially phosphorylate p38, whereas MKK4 preferentially phosphorylates JNK [166,170]. p38, which is not significantly activated via Ras-controlled pathways, is efficiently activated by MKK3 and MKK6, and can in turn activate both ATF-2 and Elk-1 [171].

The MAPK level

The activated serine/threonine protein kinases collectively referred to here as 'MAPKs' constitute a superfamily of proteins that includes the ERKs, JNKs/SAPKs and p38/RK MAP kinases. These are uniquely identified by the Thr-Xaa-Tyr dual-phosphorylation motif, where Xaa is respectively Glu, Pro and Gly for the ERKs, JNKs/SAPKs and p38/RK/CSBP. Phosphorylation of both tyrosine and threonine residues, which are found in the activation segment of the kinase domain, is essential for full kinase activity of the MAPKs [172]. (In the literature, 'MAPK' may refer either specifically to the ERKs or more generally to this superfamily.) Upon activation, the activated MAPK may be translocated to the nucleus (if it is not already there), where it can phosphorylate targeted transcription factors [e.g. c-Jun, SAP-1, Elk-1, c-Myc, activating transcription factor (ATF)-2 and the cAMP response element-binding protein (CREB)]. [SAP-1 was cloned as the serum response factor (SRF) accessory protein [173]; it is not a 'stress-activated protein' and is not activated by JNK.] The MAPKs are proline-directed in the sense that they target only serine and threonine residues that are closely followed by one or more prolines in a motif recognized preferentially by a particular MAPK. Whereas the ERKs are strongly activated by one-pass receptor tyrosine kinases (e.g. the PDGF, EGF, CSF, NGF and FGF receptors), the JNK/SAPKs are potently activated by heterotrimeric-G$\alpha\beta\gamma$-protein-coupled receptors, such as the muscarinic acetylcholine receptor, in a Ras- and PKC-independent manner that also does not involve the ERKs [174].

ERK1 (p43/44^{MAPK}) and ERK2 (p41/42^{MAPK}) have been extensively studied; they are discriminating with regard to the target proteins they phosphorylate, recognizing a proline domain that identifies the substrate and producing (usually) multiple phosphorylations on a Ser/Thr-Pro motif. ERK1 targets Elk-1, a TCF that associates with the SRF on the serum response element, whereas ERK2 exhibits a preference for c-Myc [175]. Potential cytoplasmic targets of ERK1/2 phosphorylation include cytoskeletal elements (microtubule-associated proteins, tau), various kinases [ribosomal protein S6 kinase (both p70^{s6k} and p90^{rsk}), cytoplasmic phospholipase A2, SHP-2, glycogen synthase kinase-3 (GSK-3) and protamine kinase], and upstream signalling elements (e.g. the EGF receptor, SOS, Raf-1 and MEK) [157]. The extent to which phosphorylation of upstream signalling elements reflects significant negative-feedback regulation remains to be determined. Elk-1 and SAP-1 bind to an Ets domain in the serum response element in the c-*fos* promoter along with the SRF. Serum induction of c-*fos* transcription, stimulated by lysophosphatidic acid acting through a serpentine receptor and a heterotrimeric G-protein, is the result of ERK-induced phosphorylation of these serum-response-element-

bound proteins (Elk-1 or SAP-1; SRF) [176]. Elk-1 is also phosphorylated by the JNKs, which are activated by MEKK1 in response to UV radiation, heat shock, protein synthesis inhibition and other forms of stress [177].

Mouse p38 (RK) is a kinase at the MAPK level with the unique Thr-Gly-Tyr motif as the site of Thr/Tyr phosphorylation, which is accomplished by the MKK3/RKK dual-specificity kinase [178,179]. Raingeaud et al. [170], working with HeLa and COS cells, reported that UV light, osmotic shock, LPS, TNFα and IL-1 activated JNK and p38, but not the ERKs, whereas EGF and phorbol 12-myristate 13-acetate gave maximal activation of the ERKs but only a modest activation of JNK or p38. p38 did not phosphorylate cytosolic phospholipase A2, c-Myc or c-Jun, whereas it did phosphorylate ATF-2 and small heat-shock proteins, and stimulated expression of inflammatory cytokines. There are two versions of the human p38 protein, CSBP1 and CSBP2, that are splice variants differing in an internal 25-amino-acid sequence; when expressed in yeast, the protein kinase activity of CSBP1, but not CSBP2, was increased under hyperosmolar conditions [180].

The outline of several parallel pathways leading to activation of different subsets of MAPKs is taking shape (Figure 6), although the extent to which there is cross-talk among these pathways is unknown [181–183]. In some cases apparent cross-talk can instead result from the induced secretion of a protein or hormone that acts back on the cell in a paracrine manner [184]. In U937 human leukaemic cells, transient expression of a constitutively active MEK1 stimulated SAPK as well as ERK activity by an apparent intracellular signalling process [185]. Whether or not this happens under normal signalling conditions remains to be determined. It is important to note that, usually, multiple elements are activated by a given stimulus or receptor, generating parallel signalling tracks that may or may not be self-reinforcing. One consequence of this is that loss of an individual component may have a relatively small effect. The extent to which the signal is amplified at each step is unclear. The signal is unlikely to be amplified as it passes through a multiprotein complex, for example a MAPKK–MAPK complex. Upstream (e.g. Ras, Rac) and downstream (e.g. ERKs, JNKs) elements almost certainly do amplify the signal.

The 'MAPKAPK' level

This group of MAPK-activated protein kinases consists of various kinases phosphorylated by the MAPKs. Included are upstream signalling elements whose phosphorylation may modulate the signal.

Ribosomal protein S6 kinase, p90^{rsk} (which apparently is capable of phosphorylating the ribosomal protein S6 only *in vitro*; the related p70^{s6k} does the job *in vivo*), can enter the nucleus, phosphorylate c-Fos and thus contribute to the control of transcription of genes under the control of AP-1 [157,186]. It, along with c-Myc and cytoplasmic phospholipase A2, are among the proteins phosphorylated (but not necessarily activated) by the Ras $\rightarrow$ Raf $\rightarrow$ MEK $\rightarrow$ ERK pathway in response to the activation of the T-cell receptor [187]. Phospholipase A2, activated by a single serine phosphorylation, preferentially releases arachidonic acid, a precursor of prostaglandins and leukotrienes, from the 2-position of various phospholipids. Bohm et al. [188] reported that, in normal human melanocytes stimulated with various growth factors, p90^{rsk} was likely to be responsible for the activation of the transcription factor CREB by phosphorylating it. Ginty et al. [189] made somewhat similar observations in PC12 cells and primary rat cortical neurons stimulated with NGF, although in this case the Ras-dependent protein kinase responsible for activating CREB did not appear to be p90^{rsk}. The p90^{rsk} kinase has also been called MAPKAPK-1.

The authentic S6 kinase, p70^{s6k}, is strongly but indirectly activated by PI3-K and by PLCγ1 [190]. It is considered to regulate a range of signalling pathways by phosphorylation of important target proteins in addition to the ribosomal protein S6. Stimulation of p70^{s6k} via PI3-K may be mediated by certain PKC isoforms or by p65^{PAK}; stimulation by PLC is also likely to be via PKC, a known activator of p70^{s6k}, via formation of diacylglycerol. Interestingly, there is an 85 kDa splice variant of p70^{s6k} that possesses an N-terminal addition with a nuclear localization signal that places this isoform in the nucleus. Activation of p70^{s6k} is blocked by the immunosuppressive drug rapamycin, which has a number of effects on the cell, including the suppression of translation of specific mRNAs and inhibition of NFκB activation. CDC42/Rac contributes to p70^{s6k} activation also, possibly via p65^{PAK} [190].

MAPKAPK-2 is a 50 kDa protein kinase that can be activated in various cells (KB, HeLa, PC12, monocytes and macrophages) by chemical or physical stress (sodium arsenite, heat shock, osmotic stress) and by certain cytokines and inflammatory mediators (IL-1 and LPS). It is responsible for phosphorylation of the small heat-shock proteins Hsp25/Hsp27. Phosphorylation of Hsp27 facilitates actin polymerization, probably contributing to the repair of stress-damaged actin microfilaments [191]. Activation, by phosphorylation, of any two of the three residues Thr-222, Ser-272 and Thr-334 is accomplished by the 38–40 kDa protein p38/RK, which is itself activated by an RKK [169,178,179,192]. Although *in vitro* the ERKs can phosphorylate MAPKAPK-2, this does not typically happen *in vivo*, in that activators of the Ras → Raf → MEK → ERK pathway do not stimulate phosphorylation of MAPKAPK-2 and, conversely, strong activators of p38/RK do not typically cause significant phosphorylation of the ERKs. Exceptions to this include both LPS and hyperosmolality, each of which can activate both ERK and p38 [169]. MAPKAPK-3 has also been identified as a substrate for CSBP p38 that is activated by stress, LPS, IL-1 or TNFα [193].

GSK-3 has a larger role than simply that of regulating intermediary metabolism [194]. Its control of early *Drosophila* and *Xenopus* development suggests strongly a comparable role in mammalian development. Interestingly, it also has properties of a dual-specificity protein kinase in the MAPK group [195] and it functions to complete the activation of CREB previously phosphorylated by PKA on a nearby serine residue (hierarchical phosphorylation) [196]. It can also phosphorylate multiple metabolic enzymes, cytoskeletal proteins, and nuclear factors including c-Jun, CREB and p90^{rsk} [197]. GSK-3 itself is subject to multiple levels of control, including both serine/threonine and tyrosine phosphorylation, both activating and deactivating, by PKC and MAPKAPK-1 (p90^{rsk}) [198,199]. Thus GSK-3 is a candidate for a multiplex signalling element. The insulin-stimulated inhibition of GSK-3 is mediated by protein kinase B, which is also known as AKT/Rac, a serine/threonine protein kinase distantly related to the PKC family and regulated by *D*-3-phosphorylated phosphoinositides, products of PI3-K [200,201].

Some extracellular cytokines that do not typically activate Ras nevertheless activate downstream elements of Ras-mediated signal transduction pathways. Such cytokines include IL-1 [178], interferons [202] and TNFα [109]. For example, IL-1 activates SAPK/JNK, but not ERK1 or ERK2, by a pathway that does not require Ras, Raf-1 or MEK activation [203]. David et al. [204] discovered an interaction between ERK2 and the α subunit of the interferon α/β receptor, and suggested that ERK2 may regulate the interferon → JAK → STAT signalling cascade.

TNFα stimulates a sphingomyelinase activity that generates the second messenger ceramide, which appears to act via MEK1 to activate the SAPKs/JNKs, and possibly NFκB [205,206]. IL-3 appears to require Ras to deliver certain signals (e.g. for apoptosis) but not certain other signals (e.g. for growth stimulation), at least in a mouse pro-B-cell line [207]. Interestingly, growth hormone appears to stimulate a signalling pathway mediated by JAK2 activation of ERK2 in a Ras- and Raf-dependent process [208]. Clearly, there is the potential for an immense variety of different signalling pathways.

SIGNAL DOWN-REGULATION AND TRANSCRIPTION UP-REGULATION

Receptor signalling can be terminated in several ways. Phosphorylation by serine/threonine kinases activated as a consequence of receptor activation is one (negative feedback, e.g. by PKA, PKC or one of the MAPKs), and dephosphorylation of key phosphotyrosines by phosphotyrosine phosphatases (e.g. SHP-1) is another. In contrast to this easily reversed downregulation is the more permanent shut-off resulting from ligand-induced internalization of the receptor and its degradation in lysosomes. An interesting example of signal desensitization is provided by the EGF receptor vIII transforming mutant, which is deleted for much of the extracellular domain of the receptor and undergoes spontaneous homodimerization, autophosphorylation and constitutive activation. It binds much of the cell's GRB2 and strongly activates MEK; nevertheless, ERK1 and ERK2 are activated to a lesser extent and seem to resist further activation by various agonists, apparently because of the activation of a tyrosine phosphatase that negatively regulates their activity [209].

Phosphorylation of upstream signalling elements by activated downstream kinases is one mechanism for shutting down a signalling pathway. For example, phosphorylation of SOS (on serine/threonine residues in proline-rich motifs in the C-terminal domain) by ERK (p44/p42 MAPK) inhibits its ability to activate Ras and results in separation of an *intact* SOS–GRB2 complex from the activated receptor (e.g. SHC or the EGF receptor) [210,211]. Somewhat at odds with this are the reports that ERK activation, either by an insulin-stimulated Ras/MEK pathway or by osmotic shock acting via a MEK-dependent but Ras-independent pathway, led to phosphorylation of SOS and *dissociation* of the GRB2–SOS complex, terminating the ability of SOS to promote disengagement of GDP from Ras–GDP [212,212a]. These differences could, of course, reflect differences in the input signals and the regulatory circuits activated in the different cell types in these studies. Specific activation of JNK but not ERK by anisomycin, which inhibits protein synthesis and activates the stress-activated signalling pathways, did not cause GRB2–SOS dissociation, suggesting that SOS is not a substrate for JNK [213]. In cells with a wild-type Ras requiring GEF activity, this negative-feedback regulatory circuit attenuates the Ras-activated signal transduction pathway. The signal would be more persistent in cells containing a mutant oncogenic Ras that did not require GEF activity.

The length of time a signalling pathway remains active varies with the cell type and the receptor involved. For example, PC12 cells stimulated with FGF or NGF exhibited a persistent elevation of Ras–GTP levels and a sustained activation of the ERKs, which were translocated to the nucleus, resulting ultimately in differentiation as shown by neurite outgrowth. In the same cells stimulated with EGF, Ras–GTP levels increased only transiently and ERK activation was short-lived. Besides differences in the details of just which intracellular signalling molecules are con-

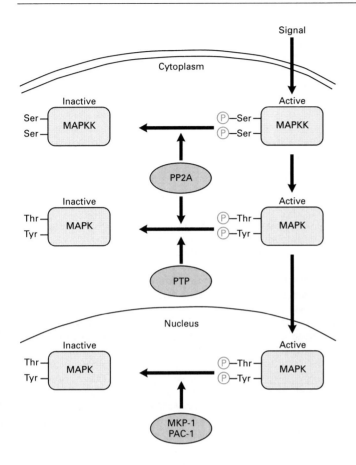

Signal

Cytoplasm

Inactive | Active

Ser—
Ser— MAPKK | Ⓟ—Ser
Ⓟ—Ser MAPKK

PP2A

Inactive | Active

Thr—
Tyr— MAPK | Ⓟ—Thr
Ⓟ—Tyr MAPK

PTP

Nucleus

Inactive | Active

Thr—
Tyr— MAPK | Ⓟ—Thr
Ⓟ—Tyr MAPK

MKP-1
PAC-1

Figure 8   Dephosphorylation of active signalling intermediates by protein phosphatases acting in the cytoplasm or the nucleus with varying specificities

PP2A and other serine/threonine phosphatases dephosphorylate phosphoserine and phosphothreonine residues as indicated. Several protein tyrosine phosphatases have the potential to dephosphorylate the pTyr moieties. PTP represents a generic protein tyrosine phosphatase. Dual-specificity protein phosphatases with a predominantly nuclear location include MKP-1 (previously known as 3CH134 and CL100) and PAC-1. (Redrawn from Biochim. Biophys. Acta, **1265** Keyse, S. M., 152–160, ©1995, with kind permission of Elsevier Science – NL, Sara Burgerhartstraat 25, 1055 KV Amsterdam, The Netherlands.)

trolled by each receptor, there is also the fact that the EGF receptor is more rapidly down-regulated than the NGF receptor through phosphorylation and internalization. Appropriate manipulation of EGF or NGF receptor levels can lead to either response, suggesting that the critical element is the intensity of ERK activation [181].

Many of the signalling elements are returned to their ground state by cytoplasmic and nuclear phosphatases (Figure 8) that remove the activating phosphates from the various intermediates [54]. A corollary of this is that inhibitors of phosphatase activity can mimic the effect of a kinase by prolonging the state of activation of a signalling intermediate. Protein serine/threonine phosphatases (e.g. PP1, PP4, PP5, PP2A, PP2B, PP2C) remove phosphates from protein phosphoserine and phosphothreonine residues [214]. Many of them are sensitive to okadaic acid. Barford [215] has reviewed structural studies of the protein phosphatases, noting their classification into four structurally distinct families and the fact that the depth of the catalytic site cleft determines whether the phosphatase will target a tyrosine phosphate or serine/threonine phosphate.

Various protein tyrosine phosphatase activities have been identified (e.g. CD45, SHP-1, SHP-2), and much current research is directed at discovering their specificities [216]. Because of the number of names given to the same protein tyrosine phosphatases, a uniform nomenclature has recently been proposed [52]. Thus SHP-1 (also known as SHP, PTP1C, SHPTP-1 and HCP) dephosphorylates many tyrosine-phosphorylated receptor proteins and has been implicated in attentuating the mitogenic response [216]. SHP-2 (also known as SYP, SHPTP-2, SHPTP-3, PTP2C and PTP1D) is a protein tyrosine phosphatase that appears to act immediately downstream of Ras as a positive mediator of growth factor signalling by a number of receptors (those for insulin, IGF-1 and EGF). Phosphatases (e.g. SHP-2) can also provide a positive signal by removing inhibitory phosphates, for example the phosphate in (unactivated) Src that obstructs an SH2 domain necessary for activity. It is not known whether SHP-2 phosphatase action itself is necessary for transmission of the signal from activated receptors such as the PDGF receptor, or whether its effect is because of its potential to associate with GRB2–SOS [217–219].

Dual-specificity protein phosphatases (e.g. MKP-1, PAC-1) are predominately nuclear proteins that dephosphorylate both phosphothreonine and phosphotyrosine residues. They exhibit considerable specificity for individual MAPKs [220]. Growth factor stimulation and many forms of physical or chemical trauma enhance the mRNA levels of these enzymes, suggesting an important role for PAC-1 and MKP-1 in modulating the signal transduction pathways initiated by mitogens and stress respectively.

In cells with wild-type Ras, activation of Ras in response to the binding of a ligand to an appropriate receptor is transient, and consequently the transmitted signal has the nature of a pulse. Importantly, the Ras-transmitted signal is also accompanied by parallel signals generated by other signalling elements mobilized by the receptor; some of these signals may serve in a negative-feedback capacity. As shown in Figure 6, the signal often results in the modification of one or more transcription factors, giving rise to changes in gene expression. These changes in gene expression are the consequence of a cell-specific integration of all the input signals, including both their intensity (i.e. number of activated receptors) and duration (which varies with the receptor activated), and although the complexity of the pathways and their cross-talk is bewildering, there do seem to be some consistent patterns emerging [221].

Oncogenic forms of Ras have been shown to activate all three of the protein phosphorylation cascades illustrated in Figure 6 [94]. In contrast to wild-type Ras, the signal from a mutant oncogenic Ras (a transformation signal) is continuous, is not accompanied by independently generated upstream signals and results in the perpetual stimulation of certain downstream signalling elements, probably giving rise to an altered relationship among the signalling intermediates and a permanent state of modification of particular downstream transcription factors [16]. As a consequence of the particular Ras mutation or the permanent stimulation of certain pathways, there may also be changes in the way Ras interacts with proteins such as Ral-GDS, PI3-K, and the GAPs, GEFs, Rafs and MEKKs. Many of the identified Ras response elements in promoters and enhancers involve either AP-1 (Fos/Jun) or Ets transcription factors, often in situations where the two binding sites overlap. The resulting changes include the increased expression of genes that are important in making cells tumorigenic and metastatic, e.g. proteinases and adhesion molecules [222,223]. It is interesting to note that in at least some cell types both Rac and Rho make important contributions to Ras transformation [224]. Finally, the fact that most of the studies that have led to our current picture of all these signalling pathways have been done with immortal cell lines must not be ignored. It remains to be determined how the immortalization process has impacted on these signalling pathways.

PERSPECTIVE: IMPLICATIONS OF MULTIPLE SIGNALLING CASCADES AND MULTIPLEX SIGNAL TRANSMISSION

When a receptor is activated, a number of signalling pathways are typically stimulated to various degrees. Receptor, non-receptor and receptor-associated protein tyrosine kinases phosphorylate themselves and specific target proteins. Other proteins are then activated by virtue of their association with tyrosine phosphate groups on these proteins, directly or via an adaptor. Heterotrimeric G-proteins coupled to serpentine receptors trigger signalling pathways through both their Gα and G$\beta\gamma$ subunits. The resulting cascades of interactions provide a self-reinforcing stability to the signal; the contribution of any one element is important to, but may not be critical to, the successful delivery of the signal. The branching of the pathways permits a degree of signal diversification, while at the same time cross-talk among pathways permits one pathway to reinforce or attenuate the signal transmitted by a second pathway. Signals initiated by different receptors are distinguished by the different mix of pathways stimulated and the intensity of their stimulation. Thus NGF stimulation of 3T3 cells expressing the human TrkA receptor inhibits proliferation, possibly via the induction of the cyclin-dependent kinase inhibitor p21$^{Cip1/WAF1}$; this occurs despite the fact that NGF is more effective at activating the p42 and p44 ERKs than is EGF, which stimulates proliferation [225]. It may be instructive that EGF binds GRB2–SOS directly, whereas TrkA acts via SHC phosphorylation and the formation of SHC–GRB2–SOS complexes [181].

A very important principle to appreciate is that most of these pathways in the cell are likely to be functioning at a modest level in a healthy cell in normal (unperturbed) conditions. The balance among the pathways results in a homoeostatic state. Perturbations of different sorts upset the balance in different ways, leading for example to cell differentiation or proliferation on the one hand, or to growth inhibition or apoptosis on the other. For example, removal of NGF from PC12 cells reduces the stimulation of the Ras → Raf → MEK → ERK1/2 pathway and tilts the balance towards net activation of the MEKK → SEK → JNK and MKK3 → p38 pathways, resulting in apoptosis [226]. Similarly, activation of the B-cell surface IgM receptor activates the Ras → Raf → MEK → ERK (but not the JNK) pathway, leading to cell activation, differentiation or apoptosis (depending upon other variables), whereas activation of the CD40 receptor (a member of the TNF receptor family that recognizes a ligand expressed on activated T-cells) activates JNK/SAPK, but not ERK, and inhibits the apoptotic response induced by anti-IgM [227]. Insulin promotes the survival of neurons in culture, perhaps by virtue of the fact that it negatively regulates p38 kinase activity [228].

The fact that some of the proteins in a pathway become multiply phosphorylated (e.g. Raf-1, MEK, GSK-3) raises the possibility that more than one signal may be transmitted simultaneously via the same pathway (Figure 9). This is known as multiplex signalling. For example, MEK is preferentially phosphorylated on Ser-218 by MEKK1, but on both Ser-218 and Ser-

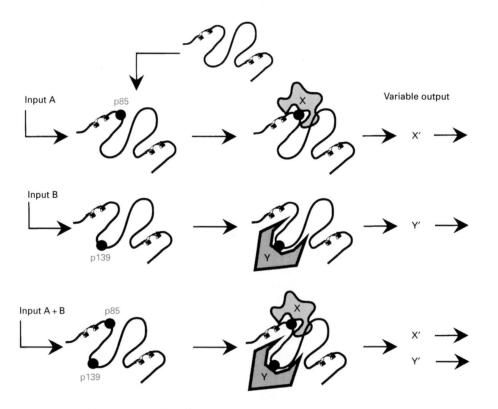

Figure 9   Model for multiplex signalling via a multiply phosphorylated protein

In this hypothetical example, one input signal leads to phosphorylation of amino acid 85, whereas input of a second signal phosphorylates residue 139. Each of the phosphorylated amino acids must interact independently with a different downstream signalling element, providing a distinct output. When both amino acids are phosphorylated, then the protein conveys both signals simultaneously.

222 by Raf-1 [4]. Is it possible that the type of signal MEK delivers is different in these two cases? As illustrated by the hypothetical example in Figure 9, when amino acid 85 is phosphorylated it interacts with and passes the signal on to protein X, whereas when residue 139 is phosphorylated it interacts with and passes a signal on to protein Y; phosphorylation of both amino acids leads to both signals being transmitted. Selective mutation of one or other of the relevant amino acids followed by determination of the type of signal transmitted would give some indication of whether different signals are being sent or not. Activating and inhibiting phosphorylations of a particular element, e.g. a transcription factor, are a form of multiplex signalling only in the sense that the second signal is a Null signal. If multiplex signalling is occurring, it adds yet another layer of complexity to the signalling process, one that will be as fascinating as it will be difficult to sort out in future research.

The comments of various colleagues, especially Ann Chambers, Mike Kiledjian, Mike Ostrowski and Ren Ping Zhou, on earlier versions of the manuscript were much appreciated. I thank Kathleen Curtis for preparing the illustrations. Because of the vast number of publications in this field, references have for the most part been restricted to representative research reports and reviews published in the last 2 years; sincere apologies are offered to those whose work has not been cited.

REFERENCES

1 Maher J., Baker, D. A., Manning, M., Dibb, N. J. and Roberts, I. A. G. (1995) Oncogene **11**, 1639–1647

2 Khosravi-Far, R. and Der, C. J. (1994) Cancer Metastasis Rev. **13**, 67–89

3 van der Geer, P., Hunter, T. and Lindberg, R. A. (1994) Annu. Rev. Cell Biol. **10**, 251–337

4 Malarkey, K., Belham, C. M., Paul, A., Graham, A., McLees, A., Scott, P. H. and Plevin, R. (1995) Biochem. J. **309**, 361–375

5 Wagner, A. C. and Williams, J. A. (1994) Am. J. Physiol. **266**, G1–G14

6 Nuoffer, C. and Balch, W. E. (1995) Annu. Rev. Biochem. **63**, 949–990

7 Rothman, J. E. (1994) Nature (London) **372**, 55–63

8 Vojtek, A. B. and Cooper, J. A. (1995) Cell **82**, 527–529

9 Chant, J. and Stowers, L. (1995) Cell **81**, 1–4

10 Ridley, A. J. (1995) Curr. Opin. Genet. Dev. **5**, 24–30

11 Kim, S.-H., Prive, G. G. and Milburn, M. V. (1993) Handb. Exp. Pharmacol. **108**, 177–194

12 Lowy, D. R. and Willumsen, B. M. (1993) Annu. Rev. Biochem. **62**, 851–891

13 Polakis, P. and McCormick, F. (1993) J. Biol. Chem. **268**, 9157–9160

14 Yoder-Hill, J., Golubic, M. and Stacey, D. W. (1995) J. Biol. Chem. **270**, 27615–27621

15 Quilliam, L. A., Khosravi-Far, R., Huff, S. Y. and Der, C. J. (1995) BioEssays **17**, 395–404

16 Lowe, P. N. and Skinner, R. H. (1994) Cell. Signalling **6**, 109–123

17 Gutmann, D. H., Geist, R. T., Wright, D. E. and Snider, W. D. (1995) Cell Growth Differ. **6**, 315–323

18 Maekawa, M., Li, S., Iwamatsu, A., Morishita, T., Yokota, K., Imai, Y., Kohsaka, S., Nakamura, S. and Hattori, S. (1994) Mol. Cell. Biol. **14**, 6879–6885

19 Cullen, P. J., Hsuan, J. J., Truong, O., Letcher, A. J., Jackson, T. R., Dawson, A. P. and Irvine, R. F. (1995) Nature (London) **376**, 527–530

20 Takai, Y., Sasaki, T., Tanaka, K. and Nakanishi, H. (1995) Trends Biochem. Sci. **20**, 227–231

21 Henkemeyer, M., Rossi, D. J., Holmyard, D. P., Puri, M. C., Mbamalu, G., Harpal, K., Shih, T. S., Jacks, T. and Pawson, T. (1995) Nature (London) **377**, 695–701

22 Vogel, K. S., Brannan, C. I., Jenkins, N. A., Copeland, N. G. and Parada, L. F. (1995) Cell **82**, 733–742

23 Hori, Y., Beeler, J. F., Sakaguchi, K., Tachibana, M. and Miki, T. (1994) EMBO J. **13**, 4776–4786

24 Huby, R. D. J., Carlile, G. W. and Ley, S. C. (1995) J. Biol. Chem. **270**, 30241–30244

25 Romero, F., Dargemont, C., Pozo, F., Reeves, W. H., Camonis, J., Gisselbrecht, S. and Fischer, S. (1995) Mol. Cell. Biol. **16**, 37–44

25a Nomanbhoy, T. K. and Cerione, R. A. (1996) J. Biol. Chem. **271**, 10004–10009

25b Chuang, T.-H., Xu, X., Kaartinen, V., Heisterkamp, N., Groffen, J. and Bokoch, G. M. (1995) Proc. Natl. Acad.Sci. U.S.A. **92**, 10282–10286

26 Cantor, S. B., Urano, T. and Feig, L. A. (1995) Mol. Cell. Biol. **15**, 4578–4584

27 Urano, T., Emkey, R. and Feig, L. A. (1996) EMBO J. **15**, 810–816

28 Park, S.-H. and Weinberg, R. A. (1995) Oncogene **11**, 2349–2355

29 Nobes, C. D. and Hall, A. (1995) Cell **81**, 53–62

30 Kozma, R., Ahmed, S., Best, A. and Lim, L. (1995) Mol. Cell. Biol. **15**, 1942–1952

31 Khosravi-Far, R., Solski, P. A., Clark, G. J., Kinch, M. S. and Der, C. J. (1995) Mol. Cell. Biol. **15**, 6443–6453

32 Heldin, C.-H. (1995) Cell **80**, 213–223

33 Ihle, J. N. (1995) Nature (London) **377**, 591–594

34 Claesson-Welsh, L. (1994) J. Biol. Chem. **269**, 32023–32026

35 Pawson, T. (1995) Nature (London) **373**, 573–580

36 Bork, P. and Margolis, B. (1995) Cell **80**, 693–694

37 van der Geer, P. and Pawson, T. (1995) Trends Biochem. Sci. **20**, 277–280

38 Zhou, M.-M., Ravichandran, K. S., Olejniczak, E. T., Petros, A. M., Meadows, R. P., Sattler, M., Harlan, J. E., Wade, W. S., Burakoff, S. J. and Fesik, S. W. (1995) Nature (London) **378**, 584–592

39 Ladbury, J. E., Lemmon, M. A., Zhou, M., Green, J., Botfield, M. C. and Schlessinger, J. (1995) Proc. Natl. Acad. Sci. U.S.A. **92**, 3199–3203

40 Feng, S., Chen, J. K., Yu, H., Simon, J. A. and Schreiber, S. L. (1994) Science **266**, 1241–1247

41 Hu, Q., Milfay, D. and Williams, L. T. (1995) Mol. Cell. Biol. **15**, 1169–1174

42 de Vries-Smits, A. M. M., Pronk, G. J., Medema, J. P., Burgering, B. M. T. and Bos, J. L. (1995) Oncogene **10**, 919–925

43 Decker, S. J. (1993) J. Biol. Chem. **268**, 9176–9179

44 Gotoh, N., Tojo, A., Muroya, K., Hashimoto, Y., Hattori, S., Nakamura, S., Takenawa, T., Yazaki, Y. and Shibuya, M. (1994) Proc. Natl. Acad. Sci. U.S.A. **91**, 167–171

45 Sato, K.-i., Sato, A., Aoto, M. and Fukami, Y. (1995) Biochem. Biophys. Res. Commun. **215**, 1078–1087

46 Burgering, B. M. T., de Vries-Smits, A. M. M., Medema, R. H., van Weeren, P. C., Tertoolen, L. G. J. and Bos, J. L. (1993) Mol. Cell. Biol. **13**, 7248–7256

47 Songyang, Z., Gish, G., Mbamalu, G., Pawson, T. and Cantley, L. C. (1995) J. Biol. Chem. **270**, 26029–26032

48 Songyang, Z. and Cantley, L. C. (1995) Trends Biochem. Sci. **20**, 470–475

49 Yao, R. and Cooper, G. M. (1995) Oncogene **11**, 1607–1614

50 McGlade, J., Brunkhorst, B., Anderson, D., Mbamalu, G., Settleman, J., Dedhar, S., Rozakis-Adcock, M., Chen, L. B. and Pawson, T. (1993) EMBO J. **12**, 3073–3081

51 Johnson, M. R., Declue, J. E., Felzmann, S., Vass, W. C., Xu, G., White, R. and Lowy, D. R. (1994) Mol. Cell. Biol. **14**, 641–645

52 Adachi, M., Fischer, E. H., Ihle, J., Imai, K., Jirik, F., Neel, B., Pawson, T., Shen, S.-H., Thomas, M., Ullrich, A. and Zhao, Z. (1996) Cell **85**, 15

53 Li, S., Janosch, P., Tanji, M., Rosenfeld, G. C., Waymire, J. C., Mischak, H., Kolch, W. and Sedivy, J. M. (1995) EMBO J. **14**, 685–696

54 Sun, H. and Tonks, N. K. (1994) Trends Biochem. Sci. **19**, 480–485

55 Peng, Z.-Y. and Cartwright, C. A. (1995) Oncogene **11**, 1955–1962

56 Chen, Q., Kinch, M. S., Lin, T. H., Burridge, K. and Juliano, R. L. (1994) J. Biol. Chem. **269**, 26602–26605

57 Schlaepfer, D. D., Hanks, S. K., Hunter, T. and van der Geer, P. (1994) Nature (London) **372**, 786–791

58 Hunter, T. (1995) Cell **83**, 1–4

59 Divecha, N. and Irvine, R. F. (1995) Cell **80**, 269–278

60 Rameh, L. E., Chen, S.-S. and Cantley, L. C. (1995) Cell **83**, 821–830

61 Nakanishi, H., Brewer, K. A. and Exton, J. H. (1993) J. Biol. Chem. **268**, 13–16

62 Liu, J.-P. (1996) Mol. Cell. Endocrinol. **116**, 1–29

63 Erpel, T. and Courtneidge, S. A. (1995) Curr. Opin. Cell Biol. **7**, 176–182

64 Jiang, H., Luo, J.-Q., Urano, T., Frankel, P., Lu, Z., Foster, D. A. and Feig, L. A. (1995) Nature (London) **378**, 409–412

65 Xie, W. and Herschman, H. R. (1995) J. Biol. Chem. **270**, 27622–27628

66 Downward, J. (1994) FEBS Lett. **338**, 113–117

67 Aronheim, A., Engelberg, D., Li, N., Al-Alawi, N., Schlessinger, J. and Karin, M. (1994) Cell **78**, 949–961

68 Xie, Y., Pendergast, A. M. and Hung, M.-C. (1995) J. Biol. Chem. **270**, 30717–30724

69 Klint, P., Kanda, S. and Claesson-Welsh, L. (1995) J. Biol. Chem. **270**, 23337–23344

70 Kavanaugh, W. M. and Williams, L. T. (1994) Science **268**, 1862–1865

71 Blaikie, P., Immanuel, D., Wu, J., Li, N., Yajnik, V. and Margolis, B. (1994) J. Biol. Chem. **269**, 32031–32034

72 Prigent, S. A., Pillay, T. S., Ravichandran, K. S. and Gullick, W. J. (1995) J. Biol. Chem. **270**, 22097–22100

73 Reference deleted

74 Strader, C. D., Fong, T. M., Graziano, M. P. and Tota, M. R. (1995) FASEB J. **9**, 745–754

75 DeVivo, M. and Iyengar, R. (1994) Mol. Cell. Endocrinol. **100**, 65–70

76 Inglese, J., Koch, W. J., Touhara, K. and Lefkowitz, R. J. (1995) Trends Biochem. Sci. **20**, 151–156

77 Burgering, B. M. T. and Bos, J. L. (1995) Trends Biochem. Sci. **20**, 18–22

78 Faure, M., Voyno-Yasenetskaya, T. A. and Bourne, H. R. (1994) J. Biol. Chem. **269**, 7851–7854

79 Crespo, P., Xu, N., Simonds, W. F. and Gutkind, J. S. (1994) Nature (London) **369**, 418–420

80 Touhara, K., Hawes, B. E., van Biesen, T. and Lefkowitz, R. J. (1995) Proc. Natl. Acad. Sci. U.S.A. **92**, 9284–9287

81 van Biesen, T., Hawes, B. E., Luttrell, D. K., Krueger, K. M., Touhara, K., Porfiri, E., Sakaue, M., Luttrell, L. M. and Lefkowitz, R. J. (1995) Nature (London) **376**, 781–784

82 Luttrell, L. M., van Biesen, T., Hawes, P. E., Koch, W. J., Touhara, K. and Lefkowitz, R. J. (1995) J. Biol. Chem. **270**, 16495–16498

83 Crespo, P., Cachero, T. G., Xu, N. and Gutkind, J. S. (1995) J. Biol. Chem. **270**, 25259–25265

84 Musacchio, A., Gibson, T., Rice, P., Thompson, J. and Saraste, M. (1993) Trends Biochem. Sci. **18**, 343–348

85 Harlan, J. E., Hajduk, P. J., Yoon, H. S. and Fesik, S. W. (1994) Nature (London) **371**, 168–170

86 Abrams, C. S., Wu, H., Zhao, W., Belmonte, E., White, D. and Brass, L. F. (1995) J. Biol. Chem. **270**, 14485–14492

87 Hyvonen, M., Macias, M. J., Nilges, M., Oschkinat, H., Saraste, M. and Wilmanns, M. (1995) EMBO J. **14**, 4676–4685

88 Yao, L., Kawakami, Y. and Kawakami, T. (1994) Proc. Natl. Acad. Sci. U.S.A. **91**, 9175–9179

89 Lemmon, M. A., Ferguson, K. M., O'Brien, R., Sigler, P. B. and Schlessinger, J. (1995) Proc. Natl. Acad. Sci. U.S.A. **92**, 10472–10476

90 Lee, S. B. and Rhee, S. G. (1995) Curr. Opin. Cell Biol. **7**, 183–189

91 Marshall, M. S. (1995) FASEB J. **9**, 1311–1318

92 Blank, J. L., Gerwins, P., Elliott, E. M., Sather, S. and Johnson, G. L. (1996) J. Biol. Chem. **271**, 5361–5368

93 Joneson, T., White, M. A., Wigler, M. H. and Bar-Sagi, D. (1996) Science **271**, 810–812

94 Waskiewicz, A. J. and Cooper, J. A. (1995) Curr. Opin. Cell Biol. **7**, 798–805

95 Elion, E. A. (1995) Trends Cell Biol. **5**, 322–327

96 Zhu, A. X., Zhao, Y., Moller, D. E. and Flier, J. S. (1994) Mol. Cell. Biol. **14**, 8202–8211

97 English, J. M., Vanderbilt, C. A., Xu, S., Marcus, S. and Cobb, M. H. (1995) J. Biol. Chem. **270**, 28897–28902

98 Newton, A. C. (1995) J. Biol. Chem. **270**, 28495–28498

99 Manser, E., Leung, T., Salihuddin, H., Zhao, Z.-S. and Lim, L. (1994) Nature (London) **367**, 40–46

100 Jakobi, R., Chen, C.-J., Tuazon, P. T. and Traugh, J. A. (1996) J. Biol. Chem **271**, 6206–6211

101 Knaus, U. G., Morris, S., Dong, H.-J., Chernoff, J. and Bokoch, G. M. (1995) Science **269**, 221–223

102 Coso, O. A., Chiariello, M., Yu, J.-C., Teramoto, H., Crespo, P., Xu, N., Miki, T. and Gutkind, J. S. (1995) Cell **81**, 1137–1146

103 Bagrodia, S., Derijard, B., Davis, R. J. and Cerione, R. A. (1995) J. Biol. Chem. **270**, 27995–27998

104 Minden, A., Lin, A., Claret, F.-X., Abo, A. and Karin, M. (1995) Cell **81**, 1147–1157

105 Qiu, R.-G., Chen, J., Kirn, D., McCormick, F. and Symons, M. (1995) Nature (London) **374**, 457–459

106 Zhang, S., Han, J., Sells, M. A., Chernoff, J., Knaus, U. G., Ulevitch, R. J. and Bokoch, G. M. (1995) J. Biol. Chem. **270**, 23934–23936

107 Wu, X., Noh, S. J., Zhou, G., Dixon, J. E. and Guan, K.-L. (1996) J. Biol. Chem. **271**, 3265–3271

108 Jelinek, T., Catling, A. D., Reuter, C. W. M., Moodie, S. A., Wolfman, A. and Weber. M. J. (1994) Mol. Cell. Biol. **14**, 8212–8218

109 Minden, A., Lin, A., McMahon, M., Lange-Carter, C., Derijard, B., Davis, R. J., Johnson, G. L. and Karin, M. (1994) Science **266**, 1719–1723

110 Lange-Carter, C. A. and Johnson, G. L. (1994) Science **265**, 1458–1561

111 Matsuda, S., Kawasaki, H., Moriguchi, T., Gotoh, Y. and Nishida, E. (1995) J. Biol. Chem. **270**, 12781–12786

112 Lander, H. M., Ogiste, J. S., Teng, K. K. and Novogrodsky, A. (1995) J. Biol. Chem. **270**, 21195–21198

113 Fialkow, L., Chan, C. K., Rotin, D., Grinstein, S. and Downey, G. P. (1994) J. Biol. Chem. **269**, 31234–31242

114 Chen, Q., Olashaw, N. and Wu, J. (1995) J. Biol. Chem. **270**, 28499–28502

115 Russell, M., Lange-Carter, C. A. and Johnson, G. L. (1995) J. Biol. Chem. **270**, 11757–11760

116 Xu, S., Robbins, D., Frost, J., Dang, A., Lange-Carter, C. and Cobb, M. H. (1995) Proc. Natl. Acad. Sci. U.S.A. **92**, 6808–6812

117 Chatani, Y., Tanimura, S., Miyoshi, N., Hattori, A., Sato, M. and Kohno, M. (1995) J. Biol. Chem. **270**, 30686–30692

118 Yamaguchi, K., Shirakabe, K., Shibuya, H., Irie, K., Oishi, I., Ueno, N., Taniguchi, T., Nishida, E. and Matsumoto, K. (1995) Science **270**, 2008–2011

119 Salmeron, A., Ahmad, T. B., Carlile, G. W., Pappin, D., Narsimhan, R. P. and Ley, S. C. (1996) EMBO J. **15**, 817–826

120 Dorow, D. S., Devereux, L., Tu, G. F., Price, G., Nicholl, J. K., Sutherland, G. R. and Simpson, R. J. (1995) Eur. J. Biochem. **234**, 492–500

121 Hirai, S., Izawa, M., Osada, S.-I., Spyrou, G. and Ohon, S. (1996) Oncogene **12**, 641–650

122 Reuter, C. W. M., Catling, A. D., Jelinek, T. and Weber, M. J. (1995) J. Biol. Chem. **270**, 7644–7655

122a Hill, C. S. and Treisman, R. (1995) EMBO J. **14**, 5037–5047

123 Jaiswal, R. K., Moodie, S. A., Wolfman, A. and Landreth, G. E. (1994) Mol. Cell. Biol. **14**, 6944–6953

124 Pritchard, C. A., Samuels, M. L., Bosch, E. and McMahon, M. (1995) Mol. Cell. Biol. **15**, 6430–6442

125 Daum, G., Eisenmann-Tappe, I., Fries, H.-W., Troppmair, J. and Rapp, U. R. (1994) Trends Biochem. Sci. **19**, 474–480

126 Schulte, T. W., Blagosklonny, M. V., Ingui, C. and Necker, L. (1995) J. Biol. Chem. **270**, 24585–24588

127 Hu, C.-D., Kariya, K.-i., Tamada, M., Akasaka, K., Shirouzu, M., Yokoyama, S. and Kataoka, T. (1995) J. Biol. Chem. **270**, 30274–30277

128 Druggan, J. K., Khosravi-Far, R., White, M. A., Der, C. J., Sung, Y.-., Hwang, Y.-W. and Campbell, S. L. (1996) J. Biol. Chem. **271**, 233–237

129 Leevers, S. J., Paterson, H. F. and Marshall, C. J. (1994) Nature (London) **369**, 411–414

130 Stokoe, D., Macdonald, S. G., Cadwallader, K., Symons, M. and Hancock, J. F. (1994) Science **264**, 1463–1467

131 Shirouzu, M., Koide, H., Fujita-Yoshigaki, J., Oshiro, H., Toyama, Y., Yamasaki, K., Fuhrman, S. A., Villacranca, E., Kaziro, Y. and Yokoyama, S. (1994) Oncogene **9**, 2153–2157

132 Marais, R., Light, Y., Paterson, H. F. and Marshall, C. J. (1995) EMBO J. **14**, 3136–3145

133 Bjorkoy, G., Overvatn, A., Diaz-Meco, M. T., Moscat, J. and Johansen, T. (1995) J. Biol. Chem. **270**, 21299–21306

134 Therrien, M., Chang, H. C., Solomon, N. M., Karim, F. D., Wassarman, D. A. and Rubin, G. M. (1995) Cell **83**, 879–888

135 Jelinek, T., Dent, P., Sturgill, T. W. and Weber, M. J. (1996) Mol. Cell. Biol. **16**, 1027–1034

136 Morrison, D. K., Heidecker, G., Rapp, U. R. and Copeland, T. D. (1993) J. Biol. Chem. **268**, 17309–17316

137 Dent, P., Jelinek, T., Morrison, D. K., Weber, M. J. and Sturgill, T. W. (1995) Science **268**, 1902–1906

138 Zheng, C.-F. and Guan, K.-L. (1994) EMBO J. **13**, 1123–1131

139 Catling, A. D., Schaeffer, H.-J., Reuter, C. W. M., Reddy, G. R. and Weber, M. J. (1995) Mol. Cell. Biol. **15**, 5214–5225

140 Li, S. and Sedivy, J. M. (1993) Proc. Natl. Acad. Sci. U.S.A. **90**, 9247–9251

141 Jamal, S. and Ziff, E. B. (1994) Oncogene **10**, 2095–2101

142 Galaktionov, K., Jessus, C. and Beach, D. (1995) Genes Dev. **9**, 1046–1058

143 Galaktionov, K., Lee, A. K., Eckstein, J., Draetta, G., Meckler, J., Loda, M. and Beach, D. (1995) Science **269**, 1575–1577

144 Hafner, S., Adler, H. S., Mischak, H., Janosch, P., Heidecker, G., Wolfman, A., Pippig, S., Lohse, M., Ueffing, M. and Kolch, W. (1994) Mol. Cell. Biol. **14**, 6696–6703

145 Erhardt, P., Troppmair, J., Rapp, U. R. and Cooper, G. M. (1995) Mol. Cell. Biol. **15**, 5524–5530

146 Yao, B., Zhang, Y., Delikat, S., Mathias, S., Basu, S. and Kolesnick, R. (1995) Nature (London) **378**, 307–310

147 Aitken, A. (1995) Trends Biochem. Sci. **20**, 95–97

148 Bonnefoy-Berard, N., Liu, Y.-C., von Willebrand, M., Sung, A., Elly, C., Mustelin, T., Yoshida, H., Ishizaka, K. and Altman, A. (1995) Proc. Natl. Acad. Sci. U.S.A. **92**, 10142–10146

149 Acs, P., Szallasi, Z., Kazanietz, M. G. and Blumberg, P. M. (1995) Biochem. Biophys. Res. Commun. **216**, 103–109

150 Muslin, A. J., Tanner, J. W., Allen, P. M. and Shaw, A. S. (1996) Cell **84**, 889–897

151 Suen, K.-L., Bustelo, X. R. and Barbacid, M. (1995) Oncogene **11**, 825–831

152 Li, S., Janosch, P., Tanji, M., Rosenfeld, G. C., Waymire, J. C., Mischak, H., Kolch, W. and Sedivy, J. M. (1995) EMBO J. **14**, 685–696

153 Michaud, N. R., Fabian, J. R., Mathes, K. D. and Morrison, D. K. (1995) Mol. Cell. Biol. **15**, 3390–3397

154 Rommel, C., Radziwill, G., Lovric, J., Noeldeke, J., Heinicke, T., Jones, D., Aitken, A. and Moelling, K. (1996) Oncogene **12**, 609–619

154a Braselmann, S. and McCormick, F. (1995) EMBO J. **14**, 4839–4848

155 Buscher, D., Hipskind, R. A., Krautwald, S., Reimann, T. and Caccarini, M. (1995) Mol. Cell. Biol. **15**, 466–475

156 Seger, R., Seger, D., Reszka, A. A., Munar, E. S., Eldar-Finkelman, H., Dobrowolska, G., Jensen, A. M., Campbell, J. S., Fischer, E. H. and Krebs, E. G. (1994) J. Biol. Chem. **269**, 25699–25709

157 Seger, R. and Krebs, E. G. (1995) FASEB J. **9**, 726–735
158 Mansour, S. J., Candia, J. M., Gloor, K. K. and Ahn, N. G. (1996) Growth Differ. **7**, 243–250
159 Mansour, S. J., Matten, W. T., Hermann, A. S., Candia, J. M., Rong, S., Fukasawa, K., Vande Woude, G. F. and Ahn, N. G. (1994) Science **265**, 966–970
160 Cowley, S., Paterson, H., Kemp, P. and Marshall, C. J. (1994) Cell **77**, 841–852
161 Sanchez, I., Hughes, R. T., Mayer, B. J., Yee, K., Woodgett, J. R., Avruch, J., Kyriakis, J. M. and Zon, L. I. (1994) Nature (London) **372**, 794–798
162 Lin, A., Minden, A., Martinetto, H., Claret, F.-X., Lange-Carter, C., Mercurio, F., Johnson, G. L. and Karin, M. (1995) Science **268**, 286–290
163 Derijard, B., Hibi, M., Wu, I.-H., Barrett, T., Su, B., Deng, T., Karin, M. and Davis, R. J. (1994) Cell **76**, 1025–1037
164 Kharbanda, S., Pandey, P., Ren, R., Mayer, B., Zon, L. and Kufe, D. (1995) J. Biol. Chem. **270**, 30278–30281
165 Saleem, A., Datta, R., Yuan, Z.-M., Kharbanda, S. and Kufe, D. (1995) Cell Growth Differ. **6**, 1651–1658
166 Yan, M., Dal, T., Deak, J. C., Kyriakis, J. M., Zon, L. I., Woodgett, J. R. and Templeton, D. J. (1994) Nature (London) **372**, 798–800
167 Derijard, B., Raingeaud, J., Barrett, T., Wu, I.-H., Han, J., Ulevitch, R. J. and Davis, R. J. (1995) Science **267**, 682–685
168 Han, J., Lee, J.-D., Jiang, Y., Li, Z., Feng, L. and Ulevitch, R. J. (1996) J. Biol. Chem. **271**, 2886–2891
169 Han, J., Lee, J.-D., Bibbs, L. and Ulevitch, R. J. (1994) Science **265**, 808–811
170 Raingeaud, J., Gupta, S., Rogers, J. S., Dickens, M., Han, J., Ulevitch, R. J. and Davis, R. J. (1995) J. Biol. Chem. **270**, 7420–7426
171 Raingeaud, J., Whitmarsh, A. J., Barrett, T., Derijard, B. and Davis, R. J. (1996) Mol. Cell. Biol. **16**, 1247–1255
172 Johnson, L. N., Noble, M. E. M. and Owen, D. J. (1996) Cell **85**, 149–158
173 Coso, O. A., Chiariello, M., Kalinec, G., Kyriakis, J. M., Woodgett, J. and Gutkind, J. S. (1995) J. Biol. Chem. **270**, 5620–5624
174 Dalton, S. and Treisman, R. (1992) Cell **68**, 597–612
175 Chuang, C.-F. and Ng, S.-Y. (1994) FEBS Lett. **346**, 229–234
176 Treisman, R. (1995) EMBO J. **14**, 4905–4913
177 Cavigelli, M., Dolfi, F., Claret, F.-X. and Karin, M. (1995) EMBO J. **14**, 5957–5964
178 Freshney, N. W., Rawlinson, L., Guesdon, F., Jones, E., Cowley, S., Hsuan, J. and Saklatvala, J. (1994) Cell **78**, 1039–1049
179 Rouse, J., Cohen, P., Trigon, S., Morange, M., Alonso-Llamazares, A., Zamanillo, D., Hunt, T. and Nebreda, A. R. (1994) Cell **78**, 1027–1037
180 Kumar, S., McLaughlin, M. M., McDonnell, P. C., Lee, J. C., Livi, G. P. and Young, P. R. (1995) J. Biol. Chem. **270**, 29043–29046
181 Marshall, C. J. (1995) Cell **80**, 179–185
182 Cobb, M. H. and Goldsmith, E. J. (1995) J. Biol. Chem. **270**, 14843–14846
183 Cano, E. and Mahadevan, L. C. (1995) Trends Biochem. Sci. **20**, 117–122
184 McCarthy, S. A., Samuels, M. L., Pritchard, C. A., Abraham, J. A. and McMahon, M. (1995) Mol. Cell. Biol. **9**, 1953–1964
185 Franklin, C. C. and Kraft, A. S. (1995) Oncogene **11**, 2365–2374
186 Blenis, J. (1993) Proc. Natl. Acad. Sci. U.S.A. **90**, 5889–5892
187 Franklin, R. A., Tordai, A., Patel, H., Gardner, A. M., Johnson, G. L. and Gelfand, E. W. (1994) J. Clin. Invest. **93**, 2134–2140
188 Bohm, M., Moellmann, G., Cheng, E., Alvarez-Franco, M., Wagner, S., Sassone-Corsi, P. and Halaban, R. (1995) Cell Growth Differ. **6**, 291–302
189 Ginty, D. D., Bonni, A. and Greenberg, M. E. (1994) Cell **77**, 713–725
190 Chou, M. M. and Blenis, J. (1995) Curr. Opin. Cell Biol. **7**, 806–814
191 Lavoie, J. N., Lambert, H., Hickey, E., Weber, L. A. and Landry, J. (1995) Mol. Cell. Biol. **15**, 505–516
192 Ben-Levy, R., Leighton, I. A., Doza, Y. N., Attwood, P., Morrice, N., Marshall, C. J. and Cohen, P. (1995) EMBO J. **14**, 5920–5930
193 McLaughlin, M. M., Kumar, S., McDonell, P. C., Van Horn, S., Lee, J. C., Livi, G. P. and Young, P. R. (1996) J. Biol. Chem. **271**, 8488–8492
194 He, X., Saint-Jeannet, J.-P., Woodgett, J. R., Varmus, H. E. and Dawid, I. B. (1995) Nature (London) **374**, 617–622
195 Wang, Q. M., Fiol, C. J., DePaoli-Roach, A. A. and Roach, P. J. (1994) J. Biol. Chem. **269**, 14566–14574
196 Fiol, C. J., Williams, J. S., Chou, C.-H., Wang, Q. M., Roach, P. J. and Andrisani, O. M. (1994) J. Biol. Chem. **269**, 32187–32193
197 Wang, Q. M., Vik, T. A., Ryder, J. W. and Roach, P. J. (1995) Biochem. Biophys. Res. Commun. **208**, 796–801
198 Welsh, G. I., Foulstone, E. J., Young, S. W., Tavare, J. M. and Proud, C. G. (1994) Biochem. J. **303**, 15–20
199 Eldar-Finkelman, H., Seger, R., Vandenheede, J. R. and Krebs, E. G. (1995) J. Biol. Chem. **270**, 987–990
200 Cross, D. A. E., Alessi, D. R., Cohen, P., Andjelkovich, M. and Hemmings, B. A. (1995) Nature (London) **378**, 785–789
201 Franke, T. F., Yang, S.-I., Chan, T. O., Datta, K., Kazlauskas, A., Morrison, D. K., Kaplan, D. R. and Tsichlis, P. N. (1995) Cell **81**, 727–736
202 Silvennoinen, O., Ihle, J. N., Schlessinger, J. and Levy, D. E. (1993) Nature (London) **366**, 583–585
203 Bird, T. A., Kyriakis, J. M., Tyshler, L., Gayle, M., Milne, A. and Virca, G. D. (1994) J. Biol. Chem. **269**, 31836–31844
204 David, M., Petricoin, III, E., Benjamin, C., Pine, R., Weber, M. J. and Larner, A. C. (1995) Science **269**, 1721–1723
205 Winston, B. W., Remigio, L. K. and Tiches, D. W. H. (1995) J. Biol. Chem. **270**, 27391–27394
206 Westwick, J. K., Bielawska, A. E., Dbaibo, G., Hannun, Y. A. and Brenner, D. A. (1995) J. Biol. Chem. **270**, 22689–22692
207 Terada, K., Kaziro, Y. and Satoh, T. (1995) J. Biol. Chem. **270**, 27880–27886
208 Winston, L. A. and Hunter, T. (1995) J. Biol. Chem. **270**, 30837–30840
209 Montgomery, R. B., Moscatello, D. K., Wong, A. J., Cooper, J. A. and Stahl, W. L. (1995) J. Biol. Chem. **270**, 30562–30566
210 Buday, L., Warne, P. H. and Downward, J. (1995) Oncogene **11**, 1327–1331
211 Rozakis-Adcock, M., van der Geer, P., Mbamalu, G. and Pawson, T. (1995) Oncogene **11**, 1417–1426
212 Waters, S. B., Holt, K. H., Ross, S. E., Syu, L.-J., Guan, K.-L., Saltiel, A. R., Koretzky, G. A. and Pessin, J. E. (1995) J. Biol. Chem. **270**, 20883–20886
212a Cherniack, A. D., Klarlaund, J. K., Conway, B. R. and Czech, M. P. (1995) J. Biol. Chem. **270**, 1485–1488
213 Chen, D., Waters, S. B., Holt, K. H. and Pessin, J. E. (1996) J. Biol. Chem. **271**, 6328–6332
214 Wera, S. and Hemmings, B. A. (1995) Biochem. J. **311**, 17–29
215 Barford, D. (1995) Curr. Opin. Struct. Biol. **5**, 728–734
216 Hunter, T. (1995) Cell **80**, 225–236
217 Milarski, K. L. and Saltiel, A. R. (1994) J. Biol. Chem. **269**, 21239–21243
218 Xiao, S., Rose, D. W., Sasaoka, T., Maegawa, H., Burke, Jr., T. R., Roller, P. P., Shoelson, S. E. and Olefsky, J. M. (1994) J. Biol. Chem. **269**, 21244–21248
219 Sawada, T., Milarski, K. L. and Saltiel, A. R. (1995) Biochem. Biophys. Res. Commun. **214**, 737–743
220 Keyse, S. M. (1995) Biochim. Biophys. Acta **1265**, 152–160
221 Wiesmuller, L. and Wittinghofer, F. (1994) Cell. Signalling **6**, 247–267
222 Chambers, A. F. and Tuck, A. B. (1993) Crit. Rev. Oncogenesis **4**, 95–114
223 Bortner, D. M., Langer, S. J. and Ostrowski, M. C. (1993) Crit. Rev. Oncogenesis **4**, 137–160
224 Qiu, R.-G., Chen, J., McCormick, F. and Symons, M. (1995) Proc. Natl. Acad. Sci. U.S.A. **92**, 11781–11785
225 Decker, S. J. (1995) J. Biol. Chem. **270**, 30841–30844
226 Xia, Z., Dickens, M., Raingeaud, J., Davis, R. J. and Greenberg, M. E. (1995) Science **270**, 1326–1331
227 Sakata, N., Patel, H. R., Terada, N., Aruffo, A., Johnson, G. L. and Gelfand, E. W. (1995) J. Biol. Chem. **270**, 30823–30828
228 Heidenreich, K. A. and Kummer, J. L. (1996) J. Biol. Chem. **271**, 9891–9894

Biochem. J. (1996) **319**, 657–667 (Printed in Great Britain)

REVIEW ARTICLE
Steroid hormone receptors and their regulation by phosphorylation

Nancy L. WEIGEL
Department of Cell Biology, Baylor College of Medicine, Houston, TX 77030, U.S.A.

The steroid/thyroid hormone receptor superfamily of ligand-activated transcription factors encompasses not only the receptors for steroids, thyroid hormone, retinoids and vitamin D, but also a large number of proteins whose functions and/or ligands are unknown and which are thus termed orphan receptors. Recent studies have highlighted the importance of phosphorylation in receptor function. Although most of the phosphorylation sites are serine and threonine residues, a few of the family members are also phosphorylated on tyrosine. Those steroid receptor family members that are bound to heat-shock proteins in the absence of ligand typically are basally phosphorylated and exhibit increases in phosphorylation upon ligand binding. Most of these sites contain Ser-Pro motifs, and there is evidence that cyclin-dependent kinases and MAP kinases (mitogen-activated protein kinases) phosphorylate subsets of these sites. In contrast, phosphorylation sites identified thus far in members of the family that bind to DNA in the absence of hormone typically do not contain Ser-Pro motifs and are frequently casein kinase II or protein kinase A sites. Phosphorylation has been implicated in DNA binding, transcriptional activation and stability of the receptors. The finding that some of the steroid receptor family members can be activated in the absence of ligand by growth factors or neurotransmitters that modulate kinase and/or phosphatase pathways underscores the role of phosphorylation in receptor function. Hence this family of transcription factors integrates signals from ligands as well as from signal transduction pathways, resulting in alterations in mRNA and protein expression that are unique to the complex signals received.

INTRODUCTION

The steroid/thyroid hormone receptors are members of a very large family of nuclear ligand-activated transcription factors that includes the steroid receptors [those for progesterone (PRs), androgen (ARs), oestrogen (ERs), glucocorticoids (GRs) and mineralocorticoids] and receptors for thyroid hormone, retinoids and vitamin D, as well as an even larger group of proteins termed orphan receptors whose ligands and/or function are as yet unknown [1,2]. These receptors play key roles both as transcriptional activators and as repressors in all aspects of biological function, including regulation of development, metabolism and reproduction.

Functionally, the receptor family can be divided into two groups based on their interactions or lack of interactions with heat-shock proteins, as well as on their tendency to form homodimers or heterodimers. Figure 1 shows a simplified model of the mechanisms of action of the two sides of the family. In the absence of hormone, steroid receptors are associated with preformed complexes of proteins which include hsp90 (heat-shock protein of 90 kDa) and hsp70 [3]. These complexes may be cytoplasmic or loosely associated with the nucleus. Studies from several laboratories suggest that, in the absence of ligand, the receptors shuttle between the cytoplasm and nucleus, with the relative amount of cytoplasmic receptor dependent upon the receptor type [4,5]. Thus in the absence of hormone the GR [6] and perhaps AR [7] appear to be predominantly cytoplasmic, whereas the PR and ER are predominantly nuclear [4,5,8]. Ligand binding leads to dissociation of the heat-shock proteins, receptor dimerization and binding to a specific steroid response element whose consensus sequence is an inverted repeat separated by three nucleotides. GRs, PRs, ARs and mineralocorticoid receptors all bind to the same consensus sequence (AGAACAnnnTGTTCT), whereas ERs bind to the sequence AGGTCAnnnTGACCT [9]. Although these consensus sequences have been identified, there are clearly other sequences to which the receptors bind alone or in combination with other proteins [10–14]. The receptor dimer then interacts with basal transcription factors [15], other DNA-binding proteins [16,17] and co-activators, resulting in transcription of the target gene.

In contrast, as shown in the lower portion of Figure 1, the thyroid hormone, retinoid, vitamin D and orphan receptors that have been studied all appear to be tightly bound to the nucleus at specific sites in the DNA in the absence of ligand. These proteins typically form heterodimers with retinoid X receptors (RXRs) [18–20] rather than homodimers, and these receptor forms recognize sequences that match the ER half site with variations in nucleotide spacing and the direction of the repeated elements (for a review of DNA binding specificity, see [13]). Ligand and DNA binding may also cause dissociation of repressor molecules [21,22], allowing association with general transcription factors, other DNA-binding proteins and/or co-activators, resulting in transcription of the target gene.

While much remains to be learned about this family of proteins, a great deal is known about steroid receptor structure and function; various aspects have been the subjects of previous reviews (for a comprehensive review of receptor structure and function, see [23]). Of special interest are the very recent studies that reveal the complexity of the regulation of receptor function and cross-talk with other transcription factors and signalling

Abbreviations used: AR, androgen receptor; EGF, epidermal growth factor; ER, oestrogen receptor; GR, glucocorticoid receptor; hsp90, heat-shock protein of 90 kDa; MAP kinase, mitogen-activated protein kinase; (c)PR, (chicken) progesterone receptor; RAR, retinoic acid receptor; RXR, retinoid X receptor; MMTV, mouse mammary-tumour virus.

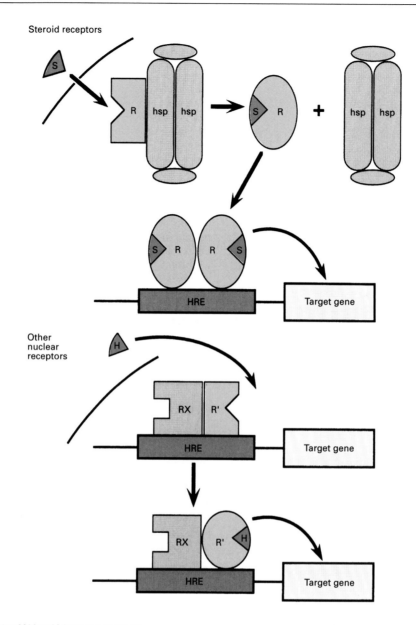

Figure 1   Activation of steroid/thyroid hormone receptors

The upper portion of the Figure depicts activation of the steroid receptor members of the family. In the absence of hormone, each receptor (R) monomer is associated with a heat-shock protein complex (hsp). These complexes may be cytoplasmic or loosely bound in the nucleus. Binding of steroid (S) causes a conformational change resulting in dissociation of the heat-shock complex and allowing dimerization of the receptor and DNA binding to a hormone response element (HRE) to produce a transcriptionally productive complex. The lower part of the Figure describes the action of the non-steroid receptors (R'). In the absence of ligand these proteins are bound to their hormone response elements, typically as heterodimers with RXRs (RX) or as homodimers, or in a few cases as monomers, and repress transcription of the target gene. Binding of hormone (H) changes the conformation so that the receptor can form an active transcription complex.

pathways. These regulatory mechanisms, which include phosphorylation, interaction with other steroid receptor family members, interactions with other transcription factors and interactions with co-activators and co-repressors, are the subject of this review.

STEROID RECEPTOR STRUCTURE AND FUNCTION

The steroid/thyroid receptor family is characterized by three regions of homology; these consist of a highly conserved DNA-binding domain that contains two Zn^{2+}-containing motifs, and two less highly conserved regions in the C-terminal region of the protein. The region N-terminal of the DNA-binding domain

exhibits the greatest variability with respect to both length and sequence similarity [24].

Shown in Figure 2 are the locations of the functional domains of a typical receptor. Although all receptors contain DNA- and hormone-binding domains, the remaining functions are a composite from studies of various receptors, and some members may lack specific functional regions. Deletion studies suggested that the portion of the receptors required to maintain ligand binding was surprisingly large (about 25 kDa) for a steroid ligand [25]. Affinity labelling studies of GRs and PRs [26,27] confirmed that the steroid interacts with amino acids that are more than 100 residues apart in the primary structure. The hormone-binding domain also contains regions important for dimerization [28]. X-

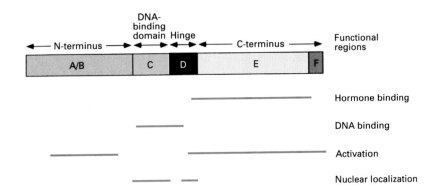

Figure 2   Location of functional regions in a typical steroid receptor

A/B represents the region N-terminal to the DNA-binding domain; C contains the DNA-binding domain; D contains the hinge region; E contains the ligand-binding domain; F is a region originally found in ER that can be deleted without eliminating hormone binding.

ray crystallographic structures of the DNA-binding domains of the GR [29] and the ER [30] revealed that this region is important both for DNA binding and for receptor dimerization.

Both the N-terminal and C-terminal regions of steroid receptors contain sequences important for trans-activation (reviewed in Sartorius et al. [31]). These were initially termed TAF-1 and TAF-2, and more recently have been renamed AF1 and AF2. The human PR B form is unique in that it contains a third activation function at the extreme N-terminus, termed AF3 [31].

Other receptor functions include a nuclear localization signal, which is in the hinge region between the DNA- and hormone-binding domains [32]. This region and the C-terminal region are important for binding of heat-shock proteins [33].

Binding of hormone initiates a series of events that result in activation or repression of target genes. Several lines of evidence demonstrate that hormone binding changes the conformation of the receptor. Studies from Gorski's laboratory initially showed that ligand binding changed the hydrophobicity of the ER, implying a major conformational change [34]. Allan et al. [35], using in vitro translated ^{35}S-labelled PRs, showed that binding of hormone caused the ligand-binding domain to become protease-resistant. Using an antibody that reacts with the C-terminal 14 amino acids of the PR, Weigel et al. [36] demonstrated that this region is accessible in the unliganded receptor, but is not recognized once the agonist, R5020, is bound. Finally, a comparison of the three-dimensional structure of the ligand-binding domain of the retinoic acid receptor (RAR) bound to ligand with the RXR in the absence of ligand suggests that a helical region moves to cover the ligand upon binding [37,38].

That ligand binding significantly alters the structure of receptors was also shown by functional studies that include the work of Smith [39], who has shown that binding of progesterone blocks reassociation of the PR with heat-shock proteins, and by the observation that, for some steroid receptors, ligand binding either is required for DNA binding [40] or enhances DNA binding [41].

The mechanism of action of antagonists has been of interest both mechanistically and clinically for many years. Antagonists are either steroidal or non-steroidal compounds that compete with agonists for binding and prevent activation of receptors. Although it was initially assumed that the binding sites for agonists and antagonists are the same, there is now evidence that they can be different, although overlapping. Vegeto et al. [42] showed that a human PR mutant lacking the C-terminal 42 amino acids no longer bound R5020, but still bound the

antagonist RU486. Moreover, RU486 acted as an agonist for this mutant. A comparison of partial protease patterns of agonist-bound versus antagonist-bound PRs revealed that a smaller limit fragment lacking the C-terminal amino acids was produced when antagonist was bound [35]. The antibody C262, which does not interact with agonist-bound receptor, does interact with antagonist-bound receptor, suggesting that different conformations are induced [36]. Many of these antagonists induce dissociation from heat-shock proteins and binding to DNA; in some cases these compounds act as partial agonists. It is thought that the conformation of the antagonist-bound receptor blocks productive interaction with one or more proteins needed to induce transcription. As described below, likely candidates for this function are the recently discovered co-activators. A second class of antagonists (e.g. ZK98299, a PR antagonist, and flutamide, an AR antagonist) fail to promote DNA binding efficiently [40,43]. One of the ER antagonists, ICI 164,384, seems to function in part by reducing the stability of the ER [44]. Thus there appear to be multiple mechanisms by which antagonists can inhibit receptor action. All antagonists identified to date are synthetic compounds. An important question yet to be answered is whether there are naturally occurring antagonists for these proteins with physiologically important functions.

REGULATION OF RECEPTOR FUNCTION BY MEANS OTHER THAN PHOSPHORYLATION

Multiple forms of steroid receptor family members as a means of regulating the response to a steroid

In most avian and mammalian species, the PR is expressed as two major DNA-binding forms, PR-A and PR-B [45–47]. The exception is the rabbit receptor, which is expressed only as the larger B form [48]. Both forms are expressed from the same gene, either from different mRNAs [49] or by alternative initiation of translation [50]. The human PR-B contains an additional 164 amino acids at the N-terminus [49], and this region contains an additional trans-activation domain [31]. Both forms are biologically active, although their relative ability to induce target genes differs [51,52]. A third form, the C receptor, which contains the C-terminal hormone-binding domain and a portion of the DNA-binding domain, was detected more recently [53].

A and B forms have recently been described for the AR [54]. In contrast with the PR, the A form of the AR is expressed at substantially lower levels than the B form [54], and its contribution to androgen action is currently being evaluated.

The GR also is expressed as two forms, α and β, that are produced by alternative splicing [55]. The α form is the classical GR and is activated by glucocorticoids. The β form differs at the C-terminus, does not bind glucocorticoids and can repress the activity of the GR α form [55]. The relative expression and importance of the β form has yet to be determined.

In addition to the variants described above, numerous splice variants, particularly of the ER, have been described [56]. For the most part, these consist of receptors lacking one of the exons found in the full-length ER. They have primarily been detected at the mRNA level, and their expression and role *in vivo* is an active subject of investigation.

Receptor variants and heterodimerization play key roles in the regulation of the receptors that do not interact with heat-shock proteins. There are multiple forms of RARs and RXRs that arise from distinct genes [57–59], as well as splice variants of these genes. There are also multiple forms of the thyroid hormone receptor [60–62].

Although some of the receptor family members that do not interact with heat-shock proteins are capable of forming homodimers, many of these proteins function as heterodimers with RXRs. Although RXRs can form homodimers, they apparently play a larger role as heterodimer partners. Thus the response of a receptor such as the vitamin D receptor will depend in part upon its RXR partner. The ligand response of the heterodimers is dependent both upon the heterodimer pair and upon the DNA response element. For example, the retinoid X ligand, 9-*cis*-retinoic acid, can inhibit the response to the ligand of the heterodimer partner, have no effect on its response, or stimulate the response [63,64]. The cellular complement of receptors, the ligands present and the DNA response elements will dictate the response of individual cells to these receptors.

Regulation of receptor function by physical or functional interactions with other proteins

Numerous studies utilizing artificial as well as natural promoters have demonstrated that the activity of a steroid receptor will in part be determined by proteins that bind to nearby or even overlapping DNA-binding sites. In the simplest case of two adjacent steroid hormone response elements, both the DNA binding of receptor dimers [65] and the resulting transcription of a target gene are greatly enhanced compared with a single steroid response element. In the complex mouse-mammary-tumour-virus (MMTV) promoter, which contains a number of near-consensus sequences as well as half sites, mutation of any of these full or half sites reduces the transcriptional activity [13]. Using a series of artificial promoters containing a single steroid response element and single sites for other transcription factors such as NF1 or SP-1, Schule et al. [16] showed that both GRs and PRs functionally synergize with these factors to enhance transcription. Other types of interactions are typified by the pfl promoter, which contains a negative glucocorticoid response element; binding of the GR can inhibit or stimulate transcription of this gene depending upon whether a Fos–Jun or Jun–Jun complex occupies the adjacent AP-1 site [66,67].

Steroid receptors interact with other transcription factors such as AP-1 [68,69] and RelA [70] independent of DNA binding, and also reciprocally affect each other's function. The less active form of the PR, PR-A, can repress the activity of PR-B [71,72]. Moreover, McDonnell's group has shown that PR-A also can repress the activity of GRs, mineralocorticoid receptors and even ERs [71,73]. Kraus et al. [74] have also reported inhibition of the ER by the PR. Because PRs and ERs bind to different response elements, this inhibition must be through a mechanism other than competition for a DNA-binding site.

Two important classes of interacting proteins, co-activators and co-repressors, have recently been described. Co-activators bind to the receptor itself, and presumably serve as bridges to the general transcription factors. Those that interact with the C-terminal portion of the receptors [75–77] have been characterized as interacting with receptors only in the presence of agonist. Co-repressors have been identified that interact with some of the non-heat-shock-binding members of the family [21,78]. The proteins that have been characterized interact with the receptor in its hinge region and typically are released upon binding of ligand.

Clearly, regulation of steroid receptor function is a complex phenomenon involving many factors and many steps. Any or all of these are targets for regulation by phosphorylation, as outlined below.

PHOSPHORYLATION OF THE STEROID/THYROID RECEPTOR SUPERFAMILY

Although the steroid/thyroid hormone receptors are a family of ligand-activated receptors, they are also phosphoproteins and their functions are regulated by phosphorylation. Studies of other transcription factors have shown that phosphorylation can play roles in nuclear translocation, DNA binding (either positive or negative), interactions with other proteins and trans-activation (for a review, see [79]). Transcription factors are frequently multiply phosphorylated and, as in the case of c-Jun, may contain phosphorylations that enhance activity as well as others that reduce activity [80]. That the phosphorylation and de-phosphorylation reactions are accomplished by multiple enzymes indicates that the cell can alter the activity of the protein in response to different signalling pathways. Although steroid receptor phosphorylation has not been as well characterized as that of some of the other transcription factors, recent studies have revealed that regulation of receptor activity by phosphorylation is complex and is involved in multiple aspects of receptor action. The finding that some of the steroid receptors can be activated by stimulation of kinase pathways in the absence of hormone (reviewed below), and that these pathways can alter responses to antagonists, confirms the importance of phosphorylation in the activity of steroid receptors.

Analysis of receptor phosphorylation

That the steroid receptors are phosphoproteins has been known for many years [81,82]. Phosphopeptide mapping studies of chicken [83], human [84,85] and rabbit [86,87] PRs showed that the proteins are multiply phosphorylated and that, in the case of the chicken [83] and human [84,88] PRs, novel sites are phosphorylated in response to hormone treatment. For the chicken [83] and human [84,88] PRs, this enhanced phosphorylation is accompanied by a change in mobility on SDS/PAGE gels. The ER [89–91] and GR [92] also exhibit enhanced phosphorylation in response to hormone treatment. Studies of the AR have yielded conflicting results. Analysis of endogenous AR in LNCaP cells showed enhanced phosphorylation of the receptor in response to androgen [93], whereas AR expressed in COS cells showed no net increase in phosphorylation per molecule, but rather a general enhancement of the stability of the receptor [7]. Sites have been identified by direct characterization of phosphorylation sites in endogenous receptors, as well as by recombinant DNA technology. In some cases the results have been consistent; in others they are conflicting.

Phosphorylation of PRs

Using ^{32}P-labelled oviduct tissue minces as a source of chicken PR (cPR), Denner et al. [83] first identified three phosphorylation sites by isolating and sequencing phosphotryptic peptides. A fourth site was identified subsequently by Poletti and Weigel [94]. Figure 3 shows the location of the four phosphorylation sites in the cPR. All four are in the region common to the PR-A and PR-B receptors and contain the consensus sequence Ser-Pro within the phosphorylation sites. Subsequent studies of other steroid receptors (see below) revealed that the majority of the phosphorylation sites in these receptors contain Ser-Pro sequences, suggesting that the proline-directed kinases, which include cyclin-dependent kinases [95], mitogen-activated protein (MAP) kinases and stress-activated protein kinases [96,97], play roles in receptor phosphorylation. As shown in Figure 3, three of the sites are in the N-terminus, which has been identified as a region important for transcriptional activation, and the fourth is in the region between the DNA-binding and hormone-binding domains (the hinge region). Two of the N-terminal sites, Ser[211] and Ser[260], are phosphorylated in the absence of hormone and their phosphorylation is increased in response to hormone treatment [83], whereas the other sites, Ser[530] and Ser[367], are detected after hormone treatment [83,94]. Of particular interest is the hinge-region site, Ser[530]. All of the steroid receptors contain either a Ser-Pro or a Thr-Pro at essentially the same position [83]. This site is also phosphorylated in the AR [98], and an analogous site has been reported in the mouse ER [99].

Phosphorylation of the cPR expressed in yeast (*Saccharomyces cerevisiae*) is indistinguishable from phosphorylation of the endogenous chicken oviduct receptor [100]. More recent studies have shown that the cPR expressed in CV1 cells is also identical with the endogenous receptor [101], so that, in this case, expression in different systems results in comparable phosphorylation patterns.

The roles of some of these sites have been examined using site-directed mutagenesis to replace the target serines with alanine. Substitution of an alanine for Ser[211] blocks the hormone-dependent change in mobility on SDS/PAGE gels [102], demonstrating that phosphorylation of this site is required for the mobility shift. A comparison of the activity of the Ala[211] mutant with wild-type receptor in cells transiently co-transfected with a receptor expression vector and a reporter showed that the activity of the mutant receptor was reduced in a cell- and promoter-specific manner [102]. The most significant reduction was observed in HeLa cells, where the mutant exhibited 25% of wild-type activity when a simple promoter consisting of two progesterone/glucocorticoid response elements and a TATA box from the E1b gene was linked to the coding region for chloramphenicol acetyltransferase. In contrast, receptor activity was reduced only slightly when the mutant was tested in CV1 cells using a complex promoter containing a portion of the thymidine kinase promoter as well as two progesterone response elements. This cell- and promoter-specific variation in activity has been observed for other steroid receptors (see below), suggesting that these sites are involved in interactions with proteins whose expression may be cell-type-specific. A model of this mechanism is presented in Figure 4.

A very different phenotype was detected when Ser[530] in the hinge region was replaced with alanine. Although the activities of the wild-type and mutant receptors were essentially identical at saturating concentrations of hormone, the activity of the mutant was several-fold less than wild-type at 0.1 nM progesterone [103]. That this decrease in activity was not a result of a change in hormone-binding affinity was shown by hormone binding assays both *in vitro* and in whole cells. This led to a model suggesting that this phosphorylation maintains the receptor in an activated state such that dissociation of hormone does not promote rebinding of the receptor to heat-shock protein complexes.

Phosphorylation of the human PR has not been characterized as thoroughly. Using peptide mapping of ^{32}P-labelled receptor isolated from T47D breast cancer cells, Sheridan et al. [84] initially showed that the phosphorylation of the human PR was substantially more complicated than that of the cPR, and that the patterns for PR-A and PR-B differed. Subsequent studies using a DNA-binding mutant suggested that some of the hormone-dependent phosphorylation of PR may also be DNA-dependent [104]. Using T47D breast cancer cells labelled with ^{32}P as a source of PR, Zhang et al. [85] identified three phosphorylation sites found only in the B form of the receptor. All reside within the N-terminal region unique to this form. Additional peptide mapping studies revealed that there are a number of basal phosphorylation sites whose phosphorylation is enhanced within a few minutes of hormone administration, and three sites that are phosphorylated slowly in response to hormone [88]. One of these is a PR-B-specific site. All five of the sites that have been identified are N-terminal to the DNA-binding domain; four contain the consensus sequence Ser-Pro, whereas the other site contains a casein kinase II consensus sequence [85]. *In vitro* phosphorylation studies suggest that one site is a casein kinase II site [85] and that three basal sites are substrates for cyclin A–CDK2 (cyclin-dependent kinase 2) complexes [105]. Based on the peptide mapping studies, as many as five additional sites are as yet unidentified.

Beck et al. [106] compared the effects of agonists and antagonists on PR phosphorylation and found that treatment with RU486 caused phosphorylation of the same hormone-dependent sites as did that with the agonist R5020. However, treatment with the pure antagonist ZK98299 failed to induce the hormone-dependent phosphorylations [106]. Hence treatment with some antagonists does not promote normal phosphorylation.

Other than a preliminary report that mutation of two potential phosphorylation sites (which have not yet been confirmed as authentic phosphorylation sites) results in reduced transcriptional activation [107], no studies of human PR phosphorylation-site mutants have been published. Consequently the roles of the many sites in the human PR remain to be determined.

Phosphorylation of GRs

Bodwell et al. [108] have identified seven phosphorylation sites in the mouse GR; all of the sites are in the region N-terminal to the DNA-binding domain, and many contain Ser-Pro sequences. One of the sites is a threonine and other sites may be phosphorylated either by casein kinase II or by other unidentified kinases. Hormone treatment increases phosphorylation at all sites, but the phosphorylation of one site, Ser[220], is preferentially increased [109]. Hu et al. [110] have shown that GRs overexpressed in Chinese hamster ovary cells undergo cell-cycle-dependent changes in phosphorylation, with basal phosphorylation highest in G2/M and hormone-dependent phosphorylation most evident in S phase. Hsu and De Franco [111] found that the relative phosphorylation of GR phosphorylation sites in cells synchronized in G2 differed from the phosphorylation in unsynchronized cells; these receptors are transcriptionally inactive, but are still capable of acting as repressors.

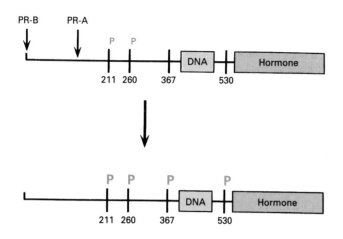

Figure 3   Location of phosphorylation sites in the cPR

The structure of the cPR and the location of the four phosphorylation sites within the receptor are shown. As indicated by the internal start for PR-A, all four sites are found in the region common to PR-A and PR-B. The presence and size of the 'P' denotes the state and extent of phosphorylation in the absence and presence of progesterone.

Mason and Housley [112] were the first to characterize the activity of steroid-receptor phosphorylation-site mutants. In these studies, the activity of mouse GRs having alanine replacing one or more of the phosphorylation sites in the receptor was measured in COS cells using MMTVCAT (a portion of the MMTV long tandem repeat-linked coding sequence for chloramphenicol acetyltransferase) as a reporter. Under these con-

ditions, none of the single mutants showed a significant alteration in activity, but a receptor lacking the five sites conserved in mouse, rat and human GRs showed a small (22%), but statistically significant, reduction in activity. More recent studies by Jewell et al. [113] have suggested that phosphorylation may affect GR stability and thus alter receptor activity.

Phosphorylation of ERs

In contrast with the studies of PRs and GRs, where individual laboratories have identified the phosphorylation sites, several laboratories have contributed to the identification of phosphorylation sites in the ER. At present, these studies have led to conflicting identifications. Auricchio's group first reported that the ER contains phosphotyrosine [114] and subsequently identified the site as Tyr[537] in the ligand-binding domain [115]. Arnold et al. [116] have also identified Tyr[537] as a phosphorylation site in the human ER. Other groups have not detected phosphotyrosine, but have detected serine phosphorylation [91,117].

Several groups have identified serine phosphorylation sites. Three separate groups studying phosphorylation of ERs transiently expressed in COS cells identified Ser[118], a site containing the consensus sequence Ser-Pro, as the major hormone-dependent phosphorylation site [90,91,118]. Mutation eliminates the change in mobility detected on SDS gels as a result of hormone treatment *in vivo* [118]. The corresponding site, Ser[122], has been identified in the mouse ER [99]. However, although Arnold et al. [119] identified Ser[118] as a phosphorylation site in the endogenous ER from MCF-7 cells, they reported that Ser[167], a casein kinase II site, was the major hormone-inducible site [120]. Lahooti et al. [99] have found that the corresponding mouse site, Ser[171], may also be phosphorylated, and reported two additional sites, Ser[156] and Ser[158], which are located in sites that contain neither Ser-Pro

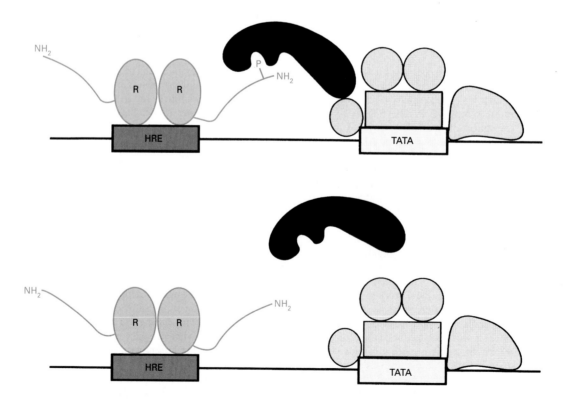

Figure 4   A potential role for Ser[211] and other receptor phosphorylation sites that exhibit cell- and/or promoter-specific changes in activity

As shown in the upper portion of the Figure, the phosphorylation induces an interaction with a protein that forms a bridge between the general transcription factors and the receptor.

sequences nor a casein kinase II consensus sequence. LeGoff et al. [91] have also identified either Ser[104] or Ser[106] (both Ser-Pro motifs) as phosphorylation sites in human ERs expressed in COS cells, and found no evidence for phosphorylation of Ser[167] or Ser[294], which corresponds to the hinge site found in the mouse ER. Although the differences between the phosphorylation patterns of human ERs expressed in COS and MCF-7 cells [91,120] may be a result of cell-specific phosphorylation, the phosphorylation of the recombinant baculovirus-expressed ER mirrors that of the MCF-7 cell ER [120]. Additional sites remain to be identified in the ER expressed in COS cells [91], as well as in the MCF-7 ER [120]. These discrepancies may be resolved once the identification of the sites is completed.

The activities of several of these sites have been examined by site-directed mutagenesis. Ali et al. [90] reported that mutation of Ser[118] to Ala reduced the activity of the ER to as little as 25% of wild type, but that the decrease was cell- and promoter-specific and in some cases was much smaller. LeGoff et al. [91] found that mutation of Ser[118] alone only marginally affected activity, but mutation of the three serines at positions 104, 106 and 118 reduced activity to about 50%. Lahooti et al. [99] reported that mutation of Ser[122] or Ser[298] had little effect on the transcriptional activity of the wild-type mouse ER, but mutation of Ser[122] in the presence of a mutated Af2 region substantially decreased the transcriptional activity.

In addition to studies showing that phosphorylation regulates transcriptional activation, others suggest that DNA binding is affected. Denton et al. [89] found that treatment of ERs with phosphatase decreased DNA binding. Arnold et al. [116] have subsequently reported that phosphorylation of Tyr[537] is important for dimerization and for DNA binding, and have also shown that phosphorylation of Ser[167] enhances DNA binding [119]. Taken together, these studies suggest that at least DNA binding and transcriptional activation of the ER are modulated by phosphorylation.

Kato et al. [121] and Arnold et al. [119] have both shown that Ser[118] can be phosphorylated in vitro by MAP kinase. Kato et al. [121] have further shown that activation of MAP kinases enhances ER phosphorylation in vivo and that this phosphorylation is on Ser[118]. Expression of ras, an intermediate in the MAP kinase pathway, stimulates the activity of wild-type ERs, but not of a Ser[118] mutant. Joel et al. [118] reported that phorbol ester treatment also stimulated phosphorylation of Ser[118]. Taken together with the in vitro evidence that Ser[167] is a casein kinase II site [120] and the finding that Tyr[537] can be phosphorylated by Src kinase in vitro [116], it is evident that the phosphorylation of the ER is regulated by multiple kinases.

Phosphorylation of ARs

The AR is phosphorylated on multiple sites, but little is known about its phosphorylation. The receptor is detected as a doublet of 110 kDa and 112 kDa on SDS/polyacrylamide gels, and it is the 112 kDa species that is phosphorylated [122]. ARs expressed in COS cells exhibit hormone-independent phosphorylation [7], whereas a specific hormone-induced peptide was detected in ARs from LNCaP cells [122]. Zhou et al. [98] have identified three phosphorylation sites in ARs expressed in COS cells. All three, Ser[81], Ser[94] and Ser[650], are in Ser-Pro motifs; these studies indicate that additional sites remain to be identified. Mutation of Ser[650] to Ala reduces the activity of the receptor by approx. 30%.

Other receptor family members

The other receptors that have been examined are also phosphoryl-

ated, but their phosphorylation patterns are different from those of the steroid receptors. The sites that have been identified to date are not Ser-Pro sites. Ser[205] (Ser[208] depending upon the numbering system) is the major hormone-dependent phosphorylation site in the vitamin D receptor [123,124]. This site is a casein kinase II site. Interestingly, when this site is mutated to glycine, the activity of the receptor is reduced slightly [125], but when it is mutated to alanine the activity is not affected [123,125]. Analysis of the alanine substitution mutant revealed that alternative phosphorylation occurred at a nearby site [123]. Co-expression of the vitamin D receptor and casein kinase II stimulated the activity of the wild-type receptor and of the alanine mutant, but not of the glycine mutant [125].

Analysis of thyroid hormone receptor phosphorylation is incomplete. Thyroid hormone receptors are phosphorylated [62,126–129] and phosphorylation enhances DNA binding [128,129]. On the other hand, DNA binding of the TR2 variant, which does not bind ligand, is negatively regulated by phosphorylation [62].

The RARs are all phosphoproteins, but a number of findings suggest that the phosphorylation patterns of the different receptors, and even of splice variants, differ. Interestingly, RARβ1 and RARβ3 show enhanced phosphorylation in response to ligand, but RARγ and RARβ2 do not [130] RARβ, in contrast with the others, contains a phosphotyrosine [130]. Ser[369] in RARα1 is phosphorylated in COS-1 cells in response to increased protein kinase A activity, and in vitro by purified protein kinase A, but is not phosphorylated in the absence of protein kinase A stimulation [131]. Additional phosphopeptides were also detected in the COS-1-expressed receptor. Finally, Lefebvre et al. [132] have shown that dephosphorylation decreases the DNA binding of both RARα and RXRα.

Little is known about the phosphorylation of orphan receptors. However, the phosphorylation of Nur77(NGFIB) (nerve growth factor-inducible protein-B) is dependent upon the activator utilized; activation by nerve growth factor produces a phosphorylation pattern distinct from that found as a result of membrane depolarization [133,134]. Davis et al. [135] have reported that one of the sites phosphorylated in vivo, Ser[354], is phosphorylated in vitro by pp90[rsk]; this site is in a region important for DNA binding and may affect the DNA binding of the protein.

STEROID RECEPTORS AND SIGNAL TRANSDUCTION PATHWAYS

The steroid receptors have classically been viewed as ligand-activated transcription factors. With the surprising discovery that modulation of kinase activity in cells can cause activation of some of the steroid receptor family members in the absence of hormone [136,137], it became apparent that there are alternative means of activating some of these receptors. Since then, many examples of cross-talk between steroid receptor pathways and other signal transduction pathways have been described.

Denner et al. [136] first reported that cPR-A, co-transfected with a reporter gene into CV1 cells, can be transcriptionally activated by treatment of the cells with 8-bromo-cAMP (an activator of protein kinase A) or with okadaic acid (an inhibitor of phosphatases 1 and 2A). Subsequent studies showed that activation of membrane-receptor-mediated signal transduction pathways by dopamine [138] or by epidermal growth factor (EGF) [139] also activated the cPR.

There is also indirect evidence that some of the endogenous mammalian PRs can be activated in the absence of progesterone, both in cells and in animals. Turgeon and Waring [140] found that a gonadotropin-releasing-hormone pathway can activate the PR in rat pituitary cells. Mani et al. [141] have shown that dopamine-induced lordosis in ovariectomized rats can be blocked by progesterone antagonists, implying that dopamine activates the PR. Despite these studies, an analysis of the human PR revealed that, although 8-bromo-cAMP stimulated the hormone-dependent activation of the PR in T47D breast cancer cells, the PR was not activated in a ligand-independent manner [142]. However, the combination of the antagonist RU486 with 8-bromo-cAMP caused receptor activation and the antagonist to act as an agonist [40,143]. In contrast, the antagonist ZK98299, which does not promote DNA binding as effectively as RU486, remained inactive in combination with 8-bromo-cAMP. This antagonist/agonist switch appears to be rather specific, in that activation of protein kinase C with phorbol esters does not induce the switch, although it does stimulate agonist-dependent transcription. Subsequent studies by Sartorius et al. [144] showed that only the B form of the receptor responds to 8-bromo-cAMP and RU486.

That the ER can be activated by signal transduction pathways in the absence of ligand was first suggested by the study of Ignar-Trowbridge et al. [137], showing that EGF treatment of ovariectomized mice resulted in nuclear translocation and altered phosphorylation of the ER, as judged by a change in mobility of the receptor on SDS/polyacrylamide gels. Moreover, EGF-stimulated uterine DNA synthesis was reduced by treatment with the anti-oestrogen ICI 164,384. Subsequent studies in transiently transfected Ishikawa endometrial adenocarcinoma cells confirmed that EGF can activate the ER and that the N-terminal portion of the receptor was more important for this activation than the C-terminus [145]. Moreover, Auricchio et al. [146] found that treatment of MCF-7 cells with vanadate, a phospho-tyrosine phosphatase inhibitor, stimulated tyrosine phosphorylation of the ER and enhanced cell growth. The stimulation of growth was blocked by ER antagonists, suggesting that vanadate acts through the ER.

Other signalling pathways also activate and/or change the phosphorylation of the ER. Rat uterine ERs can be activated in the absence of oestrogen by cAMP and by insulin-like growth factor-1, and these treatments enhance the phosphorylation of the receptor [147].

A number of studies have examined the ligand-independent activation of the human ER. This receptor can be activated by dopamine [148] and by EGF [147]. Other studies have shown that the ability to exhibit ligand-independent activation and the level of the activation is to some extent dependent on the promoter and cell context, so that in some cases little or no ligand-independent activation is detected, but there is substantial stimulation of hormone-dependent activity. In addition, these signalling pathways substantially modify the response to agonists and antagonists [149,150]. Treatment of MCF-7 breast cancer cells with activators of protein kinase A in the presence of antagonist induces transcriptional activity; the magnitude of the response is promoter-dependent [150]. Of particular interest is the finding that some mutants that cannot be activated by oestrogen alone are activated by treatment with 8-bromo-cAMP in combination with either agonists or antagonists [151].

Despite a number of attempts to demonstrate ligand-independent activation of the GR, it appears that this receptor requires a ligand for activation. Similar to the human PR, the receptor activity is stimulated by agents that activate protein kinase A and protein kinase C pathways [152]. This activation is not accompanied by changes in GR phosphorylation [152]. The GR also exhibits an antagonist-to-agonist switch in response to 8-bromo-cAMP, but not to phorbol esters [153].

Attempts to induce ligand-independent activation of the AR have led to mixed results. Some investigators have failed to detect ligand-independent activation, but have seen stimulation of hormone-dependent activity [154,155]. However, Culig et al. [156] have reported that the AR can be activated by growth factors, and that the magnitude of the response is dependent upon the growth factor as well as the promoter.

Taken together, these studies suggest that some of the steroid receptor family members are activated in the absence of ligand, but their ability to induce transcription in response to these signals will be highly dependent upon the cellular context and on the target genes.

SUMMARY

The studies summarized above demonstrate that the level of receptor protein and of its cognate hormone is only one factor in determining the final biological response. The relative abundance of the receptor relative to other receptor forms, as well as to that of other transcription factors and co-activators or co-repressors, will affect the resulting activity. In addition, the sequence, location and number of receptor-binding sites and adjacent protein-binding sites will affect the level of induction of a specific target gene within the cellular milieu. These factors presumably control the cell- and/or tissue-specific responses to hormones, and all of these interactions are potential targets for regulation by receptor phosphorylation.

Many of the phosphorylation sites in the receptors have been identified, and initial analyses indicate that at least DNA binding and transcriptional activation are modified substantially by phosphorylation. The regulation of receptors by phosphorylation is complex, and individual phosphorylations can act either to enhance or to inhibit the activity of the receptor. As with most sites in transcription factors, none of the sites in the receptors that bind ligand has been found to function as an on/off switch. That the receptors are multiply phosphorylated by several different kinases suggests that subsets of sites may subserve specific roles that are dependent upon the physiological state of the cell. Moreover, it is probable that the current methods used to assay receptor function fail to measure fully the roles of the individual phosphorylation sites. The activity of receptor mutants is typically measured in transient transfection assays, with both an artificial promoter and a receptor expression plasmid being introduced into cells that do not normally express the receptor. Since most transfection procedures are inefficient, receptors are typically overexpressed in the successfully transfected cells in order to measure their activity. This creates two potential problems. First, the higher level of receptor may partially mask changes in affinity for DNA or for other proteins. Secondly, the important interacting proteins may be in limiting quantities in these cells. That this is a problem is evident from studies of receptors that form heterodimers with the RXRs. Although most cells contain endogenous RXRs and some activity is obtained when a receptor such as the vitamin D receptor is expressed, this activity is typically much higher if additional exogenous RXR is provided.

A second limitation in evaluating fully the role of phosphorylation is the use of artificial promoters transiently introduced into cells. Under these conditions, both the context of the natural gene and the structure imposed by chromatin are lacking. As techniques for measuring steroid receptor function in a more natural context are developed, the importance of the individual phosphorylation sites will be more clearly defined.

The finding that some of the steroid receptors can be activated

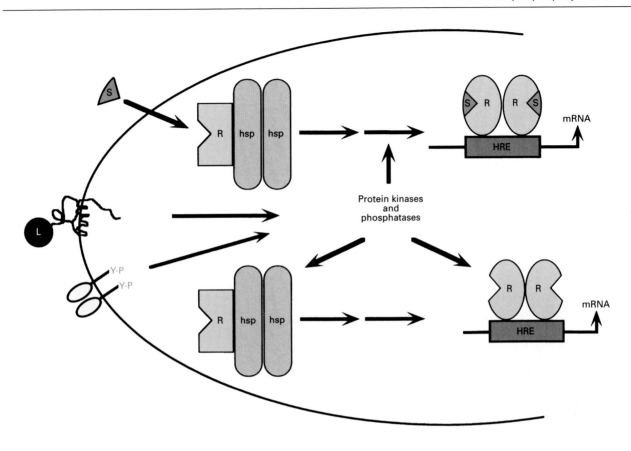

Figure 5   Steroid receptor activation

The upper portion depicts the classical steroid (S)-dependent activation pathway. The lower portion depicts signals emanating from receptors such as the dopamine receptor or growth factor receptors that alter the kinase/phosphatase balance of the cell such that the receptors are activated in the absence of ligand. L represents a ligand binding to a membrane receptor that acts through a cAMP pathway. Y-P indicates the active form of a growth-factor receptor. R, receptor; hsp, heat-shock protein complex.

in the absence of hormone has forced a re-evaluation of steroid receptor action and the interaction of steroid receptors with other signal transduction pathways. A model depicting the two pathways of steroid receptor activation is presented in Figure 5. In the upper portion is the classical hormone-dependent pathway showing that binding of ligand to receptor in the heat-shock protein complex initiates a series of steps that results in an active liganded receptor homodimer binding to DNA and inducing transcription. The lower portion shows signals generated by membrane-bound receptors such as those for dopamine or growth factors following a signal cascade that results in activation of the receptor in the absence of ligand. Although there is evidence in the case of the ER that the phosphorylation of the receptor is altered during this process, the role of this phosphorylation in the activation process remains to be determined. In addition to the receptor, the heat-shock protein complex associated with the unliganded receptor, as well as proteins that interact with receptor to produce the active transcription complex, are likely targets for these signalling pathways. Major questions yet to be answered include the mechanism by which receptors become active in the absence of ligand, and the relative importance of this pathway *in vivo*. Nonetheless, these studies highlight the importance of the interactions between signal transduction cascades and steroid hormone receptors in producing a biological response.

REFERENCES

1   Evans, R. M. (1988) Science **240**, 889–895

2   O'Malley, B. W. and Conneely, O. M. (1992) Mol. Endocrinol. **6**, 1359–1361

3   Smith, D. F. and Toft, D. O. (1993) Mol. Endocrinol. **7**, 4–11

4   DeFranco, D. B., Qi, M., Borror, K. C., Garabedian, M. J. and Brautigan, D. L. (1991) Mol. Endocrinol. **5**, 1215–1228

5   Guiochon-Mantel, A., Lescop, P., Christin-maitre, S., Loosfelt, H., Perrot-Applanat, M. and Milgrom, E. (1991) EMBO J. **10**, 3851–3859

6   Picard, D., Kumar, V., Chambon, P. and Yamamoto, K. R. (1990) Cell Regul. **1**, 291–299

7   Kemppainen, J. A., Lane, W. V., Sar, M. and Wilson, E. M. (1992) J. Biol. Chem. **267**, 968–974

8   Izhar, M., Nuchamowitz, Y. and Mirelman, D. (1982) Infect. Immun. **35**, 1110–1118

9   Beato, M., Chalepakis, G., Schauer, M. and Slater, E. P. (1989) J. Steroid Biochem. **32**, 737–747

10   Scheidereit, C., Geisse, S., Westphal, H. M. and Beato, M. (1983) Nature (London) **304**, 749–752

11   Lamian, V., Gonzalez, B. Y., Michel, F. J. and Simmen, R. C. M. (1993) J. Steroid Biochem. Mol. Biol. **46**, 439–450

12   Norris, J., Fan, D., Aleman, C., Marks, J. R., Futreal, P. A., Wiseman, R. W., Iglehart, J. D., Deininger, P. L. and McDonnell, D. P. (1995) J. Biol. Chem. **270**, 22777–22782

13   Truss, M. and Beato, M. (1993) Endocr. Rev. **14**, 459–479

14   Kasper, S., Rennie, P. S., Bruchovsky, N., Sheppard, P. C., Cheng, H., Lin, L., Shiu, R. P., Snoek, R. and Matusik, R. J. (1994) J. Biol. Chem. **269**, 31763–31769

15   Ing, N. H., Beekman, J. M., Tsai, S. Y., Tsai, M.-J. and O'Malley, B. W. (1992) J. Biol. Chem. **267**, 17617–17623

16 Schule, R., Muller, M., Kaltschmidt, C. and Renkawitz, R. (1988) Science. **242**, 1418–1420

17 Schule, R., Muller, M., Otsuka-Murakami, H. and Renkawitz, R. (1988) Nature (London) **332**, 87–90

18 Mangelsdorf, D. J., Ong, E. S., Dyck, J. A. and Evans, R. M. (1990) Nature (London) **345**, 224–229

19 Leid, M., Kastner, P., Lyons, R., Nakshatri, H., Saunders, M., Zacharewski, T., Chen, J. Y., Staub, A., Garnier, J. M., Macler, S., and Chambon, P. (1992) Cell **68**, 377–395

20 Yu, V. C., Delsert, C., Andersen, B., Holloway, J. M., Devary, O. V., Naar, A. M., Kim, S. Y., Boutin, J. M., Glass, C. K. and Rosenfeld, M. G. (1991) Cell **67**, 1251–1266

21 Horlein, A. J., Naar, A. M., Heinzel, T., Torchia, J., Gloss, B., Kurokawa, R., Ryan, A., Kamel, Y., Soderstrom, M., Glass, C. K. and Rosenfeld, M. G. (1995) Nature (London) **377**, 397–404

22 Kurokawa, R., Soderstrom, M., Horlein, A., Halachml, S., Brown, M., Rosenfeld, M. G. and Glass, C. K. (1995) Nature (London) **377**, 451–454

23 Tsai, M.-J. and O'Malley, B. W. (1994) Molecular Biology Intelligence Unit: Mechanism of Steroid Hormone Regulation of Gene Transcription, R. G. Landes Company, Austin, TX

24 Weinberger, C., Hollenberg, S. M., Rosenfeld, M. G. and Evans, R. M. (1985) Nature (London) **318**, 670–672

25 Kumar, V., Green, S., Staub, A. and Chambon, P. (1986) EMBO J. **5**, 2231–2236

26 Stromstedt, P. E., Berkenstam, A., Jornvall, H., Gustafsson, J. A. and Carlstedt-Duke, J. (1990) J. Biol. Chem. **265**, 12973–12977

27 Carlstedt-Duke, J., Stromstedt, P. E., Persson, B., Cederlund, E., Gustafsson, J.-A. and Jornvall, H. (1988) J. Biol. Chem. **263**, 6842–6846

28 Fawell, S. E., Lees, J. A., White, R. and Parker, M. G. (1990) Cell **60**, 953–962

29 Luisi, B. F., Xu, W. X., Otwinowski, Z., Freedman, L. P., Yamamoto, K. R. and Sigler, P. B. (1991) Nature (London) **352**, 497–505

30 Schwabe, J. W. R., Chapman, L., Finch, J. T. and Rhodes, D. (1993) Cell **75**, 567–578

31 Sartorius, C. A., Melville, M. Y., Hovland, A. R., Tung, L., Takimoto, G. S. and Horwitz, K. B. (1994) Mol. Endocrinol. **8**, 1347–1360

32 Guiochon-Mantel, A., Loosfelt, H., Lescop, P., Sar, S., Atger, M., Perrot-Applanat, M. and Milgrom, E. (1989) Cell **57**, 1147–1154

33 Carson-Jurica, M. A., Lee, A. T., Dobson, A. D. W., Conneely, O. M., Schrader, W. T. and O'Malley, B. W. (1989) J. Steroid Biochem. **34**, 1–9

34 Fritsch, M., Leary, C. M., Furlow, J. D., Ahrens, H., Schuh, T. J., Mueller, G. C. and Gorski, J. (1992) Biochemistry **31**, 5303–5311

35 Allan, G. F., Leng, X., Tsai, S. Y., Weigel, N. L., Edwards, D. P., Tsai, M.-J. and O'Malley, B. W. (1992) J. Biol. Chem. **267**, 19513–19520

36 Weigel, N. L., Beck, C. A., Estes, P. A., Prendergast, P., Altmann, M., Christensen, K. and Edwards, D. P. (1992) Mol. Endocrinol. **6**, 1585–1597

37 Renaud, J.-P., Rochel, N., Ruff, M., Vivat, V., Chambon, P., Gronemeyer, H. and Moras, D. (1995) Nature (London) **378**, 681–689

38 Bourguet, W., Ruff, M., Chambon, P., Gronemeyer, H. and Moras, D. (1995) Nature (London) **375**, 377–382

39 Smith, D. F. (1993) Mol. Endocrinol. **7**, 1418–1429

40 Beck, C. A., Weigel, N. L., Moyer, M. L., Nordeen, S. K. and Edwards, D. P. (1993) Proc. Natl. Acad. Sci. U.S.A. **90**, 4441–4445

41 Schauer, M., Chalepakis, G., Willmann, T. and Beato, M. (1989) Proc. Natl. Acad. Sci. U.S.A. **86**, 1123–1127

42 Vegeto, E., Allan, G. F., Schrader, W. T., Tsai, M.-J., McDonnell, D. P. and O'Malley, B. W. (1992) Cell **69**, 703–713

43 Wong, C. I., Xhou, Z. X., Sar, M. and Wilson, E. M. (1993) J. Biol. Chem. **268**, 19004–19012

44 Montano, M. M., Ekena, K., Krueger, K. D., Keller, A. L. and Katzenellenbogen, B. S. (1996) Mol. Endocrinol. **10**, 230–242

45 Schrader, W. T. and O'Malley, B. W. (1972) J. Biol. Chem. **247**, 51–59

46 Sherman, M. R., Corvol, P. L. and O'Malley, B. W. (1970) J. Biol. Chem. **245**, 6085–6096

47 Horwitz, K. B. and Alexander, P. S. (1983) Endocrinology (Baltimore) **113**, 2195–2201

48 Loosfelt, H., Logeat, F., Hai, M. T. V. and Milgrom, E. (1984) J. Biol. Chem. **259**, 14196–14202

49 Kastner, P., Krust, A., Turcotte, B., Stropp, U., Tora, L., Gronemeyer, H. and Chambon, P. (1990) EMBO J. **9**, 1603–1614

50 Conneely, O. M., Kettelberger, D. M., Tsai, M.-J., Schrader, W. T. and O'Malley, B. W. (1989) J. Biol. Chem. **264**, 14062–14064

51 Meyer, M. E., Pornon, A., Ji, J. W., Bocquel, M. T., Chambon, P. and Gronemeyer, H. (1990) EMBO J. **9**, 3923–3932

52 Tora, L., Gronemeyer, H., Turcotte, B., Gaub, M. P. and Chambon, P. (1988) Nature (London) **333**, 185–188

53 Wei, L. L., Gonzalez-Aller, C., Wood, W. M., Miller, L. A. and Horwitz, K. B. (1990) Mol. Endocrinol. **4**, 1833–1840

54 Wilson, C. M. and McPhaul, M. J. (1994) Proc. Natl. Acad. Sci. U.S.A. **91**, 1234–1238

55 Bamberger, C. M., Bamberger, A.-M., de Castro, M. and Chrousos, G. P. (1995) J. Clin. Invest. **95**, 2435–2441

56 Fuqua, S. A. W., Allred, D. C. and Auchus, R. J. (1993) J. Cell. Biochem. **17G**, 194–197

57 Glass, C. K., DiRenzo, J., Kurokawa, R. and Han, Z. H. (1991) DNA Cell Biol. **10**, 623–638

58 Leid, M., Kastner, P. and Chambon, P. (1992) Trends Biochem. Sci. **17**, 427–433

59 Mangelsdorf, D. J., Borgmeyer, U., Heyman, R. A., Zhou, J. Y., Ong, E. S., Oro, A. E., Kakizuka, A. and Evans, R. M. (1992) Genes Dev. **6**, 329–344

60 Thompson, C. C., Weinberger, C., Lebo, R. and Evans, R. M. (1987) Science **237**, 1610–1614

61 Weinberger, C., Thompson, C. C., Ong, E. S., Lebo, R., Gruol, D. J. and Evans, R. M. (1986) Nature (London) **324**, 641–646

62 Katz, D., Reginato, M. J. and Lazar, M. A. (1995) Mol. Cell. Biol. **15**, 2341–2348

63 Carlberg, C., Bendik, I., Wyss, A., Meier, E., Sturzenbecker, L. J., Grippo, J. F. and Hunziker, W. (1993) Nature (London) **361**, 657–660

64 Giguere, V. (1994) Endocr. Rev. **15**, 61–79

65 Tsai, S. Y., Tsai, M.-J. and O'Malley, B. W. (1989) Cell **57**, 443–448

66 Nicholson, R. C., Mader, S., Nagpal, S., Leid, M., Rochette-Egly, C. and Chambon, P. (1990) EMBO J. **9**, 4443–4454

67 Diamond, M. I., Miner, J. N., Yoshinaga, S. K. and Yamamoto, K. R. (1990) Science **249**, 1266–1272

68 Lucibello, F. C., Slater, E. P., Jooss, K. U., Beato, M. and Muller, R. (1990) EMBO J. **9**, 2827–2834

69 Yang-Yen, H. F., Chambard, J. C., Sun, Y. L., Smeal, T., Schmidt, T. J., Drouin, J. and Karin, M. (1990) Cell **62**, 1205–1215

70 Bailly, A., Le Page, C., Rauch, M. and Milgrom, E. (1986) EMBO J. **5**, 3235–3241

71 Vegeto, E., Shahbaz, M. M., Wen, D. X., Goldman, M. E., O'Malley, B. W. and McDonnell, D. P. (1993) Mol. Endocrinol. **7**, 1244–1255

72 Tung, L., Mohamed, M. K., Hoeffler, J. P., Takimoto, G. S. and Horwitz, K. B. (1993) Mol. Endocrinol. **7**, 1256–1265

73 McDonnell, D. P. and Goldman, M. E. (1994) J. Biol. Chem. **269**, 11945–11949

74 Kraus, W. L., Weis, K. E. and Katzenellenbogen, B. S. (1995) Mol. Cell. Biol. **15**, 1847–1857

75 Onate, S. A., Tsai, S. Y., Tsai, M.-J. and O'Malley, B. W. (1995) Science **270**, 1354–1357

76 Halachmi, S., Marden, E., Martin, G., MacKay, H., Abbondanza, C. and Brown, M. (1994) Science. **264**, 1455–1458

77 Cavailles, V., Dauvouis, S., L'Horset, F., Lopez, G., Hoare, S., Kushner, P. J. and Parker, M. G. (1995) EMBO J. **14**, 3741–3751

78 Chen, J. D. and Evans, R. M. (1995) Nature (London) **377**, 454–457

79 Hill, C. S. and Treisman, R. (1995) Cell **80**, 199–211

80 Karin, M. (1992) FASEB J. **6**, 2581–2590

81 Housley, P. R. and Pratt, W. B. (1983) J. Biol. Chem. **258**, 4630–4635

82 Dougherty, J. J., Puri, R. K. and Toft, D. O. (1984) J. Biol. Chem. **259**, 8004–8009

83 Denner, L. A., Schrader, W. T., O'Malley, B. W. and Weigel, N. L. (1990) J. Biol. Chem. **265**, 16548–16555

84 Sheridan, P. L., Evans, R. M. and Horwitz, K. B. (1989) J. Biol. Chem. **264**, 6520–6528

85 Zhang, Y., Beck, C. A., Poletti, A., Edwards, D. P. and Weigel, N. L. (1994) J. Biol. Chem. **269**, 31034–31040

86 Chauchereau, A., Cohen-Solal, K., Jolivet, A., Bailly, A. and Milgrom, E. (1994) Biochemistry **33**, 13295–13303

87 Chauchereau, A., Loosfelt, H. and Milgrom, E. (1991) J. Biol. Chem. **266**, 18280–18286

88 Zhang, Y., Beck, C. A., Poletti, A., Edwards, D. P. and Weigel, N. L. (1995) Mol. Endocrinol. **9**, 1029–1040

89 Denton, R. R., Koszewski, N. J. and Notides, A. C. (1992) J. Biol. Chem. **267**, 7263–7268

90 Ali, S., Metzger, D., Bornert, J. M. and Chambon, P. (1993) EMBO J. **12**, 1153–1160

91 LeGoff, P., Montano, M. M., Schodin, D. J. and Katzenellenbogen, B. S. (1994) J. Biol. Chem. **269**, 4458–4466

92 Orti, E., Mendel, D. B., Smith, L. I. and Munck, A. (1989) J. Biol. Chem. **264**, 9728–9731

93 Kuiper, G. G. J. M., de Ruiter, P. E., Trapman, J., Boersma, W. J. A., Grootegoed, J. A. and Brinkmann, A. O. (1993) Biochem. J. **291**, 95–101

94 Poletti, A. and Weigel, N. L. (1993) Mol. Endocrinol. **7**, 241–246

95 Morgan, D. O. (1995) Nature (London) **374**, 131–134

96 Cano, E. and Mahadevan, L. C. (1994) Trends Biochem. Sci. **20**, 117–122

97   Davis, R. J. (1994) Trends Biochem. Sci. **19**, 470–473

98   Zhou, Z. X., Kemppainen, J. A. and Wilson, E. M. (1995) Mol. Endocrinol. **9**, 605–615

99   Lahooti, H., White, R., Hoare, S. A., Rahman, D., Pappin, D. J. C. and Parker, M. G. (1995) J. Steroid Biochem. Mol. Biol. **55**, 305–313

100 Poletti, A., Conneely, O. M., McDonnell, D. P., Schrader, W. T., O'Malley, B. W. and Weigel, N. L. (1993) Biochemistry **32**, 9563–9569

101 Allgood, V. E., Weigel, N. L. and O'Malley, B. W. (1994) Endocr. Soc. Meet. 73rd, p. 631

102 Bai, W. and Weigel, N. L. (1996) J. Biol. Chem. **271**, 12801–12806

103 Bai, W., Tullos, S. and Weigel, N. L. (1994) Mol. Endocrinol. **8**, 1465–1473

104 Takimoto, G. S., Tasset, D. M., Eppert, A. C. and Horwitz, K. B. (1992) Proc. Natl. Acad. Sci. U.S.A. **89**, 3050–3054

105 Zhang, Y., Beck, C. A., Hutchens, T. W., Clement, J. P., Harper, J. W., Edwards, D. P. and Weigel, N. L. (1994) Endocr Soc. Meeting 73rd, p. 630

106 Beck, C. A., Zhang, Y., Weigel, N. L. and Edwards, D. P. (1996) J. Biol. Chem. **271**, 1209–1217

107 Takimoto, G. S. and Horwitz, K. B. (1993) Trends Endocrinol. Metab. **4**, 1–7

108 Bodwell, J. E., Orti, E., Coull, J. M., Pappin, D. J., Swift, F. and Smith, L. I. (1991) J. Biol. Chem. **266**, 7549–7555

109 Bodwell, J. E., Hu, J. M., Orti, E. and Munck, A. (1995) J. Steroid Biochem. Mol. Biol. **52**, 135–140

110 Hu, J. M., Bodwell, J. E. and Munck, A. (1994) Mol. Endocrinol. **8**, 1709–1713

111 Hsu, S.-C. and DeFranco, D. B. (1995) J. Biol. Chem. **270**, 3359–3364

112 Mason, S. A. and Housley, P. R. (1993) J. Biol. Chem. **268**, 21501–21504

113 Jewell, C. M., Webster, J. C., Bodwell, J. E., Munck, A., Sar, M. and Cidlowski, J. A. (1995) Endocrine Soc. 423

114 Migliaccio, A., Rotondi, A. and Auricchio, F. (1986) EMBO J. **5**, 2867–2872

115 Castoria, G., Migliaccio, A., Green, S., DiDomenico, M., Chambon, P. and Auricchio, F. (1993) Biochemistry **32**, 1740–1750

116 Arnold, S. F., Obourn, J. D., Jaffe, H. and Notides, A. C. (1995) Mol. Endocrinol. **9**, 24–33

117 Washburn, T., Hocutt, A., Brautigan, D. L. and Korach, K. S. (1991) Mol. Endocrinol. **5**, 235–242

118 Joel, P. B., Traish, A. M. and Lannigan, D. A. (1995) Mol. Endocrinol. **9**, 1041–1052

119 Arnold, S. F., Obourn, J. D., Yudt, M. R., Carter, T. H. and Notides, A. C. (1995) J. Steroid Biochem. Mol. Biol. **52**, 159–171

120 Arnold, S. F., Obourn, J. D., Jaffe, H. and Notides, A. C. (1994) Mol. Endocrinol. **8**, 1208–1214

121 Kato, S., Endoh, H., Masuhiro, Y., Kitamoto, T., Uchiyama, S., Sasaki, H., Masushige, S., Gotoh, Y., Nishida, E., Kawashima, H., Metzger, D. and Chambon, P. (1995) Science **270**, 1491–1494

122 Kuiper, G. G. J. M. and Brinkmann, A. O. (1995) Biochemistry **34**, 1851–1857

123 Hilliard, G. M., Cook, R. G., Weigel, N. L. and Pike, J. W. (1994) Biochemistry **33**, 4300–4311

124 Jurutka, P. W., Hsieh, J. C., MacDonald, P. N., Terpening, C. M., Haussler, C. A., Haussler, M. R. and Whitfield, G. K. (1993) J. Biol. Chem. **268**, 6791–6799

125 Jurutka, P. W., Hsieh, J.-C., Nakajima, S., Haussler, C. A., Whitfield, G. K. and Haussler, M. R. (1996) Proc. Natl. Acad. Sci. U.S.A. **93**, 3519–3524

126 Glineur, C., Bailly, A. and Ghysdael, J. (1989) Oncogene **4**, 1247–1254

127 Glineur, C., Zenke, M., Beug, H. and Ghysdael, J. (1990) Genes Dev. **4**, 1663–1676

128 Lin, K. H., Ashizawa, K. and Cheng, S. (1992) Proc. Natl. Acad. Sci. U.S.A. **89**, 7737–7741

129 Sugawara, A., Yen, P. M., Apriletti, J. W., Ribeiro, R. C. J., Sacks, D. B., Baxter, J. D. and Chin, W. W. (1994) J. Biol. Chem. **269**, 433–437

130 Rochette-Egly, C., Gaub, M. P., Lutz, Y., Ali, S., Scheuer, I. and Chambon, P. (1992) Mol. Endocrinol. **6**, 2197–2209

131 Rochette-Egly, C., Oulad-Abdelghani, M., Staub, A., Pfister, V., Scheuer, I., Chambon, P. and Gaub, M.-P. (1995) Mol. Endocrinol. **9**, 860–871

132 Lefebvre, P., Gaub, M.-P., Tahayato, A., Rochette-Egly, C. and Formstecher, P. (1995) J. Biol. Chem. **270**, 10806–10816

133 Fahrner, T. J., Carroll, S. L. and Milbrandt, J. (1990) Mol. Cell. Biol. **10**, 6454–6459

134 Hazel, T. G., Misra, R., Davis, I. J., Greenberg, M. E. and Lau, L. F. (1991) Mol. Cell. Biol. **11**, 3239–3246

135 Davis, I. J., Hazel, T. G., Chen, R.-H., Blenis, J. and Lau, L. F. (1993) Mol. Endocrinol. **7**, 953–964

136 Denner, L. A., Weigel, N. L., Maxwell, B. L., Schrader, W. T. and O'Malley, B. W. (1990) Science **250**, 1740–1743

137 Ignar-Trowbridge, D. M., Nelson, K. G., Bidwell, M. C., Curtis, S. W., Washburn, T. F., McLachlan, J. A. and Korach, K. S. (1992) Proc. Natl. Acad. Sci. U.S.A. **89**, 4658–4662

138 Power, R. F., Mani, S. K., Codina, J., Conneely, O. M. and O'Malley, B. W. (1991) Science **254**, 1636–1639

139 Zhang, Y., Bai, W., Allgood, V. E. and Weigel, N. L. (1994) Mol. Endocrinol. **8**, 577–584

140 Turgeon, J. L. and Waring, D. W. (1994) Mol. Endocrinol. **8**, 860–869

141 Mani, S. K., Allen, J. M., Clark, J. H., Blaustein, J. D. and O'Malley, B. W. (1994) Science **265**, 1246–1249

142 Beck, C. A., Weigel, N. L. and Edwards, D. P. (1992) Mol. Endocrinol. **6**, 607–620

143 Sartorius, C. A., Tung, L., Takimoto, G. S. and Horwitz, K. B. (1993) J. Biol. Chem. **268**, 9262–9266

144 Sartorius, C. A., Groshong, S. D., Miller, L. A., Powell, R. L., Tung, L., Takimoto, G. S. and Horwitz, K. B. (1994) Cancer Res. **54**, 3868–3877

145 Ignar-Trowbridge, D. M., Teng, C. T., Ross, K. A., Parker, M. G., Korach, K. S. and McLachlan, J. A. (1993) Mol. Endocrinol. **7**, 992–998

146 Auricchio, F., DiDomenico, M., Migliaccio, A., Castoria, G. and Bilancio, A. (1995) Cell Growth Differ. **6**, 105–113

147 Aronica, S. M. and Katzenellenbogen, B. S. (1993) Mol. Endocrinol. **7**, 743–752

148 Smith, C. L., Conneely, O. M. and O'Malley, B. W. (1993) Proc. Natl. Acad. Sci. U.S.A. **90**, 6120–6124

149 Cho, H. and Katzenellenbogen, B. S. (1993) Mol. Endocrinol. **7**, 441–452

150 Fujimoto, N. and Katzenellenbogen, B. S. (1994) Mol. Endocrinol. **8**, 296–304

151 Ince, B. A., Montano, M. M. and Katzenellenbogen, B. S. (1994) Mol. Endocrinol. **8**, 1397–1406

152 Nordeen, S. K., Moyer, M. L. and Bona, B. J. (1994) Endocrinology (Baltimore) **134**, 1723–1732

153 Nordeen, S. K., Bona, B. J. and Moyer, M. L. (1993) Mol. Endocrinol. **7**, 731–742

154 Ikonen, T., Palvimo, J. J., Kallio, P. J., Reinikainen, P. and Janne, O. A. (1994) Endocrinology **135**, 1359–1366

155 de Ruiter, P. E., Teuwen, R., Trapman, J., Dijkema, R. and Brinkmann, A. O. (1995) Mol. Cell. Endocrinol. **110**, R1–R6

156 Culig, Z., Hobisch, A., Cronauer, M. V., Radmayr, C., Trapman, J., Hittmair, A., Bartsch, G. and Klocker, H. (1994) Cancer Res. **54**, 5474–5478

Biochem. J. (1996) **318**, 1–14 (Printed in Great Britain)

REVIEW ARTICLE
The lipocalin protein family: structure and function

Darren R. FLOWER

Department of Physical Chemistry and BioAnalysis, Astra Charnwood, Bakewell Road, Loughborough, Leics. LE11 5RH, U.K.

The lipocalin protein family is a large group of small extracellular proteins. The family demonstrates great diversity at the sequence level; however, most lipocalins share three characteristic conserved sequence motifs, the kernel lipocalins, while a group of more divergent family members, the outlier lipocalins, share only one. Belying this sequence dissimilarity, lipocalin crystal structures are highly conserved and comprise a single eight-stranded continuously hydrogen-bonded antiparallel β-barrel, which encloses an internal ligand-binding site. Together with two other families of ligand-binding proteins, the fatty-acid-binding proteins (FABPs) and the avidins, the lipocalins form part of an overall structural superfamily: the calycins. Members of the lipocalin family are characterized by several common molecular-recognition properties: the ability to bind a range of small hydrophobic molecules, binding to specific cell-surface receptors and the formation of complexes with soluble macromolecules. The varied biological functions of the lipocalins are mediated by one or more of these properties. In the past, the lipocalins have been classified as transport proteins; however, it is now clear that the lipocalins exhibit great functional diversity, with roles in retinol transport, invertebrate cryptic coloration, olfaction and pheromone transport, and prostaglandin synthesis. The lipocalins have also been implicated in the regulation of cell homoeostasis and the modulation of the immune response, and, as carrier proteins, to act in the general clearance of endogenous and exogenous compounds.

INTRODUCTION

Members of the lipocalin protein family are typically small secreted proteins which are characterized by a range of different molecular-recognition properties: their ability to bind small, principally hydrophobic molecules (such as retinol), their binding to specific cell-surface receptors and their formation of macromolecular complexes. The lipocalins are a large and ever-expanding group of proteins exhibiting great structural and functional diversity, both within and between species. Although they have, in the past, been classified primarily as transport proteins, it is now clear that members of the lipocalin family fulfill a variety of different functions. These include roles in retinol transport, cryptic coloration, olfaction, pheromone transport, and the enzymic synthesis of prostaglandins; the lipocalins have also been implicated in the regulation of the immune response and the mediation of cell homoeostasis.

Despite common characteristics and common functions, the lipocalin family has been defined largely on the basis of sequence similarity; the constitution of the family is surveyed in Table 1: its principal members are listed together with a summary of their biochemical properties. Already the lipocalin protein family is too large a subject to be reviewed exhaustively; instead we focus, in turn, on three key aspects: sequence and structure relationships of the lipocalins, their capacity for molecular recognition, and our burgeoning understanding of function within the family.

SEQUENCE AND STRUCTURE RELATIONSHIPS

From its initial identification [1], the apparent size of the lipocalin family has grown significantly to encompass a large corpus of protein sequences. Within this the lipocalins display unusually low levels of overall sequence conservation, with pairwise sequence identity often falling below 20%, the threshold for reliable alignment. However, all lipocalins share sufficient similarity, in the form of short characteristic conserved sequence motifs, for a useful definition of family membership to be made [2,3]. This analysis shows the lipocalin protein family to be composed of a core set of quite closely related proteins, the kernel lipocalins, and a smaller number of more divergent sequences, the outlier lipocalins (see Table 1). Kernel lipocalins, which form by far the largest self-consistent subset within the whole set of related sequences, share three conserved sequence motifs (see Figure 1) [2,3], which correspond to the three main structurally conserved regions of the lipocalin fold and have proved useful in the design of primers for cloning studies [4]. The first of these three characteristic motifs is shared by all lipocalins and can be used as a diagnostic of family membership. The outlier lipocalins match no more than two of these three motifs and are more diverse, forming distinct groups at the sequence level: the α_1-acid glycoproteins (AGPs), odorant-binding proteins, and Von Ebner's-gland proteins.

Hitherto, lipocalins had only been found in eukaryotic organisms, mostly in vertebrates, although some have been identified in other phyla. This includes several examples from arthropods: butterfly insecticyanin, grasshopper lazarillo [5], cockroach Bla g 4 protein [6] and lobster crustacyanin [7]; and there is evidence to suggest that carotenoprotein lipocalins may also be present in species from the phylum Coelenterata [8]. Other calycins, FABPs and avidins are also present in both vertebrates and invertebrates, including Arthropoda and Coelentarata, but thus far there is only one example from prokaryotes, namely streptavidin. Recent reports have, for the first time, identified bacterial lipocalins [9,10]. The existence of prokaryotic lipocalins has profound implications for our under-

---

Abbreviations used: FABP, fatty-acid-binding protein; RBP, plasma retinol-binding protein; Blg, β-lactoglobulin; BBP, bilin-binding protein; MUP, major urinary protein; OBP, odorant-binding protein; NGAL, neutrophil lipocalin; VEGP, von Ebner's-gland protein; apo, apolipoprotein; PGD_2, prostaglandin D_2; APP, acute-phase protein; PP14, pregnancy protein 14; A1M, α_1-microglobulin; AGP, α_1-acid glycoprotein; MAC, membrane-attack complex; QSP, quiescence-specific protein; CHM, chicken heart mesenchymal cells; SCR, structurally conserved region.

Table 1   Properties of members of the lipocalin protein family

Members of the lipocalin protein family listed as a Table with a summary of their physical and chemical properties; the proteins listed are divided between kernel and outlier lipocalins [2,3]. Molecular masses are given in kDa. Where a property, such as glycosylation, has been shown to be present experimentally this is indicated by a +, shown to be absent by a −; otherwise, where this is unknown the value is left blank. Data are taken from references cited in the text. Well-established abbreviations of particular lipocalins, as used in the text, are given in the last column.

Protein	Subunit molecular mass	pI	No. of residues	Oligomeric state	Glycosylation	Number of disulphides	Abbreviation
Kernel lipocalins							
Retinol-binding protein	21.0	5.5	183	Monomer	−	3	RBP
Purpurin	20.0		175				PURP
Retinoic acid-binding protein	18.5	5.2	166	Monomer	−	1	RABP
α_{2u}-Globulin	18.7	5.7–6.7	162	Dimer	−	1	A2U
Major urinary protein	17.8	5.5–5.7	161	Dimer	−	1	MUP
Bilin-binding protein	19.6		173	Tetramer	−	2	BBP
α-Crustacyanin	350.0	4.3–4.7	174/181	Octamer of heterodimers	−	2/2	
Pregnancy protein 14	56.0		162	Homodimer	+		PP14
β-Lactoglobulin	18.0	5.2	162	Dimer/monomer	−	2	Blg
α_1-Microglobulin	31.0	4.3–4.8	188	Monomer + complexes	+	1	A1M
C8γ	22.0		182	Part of complex	−	1	C8γ
Apolipoprotein D	29.0–32.0	4.7–5.2	169	Dimer + complexes	+	2	ApoD
Lazarillo	45.0		183	Monomer	+	+	LAZ
Prostaglandin D synthase	27.0	4.6	168	Monomer	+	1	PGDS
Quiesence-specific protein	21.0	6.3	158			1	QSP
Neutrophil lipocalin	25.0		179	Monomer/dimer + complexes			NGAL
Choroid plexus protein	20.0		183	Monomer	−		
Outlier lipocalins							
Odorant-binding protein	37.0–40.0	4.7	159	Dimer		0	OBP
von Ebner's-gland protein	18.0	4.8–5.2	170	Dimer		1	VEGP
α_1-Acid glycoprotein	40.0	3.2	183	Monomer	+	2	AGP
Probasin	20.0	11.5	160				PBAS
Aphrodisin	17.0		151		+	2	

standing of how the family has evolved, suggesting that it is very much older and more widespread than has been supposed.

In contrast with their low conservation at the sequence level, analysis of available lipocalin crystal structures, which include plasma retinol-binding protein (RBP) [11], β-lactoglobulin (Blg) [12], insecticyanin [13], bilin-binding protein (BBP) [14,15], major urinary protein (MUP) and α_{2u}-globulin [16], odorant-binding protein (OBP) [17] and epididymal retinoic acid-binding protein [18], shows that the overall folding pattern common to the lipocalins is highly conserved. The nature of this common structure is now well described (see Figures 1 and 2) [2,11]. The lipocalin fold is a highly symmetrical all-β protein dominated by a single eight-stranded antiparallel β-sheet closed back on itself to form a continuously hydrogen-bonded β-barrel, which, in cross-section, has a flattened or elliptical shape and encloses an internal ligand-binding site. The eight β-strands of the barrel, labelled A–H (see Figure 1), are linked by a succession of +1 connections. These seven loops, labelled L1 to L7 (see Figure 1), are all typical of short β-hairpins, except loop L1; this is a large Ω loop which forms a lid folded back to close partially the internal ligand-binding site found at this end of the barrel.

Together with two other distinct families of ligand-binding proteins, the FABPs and the avidins, the lipocalins form part of the calycin protein superfamily [2,19]; see Figures 2 and 3. Like the lipocalins, both families have β-barrel structures. The FABP barrel is ten-stranded and discontinuous and that of the avidins, although eight-stranded, is less elliptical in cross-section than that of the lipocalins. Beyond an obvious functional similarity (the binding of hydrophobic ligands) these families are characterized by a similar folding pattern (an antiparallel β-barrel, with a repeated +1 topology, possessed of an internal ligand-binding site), within which large parts of their structures can be equivalenced. They are also characterized by a conserved sequence motif which corresponds to an unusual structural feature (a short 3_{10}-like helix leading into a β-strand) conserved in conformation and its location within the fold [2,19]; other than the conservation of this motif, the three families share no discernible global sequence similarity. Members of the calycin triumvirate

Figure 1   Structure of the lipocalin fold

(**a**) Characteristic features of the lipocalin fold. An unwound view of the lipocalin fold orthogonal to the axis of the barrel [2]. The nine β-strands of the antiparallel β-sheet are shown as arrows and labelled A–I. The N-terminal 3_{10}-like helix and C-terminal α-helix (labelled A1) are also marked. The hydrogen-bonded connection of two strands is indicated by a pair of dotted lines between them. Connecting loops are shown as solid lines and labelled L1–L7. The two ends of the β-barrel are topologically distinct [21]. One end has four β-hairpins (L1, L3, L5 and L7); the opening of the internal ligand-binding site is here and so is called the Open end of the molecule. The other has three β-hairpin loops (L2, L4 and L6); the N-terminal polypeptide chain crosses this end of the barrel to enter strand A via a conserved 3_{10} helix affecting closure of this end of the barrel: the Closed end of the molecule. Those parts which form the three main structurally conserved regions (SCRs) of the fold, SCR1, SCR2 and SCR3, are marked as boxes. Three sequence motifs which correspond to these SCRs are shown (MOTIF 1, MOTIF 2 and MOTIF 3). The first three sequences are from kernel lipocalins and the second three from outlier lipocalins. Note that MOTIF 1 is well conserved in all sequences, whereas the other two, particularly MOTIF 2, are only well conserved in kernel lipocalin sequences [2,3]. (**b**) The lipocalin β-barrel in cross-section β-strands are shown as triangles. Triangles pointing downwards indicate a strand direction into the plane of the paper and those pointing upwards indicate a strand direction out of the plane of the paper. The view shown, down the axis of the barrel, is orthogonal to that in (**a**). Connecting loops are shown as continuous lines. Labelling and features shown are as in (**a**). The closure of the sheet to form the lipocalin β-barrel breaks the symmetry of its elliptical cross-section, distinguishing between its two foci and suggesting a sidedness to the barrel also apparent in the location of the marked SCRs.

(a)

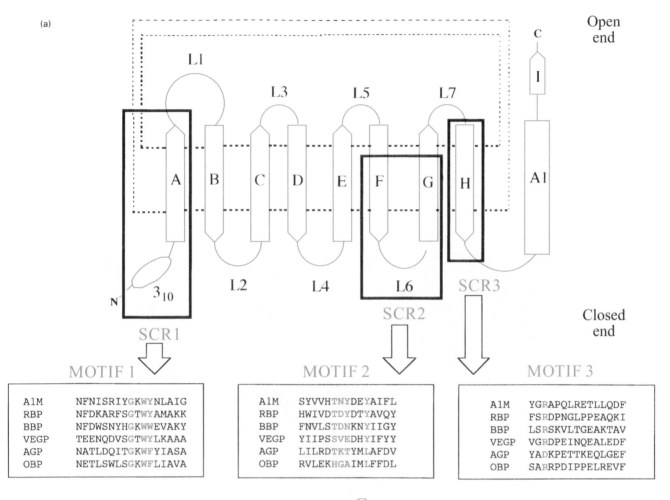

MOTIF 1

A1M	NFNISRIYGKWYNLAIG
RBP	NFDKARFSGTWYAMAKK
BBP	NFDWSNYHGKWWEVAKY
VEGP	TEENQDVSGTWYLKAAA
AGP	NATLDQITGKWFYIASA
OBP	NETLSWLSGKWFLIAVA

MOTIF 2

A1M	SYVVHTNYDEYAIFL
RBP	HWIVDTDYDTYAVQY
BBP	FNVLSTDNKNYIIGY
VEGP	YIIPSSVEDHYIFYY
AGP	LILRDTKTYMLAFDV
OBP	RVLEKHGAIMLFFDL

MOTIF 3

A1M	YGRAPQLRETLLQDF
RBP	FSRDPNGLPPEAQKI
BBP	LSRSKVLTGEAKTAV
VEGP	VGRDPEINQEALEDF
AGP	YADKPETTKEQLGEF
OBP	SARRPDIPPELREVF

(b)

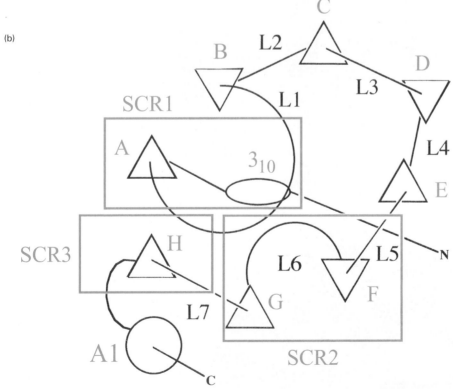

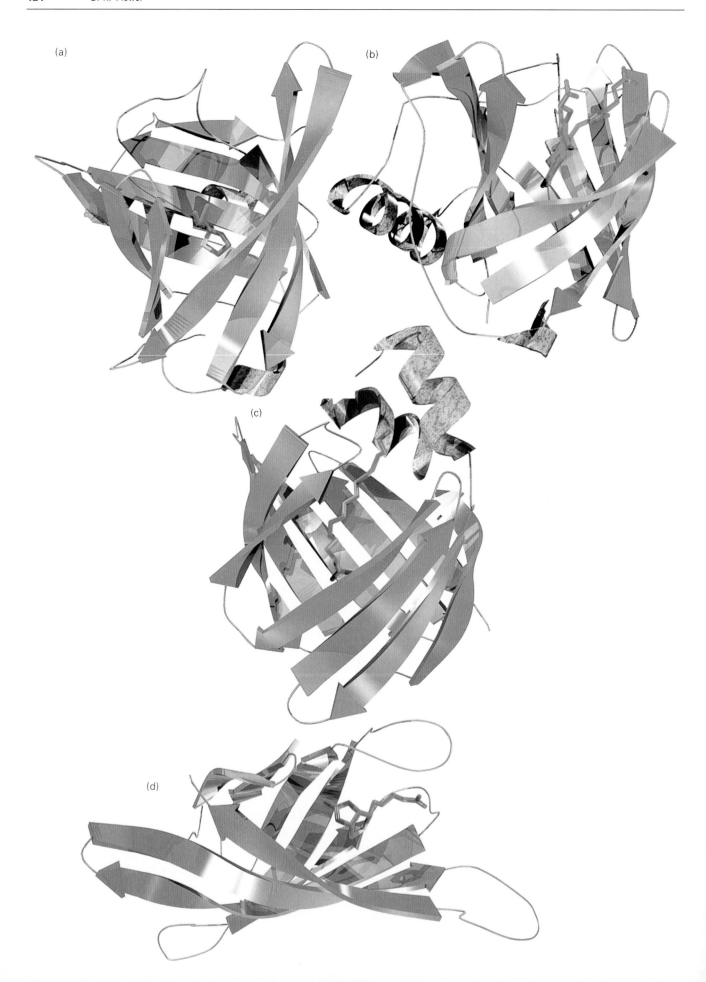

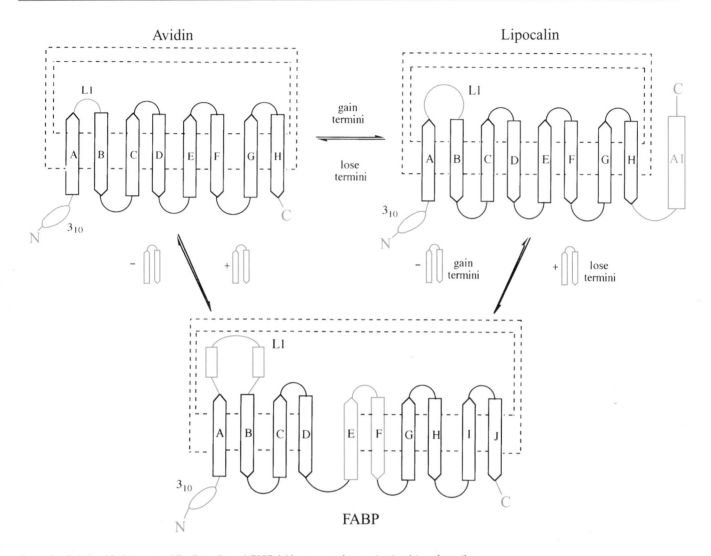

Figure 3   Relationship between avidin, lipocalin and FABP folds expressed as a structural transformation

If we imagine the removal of N- and C-terminal segments from a lipocalin structure (by cutting it at the start of the 3_{10} helix prior to strand A and at the end of strand H) and then make the necessary insertion of strands, then the resulting structure would be essentially that of an FABP. The transformation between avidin and FABP structures requires insertion of two strands into the centre of its β-barrel core and the dilation of loop L1, and from lipocalin to avidin necessitates the loss of both N- and C-terminal peptides and the truncation of loop L1. In no case is any change in topology or organization of the fold required, only the loss or gain of features and such minor structural adjustments as is normal between members of homologous families. Transforming a protein structure from any other family into either the avidin, lipocalin or FABP folds requires changes of gross conformation and topology more severe than any of these interconversions. The structural representations follow the conventions of Figure 1; β-strands are shown as arrows and labelled by letter. The N-terminal 3_{10}-like helix and C-terminal lipocalin α-helix (labelled A1) are marked. The hydrogen-bonded connection of two strands is indicated by a pair of dotted lines between them. Connecting loops are shown as continuous lines; loop L1 is labelled for each fold.

also share a distinct structural signature: an arginine or lysine residue (from the last strand of the β-barrel) which forms several hydrogen bonds with the main-chain carbonyl groups of the N-terminal 3_{10}-like helix and packs across a conserved tryptophan (from the first strand of the β-barrel) [20].

COMMON MOLECULAR-RECOGNITION PROPERTIES

Ligand binding

The lipocalins are, perhaps, best known for their binding of a remarkable array of small hydrophobic ligands. The structural features of the lipocalin fold, a large cup-shaped cavity, within

the β-barrel, and a loop scaffold at its entrance, are well adapted to the task of ligand binding: the amino acid composition of the pocket and loop scaffold, as well as its overall size and conformation, determining selectivity [21]. To accommodate ligands of different size and shape, the binding sites of different lipocalins can be quite different. For example, compare the binding mode exhibited by MUP (Figure 2a), which binds its small ligand deep within its pocket entirely enclosed by side chains, with that of BBP (Figure 2b), which binds its large and relatively hydrophilic ligand in a solvent-exposed site predominantly formed from the loop scaffold; see also Figure 4. Table 2 summarizes much of the available data concerning the binding of endogenous and exogenous ligands by members of the family. These include

Figure 2   Schematic ribbon drawings of protein structures

(**a**) MUP; (**b**) BBP; (**c**) FABP; and (**d**) avidin. Ray-traced schematic ribbon drawings of protein structures; following convention [148], β-strands are shown as smoothly curving arrows, α-helices as spiral ribbons, and loops not in secondary structures are displayed as a smoothed coil [149]. Ligand molecules are shown using a coloured all-atom representation.

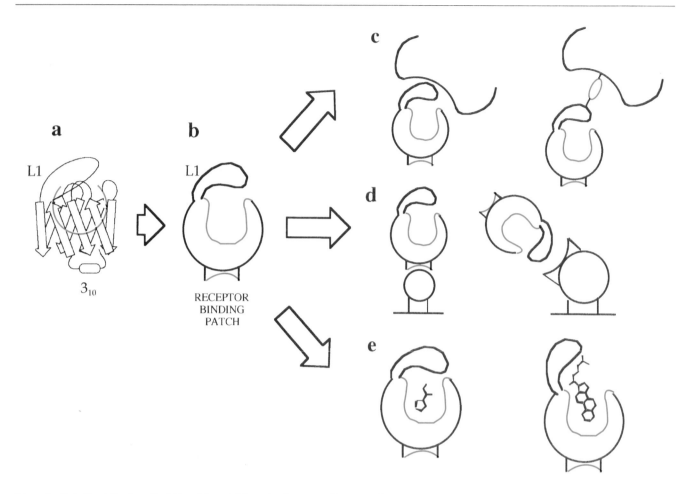

Figure 4   Polarity of the lipocalin fold and its multiple molecular recognition properties

The Figure shows a schematic representation of the lipocalin fold and visualization of the multiple interactions underlying its molecular-recognition properties. (**a**) Simplified schematic of the three-dimensional structure of the lipocalin fold; the β-strands of the lipocalin barrel are shown as arrows; the loops of the Open end of the fold are also shown, as is the N-terminal 3_{10} helix. The location of the internal ligand-binding pocket is shown as a highlighted semicircle. (**b**) Highly schematic cartoon summarizing the key structural features of the lipocalin fold and emphasizing its structural polarity; the ligand-binding pocket and hypothetical receptor-binding surface patch are highlighted. (**c**) Complexation with soluble macromolecules. The two modes exhibited by the lipocalins are shown; non-covalent association, such as between RBP and transthyretin, and covalently linked, either by disulphides (C8γ and C8α, apoD and apolipoproteins, or NGAL and gelatinase) or other groups (A1M and IgA). (**d**) Cell membrane receptor binding. Interaction with membrane proteins either via the lipocalin surface patch (Purpurin, C8γ, etc) or the loop scaffold (RBP). (**e**) Ligand-binding modes. Lipocalins bind hydrophobic ligands with a range of size and shape; extreme examples are shown, namely the binding of a small ligand, entirely enclosed within the internal pocket, and the binding of a large steroid, bound partly in the pocket and partly by the loop scaffold in an altogether more exposed manner.

molecules with critical biological functions: retinoids (retinol and retinoic acid), arachidonic acid, and various steroids. Binding of such substances may have functional significance for some of the lipocalins; however, it is important to draw a distinction between the demonstration of binding *in vitro* and the identification of endogenous ligands. The retinol binding exhibited by Blg, for example, may only reflect its general affinity for a range of different small hydrophobic molecules; retinol in milk is associated with fat globules, and retinol bound to Blg has yet to be detected. Blg also binds long-chain fatty acids and triacylglycerols. It has been shown that careful extraction procedures, which do not reduce pH or increase the ionic strength of whey, allow the isolation of Blg from milk with bound endogenous fatty acids [22,23]. The composition of these bound fatty acids, mostly palmitate and oleate, resembles that of fatty acids extracted from milk, indicating that there is no appreciable selectivity of individual fatty acids. Thus the broad selectivity of binding exhibited by some lipocalins, such as Blg or AGP, may reflect a general transport role, such as the clearance of unwanted endogenous, or exogenous, compounds. Thus the binding of many of the molecules listed in Table 2 may not have any

physiological relevance and may even give rise to pathological conditions. For example, acute exposure to many important industrial and environmental chemicals, including components of unleaded petrol, causes a toxic syndrome, known as α_{2u}-globulin nephropathy, in the kidney of adult male rats. This syndrome is characterized by an excessive accumulation, in proximal-tubule epithelial cells, of lysosomal protein droplets composed of large amounts of α_{2u}-globulin and the degeneration and necrosis of cells lining the proximal tubule [24]. Chronic exposure leads to an escalating progression of symptoms often resulting in kidney failure and death.

Receptor binding

There is experimental evidence to show that a number of lipocalins are bound by specific cell-surface receptors and may be internalized by receptor-mediated endocytosis; for example, it has been shown that, in the liver, the retinol–RBP complex is taken up by receptor-mediated endocytosis in parenchymal and stellate cells [25], probably involving potocytosis [26], whereas

Table 2   Ligand binding by members of lipocalin protein family

Ligand-binding properties of the lipocalin protein family are summarized; where the ability to bind retinol or retinoic acid has been demonstrated experimentally this is indicated by a +, or shown not to bind by a —; where this is unknown, or is uncertain, by a ? Known, or suspected, endogenous ligands are shown, as are compounds known to bind *in vitro*. Where these lists are long, for example AGP or crustacyanin, this is abbreviated; the interested reader is referred to the given references.

Protein	Retinol	Retinoic acid	Endogenous ligand	Other binding	References
RBP	+	+	Retinol	Retinal (+), retinyl acetate (+), β-ionone (+), *cis*-retinoids (—), β-carotene (—), cholesterol (—), terpenoids (—), β-lonylideneacetate (—), long-chain esters of retinol and retinoic acid (—)	[138,139]
RABP	—	+	Retinoic acid	Synthetic retinoids (—), retinal (—)	[126]
Purpurin	+	?	?		[132]
MUP	?	?	2-(s-Butyl)thiazoline, 2,3-dehydroexobrevicomin, 4-(ethyl)phenol	All (+): 2-isobutyl-3-methoxypyrazine, methyl dihydrojasmonate, (+)-methylfenchol, methyl dihydrojasmonate, thymol, 2-nonenal, β-ionone	[56,58]
A2U	+	?	Pheromones?	All (+): 2,2,4-trimethylpentane, decalin, JP-5, isophorone, 1,4-dichlorobenzene, D-limonene, dimethyl methylphosphonate, chloroethene, pentachloroethane, hexachloroethane, cyneole, 2-isobutyl-3-methoxypyrazine, methyldihydrojasmonate, (—)-methylfenchol, D-limonene oxide, 2,4,4-trimethylpentan-1-ol, 2,4,4-trimethylpentan-2-ol, 2,2,4-trimethylpentan-1-ol, α-tetralone, isophorone, α-tetralol, 2-hexanone, 2,5-dichlorophenol, phenylethylalcohol, nonen-2-al	[24,58,140]
Apo D	?	?	Progesterone/pregnenolone	Bilin (+), cholesterol (—), arachidonic acid (+)	[111,141,142]
PP14	—	—	?		[92]
Blg	+	?	Fatty acids?	All (+): stearate, palmitate, laurate, oleate, heptane, butane, pentane, iodobutane, SDS, Methyl Orange, toluene, n-octylbenzene-p-sulphonate, p-nitrophenol, p-nitrophenyl acetate, p-nitrophenyl β-glucuronide, p-nitrophenyl sulphate, p-nitrophenyl pyridoxal phosphate, heptan-2-one, octan-2-one, nonan-2-one, trifluorotoluene, hexafluorobenzene, β-ionone, haemin, protoporphyrin IX, ellipticine	[31,143,144]
BBP	?	?	Biliverdin IXγ		[82]
Crustacyanin	?	?	Astaxanthin (3,3-dihydroxy-β,β-carotene-4,4'-dione)	Range of carotenoids (+), and other molecules	[145,146]
C8γ	+	+	?		[44]
A1M	+	?	?		[147]
VEGP	+	?	?	Denatonium benzoate (+), cholesterol (—), stearate (+) palmitate (+), fatty alcohols (+), phospholipids (+), glycolipids (+)	[76,77,78,79]
A1AG	?	?	Histamine?	Extensive range of drugs and other compounds	[105]
PGDS	?	?	Prostaglandin H$_2$		[86]

AGP is endocytosed via clathrin-coated pits [26]. There is increasing evidence, from a wide variety of different tissues, that RBP binding to its target cells occurs via specific surface receptors [27,28]. A cell-surface receptor for α_1-microglobulin (A1M) has also been identified [29,30], and there is additional evidence to suggest the existence of receptors for MUP [16], Blg [12,31], and OBP [32]. Epididymal secretory protein has been shown to bind to the plasma membrane of spermatozoa [33], and may be another lipocalin to act via a specific surface receptor. It has been hypothesized that the three conserved sequence motifs characteristic of the family, which lie next to each other forming a surface patch at the Closed end of the lipocalin fold, constitute a common cell-surface receptor binding site (see Figure 4) [2,34]. However, Sivaprasadarao and colleagues showed, using mutagenesis, that specific amino acids within the Open end loop scaffold are responsible for binding to the RBP receptor [32]. Although the mechanism of retinol transfer from RBP to the cell interior remains controversial [35], it seems that retinol is taken up by target cells through a specific cell-surface receptor that recognizes RBP. Sivaprasadarao and Findlay propose that RBP remains external to the cell, with only retinol being internalized by the RBP receptor in a manner not involving receptor-mediated endocytosis [36]. In contrast, Noy and co-workers have suggested that retinol dissociates spontaneously from RBP outside the cell and undergoes passive transfer through the plasma membrane to be bound by high-affinity sites in the cytoplasm [37,38]. Thus the structural determinants of binding to different membrane-bound receptors, although still poorly understood in themselves, clearly reside in different parts of the lipocalin fold for different family members.

Macromolecular complexation

The ability of lipocalins to form complexes with soluble macromolecules is arguably their least well known molecular-recognition property [21]. In plasma, RBP is usually complexed to the protein transthyretin: RBP binds transthyretin with an association constant of 1×10^7 M; only about 4% of total plasma RBP is free, the rest being part of a transthyretin complex [39]. Transthyretin has a higher affinity for halo-RBP than apo-RBP, and its interaction with RBP is also sensitive to both ionic strength and pH; the complex dissociates at low ionic strength and is only stable between pH 5.0 and 9.0. Purpurin is the component of adherons (large extracellular multi-component macromolecular complexes present in cultured chick retina growth medium) which mediates cell–cell and cell–substratum adhesion through its interaction with a specific cell-surface receptor [40]. About 80% of all plasma apolipoprotein (apo) D exists as disulphide-linked complexes: predominantly with apoA-II in high-density lipoproteins and plasma, and mainly with apoB-100 in low-density lipoproteins and very-low-density lipoproteins [41].

More recently, it has been shown that neutrophil lipocalin (NGAL) is covalently attached to human neutrophil gelatinase (type IV collagenase) via an intermolecular disulphide [42,43], although most of the protein is secreted in an uncomplexed form. These authors propose a regulatory role for NGAL on the action of gelatinase. The lipocalin C8γ, one of three subunits of C8 [the penultimate component of the membrane attack complex (MAC) of complement], is covalently linked to C8α by an inter-molecular disulphide bridge [44]. C8γ may have a regulatory function helping to protect host cells from lysis [45]; there is evidence that C8γ is the site of interaction with the C8 binding protein, also known as homologous restriction factor, found on the cell surface of erythrocytes and other cells [46]. Half of plasma A1M exists in a free form, the rest in 1:1 covalent complexes with other macromolecules: albumin [47] and immunoglobulin A in humans [48]; fibronectin [49], via an intermolecular disulphide bridge and α_1-inhibitor-3 in the rat [50]. The components of the A1M–IgA complex are linked covalently between A1M and the C-terminal nine residues of IgA [48]. The charge heterogeneity of free A1M is due to a tightly associated chromophore also responsible for its characteristic yellow–brown fluorescence [30]. There may be more than one chromophoric group per protein [51], one of which is covalently bound to cysteine-34 [52]. The complex does not exhibit the charge heterogeneity and fluorescence of free A1M, implying that the chromophore at cysteine-34 is involved in the link. Formation of the A1M–α_1-inhibitor-3 complex abrogates the inhibitory action of the inhibitor, presumably preventing it from forming cross-links to proteinases [49,50]. It is not known whether A1M can affect its other macromolecular partners in a similar way, although its immunoglobulin complex exhibits both antibody activity and many of the biological actions of free A1M. It is tempting to suggest that A1M, and other lipocalins such as NGAL and C8γ, may down-regulate plasma protein activity by complex-formation.

The protein–protein interactions which underlie complex-formation by the lipocalins are mediated by the loop scaffold forming the Open end of the molecule (see Figure 4); the variability of loop length and conformation and the different amino acid composition shown by these loops may be the principal means by which different lipocalins are able to form macromolecular complexes with high affinity and selectivity; for example, the correct positioning of free cysteine residues enables a number of lipocalins to form disulphide-cross-linked complexes. The recently published crystal structure of the RBP–transthyretin complex [53] suggests that a more complete understanding of these processes will follow as more crystal structures, and results of mutagenesis studies, become available.

FUNCTION

The lipocalins have, in general, been classified as extracellular transport proteins. In this regard they are typified by RBP [11,35], virtually the sole retinol transporter in plasma, which binds a single all-*trans*-retinol molecule as its physiological ligand. RBP is synthesized in hepatic parenchymal cells, where the apoprotein is saturated with retinol, triggering its secretion into general circulation. 85–90 % of plasma RBP carries bound retinol, while about 96 % is complexed to another plasma protein: transthyretin. The RBP–transthyretin complex is much larger than RBP alone, preventing its loss by filtration through the

kidney glomeruli. RBP-mediated retinol transport fulfills several physiological functions [54]. First, it facilitates the transfer transport of insoluble retinol from storage sites in the liver to peripheral tissues. Secondly, RBP protects bound retinol from oxidation. Thirdly, the synthesis of RBP regulates retinol release from the liver and mediates the specificity of its uptake by target cells. Interaction of RBP with target cells is crucial to its biological function: RBP is recognized by a specific cell-surface receptor, releasing retinol, and losing its affinity for transthyretin. The resulting apo-RBP is filtered by the kidney, reabsorbed and catabolized; thus RBP carries only one retinol molecule before being degraded.

Although few members of the lipocalin family are as well characterized functionally as RBP, as we shall see there is mounting evidence that lipocalins fulfil many different and potentially significant biological functions, including roles in mediating pheromone activity, olfaction, cryptic coloration, enzymatic synthesis, immunomodulation, and the regulation of cell homoeostasis.

Pheromone activity

Rodent urine contains an unusually large amount of protein, and this phenomenon has been studied extensively in both rats and mice. MUP is the major protein component of mouse urine [55] and is expressed in several different secretory tissues of the mouse [56]. The major site of MUP synthesis is the liver; the protein is secreted by the liver into serum, where it circulates at relatively low levels before being rapidly filtered by the kidney and excreted.

Expression of MUP mRNA is under different developmental and hormonal control in different tissues. In the liver, expression of MUP is stimulated by androgens, principally testosterone, and is modulated by thyroxine and peptide growth hormones: addition of thyroxine or growth factors in the presence of testosterone increases MUP expression 150-fold [57]. However, constitutive expression of MUP has been demonstrated in the salivary and lachrymal glands [56]. The sex-dependent expression of MUP (adult male mice secrete 5–20 times as much MUP as do females) and its ability to bind a number of odorant molecules [57] is consistent with the suggestion that MUP acts as a pheromone transporter [59]; the protein may be excreted into the urine carrying a bound pheromone which is released as the urine dries and the protein denatures. This proposal is strongly supported by the work of Bacchini and colleagues, who have successfully purified MUP from mouse urine with bound ligands [60]. They identified three components from the total ligand extracted from the purified protein: the largest proportion (around 70 %) was 2-(s-butyl)thiazoline, with 2,3-dehydroexo-brevicomin and 4-(ethyl)phenol comprising minor fractions of about 15 % each. However, only about 40 % of protein contained bound ligand. The first two of these compounds are known to have pheromone activity in male rat urine, eliciting many sexually related responses in female rats. A recent report has shown that MUP, acting via the vomeronasal organ after appropriate physical contact with male mouse urine in their environment, can accelerate the onset of puberty in female mice [61]. Interestingly, this seems to be a function of the protein itself; MUP devoid of ligands, either by extraction or competitive displacement, is still active, while an organic extract containing these volatile ligands shows no activity. Moreover, a peptide corresponding to the N-terminus of MUP is also active. These results suggest that MUP is not only a carrier of pheromones, but also a pheromone in its own right.

α_{2u}-Globulin, a close homologue of MUP, is the major protein component of adult male rat urine, accounting for 30–50 % of total excreted protein [62]. As its electrophoretic mobility is similar to that of serum α_2-globulin, it was named 'α_{2u}-globulin' with the subscript 'u' denoting its origin in urine. This choice of nomenclature has proved confusing; the protein is often wrongly called α_2-microglobulin. α_{2u}-Globulin is secreted into the plasma by a number of tissues, where it circulates before being filtered through the kidney; between 20 and 50% is reabsorbed by the proximal tubule of the nephron, the rest being excreted. Hepatic parenchymal cells are the principal sites of α_{2u}-globulin synthesis in mature male rats [63], although it is also expressed, at much lower levels, in many other tissues. The physiological regulation of α_{2u}-globulin expression is under multi-hormonal control [62]: treatment with androgenic steroid hormones stimulates α_{2u}-globulin expression, with 5α-dihydrotestosterone being the most potent, while steroidal and non-steroidal oestrogens are strongly inhibitory. Combination of androgens, glucocorticoid, thyroxine, and growth hormone are required to maintain or restore normal levels of protein secretion. Although the exact physiological role of α_{2u}-globulin remains unclear, there is good circumstantial evidence that it functions in pheromone transport. This is consistent with its observed binding properties [58], its close similarity with MUP, its sex-dependent expression and the known properties of male rat urine.

Aphrodisin, another lipocalin, is the major macromolecular component of hamster vaginal discharge [64], and is secreted by vaginal tissue and the Bartholin's gland [65]. These secretions, acting via the vomeronasal organ, are known to elicit a copulatory response in male hamsters: if applied to the hind parts of an unconscious male hamster and another male hamster is encouraged to nuzzle the treated area with its snout then this hamster will reproducibly attempt an abortive copulation. By depleting these discharges of volatile components and by fractionation, it can be shown that aphrodisin is responsible for these effects, suggesting it is, like MUP, a mammalian proteinaceous pheromone [66]. Cloned aphrodisin expressed in *Escherichia coli* shows only modest activity compared with the isolated hamster protein, but is fully active when combined with organic extracts of vaginal discharge [67], suggesting that it is the protein–ligand complex which is fully functional.

Olfactory and gustatory proteins

The molecular basis of both olfaction and gustation is rather less well understood than the mechanisms underlying the other physical senses. Thus the discovery of specific proteins, usually referred to as OBPs, associated with olfactory tissue, which seem able to bind odorant molecules with high specificity, has generated considerable interest [68]; several of these proteins are lipocalins [68–70]. Two independent studies identified a soluble protein from bovine nasal mucosa which constituted about 1% of total isolated protein [71,72]. This protein binds the bell-pepper odorant 3-isobutyl-3-methoxypyrazine, and became known as pyrazine-binding protein; it is most abundant in the tubulo-acinar cells of the respiratory epithelium of nasal mucosa, but is also present in olfactory mucosa, olfactory neurons and nasal secretions [73]. Pevsner et al. cloned and sequenced a protein, homologous with bovine pyrazine-binding protein, from rat nasal mucosa, which they called rat OBP [71]. Rat OBP is localized to the lateral nasal, or Sterno's, gland, the largest of the 20 discrete rat nasal glands of the rat. Lee et al. identified, cloned and sequenced a similar protein from the olfactory tissue of the frog *Rana pipiens* which they named protein BG (Bowman's gland) [69]. Analysis of mRNA distributions showed that this protein was specific to frog olfactory tissue. It is thought that the OBPs may function by concentrating and delivering odorant molecules to their receptors.

Recently, two lipocalins, specifically expressed in the posterior and vomeronasal glands of the mouse nasal septum, have been identified and were suggested to act in the chemoreception of, as-yet-unidentified, small lipophilic pheromones [74]. One of these proteins was immunolocalized on the vomeronasal sensory epithelium, the site of primary pheromone reception, and the immunoreactivity was greatest during periods when contact between animals plays an important role in modulating behaviour.

It has been suggested that another lipocalin, highly expressed by the small acinar von Ebner's salivary glands of the tongue, but not in the secretory duct, fulfills a similar function, in salivary secretions, to that of OBPs in olfactory mucosa: the selective binding of sapid chemicals and their transport to taste receptors [75]. There is evidence that von Ebner's-gland protein (VEGP) can help to clear the bitter-tasting compound denatonium benzoate *in vivo* [76], suggesting a possible clearance function for the protein in taste reception, although it fails to bind other bitter compounds [77]. VEGP is also secreted by the lachrymal gland into tear fluid, where it has been known historically as 'tear prealbumin' [78]. Together with lysozyme and lactoferrin, VEGP forms 70–80% of total tear protein, although this decreases in diseases affecting the lachrymal gland. Tear VEGP has been suggested to enhance the bactericidal activity of lysozyme and to have an inherent antimicrobial function, perhaps through the transport of compounds with antibacterial properties [79]. VEGP has been shown to bind retinol [78], and can be co-extracted with fatty acids, particularly stearate and palmitate, fatty alcohols (including cholesterol), phospholipids and glycolipids [79]. It has been postulated that VEGP may act as a transporter of lipids, synthesized in the dorsal, or meibomian, glands of the eyelid, to the thin film they form at the air/tear-fluid interface. Interestingly, a recent report indicates that the lipocalin apoD is secreted, as a disulphide-linked dimer, by the lachrymal glands and might also be a carrier of meibomian lipids in tear fluid [80].

Coloration

A number of lipocalins act in invertebrate coloration: bilin-binding protein from the cabbage white butterfly (*Pieris brassicae*), the closely related protein insecticyanin from *Manduca sexta* (tobacco hornworm) and the lobster protein crustacyanin. Like other members of the family they bind small molecules, and gain their colorant properties from interaction with their ligands.

A variety of different bilin pigments, derived from haem breakdown products, are distributed widely in insects. Biliverdin IXγ is amongst the most common, especially in butterflies and moths [81]. These pigments are usually associated with proteins, and they contribute significantly to coloration in the epidermis or interlamellar space of the wing. Two insect bilin-binding lipocalins have been studied in great detail: insecticyanin from the tobacco hornworm [82] and BBP from *Pieris brassicae* [13,14,83]. Both are blue pigments in their complexes with biliverdin IXγ. The precise functions of these proteins is unclear; apart from their role as pigments, it has been suggested that they may function in photoreception and protection from photo-induced free radicals. Many of these pigment complexes are remarkably stable, and persist, during insect development, from their synthesis during the larval stage, through the pupal stage into adulthood.

Crustacyanin (meaning 'shell blue') is the general name given to the carotenoprotein complex found in the epicuticle, or calcified outer layer, of the lobster carapace [7,84]. It acts as the dominant pigment of the lobster shell, giving rise to its characteristic blue colour. In solution, crustacyanin exists as an equilibrium mixture between several distinct forms, differing in their physical and spectral properties [84,85]. The native, blue form (α-crustacyanin), which predominates *in vivo*, will, at low ionic strength, form α'-crustacyanin; this in turn changes to purple β-crustacyanin on standing. The $\alpha \rightleftharpoons \alpha'$ transition is favoured by low ionic strength and is reversible, while conversion into β-crustacyanin is irreversible. Native α-crustacyanin is an octamer of hetero-dimers, totalling 16 separate polypeptide chains, each dimer binding two molecules of astaxanthin, β-crustacyanin corresponding to the free heterodimer.

Prostaglandin D (PGD$_2$) synthesis

Glutathione-independent PGD$_2$ synthase (EC 5.3.99.2) is the main factor involved in synthesis of PGD$_2$ in the brain, accounting for over 90% of activity in the rat; it is responsible for catalysing the conversion of PGH$_2$ into PGD$_2$ in the presence of various thiol compounds [86,87]. PGD$_2$ is a major prostaglandin in mammalian brains, functioning in the central nervous system as both a neuromodulator and a trophic factor. The enzymic activity of PGD synthase makes it unique among the lipocalins. It is localized in the choroid plexus, meninges and oligodendrocytes, but is also a major component of cerebrospinal fluid. Immunocytochemistry indicates that the protein is associated with the rough endoplasmic reticulum and the outer nuclear membranes of rat oligodendrocytes and seems to be a peripheral membrane protein easily dissociated by detergents.

A near homologue of PGD synthase has been identified in cane-toad (*Bufo marinus*) choroid plexus, where it is the most abundant protein secreted into the cerebrospinal fluid [88]. It is also found in other areas of the brain, albeit at much lower levels, and is expressed throughout amphibian metamorphosis. The choroid plexus helps form the barrier between blood and cerebrospinal fluid, and it may be that this member of the lipocalin family helps transport lipophilic molecules across the blood/brain barrier.

Immune modulation

The plasma levels of many proteins change during the acute-phase response, a complex physiological reaction to stress and inflammatory stimulation which plays an important role in many disease states. Proteins demonstrating elevated concentrations are called positive acute-phase proteins (APPs). These include the lipocalins AGP, NGAL, pregnancy protein 14 (PP14) and A1M. Negative APPs include the lipocalin RBP. APPs are thought to have an anti-inflammatory function, preventing ongoing tissue damage, as well as other roles well-suited to lipocalins, such as the transport of factors. Another lipocalin, C8γ, is part of the membrane-attack complex of human complement.

The major protein product expressed between the late luteal phase of the menstrual cycle and the first trimester of pregnancy is a progesterone-regulated lipocalin called PP14 [89,90]; although first isolated from placenta, its principal site of synthesis is the secretory glandular epithelium of the endometrium and the decidua [91]. PP14 is also present in high concentrations in amniotic fluid, uterine luminal fluid, and at lower levels in plasma, but is not expressed by other tissues [89]. Variation in PP14 levels in these fluids reflects changes in endometrial PP14 secretion. Recently, Morrow et al. have shown that PP14 is expressed by haematopoietic cells of the megakaryocytic lineage and demonstrates potent immunosuppressive properties which

can be blocked by PP14-specific antibodies [92]. PP14 has also been reported to suppress the activity of natural killer cells and T-cells [93]. Seppala et al. recently reviewed other immunoregulatory properties of PP14 [94]. In the presence of PP14, interleukin-2 loses the ability to increase the mitogen-induced T-cell proliferation. PP14 inhibits the reactivity of mitogenic lymphocytes to phytohaemogglutinin in a dose dependent manner. The immunosuppressive effects of decidual extracts are greatly inhibited by anti-PP14 antibodies [95].

Blg, a close homologue of PP14, is the major protein component of whey from the milk of many mammals, synthesised by epithelial cells of the mammary gland regulated by prolactin and secreted into the lumen to accumulate in milk. Although an enormous amount is known about the physical and biochemical properties of Blg [31], its physiological function remains unclear. Blg may act in the delivery of free retinol to the absorptive cells of the intestine, an idea given support by evidence that Blg can enhance retinol uptake in rat intestine [96]. Thus it may be that Blg is a general carrier of nutritionally important, but insoluble, molecules between mother and child. It is also tempting to suggest some immunoregulatory function for Blg, analogous to that of PP14, but as yet there is no evidence to support this conjecture.

The lipocalin A1M is a well-studied plasma protein which has inhibitory effects on the immune system [30,97]. It is a positive APP, demonstrating enhanced hepatic expression, although plasma levels do not change significantly during inflammation. A1M suppresses antigen-induced polyclonal proliferation of cultured lymphocytes; this inhibition is partial at normal plasma concentrations and complete at 10–20 fold higher levels [98]. A1M is glycosylated by three separate chains (22% of its overall mass is carbohydrate); corresponding glycopeptides can account for most or all of the immunosuppresive activity of A1M [99]. The formation of complexes, with IgA, fibronectin and α_1-inhibitor-3, for example, may also play a role in regulating the immune system [48–50]. The increased secretion of the protein after stimulation by interleukin-6 (released by macrophages and T-lymphocytes) may be part of a negative-feedback system. A1M also inhibits the spontaneous migration of neutrophil granulocytes *in vitro* and suppresses the chemotactic attraction of granulocytes by a cytokine concentration gradient released by triggered monocytes, macrophages, and B- and T-lymphocytes [100].

Murine NGAL exhibits a 7–10-fold increase in expression in cultured mouse kidney cells infected by simian-virus 40 or other viruses [101]. In another study [102], NGAL was identified as a major secretory product of lipopolysaccharide-stimulated cultured mouse macrophages, suggesting that the protein might function in the defence against infection. Recently, NGAL has been shown to be identical with SIP24, a previously identified secretory product of quiescent mouse fibroblasts induced by serum, dexamethasone, basic fibroblast growth factor, phorbol ester, and PGF$_{2\alpha}$ [103]. Mouse plasma levels of NGAL rise, as a result of increased expression levels in the liver, in response to intramuscular turpentine injection. Tumour necrosis factor-α can regulate NGAL expression in cultured liver cells. These findings indicate that NGAL is a positive APP and may possess immunosuppressive or anti-inflammatory properties, possibly linked to its regulation of neutrophil gelatinase or other plasma protein [42]. The uterus is also a major site of NGAL synthesis, especially at parturition, when expression increases significantly, suggesting a physiological role for the protein in uterine secretions [104].

The lipocalin AGP (or orosomucoid) is an abundant plasma protein [105,106]. It is a positive APP; after induction by turpentine injection, AGP becomes one of the dominant proteins expressed by the liver and is known to accumulate at sites of inflammation. AGP has shown many immunoregulatory properties: the ability to inhibit platelet aggregation [107], an involvement in wound healing (possibly through its interaction with collagen), the inhibition of neutrophil activation, inhibition of phagocytosis, and it possesses non-specific immunosuppressive properties [105,108]. AGP has an unusually high carbohydrate content of 40 % and is consequently unusually acidic and soluble [106]; several studies show that the deglycosylated protein demonstrates little or none of AGP's immunosuppressive properties. The characteristic glycosylation of AGP changes during the acute-phase response, with concomitant effects on its immunomodulatory properties: for example acute inflammation leads to a large increase in sialyl-Lewis-X-substituted AGP, which may inhibit selectin-mediated granulocyte invasion of inflamed endothelium [109]. However, the physiological function of AGP remains unclear; it binds a very wide variety of diverse ligands; rather too diverse, in fact, for any inferences to be drawn regarding transport functions of the protein [105].

Although several lipocalins have been shown to possess immunosuppressive properties *in vitro*, it is not yet clear whether this is a property of the protein itself, perhaps mediated by complex-formation or receptor activation, results from the transport of a bound ligand, or, as most are highly glycosylated, is mediated by attached carbohydrate, as is the case for A1M and AGP. The relevance of such results to the physiological functioning of these lipocalins *in vivo* is not clear. The plasma level of AGP is monitored during pregnancy and can be used as a diagnostic and prognostic marker in a variety of disease states [105], including cancer chemotherapy, renal disfunction, myocardial infarction, arthritis, and multiple sclerosis; AGP levels also rise during infection or after surgery. Other lipocalins have also been used as biochemical markers in disease: RBP (used clinically as a marker of tubular reabsorption in the kidney) [35], A1M (the most accurate marker of tubular proteinurea) [110], PP14 (a clinical marker of endometrial status) [94], and apoD (a marker in gross cystic breast disease) [111]. However, evidence linking the observation of immunosuppression in the test tube and effects in the normal and pathological functioning of the immune system is lacking for the lipocalins; further work is need to establish this link and demonstrate their role *in vivo*. In passing, it is worth noting that several lipocalins have been shown to give rise to inflammation [6]: bovine β-lactoglobulin, rodent urinary proteins and cockroach Bla g 4 protein are all allergens, the latter specifically associated with bronchial asthma.

One of the main functions of complement, the lysis of foreign cells, is mediated by the end product of the complement cascade; the MAC, a large pore-forming multimeric protein complex which induces rapid osmotic cell lysis. Complement component C8, the penultimate component of MAC, is a large multimeric protein composed of three subunits [45], two large glycosylated proteins, C8α and C8β, and a smaller non-glycosylated protein, C8γ. The lipocalin C8γ contains three cysteine residues, one of which (cysteine-40) forms an intermolecular disulphide to C8α [44,112]. C8β is responsible for binding to the evolving MAC complex, while C8α is responsible for both membrane insertion and the C9 binding capacity of the complex. C8γ is not obligatory for C8 to function within MAC, but may be part of a regulation mechanism protecting host cells from MAC-induced lysis [45,46].

Cell regulation

Several lipocalins have been implicated in the mediation of cell regulation [113]. Among the first to be so implicated was

quiescence-specific protein (QSP). Subconfluent, actively dividing or transformed cells do not synthesize QSP. Confluent cells, including those stimulated by hormones or growth factors, express the protein in a density-dependent manner. Bedard et al. showed that QSP is among the most prevalent proteins secreted by quiescent chicken heart mesenchymal cells (CHM) cells [114]. QSP is virtually absent from CHM cells transformed by Rous-sarcoma virus or sparse cells stimulated by hormones or growth factors. However, hormone-stimulated cells grown to confluence produce significant quantities of the protein, albeit at much reduced levels compared with quiescent cells; levels of QSP expression increase with cell density. Even very dense transformed cells do not express the protein.

Chick embryo fibroblasts give similar results: a density-dependent expression of QSP at confluence but negligible protein production when sparse. Likewise, QSP secretion begins as cultured chondrocytes enter the hypertrophic phase, the last of three stages of chondrocyte differentiation [115]; its level of expression increases as the cell population matures, becoming a major secretory product with a time-dependent distribution keyed to developmental status of mesenchymal cells [115,116]. QSP expression is a specific consequence of the dense confluent state rather than of growth arrest in general, such as results from cell starvation. These properties are consistent with a protein stabilizing mature cell populations. However, the work of Nakano and Graf shows that rapidly growing v-Myb-transformed promyelocytes express high QSP [117]. However, v-Myb induces an abnormal pattern of gene expression and may abolish an inhibitory mechanism present in untransformed cells.

The lipocalin apoD is a mammalian plasma lipoprotein that has long been thought to function in cholesterol metabolism [118], although its precise physiological role remains uncertain. Most apoD is localized in some form of high-density-lipoprotein particle, representing about 5 % of high-density-lipoprotein protein, where it may help enhance lecithin:cholesterol acyltransferase activity [119]; ApoD is identical with gross-cystic-disease-fluid protein (GCDFP)-24 [111], a known progesterone/pregnenolone-binding protein, which constitutes over half the protein component of breast cyst fluid. Simard et al. [120] observed a correspondence between the steroid-induced secretion of apoD by cultured human prostate cancer cells and the inhibition of cell growth and also a higher concentration of apoD in well differentiated cells. In a related study, Provost et al. found little apoD expression by human diploid fibroblasts cells in sparse culture, while confluent quiescent cells did express the protein [121]. Cells whose growth is arrested by serum starvation also expressed the protein. Provost et al. also noted that expression levels rise as cells reach confluence. A more recent report has shown that expression of the apoD gene is induced by all-*trans*-retinoic acid in certain human breast-cancer cell lines [122]. This was accompanied by secretion of apoD, progression of cells to a more differentiated phenotype and the inhibition of cell proliferation.

Boyles et al. [123] and Spreyer et al. [124] both report a correlation between increased apoD expression and nerve regeneration consistent with the protein acting in the repair process. Boyles et al. report a 500-fold increase in apoD expression in regenerating rat, rabbit and marmoset peripheral nerves. They find apolipoprotein D is produced by astrocytes and oligodendrocytes in the central nervous system, and by neurolemmal cells and fibroblasts in the peripheral nervous system. Spreyer et al. report a 40-fold increase in apoD levels in the extracellular space of regenerating crushed rat neurons, although its expression was restricted to endoneural fibroblasts. In light of these observations, the identification of a close homologue of apoD, the grasshopper lipocalin Lazarillo, is of particular interest [5]. Lazarillo is expressed by a subset of developing neurons and is anchored to the cell surface by a glycosyl-phosphotidylinositol group. The protein is necessary for the correct navigation of growing axons in the grasshopper embryo and suggests that Lazarillo may function as a signalling molecule mediating axonal pathfinding in the developing nervous system [125].

Homologous lipocalins have been identified as composing 15–20% of the complex mixture of proteins secreted into the lumen of the epididymis from rat and lizard (*Lacerta vivapara*) [33,126]. They are known as epididymal secretory proteins and are able to bind retinoic acid, a known morphogen and regulator of gene expression, with high affinity and specificity [18]. Their synthesis is under strict androgen control and they are secreted from the principal cells of the proximal portion of the epididymis. The proteins are believed to have a significant role in the process of spermatogenesis and sperm maturation. It may be that this function is related to their ability to bind retinoic acid [126,127] and to bind to the plasma membrane of the head region of the spermatozoa [33].

Probasin, originally isolated from the nuclei of rat dorsolateral prostate epithelial cells [128], is another lipocalin implicated in cell regulation. Probasin mRNA expression, which is regulated by androgens, gives rise to both a secreted and a nuclear form of probasin [129], the relative abundance of the two forms being correlated with cell type. Probasin concentration also seems to be closely linked with cell age and state of differentiation, consistent with a role in cell regulation [128].

Purpurin, a lipocalin localized almost exclusively in neural cells of the retina [40,130], binds both glycosaminoglycan [130] and retinol [131], and is believed to function in the control of cell differentiation, adhesion, and survival [132]; experiments have shown the protein to promote the survival of cultured chick retinal nerve cells [131]. AGP has growth-promoting effects on cultured cells and has been shown to be involved in the process of nerve regeneration [133]. Apart from its inhibitory effects on the immune system, A1M also exhibits mitogenic properties [30]. Studies on mouse lymphocytes have shown an increase in thymidine uptake by B- and T-cells caused by A1M [134,135]. The protein also induces proliferation of B-lymphocytes in the absence of serum, a mitogenic effect which can be either enhanced or inhibited by other plasma components [136]. This effect seems to be a function of the C-terminus of the protein rather than its glycosylation.

The need to transfer retinoid or steroid molecules between cells in a controlled manner is a powerful argument for the involvement of extracellular carrier proteins, such as lipocalins, in mediating their effects *in vivo*, and the delivery, via endocytosis, of a lipocalin to nuclear receptors suggests a plausible mechanism of action. However, the role of lipocalins in mediating cell homoeostasis remains open to question; changes in expression correlated with changes in morphology or growth arrest may be coincidental rather than causative, and, despite good circumstantial evidence, clearly further work is required to prove their mechanistic involvement in cell regulation.

CONCLUSION

The lipocalins are a structurally and functionally diverse family of proteins; this, at least, is clear. We have seen how structural studies of the family and an increasing understanding of their molecular-recognition properties illuminates insights into their biological function. Only a few years ago, the family remained poorly understood; now many studies are showing how much more interesting and important the lipocalins are than had been

supposed. It has been the purpose of this review to highlight this change of emphasis, and so raise awareness of the lipocalin family. Roles in transport, enzymic synthesis, immunomodulation, olfaction, pheromone signalling and cell regulation have been demonstrated or inferred. Moreover, as we have seen, clinical studies have shown the practical importance of many lipocalins as biochemical markers in health and disease. For example, drug binding to AGP is important pharmacologically; with serum albumin, it represents the major site of drug binding in plasma, and affects bioavailability, particularly as AGP levels vary during different disease states [105].

However, many aspects of the family provide scope for further investigation. Greater understanding of the precise relationship between kernel and outlier lipocalins will come, particularly, from the crystal structures of outliers, as well as from analysis of protein sequences and the structure of lipocalin genes. Genomic sequencing and cloning studies will further explore the species distribution and evolutionary relationship of the lipocalins, both to each other and to other members of the calycin superfamily. Many aspects of the multiple molecular-recognition properties of the lipocalins are now increasingly well understood. However, for many members of the family the nature and extent of ligand binding is still unclear; for some their natural ligand, or ligands, remain unknown, while for others binding is, as yet, wholly uncharacterized. Likewise, the question of how widespread, among the family, is the formation of complexes, with either soluble macromolecules or membrane-bound receptors, has yet to be answered fully. The nature of both the protein–protein and protein–small molecule interactions exhibited by the family will be illuminated both by crystal structures of lipocalins complexed with macromolecules and small molecule ligands and by directly probing function through mutagenesis studies.

Future studies of the lipocalin family will aim to understand both the particular functional specialization of individual lipocalins and the possible general physiological roles of the family as a whole in the context of a growing appreciation of the general principles which underlie their common behaviour. For example, the very broad tissue distribution of many lipocalins suggests that they may fulfill either a number of tissue-specific functions or alternatively have a single, but fundamental, role, perhaps some general clearance function preventing the harmful local accumulation of exogenous, or endogenous, hydrophobic molecules [78,122]. Moreover, is the apparent multi-functionalism displayed by many members of the family real or 'artifactual'? Does it result solely from their multiple molecular-recognition properties, or does it reflect true functional adaptation? To what extent are the apparent biological properties of individual lipocalins mediated directly by the protein structure itself (ligand binding [35,83,84], complexation [35,44,50] and receptor interaction [61,136]), by post-translational modifications (glycosylation state [99,109], chromophores [51], or membrane anchors [5]), or by combination of the two, such as the formation of disulphide linked complexes [41,44,49]? Are the allergenic properties of several, evolutionarily distinct lipocalins [6] merely a co-incidental property of their individual structures, or do they result from conserved structural features, such as the ability to interact with cellular receptors, common to the family? Can the immunosuppressive and cell-regulation functions of the family be linked to macromolecular complexation, perhaps through the modulation of proteinase-mediated cell signalling [137], either by the down-regulation of protease inhibitors [50] or by the direct inhibition of proteases themselves [42]? The immunosuppressive and cell-regulatory properties of the lipocalins suggests that they might be examined as candidates for protein therapeutics. Lipocalin research has entered an exciting era; our understanding of the family has broadened and deepened in recent years; the answers to many intriguing questions about this fascinating family of proteins are now within reach.

REFERENCES

1 Pervaiz, S. and Brew, K. (1985) Science **228**, 335–337
2 Flower, D. R., North, A. C. T. and Attwood, T. K. (1993) Protein Sci. **2**, 753–761
3 Flower, D. R., North, A. C. T. and Attwood, T. K. (1991) Biochem. Biophys. Res. Commun. **180**, 69–74
4 Bartsch, S. and Tschesche, H. (1995) FEBS Lett. **357**, 255–259
5 Ganfornia, M. D., Sanchez, D. and Bastiani, M. J. (1995) Development **121**, 123–134
6 Arruda, L.K., Vailes, L. D., Hayden, M. L., Benjamin, D. C. and Chapman, M. D. (1995) J. Biol. Chem. **270**, 31196–31201
7 Keen, J. N., Caceres, I., Eliopoulos, E. E., Zagalsky, P. F. and Findlay, J. B. C. (1991) Eur. J. Biochem. **197**, 407–417
8 Zagalsky, P. F., Mummery, R. S. and Winger, L. A. (1995) Comp. Biochem. Physiol. **110B**, 385–391
9 Bishop, R. E., Penfold, S. S., Frost, L. S., Holtje, J.-V. and Weiner, J. H. (1995) J. Biol. Chem. **270**, 23097–23103
10 Flower, D. R., Sansom, C. E., Beck, M. E. and Attwood, T. K. (1995) Trends Biochem. Sci. **20**, 498–499
11 Cowan, S. W., Newcomer, M. E. and Jones, T. A. (1990) Proteins: Struct. Funct. Genet. **8**, 44–61
12 Papiz, M. Z., Sawyer, L., Eliopoulos, E. E., North, A. C. T., Findlay, J. B. C., Sivaprasadarao, R., Jones, T. A., Newcomer, M. E. and Kraulis, P. J. (1986) Nature (London) **324**, 383–385
13 Holden, H. M., Rypniewski, W. R., Law, J. H. and Rayment, I. (1987) EMBO J. **6**, 1565–1570
14 Huber, R., Schneider, M., Epp, O., Mayr, I., Messerschmidt, A., Pflugrath and Kayser, H. (1987) J. Mol. Biol. **195**, 423–434
15 Huber, R., Schneider, M., Mayr, I., Muller, R., Deutzmann, R., Suter, F., Zuber, H., Falk, H. and Kayser, H. (1987) J. Mol. Biol. **198**, 499–513
16 Bocskei, Zs., Groom, C. R., Flower, D. R., Wright, C. E., Phillips, S. E. V, Cavaggioni, A., Findlay, J. B. C. and North, A. C.T (1992) Nature (London) **360**, 186–189
17 Monaco, H. and Zanotti, G. (1992) Biopolymers **32**, 457–465
18 Newcomer, M. E. (1993) Structure **1**, 7–18
19 Flower, D. R. (1993) FEBS Lett. **333**, 99–102
20 Flower, D. R. (1995) Protein Peptide Lett. **2**, 341–346
21 Flower, D. R. (1995) J. Mol. Recognit. **8**, 185–195
22 de Villegras, M. C. D., Oria, R., Sala, F. J. and Calvo, M. (1987) Milchwissenschaft **42**, 357–358
23 Perez, M. D., de Villegras, C. D., Sanchez, L., Aranda, P., Ena, J. M. and Calvo, M. (1989) J. Biochem. (Tokyo) **106**, 1094–1097
24 Borghoff, S. J., Short, B. G. and Swenberg, J. A. (1990) Annu. Rev. Pharmacol. Toxicol. **30**, 349–367
25 Senoo, H., Strang, E., Nilsson, A., Kindberg, G. M., Berg, T., Ross, N., Norum, K. R. and Blomhoff, R. (1990) J. Lipid Res. **31**, 1229–1239
26 Malaba, L., Smeland, S., Senoo, H., Norum, K. R., Berg, T., Blomhoff, R. and Kindberg, G.M. (1995) J. Biol. Chem. **270**, 15686–15692
27 Bavik C.-O., Busch, C. and Eriksson, U. (1992) J. Biol. Chem. **267**, 23035–23042
28 Smeland, S., Bjerknes, T., Malaba, L., Eskild, W., Norum, K. R. and Blomhoff, R. (1995) Biochem. J. **305**, 419–424
29 Fernandez-Luna, J. L., Levy-Cobian, F. and Mollinedo, F. (1988) FEBS Lett. **236**, 471–474
30 Akerstrom, B. and Logdberg, L. (1990) Trends in Biochem. Sci. **15**, 240–243
31 Hambling, S. G., McAlpine, A. S. and Sawyer, L. (1992) in Advanced Dairy Chemistry, vol. 1 (Fox, P. F., ed.), pp. 140–190, Elsevier Applied Science, London
32 Sivaprasadarao, A., Boudjelal, M. and Findlay, J. B. C. (1993) Biochem. Soc. Trans. **21**, 619–622
33 Morel, L., Dufarre, J.-P. and Depeiges, A. (1993) J. Biol. Chem. **268**, 10274–10281
34 North, A. C. T. (1989) Int. J. Biol. Macromol. **11**, 56–58
35 Blaner, W. S. (1989) Endocrine Rev. **10**, 308–316
36 Sivaprasadarao, A. and Findlay, J. B. C. (1988) Biochem. J. **255**, 571–579
37 Noy, N. and Xu, Z.-J. (1990) Biochemistry **29**, 3878–3883
38 Noy, N. and Blaner, W. S. (1991) Biochemistry **30**, 6380–6386
39 Goodman, W.S (1984) in The Retinoids (Sporn, M. B., Roberts, A. B. and Goodman, W. S, eds.), vol. 2, pp. 41–88, Academic Press, New York
40 Schubert, D. and LaCorbiere, M. (1985) J. Cell Biol. **100**, 56–63
41 Blanco-Vaca, F., Via, D. P., Yang, C.-Y., Massey, J. B. and Pownall, H.J. (1992) J. Lipid Res. **33**, 1785–1795
42 Treibel, S., Blaser, J., Reinke, H. and Tschesche, H. (1992) FEBS Lett. **314**, 386–388

43 Kjeldsen, L, Johnsen, A. H., Sengelov, H. and Borregaard, N. (1993) J. Biol. Chem. **268**, 10425–10432
44 Haefliger, J.-A., Peitsch, M. C., Jenne, D. E. and Tschopp, J. (1991) Mol. Immunol. **28**, 123–131
45 Sodetz, J. M. (1988) Curr. Top. Microbiol. Immunol. **140**, 21–31
46 Hansch, G. M. (1988) Curr. Top. Microbiol. Immunol. **140**, 109–118
47 Tejler, L. and Grubb, A. O. (1976) Biochim. Biophys. Acta **439**, 82–94
48 Grubb, A., Mendez, E., Fernandez-Luna, J. L., Lopez, C., Mihaesco, E. and Vaerman, J.-P. (1986) J. Biol. Chem. **261**, 14313–14320
49 Falkenberg, C., Enghild, J. J., Thogersen, I. B., Salvesen, G. and Akerstrom, B. (1994) Biochem. J. **301**, 745–751
50 Falkenberg, C., Allhorn, M., Thogersen, I. B., Valnickova, Z., Pizzo, A. V., Salvesen, G. Akerstrom, B. and Enghild, J. J. (1995) J. Biol. Chem. **270**, 4478–4483
51 Akerstrom, B., Bratt, T. and Enghild, J.J (1995) FEBS Lett. **362**, 50–54
52 Escribano, J., Grubb, A., Calero, M. and Mendez, E. (1991) J. Biol. Chem. **266**, 15758–15763
53 Monaco, H. L., Rizzi, M. and Coda, A. (1995) Science **268**, 1039–1041
54 Blomhoff, R., Green, M. H., Berg, T. and Norum, K. R. (1990) Science **25**, 399–404
55 Finlayson, J. S., Potter, M. and Runner, R. C. (1963) J. Natl. Cancer. Inst. **31**, 91–97
56 Shaw, P. H., Held, W. A. and Hastie, M. D. (1983) Cell **32**, 755–761
57 Knopf, J. L., Gallagher, J. A. and Held, W. A. (1983) Mol. Cell Biol. **3**, 2232–2241
58 Cavaggioni, A., Findlay, J. B. C. and Tirindelli, R. (1990) Comp. Biochem. Physiol. **96B**, 513–520
59 Cavaggioni, A., Sorbi, R. T., Keen, J. N., Pappin, D. J. C. and Findlay, J. B. C. (1987) FEBS Lett. **212**, 225–228
60 Bacchini, A., Gaetani, E. and Cavaggioni, A. (1992) Experientia **48**, 419–421
61 Mucignat-Caretta, C., Caretta, A. and Cavaggioni, A. (1995) J. Physiol. (London) **486**, 517–522
62 Roy, A. K. and Neuhaus, O. W. (1966) Biochim. Biophys. Acta **127**, 82
63 Roy, A. K., Chatterjee, B., Demyan, W. F., Milin, B. S., Motwani, N. M., Nat, T. S. and Schiop, M. J. (1983) Recent Prog. Hormone Res. **39**, 425–457
64 Henzel, W. J., Rodriguez, H., Singer, A. G., Stults, J. T., Macrides, F., Agosta, W. C. and Niall, H. (1988) J. Biol. Chem. **263**, 16682–16687
65 Magert, H.-J., Hadrys, T., Cieslak, A., Groger, A., Feller, S. and Forssmann, W.-G. (1995) Proc. Natl. Acad. Sci. U.S.A. **92**, 2091–2095
66 Singer, A. G., Macrides, F., Clancy, A. N. and Agosta, W. C. (1986) J. Biol. Chem. **261**, 13323–13326
67 Singer, A. C. and Macrides, F. (1990) Chem. Senses **15**, 199–203
68 Snyder, S. H., Sklar, P. B. and Pevsner, J. (1988) J. Biol. Chem. **263**, 13971–13974
69 Lee, K.-H., Wells, R. G. and Reed, R. R. (1987) Science **235**, 1053–1056
70 Cavaggioni, A., Sorbi, R. T., Keen, J. N., Pappin, D. J. C. and Findlay, J. B. C. (1987) FEBS Lett. **212**, 225–228
71 Pevsner, J., Reed, R., Feinstein, P. G. and Snyder, S. H. (1988) Science **241**, 336–339
72 Bignetti, E., Cattaneo, P., Cavaggioni, A., Damiani, G. and Tirindelli, R. (1988) Comp. Biochem. Physiol. **90B**, 1–5
73 Pevsner, J., Trifiletti, R. R., Strittmatter, S. M. and Snyder, S. H. (1985) Proc. Natl. Acad. Sci. U.S.A. **82**, 3050–3054
74 Miyawaki, A., Matsushita, F., Ryo, Y. and Mikoshiba, K. (1995) EMBO J. **13**, 5835–5842
75 Schmale, H., Holtgreve-Grez, H. and Christiansen, H. (1990) Nature (London) **343**, 366–369
76 Kock, K., Morley, S. D., Mullins, J. J. and Schmale, H. (1994) Physiol. Behav. **56**, 1173–1177
77 Schmale, H., Ahlers, C., Blaker, M., Kock, K. and Spielman, A. I. (1993) Ciba Found. Symp. **179**, 167–185
78 Redl, B., Holzfeind, P. and Lottspeich, F. (1992) J. Biol. Chem. **267**, 20282–20287
79 Glasgow, B. J., Abduragimov, A. R., Farahbaksh, Z. T., Faull, K. F. and Hubbell, W. L. (1995) Curr. Eye Res. **14**, 363–372
80 Holzfiend, P., Mershak, P., Dieplinger, H. and Redl, B. (1995) Exp. Eye Res. **61**, 495–500
81 Kayser, H. (1985) in Comprehensive Insect Physiology, Biochemistry and Pharmacology (Kerkut, G. and Gilbert, L., eds.), vol. 10, pp. 367–415, Pergamon Press, Oxford
82 Riley, C. T., Barbeau, B. K., Keim, P. S., Kezdy, F. J., Heinrikson, R. L. and Law, J. H. (1984) J. Biol. Chem. **259**, 13159–13165
83 Zuber, H., Suter, F. and Kayser, H. (1987) Biol. Chem. Hoppe-Seyler **369**, 497–505
84 Zagalsky, P. F. (1985) Methods Enzymol. **111**, 216–247
85 Zagalsky, P. F. (1976) Pure Appl. Chem. **47**, 103–120
86 Urade, Y., Nagata, A., Suzuki, Y., Fujii, Y. and Hayaishi, O. (1989) J. Biol. Chem. **264**, 1041–1045
87 Nagata, A., Suzuki, Y., Igarashi, M., Eguchi, N., Toh, H., Urade, Y. and Hayaishi, O. (1991) Proc. Natl. Acad. Sci. U.S.A. **88**, 4020–4024

88   Achen, M. G., Harms, P. J., Thomas, T., Richardson, S. J., Wettenhall, R. E. H. and Schreiber, G. (1992) J. Biol. Chem. **267**, 23170–23174

89   Bell, S. C. (1986) Human Reprod. **1**, 129–143

90   Bell, S. C. (1986) Human Reprod. **1**, 313–318

91   Julkunen, M., Seppala, M. and Janne, O. A. (1988) Proc. Natl. Acad. Sci. U.S.A. **85**, 8845–8849

92   Morrow, D. M., Xiong, N., Getty, R. R., Ratajczak, M. Z., Morgan, D., Seppala, M., Rittinen, L., Gewirtz, A. M. and Tykocinski, M. L. (1995) Am. J. Pathol. **145**, 1485–1495

93   Okamato, N., Uchida, A., Takakura, K., Kariya, Y., Kanzaki, Y., Riitinen, L., Koistinen, R., Seppala, M. and Mori, T. (1991) Am. J. Reprod. Immunol. **26**, 137–142

94   Seppala, M., Koistinen, R. and Rutanen, E.-M. (1994) Human Reprod. **9**, 917–925

95   Bolton, A. E., Pockley, A. G., Clough, K. J., Mowles, E. A., Stoker, R. J., Westwood, O. M. R. and Chapman, M. G. (1987) Lancet **i**, 593–595

96   Said, H. M., Ong, D. E. and Singleton, J. L. (1989) Am. J. Clin. Nutr. **49**, 690–694

97   Akerstrom, B. (1992) Folia Hist. Cytobiol. **30**, 183–186

98   Logdberg, L. and Akerstrom, B. (1981) Scand. J. Immunol. **24**, 575–581

99   Akerstrom, B. and Logdberg, L. (1984) Scand. J. Immunol. **20**, 559–563

100   Mendez, E., Fernandez-Luna, J. L., Grubb, A. and Levy-Cobian (1986) Proc. Natl.Acad. Sci. U.S.A. **83**, 3050–3054

101   Hraba-Renevey, S., Turler, H., Kress, M., Salomon, C. and Weil, R. (1989) Oncogene **4**, 601–608

102   Meheus, L. H., Fransen, L. M., Raymackers, J. G., Blockx, H. A., Van Beeumen, J. J., Van Bun, S. M. and Van De Voorde, A. (1993) J. Immunol. **151**, 1535–1547

103   Liu, Q. and Nilsen-Hamilton, M. (1995) J. Biol Chem. **270**, 22565–22570

104   Kasik, J. W. and Rice, E. J. (1995) Am. J. Obstet. Gynecol. **173**, 613–617

105   Kremer, J. M. H., Wilting, J. and Janssen, L. H. M. (1988) Pharmacol. Rev. **40**, 1–47

106   Arnaud, P., Miribel, L. and Roux, A. F. (1988) Methods Enzymol. **163**, 418–431

107   Castello, M., Fiedel, B. A. and Gewurz, H. (1979) Nature (London) **281**, 677–678

108   Bennett, M. and Schmid, K. (1980) Proc. Natl. Acad. Sci. U.S.A. **77**, 6109–6113

109   De Graff, T. W., Van der Stelt, M. E., Anbergen, M. G. and van Dijk, W. (1993) J. Exp. Med. **177**, 657–666

110   Itoh, Y. and Kawai, T. (1990) J. Clin. Lab. Anal. **4**, 376–384

111   Balbin, M., Freije, J. M. P., Fueyo, A., Sanchez, L. M. and Lopez-Otin, C. (1990) Biochem. J. **271**, 803–807

112   Haefliger, J.-A., Jenne, D., Stanley, K. K. and Tschopp, J. (1987) Biochem. Biophys. Res. Commun. **149**, 750–754

113   Flower, D. R. (1994) FEBS Lett. **354**, 1–7

114   Bedard, P.-A., Yannoni, Y., Simmons, D. L. and Erikson, R. L. (1989) Mol. Cell.Biol. **9**, 1371–1375

115   Cancedda, F. D., Manduca, P., Tacchetti, C., Fossa, P., Quarto, R. and Cancedda, R. (1988) J. Cell Biol. **107**, 2455–2463

116   Cancedda, F. D., Asaro, D., Molina, F., Cancedda, R., Caruso, C., Camardella, L., Negri, A. and Ronchi, S. (1990) Biochem. Biophys. Res. Commun. **168**, 933–938

117   Nakano, T. and Graf, T. (1992) Oncogene **7**, 527–534

118   Francone, O. L., Gurakar, A. and Fielding, C. (1989) J. Biol. Chem. **264**, 7066–7072

119   Streyer, E. and Kostner, G. M. (1988) Biol. Chem. Hoppe-Seyler **369**, 924–925

120   Simard, J., Veilleux, R., de Launoit, Y., Haagensen, D. E. and Labrie, F. (1991) Cancer Res. **51**, 4336–4341

121   Provost, P. R., Marcel, Y. L., Milne, R. W., Weech, P. K. and Rassart, E. (1991) FEBS Lett. **290**, 139–141

122   Lopez-Boado, Y.-S., Toliva, J. and Lopez-Otin, C. (1994) J. Biol. Chem. **269**, 26871–26878

123   Boyles, J. K., Notterpek, L. M. and Anderson, L. J. (1990) J. Biol. Chem. **265**, 17805–17815

124   Spreyer, P., Schaal, H., Kuhn, G., Rothe, T., Unterbeck, A., Olek, K. and Muller, H. W. (1990) EMBO J. **9**, 2479–2484

125   Sanchez, D., Ganfornia, M. D. and Bastiani, M. J. (1995) Development **121**, 135–147

126   Newcomer, M. E. and Ong, D. E. (1990) J. Biol. Chem. **265**, 12876–12879

127   Brooks, D. E. (1987) Biochem. Int. **14**, 235–240

128   Matuo, Y., Nishi, N., Tanaka, Y., Muguruma, Y., Tanaka, K., Akatsuka, Y., Matsui, S.-I., Sandberg, A. and Wada, F. (1984) Biochem. Biophys. Res. Commun. **118**, 467–473

129   Spence, A. M., Sheppard, P. C., Davie, J. R., Matuo, Y., Nishi, N., McKeehan, W. L., Dodd, J. G. and Matusik, R. J. (1989) Proc. Natl. Acad. Sci. U.S.A. **86**, 7843–7847

130   Schubert, D. and LaCorbiere, M. (1985) J. Cell Biol. **101**, 1071–1077

131   Schubert, D., LaCorbiere, M. and Esch, F. (1986) J. Cell Biol. **102**, 2295–2301

132   Berman, P., Gray, P., Chen, E., Keyser, K., Ehrlich, D., Karten, H., LaCorbiere, M., Esch, F. and Schubert, D. (1987) Cell **51**, 135–142

133   Liu, H. M., Lei, H. Y. and Schmid, K. (1993) Lab. Invest. **68**, 577–583

134   Babiker-Mohamed, H., Olsen, M. L., Boketoft, A., Akerstrom, B. and Logdberg, L. (1990) Immunobiology **180**, 221–234

135   Babiker-Mohammed, H., Akerstrom, B. and Logdberg, L. (1990) Scand. J. Immunol. **32**, 37–44

136   Babiker-Mohammed, H., Forsberg, M., Olsen, M. L., Winquist, O., Logdberg, L. and Akerstrom, B. (1991) Scand. J. Immunol. **34**, 655–666

137   Hollenberg, M. D. (1996) Trends Pharmacol. Sci. **17**, 3–6

138   Cogan, U., Hopelman, M., Mokady, S. and Shinitzky, M. (1976) Eur. J. Biochem. **65**, 71–78

139   Hase, J., Kobashi, K., Nakai, N. and Onosaka, S. (1976) J. Biochem. (Tokyo) **79**, 373–380

140   Borghoff, S. J., Miller, A. B., Bowen, J. P. and Swenberg, J. A. (1991) Toxicol. Appl. Pharmacol. **107**, 228–238

141   Peitsch, M. C. and Boguski, M. S. (1990) New Biol. **2**, 197–206

142   Morais Cabral, J. H., Atkins, G. L., Sanchez, L. M., Lopez-Boado, Y. S., Lopez-Otin, C. and Sawyer, L. (1995) FEBS Lett. **366**, 53–56

143   Dodin, G., Andrieux, M. and Al Kabbani, H. (1990) Eur. J. Biochem. **193**, 697–700

144   Dufour, E., Marden, M. C. and Haertle, T. (1991) FEBS Lett. **277**, 223–226

145   Britton, G., Armitt, G. M., Lau, S. Y. M., Patel, A. K. and Shone, C. C. (1982) in Carotenoid Chemistry and Biochemistry (Britton, G. and Goodwin, T. W., eds.), pp. 237–253, Pergamon, Press, Oxford

146   Zagalsky, P. F., Eliopoulos, E. E. and Findlay, J. B. C. (1990) Comp. Biochem. Physiol. **97B**, 1–18

147   Escribano, J., Grubb, A. and Mendez, E. (1988) Biochem. Biophys. Res. Commun. **155**, 1424–1429

148   Flower, D. R. (1991) J. Mol. Graphics **9**, 257–258

149   Flower, D. R. (1995) J. Mol. Graphics **13**, 377–384

Biochem. J. (1996) **314**, 361–385 (Printed in Great Britain)

REVIEW ARTICLE

Structure–function relationships in the FK506-binding protein (FKBP) family of peptidylprolyl *cis–trans* isomerases

John E. KAY

School of Biological Sciences, University of Sussex, Brighton BN1 9QG, U.K.

1 INTRODUCTION

During protein synthesis the peptide bonds on the amino side of proline residues are formed in the *trans* conformation, but in proteins whose three-dimensional structure has been determined about 15% of these bonds are found in the alternative *cis* conformation. Enzymes with peptidylprolyl *cis–trans* isomerase (PPIase) activity catalyse this interconversion (Figure 1). PPIases were first identified in the course of studies on protein folding in Fischer's laboratory [1,2] and became the focus of intensive research when cyclophilin A (CypA), the primary target of the clinically important immunosuppressive agent cyclosporin A (CsA) [3,4], was identified as the predominant PPIase present in

the cytoplasm of mammalian cells [5,6]. The subsequent discoveries that the primary target of two alternative immunosuppressive drugs, FK506 and rapamycin, was a second but completely unrelated cytoplasmic PPIase, the FK506-binding protein (FKBP) now known as FKBP12 [7,8], and that in both PPIase families the naturally produced immunosuppressive agents bound into the proline-binding site of the PPIase to inhibit its activity, led to an explosion of interest in this class of enzymes.

Subsequent research has shown that each of these cytoplasmic PPIases is a member of a highly conserved family of enzymes found in a wide variety of prokaryotic and eukaryotic organisms (Table 1). In mammals they are not confined to lymphoid cells but are present and abundant in all tissues and most subcellular compartments [9,10]. More than a dozen different PPIases belonging to these two families are now known to be present in humans, most of them widely distributed in different tissues, and there are at least nine in the yeast *Saccharomyces cerevisiae* (Table 1).

In 1994 a new 10.1 kDa PPIase, parvulin, which is not a member of either the cyclophilin family or the FKBP family was identified in the cytoplasm of *Escherichia coli* [11–13]. It has PPIase activity comparable with that of cyclophilin and higher than that of any FKBP, but no sequence similarity to either PPIase family and no affinity for the usual cyclophilin or FKBP ligands (Table 1). Sequence similarities with domains of other larger proteins suggest that it will prove to be representative of a third family of PPIases. All three families include proteins containing just the basic PPIase domain, with signal sequences to target them to the correct subcellular compartment where

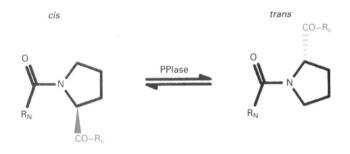

Figure 1   The *cis–trans* isomerization reaction catalysed by PPIases

R_N and R_C represent the amino acid residues on the amino and carboxyl sides of the proline residue.

Table 1   Characteristics of the three PPIase families

	FKBPs	Cyclophilins	Parvulins
Domain size (residues)	Approx. 108	Approx. 165	92
PPIase activity	− to + +	+ + to + + +	? to + + +
Distribution	Mammals	Mammals	?
	Plants	Plants	?
	Lower eukaryotes	Lower eukaryotes	?
	Prokaryotes	Prokaryotes	Prokaryotes
Natural inhibitors	Macrolides (FK506, ascomycin, rapamycin, meridamycin)	Peptides (cyclosporins)	?
Known in humans	5 (FKBP12, FKBP12A, FKBP13, FKBP25, FKBP52)	6 (CypA, CypB, CypC, CypD, Cyp40, CypNK)	?
Known in *S. cerevisiae*	4	5	−
Domains found within larger proteins	Yes	Yes	Yes
Multiple domains found	Yes	No	Yes

Abbreviations used: CsA, cyclosporin A; CypA (etc.), cyclophilin A (etc.); ER, endoplasmic reticulum; FKBP, FK506-binding protein; Mip protein, macrophage infectivity activator protein; ORF, open reading frame; PPIase, peptidylprolyl *cis–trans* isomerase; succ.AAPF.pNA, succinyl-Ala-Ala-Pro-Phe-*p*-nitroanilide (and similarly for substrates containing other amino acids); TOR, target of rapamycin; UTR, untranslated region. Members of the FKBP family are distinguished by a prefix of one or two letters to indicate the species of origin (see legend to Figure 2 for definitions).

Gene	27	37	38	43	47	55	56	57	60	82	83	88	92	100
hFKBP12	Y	F	D	R	F	E	V	I	W	A	Y	H	I	F
bFKBP12	*	*	*	*	*	*	*	*	*	*	*	*	*	*
oFKBP12	*	*	*	*	*	*	*	*	*	*	*	*	*	*
mFKBP12	*	*	*	*	*	*	*	*	*	*	*	*	*	*
rFKBP12	*	*	*	*	*	*	*	*	*	*	*	*	*	*
yFKBP12	*	*	*	*	*	Q	*	*	*	*	*	F	*	*
caFKBP12	*	*	*	*	*	Q	*	*	*	*	*	I	*	*
ncFKBP12	*	*	*	*	L	Q	*	*	*	*	*	V	*	*
scFKBP15	*	*	*	*	L	Q	*	*	*	*	*	A	*	*
nmFKBP12	*	*	*	*	L	Q	*	*	G	*	*	A	*	*
hFKBP12A	*	*	*	*	*	*	*	*	F	*	*	*	*	*
bFKBP12A	*	*	*	*	*	*	*	*	F	*	*	*	*	*
hFKBP13	*	*	*	Q	*	Q	*	*	*	G	*	A	*	*
bsFKBP13	*	*	*	*	*	Y	*	*	*	G	*	S	*	*
yFKBP13	*	*	*	*	I	R	*	*	*	*	*	V	*	*
ecFKBP16	F	A	E	N	A	S	L	S	L	*	F	P	*	*
pfFKBP16	F	V	*	K	A	N	L	L	F	*	F	P	V	*
ecFKBP22	*	V	*	S	L	S	L	*	L	*	*	E	*	*
hFKBP25	*	*	*	T	L	K	*	*	*	*	*	Q	*	*
bFKBP25	*	*	*	T	L	K	*	*	*	*	*	Q	*	*
ctFKBP25M	*	*	*	N	I	K	*	*	F	*	*	Q	L	*
lpFKBP25M	*.	*	*	T	A	Q	*	*	*	*	*	V	*	*
lmFKBP25M	*	*	*	T	A	Q	*	*	*	*	*	V	*	*
paFKBP25M	*	*	*	-	-	-	*	*	*	*	*	A	*	*
scFKBP33d1	*	*	*	*	*	M	*	*	*	G	*	Q	*	*
scFKBP33d2	*	*	*	T	Q	V	T	L	L	*	F	Q	*	*
dmFKBP39	*	*	*	-	*	*	*	*	*	*	*	A	*	*
sfFKBP46	*	*	*	-	*	*	*	*	*	*	*	S	*	*
yFKBP46	*	*	*	-	*	*	*	*	*	*	*	L	*	*
hFKBP52d1	*	*	*	*	*	*	*	*	*	*	*	S	*	*
oFKBP52d1	*	*	*	*	*	*	*	*	*	*	*	S	*	*
mFKBP52d1	*	*	*	*	*	*	*	*	*	*	*	S	*	*
hFKBP52d2	L	*	*	*	G	D	L	P	L	*	F	K	*	Y
oFKBP52d2	L	*	*	*	G	D	L	P	L	*	F	K	*	Y
mFKBP52d2	L	*	*	C	G	D	L	P	L	*	F	K	*	Y
hFKBP52d3	*	Y	E	N	Q	R	L	A	L	*	L	N	L	L
oFKBP52d3	*	Y	E	S	Q	R	L	A	L	*	L	N	L	L
mFKBP52d3	*	Y	E	G	Q	R	L	A	L	*	L	N	L	L

Figure 2   Conservation of FK506-binding residues in FKBPs and FKBP domains

Residues identified as the FK506-binding domains of hFKBP12 are compared with the corresponding residues of 37 other sequenced FKBP domains. A residue identical to the corresponding hFKBP12 residue is indicated by an asterisk, while a dash indicates that no residue clearly corresponds after alignment. Species abbreviations used are: b, *Bos taurus*; bs, *Botryllus schlosseri*; ca, *Candida albicans*; ct, *Chlamydia trachomatis*; dm, *Drosophila melanogaster*; ec, *Escherichia coli*; h, *Homo sapiens* (human); lm, *Legionella micdadei*; lp, *Legionella pneumophila*; m, *Mus musculus* (mouse); nc, *Neurospora crassa*; nm, *Neisseria meningitidis*; o, *Oryctolagus cuniculus* (rabbit); pa, *Pseudomonas aeruginosa*; pf, *Pseudomonas fluorescens*; r, *Rattus norvegicus* (rat); sc, *Streptomyces chrysomallus*; sf, *Spodoptera frugiperda*; y, *Saccharomyces cerevisiae*.

appropriate, and also cyclophilin-, FKBP- or parvulin-domain proteins containing one or more PPIase domains as part of a larger protein (Table 1). Recent studies suggest that PPIase-domain proteins, and especially proteins containing multiple PPIase domains, may be particularly common in the FKBP family.

The physiological function of the most abundant PPIases has proved surprisingly difficult to establish. Although the importance of some PPIase-domain proteins is now better understood, the specific role of their PPIase domain(s) is often less certain. The *cis–trans* isomerization of peptidylprolyl bonds has been identified as a relatively slow step in protein folding, suggesting that PPIases might play an important role in the catalysis of this process. However, some CsA and FK506 analogues that are effective PPIase inhibitors are not immunosuppressive, and even at high concentrations have little identifiable effect on the survival, growth or behaviour of mammalian cells [14–17]. In addition, experimental inactivation of genes encoding the major PPIases, and even inactivation of several PPIase genes simultaneously, may have little evident effect on the growth or behaviour of yeast and bacteria in a variety of conditions. Proteins that are universal, abundant, highly conserved and targeted by a range of microbial antibiotics, but at the same time apparently serve no essential function, present an intriguing paradox. While the possibility that the explanation may lie in the functional redundancy of the PPIases so far studied has yet to be completely excluded, some of the protocols followed inactivate the great majority of PPIase enzyme activity detectable in the cell.

1.1   Assessment of PPIase activity

PPIase activity is most commonly monitored using the spectro-photometric assay developed by Fischer et al. [1], which relies on the ability of α-chymotrypsin to release *p*-nitroaniline from peptides such as succinyl-Ala-Ala-Pro-Phe-*p*-nitroanilide (succ.AAPF.pNA) only if the bond on the N-terminal side of the proline residue is in the *trans* conformation. Addition of α-chymotrypsin thus causes very rapid release of *p*-nitroaniline from that proportion of succ.AAPF.pNA in the *trans* conformation, followed by the slower release of the remainder as the *cis* isomer spontaneously isomerizes to the *trans* form. PPIases enhance the rate of this isomerization, and thus the rate of the slow second stage of this reaction. Succ.AAPF.pNA is the synthetic peptide most commonly used for the assay of cyclophilin PPIase activity, but FKBP-family PPIases and parvulin usually have a strong preference for substrates with a bulky hydrophobic residue preceding the proline [11,18,19], so that succ.ALPF.pNA or succ.AFPF.pNA are most commonly used to assay these enzymes.

A major limitation of the assay is that in normal aqueous solution about 85 % of succ.AAPF.pNA is in the *trans* form, and attempts have been made to improve the sensitivity of the assay by identifying substrates or solvents which give a higher initial proportion of the *cis* isomer. A modification used with some success is to dissolve the peptide in LiCl and trifluoroethanol to increase the proportion in the *cis* conformation [20]. A second limitation is that the substrate concentrations necessary for maximal PPIase activity exceed the solubility of the peptides. PPIase activities are thus normally expressed as k_{cat}/K_m values.

Alternative approaches to assaying PPIase have involved the refolding of proteins such as RNase T1 [6], the use of NMR to follow the rate of interconversion of *cis* and *trans* forms [21,22], a fluorimetric assay based on greater intramolecular quenching through collision when the substrate peptide is in the *cis* conformation [23], and a spectrophotometric assay based on the different effects of the *cis* and *trans* isomers on the pK_a of the side-chain hydroxy group in -Xaa-Pro-(*m*-NO$_2$)Tyr-Xaa-peptides [24]. These methods have rarely been used for FKBP-family PPIases, due either to the lower enzymic activities of the FKBPs compared with cyclophilins or to the restricted availability of the substrates required.

1.2   Nomenclature of FKBP-family PPIases

The nomenclature used to distinguish the different members of the FKBP family has been the cause of some confusion. The usual approach, adopted here, is to add to the term FKBP a prefix of one or two letters to indicate the species of origin (see legend to Figure 2 for definitions) and a suffix to reflect the calculated molecular mass of the mature protein, where known. Thus the major human cytoplasmic FKBP becomes hFKBP12 and the recently described FKBP-domain protein from *Spodoptera frugiperda* cells is sfFKBP46.

This systematic approach rapidly encounters difficulties. Firstly, most FKBPs run anomalously slowly on SDS/PAGE, so that the suffixes ascribed to incompletely characterized family members may prove incorrect when sequencing is completed.

Thus the FKBP-domain protein initially named FKBP59 has now become FKBP52. Secondly, the fungal cytoplasmic FKBPs that are clearly homologous to hFKBP12 contain a handful of additional residues, but to name such proteins FKBP13s would suggest homology with hFKBP13, the family member present within the endoplasmic reticulum (ER). Thirdly, the recently discovered second human cytoplasmic FKBP that is distinct from hFKBP12 has an identical number of amino acids and a very similar molecular mass; it is here referred to as hFKBP12A. Fourthly, two quite distinct FKBP-domain proteins, a mammalian FKBP located primarily in the nucleus and the membrane-located Mip (macrophage infectivity potentiator) family of FKBP-domain proteins required for intracellular infectivity of some prokaryote intracellular parasites, both have very similar molecular masses of about 25 kDa. The Mip family are here distinguished by the suffix 25M, so that the *Legionella pneumophila* Mip protein becomes lpFKBP25M.

When comparing sequences between FKBP domains from different family members (as in Figure 2, which identifies the residues at the key positions in different FKBPs for interaction with FK506), the amino acid sequences of other FKBP domains are first aligned with that of hFKBP12. Residues are then assigned the number of the homologous hFKBP12 residue, with the cleaved N-terminal methionine taken as residue 1. The standard single-letter amino acid code is used when describing variant or mutated FKBPs, so that the notation A82G indicates the replacement of the alanine residue at position 82 of the hFKBP12 domain by a glycine.

2   MAMMALIAN FKBPs AND FKBP-DOMAIN PROTEINS

Mammals contain 11 currently characterized and sequenced PPIases (Table 1), but it is clear that this list is far from complete. Six of the characterized PPIases are cyclophilins. CypA is an abundant protein that accounts for the majority of PPIase activity in the cytoplasm [3,5,6]. CypB [25–27] and CypC [28,29] are secreted into the ER and are found both in the secretory pathway and on the cell surface. CypD is predicted to cross the mitochondrial membrane, where it may account for at least part of the PPIase activity found in that organelle [30–32]. The cyclophilin-domain protein Cyp40, which has 370 amino acid residues including a single cyclophilin domain near its N-terminal end, is also found in the cytoplasm, associated with the heat shock protein hsp90 in unliganded steroid receptor complexes [33]. CypA, CypB, CypD and Cyp40 are expressed in all cells and tissues studied, while the distribution of CypC is more restricted. However, a more recently characterized member of the cyclophilin family, the 1403-amino-acid cyclophilin-domain protein CypNK, in which the N-terminal signal sequence is followed by a single cyclophilin domain and then a long uncharacterized sequence, is expressed only on the surface of NK cells and appears necessary for their natural killer activity against tumour cells [34,35]. This highly selective pattern of expression is probably not unique, as antibodies specific for a peptide from the *Drosophila melanogaster* NinaA protein, a protein with a single cyclophilin domain that is present in the secretory system and is essential for the proper assembly of rhodopsin in the rhabdomeres (see section 7.1), cross-react with a 30 kDa protein present in bovine retina and cerebral cortex but not in other tissues studied [36].

There is now good evidence for a similarly complex family of FKBPs and FKBP-domain proteins in mammals. Preliminary studies showed that, in addition to FKBP12, mammalian cells contain a number of other proteins that were retained by FK506- or rapamycin-affinity matrices and then eluted with the free macrolide [37], or were recognized by antisera raised against specific hFKBP12 peptides [38]. This suggestion of the existence of a family of mammalian FKBP proteins has subsequently been confirmed by the purification and characterization of five family members. FKBP12, FKBP12A and FKBP13 contain only the basic FKBP domain, while FKBP25 and FKBP52 have one or more FKBP domains as part of a larger molecule. However, not all the proteins whose existence was suggested by the preliminary studies have yet been fully identified, and partial sequences are available for at least two further mammalian FKBP-domain proteins.

Parvulin, the prototype of the third family of PPIases, has so far been identified only in prokaryotes. No eukaryotic members of this family have yet been isolated or characterized, but a human expressed sequence tag with significant similarity to *E. coli* parvulin has been identified [13].

2.1   Mammalian FKBP12

The primary target for the immunosuppressive agents FK506 and rapamycin in human cells was first purified and identified as a cytoplasmic protein distinct from CypA in 1989 [39], and shortly afterwards it was identified as an enzyme with PPIase activity [7,8]. This protein, now called hFKBP12, has 108 amino acids including a cleaved N-terminal methionine residue [40,41], a pI of 8.9 and a molecular mass of 11.8 kDa, although it migrates on SDS/PAGE with an apparent molecular mass of 14 kDa. The sequence has no significant similarity to the cyclophilins.

hFKBP12 is resistant to heat denaturation, with heating to 56 °C for 20 min included in some purification protocols to eliminate contaminating cyclophilin PPIases [7,39], and folds and unfolds quite readily and reversibly [42]. Addition of ligand further increases the thermal stability of the protein [43]. Even when recombinant hFKBP12 is expressed in *E. coli* at high levels, i.e. up to 36 % of total *E. coli* protein, hFKBP12 is recovered from the lysed bacteria in a soluble and fully active form [40,44].

Homologous FKBP12s are present in all other mammalian species studied. Sequences have been reported for bovine (b)FKBP12 [45], murine (m)FKBP12 [46], rat (r)FKBP12 [47] and rabbit (o)FKBP12 [48]. The hFKBP12 and oFKBP12 sequences are identical, and the others show 95–97 % identity with hFKBP12 and with each other. The few substitutions are conservative and none affects the ligand-binding site (Figure 2) or the overall protein fold. Homologous proteins have been reported in the chicken, frog and goldfish [49,50].

FKBP12 is an abundant protein, accounting for up to 0.4 % of the total cytoplasmic protein in Jurkat cells [7]. It is not confined to lymphoid cells. Both the protein and its mRNA are present in the cytoplasm of all mammalian cells studied, with exceptionally high levels of expression in specific regions of the brain [8,10,41,51,52]. FKBP12 is also present, although at lower concentrations, in erythrocytes [53], with the result that virtually all circulating bioavailable FK506 or rapamycin is sequestered within these cells, a factor that complicates the clinical monitoring of therapeutically administered macrolides.

Each FKBP12 binds a single molecule of FK506 or rapamycin with high affinity, and this inactivates its PPIase activity. K_d values for ligand binding or K_i values estimated from inhibition of PPIase activity have been determined to be within the range 0.2–1.0 nM for rapamycin, FK506, ascomycin (a natural macrolide with the 21-allyl group of FK506 replaced by an ethyl group; also known as FK520) and 21-propyl-FK506 (also known as dihydroFK506) [7,54,55]. The affinity for rapamycin is about 2-fold higher than that for FK506; that for the other analogues

(a)

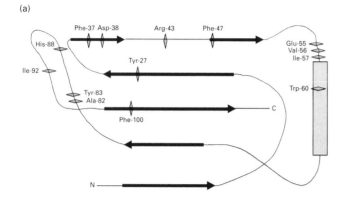

(b)

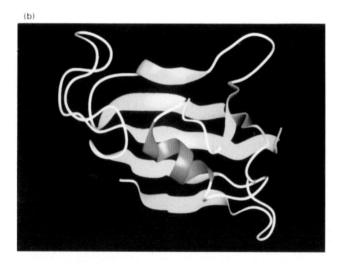

Figure 3   Structure of hFKBP12

(**a**) Topological representation of the hFKBP12 domain structure. The broad arrows indicate the strands of the β-sheet, the boxed region the α-helix and the diamonds the locations of the 14 FK506-binding residues. (**b**) Ribbon diagram to illustrate the solution structure of the hFKBP12 backbone. The strands of the β-sheet are shown in yellow and the α-helix in purple. The N-terminus is at the bottom left.

is slightly lower. The PPIase activity of hFKBP12 with the standard cyclophilin substrate, succ.AAPF.pNA, is only 1–5 % that of hCypA, but activity is increased up to 10-fold if the alanine residue preceding the proline is replaced by an amino acid with a bulky hydrophobic side chain such as leucine or phenylalanine [18,19]. The PPIase activity of FKBP12 is little affected by the pH of the incubation medium within the range pH 5–10 [55].

High-resolution structures derived from X-ray crystallography and multidimensional NMR methods have been reported for hFKBP12 alone [56,57] or complexed with FK506 [58–60], the immunosuppressive analogue ascomycin [61], the non-immunosuppressive analogue 18-hydroxyascomycin (also known as L-685,818) [62,63], rapamycin [60,64] and a synthetic cyclic peptide–FK506 hybrid molecule [65]. A detailed structure for the hFKBP12–peptide complex has been deduced using molecular dynamics and free energy perturbation methods [66].

The major hFKBP12 structural element is a five-stranded antiparallel β-sheet with topology +3, +1, −3, +1, which requires an unusual crossing of two of the intervening loops (Figure 3). One of the loops includes a short section of α-helix.

The β-sheet is twisted to give concave and convex surfaces, with the concave surface made up of hydrophobic residues that form the core of the protein and the hydrophilic residues projecting on the convex side making up much of the protein surface. The long loop containing the α-helix passes across the concave side of the β-sheet. A very similar NMR structure has been reported for uncomplexed bFKBP12 [67].

The ligand conformation changes when FK506 or other ligands are bound to hFKBP12, as has also been observed when CsA binds to hCypA [68]. In solution FK506 is predominantly in the *cis* form, but it binds to hFKBP12 in a transition state close to the *trans* configuration. Rapamycin in solution is predominantly in the *trans* form, and this may explain why it has a lower K_d than FK506 for binding to hFKBP12. These ligands bind into a hydrophobic pocket bounded by part of the convex surface of the β-sheet and three loops, including the loop containing the short α-helix. The only significant change to the hFKBP12 structure on ligand binding is decreased flexibility of two of the loops bordering the binding pocket [61,69]. The first strand of the β-sheet and the succeeding loop (which are relatively poorly conserved between FKBPs) contribute little to the hydrophobic core of the molecule and nothing at all to the binding pocket.

Fourteen hFKBP12 residues, ten of them hydrophobic, have been identified as interacting with bound FK506 (Figures 2 and 3a). All of these residues except Arg-43 and Ile-92 also interact with bound rapamycin, although there are some minor differences in the nature of the interactions. Binding involves the ligand pipecolinyl and pyranose rings and the dicarbonyl region between them, and also one edge of the cyclohexane ring. This section of the ligand, which is common to macrolides that inhibit FKBP PPIase activity, becomes buried in the hydrophobic core of the ligand–hFKBP12 complex. On binding into the hydrophobic pocket the ligand is thought to mimic a transition state of the natural peptidylprolyl substrate and form a short two-stranded antiparallel β-sheet with three amino acids in the loop immediately preceding the α-helix: residues Glu-55, Val-56 and Ile-57 [65]. The remainder of the cyclohexane ring and the rest of the macrolide ring, in which the structural differences between these FKBP ligands are located, are exposed on the surface of the ligand–hFKBP12 complexes and are available to interact with other molecules.

The gene encoding hFKBP12 is present in a single copy on chromosome 20 band p13 [70–72], although, as also noted for hCypA [73], there are additionally several processed pseudogenes. The hFKBP12 gene is spread over 24 kb, a considerable distance for a protein whose coding sequence requires only 0.3 kb. It has five exons, four of them covering the coding region. The first exon [covering the 5′ untranslated region (UTR) and the sequence to Gly-13 including the first strand of the β-sheet] and the second exon (covering residues Gly-13–Gly-29 and including the second strand of the β-sheet) are separated by an intron of only 0.1 kb. However, there is then a long 16 kb intron before the third exon, which encodes residues Gly-29–Gln-66, including the third strand of the β-sheet and the α-helix. The final coding exon encodes from Met-67 onwards, including the last two strands of the β-sheet and the long intervening loop that borders the ligand-binding pocket, but there is a final intron within the 3′ UTR. The predominant hFKBP12 mRNA is over 1500 nucleotides in length, with a short (69 nucleotide) 5′ UTR but a 3′ UTR over 1000 nucleotides long [70]. Minor hFKBP12 mRNA forms with shorter 3′ UTRs have been reported, apparently generated by the use of an alternative weak poly(A) cleavage site or by failing to splice out the 3′ UTR intron [52].

Although most mammalian FKBP12 is found free in the cytoplasm, a small proportion is associated with other proteins

Table 2   Amino acid sequence identity between human FKBP domains

| | Identity (%) | | | | | |
	hFKBP12A	hFKBP13	hFKBP25	hFKBP52 Domain 1	Domain 2	Domain 3
hFKBP12	83	43	42	49	26	7
hFKBP12A	–	38	41	47	25	8
hFKBP13	–	–	44	45	19	9
hFKBP25	–	–	–	44	24	8
hFKBP52 domain 1	–	–	–	–	29	12
hFKBP52 domain 2	–	–	–	–	–	11

in the membrane fraction. This was first demonstrated for the skeletal muscle type 1 ryanodine receptor, which is responsible for Ca^{2+} release from the sarcoplasmic reticulum into the cytoplasm during muscle contraction [48,74]. The ryanodine receptor is composed of four 565 kDa subunits associated with four FKBP12 molecules. The FKBP12 can be released from the complex by addition of FK506, rapamycin or non-immunosuppressive analogues that inhibit FKBP PPIase activity. The presence of the FKBP12 affects the probability and extent of Ca^{2+} channel opening, and decreases its sensitivity to caffeine [75–78]. The kinetics of *in vitro* exchange between free and bound FKBP12 suggest that this will be rapid *in vivo*, and FKBPs with point mutations inhibiting PPIase activity and FK506 binding can still associate with and modify the properties of the channel [78]. Bastadins synthesized by the sponge *Ianthella basta* to target this channel are antagonized by FK506, to which they have some structural resemblance. While bastadins do not themselves dissociate FKBP12 from the complex, they enhance FK506-induced dissociation [79].

Recently FKBP12 has also been reported to be a component of another major intracellular Ca^{2+} release channel, the inositol 1,4,5-trisphosphate receptor [80]. Again the FKBP12 can be released by FK506 or rapamycin, enhancing spontaneous Ca^{2+} efflux and sensitivity to inositol 1,4,5-trisphosphate. The yeast two-hybrid system has also been used to identify an association between FKBP12 and isoforms of the cytoplasmic domain of type I receptors for the transforming growth factor-β family of cytokines [47]. This interaction was also prevented by FK506. Transforming growth factor-β induces the expression of type 3 ryanodine receptors in a variety of mammalian cells.

2.2   Mammalian FKBP12A

Two recent reports have identified a novel cytoplasmic FKBP in human [81] and bovine [82] brain. Although the protein, here termed FKBP12A, migrates slightly more slowly than FKBP12 on SDS/PAGE gels, it contains the same number of amino acids (108 including the cleaved N-terminal methionine residue) and its calculated molecular mass is slightly lower. The protein is abundant in brain, but its mRNA was detected in all other tissues examined. It has rapamycin- and FK506-sensitive PPIase activity comparable with that of hFKBP12, and the K_d for FK506 was 0.55 nM, compared with 0.4 nM for hFKBP12.

The amino acid sequences of hFKBP12A and bFKBP12A are identical to each other, and have 83% and 85% identity respectively with the corresponding FKBP12s. This compares with only 38–47% identity between hFKBP12 or hFKBP12A and other human FKBP domains that are active PPIases (Table

Table 3   Amino acid sequence identity between *Saccharomyces cerevisiae* and human FKBP domains

| | Identity (%) | | |
	yFKBP12	yFKBP13	yFKBP46
yFKBP13	52	–	–
yFKBP46	45	45	–
hFKBP12	56	46	44
hFKBP12A	55	43	41
hFKBP13	44	56	35
hFKBP25	45	42	40
hFKBP52 domain 1	45	45	38

2), values comparable with their identity with FKBP domains from a variety of distantly related organisms (Tables 2 and 3). Of the 14 FK506-binding residues found in hFKBP12, 13 are retained in hFKBP12A, the exception being a W60F substitution (Figure 2).

2.3   Mammalian FKBP13

The first mammalian FKBP-family member to be identified and sequenced after FKBP12 itself was FKBP13. This was first purified from bovine thymus using a rapamycin-affinity matrix, and the N-terminal sequence determined was used to design PCR primers that enabled the hFHBP13 coding region to be identified and sequenced from a colon carcinoma cDNA library [83]. The hFKBP13 cDNA encodes a 141-amino-acid protein with a hydrophobic 21-amino-acid N-terminal signal sequence (Figure 4). The mature protein of 120 amino acids has a predicted molecular mass of 13.2 kDa, although its migration on SDS/PAGE would suggest a molecular mass of 15 kDa.

The signal sequence is predicted to cause hFKBP13 to be secreted into the ER, and it has no internal hydrophobic region suggestive of a transmembrane domain. However, hFKBP13 was not detected as being secreted from the cell. It has a C-terminal -RTEL tetrapeptide following the FKBP domain that is sufficiently homologous to the canonical -KDEL to cause its retention in the lumen of the pre-Golgi ER compartment, a prediction confirmed by subsequent experimental work [84,85]. Addition of FK506 had no evident effect on the distribution of hFKBP13 between cellular compartments.

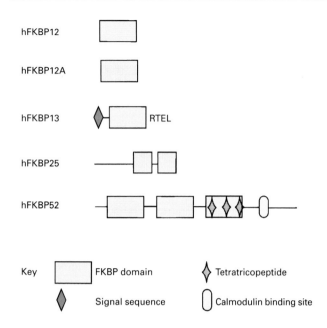

hFKBP12

hFKBP12A

hFKBP13   RTEL

hFKBP25

hFKBP52

Key   FKBP domain   ◆ Tetratricopeptide

◆ Signal sequence   Calmodulin binding site

Figure 4   Human FKBPs and FKBP-domain proteins

The FKBP domain of mFKBP13 has 100% identity with the corresponding hFKBP13 domain, although there are several differences (only 70% amino acid identity) within the signal sequence, and the C-terminal peptide is -RSEL rather than -RTEL [86]. There is 38–45% amino acid identity between hFKBP13 and other human FKBP PPIase domains (Table 2), values comparable with its identity with FKBP domains from *S. cerevisiae* (Table 3) or even from prokaryotes (e.g. 42% with the first FKBP domain of scFKBP33). Its greatest identity is with single FKBP-domain proteins containing N-terminal signal sequences from *S. cerevisiae* (yFKBP13; 56% identity) and a protochordate, the colonial tunicate *Botryllus schlosseri* (btFKBP13; 72% identity) [87]. In btFKBP13 the C-terminal -RTEL peptide is replaced by the canonical -KDEL, but yFKBP13 ends with the sequence -SAA.

Of the 14 FK506-binding residues found in hFKBP12, 10 are conserved in hFKBP13 (Figure 2). The four substitutions are R43Q (a unique, non-conservative, change), E55Q, A82G and H88A (all encountered in other active FKBP-family PPIases). The tunicate btFKBP13 retains Arg-43 but has three differences from hFKBP12 affecting three of the same residues, i.e. E55Y, A82G and H88S. The structure of the hFKBP13–FK506 complex determined by X-ray diffraction confirms that the ligand-binding pocket has a very similar conformation to that of hFKBP12 [88]. The most striking difference is in the loop crossing region. In hFKBP12 this is stabilized by inter- and intra-strand hydrogen bonds, but in hFKBP13 it is stabilized by a disulphide bond between two cysteine residues.

Like FKBP12, FKBP13 is widely expressed in all mammalian tissues examined. Although normally found at a considerably lower level of expression than FKBP12, it may become the predominant FKBP in cells such as mast cells that specialize in protein secretion and are rich in ER [89]. FKBP13 has PPIase activity that is inhibited by both FK506 and rapamycin, but its sensitivity to both drugs, especially FK506, is much lower than that of hFKBP12 [38]. The K_d for rapamycin is 3.6 nM (cf. 0.2 nM for hFKBP12), while the K_d for FK506 is 38 nM (cf. 0.4 nM), confirming the qualitative observation that bFKBP13 is

more effectively retained by rapamycin-affinity matrices than by the corresponding FK506 matrices [37].

The hFKBP13 gene is on human chromosome 11 band q13. It is a single-copy gene with no evident pseudogenes, and has six exons distributed over 3 kb of DNA [72,86]. The first exon encodes the first part of the 5′ UTR, the second exon the remainder of the 5′ UTR, the leader sequence and the first 28 residues of the FKBP domain, and the third, fourth and fifth exons residues 29–66, 66–82 and 82–94 respectively of the FKBP domain. The final exon encodes the remainder of the protein and the 3′ UTR. All of the exon–intron boundaries can be related to specific structural elements in the FKBP domain. The exon–intron boundaries at FKBP-domain residues 29 and 66 are identical in position in both hFKBP12 and hFKBP13, with the intervening exon encoding the third strand of the β-sheet and the α-helix and including nine of the 14 FK506-binding residues. The other exon–intron boundaries are not conserved between these two genes. The primary transcript of the gene is processed to give a 0.6 kb mRNA [83].

2.4   Mammalian FKBP25, an FKBP-domain protein

FKBP25, first detected as a protein in human T-cells and calf thymus that bound well to a rapamycin-affinity matrix but much less efficiently to an FK506-affinity matrix [37], was purified to homogeneity from bovine brain, thymus and spleen and sequenced almost completely by conventional means [90]. The human homologue, hFKBP25, was then isolated from cDNA libraries and fully sequenced [91–93].

Both bFKBP25 and hFKBP25 migrate as 30 kDa proteins on SDS/PAGE, but the molecular mass determined by MS or calculated from the cDNA sequence is 25.2 kDa. The hFKBP25 cDNA has a 224-amino-acid open reading frame (ORF), but the N-terminal amino group of the mature protein is blocked and cannot be identified by the Edman procedure. The FKBP domain is the C-terminal half of the molecule (Figure 4). The strongly hydrophilic N-terminal domain has no known identity with other proteins, but both CD spectroscopy and Chou–Fasman calculations predict a largely α-helical structure. It has 38% of residues with charged side chains. There is 98% identity between the N-terminal domains of hFKBP25 and bFKBP25 and 97% identity over the C-terminal FKBP domains.

The C-terminal FKBP domain differs from other mammalian FKBP domains in having an additional seven amino acids (including a KKKK motif) inserted into the loop that separates the two parts of the third strand of the β-sheet. As this loop forms one edge of the FK506-binding pocket of hFKBP12 and includes the Arg-43 residue that interacts with FK506 but not rapamycin, this may be related to the selective retention of FKBP25 on rapamycin-affinity matrices. Ten of the 14 hFKBP12 FK506-binding residues are conserved in hFKBP25 and bFKBP25 (Figure 2), the substitutions being R43T (immediately preceding the inserted loop), F47L, E55K and H88Q. All four substitutions occur, individually, in other FKBP domains known to have FK506-sensitive PPIase activity. The FKBP domain of hFKBP25 has 41–44% identity with hFKBP12, hFKBP12A, hFKBP13 and the first FKBP domain of hFKBP52 (Table 2), no higher than its 40–44% identities with *S. cerevisiae* FKBPs (Table 3). It has 34% identity with the first FKBP domain of the prokaryotic scFKBP33.

bFKBP25 has been demonstrated to have PPIase activity but, uniquely amongst FKBP PPIases, it prefers succ.AAPF.pNA to succ.ALPF.pNA as a substrate [90]. Its activity with this substrate exceeds that of hFKBP12 (although hFKBP12 has a higher activity with its preferred substrates, and only about 10% of its maximal activity with the substrate preferred by FKBP25) [90,91]. The PPIase activity is much more effectively inhibited by rapamycin (K_d 0.9 nM) than by FK506 (K_d 160 nM) [90], in agreement with the initial observation that the protein was much more readily retained by a rapamycin-affinity matrix.

Comparison of the FKBP25 sequence with those of other FKBPs suggested that the motif $KK(X)_7KK(X)_{26}KKKK$ present in the FKBP domain (the KKKK being in the inserted sequence) might serve as a potential nuclear localization signal unique to this member of the FKBP family [90,93] and that the five KXXK motifs present in the N-terminal domain and the first section of the FKBP domain might also contribute to this [91]. This prediction was confirmed by the demonstration that FKBP25 was mainly localized in the nuclear fraction in human and murine T-cells [94,95]. Unlike bFKBP12, bFKBP25 binds strongly to DNA–cellulose columns, and the N-terminal domain has been predicted to form an amphipathic helix–loop–helix that may be responsible for this [94]. The necessity for high salt concentrations for FKBP25 extraction from tissues would be compatible with a physiological association with DNA. The chromosomal location of the hFKBP25 gene has not been reported. Its transcript is processed to give a 1.8 kb mRNA [91].

Mammalian FKBP25 also contains consensus sequences for phosphorylation by casein kinase II, a protein kinase present in both the nucleus and the cytoplasm that translocates to the nucleus on cell activation and may phosphorylate a variety of nuclear and cytoplasmic substrates [90,95]. A glutathione S-transferase–hFKBP25 fusion protein when added to nuclear lysates from human or murine T-cells or HeLa cells co-precipitates with both casein kinase II and one of its principal substrates, the 100 kDa nucleolar phosphoprotein nucleolin [95]. Nucleolin is implicated in the regulation of ribosome biogenesis. At least *in vitro*, rapamycin did not inhibit the interaction of hFKBP25 with casein kinase II or nucleolin, or affect casein kinase II activity. The glutathione S-transferase–hFKBP25 fusion protein has been shown to be phosphorylated by casein kinase II when added to nuclear lysates from human or murine T-cells or HeLa cells [95], although a quantitative study using recombinant human casein kinase II found that less than 0.1% of the fusion protein became phosphorylated [96].

2.5 Mammalian FKBP52, an FKBP multidomain protein

The FKBP-domain PPIase variously called p59, FKBP59, HBI and hsp56 is referred to throughout this article as FKBP52, on the basis of its calculated molecular mass of 51.8 kDa [97]. Like most FKBP-family members it migrates on SDS/PAGE with a mobility that would predict a higher molecular mass (between 56 and 59 kDa).

The existence of a high-molecular-mass PPIase in human T-cells was first identified by its retention on FK506-and rapamycin-affinity matrices [37]. Subsequent work established that FKBP52 was identical to a protein that had already been described as a component of the unliganded glucocorticoid, androgen, oestrogen and progesterone receptor complexes in several mammalian tissues [98–100]. FKBP52, translated from a 2.4 kb mRNA, is widely expressed in human and rodent tissues, especially in tissues highly responsive to steroid hormones [97,101], and is also expressed at high levels in other vertebrate steroid-responsive tissues, such as chicken thymus and oviduct [49,102]. In chicken thymus [49] and the human lymphoblastoid IM9 cell line [100] the amount of FKBP52 present may approach or even exceed the amount of FKBP12 in the same tissue.

The hFKBP52 gene encodes a 459-amino-acid protein, including the cleaved N-terminal methionine [97]. The corre-

sponding rabbit [99] and murine [103] FKBP52s are one amino acid shorter. Each contains two, or perhaps three, identifiable FKBP domains at amino acids 31–138, 148–253 and 267–373, separated by short hydrophilic linker sequences [104]. Following the third FKBP domain is a calmodulin-binding consensus sequence, between amino acids 406 and 420 (Figure 4).

The N-terminal FKBP domain of hFKBP52 has 44–49% identity with other human FKBP PPIase domains (Table 2), and conserves 13 of the 14 FK506-binding residues of hFKBP12 (Figure 2). The second FKBP domain has only 29% amino acid identity with the first, 19–26% identity with other human FKBP domains and retains just five of the 14 FK506-binding residues. Although all of the substitutions except for F47G could be regarded as conservative, they include substitutions at residues Tyr-27, Val-56, Ile-57, Trp-60 and Phe-100 that are otherwise very strongly conserved in active PPIases. However, many other conserved residues important for the folding of the domain are retained. The third putative domain, predicted on the basis of hydrophobic cluster analysis and the conservation of short deletions also found in the second domain [104], has very limited sequence identity with other FKBP domains (Table 2). Only two of the 14 FK506-binding residues are conserved (although the majority of the substitutions appear conservative) and, of 14 glycine residues conserved in most FKBP domains, only one is preserved here. Lying largely within this putative third FKBP domain are three tetratricopeptide repeat motifs at amino acids 273–306, 322–355 and 356–389, each 34-amino-acid motif having two short amphipathic α-helices [105,106]. Comparison of the hFKBP52, oFKBP52 and mFKBP52 sequences shows 90–92% amino acid identity overall. However, the calmodulin-binding domain is completely conserved between the three, and the first FKBP domain and the tetratricopeptide repeats also appear more strongly conserved (96–97% and 97–98% identical respectively), with only 85–89% identity for the second FKBP domain and 80–84% identity for the remaining sequences. The N-terminal FKBP domain of chicken FKBP52 shows about 80% identity with the corresponding region of hFKBP52 [49].

As would be expected from its purification on FK506- and rapamycin-affinity matrices and the sequence of its N-terminal FKBP domain, mammalian FKBP52 binds both macrolides and exhibits drug-sensitive PPIase activity [97,100,107]. The PPIase specific activity of FKBP52 has been estimated at 10–90% that of hFKBP12 and, like FKBP12, it shows a preference for a bulky hydrophobic residue preceding the proline [97,107,108]. The affinity of hFKBP52 for an FK506-affinity matrix is weaker than that of hFKBP12 [98], and measured K_d and K_i values for FK506, ascomycin and rapamycin are in the 10–100 nM range [49,97,100,107]. There is a single high-affinity binding site for FK506 [100], and expression of the individual FKBP domains of oFKBP52 showed that the first FKBP domain had PPIase activity comparable with that of intact oFKBP52, that the second FKBP domain had PPIase activity significantly above background but less than 2% that of the first domain, and that the putative third domain had no detectable PPIase activity [108]. However, the second FKBP domain includes an ATP/GTP-binding site, probably in the modified FK506-binding pocket [99,104,109].

No more than a small fraction of FKBP52 is found free in mammalian cells. It was initially identified as a component of unliganded steroid hormone receptor complexes, along with the heat shock proteins hsp90 and hsp70 and a number of other proteins which are thought to be required to maintain the correct receptor conformation for steroid binding. When the steroid is available the heat shock proteins and FKBP52 dissociate from the receptor. However, stoichiometric studies have established that hsp90 and hsp70 are much more abundant than FKBP52, which is in turn more abundant than the steroid receptor. The primary association of FKBP52 is with hsp90 [110,111], mediated through its tetratricopeptide repeats [106]. It appears that the majority of cellular FKBP52 is associated with this protein, or with both hsp90 and hsp70, and that it is the hsp90 and not FKBP52 that associates directly with the steroid receptor. Stoichiometric studies suggest that only about half of the steroid receptor complexes formed in the rabbit reticulocyte lysate experimental system contain FKBP52; furthermore, receptor complexes purified using anti-FKBP52 lack another PPIase, Cyp40, that also contains tetratricopeptide repeat motifs and is associated with such complexes [110,112]. It is not at present clear whether the remaining receptor complexes contain an alternative PPIase, such as Cyp40 or FKBP54 (see section 2.6), or whether the presence of a PPIase is not essential.

ATP is not required for the FKBP52–hsp90 association but is necessary for the formation of the full steroid receptor complex [111]; hsp90 has ATPase activity [113]. The FK506-binding site of the first FKBP domain of FKBP52 appears not to be involved in the association with hsp90 or the steroid receptor, as both FKBP52–hsp90 and FKBP52–hsp90–steroid-receptor complexes can bind FK506 or rapamycin or be purified by affinity chomatography on FK506-matrices without dissociation [100,114]. However, a 63 kDa protein present in FKBP52-containing complexes and normally retained together with hFKBP52 on immunoaffinity columns was lost on preincubation of cytosol with FK506 or rapamycin, suggesting that this protein may normally be associated with the complex via the FKBP52 FK506-binding site [114].

Although the majority of FKBP52 in mammalian cell extracts is found in the cytosolic fraction, immunofluorescence of intact cells has identified prominent nuclear staining, with the same pattern of distribution as for hsp90 and steroid receptors [101,115,116]. This suggests that these components are only loosely bound within the nucleus and are released into the cytosolic fraction on cell lysis. Neither FK506 nor corticosteroids affected the staining pattern observed. However, there is also fibrillar staining of both FKBP52 and hsp90 in the cell cytoplasm, suggesting co-localization with microtubules. A microtubule-associated protein MAP-1C complex (a cytoplasmic form of dynein intermediate chains) co-absorbs with FKBP52 from cell cytoplasm, but tubulin does not [115]. Non-nucleated reticulocytes also contain oFKBP52. No nuclear localization sequence has been identified in FKBP52 itself.

In initial studies neither enhancement nor inhibition of glucocorticoid receptor function by FK506 was observed [112]. FK506 binding to FKBP52 in the untransformed glucocorticoid receptor complex had no effect on steroid binding, nor did it stabilize or destabilize the receptor–hsp90 complex *in vitro*, or affect steroid-mediated nuclear transfer of the receptor or steroid-mediated transcriptional enhancement of a reporter gene in intact cells. However, more detailed investigation showed that, although FK506 had no effect in the absence of steroid hormone or in the presence of maximally effective hormone concentrations, addition of high concentrations of FK506 or rapamycin strongly potentiated the effects of suboptimal concentrations of glucocorticoid on the expression of a glucocorticoid-inducible reporter gene in L929 cells [117]. This effect was associated with increased nuclear translocation of the glucocorticoid receptor. High macrolide concentrations (over 1μM, with an optimal effect at 10μM) were required for this effect, probably reflecting a requirement to saturate the high-affinity FKBP12 prior to any effect on FKBP52 (or other FKBP-domain proteins present in steroid receptor complexes; see section 2.6). The steroid-binding

capacity of non-activated progesterone receptors has also been reported to increase when rabbit uterus cytosol was incubated with FK506 or rapamycin [118].

Apparently similar results were obtained in yeast co-transfected with the human progesterone receptor form B and a progesterone-responsive reporter gene [119]. FK506 (10μM) alone did not induce reporter gene activity or have any effect on induction in the presence of high hormone concentrations, but it decreased 10-fold the hormone concentration needed for maximum reporter gene induction. This effect was maintained in a yeast strain deficient in yFKBP12, so was presumably mediated by the yeast homologue of FKBP52. In similar systems the authors were unable to demonstrate any effect of FK506 on the expression of androgen- or oestrogen-mediated transcription of reporter genes expressing the corresponding human receptors.

FKBP52 complexes with hsp90 and hsp70 have also been reported to be associated with a number of other proteins, including actin and a variety of protein kinases [114,120], and the activities of the protein kinases may be regulated by the association. A well-studied system is the haem-regulated eIF-2α kinase present in reticulocyte lysates which is normally retained in an inactive form by its association with this complex. In the absence of haemin or the presence of elevated levels of denatured protein, the eIF-2α kinase is displaced from the complex, becomes active and down-regulates protein synthesis at the initiation step [120–122].

FKBP52 is phosphorylated in the cell, and is found as a series of isoforms with different pI values [37,100,123]. All the isoforms bind FK506 [101]. In the hFKBP52 sequence 12 potential protein kinase phosphorylation sites have been identified: five for casein kinase I, four for casein kinase II, two for proline-dependent protein kinase and one for calmodulin-dependent protein kinase II or S6 kinase II [97]. Casein kinase II has been found associated with hsp90 in cell lysates, and this association enhances its activity. Highly purified oFKBP52 prepared by affinity chromatography (but not recombinant oFKBP52) can be phosphorylated *in vitro* by a co-purifying cation-dependent protein kinase activated by Ca^{2+}/calmodulin, but this kinase activity is not inhibited by known casein kinase II inhibitors and was unable to phosphorylate known casein kinase II substrates [109]. It is not yet clear whether the different phosphorylation states are correlated with the protein complex with which the FKBP52 is associated or with its intracellular localization.

2.6 Other mammalian FKBPs and FKBP-domain proteins

Most of the major proteins present in lysates of Jurkat T-cells or bovine thymus that are reported to be retained by FK506- or rapamycin-affinity matrices [37] have now been accounted for, the principal exception being an 80 kDa phosphoprotein present in Jurkat cells. However, several of the larger number of proteins that cross-react with antibodies specific for FKBP-domain peptides [38] remain to be explained. It seems clear that there are at least some further FKBPs and FKBP-domain-containing proteins in mammalian cells that await characterization.

Progesterone receptor complexes purified from chick oviduct have been shown to contain two distinct FKBP-domain proteins, p50 and p54, both containing peptides homologous to oFKBP52 [102,124]. Comparison of the peptide sequences and antibody cross-reactivity suggests that p50 is the chick homologue of hFKBP52 (80 % amino acid identity in the peptides isolated), while p54 is a related but distinct chick FKBP-domain protein (60 % amino acid identity). Antibodies that recognize p54 but not p50 cross-react with a previously unidentified 55 kDa protein present in steroid receptor complexes from a variety of mammalian sources. Both p50 and p54 are predominantly present in the cytosol as higher-molecular-mass complexes and dissociate from progesterone receptor complexes when the hormone is added, but p54 dissociates more readily *in vitro*. Both are present as a range of isoforms, probably reflecting different phosphorylation states. However, unlike FKBP52 and p50, p54 is retained on FK506-affinity matrices only at high salt concentrations, when the steroid receptor complexes are dissociated, suggesting both that p50 and p54 are present in different complexes and that p54 has an FK506-binding site that is not accessible in the complexed form. p54 is also more abundant in brain than in oviduct, and is selectively dissociated from the complex by *N*-ethylmaleimide. The human p54 homologue has slightly greater mobility on SDS/PAGE than FKBP52, and is detected in both nuclear and cytosolic extracts.

Hsp90 will associate with denatured proteins in the cytoplasm, and also with the apparently native or newly synthesized forms of many different proteins. In particular many protein kinases encoded by retroviral oncogenes are found complexed with hsp90 and an associated 50 kDa phosphoprotein [125]. The presence of PPIases in such complexes has not been fully investigated, but in addition to its known association with FKBP52 and the mammalian homologue of chick p54, hsp90 has been shown to form an ATP-dependent complex with CypA [113]. It has recently been reported that the v-raf protein kinase expressed in transfected rat fibroblasts and found in such a complex also contains an FKBP-family protein, identified by its ability to bind radiolabelled FK506 with a K_d of 80 nM [126]. The FK506 binding was competitively inhibited by similar concentrations of unlabelled FK506 or rapamycin, but was unaffected by CsA. Purified v-raf, hsp90 and 50 kDa phosphoprotein were all unable to bind FK506. The complexes did not contain identifiable amounts of FKBP12, FKBP52 or the rat homologue of the chicken p54 FKBP-domain protein, while the similar affinities for FK506 and rapamycin argue against FKBP13 or FKBP25. The possibility that the recently discovered FKBP12A is involved was not considered, but its K_d for FK506 would appear to be too low, suggesting that a novel FKBP-family PPIase is responsible.

Another multi-FKBP-domain protein is encoded by a gene on mouse chromosome 11 [127,127a]. This gene encodes a protein, FKBP65, with four FKBP domains, three of which retain eight or nine of the 14 FK506-binding residues of hFKBP12 and most of the other highly conserved residues. These domains contain the substitutions V56L or V56I and W60M that are not encountered in other active PPIases, but the protein has PPIase activity [127a].

Donelly and Soldin [128–130] have isolated a 52 kDa protein from cytosolic extracts of human lymphoid cells and calf thymus that is so far unique in binding CsA, FK506 and rapamycin at high affinity. The purified protein has no intrinsic protein kinase activity, but it decreased protein phosphorylation by protein kinase A, perhaps by acting as a phosphatase. The 52 kDa protein did not affect protein phosphorylation by protein kinase C or p56[lck] tyrosine kinase, and its effect on protein kinase A phosphorylation was unaffected by CsA, FK506 or rapamycin. It had very low or no PPIase activity with a range of peptide substrates. Three peptides isolated from it by proteolytic digestion and sequenced had no identified similarity with any other PPIase or any known protein.

Soldin and co-workers have also reported the purification of a low-abundance 14 kDa FK506- and rapamycin-binding protein from calf thymus cytosol that has properties quite different from those of bFKBP12 and bFKBP12A [131]. The protein binds [³H]dihydroFK506 at a single high-affinity binding site with a K_d

of 2 nM, and competition studies indicated a K_d for rapamycin of about 40 nM, distinguishing it from other FKBP-family members that invariably prefer rapamycin. The protein does not bind CsA and has no detectable PPIase activity with any substrate tested. Partial sequence information that is available excludes identity with bFKBP12 or bFKBP12A, and showed no similarity with any other known protein sequence. The cardiac (type 2) form of the ryanodine receptor also co-purifies with a protein that cross-reacts with antisera to the N-terminal sequence of FKBP12 but which migrates more slowly on SDS/PAGE, with an apparent molecular mass of 16–17 kDa [132]. The possibility that this could be FKBP12A, whose N-terminal sequence resembles that of FKBP12, has not been excluded.

3 FKBPS IN OTHER EUKARYOTES

Cyclophilins have been reported in a wide range of eukaryotes, including yeast and other fungi (section 3.1), plants (section 3.2) and a range of invertebrates (section 3.3), and cyclophilin-domain proteins are known in yeast. It is widely assumed that FKBPs will prove equally universal, although the evidence is as yet fragmentary except in yeast and fungi.

3.1 FKBPs and FKBP-domain proteins in yeast and fungi

The yeast *S. cerevisiae* has been shown to express at least five cyclophilin genes. These are a gene on chromosome IV that encodes a 162-amino-acid CypA found in the cytoplasm [133]; a gene on chromosome XIII that encodes a 182-amino-acid cyclophilin with a cleaved 20-amino-acid amphipathic leader sequence, whose mature protein is expressed in the mitochondria [134,135]; a gene on chromosome VIII encoding a 205-amino-acid cyclophilin with a 20-amino-acid hydrophobic signal sequence that enters the ER and may be secreted [136,137]; a gene on chromosome XII encoding a 225-amino-acid cyclophilin with a 22-amino-acid hydrophobic signal sequence and a C-terminal -HDEL sequence that causes its retention in the pre-Golgi ER [138]; and a gene on chromosome III encoding a 318-amino-acid protein with a cyclophilin-like domain [139]. This last protein also has an N-terminal hydrophobic signal sequence and a second hydrophobic sequence near the C-terminus that may serve as a membrane anchor. Its cyclophilin-like domain has a 12-amino-acid insert and lacks a number of residues that are normally highly conserved, including a key tryptophan residue necessary for CsA binding. In addition to these cyclophilins, *S. cerevisiae* also has two FKBPs and at least two FKBP-domain proteins: yFKBP12 expressed in the cytoplasm, yFKBP13 present within a membrane-bound compartment, yFKBP46 found in the nucleolus, and a homologue of FKBP52 complexed with hsp90 and other proteins in the cytoplasm and nucleus.

S. cerevisiae yFKBP12 is, like its mammalian equivalent, an abundant cytoplasmic protein. It is encoded by a single gene without introns on the left arm of chromosome XIV [140,141]. It has 114 amino acids and a predicted molecular mass of 12.1 kDa (Figure 5). Disregarding the additional six amino acids at the N-terminal end, it has 44–56 % amino acid identity with the human FKBP domains, the highest identity being with hFKBP12 (Table 3). Of the 14 FK506-binding residues of hFKBP12, 12 are conserved in yFKBP12, the exceptions being E55Q and H88F (Figure 2).

S. cerevisiae yFKBP12 binds FK506, rapamycin and ascomycin, can be purified on ligand-affinity matrices and has PPIase activity comparable with that of hFKBP12 [10,141,142]. The substrate specificity of yFKBP12 resembles that of hFKBP12 in preferring a hydrophobic residue preceding the proline,

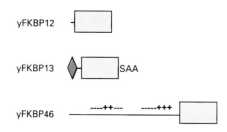

Figure 5 *S. cerevisiae* **FKBPs and FKBP-domain proteins**

Key as in Figure 4.

although it has greater tolerance for alanine and valine residues in this position [10,143]. The crystal structure of yFKBP12 complexed with FK506 is quite similar to that of hFKBP12–FK506, with the additional six residues at the N-terminal end forming a type II β-turn and a short segment that extends the β-sheet to six strands [144]. The conformations adopted by FK506 in the two complexes are similar. However, yFKBP12 (pI 4.8) is a much more acidic protein than hFKBP12 (pI 8.9), and the N-terminal amino group of the mature protein is blocked [10].

Despite its abundant expression, the yFKBP12 gene is not essential for the normal growth of *S. cerevisiae*, and addition of FK506 at concentrations below 1 μM does not affect proliferation in most strains. Proliferation is inhibited by 10–100 nM rapamycin [142,145], although it is not clear why such high concentrations are required given the reported yFKBP12 K_d values of 0.5 nM for rapamycin and 0.9 nM for FK506 [143]. Rapamycin-resistant strains are easily isolated, and in most cases such resistance is recessive in diploids and is due to the loss or instability of yFKBP12 expression [142,146]. Disruption of the yFKBP12 gene reduces FK506 binding by over 99%, showing that this is the principal high-affinity FKBP in *S. cerevisiae* [147]. Expression of hFKBP12 in such rapamycin-resistant yeast restores sensitivity to rapamycin, confirming that the two proteins are functionally equivalent [142].

The second *S. cerevisiae* single-domain FKBP, yFKBP13, is encoded by a unique gene on the right arm of chromosome IV [143,148]. The gene encodes a protein of 135 amino acids (calculated molecular mass 14.5 kDa), including a cleaved 17-amino-acid N-terminal hydrophobic signal sequence directing secretion into the ER (Figure 5). The mature protein has 118 amino acids and a calculated molecular mass of 12.4 kDa, although on SDS/PAGE it runs with an apparent molecular mass of 15 kDa. On cell fractionation over 90% of yFKBP13 is found in the membrane fraction. Its overall abundance has been estimated at 20% that of yFKBP12.

The amino acid sequence of the yFKBP13 FKBP domain shows slightly higher identity with hFKBP13 than with yFKBP12 or other human FKBP domains (Table 3). It shares with hFKBP13 the unique pair of conserved cysteine residues (Cys-13 and Cys-68). However, the C-terminal -RTEL motif present in hFKBP13 is replaced by a short -SAA extension in yFKBP13. It has PPIase activity comparable with that of yFKBP12, and a particularly strong preference for leucine or phenylalanine residues preceding the proline residue in substrate peptides. It binds rapamycin more strongly than FK506, but binds both with 10–20 times lower affinity than does yFKBP12 [143]. However, it cannot mediate the anti-proliferative action of rapamycin when yFKBP12 is not available, and mutants in which either the yFKBP13 gene or both the yFKBP13 and yFKBP12 genes are

disrupted grow normally under a wide range of conditions. Disruption of the yFKBP13 gene alone does not affect the sensitivity of *S. cerevisiae* to either 10–100 nM concentrations of rapamycin or high (1–50 μM) concentrations of FK506 [143,149].

The *S. cerevisiae* FKBP-domain protein yFKBP46, encoded by a gene on chromosome XIII, has an FKBP domain at the C-terminal end of a protein containing 411 or 413 amino acids and with a predicted molecular mass of 46.5 or 46.8 kDa [150–152] (the gene sequenced in [152] has a sequence of nine successive GAA codons in the N-terminal domain, while those in [150] and [151] have only seven). The gene is transcribed to give a 1.5 kb mRNA. The N-terminal end of the mature protein is blocked, and the region preceding the FKBP domain contains three regions rich in acidic residues and two regions rich in basic residues (Figure 5), which may account for the anomalous migration of yFKBP46 on SDS/PAGE with an apparent molecular mass of 70 kDa. This N-terminal domain makes yFKBP46 one of a number of yeast proteins that can recognize nuclear localization sequences *in vitro*, and has significant identity with mammalian nucleolin (50% identity with Chinese hamster nucleolin over residues 209–270). The protein is phosphorylated *in vivo* within the N-terminal domain, and the sequence shows a number of potential casein kinase II phosphorylation sites.

The C-terminal FKBP domain of yFKBP46 has 45% amino acid identity with yFKBP12 and yFKBP13, and 35–44% identity with human FKBP domains (Table 3). Of the 14 FK506-binding residues of hFKBP12, 12 are conserved, the exceptions being R43– and H88L (Figure 2). Purified recombinant yFKBP46 and its excised C-terminal domain have equivalent PPIase activities and prefer peptide substrates with a hydrophobic residue preceding the proline, but activity has been estimated as only 5–15% of that reported for hFKBP12 [150–152]. PPIase activity is abolished by 1 μM rapamycin and strongly inhibited by 1 μM FK506, and the protein is retained by an ascomycin-affinity matrix. Overexpression in the cytoplasm of the excised yFKBP46 FKBP domain restores the sensitivity to rapamycin of *S. cerevisiae* containing a disrupted yFKBP12 gene, although a 10-fold higher rapamycin concentration is required. Intact yFKBP46 is not effective, probably because it is not expressed in the right cellular compartment [152]. Heavy overexpression of full-length yFKBP46 or its excised N-terminal domain inhibits yeast growth, and the inhibition caused by the intact protein can be antagonized by rapamycin in yeast lacking yFKBP12.

Disruption of the yFKBP46 gene, alone or simultaneously with the yFKBP12 and yFKBP13 genes, causes no observable effect on growth under a range of conditions, mating ability, sporulation or sensitivity to rapamycin or FK506 [150–152]. Immunofluorescence shows that yFKBP46 is a relatively abundant protein localized to the nucleoli, and the difficulty in extracting it even at high NaCl concentrations suggests a tight association with other elements within the nucleolus. The nuclear localization, apparent preference for rapamycin over FK506, homology/association with nucleolin and presence of an N-terminal domain containing many charged residues might suggest parallels between yFKBP46 and mammalian FKBP25. However, Table 3 shows that the identity between the FKBP domains of yFKBP46 and hFKBP25 is lower than that between yFKBP46 and hFKBP12, and there is little evidence for conservation of other distinctive features between the two sequences.

The second FKBP-domain protein present in *S. cerevisiae* is probably the yeast homologue of hFKBP52. A rapamycin-affinity matrix retained complexes from yeast cytosol that could be eluted with free rapamycin and contained the *S. cerevisiae* homologues of hsp90, hsp70 and a number of other proteins including a 62 kDa protein tentatively identified as the hFKBP52

Table 4   Amino acid sequence identity between FKBP12s

FKBP12 prefixes used: h, human; y, *Saccharomyces cerevisiae*; ca, *Candida albicans*; nc, *Neurospora crassa*; nm, *Neisseria meningitidis*; sc, *Streptomyces chrysomallus*.

	Identity (%)				
	yFKBP12	caFKBP12	ncFKBP12	nmFKBP12	scFKBP15
hFKBP12	56	46	44	49	40
yFKBP12	–	59	43	47	42
caFKBP12	–	–	50	43	41
ncFKBP12	–	–	–	48	48
nmFKBP12	–	–	–	–	41

homologue [114]. The *S. cerevisiae* hsp90 homologue bound to an antibody-affinity matrix also retained several proteins from yeast cytosol including the band at 62 kDa, but also two bands at 53 kDa and 54 kDa recognized by polyclonal anti-hFKBP52 antibodies [113]. The proteins associated with the *S. cerevisiae* hsp90 homologue also include a probable yeast homologue of the Cyp40 also found associated with hsp90 in mammalian cells [153].

There have been few detailed studies on the presence of FKBPs and FKBP-domain proteins in other yeasts and fungi. However, FKBP12 homologues have been cloned and sequenced from the yeast *Candida albicans*, whose proliferation is strongly inhibited by rapamycin [154], and the fungus *Neurospora crassa* [155]. A 12 kDa FK506-sensitive PPIase that prefers leucine to alanine preceding the proline residue in peptide substrates has been purified from the CsA-producing fungus *Tolypocladium inflatum* [156].

The *C. albicans* caFKBP12 gene encodes a 124-amino-acid protein that has three additional residues added to the N-terminal end, two added to the C-terminal end and an extra 11 amino acids inserted into the loop between the third and fourth strands of the β-sheet. The protein has not been characterized, but it has 59% amino acid identity with yFKBP12 and 41–50% amino acid identity with other known FKBP12 homologues from prokaryotes and eukaryotes (Table 4) and conserves 12 of the 14 hFKBP12 FK506-binding residues (the substitutions being E55Q, found also in other fungal FKBP12 homologues, and H88I) (Figure 2). The *N. crassa* ncFKBP12 has six additional amino acids at the N-terminus and five at the C-terminus, but otherwise aligns perfectly with hFKBP12. It also has 41–50% identity with other prokaryotic and eukaryotic FKBP12s (Table 4) and conserves 11 of the 14 FK506-binding residues of hFKBP12 (Figure 2). The substitutions are F47L, also found in two prokaryotic FKBP homologues, E55Q and H88V. ncFKBP12 is an abundant cytoplasmic protein with FK506-sensitive PPIase activity that is 20–30% that of ncCypA with succ.AAPF.pNA or RNase T1 as substrates. Preliminary studies suggested that *N. crassa* mitochondria might also contain an FKBP [155].

Indirect evidence for the occurrence of FKBPs in other yeasts and fungi comes from reports of their sensitivity to FK506 and rapamycin [157]. Rapamycin has anti-proliferative activity at low concentrations against some yeasts, while FK506, normally inactive against yeast, is a potent inhibitor of the growth of fungi such as *Aspergillus fumigatus* and *Fusarium oxysporum*. The precedents suggest that anti-proliferative activity at such low concentrations is likely to be mediated by FKBP12.

Organisms such as *Schizosaccharomyces pombe* and *N. crassa* that are resistant to micromolar concentrations of rapamycin or FK506 (more than 100 times higher than those needed to inhibit sensitive mammalian cells) may still be affected by yet higher concentrations in the range 10–100 μM [158,159]. The need for such high concentrations to cause inhibition could be related to efficient exclusion of the inhibitor, possession of an FKBP12 with much decreased affinity for the inhibitor, or the participation of an alternative target, perhaps a member of the FKBP family with much lower affinity for the inhibitor. At least in the case of *N. crassa* it has been shown that resistance may correlate with mutation of the gene encoding ncFKBP12. However, the sensitivity of some amino acid auxotrophic strains of *S. cerevisiae* to very high (10–100 μM) concentrations of FK506 is unaffected by inactivation of the yFKBP12, yFKBP13 or yFKBP46 genes and appears to involve an alternative target whose action may be required for normal amino acid uptake. Inhibition takes several hours to develop, and is perhaps due to an effect on the maturation of amino acid transporters which contain several conserved proline residues [143,147,149,152].

3.2   Plant FKBPs

The presence of cyclophilin-family PPIases has been demonstrated in the cytoplasm, mitochondria and chloroplasts of several species of plants. CypA homologues are present in the cytoplasm in all parts of the plant, and sequences that have been reported from *Zea mays* (maize), *Lycopersicon esculentum* (tomato), *Brassica napus* [160], *Phaseolus vulgaris* (kidney bean) [161], *Arabidopsis thaliana* [162–165] and *Oryza sativa* (rice) [166] all include a plant-specific 7-amino-acid insert within the cyclophilin domain. Plant CypAs are encoded by small gene families, but, whereas in mammals only a single CypA protein is found in the cytoplasm and the additional genes are believed to be pseudogenes, in plants more than one of these genes may be functional. Four different *A. thaliana* cyclophilin cDNA sequences are reported to encode CypA homologues of 169, 172, 172 and 174 amino acids that differ as much from each other as from hCypA. Similar conclusions were reached for the two *O. sativa* sequences, which encode proteins of 172 and 179 amino acids that are both expressed in roots and leaves.

Cyclophilins specifically expressed in the chloroplast stroma may be the most abundant PPIase present in the green parts of plants. The corresponding cDNA seqences from *Vicia faba* [167] and *A. thaliana* [163] encode proteins of 248 and 268 amino acids respectively, but these include substantial cleaved N-terminal sequences, 65 amino acids long in the case of *V. faba*, that are presumed to target the protein to its final location in the stroma. These chloroplast cyclophilins lack the 7-amino-acid insert found in the plant CypAs.

By contrast, the evidence for the presence of FKBPs in plants is sparse, and no plant FKBP sequences have yet been reported. The very low level of PPIase activity detected in the mitochondrial membrane fraction from *Pisum sativum* (pea) was partially inhibited by a high concentration of rapamycin (30 μM), but also completely inhibited by a lower concentration of CsA. This same fraction contained a 25 kDa protein that cross-reacted with antibodies to yFKBP12 [168]. No evidence for the presence of FKBP-family PPIases was detected in other mitochondrial or chloroplast fractions, although the PPIase assays were performed using succ.AAPF.pNA, with which many FKBPs show low activity.

By far the most convincing evidence for the presence of FKBPs in plants has come from studies by Luan et al. [169,170], who used FK506-affinity matrices to purify FKBPs from tissues of *V. faba*. They identified proteins with apparent molecular masses of 55, 25, 18, 16 and 12 kDa that were specifically retained by the affinity matrix and eluted with FK506. The 55, 25 and 12 kDa proteins, all expressed at comparatively low levels, were suggested to be the plant homologues of FKBP52, FKBP25 and FKBP12 respectively. The 12 kDa protein had an N-terminal peptide with 35% amino acid identity with the corresponding peptide of hFKBP12. It is tempting to speculate that the low level at which this putative plant FKBP12 homologue is expressed may be related to the multiple expression of CypA isotypes.

However, the most abundant plant FKBPs were the 16 kDa FKBP, present mainly in chloroplasts and strongly up-regulated by exposure to light, and the 18 kDa FKBP, found in mitochondria in all parts of the plant. The N-terminal peptides of both had greater identity (33% for the 16 kDa protein) with hFKBP13 than with hFKBP12 [169,170]. The localization of these FKBPs within the chloroplasts and mitochondria is not yet established. All PPIase activity in chloroplast and mitochondrial stromal preparations was CsA-sensitive [163,168], but in both studies the use of succ.AAPF.pNA as substrate may have compromised detection of FKBP-catalysed activity.

3.3   FKBPs and FKBP-domain proteins in invertebrate animals and protozoa

The presence of members of the cyclophilin family has been reported in a variety of eukaryotic invertebrates and protozoa, including the insect *Drosophila melanogaster* [36,171,172], the tapeworm *Echinococcus granulosus* [173], the slime mould *Dictyostelium discoideum* [174], and the protozoa *Schistosoma japonicum* [175], *Toxoplasma gondii* [176] and *Plasmodium falciparum* [177]. However, the widespread assumption that CypA homologues are universally abundant is as yet supported by only very limited evidence from these organisms, and fully sequenced homologues are evident only in *D. melanogaster* [36] and *D. discoideum* [174]. Putative homologues in other species have ill-defined N-termini, and in *T. gondii* neither of the two major CsA-sensitive PPIases has the characteristics of CypA. These are an 18.5 kDa cyclophilin that has an N-terminal hydrophobic signal sequence, and a 20 kDa protein that may be synthesized as the C-terminal domain of a larger protein and which is found primarily in the 10 000 *g* pellet on cell fractionation [176]. The slime mould *D. discoideum* also has an ORF with identity with that of parvulin [13].

The general assumption that FKBPs and FKBP-domain proteins will also have a universal distribution is at present supported by even more limited evidence from this class of organisms. The colonial tunicate *Botryllus schlosseri* has an FKBP13 homologue, referred to in section 2.3 [87]. CsA, FK506 and rapamycin all inhibit the growth of malarial parasites *in vitro*, although quite high concentrations of FK506 and rapamycin are needed (IC$_{50}$ values 1.9 and 2.6 μM respectively) [177]. However, as the most effective cyclosporin analogue against parasite growth showed little inhibitory activity against *Plasmodium falciparum* PPIase but is a strong inhibitor of the P-glycoprotein responsible for multi-drug resistance in mammalian cells, and as no FK506-or rapamycin-sensitive PPIase activity could be detected in this organism, the involvement of PPIases in this inhibition remains unclear. The only other unambiguous evidence for the expression of characterized FKBP family members is in insects.

Sf9 cells from the insect *Spodoptera frugiperda* bind similar amounts of FK506 to human T-cells [10], although a search for an FKBP12 homologue using cDNA probes and antisera specific for mammalian FKBP12 was unsuccessful [77]. The same cell line has been shown to express an FKBP-domain protein, sfFKBP46, that has been cloned and characterized [178]. This

Table 5   Amino acid sequence identity between insect (sfFKBP46 and dmFKBP39) FKBP domains and *Saccharomyces cerevisiae* and human FKBP domains

	Identity (%)	
	sfFKBP46	dmFKBP39
sfFKBP46	–	65
yFKBP12	44	48
yFKBP13	42	47
yFKBP46	51	50
hFKBP12	43	47
hFKBP12A	47	46
hFKBP13	39	43
hFKBP25	37	36
hFKBP52 domain 1	45	50

gene encodes a 412-amino-acid protein with a calculated molecular mass of 45.8 kDa and an FKBP domain at its C-terminal end. The N-terminal domain contains strongly acidic and strongly basic regions, including a 24-amino-acid sequence containing only aspartate and glutamate residues. This domain gives it the ability to bind to DNA–cellulose, and the protein is expressed in the nucleus. However, although the most obvious potential nuclear localization sequences are in the basic regions of the N-terminal domain, a truncated sfFKBP46 that lacks the FKBP domain is expressed in the cytoplasm. sfFKBP46 binds to basic nuclear proteins, as does the N-terminal domain. It is a phosphoprotein, phosphorylated by a co-purifying Mg^{2+}-dependent protein kinase. Several potential casein kinase II phosphorylation sites are present in the sequence.

Comparison of the FKBP domain of sfFKBP46 with that of yFKBP46 shows that two of three single amino acid deletions needed to retain the alignment with hFKBP12 and the same 12 of the 14 FK506 binding residues found in hFKBP12 are conserved in both (Figure 2). sfFKBP46 binds FK506 and has PPIase activity which is reported to be comparable with that of hFKBP12 but which is inhibited only by about 50% by 5 μM FK506. The FKBP domain shows higher identity with that of yFKBP46 than with yFKBP12 or yFKBP13 (Table 5). The identity with the hFKBP25 FKBP domain is actually lower than that with other human FKBP domains.

A recently sequenced *D. melanogaster* cDNA encodes a similar but shorter protein of 357 amino acids, dmFKBP39, that also has an N-terminal domain with strongly acidic and basic regions and a C-terminal FKBP domain [179]. Its FKBP domain has 65% amino acid identity with the corresponding sfFKBP46 domain and 50% identity with that of yFKBP46. The 14 FK506-binding residues are identical to those of sfFKBP46 and yFKBP46 except at the non-conserved residue 88 (Figure 2) but, uniquely, it has a hydrophobic leucine residue at position 40. All other FKBP domains have serine, threonine or asparagine at this position, except for sfFKBP46 which has cysteine. Like sfFKBP46, dmFKBP39 has lower identity with hFKBP25 (36%) than with the other human FKBP domains (Table 5). The 1.4 kb mRNA transcript was expressed at all developmental stages and in all tissues, but at the highest levels in the early embryo.

4   FKBPS AND FKBP-DOMAIN PROTEINS IN PROKARYOTES

Three families of PPIases have been reported in prokaryotes. Two cyclophilins are present in *E. coli*, a 164-amino-acid CypA homologue in the cytoplasm [180] and a 190-amino-acid cyclophilin with a 24-amino-acid N-terminal cleaved hydrophobic signal sequence expressed in the periplasm [181,182]. Both are active PPIases but are 1000-fold less sensitive to CsA than hCypA, due mainly to the replacement of a key tryptophan residue by phenylalanine [180,182–185]. Cytoplasmic cyclophilin homologues or gene sequences predicted to encode them have also been reported in *Bacillus subtilis* [186], *Streptomyces chrysomallus* [187], *Synechococus* spp. [188], *Bacillus stearothermophilus* (where the enzyme is salt- and heat-stable) [189] and the halophilic archaeum *Halobacterium cutirubrum* [190]. Genes encoding predicted homologues of the periplasmic cyclophilin have been reported in *Salmonella typhi* [191] and *Acinetobacter calcoaceticus* [192]. The *B. subtilis*, *S. chrysomallus* and *H. cutirubrum* CypAs have been characterized and shown to be considerably more sensitive to CsA than the *E. coli* PPIases. The *S. chrysomallus* CypA retains the key tryptophan residue found in hCypA, but in the *B. subtilis* enzyme the corresponding residue is histidine. The *A. calcoaceticus* periplasmic cyclophilin gene is not essential for normal growth, at least in rich media [192].

E. coli cytoplasm also contains a distinct PPIase with a sequence indicating that it belongs to neither the cyclophilin nor the FKBP family; also, it is inhibited by neither CsA nor FK506 [11,12]. This PPIase, parvulin, is a small 10.1 kDa protein containing only 93 amino acids including a cleaved N-terminal methionine, but it appears to exist in the cytoplasm as a monomer. The sequence indicates that the protein is very hydrophilic, with 28% charged residues, and it is predicted to contain a much higher proportion of α-helix than either cyclophilins or FKBPs. The k_{cat}/K_m value for its PPIase activity is comparable with those of cyclophilins and higher than those of even the most active FKBPs, but its substrate specificity more closely resembles those of FKBPs, preferring a hydrophobic amino acid preceding the proline residue.

Comparison of the parvulin amino acid sequence with others in the databases has revealed a series of other larger proteins that contain domains with similarity to parvulin [12,13]. The closest identity (41%) was with a domain of the *B. subtilis* protein PrsA, but lower levels of similarity (20–30% amino acid identity) were found with domains in *E. coli* SurA, *Lactococcus lactis* PrtM, an unidentified *B. subtilis* ORF and the NifM gene products from a trio of nitrogen-fixing bacteria. SurA contains two parvulin-homologous domains. Amino acid alignment identified motifs that were conserved throughout this third PPIase family. PrsA, PrtM and SurA all have N-terminal signal sequences and are expressed on the outer surface of the cytoplasmic membrane or in the periplasm, whereas NifM is associated with the large nitrogenase multi-subunit complex in the bacterial cytoplasm.

Although the initial report of the hFKBP12 sequence noted strong similarity with a previously sequenced *Neisseria meningitidis* ORF [40], the few subsequent reports of bacterial FKBP12 homologues have been confined to *Neisseria* spp. [193,194] and *Streptomyces* spp. [195]. The *N. meningitidis* nmFKBP12 has 109 amino acids that align precisely to hFKBP12 except for a single insertion at the C-terminal end (Figure 6). It has 40–50% sequence identity with homologues in mammals, fungi and *Streptomyces chrysomallus* (Table 4). Most highly conserved amino acids are present in nmFKBP12, including 10 of the 14 hFKBP12 residues identified as interacting with FK506. The four differences are F47L, E55Q, A82G and H88A, all substitutions found in other known PPIases (Figure 2). The gene is expressed as an active FK506-sensitive PPIase present in the bacterial cytoplasm [193]. FKBP12s with over 97% amino acid sequence identity are present in 14 different *N. meningitidis* strains and 11 commensal *Neisseria* species, but no homologous

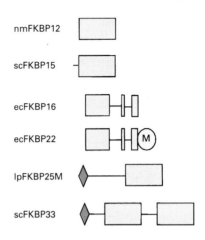

Figure 6 Prokaryotic FKBPs and FKBP-domain proteins

Key as in Figure 4.

gene was found in any of 11 *N. gonorrhoeae* strains tested or in *Moraxella catarrhalis* [194].

Cytoplasmic FKBPs slightly larger than FKBP12s have been purified from three streptomycetes, *Streptomyces chrysomallus*, *S. hygroscopicus* (which produces rapamycin) and *S. hygroscopicus* subsp. *ascomyceticus* (which produces ascomycin) [195]. *S. chrysomallus* scFKBP15 has been sequenced and contains 124 amino acids, including a 15-amino-acid N-terminal extension (Figure 6). scFKBP15 has 48% amino acid identity with ncFKBP12 and 40–42% identity with mammalian, other fungal and *N. meningitidis* FKBP12s (Table 4). Of the 14 hFKBP12 residues interacting with FK506, 11 are retained; the three substitutions (F47L, E55Q and H88A) are all identical to those found in nmFKBP12 (Figure 2), and all three streptomycete FKBPs have PPIase activity that is sensitive to both FK506 and rapamycin. Analysis of the distribution of of the FKBP inhibitor ascomycin in *S. hygroscopicus* subsp. *ascomyceticus* cultures showed that over 99.5% was secreted from the streptomycete mycelium into the medium, suggesting that the organisms producing such inhibitory macrolides have an efficient self-protection mechanism.

No simple FKBP homologue has been identified in *E. coli*. However, there is an ORF between the *E. coli lsp* and *dapB* genes that is predicted to encode a 149-amino-acid protein with some similarity to FKBP12s (ecFKBP16) [196], although it lacks the first strand of the β-sheet and contains two inserted sequences of 32 and 15 amino acids between the fourth and fifth β-sheet strands (Figure 6). Over the homologous sequences this *E. coli* ORF shows 25–31% amino acid sequence identity with mammalian, fungal and prokaryotic FKBP12s, the greatest identity being with hFKBP12. However, since only three of the 14 amino acids in hFKBP12 predicted to form the proline-binding pocket and interact with FK506 are conserved in this *E. coli* ORF and some of the substitutions are non-conservative (e.g. F37A, F47A and I57S), it is far from clear that the protein can be expected to possess PPIase activity or interact with FK506 or rapamycin (Figure 2). In addition, a G29L substitution would be predicted to insert the bulky leucine side chain into the FKBP-binding pocket [58]. No FK506- or rapamycin-sensitive PPIase has been detected in *E. coli*. An ORF predicted to encode a homologous protein is also present adjacent to the *lsp* gene in the closely related bacterium *Enterobacter aerogenes* [197].

An ORF of very similar organization to this putative *E. coli*

FKBP homologue has also been reported from *Pseudomonas fluorescens* (pfFKBP16) [198]. pfFKBP16 is predicted to encode a 150-amino-acid protein with only 36% overall identity with ecFKBP16, but with insertions and deletions of very similar length and identical position. It also retains only three of the 14 FK506-binding residues found in hFKBP12, although the substitutions are more conservative than in ecFKBP16, and it also has the G29L substitution found in ecFKBP16. The region of highest (60%) identity between ecFKBP16 and pfFKBP16 is the second insert.

E. coli have also been reported to synthesize a 196-amino-acid FKBP-domain protein encoded by the *slyD* gene (ecFKBP22) [199,200]. The N-terminal 150 amino acids encode an FKBP domain that, like the *E. coli* and *P. fluorescens* ORFs, lacks the first β-sheet strand and includes 32- and 15-amino-acid inserts at identical positions (Figure 6). This *E. coli* FKBP domain shows 29% amino acid identity with the corresponding region of ecFKBP16 and 23–28% amino acid identity with human, fungal and prokaryotic FKBP12s, the greatest identity being with *Streptomyces chrysomallus* scFKBP15. It retains seven of the 14 hFKBP12 residues that interact with FK506, and the differences are more conservative than those found in ecFKBP16 (Figure 2), but it also has the otherwise conserved Gly-29 substituted by a bulky valine residue. When the protein was overexpressed and purified no PPIase activity was detected with conventional substrates [199]. The C-terminal domain of about 50 amino acids contains an internal repeat and is extremely rich in potential metal-binding residues (15 histidine, 6 cysteine and 10 acidic residues). The purified protein is expressed in the cytoplasm. It binds Zn^{2+} and Ni^{2+} tightly with 1:1 stoichiometry and Cu^{2+} and Co^{2+} with lower affinity, but has only low affinity for Mg^{2+}, Ca^{2+}, Mn^{2+}, Fe^{2+} or Fe^{3+} [199]. Mutation of the *slyD* gene results in no obvious phenotype, except that it prevents the cell lysis normally induced by the bacteriophage ϕX174 E gene [200]. This property resides in the FKBP domain, as deletion of the C-terminal metal-binding domain does not prevent phage-induced lysis. *slyD* is no longer needed if the C-terminal region of the ϕX174 E gene is replaced by a sequence that permits its spontaneous oligomerization, suggesting that the *E. coli* FKBP-domain protein may act by facilitating this process.

A second family of FKBP-domain proteins expressed in prokaryotes is the FKBP25M family of membrane-located Mip proteins reported from some species of pathogenic bacteria that grow as intracellular parasites, such as *Legionella pneumophila*, *Legionella micdadei* and *Chlamydia trachomatis*. These prokaryotic Mip FKBP25Ms have a cleaved N-terminal signal sequence, an N-terminal domain predicted to include a long α-helix, and an FKBP domain at the C-terminal end (Figure 6). They contribute greatly to the survival of the bacteria within infected cells, probably by preventing phagosome–lysosome fusion [201,202]. The drugs FK506 and rapamycin reduce intracellular infectivity [203,204]. The Mip protein also assists *L. pneumophila* survival within the protozoa that are its natural host when growing in aquatic environments [205]. *L. pneumophila* Mip-specific probes revealed the presence of related genes in other intracellular parasites, including several other *Legionella* species, *Rochalimaea quitana* (the causative agent of trench fever), *Coxiella burnetti* (causing Q fever) and *Rickettsia* spp. [203,206].

Mip FKBP25M amino acid sequences have been reported for the 233-residue Mip of *L. pneumophila* (lpFKBP25M) [207], the 243-residue homologue from *L. micdadei* (lmFKBP25M) [206] and the 243-residue Mip protein from *Chlamydia trachomatis* (ctFKBP25M) [208]. An ORF from *Pseudomonas aeruginosa* encodes a homologous 209-residue protein (paFKBP25M) [209].

Amino acid identities between the Mip FKBP25Ms of different *L. pneumophila* strains and different *C. trachomatis* strains were very high, with over 99% overall within-species identity [208,210]. lpFKBP25M and lmFKBP25M have 84% identity in their FKBP domains and 77% identity over the remainder of the sequence if a short deletion following the signal sequence in lpFKBP25M is ignored. There is no significant similarity of the N-terminal domain to the α-helical N-terminal domain of hFKBP25. However, the FKBP domains of the other Mip proteins show between 29 and 41% amino acid identity, comparable with their similarities to hFKBP12 and the FKBP domain of hFKBP25. The FKBP domains of lpFKBP25M, lmFKBP25M and paFKBP25M all retain the same set of 10 of the 14 hFKBP12 FK506-binding residues, while ctFKBP25M retains eight of these (Figure 2). PPIase activity inhibited by high concentrations of rapamycin and FK506 has been demonstrated for lpFKBP25M and ctFKBP25M [204,211]; thus ctFKBP25M is the only known member of the FKBP family to have PPIase activity while lacking the conserved Ile-92 residue, here replaced by leucine. lpFKBP25M has typical FKBP substrate preferences but, unlike FKBP12, the active form is a homodimer, with the monomer having little or no enzymic activity [212].

The N-terminal domains of the Mip FKBP25Ms show lower levels of identity with each other (15–26% amino acid identity excluding the two *Legionella* species), although all have 25–30% residues with charged side chains and there is a clearly conserved FLX(EN)$_2$KXKXGVXXL motif preceding the FKBP domain. They are predicted to include a long continuous α-helix of up to 65 amino acids [203,206]. If the Mip FKBP25M proteins are anchored in the bacterial membrane at their N-terminal ends, this rigid helix may hold the FKBP domain away from the surface to interact with other structures. A suggested alternative function is that the long amphipathic helix may be responsible for the dimerization necessary for PPIase activity by the formation of two antiparallel helices [212].

In addition to its scFKBP15, *Streptomyces chrysomallus* has within the same operon a gene encoding a double FKBP-domain protein, scFKBP33 [213]. This protein contains 312 amino acids, including a 19-residue lipoprotein signal sequence predicted to be cleaved by signal peptidase II, leaving the protein attached to the cell membrane via a lipid linkage. Cell fractionation confirmed that the protein is expressed at a high level in the streptomycete cell membrane. It has two FKBP domains, encompassing residues 61–165 and 208–312 (Figure 6), that have 42% identity with each other. The first domain has 44% identity with scFKBP15 and 32–42% identity with the various human and yeast FKBP domains, while the second domain has in every case less similarity (32% to scFKBP15 and 25–32% to human and yeast FKBP domains). The second domain also showed non-conservative substitutions at a number of residues (e.g G20S, P46T, G52S, D61K) that are normally very strongly conserved in FKBP domains. The first FKBP domain of scFKBP33 retains 11 of the 14 hFKBP12 amino acids that interact with FK506, although one of the substitutions (E55M) is non-conservative (Figure 2). By contrast, the second FKBP domain retains only six of these amino acid residues and includes four non-conservative substitutions (R43T, F47Q, E55V and V56T). scFKBP33 shows FK506- and rapamycin-sensitive PPIase activity, with substrate specificity similar that of other FKBPs, although its specific activity is 100-fold lower than that of hFKBP12 or scFKBP15 and its inhibition requires 8-fold more rapamycin and 13 times more FK506 than for scFKBP15. Surprisingly, both domains, when expressed individually, have PPIase activity comparable with that of intact scFKBP33. The scFKBP33 protein is expressed in all growth phases, and related proteins are present in the membranes of all other streptomycetes investigated.

It is thus apparent that both FKBP12 homologues and FKBP-domain proteins are found in prokaryotes, with FKBP-domain proteins expressed both at the cell surface and in the cytoplasm. However, FKBP12 homologues may not be universal, as they are present in some species of *Neisseria* but absent in others, and only streptomycetes have so far been demonstrated to express CypA- and FKBP12-homologous PPIases simultaneously in their cytoplasm. The simultaneous expression of both CypA and parvulin in *E. coli* cytoplasm may make an FKBP12 redundant. The closest identified *E. coli* homologue, ecFKBP16, has sufficiently marked alterations in structure and key residues to make it very uncertain whether it will possess PPIase activity. In organization ecFKBP16 resembles the FKBP domain of ecFKBP22, which has been overexpressed and found not to manifest PPIase activity with conventional substrates. The Mip FKBP25M family appears so far to be restricted to bacteria living intracellularly in eukaryotic cells, and the FKBP33 family to streptomycetes. It is hardly surprising that no clear prokaryotic homologues of mammalian or yeast FKBP13 or FKBP25 have been identified, as they are targeted to intracellular compartments not found in prokaryotes.

5   SEQUENCE AND STRUCTURAL HOMOLOGIES BETWEEN FKBPs

Despite their common PPIase activity, FKBP domains have no evident sequence homologies with cyclophilins or parvulin. The hFKBP12 structure shares with cyclophilins a high proportion of β-sheet and a hydrophobic ligand-binding pocket. However, the hCypA structure, a β-barrel made up of two antiparallel four-stranded β-sheets with α-helices at each end and a variety of exposed loops and turns [214–217], is quite different from that of hFKBP12. CsA and substrate peptides are bound in a long groove on the outer surface of the β-barrel, between one face of the barrel and a loop.

The alignment of 30 available FKBP-domain sequences (12 mammalian, seven other eukaryotic and 11 prokaryotic) identifies a core of highly conserved amino acids (Figure 7). In two-thirds of the positions a single residue (or a very close homologue) occurs in more than half of the sequences studied, and 18 of the

108 residues are conserved in more than 80% of the 30 sequences. The N-terminal section of the FKBP domain, which includes the first strand of the β-sheet but contributes little to the hydrophobic core or the proline-binding pocket and is absent from ecFKBP16, pfFKBP16 and ecFKBP22, shows a lower degree of sequence conservation than the remainder of the molecule. Allowing for substitution of very similar residues (as defined in the legend to Figure 7), most mammalian FKBP domains match 82–94% of the consensus sequence, and yeast and fungal FKBP domains match 76–91%. The prokaryotic FKBP domains show greater variance, with 83% matches for nmFKBP12, 76% for scFKBP33 domain 1, 67% for ecFKBP16 and 64% for lpFKBP25M.

The only residues to be completely conserved throughout the 30 sequences are Gly-70, Gly-84 and the FK506-binding residue Phe-100. However, several other residues not involved directly in FK506 binding show extremely strong conservation; e.g. Gly-59 is replaced by Ala only in mammalian FKBP52 domain 1. The greatest variation is found among the prokaryotic members of the family, especially those with no, low or uncertain PPIase activity. If only the eukaryotic FKBP domains with good PPIase activity are considered, 16 residues are completely conserved: Gly-20, Tyr-27, Gly-29, Leu-31, Phe-37, Asp-38, Gly-52, Val-56, Ile-57, Gly-70, Tyr-83, Gly-84, Ile-92, Leu-98, Phe-100 and Leu-104. Eight of these are FK506-binding residues, and the other eight are all glycines or leucines. Gly-29 is probably conserved because any side chain here would protrude into the proline-binding pocket, while the other glycine residues are important for domain folding. The three leucine residues are all present within the strands of the β-sheet and have extensive side-chain contacts with other residues forming the hydrophobic core.

Of the 14 FK506-binding residues of hFKBP12, 10 are strongly conserved amongst FKBP domains, especially if inactive PPIase domains (ecFKBP22 and mammalian FKBP52 domain 3) or domains with uncertain or low PPIase activity (ecFKBP16, pfFKBP16, paFKBP25M, mammalian FKBP52 domain 2 and scFKBP33) are excluded (Figure 2). Residues Tyr-27, Phe-37, Asp-38, Val-56, Ile-57, Tyr-83 and Phe-100 are completely conserved in all known active PPIases. Three others have only closely similar substitutions in active PPIases: A82G in nmFKBP12, hFKBP13 and bsFKBP13; I92L in ctFKBP25M; and W60F in hFKBP12A, bFKBP12A and ctFKBP25M.

Four FK506-binding residues are less conserved. Arg-43 is

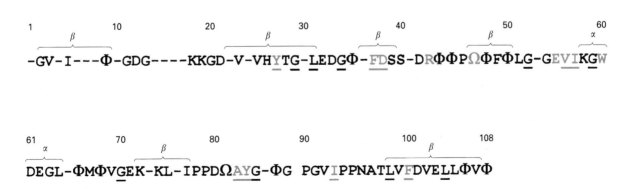

Figure 7   Consensus FKBP-domain sequence

FKBP-domain consensus sequence, derived from 30 sequences (12 mammalian but including only the first domain of FKBP52; seven other eukaryotic; 11 prokaryotic). Residues indicated (or very close homologues from the groups VIL, GA, ST, DENQ and RK) appear in at least 15 of the 30 sequences, while underlined residues appear in at least 25 out of 30. Residue numbers are by alignment with hFKBP12. The symbol Ω indicates the appearance of a hydrophobic residue in over 80% of the sequences, and the symbol Φ indicates that hydrophobic residues appear in less than 20% of sequences. Residues in hFKBP12 directly interacting with FK506 are coloured red. Residues forming part of the five strands of the β-sheet (the third strand has two sections) and the α-helix are indicated by β and α respectively.

replaced by other hydrophilic residues in hFKBP13, hFKBP25, bFKBP25 and three Mip proteins. This residue interacts with FK506 but not with rapamycin, and substitutions here may contribute to the strong preference for rapamycin of hFKBP13 and hFKBP25. Residue Phe-47 may be replaced by other hydrophobic amino acids without seriously compromising activity. Glu-55 is replaced by Gln in several active PPIases and by basic amino acids in yFKBP13, hFKBP25, bFKBP25 and ctFKBP25M. However, the E55M substitution in scFKBP33 domain 1 is the only alteration in these 14 residues that is not also observed in other active PPIases, and thus the low PPIase activity of this protein may indicate the importance of retaining a hydrophilic residue in this position. The greatest variation is observed at position 88, where the His residue found in mammalian FKBP12 and FKBP12A is substituted by a range of hydrophobic or hydrophilic residues in other active PPIases.

The importance of these FK506-binding residues has been further studied by investigating the effects of experimental substitution of individual residues on FK506 binding and PPIase activity. A Y27F substitution reduced hFKBP12 PPIase activity by 70% and binding of an FK506 analogue by 80% [78]. An identical substitution at this position is seen in ecFKBP16 and pfFKBP16. An F37Y substitution in hFKBP12 decreased PPIase activity by over 95%, although its effect on FK506 binding was much smaller [107]. A D38V substitution in hFKBP12 caused a > 90% decrease in PPIase activity and reduced FK506 sensitivity more than 50-fold [218]. A D38L substitution in the lpFKBP25M FKBP domain also decreased PPIase activity by 95% and almost abolished FK506 binding [210]. A rapamycin-resistant *S. cerevisiae* mutant expressing stable yFKBP12 proved to carry a Y83D mutation that reduced rapamycin binding by over 90% [146]. However, a more conservative Y83L substitution in hFKBP12 reduced PPIase activity with succ.ALPF.pNA by over 90% while decreasing the affinity for FK506 and rapamycin 2–3-fold [219], whereas Y83F (as found in a number of natural FKBP domains with low or uncertain activity) decreased PPIase activity by only 50% and had no significant effect on binding of an FK506 analogue [78]. Complete abolition of both PPIase activity and FK506 analogue binding was observed in hFKBP12 containing F100Y [78]. This substitution is found in mammalian FKBP52 domain 2. The importance of the conserved Tyr-27, Phe-37, Asp-38, Tyr-83 and Phe-100 residues is thus confirmed, although the nature of the replacement is evidently also important. Unfortunately, more systematic studies have been handicapped by the poor stability of some other mutant FKBP12s with mutations in key residues [146].

The cyclophilin ligand-binding site has a Trp residue that plays a key role in CsA binding but is less essential for PPIase activity; its substitution by a Phe residue accounts for the resistance of *E. coli* cytoplasmic and periplasmic cyclophilins to this drug [220,221]. In hCyp40 the equivalent Trp residue is replaced by His, and experimental substitution of this His by Trp increased the low PPIase activity and CsA affinity of this protein to values close to those found for hCypA [222]. The conserved Trp-60 residue in FKBP domains is not so critical for FK506 binding, as illustrated by the natural W60F substitution in hFKBP12A, which has an affinity for FK506 close to that of hFKBP12. However, a W60L substitution in hFKBP12 decreases PPIase activity by 90% [108]. This substitution is also observed in ecFKBP22, scFKBP33 domain 2 and domains 2 and 3 of mammalian FKBP52 (all reported to have no or very low PPIase activity) and also in ecFKBP16 (very unlikely to be an active PPIase). The non-conservative change W60H completely abolishes both PPIase activity and binding of an FK506 analogue [78].

Two of the four less conserved FK506-binding residues have also been studied by site-directed mutagenesis. The substitutions R43A, R43Q, R43I and R43K or H88A, H88F, H88L and H88V in hFKBP12 all have only minor effects on PPIase activity or FK506 binding [218,223]. Even the R43K/H88V double mutant has normal PPIase activity and only a 3-fold reduction in affinity for FK506, indicating that a range of substitutions do not compromise either enzyme activity or ligand binding. Substitutions in 20 other conserved and non-conserved hFKBP12 residues (mostly hydrophilic residues exposed on the hFKBP12 surface, and not including any of the most highly conserved residues outside the ligand-binding pocket) also had little effect on either parameter, but a G90P substitution (found naturally in caFKBP12, hFKBP13, sfFKBP46 and hFKBP52 domain 1) decreased the affinity for FK506 about 7-fold [218,223]. However, a substitution affecting a highly conserved hydrophobic residue, L31A, decreased hFKBP12 PPIase activity by 80% and binding of an FK506 analogue by 30% [78].

6   FKBP–LIGAND COMPLEXES IN IMMUNOSUPPRESSION AND SIGNAL TRANSDUCTION

When the immunosuppressive natural products CsA, FK506 and rapamycin were identified as PPIase inhibitors it was initially assumed that PPIase action might interconvert active and inactive configurations of key proteins essential for T-lymphocyte activation. However, this hypothesis was discarded when it was appreciated that, although FK506 and rapamycin both inhibited hFKBP12 and both blocked T-lymphocyte activation, they did so by quite distinct mechanisms and competitively inhibited each other's effects [54,224–226]. All ambiguity was removed by the demonstration that potent synthetic inhibitors of FKBP and cyclophilin PPIase activity were not necessarily immunosuppressive [14,15]. Sections of the PPIase ligands not involved in binding to the PPIase and normally exposed on the PPIase–ligand complex were also essential for immunosuppressive activity.

6.1   Inactivation of calcineurin

The ability of CsA and FK506 to block Ca^{2+}-dependent T- and B-lymphocyte responses was shown in 1991 to be due to the ability of CsA–CypA, CsA–CypC and FKBP12–FK506 complexes to bind to and thus inactivate a secondary target, the Ca^{2+}-activated protein serine/threonine phosphatase calcineurin [28,227]. PPIase-affinity and ligand-affinity columns alone did not retain calcineurin, but in the presence of CsA or FK506 they bound a complex of calcineurin and calmodulin. Rapamycin could not substitute for FK506 but competitively antagonized its effects, indicating a distinct secondary target for the FKBP12–rapamycin complex (Table 6). $A\alpha$ and $A\beta$ isoforms of calcineurin are both affected [228]. Both CsA and FK506 added to T-cells inhibited the protein phosphatase activity of calcineurin in the cell extracts [229].

Strong evidence has accumulated that calcineurin sequestration can account for the immunosuppressive effects of CsA and FK506. The protein phosphatase is less abundant than the PPIases, so that maximum inhibition of T-lymphocyte activation can be achieved with inhibitor concentrations insufficient to saturate all the PPIase in the cell [228,230]. For a ligand to be an effective immunosuppressant it must bind to the PPIase to form a complex that sequesters calcineurin [231–233]. Overexpression of calcineurin increases the concentration of CsA or FK506 necessary to cause inhibition [234,235].

Table 6   Inactivation of calcineurin and TOR-family phospholipid kinases

PPIase	Ligand	Secondary target	Competitor
CypA, CypB, CypC or Cyp40	CsA	Calcineurin	–
FKBP12 or FKBP12A	FK506, ascomycin or dihydroFK506	Calcineurin	Rapamycin, meridamycin or 18-hydroxyascomycin
FKBP12 (or FKBP12A?)	Rapamycin	Mammalian TOR homologue	FK506, ascomycin, dihydroFK506, meridamycin or 18-hydroxyascomycin
FKBP??	Meridamycin	??	??

One mechanism by which calcineurin is thought to promote the early phase of T-cell activation is by the dephosphorylation of the cytoplasmic precursor of a component of the T-cell-specific transcription factor NF-AT, thus enabling it to pass to the nucleus where it promotes the transcription of genes encoding interleukin-2 and other T-helper-cell-produced lymphokines required to initiate an immune response [236–239]. However, the protein phosphatase also plays additional roles in T-cells. For example, FK506 and CsA can also prevent the Ca^{2+}-mediated induction of apoptosis in T-cell hybridomas [240,241] and block Ca^{2+}-induced degranulation of T cytotoxic lymphocytes, a process not dependent on new gene expression [242]. Rapamycin antagonizes the effects of FK506 but not those of CsA.

In *in vitro* assays mammalian CypA, CypB, CypC and Cyp40 complexed with CsA can all bind calcineurin and inhibit its protein phosphatase activity [28,33,227,243]. CypB was the most potent member of the family. When over-expressed in a T-cell line both hCypA and hCypB enhanced sensitivity to CsA, but mCypC did not unless its N-terminal signal sequence was removed so that the protein was expressed in the cytoplasm [244]. However, intact hCypB was effective even when its signal sequence was retained. hCypB has a short C-terminal sequence thought to direct it to a specific ER compartment, where it co-localizes with the calcium storage protein calreticulin [26], and addition of this sequence to mCypC before overexpression made it effective in enhancing sensitivity to CsA even when its N-terminal signal sequence was retained [244]. Thus the intracellular location at which a PPIase is expressed may restrict its ability to interact with calcineurin.

Of the mammalian FKBPs and FKBP-domain proteins, only FKBP12–FK506 and FKBP12A–FK506 complexes are able to sequester calcineurin *in vitro* or *in vivo*, hFKBP13, hFKBP25 and hFKBP52 all being inactive in this respect [49,82,107,223,227,244,245]. Experimental modification of hFKBP12 has been used to investigate the basis for this. An F37Y substitution that decreased PPIase activity by over 95 % still permitted both FK506 binding and calcineurin sequestration [107], but a D38V substitution that greatly reduced FK506 binding without correspondingly affecting PPIase activity also prevented calcineurin sequestration [218]. Other mutations, such as R43I, R43K, R43A or R43Q, and especially the double substitution R43K/H88V, did not affect either PPIase activity or FK506 binding but strongly inhibited calcineurin sequestration [218,223]. Structural studies using NMR showed that, although FK506 still bound tightly to the double mutant, its position in the binding pocket was shifted, with the position of the surface-exposed sections of the FK506 molecule shown by analogue studies to be important for calcineurin sequestration being most affected [246].

A similar approach exploited the differences between hFKBP12 and hFKBP13, by exchanging short structural segments near the binding pocket between the two and determining the ability of the chimeras produced to sequester calcineurin [223,247]. Exchanging residues 41–45 (part of the loop between the two segments of the third strand of the β-sheet that forms one lip of the binding pocket and includes residue Arg-43) with the corresponding hFKBP13 sequence had little effect on calcineurin sequestration, but exchanging residues 85–92 (part of a loop that forms another lip) conferred calcineurin-binding activity on hFKBP13 and almost eliminated it from hFKBP12. A similar effect could be achieved by mutating just two residues at positions 90 and 91, from the Gly-Ile found in hFKBP12 to the Pro-Lys found in hFKBP13. An I91K substitution in hFKBP12 had little effect on PPIase activity or FK506 binding but was itself sufficient to cause a substantial reduction in calcineurin inhibition.

Residues Ile-91 and Phe-37 together with part of the FK506 molecule combine to form a hydrophobic groove on the surface of the hFKBP12–FK506 complex, but in the hFKBP13–FK506 complex the charged Lys-91 side chain covers this groove and destroys its hydrophobic character [88], suggesting that this groove plays a crucial role in the interaction with calcineurin. Gly-Ile, Gly-Leu or Gly-Val are present at residues 90/91 in almost all eukaryotic and prokaryotic FKBP12 homologues, including ecFKBP16 and pfFKBP16, and also in hFKBP12A. The only exceptions are caFKBP12, which has Pro-Ile, and scFKBP15, which has an inserted residue here to give Gly-Gly-Lys. Residues at the corresponding positions are Gly-Val in yFKBP13, Gly-Ile in yFKBP46 and Pro-Val in sfFKBP46. However, hFKBP13, bsFKBP13 and FKBP domain 1 of three mammalian FKBP52s have Pro-Lys, while hFKBP25 has Asp-Ala-Lys (again with an inserted amino acid) at the corresponding position, suggesting a common explanation for the failure of this latter group to sequester calcineurin. However, the presence of a Gly-Ile-homologous sequence at these positions is not necessarily sufficient to ensure interaction with calcineurin, as yFKBP13 (which has Gly-Val) is inactive in this respect [143,248].

CypA, CypB, FKBP12, FKBP12A and calcineurin are all widely distributed in mammalian cells, with both the PPIases and calcineurin particularly abundant in some regions of the brain [51]. The relatively selective immunosuppressive effects of CsA and FK506 in therapeutic use are thus difficult to explain. However, the marked similarity between the major clinical side-effects of CsA and FK506, and the failure of extensive attempts to identify analogues of CsA and FK506 that lack these side-effects while retaining effective immunosuppression, suggest that these too may be due to calcineurin inhibition. Actions of CsA and FK506 as varied as inhibition of antigen-induced degranulation of mast cells [89], inhibition of Ca^{2+}-mediated activation of Na^{+}/K^{+}-ATPase activity in kidney tubules [249,250], inhibition of depolarization-induced glucagon gene transcription in islet cells [251], inhibition of the nitric oxide-mediated elevation in cGMP levels involved in glutamate neurotoxicity [252], and the neuroprotective effects of FK506 in focal cerebral ischaemia [253] all appear secondary to calcineurin sequestration. Induced cytokine synthesis in murine mast cells deficient in FKBP12 (but with high levels of FKBP13) is inhibited by CsA but not by FK506, and the response to FK506 can be recovered when the cells are transfected with hFKBP12 [254,255], confirming that here too FKBP12 is the principal FKBP involved.

This mechanism of sequestration of calcineurin by PPIase–ligand complexes appears to be conserved in very different eukaryotes. The yeast *S. cerevisiae* contains protein phosphatases that are inactivated by yFKBP12–FK506 or yCypA–CsA and are homologous to mammalian calcineurin [248]. A minor difference is that the FK506 analogue 18-hydroxyascomycin (L-685,818) does not inhibit bovine calcineurin when complexed with hFKBP12, but is active when complexed with yFKBP12 [144]. Gene disruption shows that calcineurin is not required for normal *S. cerevisiae* growth, but is needed for recovery from α-pheromone-induced proliferation arrest [248,256] or growth in the presence of high LiCl or NaCl concentrations or at high pH [257,258]. In mutant *S. cerevisiae* lacking an incompletely characterized transmembrane protein, calcineurin becomes essential for normal growth [258–260]. In any of these situations when calcineurin is needed, growth is inhibited by either FK506 or CsA, providing that yFKBP12 or yCypA respectively is available, but disruption of these genes eliminates sensitivity to the corresponding ligand. Calcineurin becomes inhibitory when *S. cerevisiae* mutants defective in the system responsible for pumping excess Ca^{2+} ions from the cytoplasm into the vacuole are grown in high Ca^{2+}, and in this situation either disruption of calcineurin genes or addition of CsA or FK506 can restore normal growth [261].

CsA and FK506 also mimic the effects of disruption of a calcineurin-homologous gene in the fission yeast *Schizosaccharomyces pombe* [158], where growth is normal at 33 °C but cytokinesis is abnormal at 22 °C. In the plant *Vicia faba*, stomatal guard cell extracts have been shown to contain an okadaic acid-resistant and Ca^{2+}-activated protein phosphatase that is inhibited by CsA or added FK506–FKBP12; CsA or FK506–FKBP12 blocked Ca^{2+}-induced inactivation of K^+ channels [169]. This suggests that the sensitivity of calcineurin to both families of PPIase–ligand complexes has been conserved, although the amount of endogenous plant FKBP12 may be inadequate. The major *V. faba* chloroplast cyclophilin has also been shown to inactivate bovine calcineurin *in vitro* in the presence of CsA [167]. Thus the sensitivity of calcineurin-family protein phosphatases to inhibition by both FKBP- and cyclophilin-family PPIases complexed with ligands secreted by micro-organisms appears to be conserved in very different eukaryotes. Little information is available for the prokaryotic PPIases. Wild-type *E. coli* periplasmic cyclophilin, which has very low affinity for CsA, also has no detectable effect on the activity of calcineurin in the presence of CsA. However, the F112W mutant binds CsA with an affinity only 10-fold lower than hCypA to give a complex capable of interacting with calcineurin with an affinity 4-fold lower than that of hCypA–CsA [185].

This remarkable conservation of the ability of PPIase–ligand complexes of the two main PPIase familes to interact with calcineurin and the production of active ligands by a range of different soil micro-organisms raises the possibility of a physiological association between PPIases and calcineurin, perhaps involved in regulating the activity of the phosphatase and/or mediated by an unidentified endogenous ligand. Although several studies have concluded that neither cyclophilins nor FKBPs retain detectable amounts of calcineurin on affinity chromatography or inhibit its phosphatase activity in the absence of ligand, a weak interaction between yeast calcineurin and yFKBP12 and yCypA, and weak inhibition of bovine calcineurin phosphatase activity by yFKBP12 (but not yCypA), have been claimed [262]. The same authors recorded a weak yet significant interaction between yFKBP12 and murine calcineurin Aα1 using the sensitive yeast two-hybrid system. This interaction was greatly strengthened by addition of FK506, and was not seen if yFKBP12 was replaced by yFKBP13. The ligand-independent interaction differed from that seen in the presence of FK506 in that it was not impaired when yFKBP12 carrying mutations known to interfere with the FKBP12–FK506–calcineurin interaction was used, and it was not dependent on the presence of the calcineurin B subunit. Yeast strains lacking yFKBP12 (but not those lacking yCypA) were also found to have significantly enhanced recovery from α-pheromone-induced growth arrest and LiCl resistance, implying that yFKBP12 (but not yCypA) might antagonize endogenous calcineurin activity. It remains to be established whether this activity is unique to yFKBP12, but in another study using the yeast two-hybrid system an interaction between hFKBP12 and human calcineurin A was found to be completely dependent on the addition of FK506 [263].

6.2 Inactivation of TOR (target of rapamycin)-family phospholipid kinases

Inhibition of T-lymphocyte activation by the FKBP inhibitor rapamycin acts at a different and later step in the activation process from that affected by FK506, and its action at this stage is competitively antagonized by FK506 or non-immunosuppressive PPIase inhibitors [54,224–226], suggesting that FKBP–rapamycin complexes have a distinct secondary target. The first clue to its nature came from genetic studies in *S. cerevisiae*, which also shows rapamycin-sensitive proliferation dependent on the presence of yFKBP12 (see Section 3.1). Rapamycin-resistant *S. cerevisiae* can easily be selected, and whereas in most cases this is due to loss of yFKBP12 expression (a recessive mutation), a minority of resistant mutants show a dominant mutation in one of two alternate genes, named *TOR1* and *TOR2* [145].

S. cerevisiae TOR1 is a non-essential gene located on chromosome X, while *TOR2* is essential and is located on chromosome XI. *TOR1*, even when overexpressed, cannot substitute for *TOR2*. They encode large proteins of 2470 and 2474 amino acids with predicted molecular masses of 281 and 282 kDa respectively [264–267]. The proteins encoded by *TOR1* and *TOR2* include C-terminal domains that have significant (over 20%) identity with domains present in the yeast phosphatidylinositol 3-kinase VPS34 and the catalytic p110 subunits of mammalian phosphatidylinositide 3-kinases. It is not known whether the TOR1 and TOR2 proteins are themselves phosphatidylinositol or phosphatidylinositide 3-kinases, but VPS34-deficient yeast have been reported to lack the former activity. TOR1 and TOR2 proteins have 67% amino acid identity overall, rising to over 80% in their phospholipid-kinase-homologous domains. Exchanging these two domains between constructs shows that they are functionally equivalent. The N-terminal domain of TOR1 includes a possible leucine zipper and two helix–turn–helix motifs, but only one of the latter is clearly conserved in TOR2. Both proteins contain a possible Ca^{2+}-dependent lipid-binding domain and numerous potential phosphorylation sites. These include potential protein kinase C sites at Ser-1972 in TOR1 and Ser-1975 in TOR2 that are mutated to other residues in most rapamycin-resistant *S. cerevisiae* TOR alleles. Using the yeast two-hybrid system it has been demonstrated that in the presence of rapamycin there is a direct interaction between hFKBP12 and a 196-amino-acid fragment of yeast TOR2 spanning this key serine residue, but that the interaction is lost if the Ser-1975 residue is replaced by an arginine [268]. There is thus now compelling evidence that TOR1 and TOR2 are the direct secondary targets of yFKBP12–rapamycin complexes (Table 6), and that the interaction requires this key serine residue, perhaps in a phosphorylated state. The rapamycin-resistant TOR mutants are presumed to retain functional activity while evading sequestration.

In mammalian cells rapamycin rapidly causes the dephosphorylation and thus inactivation of p70 S6 kinase, due to an effect on an upstream signalling pathway [269–273]. Inhibition either of this enzyme or of other regulatory enzyme(s) activated by the same signalling pathway may result in sensitive cells, such as newly activated T-cells, failing to enter S phase, although many other mammalian cells may continue to proliferate almost normally.

Recently a human [274] and rat [275,276] TOR homologue, named RAFT1, FRAP or mTOR and present in many different tissues, has been identified as a direct secondary target interacting with mammalian FKBP12–rapamycin complexes by selective retention on FKBP12-affinity matrices in the presence of rapamycin. This 2549-amino-acid protein has a predicted molecular mass of 289 kDa, about 45% and 42% amino acid identity with yeast TOR2 and TOR1, and retains the C-terminal phospholipid-kinase-homologous domain, including a serine (Ser-2035) that is equivalent to the key TOR serine residues. A 35 kDa protein was also selectively retained in one study [275]. The mammalian TOR homologue was not retained if PPIase-inhibitory but non-immunosuppressive rapamycin analogues were used [274] or if FKBP12 was replaced by FKBP25 [275], and it was recovered in much lower amounts from extracts of a rapamycin-resistant murine T-cell line that contained normal mFKBP12 [276]. The yeast two-hybrid system has been used to confirm that hFKBP12 interacted directly with a 133-amino-acid fragment of the mouse TOR homologue spanning Ser-2035, but that the interaction was lost if the critical serine residue was mutated [263]. The role played by the mammalian TOR homologue in the activation of p70 S6 kinase remains to be established, but the homologous phosphatidylinositide 3-kinase has been implicated in this pathway [273]. When a mouse cDNA library was used to generate potential FKBP12–rapamycin target proteins, 21 of 28 clones that showed a rapamycin-dependent interaction with FKBP12 encoded TOR-homologous sequences. The other seven encoded six different unique sequences, as yet uncharacterized [263], raising the possibility that other secondary targets for FKBP12–rapamycin may exist in mammalian cells.

It is currently uncertain whether the interaction of PPIases with calcineurin and TOR homologues plays any role in the normal physiological regulation of the intracellular signalling pathways they modulate, or whether they represent the entire range of PPIase–ligand targets that have evolved in nature to exploit this mechanism. A suspicion that alternative targets may exist is raised by the discovery of meridamycin, a novel 27-membered macrolide produced by *Streptomyces hygroscopicus* strain F-004,081 [277]. This natural product has an FKBP-binding region identical to that of rapamycin, and binds to hFKBP12 competitively with FK506 and rapamycin to antagonize their effects, in the same manner as synthetic non-immunosuppressive analogues such as 506BD. It is not itself immunosuppressive, but it seems unlikely that it is produced to no purpose or that FKBP PPIase inhibition *per se* would be sufficient to justify its existence. Potential intracellular target systems are being sought.

7 PHYSIOLOGICAL ROLES OF FKBPs AND FKBP-DOMAIN PROTEINS

7.1 Evidence from inhibitor and gene disruption studies

Given their widespread occurrence, abundance and strong sequence conservation, inhibition or gene disruption of the major PPIases has surprisingly little evident effect on cell survival, growth or behaviour. The only known consequence of the addition of sufficient concentrations of the synthetic non-immunosuppressive FKBP inhibitors to mammalian cells to completely inhibit all FKBP12 PPIase activity is their competitive antagonism of the effects of FK506 and rapamycin [14,16,17].

The effects of inactivation or experimental disruption of FKBP genes have been most widely studied in micro-organisms. As noted in Section 3.1, disruption of the *S. cerevisiae* yFKBP12, yFKBP13 and yFKBP46 genes, individually, simultaneously or in combination with the yCypA and/or yCypB genes, has little effect on growth under a range of conditions, although one study reported that yFKBP12 disruption increased the cell doubling time by 15–30% [141]. The most dramatic effect reported was that sensitivity to rapamycin was dependent on yFKBP12. *N. crassa* lacking ncFKBP12 also grew normally at low and high temperatures and carried out the sexual life cycle successfully [159]. Some *Neisseria* appear to lack an FKBP12 homologue (see Section 4). The situation for FKBPs so far studied thus resembles that for CypA in *S. cerevisiae* and *N. crassa*, where loss of gene function had no evident effect except that the mutants became resistant to the inhibitory effects of CsA [278].

One member of the cyclophilin family that does have a well-defined function is the product of the *Drosophila melanogaster* gene *ninaA*. This encodes a 237-amino-acid member of the cyclophilin family that is expressed at a high level specifically in photoreceptor cells [36,171,172]. Its cyclophilin domain is preceded by a hydrophobic signal sequence and followed by a 46-amino-acid domain that includes a hydrophobic membrane anchor and a very short (7-amino-acid) cytoplasmic domain. The membrane anchor causes its retention in the ER and in vesicles transporting rhodopsin to the rhabdomeres. Mutant flies with defective *ninaA* genes synthesize normal amounts of rhodopsin, but less than 10% of the Rh 1 and Rh 2 opsins are successfully transported to the rhabdomeres; the majority remain in the grossly distorted ER and are eventually degraded, although other secreted proteins expressed simultaneously, including Rh 3 and Rh 4 opsins, can be transported and assembled normally [36,279]. *ninaA* gene function requires an intact cyclophilin domain, the presence of the membrane anchor and even the presence of the C-terminal seven amino acids that remain in the cytoplasm [280]. The NinaA protein forms a stable complex with the Rh 1 opsin *in vivo*, and the amount of rhodopsin successfully transported to the rhabdomeres reflects the level of NinaA expression, suggesting that NinaA acts as a chaperone or carrier. NinaA–rhodopsin complex-formation is abolished both by mutations within the cyclophilin domain and by deletion of the short C-terminal cytoplasmic sequence [281]. The only FKBPs for which inhibition or gene disruption suggests comparable importance are the FKBP-domain Mip proteins of *Legionella pneumophila* and *Chlamydia trachomatis*, which play a role in establishing intracellular infectivity (see Section 4).

7.2 Role of FKBPs in protein folding

In protein synthesis, proline residues are inserted into proteins in the *trans* configuration, and the isomerization of specific peptidylprolyl bonds into the *cis* conformation is, together with the formation of disulphide bonds between cysteine residues, one of the slower steps in protein folding [282,283]. Mammalian CypA has been shown to accelerate the isomerization of some (but not all) peptidylprolyl bonds and thus promote the *in vitro* refolding of some (but not all) proteins, including immunoglobulin chains [284,285], RNase T1 [6,286], yeast iso-2 cytochrome *c* [287] and carbonic anhydrase [288–291]. All effects of CypA are abolished by CsA. Cyclophilins from a variety of other eukaryotic and prokaryotic species [184,292,293] and the excised cyclophilin domain from hCypNK [294] have been shown to have similar activity. The *cis–trans* isomerization of the peptide hormone bradykinin [295] and of one of two peptidylprolyl bonds in calcitonin [296] was also increased by mammalian CypA. The regulation of rhodopsin assembly by the *D. melanogaster* NinaA protein would be compatible with NinaA participation in its folding in the secretory system. Two genes with sequence similarities suggesting that they may be parvulin-domain proteins may also be involved in the folding of proteins secreted into the periplasm [13]. PrsA mutants from *Bacillus subtilis* have decreased levels of exoproteins, suggesting that this protein prevents exoprotein misfolding, and the absence of PrtM from *Lactococcus lactis* leads to a cell wall serine proteinase being produced as an inactive precursor. The sequence of another parvulin-domain protein, the *E. coli* SurA protein that is essential for survival in the stationary phase, also predicts a periplasmic protein.

Not surprisingly, CypA PPIase activity with protein substrates is considerably lower than with the simple peptides normally used for laboratory assays, and some peptidylprolyl bonds appear to be inaccessible to the PPIases. FKBPs, which have lower enzymic activity and more restricted specificity with peptide substrates, have only occasionally been used in protein folding studies. However, ncFKBP12 has been shown to accelerate the folding of RNase T1 [155], hFKBP12 the folding of carbonic anhydrase [290] and an unspecified FKBP12 the folding of antibody Fab fragments [285]. In the two latter cases FKBP12 was compared with CypA and found to be less efficient.

Information available about the importance of FKBPs in protein folding in the intact cell is even more limited. CsA and FK506 have been shown to reduce folding of bacterial luciferases after their synthesis in reticulocyte lysates, with a combination of the two more being effective than either alone [297]. However, the maximum effect requires 30 μM CsA and as much as 100 μM FK506, suggesting that it is unlikely to be mediated by oFKBP12 or oFKBP52. The effect of FK506 was reduced when excess CypA was added. CsA has been reported to slow the secretion of transferrin (but not of several other proteins) from hepatocytes [298], the formation of triple-helical collagen by fibroblasts [299] and the expression of functional mammalian homo-oligomeric ligand-gated ion channels (which include several conserved proline residues) in *Xenopus laevis* oocytes [300]. These are all proteins that are secreted from the cytoplasm into the ER, and in each case correct protein maturation rather than synthesis was affected, suggesting that a PPIase of the cyclophilin family present in the secretory system, such as CypB, CypC or a NinaA homologue, might be responsible. Two of these studies also investigated the effects of FK506 and rapamycin [298,300] and both found them to be inactive, suggesting that cyclophilins may play a more important role than FKBPs in this context. However, in both *E. coli* and the related *Enterobacter aerogenes*, FKBP16 is part of the same operon as the *lsp* gene which encodes the prolipoprotein signal peptidase II involved in maturation of some proteins passing into the periplasm [196,197], which might suggest that it has a related function. However, the FKBP16s do not themselves have a signal sequence and would thus be expected to be expressed in the cytoplasm.

7.3 FKBPs as stress proteins

It has been suggested that when assisting the refolding of denatured carbonic anhydrase CypA may act as a chaperone, preventing aggregation, as well as a PPIase [289]. However, in more recent studies with the same system, addition of CypA after the unfolded protein had reformed the molten globule state still accelerated the slow step in refolding and gave the mature protein in equal yield, suggesting that the PPIase action was of primary importance [290]. There have been reports from a variety of organisms that at least some PPIases from both the cyclophilin and FKBP families may help to counteract stress conditions such as heat shock that may lead to an increase in the amount of denatured protein present.

In the yeast *S. cerevisiae*, growth at elevated temperatures increases the expression of yCypA, yCypB and yFKBP13, and disruption of the yCypA or yCypB genes increases temperature sensitivity [301,302]. A similar increase in yFKBP13 expression resulted from increasing the levels of unfolded protein within the secretory system by inhibition of glycosylation [302]. Yeasts with the mitochondrial cyclophilin gene disrupted grow equally well at normal and high temperature on most substrates but have a specific inability to utilize lactate when grown at 37 °C [135]. However, expression of yFKBP12 appears to be unaffected by stress [301], and yeasts triply deficient in yFKBP12, yFKBP13 and yFKBP46 survive and grow normally at elevated temperatures, as well as in more normal conditions [152].

In plants both cytoplasmic and chloroplast cyclophilins have been reported to be up-regulated by a range of stress-inducing treatments including heat and cold shock, wounding, osmotic stress and exposure to toxic chemicals [161,167,303]. No information is yet available about plant FKBPs. The only mammalian PPIase clearly shown to be up-regulated by heat shock is FKBP52, which has its synthesis increased to a comparable extent to that of the classical heat-shock proteins hsp90 and hsp70 with which it associates [304]. However, treatment of canine kidney cells with agents that lead to the accumulation of misfolded protein in the ER enhances the expression of FKBP13 mRNA along with the mRNAs for other ER chaperones [85], and expression of CypA in the rat hippocampus has been shown to be up-regulated after induced limbic seizures [305].

The up-regulation of PPIases by stress treatments expected to increase protein denaturation is strong, if circumstantial, evidence that they may play a role in maintaining other proteins in their native state. They could do this either by promoting refolding via their catalytic function or by themselves acting as chaperones. However, much more work will be required before it can be concluded that this is a general feature of PPIases, and the evidence presently available would suggest that the major cytoplasmic FKBP, FKBP12, is not affected. Even the PPIases found in the secretory system are not invariably up-regulated under stress conditions, as in *E. coli* expression of the periplasmic cyclophilin is not induced by heat shock, and deletion of the gene does not affect survival at high temperatures [306].

7.4 FKBPs as components of multi-subunit protein assemblies

Although the major cytoplasmic PPIases, CypA and FKBP12, exist predominantly as free monomers, there is now a striking number of examples of PPIases or PPIase-domain proteins being found in a stable or dynamic functional association with other proteins whose properties may thereby be modulated. If the FKBP interacts with a proline residue of its partner so as to maintain the peptidylprolyl bond in a transition state, the interaction would be expected to affect the conformation of the partner protein. The most studied example, the interaction of mammalian FKBP52 with hsp90 and thus with steroid hormone receptor complexes, may be atypical in that the FKBP52–hsp90 interaction is mediated by the FKBP52 tetratricopeptide repeat domain rather than by its PPIase domain, and thus does not involve the proline-binding site. Even here the additional binding of FK506, which may mimic the role of a second natural partner, appears to increase the sensitivity of the receptor–hsp90–FKBP52 complex to the hormone. Thus FKBP-domain proteins in which the non-FKBP domain includes an additional protein interaction site or multi-FKBP-domain proteins may interact as adaptors with two or more different partners to give rise to a complex protein assembly.

When the FKBP–partner interaction is dynamic and via the proline-binding site, ligands such as FK506 would be expected to competitively inhibit the interaction and thus reverse the effect of the FKBP. Non-immunosuppressive analogues of FK506 that bind to the PPIase active site would be expected to mimic these ligand effects. However, high concentrations of the ligands may be necessary to neutralize the high concentrations of PPIase present, especially when the PPIase partner in the complex has a lower affinity for the ligand than an abundant competitor, such as FKBP12. An increasing number of FKBP–partner or cyclophilin–partner interactions, such as those between mammalian FKBP12 and the ryanodine receptor or transforming growth factor-β receptor (see section 2.1), between hCypA or hCypB and the HIV-1 Gag protein [307,308] and between hCypC and its 77 kDa partner [29], appear to meet these criteria. The parvulin-homologue NifM is also part of the bacterial nitrogenase multi-subunit complex. The evident potential for physiological regulation of such PPIase–partner interactions by endogenous ligands remains at present speculative.

7.5   Is the PPIase activity of FKBPs important?

The PPIase activity of hFKBP12 is, even with its preferred peptide substrates, considerably lower than that of the abundant hCypA present in the same cellular compartment. Other members of the FKBP family have PPIases activities that are much lower or even undetectable with peptide substrates. While it is possible that FKBPs might be more active with physiological protein substrates or be able to catalyse isomerization at peptidylprolyl bonds inaccessible to cyclophilins, little evidence to support this has emerged from studies to date.

An alternative hypothesis is that the primary purpose of this family of PPIases, and perhaps the other two families also, is to participate in multi-subunit assemblies with partner proteins by interactions at proline residues and, by holding the peptidylprolyl bond in a transition state between the *cis* and *trans* conformations, to modulate partner activity. Known activities dependent on PPIases, such as the NinaA requirement for correct secretion of rhodopsin and its assembly into rhabdomeres, or for Mip proteins for the establishment of intracellular infectivity, could be accounted for by such a mechanism. The PPIase activity might simply result from the partner being released from the PPIase complex in either of the two possible conformations.

8   CONCLUSIONS

More than 30 FKBP-domain sequences have now been reported from a wide variety of eukaryotes and prokaryotes, and it is clear that there are many more to be discovered. FKBP12 homologues are particularly widespread, although they may prove less universal than CypAs in plants, lower eukaryotes and prokaryotes. Many FKBP genes appear dispensable under ex-

perimental conditions, and targeting by microbial antibiotics may have led to their loss, just as it has presumably led to the selection of *E. coli* cyclophilins resistant to CsA. In mammals there are at least seven different FKBPs or FKBP-domain proteins, most if not all generally expressed in the body, and this list is unlikely to prove complete. The yeast *S. cerevisiae* has at least four members of the family and some prokaryotes at least two, suggesting that such multiplicity will prove common. Members of the family are found together with cyclophilins (and sometimes also parvulins) in the cytoplasm, in the secretory system of animal cells, in the periplasm of at least some prokaryotes and in the chloroplasts and mitochondria of plants. The evidence presently available suggests that FKBPs predominate in nuclei, although the relationship between the mammalian nuclear FKBP25s and the larger nucleolar FKBPs with amphipathic N-terminal domains found in lower eukaryotes is uncertain. Conversely, cyclophilins are abundant in mitochondria, where there is at present only limited evidence for FKBPs.

Sequence and structural studies have helped to define the key elements of the FKBP domain. Over two-thirds of the residues are conserved in at least half of 30 sequences aligned, allowing increasingly confident prediction of the PPIase and FK506-binding activities of new FKBPs. PPIase activities vary widely, but even the best FKBPs are less active than CypA and parvulin and some have little or no activity. Conversely PPIase-domain proteins, especially those with multiple PPIase domains, appear more common in the FKBP family. It is suggested here that PPIase activity with peptide substrates may be a consequence of binding peptidylprolyl ligands in a transition state, and that the physiological role of FKBPs may be their interaction with one or more additional polypeptides to form multi-subunit assemblies in which the FKBP modulates the activities of other elements.

Both cyclophilins and FKBPs have been targeted by natural products secreted, presumably as antibiotics, by soil microorganisms. Cyclosporins produced by several fungi form a complex with many cyclophilins to sequester the protein phosphatase calcineurin. Some macrolides produced by streptomycetes complex with FKBP12 or FKBP12A (but not most other FKBPs) and also sequester calcineurin, which may interact weakly with FKBP12 and CypA even in the absence of such a ligand. Calcineurin is the only known secondary target for cyclophilin–ligand complexes, but other naturally produced macrolides may complex with FKBP12 to sequester at least one additional secondary target (and meridamycin suggests that there may be more). It remains unclear whether there are endogenous ligands, perhaps containing peptidylprolyl homologues, that regulate cyclophilin- or FKBP-domain interactions. However, the principles revealed by the study of PPIase-mediated regulation of intracellular signalling may be amenable to experimental or pharmaceutical exploitation.

I thank Yvonne Morgan, Malcolm Smith, Terry Gaymes, Mandy Fenton, Vicky Frost, Anne Wellhausen, Nishith Patel, Andrew Smith, David Gilfoyle and Mike Rau for valuable discussions, and the BBSRC Intracellular Signalling Initiative for supporting my research in this area.

REFERENCES

1   Fischer, G., Bang, H. and Mech, C. (1984) Biomed. Biochim. Acta **43**, 1101–1111
2   Fischer, G. and Bang, H. (1984) Biochim. Biophys. Acta **828**, 39–42
3   Handschumacher, R. E., Harding, M. W., Rice, J., Drugge, R. J. and Speicher, D. W. (1984) Science **226**, 544–547
4   Harding, M. W., Handschumacher, R. E. and Speicher, D. W. (1986) J. Biol. Chem. **261**, 8547–8555
5   Takahashi, N., Hayano, T. and Suzuki, M. (1989) Nature (London) **337**, 473–475
6   Fischer, G., Wittmann-Liebold, B., Lang, K., Kiefhaber, T. and Schmid, F. X. (1989) Nature (London) **337**, 476–478

7 Siekierka, J. J., Hung, S. H. Y., Poe, M., Lin, C. S. and Sigal, N. H. (1989) Nature (London) **341**, 755–757

8 Harding, M. W., Galat, A., Uehling, D. E. and Schreiber, S. L. (1989) Nature (London) **341**, 758–760

9 Koletsky, A. J., Harding, M. W. and Handschumacher, R. E. (1986) J. Immunol. **137**, 1054–1059

10 Siekierka, J. J., Wiederrecht, G., Greulich, H., Boulton, D., Hung, S., Cryan, J., Hodges, P. and Sigal, N. (1990) J. Biol. Chem. **265**, 21011–21015

11 Rahfeld, J.-U., Schierhorn, A., Mann, K. and Fischer, G. (1994) FEBS Lett. **343**, 65–69

12 Rahfeld, J.-U., Rucknagel, K. P., Schelbert, B., Ludwig, B., Hacker, J., Mann, K. and Fischer, G. (1994) FEBS Lett. **352**, 180–184

13 Rudd, K. E., Sofia, H. J., Koonin, E. V., Plunkett, G., Lazar, S. and Rouviere, P. R. (1995) Trends Biochem. Sci. **20**, 12–14

14 Bierer, B. E., Somers, P. K., Wandless, T. J., Burakoff, S. J. and Schreiber, S. L. (1990) Science **250**, 556–559

15 Sigal, N. H., Dumont, F., Durette, P., Siekierka, J. J., Peterson, L., Rich, D. H., Dunlap, B. E., Staruch, M. J., Melino, M. R., Koprak, S. L., Williams, D., Witzel, B. and Pisano, J. M. (1991) J. Exp. Med. **173**, 619–628

16 Dumont, F. J., Staruch, M. J., Koprak, S. L., Siekierka, J. J., Lin, C. S., Harrison, R., Sewell, T., Kindt, V. M., Beattie, T. R., Wyvratt, M. and Sigal, N. H. (1992) J. Exp. Med. **176**, 751–760

17 Ocain, T. D., Longhi, D., Steffan, R. J., Caccese, R. G. and Sehgal, S. N. (1993) Biochem. Biophys. Res. Commun. **192**, 1340–1346

18 Albers, M. W., Walsh, C. T. and Schreiber, S. L. (1990) J. Org. Chem. **55**, 4984–4986

19 Harrison, R. K. and Stein, R. L. (1990) Biochemistry **29**, 3813–3816

20 Kofron, J. L., Kuzmic, P., Kishore, V., Colon-Bonilla, E. and Rich, D. H. (1991) Biochemistry **30**, 6127–6134

21 Hsu, V. L., Handschumacher, R. E. and Armitage, I. A. (1990) J. Am. Chem. Soc. **112**, 6745–6747

22 Hubner, D., Drakenberg, T., Forsen, S. and Fischer, G. (1991) FEBS Lett. **284**, 79–81

23 Garcia-Echeverria, C., Kofron, J. I., Kuzmic, P., Kishore, V. and Rich, D. H. (1992) J. Am. Chem. Soc. **114**, 2758–2759

24 Garcia-Echeverria, C., Kofron, J. I., Kuzmic, P. and Rich, D. H. (1993) Biochem. Biophys. Res. Commun. **191**, 70–75

25 Price, E. R., Zydowsky, L. D., Jin, M. J., Baker, C. H., McKeon, F. D. and Walsh, C. T. (1991) Proc. Natl. Acad. Sci. U.S.A. **88**, 1903–1907

26 Arber, S., Krause, K.-H. and Caroni, P. (1992) J. Cell Biol. **116**, 113–125

27 Price, E. R., Jin, M., Lim, D., Pati, S., Walsh, C. T. and McKeon, F. D. (1994) Proc. Natl. Acad. Sci. U.S.A. **91**, 3931–3935

28 Friedman, J. and Weissman, I. (1991) Cell **66**, 799–806

29 Friedman, J., Trahey, M. and Weissman, I. (1993) Proc. Natl. Acad. Sci. U.S.A. **90**, 6815–6819

30 Bergsma, D. J., Eder, C., Gross, M., Kersten, H., Sylvester, D., Appelbaum, E., Cusimano, D., Livi, G. P., McLaughlin, M. M., Kasyan, K., Porter, T. G., Silverman, C., Dunnington, D., Hand, A., Prichett, W. P., Bossard, M. J., Brandt, M. and Levy, M. A. (1991) J. Biol. Chem. **266**, 23204–23214

31 Connern, C. P. and Halestrap, A. P. (1992) Biochem. J. **284**, 381–385

32 Kay, J. E. (1992) Biochem. J. **288**, 1074–1075

33 Kieffer, L. J., Seng, T. W., Li, W., Osterman, D. G., Handschumacher, R. E. and Bayney, R. M. (1993) J. Biol. Chem. **268**, 12303–12310

34 Anderson, S. K., Gallinger, S., Roder, J., Frey, J., Young, H. A. and Ortaldo, J. R. (1993) Proc. Natl. Acad. Sci. U.S.A. **90**, 542–546

35 Giardina, S. L., Anderson, S. K., Sayers, T. J., Chambers, W. H., Palumbo, G. A., Young, H. A. and Ortaldo, J. R. (1995) J. Immunol. **154**, 80–87

36 Stamnes, M. A., Shieh, B.-H., Chuman, L., Harris, G. L. and Zuker, C. S. (1991) Cell **65**, 219–227

37 Fretz, H., Albers, M. W., Galat, A., Standaert, R. F., Lane, W. S., Burakoff, S. J., Bierer, B. E. and Schreiber, S. L. (1991) J. Am. Chem. Soc. **113**, 1409–1411

38 Rosborough, S. L., Fleming, M., Nelson, P. A., Boger, J. and Harding, M. W. (1991) Transplant. Proc. **23**, 2890–2893

39 Siekierka, J. J., Staruch, M. J., Hung, S. H. Y. and Sigal, N. H. (1989) J. Immunol. **143**, 1580–1583

40 Standaert, R. F., Galat, A., Verdine, G. L. and Schreiber, S. L. (1990) Nature (London) **346**, 671–674

41 Maki, N., Sekiguchi, F., Nishimaki, J., Miwa, K., Hayano, T., Takahashi, N. and Suzuki, M. (1990) Proc. Natl. Acad. Sci. U.S.A. **87**, 5440–5443

42 Egan, D. A., Logan, T. M., Liang, H., Matayoshi, E., Fesik, S. W. and Holzman, T. F. (1993) Biochemistry **32**, 1920–1927

43 Marquis-Omer, D., Sanyal, G., Volkin, D. B., Marcy, A. I., Chan, H. K., Ryan, J. A. and Middaugh, C. R. (1991) Biochem. Biophys. Res. Commun. **179**, 741–748

44 Pilot-Matias, T. J., Pratt, S.D. and Lane, B. C. (1993) Gene **128**, 219–225

45 Mozier, N. M., Zurcher-Neely, H. A., Guido, D. M., Mathews, W. R., Heinrikson, R. L., Fraser, E. D., Walsh, M. P. and Pearson, J. D. (1990) Eur. J. Biochem. **194**, 19–23

46 Nelson, P. A., Lippke, J. A., Murcko, M. A., Rosborough, S. L. and Peattie, D. A. (1991) Gene **109**, 255–258

47 Wang, T., Donahoe, P. K. and Zervos, A. S. (1994) Science **265**, 674–676

48 Jayaraman, T., Brillantes, A.-M., Timerman, A. P., Fleischer, S., Erdjument-Bromage, H., Tempst, P. and Marks, A. R. (1992) J. Biol. Chem. **267**, 9474–9477

49 Yem, A. W., Reardon, I. M., Leone, J. W., Heinrikson, R. L. and Deibel, M. R. (1993) Biochemistry **32**, 12571–12576

50 Freund, E. A., Timerman, A. P. and Fleischer, S. (1994) Biophys. J. **66**, A413

51 Steiner, J. P., Dawson, T. M., Fotuhi, M., Glatt, C. E., Snowman, A. M., Cohen, N. and Snyder, S. H. (1992) Nature (London) **358**, 584–587

52 Peattie, D. A., Hsiao, K., Benasutti, M. and Lippke, J. A. (1994) Gene **150**, 251–257

53 Kay, J. E., Sampare-Kwateng, E., Geraghty, F. and Morgan, G. Y. (1991) Transplant. Proc. **23**, 2760–2762

54 Bierer, B. E., Mattila, P. S., Standaert, R. F., Herzenberg, L. A., Burakoff, S. J., Crabtree, G. and Schreiber, S. L. (1990) Proc. Natl. Acad. Sci. U.S.A. **87**, 9231–9235

55 Park, S. T., Aldape, R. A., Futer, O., DeCenzo, M. T. and Livingston, D. J. (1992) J. Biol. Chem. **267**, 3316–3324

56 Michnick, S. W., Rosen, M. K., Wandless, T. J., Karplus, M. and Schreiber, S. L. (1991) Science **252**, 836–839

57 Rosen, M. K., Michnick, S. W., Karplus, M. and Schreiber, S. L. (1991) Biochemistry **30**, 4774–4789

58 Van Duyne, G. D., Standaert, R. F., Karplus, P. A., Schreiber, S. L. and Clardy, J. (1991) Science **252**, 839–842

59 Lepre, C. A., Thomson, J. A. and Moore, J. M. (1992) FEBS Lett. **302**, 89–96

60 Van Duyne, G. D., Standaert, R. F., Karplus, P. A., Schreiber, S. L. and Clardy, J. (1993) J. Mol. Biol. **229**, 105–124

61 Meadows, R. P., Nettesheim, D. G., Xu, R. X., Olejniczak, E. T., Petros, A. M., Holzman, T. F., Severin, J., Gubbins, E., Smith, H. and Fesik, S. W. (1993) Biochemistry **32**, 754–765

62 Petros, A. M., Kawai, M., Luly, J. R. and Fesik, S. W. (1992) FEBS Lett. **308**, 309–314

63 Becker, J. W., Rotonda, J., McKeever, B. M., Chan, H. K., Marcy, A. I., Wiederrecht, G., Hermes, J. D. and Springer, J. P. (1993) J. Biol. Chem. **268**, 11335–11339

64 Van Duyne, G. D., Standaert, R. F., Schreiber, S. L. and Clardy, J. (1991) J. Am. Chem. Soc. **113**, 7433–7434

65 Ikeda, Y., Schultz, L. W., Clardy, J. and Schreiber, S. L. (1994) J. Am. Chem. Soc. **116**, 4143–4144

66 Orozco, M., Tirado-Rives, J. and Jorgensen, W. L. (1993) Biochemistry **32**, 12864–12874

67 Moore, J. M., Peattie, D. A., Fitzgibbon, M. J. and Thomson, J. A. (1991) Nature (London) **351**, 248–250

68 Theriault, Y., Logan, T. M., Meadows, R., Yu, L., Olejniczak, E. T., Holzman, T. F., Simmer, R. L. and Fesik, S. W. (1993) Nature (London) **361**, 88–91

69 Cheng, J.-Y., Lepre, C. A. and Moore, J. M. (1994) Biochemistry **33**, 4093–4100

70 DiLella, A. G. and Craig, R. J. (1991) Biochemistry **30**, 8512–8517

71 DiLella, A. G. (1991) Biochem. Biophys. Res. Commun. **179**, 1427–1433

72 DiLella, A. G., Hawkins, A., Craig, R. J., Schrieber, S. L. and Griffin, C. A. (1992) Biochem. Biophys. Res. Commun. **189**, 819–823

73 Haendler, B. and Hofer, E. (1990) Eur. J. Biochem. **190**, 477–482

74 Collins, J. H. (1991) Biochem. Biophys. Res. Commun. **178**, 1288–1290

75 Timerman, A. P., Ogunbunmi, E., Freund, E., Wiederrecht, G., Marks, A. R. and Fleischer, S. (1993) J. Biol. Chem. **268**, 22992–22999

76 Mayrleitner, M., Timerman, A. P., Wiederrecht, G. and Fleischer, S. (1994) Cell Calcium **15**, 99–108

77 Brillantes, A.-M. B., Ondrias, K., Scott, A., Kobrinsky, E., Ondriasova, E., Moschella, M. C., Jayaraman, T., Landers, M., Ehrlich, B. E. and Marks, A. R. (1994) Cell **77**, 513–523

78 Timerman, A. P., Wiederrecht, G., Marcy, A. and Fleischer, S. (1995) J. Biol. Chem. **270**, 2451–2459

79 Mack, M. M., Molinski, T. F., Buck, E. D. and Pessah, I. N. (1994) J. Biol. Chem. **269**, 23236–23249

80 Cameron, A. M., Steiner, J. P., Sabatini, D. M., Kaplin, A. I., Walensky, L. D. and Snyder, S. H. (1995) Proc. Natl. Acad. Sci. U.S.A. **92**, 1784–1788

81 Arakawa, H., Nagase, H., Hayashi, N., Fujiwara, T., Ogawa, M., Shin, S. and Nakamura, Y. (1994) Biochem. Biophys. Res. Commun. **200**, 836–843

82 Sewell, T. J., Lam, E., Martin, M. M., Leszyk, J., Weidner, J., Calaycay, J., Griffin, P., Williams, H., Hung, S., Cryan, J., Sigal, N. H. and Wiederrecht, G. J. (1994) J. Biol. Chem. **269**, 21094–21102

83 Jin, Y.-J., Albers, M. W., Lane, W. S., Bierer, B. E., Schreiber, S. L. and Burakoff, S. L. (1991) Proc. Natl. Acad. Sci. U.S.A. **88**, 6677–6681

84 Nigam, S. K., Jin Y.-J., Jin, M.-J., Bush, K. T., Bierer, B. E. and Burakoff, S. J. (1993) Biochem. J. **294**, 511–515

85   Bush, K. T., Hendrickson, B. A. and Nigam, S. K. (1994) Biochem. J. **303**, 705–708
86   Hendrickson, B. A., Zhang, W., Craig, R. J., Jin, Y.-J., Bierer, B. E., Burakoff, S. and DiLella, A. G. (1993) Gene **134**, 271–275
87   Pancer, Z., Gershon, H. and Rinkevich, B. (1993) Biochem. Biophys. Res. Commun. **197**, 973–977
88   Schultz, L. W., Martin, P. K., Liang, J., Schreiber, S. L. and Clardy, J. (1994) J. Am. Chem. Soc. **116**, 3129–3130
89   Hultsch, T., Albers, M. W., Schreiber, S. L. and Hohman, R. J. (1991) Proc. Natl. Acad. Sci. U.S.A. **88**, 6229–6233
90   Galat, A., Lane, W. S., Standaert, R. F. and Schreiber, S. L. (1992) Biochemistry **31**, 2427–2434
91   Jin, Y. J., Burakoff, S. J. and Bierer, B. E. (1992) J. Biol. Chem. **267**, 10942–10945
92   Hung, D. T. and Schreiber, S. L. (1992) Biochem. Biophys. Res. Commun. **184**, 733–738
93   Wiederrecht, G., Martin, M. M., Sigal, N. H. and Siekierka, J. J. (1992) Biochem. Biophys. Res. Commun. **185**, 298–303
94   Riviere, S., Menez, A. and Galat, A. (1993) FEBS Lett. **315**, 247–251
95   Jin, Y. J. and Burakoff, S. L. (1993) Proc. Natl. Acad. Sci. U.S.A. **90**, 7769–7773
96   Shi, Y., Brown, E. D. and Walsh, C. T. (1994) Proc. Natl. Acad. Sci. U.S.A. **91**, 2767–2771
97   Peattie, D. A., Harding, M. W., Fleming, M. A., DeCenzo, M. T., Lippke, J. A., Livingston, D. J. and Benasutti, M. (1992) Proc. Natl. Acad. Sci. U.S.A. **89**, 10974–10978
98   Yem, A. W., Tomasselli, A. G., Heinrikson, R. L., Zurcher-Neely, H., Ruff, V. A., Johnson, R. A. and Deibel, M. R. (1992) J. Biol. Chem. **267**, 2868–2871
99   Lebeau, M.-C., Massol, N., Herrick, J., Faber, L. E., Renoir J.-M., Radanyi, C. and Baulieu, E.-E. (1992) J. Biol. Chem. **267**, 4281–4284
100 Tai, P.-K. K., Albers, M. W., Chang, H., Faber, L. E. and Schreiber, S. L. (1992) Science **256**, 1315–1318
101 Ruff, V. A., Yem, A. W., Munns, P. L., Adams, L. D., Reardon, I. M., Deibel, M. R. and Leach, K. L. (1992) J. Biol. Chem. **267**, 21285–21288
102 Smith, D. F., Baggenstoss, B. A., Marion, T. N. and Rimerman, R. A. (1993) J. Biol. Chem. **268**, 18365–18371
103 Schmitt, J., Pohl, J. and Stunnenberg, H. G. (1993) Gene **132**, 267–271
104 Callebaut, I., Renoir, J.-M., Lebeau, M.-C., Massol, N., Burny, A., Baulieu, E.-E. and Mornon, J.-P. (1992) Proc. Natl. Acad. Sci. U.S.A. **89**, 6270–6274
105 Ratajczak, T., Carrello, A., Mark, P. J., Warner, B. J., Simpson, R. J., Moritz, R. L. and House, A. K. (1993) J. Biol. Chem. **268**, 13187–13192
106 Radanyi, C., Chambraud, B. and Baulieu, E.-E. (1994) Proc. Natl. Acad. Sci. U.S.A. **91**, 11197–11201
107 Wiederrecht, G., Hung, S., Chan, H. C., Marcy, A., Martin, M., Calaycay, J., Boulton, D., Sigal, S., Kincaid, R. L. and Siekierka, J. J. (1992) J. Biol. Chem. **267**, 21753–21760
108 Chambraud, B., Rouviere-Fourmy, N., Radanyi, C., Hsiao, K., Peattie, D. A., Livingston, D. J. and Baulieu, E.-E. (1993) Biochem. Biophys. Res. Commun. **196**, 160–166
109 Le Bihan, S., Renoir, J.-M., Radanyi, C., Chambraud, B., Joulin, V., Catelli, M.-G. and Baulieu, E.-E. (1993) Biochem. Biophys. Res. Commun. **195**, 600–607
110 Johnson, J. L. and Toft, D. O. (1994) J. Biol. Chem. **269**, 24989–24993
111 Czar, M. J., Owens-Grillo, J. K., Dittmar, K. D., Hutchison, K. A., Zacharek, A. M., Leach, K. L., Deibel, M. R. and Pratt, W. B. (1994) J. Biol. Chem. **269**, 11155–11161
112 Hutchison, K. A., Scherrer, L. C., Czar, M. J., Ning, Y., Sanchez, E. R., Leach, K. L., Deibel, M. R. and Pratt, W. B. (1993) Biochemistry **32**, 3953–3957
113 Nadeau, K., Das, A. and Walsh, C. T. (1993) J. Biol. Chem. **268**, 1479–1487
114 Tai, P.-K. K., Chang, H., Albers, M. W., Schreiber, S. L., Toft, D. O. and Faber, L. E. (1993) Biochemistry **32**, 8842–8847
115 Czar, M. J., Owens-Grillo, J. K., Yem, A. W., Leach, K. L., Deibel, M. R., Welsh, M. J. and Pratt, W. B. (1994) Mol. Endocrinol. **8**, 1731–1741
116 Perrot-Applanat, M., Cibert, C., Geraud, G., Renoir, J.-M. and Baulieu, E.-E. (1995) J. Cell Sci. **108**, 2037–2051
117 Ning, Y.-M. and Sanchez, E. D. (1993) J. Biol. Chem. **268**, 6073–6076
118 Renoir, J. M., Radanyi, C. and Baulieu, E.-E. (1992) C. R. Acad. Sci. Paris **315**, 421–428
119 Tai, P.-K. K., Albers, M. W., McDonnell, D. P., Chang, H., Schreiber, S. L. and Faber, L. E. (1994) Biochemistry **33**, 10666–10671
120 Matts, R. L., Xu, Z. Y., Pal, J. K. and Chen, J. J. (1992) J. Biol. Chem. **267**, 18160–18167
121 Matts, R. L. and Hurst, R. (1992) J. Biol. Chem. **267**, 18168–18174
122 Matts, R.L., Hurst, R. and Xu, Z. Y. (1993) Biochemistry **32**, 7323–7328
123 Sanchez, E. R., Faber, L. E., Henzel, W. J. and Pratt, W. B. (1990) Biochemistry **29**, 5145–5152
124 Smith, D. F., Albers, M. W., Schreiber, S. L., Leach, K. L. and Deibel, M. R. (1993) J. Biol. Chem. **268**, 24270–24273
125 Hartson, S.D. and Matts, R. L. (1994) Biochemistry **33**, 8912–8920
126 Stancato, L. F., Chow, Y.-H., Owens-Grillo, J. K., Yem, A. W., Deibel, M. R., Jove, R. and Pratt, W. B. (1994) J. Biol. Chem. **269**, 22157–22161
127 Simek, S. L., Kozak, C. A., Winterstein, D., Hegamyer, G. and Colburn, N. H. (1993) Genomics **18**, 407–409
127a Coss, M. C., Winterstein, D., Sowder, R. C. and Simek, S. L. (1995) J. Biol. Chem. **270**, 29336–29341
128 Donnelly, J. G. and Soldin, S. L. (1991) Transplant. Proc. **23**, 2887–2890
129 Palaszynski, E. W., Donnelly, J. G. and Soldin, S. J. (1991) Clin. Biochem. **24**, 63–70
130 Donnelly, J. G. and Soldin, S. L. (1994) Clin. Biochem. **27**, 367–372
131 Murthy, J. N., Chen, Y. and Soldin, S. J. (1994) Clin. Biochem. **27**, 357–365
132 Timerman, A. P., Jayaraman, T., Wiederrecht, G., Onoue, H., Marks, A. R. and Fleischer, S. (1994) Biochem. Biophys. Res. Commun. **198**, 701–706
133 Haendler, B., Keller, R., Hiestand, P. C., Kocher, H. P., Wegmann, G. and Movva, N. R. (1989) Gene **83**, 39–46
134 McLaughlin, M. M., Bossard, M. J., Koser, P. L., Cafferkey, R., Morris, R. A., Miles, L. M., Strickler, J., Bergsma, D. J., Levy, M. A. and Livi, G. P. (1992) Gene **111**, 85–92
135 Davis, E. S., Becker, A., Heitman, J., Hall, M. N. and Brennan, M. B. (1992) Proc. Natl. Acad. Sci. U.S.A. **89**, 11169–11173
136 Koser, P. L., Sylvester, D., Livi, G. P. and Bergsma, D. J. (1990) Nucleic Acids Res. **18**, 1643
137 Zydowsky, L. D., Ho, S. I., Baker, C. H., McIntyre, K. and Walsh, C. T. (1992) Protein Sci. **1**, 961–969
138 Frigerio, G. and Pelham, H. R. B. (1993) J. Mol. Biol. **233**, 183–188
139 Franco, L., Jimenez, A., Demolder, J., Molemans, F., Fiers, W. and Contreras, R. (1992) Yeast **7**, 971–979
140 Wiederrecht, G., Brizuela, L., Elliston, K., Sigal, N. H. and Siekierka, J. J. (1991) Proc. Natl. Acad. Sci. U.S.A. **88**, 1029–1033
141 Heitman, J., Movva, N. R., Hiestand, P. C. and Hall, M. N. (1991) Proc. Natl. Acad. Sci. U.S.A. **88**, 1948–1952
142 Koltin, Y., Faucette, L., Bergsma, D. J., Levy, M. A., Cafferkey, R., Koser, P. L., Johnson, R. K. and Livi, G. P. (1991) Mol. Cell. Biol. **11**, 1718–1723
143 Nielsen, J. B., Foor, F., Siekierka, J. J., Hsu, M.-J., Ramadan, N., Morin, N., Shafiee, A., Dahl, A. M., Brizuela, L., Chrebet, G., Bostian, K. A. and Parent, S. A. (1992) Proc. Natl. Acad. Sci. U.S.A. **89**, 7471–7475
144 Rotonda, J., Burbaum, J. J., Chan, H. K., Marcy, A. I. and Becker, J. W. (1993) J. Biol. Chem. **268**, 7607–7609
145 Heitman, J., Movva, N. R. and Hall, M. N. (1991) Science **253**, 905–909
146 Koser, P. L., Eng, W.-K., Bossard, M. J., McLaughlin, M. M., Cafferkey, R., Sathe, G. M., Faucette, L., Levy, M. A., Johnson, R. K., Bergsma, D. J. and Livi, G. P. (1993) Gene **129**, 159–165
147 Brizuela, L., Chrebet, G., Bostian, K. A. and Parent, S. A. (1991) Mol. Cell. Biol. **11**, 4616–4626
148 Partaledis, J. A., Fleming, M. A., Harding, M. A. and Berlin, V. (1992) Yeast **8**, 673–680
149 Heitman, J., Koller, A., Kunz, J., Henriquez, R., Schmidt, A., Movva, N. R. and Hall, M. N. (1993) Mol. Cell. Biol. **13**, 5010–5019
150 Shan, X., Xue, Z. and Melese, T. (1994) J. Cell Biol. **126**, 853–862
151 Manning-Kreig, U. C., Henriquez, R., Cammas, F., Graff, P., Gaveriaux, S. and Movva, N. R. (1994) FEBS Lett. **352**, 98–103
152 Benton, B. M., Zang, J.-H. and Thorner, J. (1994) J. Cell Biol. **127**, 623–639
153 Chang, H.-C. J. and Lindquist, S. (1994) J. Biol. Chem. **269**, 24983–24988
154 Ferrara, A., Cafferkey, R. and Livi, G. P. (1992) Gene **113**, 125–127
155 Tropschug, M., Wachter, E., Mayer, S., Schonbrunner, E. R. and Schmid, F. X. (1990) Nature (London) **346**, 674–677
156 Lee, C., Hoffmann, K. and Zocher, R. (1992) Biochem. Biophys. Res. Commun. **182**, 1282–1287
157 High, K. P. (1994) Transplantation **57**, 1689–1700
158 Yoshida, T., Toda, T. and Yanagida, M. (1994) J. Cell Sci. **107**, 1725–1735
159 Barthelmess, I. B. and Tropschug, M. (1993) Curr. Genet. **23**, 54–58
160 Gasser, C. S., Gunning, D. A., Budelier, K. A. and Brown, S. M. (1990) Proc. Natl. Acad. Sci. U.S.A. **87**, 9519–9523
161 Marivet, J., Frendo, P. and Burkard, G. (1992) Plant Sci. **84**, 171–178
162 Bartling, D., Heese, A. and Weiler, E. W. (1992) Plant Mol. Biol. **19**, 529–530
163 Lippuner, V., Chou, I. T., Scott, S. V., Ettinger, W. F., Theg, S. M. and Gasser, C. S. (1994) J. Biol. Chem. **269**, 7863–7868
164 Hayman, G. T. and Miernyk, J. A. (1994) Biochim. Biophys. Acta **1219**, 536–538
165 Saito, T., Ishiguro, S., Ashida, H., Kawamukai, M., Matsuda, H., Ochiai, H. and Nakagawa, T. (1995) Plant Cell Physiol. **36**, 377–382
166 Buchholz, W. G., Harris-Haller, L., DeRose, R. T. and Hall, T. C. (1994) Plant Mol. Biol. **25**, 837–843
167 Luan, S., Lane, W. S. and Schreiber, S. L. (1994) Plant Cell **6**, 885–892

168 Breiman, A., Fawcett, T. W., Ghirardi, M. L. and Mattoo, A. K. (1992) J. Biol. Chem. **267**, 21293–21296

169 Luan, S., Li, W., Rusnak, F., Assman, S. M. and Schreiber, S. L. (1993) Proc. Natl. Acad. Sci. U.S.A. **90**, 2202–2206

170 Luan, S., Albers, M. W. and Schreiber, S. L. (1994) Proc. Natl. Acad. Sci. U.S.A. **91**, 984–988

171 Shieh, B.-H., Stamnes, M. A., Seavallo, S., Harris, G. L. and Zuker, C. S. (1989) Nature (London) **338**, 67–70

172 Schneuwly, S., Shortridge, R. D., Larrivee, D. C., Ono, T., Ozaki, M. and Pak, W. L. (1989) Proc. Natl. Acad. Sci. U.S.A. **86**, 5390–5394

173 Lightowlers, M. W., Haralambous, A. and Rickard, M. D. (1989) Mol. Biochem. Parasitol. **36**, 287–290

174 Barisic, K., Mollner, S., Noegel, A. A., Gerisch, G. and Segall, J. E. (1991) Dev. Genet. **12**, 50–53

175 Argaet, V. P. and Mitchell, G. F. (1992) J. Parasitol. **78**, 660–664

176 High, K. P., Joiner, K. A. and Handschumacher, R. E. (1994) J. Biol. Chem. **269**, 9105–9112

177 Bell, A., Wernli, B. and Franklin, R. M. (1994) Biochem. Pharmacol. **48**, 495–503

178 Alnemri, E. S., Fernandes-Alnemri, T., Pomerenke, K., Robertson, N. M., Dudley, K., DuBois, G. C. and Litwack, G. (1994) J. Biol. Chem. **249**, 30828–30834

179 Theopold, U., Dal Zotto, L. and Hultmark, D. (1995) Gene **156**, 247–251

180 Hayano, T., Takahashi, N., Kato, S., Maki, N. and Suzuki, M. (1991) Biochemistry **30**, 3041–3048

181 Kawamukai, M., Matsuda, H., Fujii. W., Utsumi. R. and Komano, T. (1989) J. Bacteriol. **171**, 4525–4529

182 Liu, J. and Walsh, C. T. (1990) Proc. Natl. Acad. Sci. U.S.A. **87**, 4028–4032

183 Liu, J., Chen, C.-M. and Walsh, C. T. (1991) Biochemistry **29**, 3813–3816

184 Compton, L. A., Davis, J. M., MacDonald, J. R. and Bachinger, H. P. (1992) Eur. J. Biochem. **206**, 927–934

185 Fejzo, J., Etzkorn, F. A., Clubb, R. T., Shi, Y., Walsh, C. T. and Wagner, G. (1994) Biochemistry **33**, 5711–5720

186 Herrler, M., Bang, H. and Marahiel, M. A. (1994) Mol. Microbiol. **11**, 1073–1083

187 Pahl, A., Uhlein, M., Bang, H., Schlumbohn, W. and Keller, U. (1992) Mol. Microbiol. **6**, 3551–3558

188 Hassidim, M., Schwarz, R., Lieman-Hurwitz, J., Marco, E., Ronen-Tarazi, M. and Kaplan, A. (1992) Plant Physiol. **100**, 1982–1986

189 Kim, D. J., Morikawa, M., Takagi, M. and Imanaka, T. (1995) J. Ferment. Bioeng. **79**, 87–94

190 Nagashima, K., Mitsuhashi, S., Kamino. K. and Maruyama, T. (1994) Biochem. Biophys. Res. Commun. **198**, 466–472

191 Tran, P. V., Bannor, T. A., Doktor, S. Z. and Nichols, B. P. (1990) J. Bacteriol. **172**, 397–410

192 Kok, R. G., Christoffels, V. M., Vosman, B. and Hellingwerf, K. J. (1994) Biochim. Biophys. Acta **1219**, 601–606

193 Sampson, B. A. and Gotschlich, E. C. (1992) Proc. Natl. Acad. Sci. U.S.A. **89**, 1164–1168

194 McAllister, C. F. and Stephens, D. S. (1993) Mol. Microbiol. **10**, 13–23

195 Pahl, A. and Keller, U. (1992) J. Bacteriol. **174**, 5888–5894

196 Bouvier, J. and Stragier, P. (1990) Nucleic Acids Res. **19**, 180

197 Isaki, L., Kawakami, M., Beers, R., Hom, R. and Wu, H. C. (1990) J. Bacteriol. **172**, 469–472

198 Isaki, L., Beers, R. and Wu, H. C. (1990) J. Bacteriol. **172**, 6512–6517

199 Wulfing, C., Lombardero, J. and Pluckthun, A. (1994) J. Biol. Chem. **269**, 2895–2901

200 Roof, W. D., Horne, S. M., Young, K. D. and Young, R. (1994) J. Biol. Chem. **269**, 2902–2910

201 Cianciotto, N. P., Eisenstein, B. I., Mody, C. H., Toews, G. B. and Engleberg, N. C. (1989) Infect. Immun. **57**, 1255–1262

202 Lundemose, A. G., Rouch, D. A., Birkelund, S., Christiansen, G. and Pearce, J. H. (1992) Mol. Microbiol. **6**, 2539–2548

203 Hacker, J. and Fischer, G. (1993) Mol. Microbiol. **10**, 445–456

204 Lundemose, A. G., Kay, J. E. and Pearce, J. H. (1993) Mol. Microbiol. **7**, 777–783

205 Cianciotto, N. P. and Fields, B. S. (1992) Proc. Natl. Acad. Sci. U.S.A. **89**, 5188–5191

206 Bangsborg, J. M., Cianciotto, N. P. and Hindersson, P. (1991) Infect. Immun. **59**, 3836–3840

207 Engleberg, N. C., Carter, C., Weber, D. R., Cianciotto, N. P. and Eisenstein, B. I. (1989) Infect. Immun. **57**, 1263–1270

208 Lundemose, A. G., Rouch, D. A., Birkelund, S., Christiansen, G. and Pearce, J. H. (1992) Mol. Microbiol. **6**, 2539–2548

209 Kato, J., Chu, L., Kitano, K., DeVault, J. D., Kimbara, K., Chahrabarty, A. M. and Misra, T. K. (1989) Gene **84**, 31–38

210 Ludwig, B., Rahfeld, J., Schmidt, B., Mann, K., Wintermeyer, E., Fischer, G. and Hacker, J. (1994) FEMS Microbiol. Lett. **118**, 23–30

211 Fischer, G., Bang, H., Ludwig, B., Mann, K. and Hacker, J. (1992) Mol. Microbiol. **6**, 1375–1383

212 Schmidt, B., Rahfeld, J., Schierhorn, A., Ludwig, B., Hacker, J. and Fischer, G. (1994) FEBS Lett. **352**, 185–190

213 Pahl, A. and Keller, U. (1994) EMBO J. **13**, 3472–3480

214 Kallen, J., Spitzfaden, D., Zurini, M. G. M., Wider, G., Widmer, H., Wuthrich, K. and Walkinshaw, M. D. (1991) Nature (London) **353**, 276–279

215 Theriault, Y., Logan, T. M., Meadows, R., Yu, L., Olejniczak, E. T., Holzman, T. F., Simmer, R. L. and Fesik, S. W. (1993) Nature (London) **361**, 88–91

216 Mikol, V., Kallen, J., Pflugl, G. and Walkinshaw, M. D. (1993) J. Mol. Biol. **234**, 1119–1130

217 Ke, H. M., Mayrose, D., Belshaw, P. J., Alberg, D. G., Schreiber, S. L., Chang, Z. Y., Etzkorn, F. A., Ho, S. and Walsh, C. T. (1994) Structure **2**, 33–44

218 Aldape, R. A., Futer, O., DeCenzo, M. T., Jarrett, B. P., Murcko, M. A. and Livingston, D. J. (1992) J. Biol. Chem. **267**, 16029–16032

219 Bossard, M. J., Bergsma, D. J., Brandt, M., Livi, G. P., Eng, W.-K., Johnson, R. K. and Levy, M. A. (1994) Biochem. J. **297**, 365–372

220 Liu, J., Chen, C. and Walsh, C. T. (1991) Biochemistry **30**, 2306–2310

221 Bossard, M. J., Koser, P. L., Brandt, M., Bergsma, D. J. and Levy, M. A. (1991) Biochem. Biophys. Res. Commun. **176**, 1142–1148

222 Hoffman, K., Kakalis, L. T., Anderson, K. S., Armitage, I. M. and Handschumacher, R. E. (1995) Eur. J. Biochem. **229**, 188–193

223 Yang, D., Rosen, M. K. and Schreiber, S. L. (1993) J. Am. Chem. Soc. **115**, 819–820

224 Dumont, F. J., Staruch, M. J., Koprak, S. L., Melino, M. R. and Sigal, N. H. (1990) J. Immunol. **144**, 251–258

225 Dumont, F. J., Melino, M. R., Staruch, M. J., Koprak, S. L., Fischer, P. A. and Sigal, N. H. (1990) J. Immunol. **144**, 1418–1424

226 Kay, J. E., Kromwel, L., Doe, S. E. A. and Denyer, M. (1991) Immunology **72**, 544–549

227 Liu, J., Farmer, J. D., Lane, W. S., Friedman, J., Weissman, I. and Schreiber, S. L. (1991) Cell **66**, 807–815

228 Mukai, H., Kuno, T., Chang, C.-D., Lane, B., Luly, J. R. and Tanaka, C. (1993) J. Biochem. (Tokyo) **113**, 292–298

229 Fruman, D. A., Klee, C. B., Bierer, B. E. and Burakoff, S. J. (1992) Proc. Natl. Acad. Sci. U.S.A. **89**, 3686–3690

230 Asami, M., Kuno, T., Mukai, H. and Tanaka, C. (1993) Biochem. Biophys. Res. Commun. **192**, 1388–1394

231 Liu, J., Albers, M. W., Wandless, T. J., Luan, S., Alberg, D. G., Belshaw, P. J., Cohen, P., Mackintosh, C., Klee, C. B. and Schreiber, S. L. (1992) Biochemistry **31**, 3896–3901

232 Kawai, M., Lane, B. C., Hsieh, G. C., Mollison, K. W., Carter, G. W. and Luly, J. R. (1993) FEBS Lett. **316**, 107–113

233 Nelson, P. A., Akselband, Y., Kawamura, A., Su, M., Tung, R. D., Rich, D. H., Kishore, V., Rosborough, S. L., DeCenzo, M. T., Livingston, D. J. and Harding, M. W. (1993) J. Immunol. **150**, 2139–2147

234 O'Keefe, S. J., Tamura, J., Kincaid, R. L., Tocci, M. J. and O'Neill, E. A. (1992) Nature (London) **357**, 692–694

235 Clipstone, N. A. and Crabtree, G. R. (1992) Nature (London) **357**, 695–697

236 Flanagan, W. M., Corthesy, B., Bram, R. J. and Crabtree, G. R. (1991) Nature (London) **352**, 803–807

237 McCaffrey, P. G., Perrino, B. A., Soderling, T. R. and Rao, A. (1993) J. Biol. Chem. **268**, 3747–3752

238 Jain, J., McCaffrey, P. G., Miner, Z., Kerppola, T. K., Lambert, J. N., Verdine, G. L., Curran, T. and Rao, A. (1993) Nature (London) **365**, 352–355

239 Northrop, J. P., Ullman, K. S. and Crabtree, G. R. (1993) J. Biol. Chem. **268**, 2917–2923

240 Fruman, D. A., Mather, P. E., Burakoff, S. J. and Bierer, B. E. (1992) Eur. J. Immunol. **22**, 2513–2517

241 Yazdanbakhsh, K., Choi, J.-W., Li, Y., Lau, L. F. and Choi, Y. (1995) Proc. Natl. Acad. Sci. U.S.A. **92**, 437–441

242 Dutz, J. P., Fruman, D. A., Burakoff, S. J. and Bierer, B. E. (1993) J. Immunol. **150**, 2591–2598

243 Swanson, S. K.-H., Born, T., Zydowsky, L. D., Cho, H., Chang, H. Y., Walsh, C. T. and Rusnak, F. (1992) Proc. Natl. Acad. Sci. U.S.A. **89**, 3741–3745

244 Bram, R., Hung, D., Martin, P., Schreiber, S. and Crabtree, G. (1993) Mol. Cell. Biol. **13**, 4760–4769

245 Lebeau, M.-C., Myagkikh, I., Rouviere-Fourmy, N., Baulieu, E. E. and Klee, C. B. (1994) Biochem. Biophys. Res. Commun. **203**, 750–755

246 Lepre, C. A., Pearlman, D. A., Cheng, J.-W., DeCenzo, M. T., Livingston, D. J. and Moore, J. M. (1994) Biochemistry **33**, 13571–13580

247 Rosen, M. K., Yang, D., Martin, P. K. and Schreiber, S. L. (1993) J. Am. Chem. Soc. **115**, 821–822

248 Foor, F., Parent, S. A., Morin, N., Dahl, A. M., Ramadan, N., Chrebet, G., Bostian, K A. and Nielsen, J. B. (1992) Nature (London) **360**, 682–684

249 Aperia, A., Ibarra, F., Svensson, L.-B., Klee, C. and Greengard, P. (1992) Proc. Natl. Acad. Sci. U.S.A. **89**, 7394–7397

250 Lea, J. P., Sands, J. M., McMahon, S. J. and Tumlin, J. A. (1994) Kidney Int. **46**, 647–652

251 Schwaninger, M., Blume, R., Oetjen, E., Lax, G. and Knepel, W. (1993) J. Biol. Chem. **268**, 23111–23115

252 Dawson, T. M., Steiner, J. P., Dawson, V. L., Dinerman, J. L., Uhl, G. R. and Snyder, S. H. (1993) Proc. Natl. Acad. Sci. U.S.A. **90**, 9808–9812

253 Sharkey, J. and Butcher, S. P. (1994) Nature (London) **371**, 336–339

254 Kaye, R. E., Fruman, D. A., Bierer, B. E., Albers, M. W., Zydowsky, L. D., Ho, S. I., Jin, Y.-J., Castells, M. C., Schreiber, S. L., Walsh, C. T., Burakoff, S. J., Austen, K. F. and Katz, H. R. (1992) Proc. Natl. Acad. Sci. U.S.A. **89**, 8542–8546

255 Fruman, D. A., Bierer, B. E., Benes, J. E., Burakoff, S. J., Austen, K. F. and Katz, H. R. (1995) J. Immunol. **154**, 1846–1851

256 Cyert, M. S. and Thorner, J. (1992) Mol. Cell. Biol. **12**, 3460–3469

257 Nakamura, T., Liu, Y., Hirata, D., Namba, H., Harada, S., Hirokawa, T. and Miyakawa, T. (1993) EMBO J. **12**, 4063–4071

258 Breuder, T., Hemenway, C. S., Movva, N. R., Cardenas, M. E. and Heitman, J. (1994) Proc. Natl. Acad. Sci. U.S.A. **91**, 5372–5376

259 Parent, S. A., Nielsen, J. B., Morin, N., Chrebet, G., Ramadan, N., Dahl, A. M., Hsu, M.-J., Bostian, K. A. and Foor, F. (1993) J. Gen. Microbiol. **139**, 2973–2984

260 Eng, W.-K., Faucette, L., McLaughlin, M. M., Cafferkey, R., Koltin, Y., Morris, R. A., Young, P. R., Johnson, R. K. and Livi, G. P. (1994) Gene **151**, 61–71

261 Cunningham, K. W. and Fink, G. R. (1994) J. Cell Biol. **124**, 351–363

262 Cardenas, M. E., Hemenway, C., Muir, R. S., Ye, R., Fiorentino, D. and Heitman, J. (1994) EMBO J. **13**, 5944–5957

263 Chiu, M. I., Katz, H. and Berlin, V. (1994) Proc. Natl. Acad. Sci. U.S.A. **91**, 12574–12578

264 Kunz, J., Henriquez, R., Schneider, U., Deuter-Reinhard, M., Movva, N. R. and Hall, M. N. (1993) Cell **73**, 585–596

265 Cafferkey, R., Young, P. R., McLaughlin, M. M., Bergsma, D. J., Koltin, Y., Sathe, G. M., Faucette, L., Eng, W.-K., Johnson, R. K. and Livi, G. P. (1993) Mol. Cell. Biol. **13**, 6012–6023

266 Cafferkey, R., McLaughlin, M. M., Young, P. R., Johnson, R. K. and Livi, G. P. (1994) Gene **141**, 133–136

267 Helliwell, S. B., Wagner, P., Kunz, J., Deuter-Reinhard, M., Henriquez, R. and Hall, M. N. (1994) Mol. Biol. Cell **5**, 105–118

268 Stan, R., McLaughlin, M. M., Cafferkey, R., Johnson, R. K., Rosenberg, M. and Livi, G. P. (1994) J. Biol. Chem. **269**, 32027–32030

269 Chung, J., Kuo, C. J., Crabtree, G. R. and Blenis, J. (1992) Cell **69**, 1227–1236

270 Price, D. J., Grove, J. R., Calvo, V., Avruch, J. and Bierer, B. E. (1992) Science **257**, 973–977

271 Terada, N., Franklin, R. A., Lucas, J. J., Blenis, J. and Gelfand, E. W. (1993) J. Biol. Chem. **268**, 12062–12068

272 Calvo, V., Wood, M., Gjertson, C., Vik, T. and Bierer, B. E. (1994) Eur. J. Immunol. **24**, 2664–2671

273 Monfar, M., Lemon, K. P., Grammer, T. C., Cheatham, L., Chung, J., Vlahos, C. J. and Blenis, J. (1995) Mol. Cell. Biol. **15**, 326–337

274 Brown, E. J., Albers, M. W., Shin, T. B., Ichikawa, K., Keith, C. T., Lane, W. S. and Schreiber, S. L. (1994) Nature (London) **369**, 756–758

275 Sabatini, D. M., Erdjument-Bromage, H., Lui, M., Tempst, P. and Snyder, S. H. (1994) Cell **78**, 35–43

276 Sabers, C. J., Martin, M. M., Brunn, G. J., Williams, J. M., Dumont, F. J., Wiederrecht, G. and Abraham, R. T. (1995) J. Biol. Chem. **270**, 815–822

277 Salituro, G. M., Zink, D. L., Dahl, A., Nielsen, J., Wu, E., Huang, L., Kastner, C. and Dumont, F. J. (1995) Tetrahedron Lett. **36**, 997–1000

278 Tropschug, M., Barthelmess, I. B. and Neupert, W. (1989) Nature (London) **342**, 953–955

279 Colley, N. J., Baker, E. K., Stamnes, M. A. and Zuker, C. S. (1991) Cell **67**, 255–263

280 Ondek, B., Hardy, R. W., Baker, E. K., Stamnes, M. A., Shieh, B.-H. and Zuker, C. S. (1992) J. Biol. Chem. **267**, 16460–16466

281 Baker, E. K., Colley, N. J. and Zuker, C. S. (1994) EMBO J. **13**, 4886–4895

282 Gething, M.-J. and Sambrook, J. (1992) Nature (London) **355**, 33–45

283 Schmid, F. X., Mayr, L. M., Mucke, M. and Schonbrunner, E. R. (1993) Adv. Protein Chem. **44**, 25–66

284 Lang, K., Schmid, F. X. and Fischer, G. (1987) Nature (London) **329**, 268–270

285 Lilie, H., Rudolph, R. and Buchner, J. (1995) J. Mol. Biol. **248**, 190–201

286 Kiefhaber, T., Quaas, R., Hahn, U. and Schmid, F. X. (1990) Biochemistry **29**, 3061–3070

287 Veeraghavan, S. and Nall, B. T. (1994) Biochemistry **33**, 687–692

288 Fransson, C., Freskgard, P.-O., Herbertsson, H., Johansson, A., Jonasson, P., Martensson, L.-G., Svensson, M., Jonsson, B.-H. and Carlsson, U. (1992) FEBS Lett. **296**, 90–94

289 Freskgard, P.-O., Bergenhem, N., Jonsson, B.-H., Svensson, M. and Carlsson, U. (1992) Science **258**, 466–468

290 Kern, G., Kern, D., Schmid, F. X. and Fischer, G. (1994) FEBS Lett. **348**, 145–148

291 Kern, G., Kern, D., Schmid, F. X. and Fischer, G. (1995) J. Biol. Chem. **270**, 740–745

292 Schonbrunner, E. R., Mayer, S., Tropschug, M., Fischer, G., Takahashi, N. and Schmid, F. X. (1991) J. Biol. Chem. **266**, 3630–3635

293 Frech, C. and Schmid, F. X. (1995) J. Biol. Chem. **270**, 5367–5374

294 Rinfret, A., Collins, C., Menard, R. and Anderson, S. K. (1994) Biochemistry **33**, 1668–1673

295 London, R. E., Davis, D. G., Vavrek, R. J., Stewart, J. M. and Handschumacher, R. E. (1990) Biochemistry **29**, 10298–10302

296 Kern, D., Drakenberg, T., Wikstrom, M., Forsen, S., Bang, H. and Fischer, G. (1993) FEBS Lett. **323**, 198–202

297 Kruse, M., Brunke, M., Escher, A., Szalay, A. A., Tropschug, M. and Zimmermann, R. (1995) J. Biol. Chem. **270**, 2588–2594

298 Lodish, H. F. and Kong, N. (1991) J. Biol. Chem. **266**, 14835–14838

299 Steinmann, B., Bruckner, P. and Superti Furga, A. (1991) J. Biol. Chem. **266**, 1299–1303

300 Helekar, S. A., Char, D., Neff, S. and Patrick, J. (1994) Neuron **12**, 179–189

301 Sykes, K., Gething, M. J. and Sambrook, J. (1993) Proc. Natl. Acad. Sci. U.S.A. **90**, 5853–5857

302 Partaledis, J. A. and Berlin, V. (1993) Proc. Natl. Acad. Sci. U.S.A. **90**, 5450–5454

303 Marivet, J., Margis-Pinheiro, M., Frendo, P. and Burkard, G. (1994) Plant Mol. Biol. **26**, 1181–1189

304 Sanchez, E. R. (1990) J. Biol. Chem. **265**, 22067–22070

305 Yount, G. L., Gall, C. M. and White, J. D. (1992) Mol. Brain Res. **14**, 139–142

306 Norregaard-Madsen, M., Mygind, B., Pedersen, R., Valentin-Hansen, P. and Sogaard-Andersen, L. (1994) Mol. Microbiol. **14**, 989–997

307 Luban, J., Bossolt, K. L., Franke, E. K., Kalpana, G. V. and Goff, S. P. (1993) Cell **73**, 1067–1078

308 Steinkasserer, A., Harrison, R., Billich, A., Hammerschmid, F., Werner, G., Wolff, B., Peichl, P., Palfi, G., Schnitzel, W., Mylnar, E. and Rosenwirth, B. (1995) J. Virol. **69**, 814–824

Biochem. J. (1996) **316**, 1–11 (Printed in Great Britain)

REVIEW ARTICLE
Collagen fibril formation

Karl E. KADLER*§, David F. HOLMES*, John A. TROTTER† and John A. CHAPMAN‡
*Wellcome Trust Centre for Cell-Matrix Research, School of Biological Sciences, and ‡Department of Medical Biophysics, University of Manchester, Stopford Building 2.205, Oxford Road, Manchester M13 9PT, U.K., and †University of New Mexico School of Medicine, Department of Anatomy, Albuquerque, NM 87131, U.S.A.

Collagen is most abundant in animal tissues as very long fibrils with a characteristic axial periodic structure. The fibrils provide the major biomechanical scaffold for cell attachment and anchorage of macromolecules, allowing the shape and form of tissues to be defined and maintained. How the fibrils are formed from their monomeric precursors is the primary concern of this review. Collagen fibril formation is basically a self-assembly process (i.e. one which is to a large extent determined by the intrinsic properties of the collagen molecules themselves) but it is also sensitive to cell-mediated regulation, particularly in young or healing tissues. Recent attention has been focused on 'early fibrils' or 'fibril segments' of ∼ 10 μm in length which appear to

be intermediates in the formation of mature fibrils that can grow to be hundreds of micrometres in length. Data from several laboratories indicate that these early fibrils can be unipolar (with all molecules pointing in the same direction) or bipolar (in which the orientation of collagen molecules reverses at a single location along the fibril). The occurrence of such early fibrils has major implications for tissue morphogenesis and repair. In this article we review the current understanding of the origin of unipolar and bipolar fibrils, and how mature fibrils are assembled from early fibrils. We include preliminary evidence from invertebrates which suggests that the principles for bipolar fibril assembly were established at least 500 million years ago.

INTRODUCTION

Collagen is distinct from other proteins in that the molecule comprises three polypeptide chains (α-chains) which form a unique triple-helical structure. For the three chains to wind into a triple helix they must have the smallest amino acid, glycine, at every third residue along each chain. Each of the three chains therefore has the repeating structure Gly-Xaa-Yaa, in which Xaa and Yaa can be any amino acid but are frequently the imino acids proline and hydroxyproline. More than 20 genetically distinct collagens exist in animal tissues. Collagen types I, II, III, V and XI self-assemble into D-periodic cross-striated fibrils [1–4] (Figure 1) (where $D = 67$ nm, the characteristic axial periodicity of collagen) and collectively are the most abundant collagens in vertebrates. The fibril-forming collagen molecules consist of an uninterrupted triple helix of approx. 300 nm in length and 1.5 nm in diameter flanked by short extrahelical telopeptides. The telopeptides, which do not have a repeating Gly-Xaa-Yaa structure and do not adopt a triple-helical conformation, account for 2% of the molecule and are critical for fibril formation (see below).

Type I collagen {[α1(I)]₂α2(I)} is found throughout the body except in cartilaginous tissues. It is also synthesized in response to injury and in the fibrous nodules formed in the sequelae of fibrotic disease. Type II collagen {[α1(II)]₃} is found in cartilage, developing cornea and vitreous humour. These major collagen fibrils are almost certainly not formed from just one collagen type but instead are co-polymers of two or more fibril-forming collagens. Type III collagen {[α1(III)]₃} is found in the walls of arteries and other hollow organs and usually occurs in the same fibril with type I collagen. Type V collagen [α1(V), α2(V), α3(V)] and type XI collagen [α1(XI), α2(XI), α3(XI)] are minor components of tissue and occur as heterotypic fibrils with type I and type II collagen respectively (for a review of collagen distribution, see [5]).

Much of what is known about collagen fibril assembly has

resulted from studies of the type-I-collagen-containing fibrils in tendon and skin and from studies in which fibrils are reconstituted *in vitro* from purified type I collagen. Therefore, out of necessity, the present review is concerned primarily with the assembly of

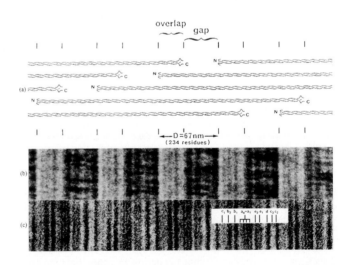

Figure 1 Axial structure of D-periodic collagen fibrils

(**a**) Schematic representation of the axial packing arrangement of triple-helical collagen molecules in a fibril, as derived from analysis of the positive (**c**) and negative (**b**) staining patterns. (**b**) Collagen fibril negatively stained with sodium phosphotungstic acid (1%, pH 7). The fibril is from a gel of fibrils reconstituted from acetic-acid-soluble calf-skin collagen. The repeating broad dark and light zones are produced by preferential stain penetration into regions of lowest packing (the gap regions). (**c**) Similar fibril positively stained with phosphotungstic acid (1%, pH 3.4) and then uranyl acetate (1%, pH 4.2). The darkly staining transverse bands are the result of uptake of electron-dense heavy-metal ions from the staining solutions on to charged residue side groups of collagen. For a detailed explanation of the band assignments and analysis, see [1].

Abbreviations used: D, the axial periodicity of collagen fibrils (= 67 nm); pNcollagen, procollagen containing the N-propeptides and lacking the C-propeptides; pCcollagen, procollagen containing the C-propeptides and lacking the N-propeptides; DPS III, D-periodic symmetrical banding type III.
§ To whom correspondence should be addressed.

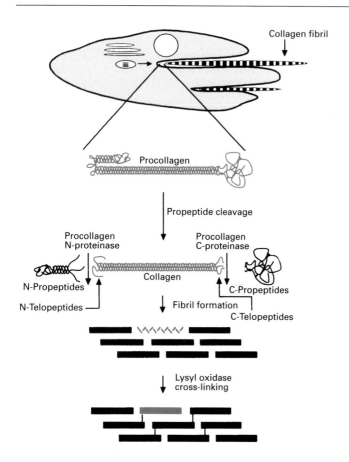

Figure 2   Extracellular events in the synthesis of fibrillar collagens

Procollagen consists of a 300-nm-long triple-helical domain (comprised of three α-chains each of approx. 1000 residues) flanked by a trimeric globular C-propeptide domain (the right-hand side of the diagram) and a trimeric N-propeptide domain (the left-hand side of the diagram). Procollagen is secreted from cells and is converted into collagen by the removal of the N- and C-propeptides by procollagen N-proteinase and procollagen C-proteinase respectively. The collagen generated in the reaction spontaneously self-assembles into cross-striated fibrils that occur in the extracellular matrix of connective tissues. The fibrils are stabilized by covalent cross-linking, which is initiated by oxidative deamination of specific lysine and hydroxylysine residues in collagen by lysyl oxidase. The process is shown occurring in cell-surface crypts according to the model generated by Birk and co-workers (see text for references).

type I collagen into fibrils. Other collagens will be mentioned only with regard to how their assembly into fibrils differs from that of type I collagen and how they influence or participate in fibril formation.

The assembly of collagen molecules into fibrils is an entropy-driven process, similar to that occurring in other protein self-assembly systems, such as microtubules, actin filaments and flagella (for a review, see [6]). These processes are driven by the loss of solvent molecules from the surface of protein molecules and result in assemblies with a circular cross-section, which minimizes the surface area/volume ratio of the final assembly. Although the broad principles of collagen fibril self-assembly are generally accepted, less is known about the molecular mechanisms of the assembly process.

A fundamental feature of fibril-forming collagens is that they are synthesized as soluble procollagens (Figure 2), which are converted into collagens by specific enzymatic cleavage of terminal propeptides by the procollagen metalloproteinases. Without these proteinases the synthesis of collagen fibrils would not occur. A suitable cell-free system of assembling fibrils has been developed in which procollagen is sequentially cleaved with the purified

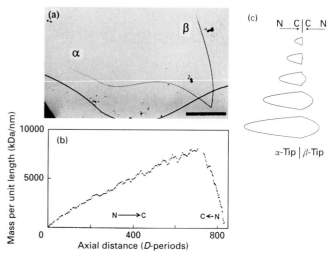

Figure 3   Bipolar fibrils formed *in vitro* by cleavage of purified pCcollagen with procollagen C-proteinase

(**a**) Transmission electron micrograph of a positively stained collagen fibril generated by cleavage of pCcollagen (50 μg/ml) with the C-proteinase (50 units/ml) at 37 °C. The fibril displays fine (α-tip) and coarse (β-tip) ends. Scale bar = 1 μm. (**b**) Axial mass distribution of an entire unstained fibril similar to the one shown in (**a**). The fibril shows a near-linear axial mass distribution of the two tips, with no evidence of a limiting diameter. Arrows show orientations of collagen molecules within the fibril; E/S [enzyme units/substrate mass (μg)] = 50:50. (**c**) Schematic representation of the growth of a bipolar fibril in the cell-free system. The model shows a two-stage model, as indicated by light-microscope observations [8], in which growth occurs first from a pointed tip (the α-tip) and additional growth occurs from the blunt end after the formation of a second pointed tip (β-tip) for growth in the opposite direction. The tip profiles are shown as sections of parabolas, consistent with the linear axial mass distributions of a fibril with a circular cross-section.

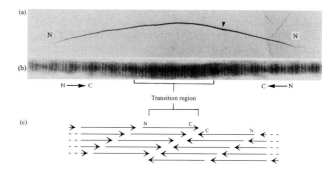

Figure 4   Transmission electron micrograph of a positively stained bipolar fibril from 18-day chick embryo metatarsal tendon

(**a**) The fibril is 10.5 μm long and shows a polarity reversal 3.5 μm from one end (arrowhead). (**b**) Enlargement of the banding pattern at the polarity reversal region (brace). The braced region shows four *D*-periods where the staining pattern illustrates molecules in anti-parallel arrangement. The centre two *D*-periods show a symmetrical pattern with two axial planes of mirror symmetry. These mirror planes occur between the **d** and **c2** staining bands and in the vicinity of the **a3** band. (**c**) Analysis of the staining pattern in the transition region indicates an anti-parallel arrangement of molecules. The schematic representation shows that the axial extent of the transition region is about four *D*-periods. This corresponds to the minimal distance possible to achieve polarity reversal and to maintain the *D*-periodicity concomitantly.

procollagen metalloproteinases to generate collagen *de novo* [6–9]. Fibrils generated in the system initially have a near-paraboloidal pointed tip [10] and a blunt end, and growth is exclusively from the pointed tip [11]. As growth proceeds, the blunt end becomes a new pointed tip for growth in the other

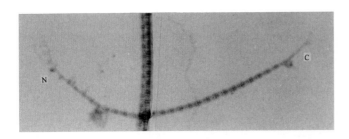

Figure 5   Transmission electron micrograph of a positively stained unipolar collagen fibril from 18-day chick embryo metatarsal tendon

The unipolar fibril is positioned from left to right and is seen crossing a larger cross-banded fibril (running from top to bottom). The unipolar fibril is 2 μm in length and shows N- and C-terminal tips with no polarity reversal.

direction [12]. Furthermore, the two pointed tips each have collagen molecules oriented with N-termini closest to the fibril end [11] (Figure 3). Thus the fibrils are N–N bipolar, in which a switch in molecular orientation occurs at a region along the fibril. These features had previously not been seen in fibrils formed *in vivo* and initially appeared to be artefacts arising from contaminants in the preparations of procollagen or the procollagen metalloproteinases. However, subsequent work has shown that bipolar fibrils having two N-terminal paraboloidal tips occur in developing chick tendon [13] (Figure 4). Thus the cell-free system had accurately predicted several fundamental features of the assembly of collagen fibrils, including the paraboloidal shape of the tip and the occurrence of N–N bipolars. Recent data from our laboratories have shown that fibrils in developing chick tendon have features additional to those of fibrils formed in the cell-free system. For example, some fibrils are unipolar, having molecules pointing exclusively in one direction, in which case fibrils have a C-terminal and an N-terminal end [13] (Figure 5).

UNIPOLAR FIBRIL FORMATION FROM ACID-SOLUBLE COLLAGEN

Collagen may be extracted from several tissues into neutral salt buffers or, with greater yield, into dilute acidic solutions [14]. Typical acetic acid extracts of skin and tendon yield milligram quantities of type I collagen, mainly in the form of monomers but also including variable amounts of cross-linked components (dimers, trimers and some higher components). Preparations may also vary in respect of the intactness of the proteinase-susceptible, non-helical, telopeptide regions of the molecule. Such preparations, when neutralized and warmed to temperatures between 20 and 34 °C, produce a gel of *D*-periodic fibrils over the course of several hours (Figure 6). At 34 °C fibril diameters are typically in the range 20–70 nm. Lower temperatures generally result in broader fibrils, with diameters of up to 200 nm found at 20 °C [15,16]. Samples of these final gels show a meshwork of very long fibrils in which ends are not observed (Figure 6). The rate of assembly of fibrils can be monitored by measuring turbidity which, to a close approximation, is proportional to the amount of fibrillar material formed [14–17]. A typical near-sigmoidal plot shows three regions: a lag region, a growth region and a plateau. Diameter measurements on fibrils obtained during the time course of assembly have demonstrated that a limiting fibril diameter distribution occurs when about 20 % of the collagen molecules have assembled into fibrils, suggesting that the latter stages of assembly must be at the ends of existing fibrils [17].

Fibrils formed from acid-soluble collagen are unipolar, *D*-periodic and have two smoothly tapered ends. Early fibrils, ranging in length from 1 to 20 μm, are observed at the end of the lag phase and in the early growth phase (Figure 6). Such early fibrils showed a well defined shape under particular solution conditions, with the occurrence of a 'limiting early fibril' of about 90 *D*-periods (6 μm) in length and with a maximal cross-section containing about 160 molecules (Figure 6) [18,19]. Such observations imply a greater level of growth control in the self-assembly of these fibrils than is indicated from observations on the final fibril gel.

Other workers [20], however, have reported the occurrence of non-banded filaments (of diameter in the range 10–20 nm) during this early phase of fibril assembly, and concluded that the final banded fibrils are formed by lateral fusion of the first formed filaments. These apparently conflicting observations of the assembly pathways were found to be due to differences in the method used to initiate fibril formation, rather than to differences in collagen preparation or solution conditions [19]. The same sample of collagen could show different aggregation states depending on the order of warming and neutralizing of the solution. The occurrence of non-banded filaments required the solution to go through a cold neutral step. The molecular mechanism leading to these different assembly routes is likely to involve changes in the conformation of the telopeptides that accompany fibril assembly (see below).

Initial oligomer formation in fibril assembly has been studied by photon correlation spectroscopy of solutions [21–26] or by electron microscopy of rotary-shadowed samples adsorbed to support films [27,28]. Studies have included experiments starting from either near-monomeric preparations of lathyritic collagen or monomer fractions of rat-tail tendon collagen. (Lathyritic collagen is obtained from animals fed 2-aminopropionitrile, which inhibits the enzyme lysyl oxidase, and consequently the animals have collagen with a much decreased cross-linking capacity.) In both solution and electron microscope studies the formation of a 4*D*-staggered dimer has been identified as a preferred initial aggregation step. Some solution studies have indicated a second stage of assembly involving the lateral aggregation of dimers and trimers into oligomers [25,26]. These species cannot be definitively assigned to a specific part of the early fibril assembly pathway such as nucleation or propagation. Some key intermediates may not accumulate in solution, whereas other abundant species may not be true intermediates contributing to the final fibril. The difficulty in experimentally determining a nucleating and accreting species is pronounced if the acid-extracted collagen is warmed prior to being neutralized to initiate fibril growth. In this method of initiating fibril formation there is a rapid onset of heterogeneity of aggregate size, with both dimers and early fibrils (containing 10^3–10^4 molecules) present at the end of the lag phase [19].

Fourier-transform IR spectroscopy suggests that conformational changes occur in the collagen molecule during assembly into fibrils [29]. Changes in the carbonyl group spectrum (amide I; 1700 to 1600 cm^{-1}) were evident in the 22–26 °C temperature range, under fibril-forming conditions, which led to the hypothesis that the triple helix of the semi-flexible collagen molecule is actually perfected during the lag phase, facilitating nucleation and intermolecular interaction. Spectra were also obtained in the amide II and III regions. Further spectral changes after fibrils had formed showed that the molecules are once again distorted as they are bent to fit within the fibrils.

Partial loss of the telopeptides of the collagen molecule has major effects on fibril growth [30,31]. These include loss of diameter uniformity, loss of unidirectional packing and changes

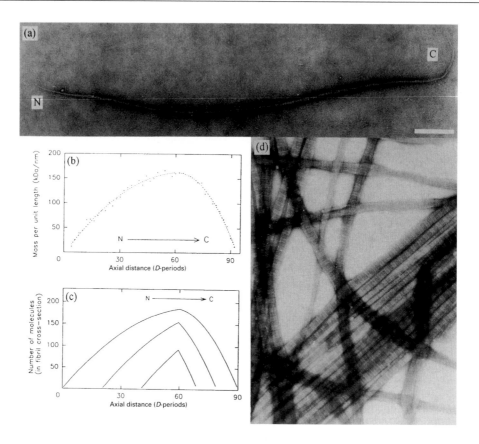

Figure 6   Unipolar fibrils formed *in vitro* by reconstitution from acetic-acid-soluble calf-skin collagen

(**a**) Transmission electron micrograph of a negatively stained unipolar fibril displaying an N- and a C-terminal tip. The fibril was sampled when approx. 1% of the collagen had assembled into fibrils. Bar = 300 nm. (**b**) Axial mass distribution of an unstained early fibril. Mass determination was by quantitative scanning transmission electron microscopy. (**c**) Set of growth curves obtained by averaging axial mass distributions similar to that shown in (**b**). The slopes of the N- and C-ends of the fibril remain constant with fibril elongation. Analysis of the slopes of the axial mass distributions indicates an increase of five collagen molecules per *D*-period at the N-tip and 10 molecules per *D*-period at the C-tip. (**d**) Negatively stained sample of the final fibril gel shown at the same magnification as (**a**). Note that the final fibrils are larger in diameter than early fibrils and that the ends of the fibrils are not observed.

in the fibril assembly pathway, depending on the extent of removal of each of the N- and C-telopeptides. Experimental approaches include exposure of the collagen solution to pepsin with partial removal of both telopeptides [32,33], or selective degradation of the N-telopeptide or of the C-telopeptide with, respectively, leucine aminopeptidase and carboxypeptidase [30,31]. Loss of the N-telopeptides has been linked with the formation of *D*-periodic symmetrical fibrils with molecules in anti-parallel contact, while loss of part of the C-telopeptides has been associated with the formation of *D*-periodic tactoids. Complete removal of both telopeptides prevents the formation of fibrils, assembly being limited to the formation of small non-banded fibrous aggregates. The experimental data have been interpreted in terms of a simple model where the N-telopeptide is critical for the formation of the polarized 4*D*-staggered dimers that occurs as an early stage of assembly, and the C-telopeptide has a dual role, promoting a lateral accretion of linear aggregates as well as participating in the formation of the early linear assemblies [31].

Electron optical data [34] and X-ray data [35] both indicate that the N-telopeptides are axially contracted when the collagen molecules are assembled into fibrils. The X-ray data predict a mean residue spacing of 0.7 *h*, where *h* is the axial spacing of residues (= 0.286 nm) in the triple helix. Other experimental evidence suggests that the condensed structure is explained by a hairpin conformation of the telopeptides. Thus NMR studies on

N-telopeptides in solution indicate the occurrence of *β*-folds and flexible hinge regions [36,37], and sequence information points to a hairpin loop conformation for N-telopeptides [32]. Rotary shadowing of individual procollagen molecules and mass mapping of assembled pNcollagen (i.e. procollagen containing the N-propeptides) molecules confirms that the N-terminal ends are in a bent-back conformation [38–41].

NMR studies of synthetic peptides show no preferred secondary structure of the C-telopeptides [42]. However, X-ray data do indicate that the C-telopeptides are axially contracted with a mean residue spacing of 0.5 *h* when the collagen molecules are assembled into fibrils. Sequence analysis suggests that the telopeptides form a hydrophobic cluster [31]. The absence of structure when the telopeptides are free in solution suggests that the contracted conformation of the C-telopeptides may only occur when molecules are in close association with neighbouring collagen molecules in a fibril.

BIPOLAR FIBRIL FORMATION BY CLEAVAGE OF PROCOLLAGEN CONTAINING THE C-PROPEPTIDES (PCCOLLAGEN) WITH PROCOLLAGEN C-PROTEINASE

Fibrils formed by neutralizing and warming of solutions of extracted collagen do not usually have the same diameters as the fibrils from which the collagen was extracted. Thus cell-mediated control must be exerted over the self-assembly process *in vivo*.

Evidence suggests that N- and C-propeptides of procollagen may play a role in such diameter regulation (see below for references). For these and other reasons, a cell-free system of fibril assembly was developed whereby procollagen is sequentially cleaved to collagen by the procollagen metalloproteinases. Fibrils formed by cleavage of purified type I pCcollagen with procollagen C-proteinase are exclusively bipolar. Early studies showed that fibrils formed by cleavage of pCcollagen with C-proteinase at low temperatures (29–32 °C) have sufficiently large diameters to be visualized by dark-field light microscopy. The earliest seen aggregates are needle-like, with a pointed end and a blunt end. Growth, observed by time-lapse photography, occurs exclusively at pointed tips. Growth at a blunt end only occurs at high pCcollagen concentrations and, at temperatures above 34 °C, by the appearance of a spear-like projection from the blunt end (as shown schematically in Figure 3), with the projection becoming a new pointed tip for growth in the opposite direction. Smaller-diameter fibrils formed at 37 °C have been shown by electron microscopy to have two smoothly pointed tips. Each tip has collagen molecules oriented with their N-termini directed towards the tip, indicating that a reversal in molecular polarity must exist within the central body of the fibril. Furthermore, these bipolar fibrils are usually shape-asymmetrical and have a fine (α-) tip and a coarse (β-) tip. The tips have linear axial mass distributions, consistent with a paraboloidal shape (Figure 3). In addition, the fineness of a tip (slope of the axial mass distribution) appears to be set at an early stage of growth, such that subsequent fibril growth preserves the shape of the tips. Recent work has shown that the level of C-proteinase has a major effect on the shape of the fibrils in this cell-free system (D. F. Holmes, R. B. Watson, J. A. Chapman and K. E. Kadler, unpublished work). At a constant initial concentration of pCcollagen, increased levels of C-proteinase led to increased slopes of the axial mass distributions of α-tips. Increasing the C-proteinase levels also resulted in a reduced shape asymmetry, yielding fibrils that were nearly shape-symmetrical. It cannot be assumed that all the features of fibril growth *in vitro* are applicable to the process *in vivo*. However, the recent observation of both unipolar and bipolar early fibrils occurring *in vivo* has given encouragement to the simple working model that some aspects of fibril assembly in tissues can be understood as an intrinsic self-assembly process governed by the interaction properties of the collagen monomer. Since early fibrils formed by reconstitution from acid-extracted collagen are exclusively unipolar [43] and those formed by cleavage of pCcollagen with C-proteinase are exclusively bipolar, it seems possible that the occurrence of bipolar fibrils is, in part, dependent on the initial concentration of pCcollagen at the onset of fibril assembly.

These studies raise a number of questions. First, do type II, III, V and XI collagens exhibit similar assembly properties? Limited studies *in vitro* suggest that recombinant type II collagen [44] assembles into thin fibrils arranged in network-like structures [45]; the rate of assembly is lower than that observed for type I collagen under the same experimental conditions [45], and the fibrils formed are C–C bipolar fibrils in which most of the central shaft consists of molecules in an anti-parallel orientation [4,45a]. Fibrils formed from type III and V collagen appear to have the usual parallel orientation of collagen molecules seen in type I collagen fibrils, but no information is available on the occurrence of unipolar or bipolar fibrils of these collagen types.

Secondly, what factors determine the shape of fibrils? Do fibrils grow at equilibrium with a shape of minimum free energy or is, as seems more likely, their shape determined by kinetic factors? Computer-generated simulations of fibril growth have been used to seek answers to these questions. Two models have been proposed for a kinetically determined growth process. One is an interface-controlled growth model where the rate of accretion on to the fibril surface is limited by transition processes on that surface; the shape of the fibril would be determined by the existence of different types of binding site with different accretion rates [46]. In another model, fibrils form by diffusion-limited aggregation [47,48]. Both models can account for tips with linear axial mass distributions. The model suggested by Silver et al. [46] adopts a set of growth rules for the formation of a nucleus and the subsequent addition of monomers to the fibril surface. Several growth steps can be defined involving axial growth of microfibril units and lateral growth which generates additional microfibrils. The model proposed by Parkinson et al. [47,48] involves a general scheme of assembly of rod-shaped molecules where the rate of fibril assembly is limited by the rate of diffusion of monomer to the fibril surface. Such a process was shown to generate fibrils with the same structural features as fibrils formed by cleavage of pCcollagen with C-proteinase *in vitro*, in that they had no constant diameter and the tips showed linear axial mass distributions. The combination of these advanced computer modelling methods together with the experimental capability of obtaining precise size and shape data on fibrils (by quantitative scanning transmission electron microscopy) provides a new route to understanding the underlying principles of self-assembly of collagen fibrillar structures.

FIBRIL ASSEMBLY AND POLARITY *IN VIVO*

A remarkable feature of fibrils formed *in vivo* (see Figures 4 and 5) is that not all are unipolar (i.e. with the molecular polarity unchanged throughout the entire length of the fibril). In 18-day chick embryos about half of the early fibrils are bipolar and half are unipolar. Analysis of the positive stain pattern throughout the entire length of the bipolar fibrils reveals the axial zone of molecular polarity reversal to be highly localized [13]. The transition region contains molecules in anti-parallel contact, of little more than 4 D-periods in extent. Analysis of the staining pattern in the transition region demonstrated that the anti-parallel arrangement of the molecules is such that the C-termini of oppositely directed molecules are in axial register. A similar anti-parallel contact is seen in a D-periodic symmetrical fibril (where molecules are in anti-parallel arrangement throughout the length of the fibril) formed *in vitro* from type I collagen (D-periodic symmetrical banding type III; DPS-III) [49,50]. Furthermore, this arrangement allows for axial continuity of the intermolecular cross-links in fibrils. As described below, fibrils from echinodermis are also N–N bipolar, with a centrally located molecular switch region in which the transition region is restricted to approx. 10 D-periods [51].

The mechanical and physical properties of a tissue depend on a hierarchical spatial arrangement of collagen fibrils, whether unipolar or bipolar. Thus narrow fibrils (~ 20 nm) in precise orthogonal array occur in the cornea, where optical transparency is important, whereas mature tendon, where high tensile strength is paramount, displays a high density ($\sim 10^8$ fibrils/mm^3) of large-diameter fibrils (~ 500 nm) arranged in parallel bundles. Experiments by Birk and co-workers using electron microscopy of serial transverse sections of embryonic chick tendon have shown that cells can have considerable control over the shape of the pericellular space where fibrils occur. Early fibrils are deposited in extracellular compartments formed between cellular protrusions close to the cell body [52]. Further away from the cell, such compartments (containing a few fibrils) anastomose to generate larger spaces containing bundles of fibrils. In 14-day chick tendon, entire fibrils showing smoothly tapered ends can be

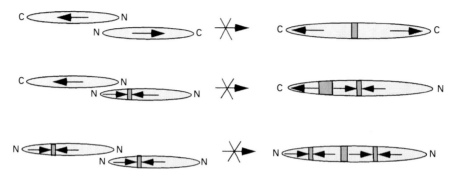

Possible fusions consistent with experimental data:

Fusions inconsistent with experimental data:

Figure 7   Models of fibril fusion

The fusion of fibrils is modelled using the assumption that fibrils can only have either two N-ends or one N-end and one C-end. Extended or multiple regions of asymmetrical orientation of collagen molecules were not allowed. There is the possibility that two unipolar fibrils pointing in opposite directions and with limited overlap could fuse to generate a bipolar fibril with a molecular switch region limited to approx. four *D*-periods (not shown). Arrows indicate molecular polarity within a fibril. Pink boxes indicate regions of polarity reversal.

found within such bundles [52,53]. Moreover, the tips of fibrils are of different taper, giving an overall shape asymmetry to the fibril. The fibrils are bipolar, and it has been proposed by Birk and co-workers that end-to-end fusion of these early bipolar fibrils could contribute to fibril elongation, and that their side-by-side fusion may account for the observed increase in fibril diameter during tissue development [54]. Fibril fusion, it is suggested, may also be influenced by the presence of surface-bound molecules of decorin [54,55].

End-to-end fusion of bipolar fibrils of type I collagen would seem, however, to be inconsistent with existing experimental data (Figure 7). Bipolar fibrils of type I collagen observed hitherto have invariably displayed two N-ends, i.e. none with two C-ends have been seen, either *in vivo* or *in vitro*. Moreover, the switch region (the polarity transition region in which molecules occur in anti-parallel array) is of limited axial extent and fibrils with multiple switch regions have not been found. Unipolar fibrils can fuse in a number of ways that would accord with the experimental data. For example, two unipolar fibrils with molecules oriented N- to C- from left to right could fuse in side-by-side register (to result in a thicker fibril), or in end-on-end register (giving a longer fibril). End-on-end fusion of two oppositely directed unipolar fibrils (i.e. one fibril oriented N- to C- from left to right and the other from right to left) would also seem to be permitted (yielding a bipolar fibril). Side-by-side fusion of two oppositely directed unipolar fibrils is, however, disallowed because this would result in axially extended regions of anti-parallel axial packing of collagen molecules, which is not seen *in vivo* or *in vitro*.

The need to satisfy these criteria introduces severe limitations on the participation of bipolar fibrils in fibril fusion. In only one special case, where the molecular switch regions of two neighbouring bipolar fibrils are perfectly aligned, is the fusion of two bipolar fibrils allowed. In all other combinations, the fusion of two bipolar fibrils would result in extended or multiple regions of anti-parallel molecular packing within the resultant fibril. In a mixed population of unipolar and bipolar fibrils, end-to-end fusion is allowed but would have the effect of decreasing the unipolar population, resulting in a population enriched in bipolar fibrils, unable to fuse further. Could a possible function for bipolar fibrils be to limit the extent of fibril fusion?

DIAMETER MEASUREMENTS OF FIBRILS FORMED *IN VIVO*

Extensive diameter measurements have been made on fibrils in a large range of tissues, mainly from electron micrographs of fibril transverse sections. Diameter distributions are typically narrow and unimodal during early stages of development, but the mean diameters and spread of diameters increase with maturation of the tissue. Parry and Craig [56] have reported the occurrence of preferred diameters of multiples of 8 nm. This conclusion was based mainly on the mean diameters from unimodal distributions obtained from different tissues at various stages of development in a large number of species. In addition, occasional observations have been made of fibril diameters showing a multi-modal distribution with peaks at multiples of 8 nm [57]. Such diameter measurements have been made on dehydrated, plastic-embedded samples where shrinkage is typically 27% , as indicated by comparative electron microscopy and X-ray diffraction data [57–59]. The corrected value for the preferred diameter increment is then 11 nm, resulting in fibrils of diameter 22, 33, 44, 55 nm, etc. [60,60a].

Studies have located partially processed type I collagen molecules retaining N-propeptides at the surface of fibrils [61–64], and this has led to the suggestion that uncleaved N-propeptides at the fibril surface limit further accretion of collagen molecules on to growing fibrils [65]. A quantitative model of diameter limitation that predicts the reported quantification of diameters has been proposed by Chapman [60,60a]. The model is based on the transient retention of N-propeptides. In this model the N-propeptides are constrained to the surface of the growing fibril, leading to an initial limiting diameter, when N-propeptides become close-packed, of about 22 nm (based on a lateral separation of between 1.8 and 2.0 nm between N-propeptide units). Subsequent cleavage of surface-located N-propeptides and accretion of a further round of pNcollagen molecules leads to a second limiting diameter of about 33 nm. Repetition of this cycle of deposition and cleavage yields fibrils with preferred diameters of multiples of 11 nm. Variants of this type of diameter-limitation process could involve other surface-bound macromolecules. An alternative mechanism of diameter limitation involving surface-associated small proteoglycans has been proposed [66–72]. Here the glycosaminoglycan chains are extended around the circumference in complexes and act as a molecular ruler to define the lateral extent of the fibril.

Evidence from immunoelectron microscopy and biochemical studies suggests that collagen fibrils *in vivo* are composed of several collagen types, partially processed collagens and other macromolecules, including proteoglycans. For example, it has been suggested that dermal collagen fibrils are hybrids of type I and type III collagens and that the type III collagen is located at the periphery of the fibril. The surface location of type III collagen implies that this collagen may have a different role than type I collagen and may have a regulatory function in fibrillogenesis [73]. Double immunoelectron microscopy of foetal skin using an antibody directed against the N-propeptide of type III procollagen and another directed against the N-propeptide of type I procollagen revealed labelling of type I and III pNcollagens on the same thin (20–30 nm) fibrils. Larger fibrils (90–100 nm) were coated with type III collagen and type III pNcollagen but not with type I pNcollagen. The type I N-propeptide was present on thin fibrils only at restricted locations in adult skin. Immunoblotting of skin extracts revealed the presence of both type III pNcollagen and type I pNcollagen in foetal skin, but only type III pNcollagen in adult skin. These data were interpreted to suggest that type I and type III collagens form hybrid fibrils and that type III pNcollagen is added to mature fibrils [74]. The data did not, however, exclude the possibility that type III pNcollagen can be deposited on preformed type I collagen fibrils after the fibrils are assembled.

Experiments *in vitro* in which mixtures of type III pNcollagen and type I collagen were generated simultaneously by enzymic cleavage of precursor forms of the proteins suggested that type III pNcollagen forms true co-polymers with type I collagen [75]. In addition, the results demonstrated that co-polymerization of type III pNcollagen with type I collagen generated fibrils that were thinner than fibrils generated under the same conditions from type I collagen alone. Of particular interest is the finding that co-polymers of type III pNcollagen and type I collagen are apparently circular in outline and are not like the hieroglyphic fibrils formed from mixtures of type I pNcollagen and type I collagen. A further important difference was that the co-polymerization increased the concentration in solution at equilibrium of type I pNcollagen, whereas co-polymerization with type I collagen decreases the concentration in solution of type III

pNcollagen [76]. These differences imply important biological roles for types I and III pNcollagen.

As a further example of the occurrence of hybrid fibrils, type V collagen has been immunolocalized to the type-I-collagen-containing fibrils and as fine-diameter fibrils in the corneal stroma [77,78]. Corneal stroma has a higher concentration of type V collagen than do other type-I-collagen-containing tissues with large-diameter fibrils. This has led to the suggestion that co-polymerization of type V and type I collagens may limit the diameter of the fibrils formed. It is noteworthy that type V collagen contains a small pepsin-sensitive N-terminal domain, presumably part or whole of the N-propeptide of the molecule. When type I collagen was mixed with type V collagen, the mean fibril diameter decreased with increasing type V/I collagen ratio [78]. Moreover, the N-terminal domain of type V collagen was needed for this regulatory effect, and without the domain little diameter-regulating ability was observed.

Type IX, XII, XIV and XIX collagens are members of a subfamily of fibril-associated collagens with interrupted triple helices (FACITs) which are found at the surfaces of, and may participate in the formation of, collagen fibrils. Type IX collagen associates with the surfaces of the narrow-diameter fibrils in cartilage, vitreous and developing cornea, where it is covalently bound to chains of type II and XI collagens [79]. It has been proposed, based on immunolocalization using monospecific antibodies, that type XI collagen forms a small cylindrical core around which molecules of type II collagen are arranged. Type IX collagen appears to be located exclusively at the fibril surface, where it may act as a molecular linker between collagen fibrils and macromolecules in the extracellular matrix. Type XII and XIV collagens have very large N-terminal domains. These molecules have been implicated in modulating the deformability of the extracellular matrix, as they are localized near the surface of banded collagen fibrils and they can mediate interactions between fibrils *in vitro* [80,81].

FIBRILS IN INVERTEBRATES

Collagen fibrils have been major constituents of the connective tissues of animals since the emergence of multicellularity. Because invertebrates comprise approximately 95 % of animal species, it might be expected that their collagens would form a great variety of fibrillar aggregates. Indeed, a number of different aggregate forms of fibrillar collagens have been described, and the subject has been reviewed [82–85]. However, the aggregation of triple-helical collagen molecules into unbranched, quasi-crystalline fibrils has been the dominant mode of fibril organization since the sponges first evolved.

As in vertebrates, the fibrillar aggregates in sponges are stabilized by covalent cross-links, but these cross-links are of unknown chemistry [86]. In other animals tyrosine-, quinone-, aldimine-, ketoimine- and cysteine-derived cross-links exist [87]. An obvious extrapolation of these observations is that the pathways for the secretion and extracellular assembly of collagen molecules into fibrils and the processing enzymes required for converting the insoluble aggregates into mechanically and chemically stable structures have, from the beginning, been essential components of collagenous matrices.

From their first appearance in sponges, collagen fibrils have been associated with other macromolecules, and have been organized into fibre-reinforced composite tissues consisting of stiff and strong collagen fibrils in an isotropic matrix dominated by proteoglycans and water. As is true of all fibrous composites, the mechanical properties of collagenous matrices are determined by the physical–chemical properties, concentrations, spatial

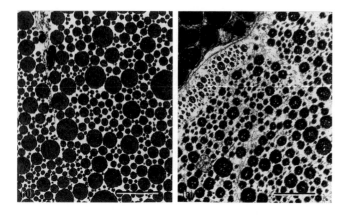

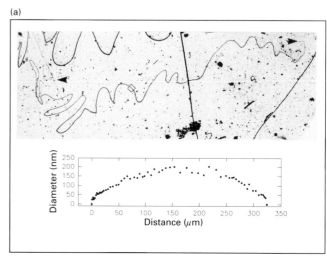

Figure 8 Transmission electron micrographs of echinoderm collagenous tissues

(**a**) Sea urchin ligament. Fibrils have approximately circular cross-sections and uniform electron density. Diameters are highly variable. (**b**) Sea cucumber dermis. Fibrils are somewhat irregular in outline and possess internal electron-lucent regions. Scale marker = 0.5 μm. Reproduced from [92], with permission.

arrangements and interactions of the separate constituents. The phylum *Echinodermata* has evolved collagenous tissues with mutable mechanical properties that make them interesting from the perspective of their unique physiology, as well as providing a unique opportunity to study collagen fibril structure and inter-fibrillar interactions. The connective tissues of these animals are mechanically regulated by the secretions of resident neuro-secretory cells, and change rapidly from stiff and strong to weak and compliant states [88,89]. These mechanical changes are thought to be due to changes in the strength of associations between adjacent fibrils.

The organization of collagen fibrils in the spine ligaments of sea urchins resembles that in vertebrate ligaments; the relaxed spine ligaments even have a crimp pattern. The fibrils are allowed to slide when the ligament is plastic, and are strongly inter-connected when it is elastic (stiff) [90,91]. Electron micrographs of *Eucidaris tribuloides* ligaments show round fibrils with a uniform interior (Figure 8). Their diameter distribution is unimodal; the range is 20–750 nm, with a modal value of 100 nm [92]. The diameter distribution is not unlike the distributions seen in vertebrate tissues composed of type I collagen fibrils, such as tendon and ligament [93] and, in fact, the isolated collagen resembles type I collagen [94]. The fibrils in the dermis of the sea cucumber *Cucumaria frondosa* are basically circular but with irregular perimeters and with numerous electron-lucent interiors [92], which is true of some other echinoderm fibrils as well [93,95,96]. They also have a wide range of diameters in tissue sections, up to a maximum of about 400 nm [92]. The isolated collagen is an $\alpha 1$ trimer, with no compositional or solubility similarities to vertebrate type I collagen, and with no immuno-logical similarity either to vertebrate type I collagen or to *E. tribuloides* collagen [97].

The collagenous tissues of several different echinoderms have been disaggregated into suspensions of separate intact fibrils [51,92,97–99]. The fibrils have tapered ends (Figure 9). Those isolated from the spine ligaments of *E. tribuloides* and the dermis of *C. frondosa* have lengths of 30 μm–1.2 mm and 39–436 μm respectively [92,99]. Surprisingly, the aspect ratio (length/diameter) is constant in the fibrils from each species, regardless of length: about 2000 in *C. frondosa* fibrils and about 2500 in *E. tribuloides* fibrils, where the diameter was measured at the

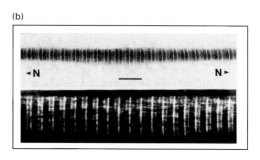

Figure 9 Isolated collagen fibrils from sea urchin ligament and sea cucumber dermis

(**a**) Electron micrograph of an entire positively stained sea urchin fibril. The ends of the 170-μm-long fibril are marked by arrowheads. The central region of the fibril is boxed. The scatter plot shows the relationship between diameter and length along a similarly prepared sea cucumber fibril. (**b**) The central regions of a positively stained sea cucumber fibril (top) and a negatively stained sea urchin fibril (bottom) are shown. At the midpoint of these fibrils (indicated by the bar), as in all others examined, the molecular polarity undergoes a reversal (compare with fibril in Figure 4). Reproduced from [51], with permission.

midpoint of the fibrils. The constant aspect ratio means that growing fibrils must increase their diameter in proportion to their increase in length, with the result that fibrils of all lengths have the same shape. This observation is all the more interesting when it is considered in the light of the fact that the fibrils are not cylindrical, but are rather symmetrically spindle-shaped, which may be related to their role as the reinforcing fibres in a discontinuous composite material [51,92,99].

Free fibrils are isolated from *C. frondosa* dermis in solutions of both high and low ionic strength [100]. Similarly, biomechanical tests on *C. frondosa* dermis have shown that its stiffness is many times lower at both low and high ionic strengths than it is at intermediate values (J. A. Trotter, J. P. Salgado and T. J. Koob, unpublished work). The isolated fibrils form aggregates *in vitro* when the ionic strength is in the physiological range. These experiments indicate that the fibrils interact with one another *in vivo* and *in vitro* largely by electrostatic interactions.

The fibrils of echinoderms have sulphated glycosaminoglycan moieties associated with their surfaces. The presence and locations of these moieties have been identified using Cupro-meronic or Cuprolinic Blue, or Ruthenium Red [97,99,101–103]. In sea urchin and sea cucumber fibrils the precipitates are associated with the middle of the gap zone, where charged amino

Table 1 Summary of the growth characteristics of early collagen fibrils

Abbreviation: AMD, axial mass distribution. See the text for further details.

Fibril assembly characteristics	Developing vertebrate tendon (*in vivo* 1)	Acid-soluble type I collagen (*in vitro* 1)	Type I pCcollagen plus C-proteinase (*in vitro* 2)	Echinoderm ligament and dermis (*in vivo* 2)
Linear AMDs of fibril tips	+	+	+	+
Tip shape set at early stage	?	+	+	−
Limiting diameter at early stage	+	+	−	−
Unipolar fibrils	+	+	−	−
Bipolar fibrils (N–N)	+	−	+	+

acids are concentrated in bands corresponding to the **d**/**e** bands of vertebrate fibrils [97,99]. In crinoid fibrils there are apparently two glycosaminoglycan sites per *D*-period, but the banding pattern has not yet been well defined for these fibrils [104].

The proteoglycans on the surfaces of echinoderm fibrils could be involved in the control of fibril shape and size, as is postulated to occur in vertebrate type I collagen fibrils. They could also be involved in the electrostatic interactions that determine the stiffness of echinoderm tissues, although definitive evidence for this role is lacking. It has recently been shown that the salt-dependent aggregation of *C. frondosa* fibrils depends on the presence of a protein that is bound to the isolated fibrils [100]. Fibrils that lack this protein do not aggregate in univalent salt solutions. The protein has been purified and partially characterized. Full characterization, including the definition of its binding sites on fibrils and the mechanism by which it induces fibril aggregation, should prove useful. This protein is the first to have been isolated that plays a crucial role in creating the fibril–fibril interactions that convert an array of individual fibrils into a mechanically functional tissue.

When compared with vertebrate fibrils, echinoderm collagen fibrils have greatly reduced intensities in their **a3** and **b2** bands, and greatly increased intensity in the **c3** band [92]. It is not known whether these changes in charge density are related to fibril functions, for example the binding of the aggregating protein mentioned above. The similarities between vertebrate and echinoderm collagen molecules, including their segment long spacing and fibril-positive and -negative stain patterns, allowed the banding pattern of echinoderm fibrils to be related to the orientation of molecules in the fibril [92]. This information was used to analyse the molecular orientation in the symmetrically spindle-shaped echinoderm fibrils [51]. Ultrastructural analyses of many fibrils from both sea urchins and sea cucumbers showed that they are all composed of molecules that have their N-terminal ends oriented towards the nearest fibril end. It follows that the centre of each fibril contains a region in which the polarity is reversed. In fibrils from both species this region contains between one and three *D*-periods in which the banding pattern is symmetrical, and is identical to the DPS-III [50] banding pattern that occurs when vertebrate collagens are induced to form fibrils under certain conditions, such as elevated phosphate concentrations (as discussed above). This pattern results from the alignment of equal numbers of anti-parallel molecules in an arrangement that brings the C-telopeptides of anti-parallel molecules into register, where they can both participate in trivalent cross-links with the same helical hydroxylysine residue. In contrast, the N-telopeptides are not in register, and thus could not participate in cross-links between anti-parallel molecules. On either side of the symmetrical *D*-periods are transitional zones of about three *D*-periods wide, in which

the DPS-III pattern changes into the normal one. These fibrils from adult echinoderms thus resemble those isolated from embryonic chick tendons [13,54].

Although it is clear that equal numbers of molecules of each polarity are present in the DPS-III regions, it is not clear whether the total number of molecules in that region is different from the number in adjacent regions. This question is being approached using mass determinations, but no definitive answers yet exist. The answers are required in order to produce a realistic model of the molecular packing in that region.

As was noted above, all the fibrils from sea urchins and sea cucumbers are similarly spindle-shaped, regardless of their lengths. Preliminary results have indicated that the fibrils have linear axial mass distributions, and are therefore similar to the ends of fibrils isolated from foetal chicken tendons and the fibrils formed *in vitro* from vertebrate type I pCcollagen using C-proteinase to control assembly. This means that the fibrils have an essentially paraboloidal shape, with a plane of mirror symmetry through their centres. The observation that all echinoderm fibrils, regardless of size, are symmetrical about a plane through the central anti-parallel *D*-period suggests that the assembly of echinoderm bipolars is different from that of vertebrate bipolars. A specific model to account for this growth process has yet to be developed.

CONCLUSIONS

The evolution of multicellular organisms depended on the formation of an extracellular matrix. Collagen fibrils, which can be several millimetres in length, have played a central role in extracellular matrices for at least 500 million years. Current evidence suggests that collagen fibril formation is intrinsically a self-assembly process but that considerable cellular control is exerted on the process *in vivo*. In this review we have surveyed four different systems of fibril formation, two *in vivo* and two *in vitro* (Table 1). The summary shows that fibrils, whether they occur *in vivo* or *in vitro*, all have growing tips with linear axial mass distributions. In the two systems *in vitro* the tip shape is established at an early time in the assembly pathway and remains unchanged during growth of the fibrils. Our understanding of the corresponding behaviour *in vivo* is incomplete and must await further quantitative studies.

A new observation is that early fibrils in developing tissues can be either unipolar or bipolar. This finding has major implications for our understanding of fibril fusion in young tissues and in healing wounds. Models based on existing experimental data suggest that fusion of a bipolar fibril to a bipolar fibril is unlikely to occur and that the presence of such fibrils may impose a limit on fibril fusion. The exclusivity of short bipolar fibrils in a cell-free system in which pCcollagen is cleaved with procollagen C-proteinase suggests that the concentration of pCcollagen during early fibril assembly may contribute to the formation of bipolar fibrils. Conversely, long and exclusively unipolar fibrils form when pCcollagen is absent during early fibril formation of acid-soluble collagen *in vitro*.

Their occurrence not only in vertebrates but also in invertebrates suggests that bipolar fibrils could be widespread throughout the animal kingdom. The observation that invertebrate bipolar fibrils are centro-symmetrical (with the molecular switch region at the midpoint of the fibril), whereas those in chick tendon are shape-asymmetrical, implies that the mechanism of assembly of bipolar fibrils could be different in vertebrates and invertebrates. Nevertheless, the occurrence of bipolar fibrils in organisms as diverse as echinoderms and chickens points to a common, but as yet unknown, physiological function of bipolar fibrils.

FUTURE PROSPECTS

A major direction of collagen fibril research will be to understand, at the molecular level, the principles governing the self-assembly of unipolar and bipolar fibrils, to understand how cells regulate this process, to learn how the deposition of early collagen fibrils is orchestrated in embryonic tissues, to understand the processes by which early fibrils are converted into larger fibrils in newly forming tissues, and to understand the role of other collagens and macromolecules in these processes. A number of short-term goals have been set, and these include the understanding of the occurrence of bipolar fibrils in tissues and in healing wounds, and the identification of cell-mediated factors in determining collagen fibril size and shape. The longer-term objective must be to place collagen fibril assembly in a broader context involving gene regulation, cytokine induction and cell-matrix signalling events that underpin tissue patterning and the generation of the body form. A recent observation has been that the gene for procollagen C-proteinase [105] is the same as that for bone morphogenic protein-1, which is related to the patterning gene *tolloid*. Thus the identification of a proteinase which is involved in collagen fibril formation, in determining the shape of collagen fibrils *in vitro*, and which also plays a major role in pattern formation in the early embryo, begins to provide new insights into the relationship between extracellular matrix deposition and vertebrate development.

D.F.H. and K.E.K. are supported by a grant from the Wellcome Trust (19512). J.A.T.'s research was supported by grants from the National Science Foundation and the Office of Naval Research. We thank Dr. Rod Watson for providing Figure 2. The electron microscope work in the laboratories of D.F.H. and K.E.K. was carried out in the Electron Microscope Unit, School of Biological Sciences, University of Manchester.

REFERENCES

1 Chapman, J. A., Tzaphlidou, M., Meek, K. M. and Kadler, K. E. (1990) Electron Microsc. Rev. **3**, 143–182
2 Veis, A. and George, A. (1994) in Extracellular Matrix Assembly and Structure (Yurchenco, P. D., Birk, D. E. and Mecham, R. P., eds.), pp. 15–45, Academic Press, New York
3 Prockop, D. J. and Hulmes, D. J. S. (1994) in Extracellular Matrix Assembly and Structure (Yurchenco, P. D., Birk, D. E. and Mecham, R. P., eds.), pp. 47–90, Academic Press, New York
4 Prockop, D. J. and Kivirikko, K. I. (1995) Annu. Rev. Biochem. **64**, 403–433
5 Kadler, K. E. (1995) Protein Profile **2**, 491–619
6 Kadler, K. E., Hojima, Y. and Prockop, D. J. (1987) J. Biol. Chem. **262**, 15696–15701
7 Miyahara, M., Njieha, F. K. and Prockop, D. J. (1982) J. Biol. Chem. **257**, 8442–8448
8 Miyahara, M., Hayashi, K., Berger, J., Tanzawa, K., Njieha, F. K., Trelstad, R. L. and Prockop, D. J. (1984) J. Biol. Chem. **259**, 9891–9898
9 Kadler, K. E., Hojima, Y. and Prockop, D. J. (1988) J. Biol. Chem. **263**, 10517–10523
10 Holmes, D. F., Chapman, J. A., Prockop, D. J. and Kadler, K. E. (1992) Proc. Natl. Acad. Sci. U.S.A. **89**, 9855–9859
11 Kadler, K. E., Hojima, Y. and Prockop, D. J. (1990) Biochem. J. **268**, 339–343
12 Kadler, K. E., Hulmes, D. J. S., Hojima, Y. and Prockop, D. J. (1990) Ann. N. Y. Acad. Sci. **580**, 214–224
13 Holmes, D. F., Lowe, M. P. and Chapman, J. A. (1994) J. Mol. Biol. **235**, 80–83
14 Gross, J. and Kirk, D. (1958) J. Biol. Chem. **233**, 355–360
15 Wood, G. C. and Keech, M. K. (1960) Biochem. J. **75**, 588–598
16 Wood, G. C. and Keech, M. K. (1960) Biochem. J. **75**, 598–605
17 Bard, J. B. and Chapman, J. A. (1973) Nature New Biol. **246**, 83–84
18 Holmes, D. F. and Chapman, J. A. (1979) Biochem. Biophys. Res. Commun. **87**, 993–999
19 Holmes, D. F., Capaldi, M. J. and Chapman, J. A. (1986) Int. J. Biol. Macromol. **8**, 161–166
20 Williams, B. R., Gelman, R. A. Poppke, D. C. and Piez, K. A. (1978) J. Biol. Chem. **253**, 6578–6585
21 Gelman, R. A. and Piez, K. A. (1980) J. Biol. Chem. **255**, 8098–8102
22 Silver, F. H. (1982) Collagen Relat. Res. **2**, 219–229
23 Silver, F. H. (1983) Collagen Relat. Res. **3**, 167–179
24 Silver, F. H. and Trelstad, R. L. (1979) J. Theor. Biol. **81**, 515–526
25 Silver, F. H., Langley, K. H. and Trelstad, R. L. (1979) Biopolymers **18**, 2523–2535
26 Silver, F. H. (1981) J. Biol. Chem. **256**, 4973–4977
27 Kadler, K. E. and Chapman, J. A. (1985) Ann. N. Y. Acad. Sci. **460**, 456–460
28 Ward, N. P., Hulmes, D. J. and Chapman, J. A. (1986) J. Mol. Biol. **190**, 107–112
29 George, A. and Veis, A. (1991) Biochemistry **30**, 2372–2377
30 Leibovich, S. J. and Weiss, J. B. (1970) Biochim. Biophys. Acta **214**, 445–454
31 Capaldi, M. J. and Chapman, J. A. (1982) Biopolymers **21**, 2291–2313
32 Helseth, D. L., Lechner, J. H. and Veis, A. (1979) Biopolymers **18**, 3005–3014
33 Helseth, D. L. J. and Veis, A. (1981) J. Biol. Chem. **256**, 7118–7128
34 Meek, K. M., Chapman, J. A. and Hardcastle, R. A. (1979) J. Biol. Chem. **254**, 10710–10714
35 Hulmes, D. J., Miller, A., White, S. W. and Doyle, B. B. (1977) J. Mol. Biol. **110**, 643–666
36 Otter, A., Kotovych, G. and Scott, P. G. (1989) Biochemistry **28**, 8003–8010
37 Otter, A., Scott, P. G. and Kotovych, G. (1988) Biochemistry **27**, 3560–3567
38 Mould, A. P., Hulmes, D. J. S., Holmes, D. F., Cummings, C., Sear, C. H. and Chapman, J. A. (1990) J. Mol. Biol. **211**, 581–594
39 Holmes, D. F., Mould, A. P. and Chapman, J. A. (1991) J. Mol. Biol. **220**, 111–123
40 Watson, R. B., Wallis, G. A., Holmes, D. F., Viljoen, D., Byers, P. H. and Kadler, K. E. (1992) J. Biol. Chem. **267**, 9093–9100
41 Holmes, D. F., Watson, R. B., Steinmann, B. and Kadler, K. E. (1993) J. Biol. Chem. **268**, 15758–15765
42 Liu, X. H., Scott, P. G., Otter, A. and Kotovych, G. (1990) J. Biomol. Struct. Dyn. **8**, 63–80
43 Haworth, R. A. and Chapman, J. A. (1977) Biopolymers **16**, 1895–1906
44 Fertala, A., Sieron, A. L., Ganguly, A., Li, S. W., AlaKokko, L., Anumula, K. R. and Prockop, D. J. (1994) Biochem. J. **298**, 31–37
45 Fertala, A., Sieron, A. L., Hojima, Y., Ganguly, A. and Prockop, D. J. (1994) J. Biol. Chem. **269**, 11584–11589
45a Fertala, A., Holmes, D. F., Kadler, K. E., Sieron, A. L. and Prockop, D. J. (1996) J. Biol. Chem., in the press
46 Silver, D., Miller, J., Harison, R. and Prockop, D J. (1992) Proc. Natl. Acad. Sci. U.S.A. **89**, 9860–9864
47 Parkinson, J., Kadler, K. E. and Brass, A. (1994) Phys. Rev. E. Stat. Phys. Plasmas Fluids Relat. Interdiscip. Top. **50**, 2963–2966
48 Parkinson, J., Kadler, K. E. and Brass, A. (1995) J. Mol. Biol. **247**, 823–831
49 Doyle, B. B., Hukins, D. W. L., Hulmes, D. J. S., Miller, A. and Woodhead-Galloway, J. (1975) J. Mol. Biol. **91**, 79–99
50 Bruns, R. R. (1976) J. Cell Biol. **68**, 521–538
51 Thurmond, F. A. and Trotter, J. A. (1994) J. Mol. Biol. **235**, 73–79
52 Birk, D. E. and Trelstad, R. L. (1986) J. Cell Biol. **103**, 231–240
53 Birk, D. E., Zycband, E. I., Winkelmann, D. A. and Trelstad, R. L. (1989) Proc. Natl. Acad. Sci. U.S.A. **86**, 4549–4553
54 Birk, D. E., Nurminskaya, M. V. and Zycband, E. I. (1995) Dev. Dyn. **202**, 229–243
55 Yu, L., Cummings, C., Sheehan, J. K., Kadler, K. E., Holmes, D. F. and Chapman, J. A. (1993) in Dermatan Sulphate Proteoglycans (Scott, J. E., ed.), pp. 183–188, Portland Press, London
56 Parry, D. A. D. and Craig, A. S. (1979) Nature (London) **282**, 213–215
57 Merrilees, M. J., Tiang, K. M. M. and Scott, L. (1987) Connect. Tissue Res. **16**, 237–257
58 Eikenberry, E. F., Brodsky, B. and Parry, D. A. D. (1982) Int. J. Biol. Macromol. **4**, 322–328

59 Eikenberry, E. F., Craig, A. S. and Parry, D. A. D. (1982) Int. J. Biol. Macromol. **4**, 393–398

60 Chapman, J. A. (1989) Biopolymers **28**, 1367–1382

60a Chapman, J. A. (1989) Biopolymers **28**, 2201–2205

61 Fleischmajer, R., Olsen, B. R., Timpl, R., Perlish, J. S. and Lovelace, O. (1983) Proc. Natl. Acad. Sci. U.S.A. **80**, 3354–3358

62 Fleischmajer, R., Perlish, J. S. and Olsen, B. R. (1987) Cell Tissue Res. **247**, 105–109

63 Fleischmajer, R., Perlish, J. S. and Olsen, B. R. (1987) J. Invest. Dermatol. **89**, 212–215

64 Fleischmajer, R., Perlish, J. S. Timpl, R. and Olsen, B. R. (1988) J. Histochem. Cytochem. **36**, 1425–1432

65 Hulmes, D. J. S. (1983) Collagen Relat. Res. **3**, 317–321

66 Scott, J. E. (1980) Biochem. J. **187**, 887–891

67 Scott, J. E. (1984) Biochem. J. **218**, 229–233

68 Scott, J. E. and Orford, C. R. (1981) Biochem. J. **197**, 213–216

69 Scott, J. E. and Parry, D. A. D. (1992) Int. J. Biol. Macromol. **14**, 292–293

70 Scott, J. E. and Haigh, M. (1985) Biosci. Rep. **5**, 765–774

71 Scott, J. E. and Haigh, M. (1988) Biochem. J. **253**, 607–610

72 Scott, J. E., Orford, C. R. and Hughes, E. W. (1981) Biochem. J. **195**, 573–581

73 Fleischmajer, R., MacDonald, E. D., Perlish, J. S., Burgeson, R. E. and Fisher, L. W. (1990) J. Struct. Biol. **105**, 162–169

74 Fleischmajer, R., Perlish, J. S., Burgeson, R. E., Shaikh Bahai, F. and Timpl, R. (1990) Ann. N. Y. Acad. Sci. **580**, 161–175

75 Romanic, A. M., Adachi, E., Kadler, K. E., Hojima, Y. and Prockop, D. J. (1991) J. Biol. Chem. **266**, 12703–12709

76 Romanic, A. M., Adachi, E., Hojima, Y., Engel, J. and Prockop, D. J. (1992) J. Biol. Chem. **267**, 22265–22271

77 Birk, D. E., Fitch, J. M. and Linsenmayer, T. F. (1986) Invest. Opthalmol. Vis. Sci. **27**, 1470–1477

78 Birk, D. E., Fitch, J. M., Babiarz, J. P., Doane, K. J. and Linsenmayer, T. F. (1990) J. Cell Sci. **95**, 649–657

79 Wu, J. J., Woods, P. E. and Eyre, D. R. (1992) J. Biol. Chem. **267**, 23007–23014

80 Nishiyama, T., McDonough, A. M., Bruns, R. R. and Burgeson, R. E. (1994) J. Biol. Chem. **269**, 28193–28199

81 Watt, S. L., Lunstrum, G. P., McDonough, A. M., Keene, D. R., Burgeson, R. E. and Morris, N. P. (1992) J. Biol. Chem. **267**, 20093–20099

82 Bairati, A. and Garrone, R. (1985) Biology of Invertebrate and Lower Vertebrate Collagens, Plenum Press, New York

83 van der Rest, M. and Garrone, R. (1991) FASEB J. **5**, 2814–2823

84 Gross, J. (1985) in Biology of Invertebrate and Lower Vertebrate Collagens (Bairati, A. and Garrone, R., eds.), pp. 1–28, Plenum Press, New York

85 Har-El, R. and Tanzer, M. (1993) FASEB J. **7**, 1115–1123

86 Garrone, R. (1985) in Biology of Invertebrate and Lower Vertebrate Collagens (Bairati, A. and Garrone, R., eds.), pp. 157–176, Plenum Press, New York

87 Tanzer, M. L. (1985) in Biology of Invertebrate and Lower Vertebrate Collagens (Bairati, A. and Garrone, R., eds.), pp. 65–72, Plenum Press, New York

88 Motokawa, T. (1984) Biol. Rev. **59**, 255–270

89 Wilkie, I. C. (1984) Mar. Behav. Physiol. **11**, 1–34

90 Smith, D. S., Wainwright, S. A., Baker, J. and Cayer, M. L. (1981) Tissue Cell **13**, 299–320

91 Hidaka, M. and Takahashi, K. (1983) J. Exp. Biol. **103**, 1–14

92 Trotter, J. A., Thurmond, F. A. and Koob, T. J. (1994) Cell Tissue Res. **275**, 451–458

93 Parry, D. A. D. and Craig, A. S. (1984) in Ultrastructure of the Connective Tissue Matrix (Ruggeri, A. and Motta, P. M., eds.), pp. 34–64, Martinus Nijhoff, Boston

94 Trotter, J. A. and Koob, T. J. (1994) Comp. Biochem. Physiol. **107B**, 125–134

95 Motokawa, T. (1982) Galaxea **1**, 55–64

96 Bailey, A. J. (1985) in Biology of Invertebrate and Lower Vertebrate Collagens (Bairati, A. and Garrone, R., eds.), pp. 369–388, Plenum Press, New York

97 Trotter, J. A., Lyons-Levy, G., Thurmond, F. A. and Koob, T. J. (1995) Comp. Biochem. Physiol. **112A**, 463–478

98 Matsumara, T., Shinmei, M. and Nagai, Y. (1973) J. Biochem. (Tokyo) **73**, 155–162

99 Trotter, J. A. and Koob, T. J. (1989) Cell Tissue Res. **258**, 527–539

100 Trotter, J. A., Lyons-Levy, G., Luna, D., Koob, T. J., Keene, D. R. and Atkinson, M. A. L. (1996) Matrix Biol., in the press

101 Bailey, A. J., Gathercole, L. J., Duglosz, J., Keller, A. and Voyle, C. A. (1982) Int. J. Biol. Macromol. **4**, 329–334

102 Scott, J. E. (1988) Biochem. J. **252**, 313–323

103 Kariya, Y., Watabe, S., Ochial, Y., Murata, K. and Hashimoto, K. (1990) Connect. Tissue Res. **25**, 149–159

104 Erlinger, R., Welsch, U. and Scott, J. E. (1993) J. Anat. **183**, 1–11

105 Kessler, E., Takahara, K., Biniaminov, L., Brusel, M. and Greenspan, D. S. (1996) Science **271**, 360–362

Biochem. J. (1996) **319**, 1–8 (Printed in Great Britain)

REVIEW ARTICLE

Cation–π bonding and amino–aromatic interactions in the biomolecular recognition of substituted ammonium ligands

Nigel S. SCRUTTON*‡ and Andrew R.C. RAINE†
*Department of Biochemistry, University of Leicester, Adrian Building, University Road, Leicester LE1 7RH, and †Cambridge Centre for Molecular Recognition, Department of Biochemistry, University of Cambridge, Tennis Court Road, Cambridge CB2 1QW, U.K.

Cation–π bonds and amino–aromatic interactions are known to be important contributors to protein architecture and stability, and their role in ligand–protein interactions has also been reported. Many biologically active amines contain substituted ammonium moieties, and cation–π bonding and amino–aromatic interactions often enable these molecules to associate with proteins. The role of organic cation–π bonding and amino–aromatic interactions in the recognition of small-molecule amines and peptides by proteins is an important topic for those involved in structure-based drug design, and although the number of structures determined for proteins displaying these interactions is small, general features are beginning to emerge. This review explores the role of cation–π bonding and amino–aromatic interactions in the biological molecular recognition of amine ligands. Perspectives on the design of ammonium-ligand-binding sites are also discussed.

INTRODUCTION

Amines are an extensive group of small molecules that are widespread in biology. They include the large and growing family of synthetic therapeutic and abused drugs, as well as several physiological amines (e.g. histamine, catecholamines and acetylcholine), peptides and 'protein elements' involved in a variety of signalling processes. Other physiological amines of direct clinical relevance include creatinine (a marker of kidney dysfunction) and creatine (released following myocardial infarction and in certain degenerative muscle diseases). Given the widespread occurrence and use of substituted ammonium compounds in biology and medicine, a study of their recognition by target receptor proteins is essential to our understanding of small-molecule-receptor signalling at the atomic level. A fuller understanding of the interactions of these compounds with proteins will impact directly on the discipline of structure-based drug design. By identifying generic features in the molecular recognition of ammonium groups by proteins, a more focused approach to the design of drugs and new target proteins will result.

Recent years have witnessed a steady growth in the number of determined structures of proteins and protein–ligand complexes. As data accumulate, it is becoming increasingly apparent that proteins can recognize ammonium cations and amino groups via interactions with aromatic side chains. Commensurately, there has been renewed interest in these interactions from both the chemical and biological viewpoints [1]. Two types of interaction are recognized: (i) the interaction of an organic cation (e.g. a protonated or quaternary amine bearing a formal positive charge) with aromatic side chains, termed a cation–π interaction, and (ii) the interaction of an amino group bearing a $\delta(+)$ charge with aromatic side chains, here termed an amino–aromatic interaction. Cation–π and amino–aromatic interactions are prominent in protein–small-molecule recognition, ranging from the interaction of receptors with transmitters/drugs to the recognition of substrates by enzymes, and they also contribute to the stability of the folded protein.

This review sets out to illustrate the importance of amino–aromatic/cation–π interactions in biological molecular recognition, and to heighten awareness of the need to involve aromatic residues in the design of substituted ammonium ligand-binding sites rather than relying on the more conventional ion-pair interactions seemingly favoured by protein engineers. Following brief comments on cation–π and amino–aromatic interactions in chemical systems and within protein structures, the emphasis is a structural one focusing on protein–protein, protein–peptide and protein–small-ligand associations that employ cation–π/amino–aromatic interactions. Comments on the design of ammonium ligand-binding sites are also included.

CHEMICAL SYSTEMS AND INTERACTIONS WITHIN PROTEINS

Cation–π bonding in chemical systems

Chemical studies of cation–π bonding were reported in 1986 as unconventional 'ionic' bonding between substituted ammonium ligands and π-donors, as seen in the clustering reactions of NH_4^+ and $MeNH_3^+$ with benzene derivatives [2]. Experimental results and *ab initio* calculations indicated that the interaction is predominantly electrostatic and that there is no π donation into the bond. Interaction energies were found to range from approx. 42 to 92 kJ·mol⁻¹ (10 to 22 kcal·mol⁻¹). For interactions of NH_4^+ with π-dimers (benzene or fluorobenzene), two of the four NH_4^+ hydrogens project towards the aromatic rings [2]. More recently, a combined quantum mechanical and molecular mechanical Monte Carlo simulation of the binding of the tetra-methylammonium ion and benzene in water suggested that

---

Abbreviation used: SH2 domain, *src* homology 2 domain.
‡ To whom correspondence should be addressed.

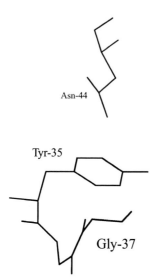

Figure 1   The amino–aromatic interaction identified by Burley and Petsko [8] in bovine pancreatic trypsin inhibitor

The backbone HN of Gly-37 and the side-chain NH₂ of Asn-44 interact with opposite faces of the aromatic ring of Tyr-35. The co-ordinates were taken from the Brookhaven Protein Databank, entry 4PTI. This and all other molecular drawings were made with the Molscript program [60].

organic-cation–π bonds are more stable than ion-pair interactions between tetra-alkylammonium ions and anionic residues in aqueous solution [3], and calculations reveal that the tetra-methylammonium ion binds to benzene with an association constant of approx. 0.8 M^{-1}. In purely chemical systems, molecular recognition by cation–π interactions is now enjoying widespread study. Various synthetic host–guest systems displaying cation–π bonding [4,5], and cationic transition-state stabilization via cation–π interactions, have been reported [6,7]. The stabilization energies that cation–π interactions contribute in the biological context are expected to be less than those for the purely chemical systems; this is a necessary consequence of the fact that stabilization in proteins is more likely to be contributed by different, and fewer, aromatic residues and is also affected by geometrical constraints and the properties of the protein interior. Nonetheless, the cation–π bond is a favourable interaction in biology, and is used by proteins that interact with protonated or substituted ammonium moieties.

Interactions within proteins

For some time, cation–π/amino–aromatic interactions within protein structures have been recognized as important associations that contribute to the protein structure. In an early study, Burley and Petsko [8] performed a geometric analysis of 33 high-resolution [2 Å (0.2 nm) resolution or better] protein structures. They were able to show a statistical preference for positively charged or δ(+) amino groups of lysine, arginine, asparagine, glutamine and histidine to pack within 6 Å of the centroids of phenylalanine, tyrosine and tryptophan. In this location, the amino groups make contact with the δ(−) π-electrons of the aromatic rings in these residues. The amino groups and the aromatic side chains are preferentially separated by between 3.4 Å and 6 Å, and they avoid contact with the δ(+) edge of the aromatic ring. An early, and now seminal, example of an amino–aromatic interaction within protein molecules is seen in bovine pancreatic trypsin inhibitor, where the aromatic ring of

residue Tyr-35 is sandwiched between the peptide amide proton of Gly-37 and the primary amide proton of Asn-44 (Figure 1; [9,10]). The interaction is seen both in the crystal structure of bovine pancreatic trypsin inhibitor and in those structures calculated from NMR data. In the latter, the chemical shifts of the Asn-44 primary amide proton and the peptide amide proton of Gly-37 are shifted upfield by the ring current of the Tyr-35 side chain [10]. The abnormal chemical shifts confirm that the protons interact with the π-electron cloud of Tyr-35 [10]. Amino–aromatic interactions involving amide protons of asparagine residues have also been observed in the crystal structure of haemoglobin bound to the drug bezafibrate [11]. In this case, an amide proton of Asn-108 from the β_1 subunit of haemoglobin is positioned to make an unconventional hydrogen bond with the π-electrons of the first aromatic nucleus of bezafibrate.

Burley and Petsko's early analysis of amino–aromatic interactions within protein molecules has been extended by Thornton and co-workers [12,13]. Their analysis focused on sp² hybridized nitrogen atoms (i.e. no formal positive charge) within protein structures. Through an analysis of 55 high-resolution structures, the results of Burley and Petsko were confirmed in that a statistical preference for the positioning of δ(+) amino groups within 6 Å of the centroids of aromatic side chains exists. Additionally, Mitchell and co-workers [13] were able to define two types of amino–aromatic interaction, i.e. those in which the sp² nitrogen atoms form stacking interactions with the aromatic rings and, secondly, those geometries which give rise to amino–aromatic hydrogen bonds, as proposed by Levitt and Perutz [14] and subsequently elaborated [15]. The former interaction is preferred by a factor of about 2.5:1, even though *ab initio* calculations of the gas-phase interaction energies for model systems favour the amino–aromatic hydrogen bond over stacking interactions. The reason for this disparity between gas-phase behaviour and interactions formed within protein structures is that, in stacked geometries, the nitrogen-containing groups form conventional (non-amino–aromatic) hydrogen bonds with neighbouring groups in the protein or solvent; these additional hydrogen bonds, which are absent from model gas-phase studies, are sufficient in the majority of cases to stabilize the stacked geometry within protein molecules. A similar analysis of the stacking interactions of arginine and aromatic side chains within protein structures has also been advanced by Flocco and Mowbray [16].

MOLECULAR RECOGNITION OF SMALL LIGANDS AND SUBSTRATES

Acetylcholinesterase: an archetype for cation–π bonding in biology?

Acetylcholinesterase is often considered as the foremost example of cation–π bonding in biological molecular recognition. In its interaction with acetylcholine, it serves as an excellent model for the recognition of quaternary amines by proteins. Early kinetic, spectroscopic and chemical modification studies [17] suggested that the active site of acetylcholinesterase is divided into two subsites: the 'esteratic' site (the site of bond breaking/making) and the 'anionic' (choline binding) site. The 'anionic' site is a misnomer, as this site is in fact uncharged and lipophilic. The molecular detail of acetylcholinesterase was revealed following the determination of the crystal structure of the enzyme from *Torpedo californicans* [18]. A structure for the enzyme–substrate complex is not available, but the details of substrate binding can be extrapolated from the structure of the enzyme alone [18] and those of the enzyme complexed with tacrine, edrophonium and decamethonium [19].

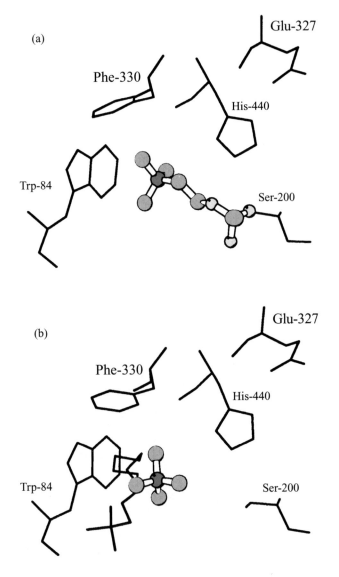

(a)

Glu-327

Phe-330

His-440

Trp-84

Ser-200

(b)

Glu-327

Phe-330

His-440

Trp-84

Ser-200

Figure 2 (a) Proposed binding of acetylcholine to acetylcholinesterase, and (b) binding of decamethonium to acetylcholinesterase

(a) Acetylcholine has been built into the crystal structure of unliganded *Torpedo californicans* acetylcholinesterase (modelled structure taken from Brookhaven entry 1ACE). The $N(CH_3)_3$ group is positioned to make a classic cation–π bond with the side chain of Trp-84, and is a similar distance away from the ring of Phe-330. Ser-200, His-440 and Glu-327 form a catalytic triad that is the mirror image of those seen in serine proteinases. (b) In the structure of acetylcholinesterase complexed with decamethonium (modelled structure taken from Brookhaven entry 1ACL), one of the terminal $N(CH_3)_3$ groups of decamethonium binds in the same position as the proposed choline-binding site in (a).

Access to the active site of acetylcholinesterase is via a deep and narrow gorge lined by 14 aromatic residues (making up about 40% of the surface of the gorge) and other residues. The gorge is 20 Å in length and the residues comprising the surface of the gorge are highly conserved in acetylcholinesterases from different species. Manual docking of the substrate acetylcholine at the base of the gorge reveals the esteratic and choline-binding sites. Associated with the esteratic site, a catalytic triad and putative oxyanion hole have been identified, and modelling of the choline moiety of acetylcholine suggests that it forms a cation–π bond with Trp-84 in the 'anionic' site (Figure 2). Perhaps the most remarkable feature of acetylcholinesterase is the preponderance of aromatic residues in the active-site gorge.

The chemical character of the gorge leads to the question of its function in contributing to the rapid rate of substrate binding and catalysis. Sussman and colleagues suggested two mechanisms by which the on-rate for ligand binding might be increased [18]. First, the high hydrophobicity of the gorge produces a low dielectric constant in the gorge. The effect is to enhance the effective local charge contributed by the small number of acidic groups in the vicinity of the gorge, which electrostatically 'steer' substrate to the active site. In the second scenario, the aromatic lining acts as a series of low-affinity sites for the substrate (in particular, the choline moiety), and guides the trapped substrate to the active site. Because of the reduction-in-dimensionality, the rate of substrate binding is increased. Relatively weak cation–π interactions may, therefore, have a major role to play in directing the substrate towards the productive enzyme–substrate complex, whereas stronger cation–π bonding is presumably responsible for binding the choline moiety of acetylcholine in the enzyme–substrate complex. Given the wealth of cation–π interactions found in acetylcholinesterase, the enzyme no doubt will remain a principal target for investigating these interactions in biological macromolecules. Interestingly, chemical modification studies of the nicotinic acetylcholine receptor have also suggested that aromatic residues are located in the acetylcholine-binding site in this molecule [20,21].

Substituted 'small-molecule' ammonium cations

Cation–π bonding of quaternary ammonium ligands has been visualized in other structurally determined protein molecules. The McPC603 Fab, a mouse myeloma IgA (κ) that binds phosphocholine, was the first determined protein structure found to bind a quaternary ammonium moiety through the use of aromatic residues [22,23]. The phosphocholine hapten is in contact with only a small number of residues; these include side chains from the three heavy-chain hypervariable regions and one light-chain region. Glu-35 (heavy chain), Tyr-100 (light chain) and Trp-107 (heavy chain) form the choline-binding site. The first residue makes a hydrogen bond with the phenolic hydroxy group of Tyr-100L and in so doing orients the side chain of Tyr-100L to make a cation–π bond with the choline group of the hapten (Figure 3). Trp-107 also makes a good cation–π bond with the quaternary ammonium group. The quaternary ammonium-binding site is revealed directly in the crystal structure of the McPC603 protein–ligand complex, and the protein is therefore an excellent model for investigating the structure of quaternary ammonium-binding sites. But what of tertiary and secondary ammonium ligands? Many bioactive amines contain these groups, and consequently structural models of binding sites for these molecules are also required.

The simple tertiary and secondary amines tri- and di-methylamine are bound by the bacterial proteins trimethylamine dehydrogenase and dimethylamine dehydrogenase respectively. Both proteins are found in methylotrophic bacteria, where they are responsible for the ability of these bacteria to grow on trimethylamine and dimethylamine as the sole carbon source [24]. The crystal structure of trimethylamine dehydrogenase has been determined at 2.4 Å resolution [25] and recently refined at 1.7 Å (S. A. White and F. S. Mathews, unpublished work). More importantly, the structures of trimethylamine dehydrogenase in complex with the substrate inhibitor tetramethylammonium chloride or the substrate trimethylamine (as the protonated ammonium cation trimethylammonium) have demonstrated that these molecules are bound by cation–π bonding in an 'aromatic bowl' [26,27] (Figure 4). The bowl, comprising the three residues Tyr-60, Trp-264 and Trp-355, positions trimethylamine in the

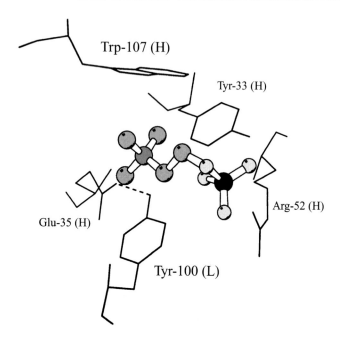

Figure 3   Phosphocholine-binding site of the McPC603 Fab

Phosphocholine is shown in ball-and-stick representation, with nitrogen dark pink and phosphorus black, while the protein residues forming the binding site are in stick representation. The choline group of the hapten is in contact with the aromatic side chains of Trp-107 (heavy chain; H) and Tyr-100 (light chain; L). Tyr-33 (H) and Arg-52 (H) interact with the phosphate. Co-ordinates were taken from Brookhaven entry 2MCP.

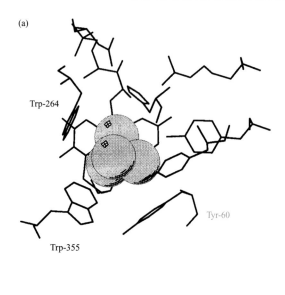

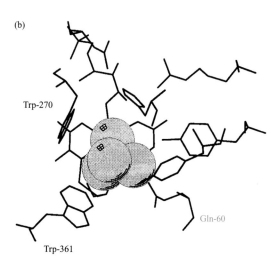

Figure 4   (a) Substrate-binding bowl in trimethylamine dehydrogenase, with tetramethylammonium (an inhibitor) bound, and (b) proposed mode of binding of the closely related dimethylamine dehydrogenase

(a) The bowl is formed by the aromatic side chains of Trp-264, Trp-355 and Tyr-60, which form cation–π bonds with the methyl groups of the inhibitor or, when the substrate trimethylamine is bound, with the three methyl groups of the substrate. (b) The only active-site difference between this enzyme and trimethylamine dehydrogenase is the replacement of Tyr-60 by a Gln. It is predicted that Trp-264 and Trp-355 will still form cation–π bonds with the two remaining methyl groups on the substrate, with Gln-60 forming a hydrogen bond with the N-H of dimethylamine.

active site so that the substrate is located close to the N-5 atom of the enzyme-bound FMN. During catalysis, the N-5 atom is involved in the formation of a covalent enzyme–substrate intermediate [28]. In trimethylamine dehydrogenase, tertiary ammonium cations are bound in a similar fashion to that previously described for quaternary ammonium ions, and the enzyme's aromatic bowl can therefore serve as a model structure for capturing tertiary ammonium groups in proteins.

Biochemically, dimethylamine dehydrogenase is closely related to trimethylamine dehydrogenase [29], and this similarity is supported by the very high degree of sequence identity (63%) between the two proteins [30,31]. On the basis of this sequence identity, a model for the structure of dimethylamine dehydrogenase has been constructed using the crystal co-ordinates of trimethylamine dehydrogenase [27], to identify the structural changes that direct the binding of dimethylamine in dimethylamine dehydrogenase. The model of dimethylamine dehydrogenase revealed that the active sites of the two proteins are almost completely identical. Those residues involved in demethylation of substrate are totally conserved; the only change is the exchange of Tyr-60 in trimethylamine dehydrogenase (a residue of the substrate-binding aromatic bowl) for a glutamine residue in dimethylamine dehydrogenase [27,31]. On placing dimethylamine in the active site of the model, Gln-60 is ideally positioned to make a conventional hydrogen bond from the side-chain amide carbonyl to the NH hydrogen of dimethylamine (Figure 4), suggesting that this single residue change is responsible for the switch in ammonium cation specificity between the two dehydrogenases. The two remaining methyl groups of dimethylamine are positioned to make cation–π bonds with the two tryptophan residues in the same way as they do in trimethylamine dehydrogenase. In trimethylamine dehydrogenase there is no requirement for the specific orientation of substrate in the aromatic bowl; trimethylamine has C-3 rotational symmetry, and any of the methyl groups can be oxidized during catalysis. In dimethylamine dehydrogenase the substrate is bound in a specific way, and this specificity is acquired through the provision of a conventional hydrogen bond to the substrate, ensuring that one of the methyl groups can be oxidized during catalysis.

Good models for cation–π bonding of primary ammonium cations by proteins are still lacking, but the hope is that bacterial methylamine dehydrogenase will prove to be a good example. Crystal structures exist for two methylamine dehydrogenases, one for the *Paracoccus denitrificans* enzyme [32] and the other for the *Thiobacillus versutus* enzyme [33,34]. The active-site regions

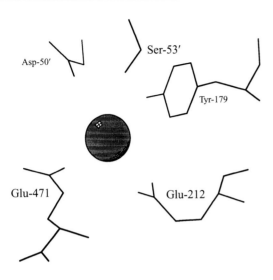

Figure 5   Ammonium-binding site in glutamine synthetase

An NH_4^+ ion has been built into the co-ordinates of Brookhaven entry 2GLS in the position at which electron density was seen for caesium and thallium ions. No energy minimization was performed after placing the ammonium ion into the ammonium-binding site of glutamine synthetase. See the text for further details.

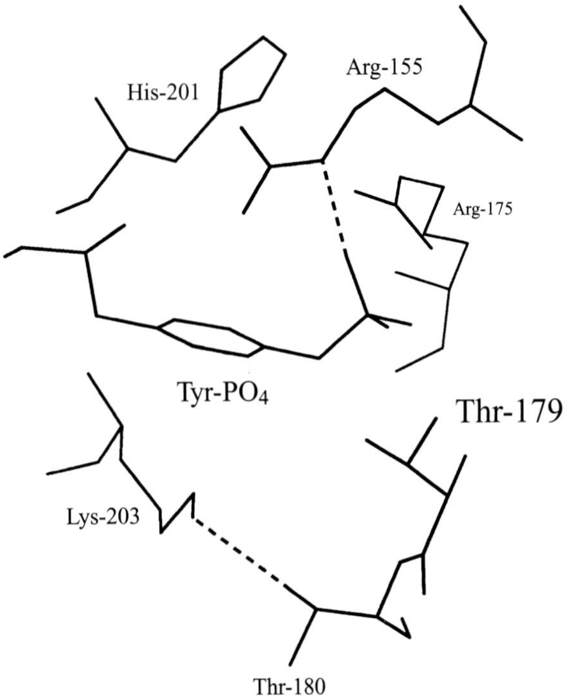

Figure 6   Detail of the binding of phosphotyrosine to the SH2 domain of the v-*src* oncogene product

One of the side-chain NH_2 groups of Arg-155 and the side-chain NH_3^+ of Lys-203 interact with the aromatic ring of the phosphotyrosine. Arg-175, Arg-155 and possibly Thr-179 interact with the phosphate group. Co-ordinates were taken from Brookhaven entry 1SHA.

of both are spatially defined, but to date only complete sequence information is available for the enzyme from *Paracoccus denitrificans* [35]. A recent refinement of this enzyme at 1.75 Å resolution indicates that the side chains of a phenylalanine (Phe-42; large subunit) and a tyrosine (Tyr-119; small subunit) residue are located in the conjectured methylamine-binding site along with an aspartate residue (F. S. Mathews, personal communication). These amino acid residues are close to the O-6 atom of the tryptophan tryptophylquinone redox cofactor of the enzyme, where the substrate forms a covalent intermediate in catalysis. A structure for methylamine dehydrogenase solved with bound substrate is lacking, but the presence of two aromatic side chains in the methylamine-binding site is highly suggestive that cation–π bonding is responsible (at least in part) for substrate recognition.

Recently, the ammonium ion-binding site has been localized in the crystal structure of glutamine synthetase [36]. Assignment of the site was made possible by performing crystal soaks of glutamine synthetase with caesium chloride and thallium acetate, the cations residing in the ammonium ion-binding site. The co-ordination shell for the ammonium ion comprises oxygen atoms donated from the γ-carboxylate of the substrate glutamate, the side–chain carboxylates of two acidic residues (Glu-212 and Asp-50'), the hydroxy group of Ser-53' and the aromatic side chain of Tyr-179 (Figure 5). Although, in this analysis, electron densities for thallium and caesium ions were observed in the difference-density maps of glutamine synthetase, the implication is that Tyr-179 forms a cation–π interaction with the physiological substrate, the ammonium ion. Therefore, as seen for dimethylamine dehydrogenase, a mixture of cation–π bonding and more conventional bonding may well be responsible for the binding of ammonium ions in glutamine synthetase.

PROTEIN–PROTEIN AND PROTEIN–PEPTIDE RECOGNITION

Recognition of phosphotyrosine-containing peptides/proteins by the v-*src* oncogene product

Cation–π and amino–aromatic interactions are observed in protein–protein and protein–peptide recognition, and perhaps

the most elegant example in this category is the recognition of tyrosine-phosphorylated peptides by the *src* homology 2 (SH2) domain of the v-*src* oncogene product. Tyrosine kinases occupy a pivotal position in signal transduction pathways [37] and, following phosphorylation of tyrosine residues, signals are transmitted by recognition of the phosphorylated tyrosines by downstream proteins containing SH2 domains. These domains reside in a variety of cytosolic non-receptor tyrosine kinases and other proteins [38]. An understanding of the molecular details of the interaction between phosphorylated-tyrosine-containing proteins and SH2 domains is, therefore, central to our understanding of signal transduction mechanisms in the cell.

Crystal structures of the v-*src* oncogene product complexed with two tyrosine-phosphorylated peptides have been solved at 1.5 Å and 2.0 Å resolution by X-ray crystallography [39]. The SH2 domain comprises an antiparallel β-sheet flanked by two α-helices, and the overall fold resembles a flattened hemisphere the face of which provides the surface for peptide binding. The phosphotyrosine residue of the bound peptide in each structure is located in a small cleft formed by residue Arg-155, the side chains of Lys-203 and His-201 and a phosphate-binding loop. Three positively charged residues (Arg-155, Arg-175 and Lys-203) are located in the phosphotyrosine-binding site (Figure 6). Arg-175 forms a conventional ion-pair interaction with two of the oxygens of phosphate. The remaining two residues (Arg-155 and Lys-203) interact with the aromatic nucleus of the phosphotyrosine. In addition to forming an optimal interaction with the aromatic ring of the phosphotyrosine, Arg-155 also forms a hydrogen bond with the phosphate of the same residue. Lys-203 is located on the opposite face of the phosphotyrosine, and is tethered by a hydrogen bond from Thr-180. In this position the

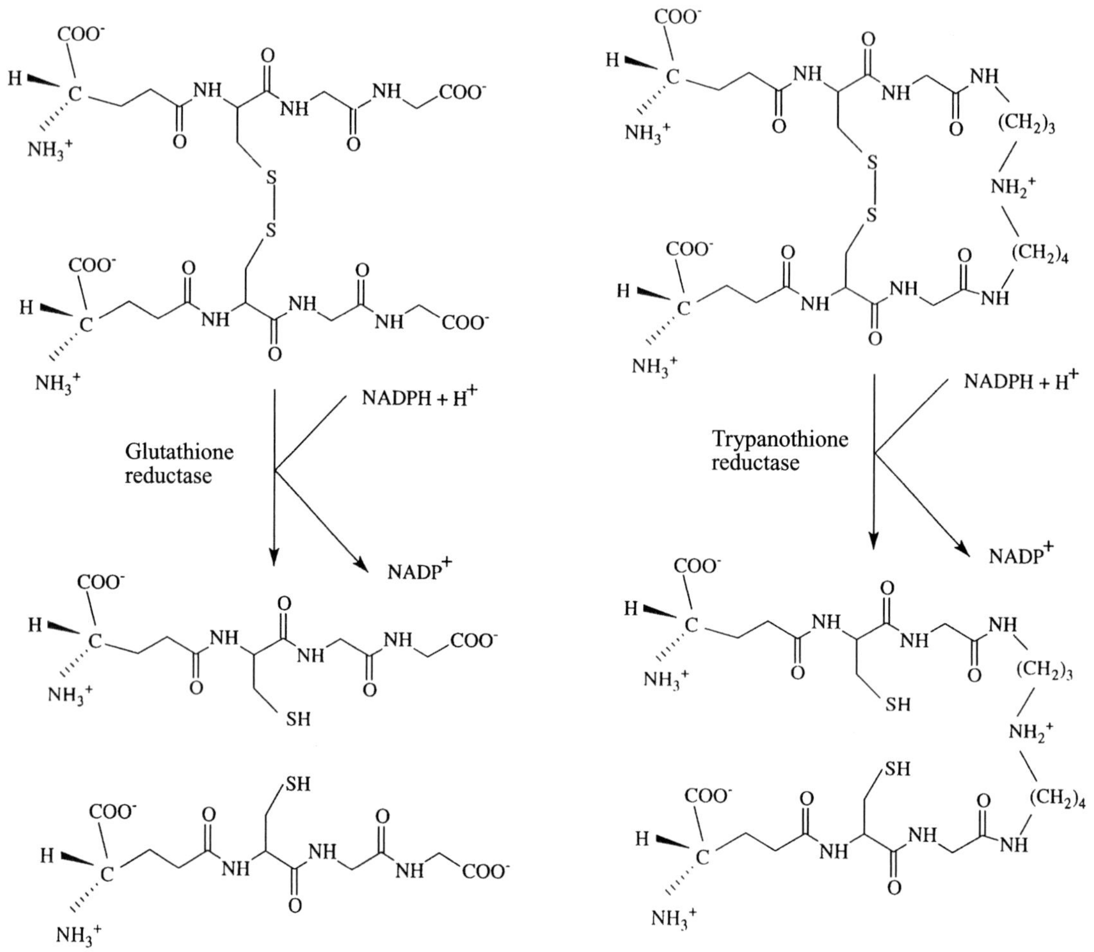

Figure 7   Reactions and substrates of glutathione reductase and trypanothione reductase

See the text for details.

terminal amino group makes no interaction with the phosphate, but forms a cation–π interaction with the aromatic side chain of the phosphorylated tyrosine residue.

The mode of binding offers a simple explanation for the selection of phosphotyrosine residues and the rejection of phosphoserine and phosphothreonine. In the latter cases, not only is the phosphate group attached to shorter side chains, which would not project sufficiently into the binding pocket to interact with those residues that form conventional hydrogen bonds with the phosphate (Arg-175, Glu-178 and Thr-179), but cation–π interactions would not form. The structural features observed in the binding of phosphotyrosine peptides are expected to hold for phosphotyrosine-containing proteins, although the contact between the SH2 domain and phosphotyrosine-containing protein is probably more extensive. Indeed, potential binding sites for residues beyond the phosphotyrosine residue (the N-terminal residue in the peptides used for the structural analysis) have been identified on the surface of the SH2 domain [39] and modelled for various SH2-domain–phosphotyrosine-peptide interactions [40]. The determined structures of other SH2-domain–phosphotyrosine-peptide complexes indicate that cation–π bonding is an important element in the recognition of the phosphorylated tyrosine residue [41–44], although this type of interaction is not a necessary hallmark of SH2-domain–phosphotyrosine complex formation [45].

Trypanothione reductase and the recognition of spermidine-linked peptides

Trypanothione reductases and glutathione reductases are related enzymes belonging to the family of disulphide oxidoreductases [46]. The enzymes are active as homodimers, using FAD as cofactor and NADPH as an electron donor. The function of trypanothione reductase is to reduce N^1,N^8-bis(glutathionyl)-spermidine (trypanothione) in trypanosomal parasites, the causative agents of African sleeping sickness, Chagas' disease and leishmaniasis. Glutathione reductase, an almost ubiquitous enzyme, is responsible for reducing disulphide-linked glutathione (Figure 7). Trypanothione reductase is unique to trypanosomal parasites and, consequently, this enzyme has become an attractive potential drug target to combat trypanosomal infections [47,48].

The structure, specificity and mechanism of human glutathione reductase have been well characterized [49–51], and the enzyme is highly specific for glutathione. In the rational design of anti-trypanosomal drugs that bind in the active site of trypanothione

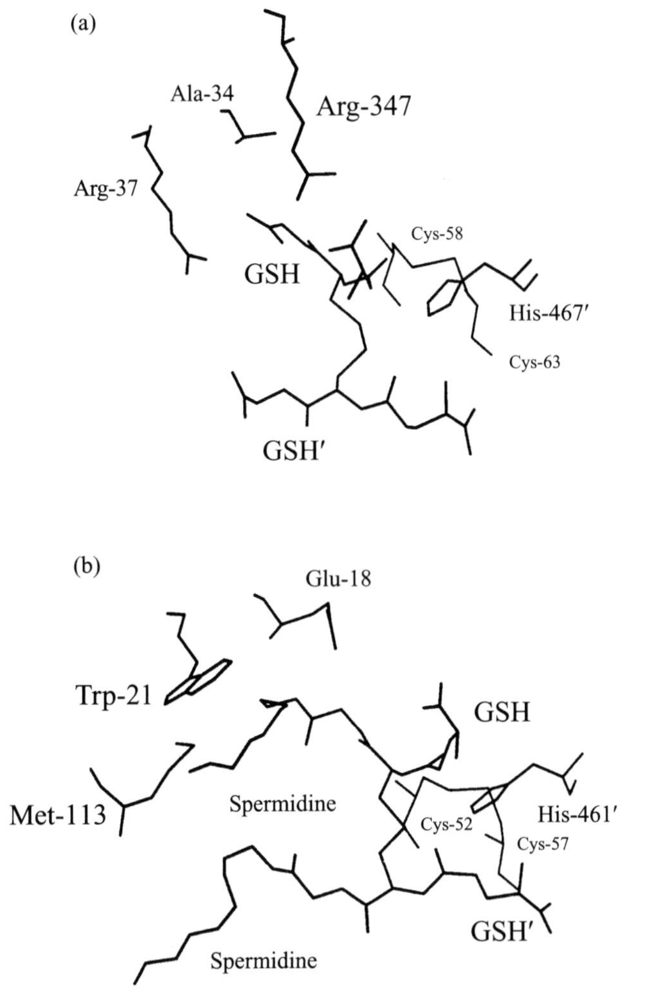

(a)

Ala-34   Arg-347

Arg-37

GSH

Cys-58

His-467'

Cys-63

GSH'

(b)

Glu-18

Trp-21

GSH

Met-113   Spermidine

Cys-52   His-461'

Cys-57

GSH'

Spermidine

Figure 8   Detail of the binding of (a) glutathione disulphide to human glutathione reductase, and (b) N¹-glutathionyl spermidine disulphide to *Crithidia fasiculata* trypanothione reductase

(**a**) The catalytic disulphide bridge (Cys-58–Cys-63) and His-467' (from the other subunit of the dimer) are shown, as are the residues making contact with the other end of the substrate. Co-ordinates taken from Brookhaven entry 1GRA. (**b**) Equivalent view to that of (**a**) above. The equivalent disulphide bridge (Cys-52–Cys-57) and histidine (His-461') are shown. Met-113 and Trp-21 form a 'hydrophobic patch' where the spermidine bridge of trypanothione would be expected to bind, with the side chain of Trp-21 making a cation–π bond with the cationic ammonium group of the spermidine bridge. Co-ordinates taken from Brookhaven entry 1TYP.

active site of human glutathione reductase. A glutamate residue (Glu-18) is also located in this region of trypanothione reductase, where it may form a favourable ion-pair interaction with the cationic ammonium group of the spermidine bridge. Trp-21 is ideally positioned to make a cation–π interaction with the ammonium group of trypanothione and is conjectured to stabilize the enzyme–substrate complex [52,53].

The observations made in the crystallographic analyses are in accord with earlier mutagenesis experiments on human [55] and *Escherichia coli* [54] glutathione reductases and on *T. congolense* trypanothione reductase [56], in which attempts were made to switch the disulphide specificity of glutathione reductase to function with trypanothione, and vice versa. In each case, the targeted residues were identified by sequence alignment of trypanothione reductase and glutathione reductase and by reference to the available high-resolution structure of human glutathione reductase. A double mutant of trypanothione reductase, in which Trp-21 was replaced by arginine and Glu-18 was replaced by alanine, acquired moderate activity as a glutathione reductase [56]. When the equivalent arginine and alanine residues were exchanged in human glutathione reductase for the naturally occurring tryptophan and glutamate residues of trypanothione reductase, the enzyme was found to discriminate against glutathione and possess significant activity with trypanothione [55], thus demonstrating the importance of these two residues in stabilizing the enzyme–substrate complex. In *E. coli* glutathione reductase, by introducing the same tryptophan and glutamate residues and an additional asparagine residue (at position 22 in the *E. coli* sequence), glutathione reductase activity is effectively abolished, but the mutant enzyme is capable of reducing trypanothione, with a selectivity coefficient (k_{cat}/K_m) 10% of that seen for wild-type trypanothione reductase [54]. Naturally, the cation–π bond and ion-pair interaction made by Trp-21 and Glu-18 respectively with the ammonium cation of the spermidine bridge of trypanothione are important discriminating interactions in substrate binding, and these residues will no doubt become a focal point in the future rational design of drugs against trypanosomal infections.

Design of ammonium-ligand-binding sites

From the available structural data, it is apparent that cation–π bonding and amino–aromatic interactions, often in combination with other more conventional interactions, can facilitate the recognition of substituted ammonium ligands, peptides and proteins by receptor protein molecules. Structural models now exist for quaternary, tertiary and secondary ammonium groups, for the ammonium ion itself and for selected peptide–protein interactions. A model for primary ammonium cations will be forthcoming in the structure of methylamine dehydrogenase. By accepting that cation–π and amino–aromatic interactions should be considered in the design of ammonium-ligand-binding sites, the question remains of how to achieve this recognition in peptide/protein design. Given the limited structural data for ammonium-ligand–protein complexes, rational design strategies seem inappropriate, especially given the generally poor record these approaches have acquired in recent years. The use of phage-display technology [57] for isolating ammonium-ligand-binding peptides seems a more attractive proposition. Ammonium-ligand-binding peptides might find use as generic clathrates for drugs and small molecules that contain ammonium groups.

 The design of ammonium-ligand-binding sites in protein molecules is perhaps more challenging than the synthesis of peptide clathrates for ammonium groups. Recently, Gold and

reductase, any effects on human glutathione reductase must also be taken into account. An understanding of the molecular details of the trypanothione-binding site in trypanothione reductase in relation to the structure of the glutathione-binding site in glutathione reductase is, therefore, central to a programme of rational drug design. With this in mind, the crystallographic structure of trypanothione reductase from *Crithidia fasiculata* [52] and the structure of the enzyme complexed with N¹-glutathionyl spermidine disulphide [53] have been determined. The position occupied by trypanothione in the active site of trypanothione reductase has also been modelled [54]. Trypanothione carries a cationic ammonium group on the spermidine bridge linking the carboxylate groups of the glycine residues (Figure 7). Trp-21 and Met-113 in trypanothione reductase form a non-polar patch in the vicinity of the spermidine bridge of trypanothione (Figure 8); these residues are absent from the colleagues [58] reported a modest improvement (up to 25-fold) in the specificity of *Bacillus stearothermophilus* L-lactate dehydrogenase for oxo acids containing ammonium groups. Analysis of the crystal structure of wild-type lactate dehydrogenase prompted these workers to exchange Gln-102 in the wild-type protein for acidic aspartate and glutamate residues, in the hope that the side-chain carboxylates would form ion-pair interactions with the ammonium groups on the new substrates. The mutant enzymes still show poor Michaelis constants (in the millimolar region) for the oxo acids bearing ammonium groups. From the evidence presented in this review, one might conjecture that improved catalysis would be realized through the provision of cation–π bonds with the new substrates rather than conventional ion-pair interactions. In this regard, the forced evolution of substrate specificity for *B. stearothermophilus* L-lactate dehydrogenase developed by Holbrook and co-workers might be a more appropriate route to follow. Interestingly, Holbrook has isolated a range of mutant lactate dehydrogenases that operate effectively with phenylpyruvate (a poor substrate for the wild-type enzyme). In one of these mutants, residues 101 and 102 are changed to proline and lysine respectively ([59]; J. J. Holbrook, personal communication) and the enzyme displays about a 10-fold improvement in K_m for phenylpyruvate compared with the wild-type enzyme. In this case, it is tempting to speculate that the side-chain ϵ-amino group of Lys-102 (the same residue position targeted by Gold and colleagues) might form a cation–π bond with the phenyl group of the new substrate, and the crystal structure for this mutant enzyme is therefore eagerly awaited. As the ground-breaking work on lactate dehydrogenase suggests, laboratory-based evolution may prove to be an attractive method for 'engineering' of new ammonium ligand specificities into existing protein scaffolds.

We thank Professor J.J. Holbrook and Professor F.S. Mathews for permission to discuss unpublished work from their laboratories. Some of the work discussed in this review on tertiary and secondary ammonium ligand binding is funded by the Biotechnology and Biological Sciences Research Council (N.S.S.) and the Royal Society (N.S.S.).

REFERENCES

1 Dougherty, D.A. (1996) Science **271**, 163–168
2 Deakyne, C.A. and Meot-Ner, M. (1986) J. Am. Chem. Soc. **107**, 474–479
3 Gao, J., Chou, L.W. and Auerbach, A. (1993) Biophys. J. **65**, 43–47
4 Shepodd, T.J., Petti, M.A. and Dougherty, D.A. (1986) J. Am. Chem. Soc. **108**, 6085–6087
5 Kearney, P.C., Mizoue, L.S., Kumpf, R.A., Forman, J.E., McCurdy, A. and Dougherty, D.A. (1993) J. Am. Chem. Soc. **115**, 9907–9919
6 Stauffer, D.A., Barrans, R.E. and Dougherty, D.A. (1990) Angew. Chem. Int. Ed. Engl. **29**, 915–917
7 McCurdy, A., Jimenez, L., Stauffer, D.A. and Dougherty, D.A. (1992) J. Am. Chem. Soc. **114**, 10314–10321
8 Burley, S.K. and Petsko, G.A. (1986) FEBS Lett. **203**, 139–143
9 Wlodawer, A., Walter, J., Huber, R. and Sjorn, L. (1984) J. Mol. Biol. **180**, 301–329
10 Tuchsen, E. and Woodward, C. (1987) Biochemistry **26**, 1918–1925
11 Perutz, M.F., Fermi, G., Abraham, D.J., Poyart, C. and Bursaux, E. (1986) J. Am. Chem. Soc. **108**, 1064–1078
12 Singh, J. and Thornton, J.M. (1990) J. Mol. Biol. **211**, 595–615
13 Mitchell, J.B.O., Nandi, C.L., McDonald, I.K., Thornton, J.. and Price, S.L. (1994) J. Mol. Biol. **239**, 315–331
14 Levitt, M. and Perutz, M.F. (1988) J. Mol. Biol. **201**, 751–754
15 Perutz, M.F. (1993) Philos. Trans. R. Soc. London A **345**, 105–112
16 Flocco, M.M. and Mowbray, S.L. (1994) J. Mol. Biol. **235**, 709–717
17 Nachmansohn, D. and Wilson, I.B. (1951) Adv. Enzymol. **12**, 259–339
18 Sussman, J.L., Harel, M., Frolow, F., Oefner, C., Goldman, A., Toker, L. and Silman, I. (1991) Science **253**, 872–879
19 Harel, M., Schalk, I., Ehret-Sabattier, L., Bouet, F., Goeldner, H., Hirth, C., Axelsen, P., Silman, I. and Sussman, J. (1993) Proc. Natl. Acad. Sci. U.S.A. **90**, 9031–9035
20 Dennis, M., Giraudat, J., Kotzybahibert, F., Goeldner, M., Hirth, C., Chang, J.Y., Lazure, C., Chretien, M. and Changeux, J.P. (1988) Biochemistry **27**, 2346–2357
21 Galzi, J.L., Revah, F., Black, D., Goeldner, H., Hirth, C. and Changeux, J.P. (1990) J. Biol. Chem. **265**, 10430–10437
22 Davis, D.R. and Metzger, H. (1983) Annu. Rev. Immunol. **1**, 87–117
23 Satow, Y., Cohen, G.H., Padlan, E.A. and Davis, D.R. (1986) J. Mol. Biol. **190**, 593–604
24 Steenkamp, D.J. and Mathews, F.S. (1992) in Chemistry and Biochemistry of Flavoenzymes (Muller, F., ed.), vol. 2, pp. 395–423, CRC Press, Boca Raton
25 Lim, L.W., Shamala, N., Mathews, F.S., Steenkamp, D.J., Hamlin, R. and Xuong, N. (1986) J. Biol. Chem. **261**, 15140–15146
26 Bellamy, H.D., Lim, L.W., Mathews, F.S. and Dunham, W.R. (1989) J. Biol. Chem. **264**, 11887–11892
27 Raine, A.R.C., Yang, C.-C., Packman, L.C., White, S.A., Mathews, F.S. and Scrutton, N.S. (1995) Protein Sci. **4**, 2625–2628
28 Rohlfs, R.J. and Hille, R. (1994) J. Biol. Chem. **269**, 30869–30879
29 Kasprzak, A.A., Papas, E.J. and Steenkamp, D.J. (1983) Biochem. J. **211**, 535–541
30 Boyd, G., Mathews, F.S., Packman, L.C. and Scrutton, N.S. (1992) FEBS Lett. **308**, 271–276
31 Yang, C.-C., Packman, L.C. and Scrutton, N.S. (1995) Eur. J. Biochem. **232**, 264–271
32 Chen, L., Mathews, F.S., Davidson, V.L., Huizinga, E.G., Vellieux, F.M.D. and Hol, W.G.J. (1992) Proteins Struct. Funct. Genet. **14**, 288–299
33 Vellieux, F.M.D., Huitema, F., Groendijk, H., Kalk, K.H., Frank, J., Jongejan, J.A., Duine, J.A., Petratos, K., Drenth, J. and Hol, W.G.J. (1989) EMBO J. **8**, 2171–2178
34 Vellieux, F.M.D., Kalk, K.H., Drenth, J. and Hol, W.G. (1990) Acta Crystallogr. **B46**, 806–823
35 Chistersderov, A.Y., Boyd, J., Mathews, F.S. and Lidstrom, M.E. (1992) Biochem. Biophys. Res. Commun. **184**, 1181–1189
36 Liaw, S.-H., Kuo, I. and Eisenberg, D. (1995) Protein Sci. **4**, 2358–2365
37 Cantley, L.C., Auger, K.R., Carpenter, C., Duckworth, B., Graziani, A., Kapeller, R. and Soltoff, S. (1991) Cell **64**, 281–302
38 Koch, C.A., Anderson, D., Moran, M.F., Ellis, C. and Pawson, T. (1991) Science **252**, 668–674
39 Waksman, G., Kominos, D., Robertson, S.C., Pant, N., Baltimore, D., Birge, R., Cowburn, D., Hanafusa, H., Mayer, B.J., Overduin, M., Resh, M.D., Rios, C.B., Silverman, L. and Kuriyan, J. (1992) Nature (London) **358**, 646–653
40 Zvelebil, M.J.J.M., Panayotou, G., Linacre, J. and Waterfield, M.D. (1995) Protein Eng. **8**, 527–533
41 Eck, M.J., Shoelson, S.E. and Harrison, S.C. (1993) Nature (London) **362**, 87–91
42 Waksman, G., Shoelson, S.E., Pant, N., Cowburn, D. and Kuriyan, D. (1993) Cell **72**, 779–790
43 Narula, S.S., Yuan, R.W., Adams, S.E., Green, O.M., Green, J., Philips, T.B., Zydowsky, L.D., Botfield, M.C., Hatada, M., Laird, E.R., Zoller, M.J., Karas, J.L. and Delgarno, D.C. (1995) Structure **3**, 1061–1073
44 Zhou, M.-M., Meadows, R.R., Logan, T.M., Yoon, H.S., Wade, W.S., Ravichandran, K.S., Burakoff, S.J. and Fesik, S.W. (1995) Proc. Natl. Acad. Sci. U.S.A. **92**, 7784–7788
45 Hatada, M.H., Lu, X.D., Laird, E.R., Green, J., Morgenstern, J.P., Lou, M.Z., Marr, C.S., Phillips, T.B., Ram, M.K., Theriault, K., Zoller, M.J. and Karas, J.L. (1995) Nature (London) **377**, 32–38
46 Williams, Jr., C.H. (1992) in Chemistry and Biochemistry of Flavoenzymes (Muller, F., ed.), vol. 3, pp. 121–211, CRC Press, Boca Raton
47 Fairlamb, A.H. (1989) Parasitology **99S**, 93–112
48 Ponasik, J.A., Strickland, C., Faerman, C., Savvides, S., Karplus, P.A. and Ganem, B. (1995) Biochem. J. **311**, 371–375
49 Karplus, P.A. and Schulz, G.E. (1987) J. Mol. Biol. **195**, 701–729
50 Karplus, P.A. and Schulz, G.E. (1989) J. Mol. Biol. **210**, 163–180
51 Karplus, P.A., Pai, E.F. and Schulz, G.E. (1989) Eur. J. Biochem. **178**, 693–703
52 Hunter, W.N., Bailey, S., Habash, J., Harrop, S.J., Helliwell, J.R., Aboagye-Kwarteng, T., Smith, K. and Fairlamb, A.H. (1992) J. Mol. Biol. **227**, 322–333
53 Bailey, S., Smith, K., Fairlamb, A.H. and Hunter, W.N. (1993) Eur. J. Biochem. **213**, 67–75
54 Henderson, G.B., Murgolo, N.J., Kuriyan, J., Osapay, K., Kominos, D., Berry, A., Scrutton, N.S., Hinchliffe, N.W., Perham, R.N. and Cerami, A. (1991) Proc. Natl. Acad. Sci. U.S.A. **88**, 8769–8773
55 Bradley, M., Bucheler, U.S. and Walsh, C.T. (1991) Biochemistry **30**, 6124–6127
56 Sullivan, F.X., Sobolov, S.B., Bradley, M. and Walsh, C.T. (1991) Biochemistry **30**, 2761–2767
57 Scott, J.K. and Smith, G.P. (1990) Science **249**, 386–390
58 Hogan, J.K., Pittol, C.A., Jones, B. and Gold, M. (1995) Biochemistry **34**, 4225–4230
59 Hawrani, A.E. (1995) Ph.D. Thesis, University of Bristol, U.K.
60 Kraulis, P.J. (1991) J. Appl. Crystallogr. **24**, 946–950

Biochem. J. (1996) **313**, 17–29 (Printed in Great Britain)

REVIEW ARTICLE
Damage to DNA by reactive oxygen and nitrogen species: role in inflammatory disease and progression to cancer

Helen WISEMAN* and Barry HALLIWELL†
*Department of Nutrition and Dietetics, King's College London, Campden Hill Road, London W8 7AH, and †Pharmacology Group, King's College London, Manresa Road, London SW3 6LX, U.K.

INTRODUCTION

It is increasingly proposed that reactive oxygen species (ROS) and reactive nitrogen species (RNS) play a key role in human cancer development [1–6], especially as evidence is growing that antioxidants may prevent or delay the onset of some types of cancer (reviewed in [7,8]). ROS is a collective term often used by biologists to include oxygen radicals [superoxide ($O_2^{•-}$), hydroxyl ($OH^•$), peroxyl ($RO_2^•$) and alkoxyl ($RO^•$)] and certain non-radicals that are either oxidizing agents and/or are easily converted into radicals, such as HOCl, ozone (O_3), peroxynitrite ($ONOO^-$), singlet oxygen ($^1O_2$) and H_2O_2. RNS is a similar collective term that includes nitric oxide radical ($NO^•$), $ONOO^-$, nitrogen dioxide radical ($NO_2^•$), other oxides of nitrogen and products arising when $NO^•$ reacts with $O_2^{•-}$, $RO^•$ and $RO_2^•$. 'Reactive' is not always an appropriate term; H_2O_2, $NO^•$ and $O_2^{•-}$ react quickly with very few molecules, whereas $OH^•$ reacts quickly with almost anything. $RO_2^•$, $RO^•$, HOCl, $NO_2^•$, $ONOO^-$ and O_3 have intermediate reactivities. ROS and RNS have been shown to possess many characteristics of carcinogens [4] (Figure 1). Mutagenesis by ROS/RNS could contribute to the initiation of cancer, in addition to being important in the promotion and progression phases. For example, ROS/RNS can have the following effects.

(1) Cause structural alterations in DNA, e.g. base pair mutations, rearrangements, deletions, insertions and sequence amplification. $OH^•$ is especially damaging, but $^1O_2$, $RO_2^•$, $RO^•$, HNO_2, O_3, $ONOO^-$ and the decomposition products of $ONOO^-$ are also effective [9–13]. ROS can produce gross chromosomal alterations in addition to point mutations and thus could be involved in the inactivation or loss of the second wild-type allele of a mutated proto-oncogene or tumour-suppressor gene that can occur during tumour promotion and progression, allowing expression of the mutated phenotype [4].

(2) Affect cytoplasmic and nuclear signal transduction pathways [14,15]. For example, H_2O_2 (which crosses cell and organelle membranes easily) can lead to displacement of the inhibitory subunit from the cytoplasmic transcription factor nuclear factor κB, allowing the activated factor to migrate to the nucleus [14]. Nitration of tyrosine residues by $ONOO^-$ may block phosphorylation.

(3) Modulate the activity of the proteins and genes that respond to stress and which act to regulate the genes that are related to cell proliferation, differentiation and apoptosis [4,14–17]. For example, H_2O_2 can stimulate transcription of c-*jun*

[18] and can activate mitogen-activated protein kinase in NIH 313 cells [19].

CHEMISTRY OF DNA DAMAGE

The endogenous reactions that are likely to contribute to ongoing DNA damage are oxidation, methylation, depurination and deamination [1,2]. Nitric oxide or, more likely, reactive products derived from it, such as $NO_2^•$, $ONOO^-$, N_2O_3 and HNO_2, are mutagenic agents, with the potential to produce nitration, nitrosation and deamination reactions on DNA bases [3,6]. Methylation of cytosines in DNA is important for the regulation of gene expression, and normal methylation patterns can be altered during carcinogenesis [20]. Conversion of guanine to 8-hydroxyguanine (Figure 2), a frequent result of ROS attack [9,10,21], has been found to alter the enzyme-catalysed methylation of adjacent cytosines [20], thus providing a link between oxidative DNA damage and altered methylation patterns.

The chemistry of DNA damage by several ROS has been well characterized *in vitro* [9,11–13,21], although more information is needed about the changes produced by $RO_2^•$, $RO^•$, O_3, $ONOO^-$ and several of the RNS. Different ROS affect DNA in different ways, e.g. $O_2^{•-}$ and H_2O_2 do not react with DNA bases at all [9,10]. $OH^•$ generates a multiplicity of products from all four DNA bases and this pattern appears to be a diagnostic 'fingerprint' of $OH^•$ attack [10]. By contrast $^1O_2$ selectively attacks guanine [13,22]. The most commonly produced base lesion, and the one most often measured as an index of oxidative DNA damage, is 8-hydroxyguanine (8-OHG). It is sometimes measured as the nucleoside, 8-hydroxydeoxyguanosine (8-OHdG) [2,23]. These assay methods have been reviewed in detail [9,10,23,24]. Figure 2 shows the structures of the products of ROS attack on DNA.

Damage to DNA by ROS/RNS appears to occur naturally, in that low steady-state levels of base damage products have been detected in nuclear DNA from human cells and tissues [2,23–26]. The pattern of damage to the purine and pyrimidine bases (Figure 2) suggests that at least some of the damage occurs by $OH^•$ attack, suggesting that $OH^•$ is formed in the nucleus *in vivo* [24].

MITOCHONDRIAL DNA DAMAGE

ROS/RNS can also damage mitochondrial DNA, and such damage has been suggested to be important in several human

---

Abbreviations used: ROS, reactive oxygen species; RNS, reactive nitrogen species; $O_2^{•-}$, superoxide radical; $OH^•$, hydroxyl radical; $RO_2^•$, peroxyl radical; $RO^•$, alkoxyl radical; O_3, ozone; $ONOO^-$, peroxynitrite; $^1O_2$, singlet oxygen; $NO^•$, nitric oxide radical; $NO_2^•$, nitrogen dioxide radical; GC-MS, gas chromatography/MS; 8-OHG, 8-hydroxyguanine; 8-OHdG, 8-hydroxydeoxyguanosine; PARP, poly(ADP-ribose) polymerase; AP, apurinic/apyrimidinic; XP, xeroderma pigmentosum; SOD, superoxide dismutase; MnSOD, manganese-containing SOD; IBD, inflammatory bowel disease; PMA, phorbol 12-myristate 13-acetate; iNOS, inducible form of nitric oxide synthase; 5-ASA, 5-aminosalicylic acid.

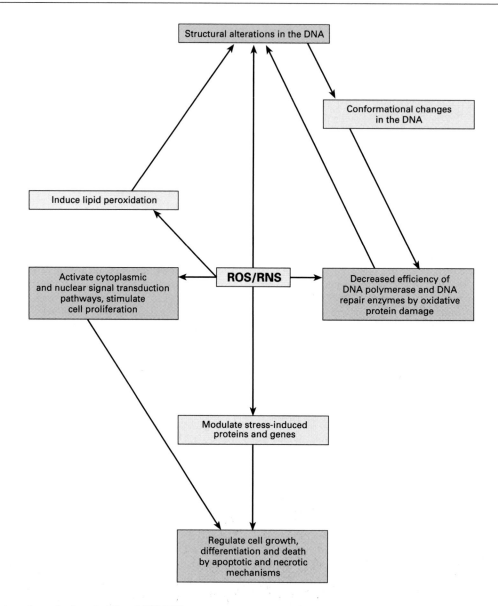

Figure 1   Potential carcinogenic characteristics of ROS/RNS

Key processes are shown in red. Aldehyde end-products of lipid peroxidation can bind to DNA and are potentially mutagenic [213].

diseases and in the aging process [27,28]. The free-radical theory of aging postulates that aging is caused by free-radical reactions and that life expectancy can be increased by nutritious low calorific diets supplemented by free-radical inhibitors [27]. Indeed, it has been claimed that mechanisms of aging and life span shortening by enhanced calorific intake are associated with increased oxidative damage resulting from associated changes in mitochondrial ROS production [29].

Mitochondria are often said to be the most important intracellular source of ROS, but it is hard to be sure of this [30]. However, it seems very likely that the mitochondrial electron transport chain generates ROS *in vivo* [31,32] and that mitochondrial DNA is damaged by them. Indeed, oxidative DNA base damage (measured as 8-OHdG) has been detected in mitochondrial DNA at steady-state levels several-fold higher than in nuclear DNA [26,28,33]. Which ROS or RNS are responsible has not yet been elucidated. This apparent increased net oxidative damage in mitochondrial DNA compared with

nuclear DNA could be because of the proximity of mitochondrial DNA to ROS generated during electron transport, the lack of histone proteins to protect the DNA against attack, or inefficient repair, so that base damage accumulates to higher levels. Intermediate radicals formed during lipid peroxidation, as well as end-products of peroxidation (Figure 1), can also attack DNA [34] and have been suggested to damage mitochondrial DNA, which is in close proximity to the mitochondrial inner membrane [35].

Oxidative damage by all the above mechanisms could contribute to the deletions and mutations in mitochondrial DNA that accumulate with age at a higher rate than in nuclear DNA [36]. Damage to mitochondrial DNA could play a role in neurodegenerative diseases: mitochondrial deletions and increased steady-state mitochondrial oxidative DNA damage (measured as elevations in 8-OHdG) have been reported in Alzheimer's disease [37]. Increased mitochondrial DNA damage in tissue from atherosclerotic hearts has also been reported [38].

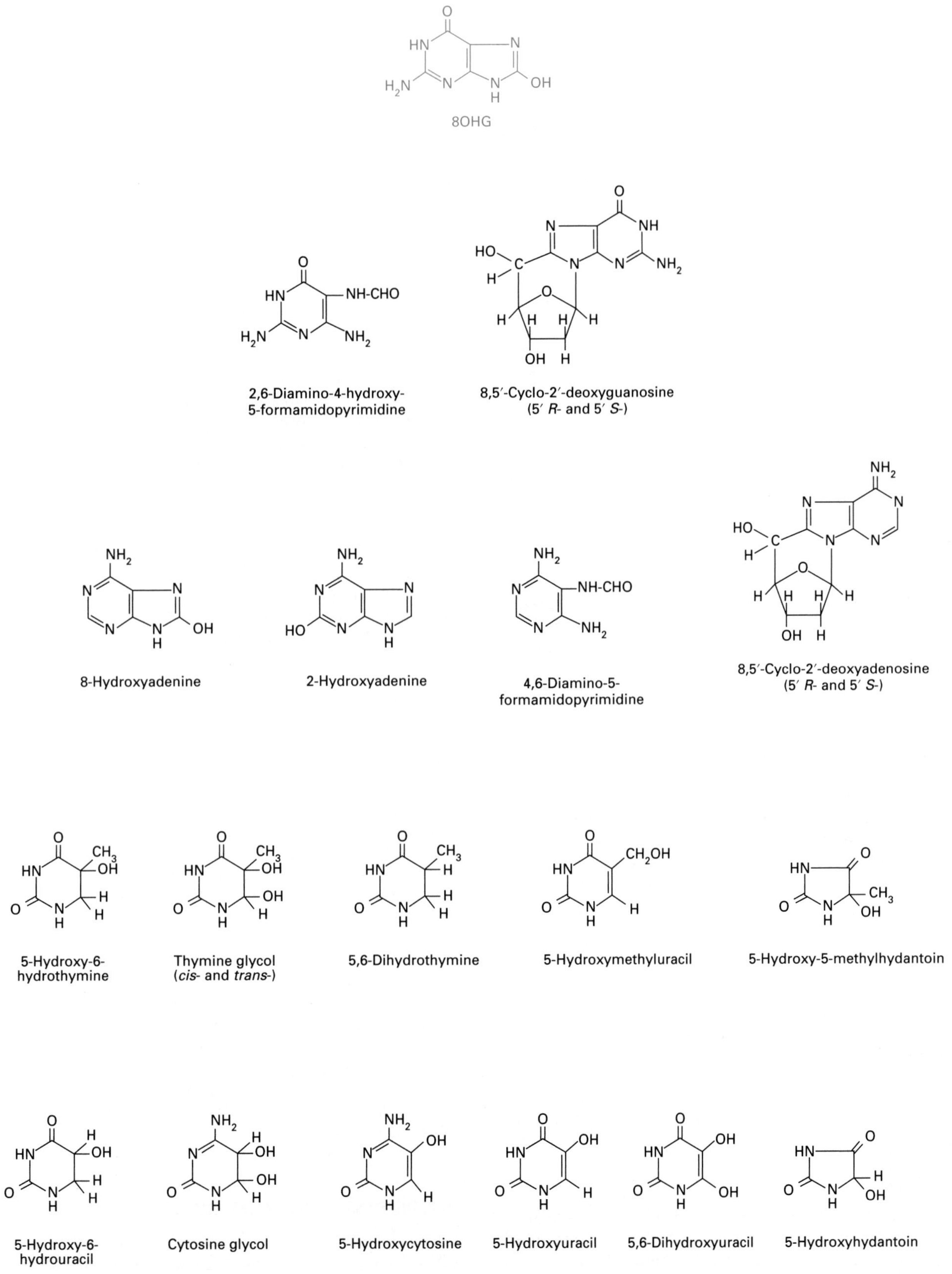

8OHG

2,6-Diamino-4-hydroxy-
5-formamidopyrimidine

8,5'-Cyclo-2'-deoxyguanosine
(5' R- and 5' S-)

8-Hydroxyadenine

2-Hydroxyadenine

4,6-Diamino-5-
formamidopyrimidine

8,5'-Cyclo-2'-deoxyadenosine
(5' R- and 5' S-)

5-Hydroxy-6-
hydrothymine

Thymine glycol
(cis- and trans-)

5,6-Dihydrothymine

5-Hydroxymethyluracil

5-Hydroxy-5-methylhydantoin

5-Hydroxy-6-
hydrouracil

Cytosine glycol

5-Hydroxycytosine

5-Hydroxyuracil

5,6-Dihydroxyuracil

5-Hydroxyhydantoin

Figure 2   Structures of modified DNA bases

8-OHG is shown in red.

SOURCES OF OH·

The pattern of damage to DNA bases in nuclear DNA suggests that OH· attack occurs *in vivo* (no data are available as yet on whether the same is true for mitochondrial DNA). How could OH· be formed to attack DNA in the nucleus and mitochondria? If OH· is attacking DNA, it must be produced very close to the DNA since it is so reactive that it cannot diffuse from its site of formation [11,30]. Background radiation may be one source [11], but radiation-generated OH· is formed over the whole cell and only a small fraction would be likely to hit DNA [39]. Other sources of OH· include reaction of O_2^- with HOCl [40] and reaction of nitric oxide radical (NO·) with $O_2^{·-}$. NO· reacts very quickly with $O_2^{·-}$ to give $ONOO^-$ [41] and with $RO_2^·$ and $RO^·$ to give organic peroxynitrites, which appear much more stable than $ONOO^-$ [42]. $ONOO^-$ itself is probably directly damaging to DNA bases (e.g. by deamination and nitration of guanine residues) and, at physiological pH, it decomposes into a range of toxic products, including species identical with (or closely resembling) $NO_2^·$, OH· and NO_2^+ [43–45].

By far the greatest interest has been, however, in Fenton chemistry as a source of OH· [30,46]. The question of whether transition metal ions that can convert H_2O_2 into OH· (e.g. iron and copper ions) really are in close proximity to DNA *in vivo* is clearly a critical one. Although iron and copper appear to be present in the nucleus [10,46] and may easily be released from non-haem iron proteins in mitochondria (perhaps as a result of attack by ROS [47]), it remains to be established how they could reach the DNA. Iron and copper ions are normally carefully sequestered by the body, into proteins such as ferritin, transferrin, caeruloplasmin and metallothionein [48]. DNA is a powerful chelating agent for transition metal ions, however, and oxidative stress may cause the release of intracellular iron and/or copper ions into forms that could then bind to DNA [10]. Thus $O_2^{·-}$ releases some iron from ferritin [49], H_2O_2 can release iron from haem proteins [50] and $ONOO^-$ releases copper from caeruloplasmin [51].

DNA-associated copper ions in cells might also react with phenolic compounds to produce ROS and electrophilic phenolic intermediates [52–54]. This interaction could cause a range of DNA lesions including base modifications, strand breaks and phenol adducts to the DNA bases, all of which might contribute to the carcinogenicity of certain phenolic compounds. Phenolic compounds that cause DNA damage in the presence of copper ions include 2-hydroxyoestradiol, 2-methoxyoestradiol, diethylstilboestrol, butylated hydroxytoluene, butylated hydroxyanisole, L-DOPA, dopamine, ferulic acid and caffeic acid [52–54]. However, phenols have complex pro- and anti-oxidant effects *in vitro*, depending on the assay system used, and it is often hard to predict their net effect *in vivo* [55]. For example, many synthetic and dietary polyphenols (including quercetin, catechin, gallic acid ester and caffeic acid ester) can protect mammalian and bacterial cells from the cytotoxicity induced by peroxides such as H_2O_2 [56].

DNA REPAIR

DNA damage can be repaired by the action of a series of enzymes (reviewed in [57]). However, DNA from human cells and tissues contains low levels of DNA base damage products [2,24,25,58–61], suggesting that these enzymes do not achieve complete removal of modified bases, perhaps because they operate at close to maximum capacity *in vivo*. In agreement with this, the steady-state levels of one or more base damage products have been observed to increase in a number of chronic inflammatory diseases accompanied by increased ROS/RNS production, including hepatitis [62] and rheumatoid arthritis [58]. They are also increased in DNA isolated from cancerous tumour biopsies of human lung, colon, kidney, breast, liver and bladder (e.g. see [59–62]) and from benign prostatic hyperplasia tissues [63]. Of course, a rise in DNA base damage products could be due either to increased oxidative damage and/or to decreased repair activity; the increased damage to DNA in inflammation is presumably due to increased ROS/RNS production, often by the activation of phagocytes [10,17].

DNA glycosylases exist for the repair of several DNA base lesions, including oxidized, methylated and deaminated bases. A repair system for the abasic [apurinic/apyrimidinic (AP)] sites produced by spontaneous depurination also exists. Areas of current interest include the role of poly(ADP-ribose) polymerase (PARP) in the rejoining of DNA strand breaks, including those induced by ROS [64,65], and the fact that repair of oxidative DNA damage is defective in xeroderma pigmentosum (XP) cells [65,66]. Human XP is a genetic disorder with an autosomal recessive mode of inheritance, and there are seven genetic complementation groups (A–G) [67]. The defective gene products in these groups are involved in nucleotide excision repair, particularly in damage recognition and incision processes. Thus the XPA protein recognizes and binds to damaged DNA, whereas XPB and XPD are involved in DNA unwinding to facilitate the removal of the faulty base [67]. Normal repair can be restored by mixing two XP cell extracts derived from different complementation groups [65]. The XPG protein is an endonuclease that plays a direct role in making one of the incisions required to excise a damaged base [67]. It seems likely that the products of RNS- and ROS-induced damage may accumulate in XP cells; this could contribute to the neurological deterioration and increased occurrence of cancer observed in XP patients [65,66], an illustration of the importance of DNA repair processes. Defective DNA repair is also responsible for one of the most common cancers, hereditary non-polyposis colon cancer [67].

DNA damage by ROS/RNS can cause multiple lesions, including single and double strand breaks, AP sites and modified pyrimidines and purines. Repair of these lesions occurs primarily by base excision repair, although nucleotide excision repair may also be involved. Recognition of 'spontaneous' [hydrolysis of the base–sugar (glycosylic) bond] and ROS/RNS-generated AP sites may be carried out by various AP endonucleases with different specificities, and this can be used to differentiate between different types of AP site [68].

DNA repair enzymes have usually been purified by assaying their ability to act upon a single specific base lesion. However, recent studies using gas chromatography/MS (GC-MS) have investigated the ability of enzymes to repair DNA containing a wide range of lesions, and have shown that they can sometimes have a broader specificity than expected. For example, *Escherichia coli* endonuclease III can excise several thymine- and cytosine-derived lesions, e.g. 5-hydroxy-2′-deoxycytidine, from DNA [57,69,70].

MEASUREMENT OF OXIDATIVE DNA DAMAGE

In principle, there are two types of measurement of oxidative DNA damage. Steady-state damage is measured when DNA isolated from human cells and tissues is analysed: it is the balance between damage and repair. It is worth mentioning that the measurement of baseline levels of modified DNA bases, although very important, does not provide information as to whether this damage is in active genes or in quiescent DNA. It does, however, seem likely that 'exposed' DNA could be more sensitive to damage by ROS/RNS than that packaged into condensed chromatin.

It is also useful to have an index of total DNA damage (i.e. that which has occurred but has been repaired) and this has been attempted in humans by analysis of urine. Several base damage products are excreted in urine [2,71–73] but the one most exploited is 8-OHdG because it can be assayed using HPLC with electrochemical detection [2,23]. 8-OHG is sometimes measured, but the 8-OHG content of urine is affected by the diet (cooking foods oxidizes their DNA and proteins, just as it oxidizes their lipids) and it can also arise by oxidative damage to RNA. By contrast, the 8-OHdG content of urine is thought not to be affected by the diet since nucleosides are not absorbed from the gut. However, some or all of the 8-OHdG measured in urine may come not from DNA, but from the deoxyGTP precursor pool [74]. An enzyme is believed to hydrolyse dGTP containing oxidized guanine to prevent its incorporation into DNA [74,75]. The activity of this enzyme may vary between different human cells, perhaps contributing to an explanation of variable mutation rates in different tissues exposed to oxidative stress [76]. However, if some or all of the 8-OHdG in urine comes from oxidized dGTP, it follows that we urgently need better markers of total body oxidative DNA damage.

8-OHG and 8-OHdG are the products most frequently measured in isolated DNA as an indicator of oxidative DNA damage. The former can be released from DNA by acid hydrolysis, whereas enzymic hydrolysis liberates 8-OHdG. The methods commonly used for their analysis also raise some questions. Measurement of 8-OHdG by HPLC with electrochemical detection is a highly sensitive method [23]. One alternative is GC-MS with selected ion monitoring, which can measure a wide spectrum of modified (methylated, oxidized, deaminated, etc.) DNA bases [9,10,71]. Both methods are sufficiently sensitive to measure steady-state levels of oxidative base damage in human cells and tissues, but more comparisons of these methods are needed. HPLC can underestimate the amount of 8-OHdG in DNA if the enzymic hydrolysis is not completely efficient; the efficiency of the exonucleases and endonucleases used to hydrolyse the DNA may be diminished by oxidative modification of the bases [24,77]. By contrast, GC-MS might overestimate base damage products if they are generated artifactually during derivatization procedures [78]. Dizdaroglu (e.g. [79]) has used stable isotope dilution MS to quantify DNA damage.

Many additional methods have been described, e.g. GC-MS has been used to analyse thymine glycol residues [80], uracil [81] and malondialdehyde–guanine adducts [82]. Analysis of uracil in DNA by GC-MS following its removal by uracil DNA glycosylase has been used to demonstrate that inhibition of folic acid metabolism induces uracil accumulation in DNA [81]. We have used HPLC after acid hydrolysis of DNA to measure several base damage products, including 8-OHG [83]. HPLC coupled to ^{32}P post-labelling has been used to measure 8-OHdG adducts in the peripheral blood of human subjects exposed to ionizing radiation [84]. Indeed, ^{32}P post-labelling has been used to measure several DNA adducts [85,86].

Different approaches to measuring oxidative base damage have been reported [87,88]. One of these [87] utilizes the ability of endonuclease III to make breaks in DNA at sites of base damage; the breaks are then measured by single-cell gel electrophoresis. For example, normal human lymphocytes *in vitro* were found to contain several hundred endonuclease III-sensitive sites per cell. By contrast, no endonuclease III-sensitive sites were found in HeLa cells, perhaps reflecting less oxidative damage and/or more efficient repair processes in these cells as compared with lymphocytes.

A rigorous comparison and standardization of these various methods is clearly needed (discussed in [24]). Another problem to be considered is the possibility that DNA is oxidatively damaged during its isolation, particularly if phenol-based methods are used [89]. However, the rigorous control of isolation procedures and avoidance of phenol (e.g. by isolation of chromatin for GC-MS analysis [90]) in many laboratories does not decrease oxidative DNA base damage to zero [24,58,59,87,91], strongly supporting the view that there is a low steady-state DNA damage level *in vivo*. Indeed, the fact that an extensive system of repair enzymes exists (see above) supports this view.

HOW COULD DNA DAMAGE BY ROS/RNS CAUSE MUTATION AND CANCER?

ROS/RNS can cause DNA base changes, strand breaks, damage to tumour-suppressor genes and enhanced expression of proto-oncogenes [4,17], and oxidative stress has been shown to induce malignant transformation of cells in culture [92,93]. However, the development of human cancer depends on many other factors, including the extent of DNA damage (excessive DNA damage can cause cell suicide by activation of PARP; see below), antioxidant defences, DNA repair systems, the efficiency of removal of oxidized nucleosides (e.g. oxo-dGTP) before they are incorporated into DNA, and the cytotoxic effects of ROS in large amounts (a dead cell will not lead to cancer) as well as their growth-promoting effects in small amounts [15,94]. For example, the proliferative responses of Syrian hamster embryo fibroblasts to $O_2^{·-}$ have been found to depend on tumour-suppressor gene function, and low levels of $O_2^{·-}$ can enhance cell growth [95]. Persistent excessive oxidative DNA damage may transmit signals to other cellular components (including the tumour-suppressor protein p53) and may trigger apoptosis. In addition, DNA damage (including ROS/RNS-induced DNA strand breaks) can result in p53 accumulation in the nucleus, which may arrest cell growth at the G_1/S border in an attempt to allow repair of DNA lesions before replication. Roles for ROS in apoptosis are frequently suggested (although fiercely contested, e.g. see [96]); the anti-apoptotic gene *bcl-2* decreases the overall cellular production of ROS [16] and ROS can trigger apoptosis [16,97,98]. This does not mean that ROS are essential mediators of apoptosis, of course. The *bcl-2* gene is the prototype of a family of genes that inhibit apoptosis and was originally cloned from the t(14;18) translocation breakpoint found in follicular B-cell lymphomas. Unfortunately, the *bcl-2* gene is also an oncogene and the oncogenic properties of the *bcl-2* protein could relate to its ability to prevent 'useful' apoptotic cell death. The *bcl-2* protein has a C-terminal membrane anchor and is localized to the nuclear, endoplasmic reticulum and mitochondrial outer membranes; it may protect cells by inhibiting ROS-induced lipid peroxidation [16,98,99]. Reasoning from first principles, this could be a direct antioxidant effect of *bcl-2*, and/or it could occur if *bcl-2* up-regulates endogenous antioxidant defences. Thus, while a high cellular antioxidant capacity tends to protect DNA from oxidative damage and related mutagenesis, antioxidant activity may also (ironically) protect 'initiated' cells from ROS-mediated killing. Down-regulation of the *bcl-2* gene by the p53 tumour-suppressor protein in human breast cancer cells has been reported [100]. However, *bcl-2* can rescue cells from apoptosis caused by non-oxidative events, so it does not simply act by antioxidant mechanisms [96,98,99].

Which of the multiple different types of chemical alteration in DNA caused by ROS/RNS are relevant to cancer development? Recent studies have attempted to identify mutations that are caused by ROS/RNS damage to DNA, with the aim of ascertaining their association (if any) with cancer [4].

Damage to DNA by ROS, as measured in a single-stranded DNA, *E.coli*-based, forward mutation assay, was found to induce a wide spectrum of mutations, which depended not only on the ROS used but also on the DNA replication apparatus that encountered the lesion [5]. The most frequent mutations found in this system were C to T transitions. However, mutations arising from C to T transitions are not diagnostic for mutagenesis by ROS because they can be caused by DNA polymerase errors and by DNA damage arising from the action of other genotoxic agents [5]. Incubation of DNA with ROS-generating systems such as $FeSO_4/O_2$ or $CuCl_2/H_2O_2$ resulted in the formation of mainly C to T, G to T and G to C substitutions in the iron system; and C to T and G to T substitutions and CC to TT tandem double transitions in the copper ion system [101]. Each of these mutation patterns was clustered in 'hot spots' that are characteristic for each system; some of this effect might relate to different binding of the two metal ions to DNA, so that ROS are directed to particular DNA base sequences [5]. For example, copper ions bind preferentially to GC-rich sequences [102].

There are several different pathways leading from initial DNA base damage by ROS/RNS to subsequent mutation (Figure 3). The first (and probably the simplest) is the chemical modification of DNA bases causing a change in their hydrogen bonding specificity, e.g. 8-OHG, thymine glycol and 2-hydroxyadenine [101,103,104]. In addition, 8-hydroxyadenine, ring-opened purines and a number of pyrimidine fragmentation products can block replication in *E. coli* and could thus be mutagenic [105,106]. Singlet-oxygen-induced DNA damage is targeted selectively at guanine residues [22,107] and the G to T transversion mutations induced may be generated by an error-prone bypass of damaged G [108]. Oxidation products of cytosine (5-hydroxycytosine and 5-hydroxyuracil) exhibit sequence-context-dependent mispairing *in vitro* resulting in C to T transitions and C to G transversions [109]. Although it is not obvious which modified DNA bases are responsible for the mutations that can be introduced by DNA polymerases α and β, it seems likely that the C to T and C to A

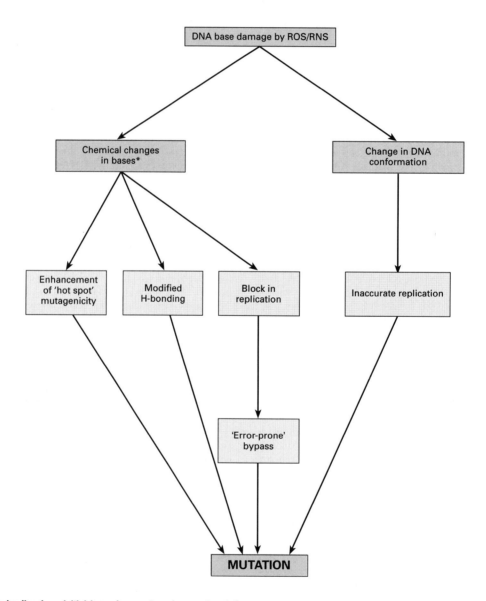

Figure 3   Pathways leading from initial base damage to subsequent mutation

Key processes are shown in red. *See Figure 2.

substitutions observed in the *E. coli* system result from the oxidative modification of cytosines [5]. DNA polymerase is known to be sensitive to damage-induced errors at guanines.

The contribution of oxidative damage to polymerase-specific 'hot spots', which is a likely major contributor to DNA polymerase-mediated mutagenesis, is a second possible mechanism. In studies on the *lacZ* gene, T to G transversions at positions +70 and +103 were observed when polymerase was used to copy an undamaged DNA template [110]. A 6–10-fold increase in the frequency of transversions at both sites was caused by iron-ion-mediated oxidative damage, and an increased frequency of mutation at position +103 only was produced by copper-ion-dependent damage [110]. The 'background' mutations might also be caused by endogenous oxidative damage. Perhaps also the sequences used were inherently mutagenic for polymerase and could thus enhance the miscoding potential of oxidatively induced adjacent lesions [5].

A third mechanism is linked to a conformational change in the DNA template that diminishes the accuracy of replication by DNA polymerases [110]. Evidence for this mechanism has been provided by the finding that iron-ion-dependent damage to M13mp2 double-stranded circular DNA produces a high frequency of mutants that possess substitutions at positions +95 and +103 on the *lacZ* gene. The frequency of this double mutation in the polymerase spectrum is four orders of magnitude higher than would be expected from two independent mutational events, and it has been suggested [110] that oxidative damage to the DNA template alters the pausing pattern for DNA polymerase at the two sites of mutation by changing the DNA structure. The frequency of mutation is also greatly increased by exposure of the DNA to ROS, probably by mechanisms involving rearrangements of the nascent strand-template. Conformational changes that cause an increase in the pausing of the DNA polymerase or a decrease in processivity could enhance such rearrangements. Although direct studies of the effects of base modifications on DNA conformation are just beginning, it is known that many oxidized bases are non-planar and could change local DNA structure [5].

Human relevance of bacterial studies

Studies of bacterial mutation depend on the so-called SOS response (which includes DNA repair enzymes) and the nature of the bacterial DNA replication system. Such studies can only provide indications about the mutations likely to be caused in human cells by ROS/RNS. Indeed, studies with mammalian DNA polymerases have shown that DNA damage by ROS can yield mutations different from those observed in *E. coli* [110]. Nevertheless, studies using bacterial systems have provided useful information, including the suggestion that tandem CC to TT double substitutions, induced by ROS from many different sources including iron(II), copper(I) and nickel(II) ions and γ radiation [111,112], may be specific markers for oxidative DNA damage, at least in cells not exposed to UV light (which can also induce this mutation [113]). Nickel(II) is a known human carcinogen which has been shown to increase oxidative DNA base damage in rats [114] and to induce the above 'signature mutation' for ROS damage in an *E. coli* forward mutation assay [112].

The *p53* tumour-suppressor gene and the *ras* family of proto-oncogenes are known to be important cancer-related genes. It is possible that ROS may be the mutagens involved in some of the mutations observed in these genes [4,17,115], e.g. the G to T transversion often found in Ki-*ras* and H-*ras* in non-melanoma skin tumours could be produced by misreplication of the 8-

OHdG lesions induced by ROS (although GT tranversions may be generated by other mutagenic mechanisms). G to T transversions have also been observed in *p53* codons in hepatocellular carcinoma and smoking-related lung carcinoma, and again they might arise by the actions of ROS. In colorectal cancer ROS may yet again be involved, as around 95 % of the mutations at *p53* hot spot codons in these tumours are C to T and G to A transitions, and these base-pair changes are often produced by the deamination of 5-methylcytosine, which is enhanced by ROS and RNS [4].

Indirect mutagenicity by ROS/RNS: damage to proteins and lipids

ROS/RNS-induced mutations could arise not only from DNA damage, but also from protein damage. Protein damage is a major consequence of excess ROS generation *in vivo* [116] and damage to DNA polymerases could alter their fidelity. It has been suggested that an alteration in the conformation of DNA polymerase could explain the frequency of close-proximity double mutations that occur secondarily to a wide range of genetic stresses [5,117]. Several RNS, e.g. ONOO⁻ and NO₂·, can attack proteins, nitrating aromatic amino acid residues and possibly affecting their ability to participate in signal transduction mechanisms [43,44,118]. Oxidative protein damage could also affect the activity of DNA repair enzymes.

Another possible mutagenic effect of ROS involves their attack on lipids, to initiate lipid peroxidation (Figure 2). Peroxides can decompose to a range of mutagenic carbonyl products [34]. For example, 4-hydroxynonenal is genotoxic to lymphocytes and hepatocytes and also disrupts gap-junction communications in cultured endothelial cells [119]. In a study on a baby hamster kidney cell line (BHK-21/C13) and its polyoma-virus-transformed malignant counterpart (BHK-21/PyY cells) the level of lipid peroxidation (as measured by HPLC of malondialdehyde) was higher in transformed cells than in non-transformed cells, suggesting that the level of lipid peroxidation is increased in the malignant state [120]. By contrast, earlier work claimed that susceptibility to lipid peroxidation is decreased in malignant hepatoma cells (e.g. Novikoff and Yoshida ascites hepatoma cells) (reviewed in [34]) and this was found to be directly related to their extent of dedifferentiation [121]. It is of related interest that regenerating rat liver shows increased steady-state levels of 8-OHdG in nuclear DNA [122] and that partially hepatectomized rats show increased urinary excretion of thymine glycols [123].

These findings illustrate the well-known observation that cell transformation alters cell responsiveness to oxidative stress. Changes in antioxidant defence enzymes such as superoxide dismutase (SOD) have been widely described in cancerous cells, although the most consistent reports seem to be of falls in Mn-containing SOD (MnSOD) (the mitochondrial enzyme) activity [124]. This change may be significant in relation to the malignant phenotype, since radiation-induced neoplastic transformation in mouse C3H 10T1/2 cells (a cell line susceptible to transformation by ROS [93]) was decreased by transfection of the cells with a gene encoding human MnSOD [125]. Similar results have been reported in human melanoma cells [126] and a mouse fibrosarcoma cell line [127].

'Slippery' DNA

Another area worthy of consideration in the ROS/RNS context is that of misalignment mutagenesis ('slippery DNA'). The human genome contains many sequences in which 1–6-nucleotide motifs are tandemly repeated a number of times. The contraction and expansion of these 'microsatellites' is associated with cancer (e.g. mono-, di- and tri-nucleotide repeats are unstable in colon cancer cells and this instability has been linked to a gene on chromosome 2) and several genetic diseases, including myotonic dystrophy and Huntington's disease [128]. The gain or loss of repetitive DNA sequences may be caused by errors that result from strand slippage during DNA replication remaining uncorrected because of defective post-replication heteroduplex repair [129]. The instability of these repeats is associated with DNA polymerases slipping during replication, and some types of colorectal cancer may reflect mutations in genes involved in DNA mismatch repair [129]. Theoretically, ROS/RNS should be capable of accelerating such changes [129].

PARP

The nuclear enzyme PARP modifies proteins (including itself; automodification) by the attachment of poly(ADP-ribose) polymers. PARP is involved in base excision repair but not in nucleotide excision repair, which requires the formation of a multiprotein repair complex before DNA incision that may prevent the binding of PARP to DNA breaks [64,65]. The repair of DNA single strand breaks induced by the antibiotic bleomycin [65] is totally dependent on the activation of PARP, and the repair of modified bases generated by alkylating agents is also promoted by PARP activation. In base excision repair, damaged bases are eliminated by DNA glycosylases, breaks are then induced at the AP sites by AP endonuclease and this is followed by the excision of deoxyribose phosphate, DNA synthesis and finally ligation. The process of DNA break rejoining initiated by AP endonuclease is aided by PARP, which may have a structural role in chromatin conformation and may provide temporary protection for the DNA breaks during the initial stage of the recombination and repair processes. Excess activation of PARP can kill cells by depleting NAD^+ and preventing energy production [130], and this may be one mechanism by which cells with excess DNA damage are eliminated (see above).

ROLE OF INFLAMMATION IN INCREASED DNA DAMAGE, MUTATION AND CANCER

The cumulative risk of cancer increases with approximately the fourth power of age [2]. This is true for both short-lived species such as mice, where about 30 % have cancer by the end of their 2–3-year life span, and in long-lived species such as humans, where approx. 30 % have cancer by age 85. Much cancer can thus be considered as a degenerative disease of old age and it is frequently suggested that this is related to the effects of continuous damage over a life span by ROS and RNS [1,2]. For example, prostate cancer is most prevalent among elderly males and has been suggested to be associated with endogenous cellular processes, in particular ROS generation [5].

Inflammation can accelerate the development of cancer [3,131]. Many sources of inflammation are effective, including that caused by viral, bacterial and parasitic infections. In colon cancer, predisposing sources of chronic inflammation include ulcerative colitis and infection with the parasite *Schistosoma japonicum* [28,62,92].

However, the link between cancer and inflammation is by no means a simple one. One chronic inflammatory disease in which patients suffer oxidative stress is rheumatoid arthritis. There is increased damage by ROS [132,133] (and probably also by RNS [134]) to lipids, proteins and DNA. Rheumatoid arthritis patients also show increased urinary excretion of 8-OHdG [135]. Increased DNA damage, and increased susceptibility to killing by H_2O_2, have also been reported in lymphocytes from patients with autoimmune diseases: lymphocyte DNA from patients with rheumatoid arthritis, systemic lupus erythematosus, vasculitis

and Behcet's disease contained significantly more 8-OHdG than that from healthy controls [58].

Despite all this, there is no clear evidence that rheumatoid arthritis patients develop cancers at an increased rate, certainly not at the most intense site of oxidative stress, the inflamed joint. Perhaps this is linked to the nature of the cells of the synovium. Synovium seems to be a hostile environment for neoplastic cells [136], although one might argue that the excessive synovial cell proliferation in rheumatoid arthritis patients might be related to the growth-promoting properties of ROS.

Despite the apparent anomaly of rheumatoid arthritis, there is considerable evidence that ROS/RNS are somehow involved in the link between chronic inflammation and cancer [3,28,62,92,131]. A notable activity of tumour promoters is their ability to recruit inflammatory cells and to stimulate them to generate ROS/RNS. Indeed, there is a strong relationship between the capacity of tumour promoters to stimulate inflammatory cells to release ROS/RNS and their capacity to promote tumours [131,137,138]. Genetic damage and neoplastic transformation have been demonstrated in cells co-cultured *in vitro* with activated phagocytes [92]. The genotoxic effects observed in these cells include the formation of DNA strand breaks [138], sister chromatid exchange [139] and mutations [140]. The DNA base modifications observed in cells co-cultured with phorbol 12-myristate 13-acetate (PMA)-activated inflammatory cells were characteristic of attack by OH˙ [141]. Presumably phagocyte products such as H_2O_2 penetrated to the nucleus and were converted into OH˙ by reaction with localized transition metal ions.

Tumour promotion can be inhibited in animal models by the use of agents that can inhibit the phagocyte respiratory burst, including certain antioxidants, as well as steroids and retinoids [131,137]. Increased levels of oxidatively modified DNA bases (thymidine glycol, 5-hydroxymethyl-2'-deoxyuridine and 8-OHdG [142]) have been induced in the skin of mice by topical PMA exposure. 8-OHdG has also been identified in the epidermis of nude mice exposed to near-UV [143]. The production of ROS *in vivo* following the application of phorbol esters to mouse skin requires two applications; each triggers a separate biochemical event, priming and activation [144]. Both of these events can be triggered by tumour-promoting phorbol esters but not by non-promoting ones. Priming events include the recruitment of inflammatory cells: PMA recruits neutrophils and activates them to produce ROS/RNS. The inhibitor of phorbol ester tumour promotion, retinoic acid, inhibits activation but not priming and similar results are found for several phenolic antioxidants [142,144]. This suggests that retinoic acid, usually considered to inhibit promotion by altering gene expression in transformed cells, might act additionally to exert indirect antioxidant effects under certain circumstances, at least when applied topically to mouse epidermis. Similarly, 13-*cis*-retinoic acid inhibits X-ray-induced skin cancer in XP patients (see above), where it has been proposed to exert antioxidant effects [145].

Inflammatory cells may also increase DNA damage by activating pro-carcinogens to DNA-damaging species, e.g. neutrophils can activate aromatic amines, aflatoxins, oestrogens, phenols and polycyclic aromatic hydrocarbons by ROS-dependent mechanisms [131,146]. RNS can generate carcinogenic nitrosamines [3,147,148], e.g. nitrosation of morpholine has been reported in immunostimulated rats [149].

Some examples

Schistosomiasis

The schistosomiasis model has been used to study the interrelationship between inflammation, oxidative DNA damage, chromosomal instability and dysregulated cell proliferation [131]. Infection with *Schistosoma haematobium* produces chronic bladder inflammation and is associated with increased cancer at this site. Indeed, infected individuals show elevated levels of genetic damage in their bladders, as measured by the exfoliated cell micronucleus test [131,150], and micronucleus frequency is decreased by treatments that kill the parasite. It is possible that clones of cells in these patients develop an inherited altered ability to repair oxidative DNA damage and thus an increased sensitivity to the ROS/RNS produced by activated inflammatory cells. Alterations in chromosome 11 are common in bladder cancer [151] and loci on this chromosome may be involved in controlling the level of chromosomal breakage caused by oxidative DNA damage. This sensitivity to oxidative stress is apparently not due to a difference in single-strand DNA breakage or repair [152]. In addition, an abnormally high frequency of chromatid breaks and gaps has been reported when human tumours (of many different tissue origins and/or histopathology) are X-irradiated during the G_2 phase of the cell cycle; insertion of a normal chromosome 11 decreased radiation-induced damage, possibly by restoration of a defective repair process (see [153]). Bladder carcinoma cells have been shown to be sensitive to micronucleus induction by tumour-promoter-activated neutrophils and protection is possible via the insertion of a normal chromosome 11 [131]. Again, this is thought to restore the defective DNA repair process [131]. It is of related interest that infection with the pro-inflammatory bacterium *Helicobacter pylori* appears to be an important risk factor for stomach ulcers, gastritis and possibly stomach cancer [154].

Lung cancer

Lung cancer is a frequent cause of death, and most cases are linked to smoking. Cessation of smoking leads to a rapid decrease in the risk of lung cancer, suggesting that a series of smoking-related events is required for cancer development. Cigarette smoke is rich in carcinogens such as nitrosamines, acrolein and carcinogenic hydrocarbons, but ROS/RNS may also contribute to cancer development, since smoke is rich in ROS and oxides of nitrogen [118,155]. Higher levels of oxidative DNA base damage have been reported in lung cancer tissue compared with surrounding normal tissue [156] and in cells exposed to cigarette smoke [157]. Additionally, a 4–10-fold elevation of urinary 8-OHdG excretion has been found in smokers [72,73].

Exposure to asbestos is a major risk factor for mesothelioma; asbestos-induced chronic inflammation and resulting DNA damage may contribute [158–160]. Crocidolite (one of the most carcinogenic types of asbestos) induces release of ROS from neutrophils and macrophages, and increased 8-OHdG levels in the DNA of a human promyelocytic leukaemia cell line (HL60) [159]. Furthermore, exposure of rat pleural mesothelial cells to crocidolite and chrysotile fibres resulted in DNA damage and cell toxicity that was partly due to ROS [160].

Liver cancer

In both Asia and Africa, hepatocellular carcinoma is a major cause of mortality. Primary hepatoma in these countries is often associated with chronic infection with hepatitis viruses B or C [161], or ingestion of aflatoxin [162]. In primary hepatocellular carcinoma, aflatoxin exposure often results in mutations involving codon 249 of the *p53* gene [163]. Indeed, aflatoxin frequently produces G to T transversions, and this is the predominant substitution at codon 249 in *p53* found in aflatoxin-associated tumours. This transversion can also be produced by

oxidative damage [115]. Chronic hepatitis [164] is associated with the presence of inflammatory cells, presumably generating ROS and RNS. Indeed, increased levels of 8-OHdG have been detected in DNA from livers with chronic hepatitis [62,165].

Inflammatory bowel disease (IBD)

IBD is the general name given to a series of chronic inflammatory diseases of the gastrointestinal tract, including ulcerative colitis and Crohn's disease. ROS are formed in excess in IBD and are likely to play an important role not only in the pathogenesis of IBD [166,167] but also in the increased risk of cancer seen in certain IBDs. An elevated production of ROS has been shown using colorectal biopsy specimens [168]. Further evidence consistent with damage by ROS is provided by the increase in lipid peroxides in rectal biopsy specimens from patients with active ulcerative colitis, and the reports of low levels of GSH, SOD and glutathione peroxidase in patients with active IBD [166,169,170]. The main sources of ROS in the gut are probably phagocytes, which accumulate in the mucosa of patients with IBD and generate ROS (and presumably RNS) upon activation. A marked increase in the activity of the inducible form of nitric oxide synthase (iNOS) in the inflamed colonic mucosa from patients with ulcerative colitis has been reported [171,172]. By contrast, there was no increase in iNOS activity in the inflamed colonic mucosa from patients with active Crohn's disease, even though the extent of inflammation was similar [171]. In addition, nitrotyrosine (a putative chemical marker for the formation of $ONOO^-$ [43,173]) has been detected by an immunohistochemical stain in an animal model of chronic gut inflammation and found to co-localize with NOS [174]. This suggests that both $NO^{\cdot}$ and $O_2^{\cdot-}$ are formed *in vivo*, and undergo reaction to give $ONOO^-$. Furthermore, intracolonic instillation of $ONOO^-$ is pro-inflammatory in a rat colitis model [175].

Another important ROS in IBD may be HOCl. HOCl, produced by the enzyme myeloperoxidase from activated neutrophils, may attack membrane proteins directly (e.g. by oxidizing -SH groups, destroying methionine and chlorinating aromatic amino acid residues), or indirectly by the formation of chloramines. Both HOCl and chloramines can stimulate colonic secretion [176]. A role for ROS has also been described in the stimulation of colonic mucosal proliferation by bile salts [94]. It has been proposed that some of the drugs effective in the treatment of IBD may act by scavenging HOCl and other ROS [167,177].

What is the evidence for increased oxidative DNA damage in IBD? DNA from colon biopsies from patients with ulcerative colitis had significantly increased levels of 8-OHG, 2-hydroxyadenine, 8-hydroxyadenine and 2,6-diamino-5-formamidopyrimidine as measured by GC-MS/selected ion monitoring (H. Wiseman, M. Dizdaroglu and B. Halliwell, unpublished work). These lesions, suggestive of $OH^{\cdot}$ attack, could signify increased DNA damage and/or decreased repair. The constitutive and ROS-induced activity of PARP has been shown to be decreased in patients with IBD and colon cancer [178].

Breast cancer

Oxidative DNA damage may be involved in the development of breast cancer. Increased steady-state levels of DNA base damage, with a pattern characteristic of $OH^{\cdot}$ attack, have been reported in DNA from invasive ductal carcinoma [60,179]. One study found a 9-fold increase in 8-OHG, 8-hydroxyadenine and 2,6-diamino-4-hydroxy-5-formamidopyrimidine in DNA from invasive ductal carcinoma compared with control tissue [60,179].

Whether this is due to decreased DNA repair and/or increased oxidative DNA damage is uncertain. DNA damage by ROS is also implicated in inflammatory breast disease [180], where malignant progression can occur.

PROSPECTS FOR DIETARY AND DRUG ANTIOXIDANT INTERVENTION IN THE PREVENTION OF OXIDATIVE DNA DAMAGE AND HUMAN DISEASE

Dietary antioxidants

Cells have multiple antioxidant defences to protect themselves against ROS. These protective mechanisms are not present in excess; if they were, oxidative damage would not occur and repair mechanisms would not be required. Instead, oxidative damage occurs continuously in the human body (Table 1). Fortunately, enzymic and some other (e.g. metallothionein, caeruloplasmin and haptoglobin) antioxidant defences are often inducible in response to oxidative damage [30].

We also obtain several antioxidants from the diet. Indeed, the consumption of fruit, grains and vegetables, which are the main sources of these antioxidants, is of importance in protecting against oxidative damage and resulting disease [7,8,181–184]. Intake of fresh fruit and vegetables appears to be inversely correlated with cancer of the stomach, pancreas, oral cavity and oesophagus, and to a lesser extent of the breast, cervix, rectum and lung [183,184], and emphasis has been placed on the protective role of ascorbate. Indeed, there is evidence that ascorbate can react with, and/or inhibit the formation of, carcinogenic N-nitroso compounds such as N-nitrosamines [185,186]. Vitamin C supplementation has been reported to decrease mucosal DNA damage, as measured by a ^{32}P post-labelling assay, in 28 of 43 patients attending a gastric follow-up clinic [187]. In patients with normal gastric mucosa, treatment with vitamin C resulted in an elevation of intragastric ascorbate in all cases, whereas in patients with chronic atrophic gastritis the effect was variable [187]. These data support epidemiological evidence suggesting that vitamin C may exert a protective effect against the development of gastric cancer. Very low vitamin C intakes have been associated with elevated levels of 8-OHdG in sperm DNA [188]. However, ascorbate supplementation did not affect tumour development in chemically induced bladder, mammary or colon cancers in rodents (reviewed in [7]). Therefore the extent of the benefits of ascorbate in human cancer prevention remain to be ascertained.

A sufficient level of dietary antioxidants has been suggested to be achievable by the intake of a minimum of three servings of vegetables and two of fruit per day. Furthermore, dietary supplementation of this daily intake of fruit and vegetables with

moderate amounts of the relatively inexpensive and non-toxic vitamins ascorbate and α-tocopherol may also be desirable in some population groups, such as smokers [8,181,183,184,189]. In addition to antioxidants, fruit and vegetables contain many vital micronutrients that may be protective. These include folic acid, which is required for the synthesis of DNA precursors, and niacin, which is required for the NAD$^+$ used by PARP. Many carotenoids, including β-carotene, can be metabolized to vitamin A (retinol), and the antioxidant activity of carotenoids is thought to be particularly directed against $^1O_2$ [190–193].

However, it would be naive in the extreme to assume that the protective effects of fruits and vegetables are related only to their antioxidant content; among the many other potentially protective substances present are anti-angiogenesis factors, inducers of carcinogen-removing enzymes, fibre and phytates [194]. Diets rich in fruits and vegetables are often low in fat, which could contribute to their anti-cancer effect [195]. Such diets are also often low in iron; high body iron levels are (controversially) associated with increased risk of cancer [196–199]. Furthermore, in a prospective study of the intake of vitamins C, E and A and the risk of breast cancer, it was found that large intakes of vitamins C or E were not protective to women [200]. There is an urgent need to investigate to what extent dietary changes can decrease steady-state and total-body oxidative DNA damage in humans (which relates to the methodological questions discussed earlier) and, if so, what is the optimal intake of, for example, fruit, vegetables or antioxidant supplements. The rapid development of accurate assays for measuring oxidative damage to DNA, lipids and proteins in the human body should help to make this possible (reviewed in [181]).

Drug antioxidants

Several drugs in current clinical use may exert some antioxidant effects in vivo [201]. Tamoxifen, which is widely used in the treatment of breast cancer and is being investigated as a prophylactic treatment for this disease, may exert antioxidant effects additional to its anti-oestrogenic properties [202]. Thus it has been reported to suppress H_2O_2 production by human neutrophils [203–205]. The tamoxifen metabolite 4-hydroxy-tamoxifen is a more powerful inhibitor of lipid peroxidation than tamoxifen in lipoproteins [206] and in the nuclear membrane [207].

Another drug that may act as a free-radical scavenger in vivo is sulphasalazine and its metabolites, used in the treatment of IBD. Sulphasalazine is converted by colonic bacteria into 5-aminosalicylic acid (5-ASA), which has a number of antioxidant properties. It is an excellent scavenger of several ROS, especially HOCl [177,208,209]. The finding, in IBD patients treated with sulphasalazine, of 5-ASA-derived products identical to those formed when this compound reacts with OH· in vitro, suggests that ROS scavenging by 5-ASA is a significant mechanism of action in vivo [210,211].

CONCLUSION AND FUTURE PROSPECTS

We need oxygen in order to survive, but the constant assault on our DNA by ROS/RNS may lead to cancer development [1,2]. Even phagocyte ROS/RNS production, useful in the short term as a defence against infection, may harm us in the long term, and certainly harms patients with chronic inflammatory diseases. Understanding the mechanisms by which these chemical changes relate to alterations in gene expression and the development of cancer, and how they can be affected by drug treatment and/or dietary changes, requires a combination of expertise in molecular

Table 1   Evidence for ROS/RNS-mediated damage in vivo

Target of damage	Evidence
DNA	Urinary excretion of damaged DNA bases; low baseline levels of damaged DNA bases in DNA isolated from human cells
Lipid	Lipid peroxidation in atherosclerotic lesions; presence of peroxidation end-products in plasma and urine
Uric acid	Damaged by ROS to form products (including allantoin) found in human body fluids; levels increase during oxidative stress
Protein	Protein carbonyls and o-tyrosine formed from ROS attack and nitrosothiols/nitro-aromatics from RNS attack; low levels of some of these products can be detected in human tissues and body fluids and may increase during oxidative stress

biology and analytical biochemistry. As an example we have little information on how oxidative base modification affects the PCR. Thus amplification of oxidized DNA (e.g. ancient DNA [212]) may produce misleading results. Correlations of structural chemistry and analytical methodology with changes in gene expression should lead to valuable new concepts and, hopefully, novel ways of preventing cancer.

We are grateful to the Cancer Research Campaign, the Medical Research Council and MAFF for research support.

REFERENCES

1 Totter, J. R. (1980) Proc. Natl. Acad. Sci. U.S.A. **77**, 1763–1767
2 Ames, B. N. (1989) Free Radical Res. Commun. **7**, 121–128
3 Oshima, H. and Bartsch, H. (1994) Mutat. Res. **305**, 253–264
4 Cerutti, P. (1994) Lancet **344**, 862–863
5 Feig, D. I., Reid, T. M. and Loeb, L. A. (1994) Cancer Res. (Suppl.) **54**, 1890s–1894s
6 Routledge, M. N., Wink, D. A., Keefer, L. K. and Dipple, A. (1994) Chem. Res. Toxicol. **7**, 628–632
7 Byers, T. and Perry, G. (1992) Annu. Rev. Nutr. **12**, 139–159
8 Diplock, A. (1994) Mol. Aspects Med. **15**, 293–376
9 Dizdaroglu, M. (1993) in DNA and Free Radicals (Halliwell, B. and Aruoma, O. I., eds.), pp. 19–39, Ellis Horwood, Chichester
10 Halliwell, B. and Aruoma, O. I. (1991) FEBS Lett. **281**, 9–19
11 von Sonntag, C. (1987) The Chemical Basis of Radiation Biology, Taylor & Francis, London
12 Steenken, S. (1989) Chem. Rev. **89**, 503–520
13 Epe, B (1993) in DNA and Free Radicals (Halliwell, B. and Aruoma, O. I., eds.), pp. 41–65, Ellis Horwood, Chichester
14 Schreck, R., Albermann, K and Bauerle, P. A. (1992) Free Radical Res. Commun. **17**, 221–237
15 Burdon, R. H., Alliangana, D. and Gill, V. (1995) Free Radical Biol. Med. **18**, 775–794
16 Sarafian, T. A. and Bredesen, D. E. (1994) Free Radical Res. **21**, 1–8
17 Jackson, J. H. (1994) Environ. Health Perspect. **102** (suppl. 10), 155–158
18 Rao, G. N., Lessegue, B., Griendling, K. K. and Alexander, R. W. (1993) Oncogene **8**, 2759–2764
19 Stevenson, M. A., Pollock, S. S., Coleman, C. N. and Calderwood, S. K. (1994) Cancer Res. **54**, 12–15
20 Weitzman, S. A., Turk, P. W., Howard-Milkowski, D. and Kozlowski, K. (1994) Proc. Natl. Acad. Sci. U.S.A. **91**, 1261–1264
21 Box, H. C., Freund, H. G., Budzinski, E., Wallace, J. C. and Maccubbin, A. E. (1995) Radiat. Res. **141**, 91–94
22 Van den Akker, E., Lutgerink, J. T., Lafleur, M. V. M., Joenje, H. and Retel, J. (1994) Mutat. Res. **309**, 45–52
23 Floyd, R. A., Watson, J. J., Wong, P. K., Altmiller, D. H., and Rickard, R. C. (1986) Free Radical Res Commun. **1**, 163–172
24 Halliwell, B. and Dizdaroglu, M. (1992) Free Radical Res. Commun. **16**, 75–87
25 Musarrat, J. and Wani, A. A. (1994) Carcinogenesis **15**, 2037–2043
26 Richter, C. (1992) Mutat. Res. **275**, 249–255
27 Harman, D. (1992) Mutat. Res. **275**, 257–266
28 Shigenaga, M. K., Hagen, T. M. and Ames, B. N. (1994) Proc. Natl. Acad. Sci. U.S.A. **91**, 10771–10778
29 Sohal, R. S., Ku, H. H., Agarwal, S., Forster, M. J. and Lal, H. (1994) Mech. Ageing Dev. **74**, 121–133
30 Halliwell, B. and Gutteridge, J. M. C. (1989) Free Radicals in Biology and Medicine, 2nd edn., Clarendon Press, Oxford
31 Ambrosio, G., Zweier, J. L., Duilio, C., Kuppusamy, P., Santoro, G., Elia, P. P., Tritto, I., Cirillo, P., Condorelli, M., Chiariello, M. and Flaherty, J. T. (1993) J. Biol. Chem. **268**, 18532–18541
32 Guidot, D. M., McCord, J. M., Wright, R. M. and Repine, J. E. (1993) J. Biol. Chem. **268**, 26699–26703
33 Agarwal, S. and Sohal, R. S. (1994) Proc. Natl. Acad. Sci. U.S.A. **91**, 12332–12335
34 Cheeseman, K. (1993) in DNA and Free Radicals (Halliwell, B. and Aruoma, O. I., eds.), pp. 109–144, Ellis Horwood, Chichester
35 Hruszkewycz, A. M. (1992) Mutat. Res. **275**, 234–248
36 Arnheim, N. and Cortopassi, G. (1992) Mutat. Res. **275**, 157–167
37 Mecocci, P., MacGarvey, U. and Beal, M. F. (1994) Ann. Neurol. **36**, 747–751
38 Corral-Debrinski, M., Shoffner, J. M., Lott, M. T. and Wallace, D. C. (1992) Mutat. Res. **275**, 169–180
39 Nackerdien, Z., Olinski, R. and Dizdaroglu, M. (1992) Free Radical Commun. **16**, 259–273
40 Candeias, L. P., Patel, K. B., Stratford, M. R. L. and Wardman, P. (1993) FEBS Lett. **333**, 151–153
41 Huie, R. E. and Padmaja, S. (1993) Free Radical Res. Commun. **18**, 195–199
42 Padmaja, S. and Huie, R. E. (1993) Biochem. Biophys. Res. Commun. **195**, 539–544
43 Beckman, J. S., Chen, J., Ischiropoulos, H. and Crow, J. P. (1994) Methods Enzymol. **233**, 229–240
44 Yermilov, V., Rubio, J., Becchi, M., Friesen, M. D., Pignatelli, B. and Ohshima, H. (1995) Carcinogenesis **16**, 2045–2050
45 Rojas-Walker, T., de Tamir, S., Ji, H., Wishnok, J. S. and Tannenbaum, S. R. (1995) Chem. Res. Toxicol. **8**, 473–477
46 Nascimento, A. L. T. O. and Meneghini, R. (1995) Carcinogenesis **16**, 1335–1338
47 Flint, D. H., Smyk-Randall, E., Tuminello, J. F., Draczynska-Lusiak, B. and Brown, O. R. (1993) J. Biol. Chem. **268**, 25547–25552
48 Halliwell, B and Gutteridge, J. M. C. (1990) Arch. Biochem. Biophys. **280**, 1–8
49 Bolann, B. J. and Ulvik, R. J. (1990) Eur. J. Biochem. **193**, 899–904
50 Gutteridge, J. M. C. (1986) FEBS Lett. **201**, 291–295
51 Swain, J. A., Darley-Usmar, V. and Gutteridge, J. M. C. (1994) FEBS Lett. **342**, 49–52
52 Li, Y., Trush, M. A. and Yager, J. D. (1994) Carcinogenesis **15**, 1421–1427
53 Li, Y. and Trush, M. A. (1994) Cancer Res. (Suppl.) **54**, 1895s–1898s
54 Spencer, J. P. E., Jenner, A., Aruoma, O. I., Evans, P. J., Kaur, H., Dexter, D. T., Jenner, P., Lees, A. J., Marsden, C. D. and Halliwell, B (1994) FEBS Lett. **353**, 246–250
55 Halliwell, B. (1990) Free Radical Res. Commun. **9**, 1–32
56 Nakayama, T. (1994) Cancer Res. (Suppl.) **54**, 1991s–1993s
57 Demple, B. and Harrison, L. (1994) Annu. Rev. Biochem. **63**, 915–948
58 Bashir, S. Harris, G., Denman, M. A., Blake, D. R. and Winyard, P. G. (1993) Ann. Rheum. Dis. **52**, 659–666
59 Adachi, S., Zeisig, M. and Moller, L. (1995) Carcinogenesis **16**, 253–258
60 Malins, D. C. and Haimanot, R. (1991) Cancer Res. **51**, 5430–5432
61 Jaruga, P., Zastawny, T. H., Skokowski, J., Dizdaroglu, M. and Olinski, R. (1994) FEBS Lett. **341**, 59–64
62 Hagen, T. M., Huang, S., Curnutte, J., Fowler, P., Martinez, V., Wehr, C. M., Ames, B. N. and Chisari, F. V. (1994) Proc. Natl. Acad. Sci. U.S.A. **91**, 12808–12812
63 Olinski, R., Zastawny, T. H., Foksinski, M., Barecki, A. and Dizdaroglu, M. (1995) Free Radical Biol. Med. **18**, 807–813
64 Satoh, M. S., Poirier, G. G. and Lindahl, T. (1993) J. Biol. Chem. **268**, 5480–5487
65 Satoh, M. and Lindahl, T. (1994) Cancer Res. (Suppl.) **54**, 1899s–1901s
66 Satoh, M. S., Jones, C. J., Wood, R. D. and Lindahl, T. (1993) Proc. Natl. Acad. Sci. U.S.A. **90**, 6335–6339
67 Marx, J. (1994) Science **266**, 728–730
68 Haring, M., Rudiger, H., Demple, B., Boiteux, S. and Epe, B. (1994) Nucleic Acids Res. **22**, 2010–2015
69 Hatahet, Z., Kow, Y. W., Purmal, A. A., Cunningham, R. P. and Wallace, S. S. (1994) J. Biol. Chem. **269**, 18814–18820
70 Dizdaroglu, M., Laval, J. and Boiteux, S. (1993) Biochemistry **32**, 12105–12111
71 Stillwell, W. G., Xu, H. X., Adkins, J. A., Wishnok, J. S. and Tannenbaum, S. R. (1989) Chem. Res. Toxicol. **2**, 94–99
72 Loft, S., Vistisen, K., Ewertz, M., Tjonnelabnd, A., Overvad, K.and Poulsen, H. E. (1992) Carcinogenesis **13**, 2241–2247
73 Loft, S., Astrup, A., Buemann, B. and Poulsen, H. E. (1993) FASEB J. **8**, 534–537
74 Sakumi, K., Furuichi, M., Tsuzuki, T., Kakuma, T., Kawabata, S. I., Maki, H and Sekiguchi, M. (1993) J. Biol. Chem. **268**, 23524–23530
75 Mo, J. Y., Maki, H. and Sekiguchi, M. (1992) Proc. Natl. Acad. Sci. U.S.A. **89**, 11021–11025
76 Wani, G. and D'Ambrosio, S. M. (1995) Carcinogenesis **16**, 277–283
77 Turk, P. W. and Weitzman, S. A. (1995) Free Radical Res. **23**, 255–258
78 Hamberg, M. and Zhang, L. Y. (1995) Anal. Biochem. **229**, 336–344
79 Dizdaroglu, M. (1993) FEBS Lett. **315**, 1–6
80 Markey, S. P., Markey, C. J., Wang, T. C. L. and Rodriguez, J. B. (1993) J. Am. Soc. Mass Spectrom. **4**, 336–342
81 Blount, B. C. and Ames, B. N. (1994) Anal. Biochem. **219**, 195–200
82 Chaudhary, A. K., Nokubo, M., Marnett, L. J. and Blair, I. A. (1994) Biol. Mass Spectrom. **23**, 457–464
83 Wiseman, H., Kaur, H. and Halliwell, B. (1995) Cancer Lett. **93**, 113–120
84 Wilson, V. L., Taffe, B. G., Shields, P. G., Povey, A. C. and Harris, C. C. (1993) Environ. Health Perspect. **99**, 261–263
85 Carmichael, P. L., She, M. N. and Phillips, D. H. (1992) Carcinogenesis **13**, 1127–1135
86 Marnett, L. J. and Burcham, P. C. (1993) Chem. Res. Toxicol. **6**, 771–785
87 Collins, A. R., Duthie, S. J. and Dobson, V. L. (1993) Carcinogenesis **14**, 1733–1735
88 Frenkel, K., Karkoszka, J., Cohen, B., Baranski, B., Jakubowski, M., Cosma, G., Taioli, E. and Toniolo, P. (1994) Environ. Health Perspect. **102** (suppl. 3), 221–225
89 Claycamp, H. G. (1992) Carcinogenesis **13**, 1289–1292

90 Dizdaroglu, M., Rao, G., Halliwell, B. and Gajewski, E. (1991) Arch. Biochem. Biophys. **285**, 317–324

91 Harris, G., Bashir, S., and Winyard, P. G. (1994) Carcinogenesis **15**, 411–413

92 Weitzman, S. A. and Gordon, L. I. (1990) Blood **76**, 655–663

93 Zimmerman, R. and Cerutti, P. (1984) Proc. Natl. Acad. Sci. U.S.A. **81**, 2085–2087

94 Craven, P. A., Pfanstiel, T. and Rubertis, F. R. (1986) J. Clin. Invest. **77**, 850–859

95 Nicotera, T. M., Privalle, C., Wang, T. C., Oshimura, M. and Barrett, J. C. (1994) Cancer Res. **54**, 3884–3888

96 Jacobson, M. D. and Raff, M. C. (1995) Nature (London) **374**, 814–816

97 Slater, A. F. G., Nobel, S. I., Maellaro, E., Bustamante, J., Kimland, M. and Orrenius, S. (1995) Biochem. J. **306**, 771–778

98 Buttke, T. M. and Sandstrom, P. A. (1995) Free Radical Res. **22**, 389–397

99 Hockenbery, D. M., Oltvai, Z. N., Yin, X. M., Milliman, C. L. and Korsmeyer, S. J. (1993) Cell **75**, 241–251

100 Halder, S., Negrini, M., Monne, M., Sabbioni, S. and Croce, C. M. (1994) Cancer Res. **54**, 2095–2097

101 Reid, T. M., Feig, D. I. and Loeb, L. A. (1994) Environ. Health Perspect. **102** (suppl. 3), 57–61

102 Geierstanger, B. H., Kagawa, T. F., Chen, S. L., Quigley, G. J. and Ho, P. S. (1991) J. Biol. Chem. **266**, 20185–20191

103 Basu, A. K., Loechler, E. L., Leadon, S. A. and Essigmann, J. M. (1989) Proc. Natl. Acad. Sci. U.S.A. **86**, 7677–7681

104 Kamiya, H., Ueda, T., Ohgi, T., Matsukage, A. and Kasai, H. (1995) Nucleic Acids Res. **23**, 761–766

105 Evans, J., Maccabee, M., Hatahet, Z., Courcele, J., Bockrath, R., Ide, H. and Wallace, S. (1993) Mutat. Res. **299**, 147–156

106 Maccabee, M., Evans, J. S., Glackin, M., Hatahet, Z. and Wallace, S. S. (1994) J. Mol. Biol. **236**, 514–530

107 Costa de Oliveira, R., Ribeiro, D., Nigro, R. G., Di Mascio, P. and Menck, C. F. M. (1992) Nucleic Acids Res. **20**, 4319–4323

108 Ribeiro, D. T., Costa de Oliveira, R., Di Mascio, P. and Menck, C. F. M. (1994) Free Radical Res. **21**, 75–83

109 Purmal, A. A., Kow, Y. W. and Wallace, S. S. (1994) Nucleic Acids Res. **22**, 72–88

110 Feig, D. I. and Loeb, L. A. (1993) Biochemistry **32**, 4466–4473

111 Reid, T. M. and Loeb, L. A. (1993) Proc. Natl. Acad. Sci. U.S.A. **90**, 3905–3907

112 Tkeshelashvili, L. K., Reid, T. M., McBride, T. J. and Loeb, L. A. (1993) Cancer Res. **53**, 4172–4174

113 Brash, D., Rudolph, J., Simon, J., Lin, A., Mckenna, G., Baden, H., Halperin, A. and Ponten, J. (1991) Proc. Natl. Acad. Sci. U.S.A. **88**, 10124–10128

114 Kasprzak, K., Diwasn, B. A., Rice, J. M., Misra, M., Riggs, C. W. Olinski, R. and Dizdaroglu, M (1992) Toxicology **6**, 809–815

115 Hussain, S. P., Aguilar, F., Amstad, P. and Cerutti, P. (1994) Oncogene **9**, 2277–2281

116 Stadtman, E. R. (1993) Annu. Rev. Biochem. **62**, 797–821

117 Madzak, C. and Sarasin, A. (1991) J. Mol. Biol. **218**, 667–673

118 Eiserich, J. P., Vossen, V., O'Neill, C. A. Halliwell, B., Cross, C. E. and Van der Vliet, A. (1994) FEBS Lett. **353**, 53–56

119 Esterbauer, H. (1993) Am. J. Clin. Nutr. (Suppl.) **57**, 779s–786s

120 Goldring, C. E. P., Rice-Evans, C. A., Burdon, R. H., Rao, R. and Diplock, A. T. (1993) Arch. Biochem. Biophys. **303**, 429–435

121 Slater, T. F., Cheeseman, K. H. and Benedetto, C. (1990) Biochem. J. **265**, 51–59

122 Adachi, S., Kawamura, K. and Takemoto, K. (1994) Carcinogenesis **15**, 537–543

123 Tsuchiya, T., Ichikawa, N., Nagao, T. and Uchida, H. (1994) Int. Hepatol. Commun. **2**, 227–229

124 Oberley, L. W. and Buettner, G. R. (1979) Cancer Res. **39**, 1141–1149

125 St Clair, D. K., Wan, X. S., Oberley, T. D., Muse, K. E. and St Clair, W. H. (1992) Mol. Carcinogenesis **6**, 238–242

126 Church, S. L., Grant, J. W., Ridnour, L. A., Oberley, L. W., Swanson, P. E., Meltzer, P. S. and Trent, J. M. (1993) Proc. Natl. Acad. Sci. U.S.A. **90**, 3113–3117

127 Safford, S. E., Oberley, T. D., Urano, M. and St Clair, D. K. (1994) Cancer Res. **54**, 4261–4265

128 Kunkel, T. A. (1993) Nature (London) **365**, 207–208

129 Strand, M., Prolla, T. A., Liskay, R. M. and Petes, T. D. (1993) Nature (London) **365**, 274–276

130 Heller, B., Wang, Z. Q., Wagner, E. F., Radons, J., Bürkle, A., Fehsel, K., Burkart, V. and Kolb, H. (1995) J. Biol. Chem. **270**, 11176–11180

131 Rosin, M. P., Anwar, W. A. and Ward, A. J. (1994) Cancer Res. (Suppl.) **54**, 1929s–1933s

132 Halliwell, B. (1995) Ann. Rheum. Dis. **54**, 505–510

133 Merry, P., Winyard, P. G., Morris, C. J., Grootveld, M. and Blake, D. R. (1989) Ann. Rheum. Dis. **48**, 864–870

134 Kaur, H. and Halliwell, B. (1994) FEBS Lett. **350**, 9–12

135 Lunec, J., Herbert, K., Blount, S., Griffiths, H. R. and Emery, P. (1994) FEBS Lett. **348**, 131–138

136 Gardener, D. L., Fitzmaurice, R. L. and Stachan, M. (1995) Ann. Rheum. Dis., in the press

137 Frenkel, K. (1992) Pharmacol. Ther. **53**, 127–166

138 Shacter, E., Beecham, E. J., Covey, J., Kohn, K. W. and Potter, M. (1988) Carcinogenesis **9**, 2297–2304

139 Weitberg, A. B. (1989) Mutat. Res. **224**, 1–4

140 Yamashina, K., Miller, B. E. and Heppner, G. H. (1986) Cancer Res. **46**, 2396–2401

141 Dizdaroglu, M., Olinski, R., Doroshow, J. H. and Akman, S. A. (1993) Cancer Res. **53**, 1269–1272

142 Wei, H. and Frenkel, K. (1992) Cancer Res. **52**, 2298–2303

143 Hattori-Nakakuki, Y., Nishigori, C., Okamoto, K. Imamura, S., Hiai, H. and Toyokuni, S. (1994) Biochem. Biophys. Res. Commun. **201**, 1132–1139

144 Marnett, L. J. and Ji, C. (1994) Cancer Res. (Suppl.) **54**, 1886s–1889s

145 Sanford, K. K., Parshad, R., Price, F. M., Tarone, R. E. and Kraemer, K. H. (1992) J. Clin. Invest. **90**, 2069–2074

146 Trush, M. A., Twerdok, L. E. and Esterline, R. L. (1990) Xenobiotica **20**, 925–932

147 Marletta, M. A. (1988) Chem. Res. Toxicol. **1**, 249–257

148 Grisham, M. B., Ware, K., Gilleland, H. E., Abell, C. L. and Yamada, T. (1992) Gastroenterology **103**, 1260–1266

149 Leaf, C. D., Wishnok, J. S. and Tannebaum, S. R. (1991) Carcinogenesis **12**, 537–539

150 Rosin, M. P. (1992) Mutat. Res. **267**, 265–276

151 Hopman, A. H. N., Moesker, O., Smeets, W. A. G. B., Pauwels, R. P. E., Vooijs, G. P. and Ramaekers, F. C. S. (1991) Cancer Res. **51**, 644–651

152 Ward, A. J., Oliver, P. L., Burr, A. H. and Rosin, M. P. (1993) Mutat. Res. **294**, 299–308

153 Parshad, R., Price, F. M., Oshimura, M., Barett, J. C., Satoh, H., Weissman, B. E., Stanbridge, E. J. and Sanford, K. K. (1992) Hum. Genet. **88**, 524–528

154 The EUROGAST Study Group (1993) Lancet **341**, 1359–1362

155 Pryor, W. A. and Stone, K. (1993) Ann. N.Y. Acad. Sci. **686**, 12–28

156 Olinski, R., Zastawny, T., Budzbin, J., Skokowski, J., Zegarski, W. and Dizdaroglu, M. (1992) FEBS Lett. **309**, 193–198

157 Kiyosawa, H., Suko, M., Okudaira, H., Murata, K., Miyamoto, T., Chung, M. H., Kasai, H. and Nishimura, S. (1990) Free Radical Res. Commun. **11**, 23–27

158 Leanderson, P. and Tageson, C. (1993) in DNA and Free Radicals (Halliwell, B. and Aruoma, O. I., eds.), pp. 293–314, Ellis Horwood, Chichester

159 Takeuchi, T. and Morimoto, K. (1994) Carcinogenesis **15**, 635–639

160 Dong, H. Y., Buard, A., Renier, A., Levy, F., Saint-Etienne, L. and Jaurand, M. C. (1994) Carcinogenesis **15**, 1251–1255

161 Blumberg, B. S., Larouze, B., London, W. T., Werner, B., Hesser, J. E., Millman, I., Saimot, G. and Payet, M. (1975) Am. J. Pathol. **81**, 669

162 Ross, R. K., Yuan, J. M., Yu, M. C., Wogan, G. S., Qian, G. S., Tu, J. T., Groopman, J. D., Gao, Y. T. and Henderson, B. E. (1992) Lancet **339**, 943–946

163 Hsu, I. C., Metcalf, R. A., Sun, T., Welsh, J. A., Wang, N. J. and Harris, C. C. (1991) Nature (London) **350**, 429–431

164 Stein, W. D. (1991) Adv. Cancer Res. **56**, 161–212

165 Shimoda, R., Nagashima, M., Sakamoto, M., Yamaguchi, N., Hirohashi, S., Yokota, J. and Kasai, H. (1994) Cancer Res. **54**, 3171–3172

166 Simmonds, N. J. and Rampton, D. S. (1993) Gut **34**, 865–868

167 Grisham, M. B. (1994) Lancet **344**, 859–861

168 Simmonds, N. J., Allen, R. E., Stevens, T. R. J., van Someren, R. N. M., Blake, D. R. and Rampton, D. S. (1992) Gastroenterology **103**, 186–196

169 Inauen, W., Bilzer, M., Rowedder, E., Halter, F. and Laureburg, B. H. (1988) Gastroenterology **94**, A199

170 Berger, S. J., Gosky, D., Zborowska, E., Willson, J. K. V. and Berger, N. A. (1994) Cancer Res. **54**, 4077–4083

171 Boughton-Smith, N. K., Evans, S. M., Hawkey, C. J., Cole, A. T., Balsitis, A. T. and Whittle, B. J. R. (1993) Lancet **342**, 338–340

172 Boughton-Smith, N. K. (1994) J. R. Soc. Med. **87**, 312–314

173 Van der Vliet, A., Smith, D., O'Neill, C. A., Kaur, H., Darley-Usmar, V., Cross, C. E. and Halliwell, B. (1994) Biochem. J. **303**, 295–301

174 Miller, M. J. S., Sadowska-Krowicka, H., Zhang, X. J. and Clark, D.A (1994) FASEB J. **8**, A2100

175 Rachmilewitz, D., Stamler, J. S., Karmeli, F., Mullins, M. E., Singel, D. J., Loscalzo, J., Xavier, R.J and Podolsky, D. K. (1993) Gastroenterology **105**, 1681–1688

176 Tamai, H., Kachur, J. F., Baron, D. A., Grisham, M. B. and Gaginella, T. S. (1991) J. Pharmacol. Exp. Ther. **257**, 887–894

177 Aruoma, O. I., Wasil, M., Halliwell, B., Hoey, B. M. and Butler, J. (1987) Biochem. Pharmacol. **36**, 3739–3742

178 Markowitz, M. M., Rozen, P., Pero, R. W., Tobi, M. and Miller, D. G. (1988) Gut **29**, 1680–1686

179 Malins, D. C., Holmes, E. H., Polissar, N. J. and Gunselman, S. J. (1993) Cancer **71**, 3036–3043

180 Jaiyesimi, I. A., Buzdar, A. U. and Hortobagyi, G. (1992) J. Clin. Oncol. **10**, 1014–1024

181 Halliwell, B. (1994) Nutr. Rev. **52**, 253–265

182 Gey, K. F. (1995) J. Nutr. Biochem. **6**, 206–236

183 Block, G. (1992) Nutr. Rev. **50**, 207–213

184 Block, G., Patterson, B. and Subar, A. (1992) Nutr. Cancer **18**, 1–29

185 Licht, W. R., Tannenbaum, S. R. and Deen, W. M. (1988) Carcinogenesis **9**, 365–372

186 Mirvish, S. S. (1994) Cancer Res. (Suppl.) **54**, 1948s–1951s

187 Dyke, G. W., Craven, J. L., Hall, R. and Garner, R. C. (1994) Carcinogenesis **15**, 291–295

188 Fraga, C. G., Motchnik, P. A., Shigenaga, M. J. K., Helbock, H. J., Jacob, R. A. and Ames, B. N. (1991) Proc. Natl. Acad. Sci. U.S.A. **88**, 11003–11006

189 Gutteridge, J. M. C. and Halliwell, B. (1994) Antioxidants in Nutrition, Health and Disease, Oxford University Press, Oxford

190 Palozza, P. and Krinsky, N. I. (1992) Arch. Biochem. Biophys. **297**, 184–187

191 Sies, H., Stahl, W. and Sundquist, A. R. (1992) Ann. N. Y. Acad. Sci. **669**, 7–20

192 Bertram, J. S. (1993) Ann. N. Y. Acad. Sci. **686**, 161–176

193 Krinsky, N. I. (1993) Ann. N.Y. Acad. Sci. **686**, 229–242

194 Graf, E. and Eaton, J. W. (1993) Nutr. Cancer **19**, 11–19

195 Willett, W. (1994) Science **264**, 532–537

196 Stevens, R. G., Jones, Y., Micozzi, M. and Taylor, P. R. (1988) N. Engl. J. Med. **319**, 1047–1052

197 Smith, A. G., Francis, J. E. and Carthew, P. (1990) Carcinogenesis **11**, 437–444

198 Weinberg, E. D. (1993) J. Trace Elem. Exp. Med. **6**, 117–123

199 Nelson, R. L., Davis, F. G., Sutter, E., Sobin, L. H., Kikendall, J. W. and Bowen P. (1994) J. Natl. Cancer Inst. **86**, 455–460

200 Hunter, D. J., Manson, J. E., Colditz, G. A., Stampfer, M. J., Rosner, B., Hennekens, C. H., Speizer, F. E. and Willett, W. C. (1993) N. Engl. J. Med **329**, 234–240

201 Halliwell, B. (1991) Drugs **42**, 569–605

202 Wiseman, H. (1994) Tamoxifen: Molecular Basis of Use in Cancer Treatment and Prevention, John Wiley and Sons, Chichester

203 Lim, J. S., Frenkel, K. and Troll, W. (1992) Cancer Res. **52**, 4969–4972

204 Wei, H. and Frenkel, K. (1993) Carcinogenesis **14**, 1195–1201

205 Troll, W., Lim, J. S. and Frenkel, K. (1994) ACS Symp. Ser. **547**, 116–121

206 Wiseman, H., Paganga, G., Rice-Evans, C. and Halliwell, B. (1993) Biochem. J. **292**, 635–638

207 Wiseman, H. and Halliwell, B. (1994) Free Radical Biol. Med. **17**, 485–488

208 Grisham, M. B. (1990) Biochem. Pharmacol. **39**, 2060–2063

209 Yamada, T., Volkmer, B. S. and Grisham, M. B. (1990) Can. J. Gastroenterol. **4**, 295–302

210 Ahnfelt-Ronne, I. and Nielsen, O. H. (1987) Agents Actions **21**, 191–194

211 Ahnfelt-Ronne, I., Nielsen, O. H., Christensen, A., Langholtz, E. and Reis, P. (1990) Gastroenterology **98**, 1162–1169

212 Brown, T. A. and Brown, K. A. (1994) BioEssays **16**, 719–726

213 Benamira, M., Johnson, K., Chaudhary, A., Bruner, K., Tibbetts, C. and Marnett, L. J. (1995) Carcinogenesis **16**, 93–99

Biochem. J. (1996) **313**, 697–710 (Printed in Great Britain)

REVIEW ARTICLE
The role of cellular hydration in the regulation of cell function*

Dieter HÄUSSINGER

Medizinische Universitätsklinik, Heinrich Heine Universität, Moorenstrasse 5, D-40225 Düsseldorf, Germany

INTRODUCTION

The cellular hydration state is dynamic and changes within minutes under the influence of aniso-osmolarity, hormones, nutrients and oxidative stress. This occurs despite the activity of potent mechanisms for cell volume regulation, which have been observed in virtually all cell types studied so far. These volume-regulatory mechanisms are apparently not designed to maintain absolute cell volume constancy; rather, they act as dampeners in order to prevent excessive cell volume deviations which would otherwise result from cumulative substrate uptake. On the other hand, these volume-regulatory mechanisms can even be activated in the resting state by hormones, and by this means changes in cell hydration are created. Most importantly, small fluctuations of cell hydration, i.e. of cell volume, act as a separate and potent signal for cellular metabolism and gene expression. Accordingly, a simple but elegant method is created for the adaptation of cell function to environmental challenges. In liver, cell swelling and shrinkage lead to certain opposite patterns of cellular metabolic function. Apparently, hormones and amino acids can trigger these patterns by altering cell volume. Thus cell volume homeostasis does not simply mean volume constancy, but rather the integration of events which allow cell hydration to play its physiological role as a regulator of cell function (for reviews see [1–4]). The interaction between cellular hydration and cell function has been most extensively studied in liver cells, but evidence is increasing that regulation of cell function through alterations of cell hydration also occurs in other cell types. This review will largely refer to hepatocytes, but when appropriate other cell types will also be considered. Regulation of mitochondrial function by hormone-induced changes of matrix volume has been established in the past (for reviews see [5,6]); this aspect will only be covered briefly. For further details, the reader is referred to recent surveys [2,4,7,8].

MECHANISMS OF CELL VOLUME CONTROL

Aniso-osmotic exposure of cells has been widely used to study cell volume regulation. Owing to the high water permeability of cell membranes and their inability to withstand significant hydrostatic pressure gradients, net water movements across the plasma membrane are driven almost exclusively by the osmotic gradient. In fact, when cells are suddenly exposed to hypo-osmotic media, they initially swell like more or less perfect osmometers, but within minutes they regain almost their original cell volume. This behaviour has been labelled regulatory cell volume decrease (RVD). Conversely, upon sudden exposure to hyperosmotic media, the cells shrink like osmometers, but display within minutes a regulatory volume increase (RVI), which brings back cell volume largely (but not completely) to the starting level. The mechanisms responsible for RVD and RVI may differ among different cell types and species, but in general involve the activation of ion transport systems in the plasma membrane. In

addition, some cell types augment RVD and RVI by the release or accumulation of organic osmolytes.

Ionic mechanisms

The major cell-type-specific ionic mechanisms of cell volume regulation have been reviewed extensively in the past [2,3,7–13] and are schematically summarized in Figure 1. In rat liver, RVD is largely achieved by the release of cellular potassium, chloride and bicarbonate [9,14–19] following activation of Ba^{2+}- and quinidine-sensitive potassium channels in parallel with anion channels. It is unclear which intracellular signals activate these volume-regulatory responses in liver; however, tyrosine phosphorylation was recognized as an essential step in the RVD response in the human intestinal 407 cell line [20]. Hepatocyte swelling may open stretch-activated non-selective cation channels, which allow passage of Ca^{2+} into the cell [21]. The increase of intracellular Ca^{2+} then may activate Ca^{2+}-sensitive K^+ channels [22]. However, there is controversy as to whether hypo-osmotic hepatocyte swelling increases intracellular Ca^{2+} [23–25]. Thus Ca^{2+} activation of K^+ channels in liver may not be the only mechanism allowing RVD; indeed, K^+ channels have also been described in isolated hepatocytes which do not require an increase of intracellular Ca^{2+} [26]. It is conceivable that these K^+ channels are directly activated by cell membrane stretching, as is observed in rat proximal tubules [27]. A swelling-induced chloride conductance regulatory protein (termed pI_{Cln}) has been identified recently in Madin–Darby canine kidney (MDCK) and cardiac cells [28]. This protein is abundant in the cytosol and bound to cytosolic proteins such as actin, and may provide a link between cell swelling and opening of volume-regulatory Cl^- channels.

RVI in liver is achieved, at least in part, by parallel activation of Na^+/H^+ exchange and Cl^-/HCO_3^- exchange (Figure 1). Hyperosmotic exposure of perfused liver stimulates amiloride- and ouabain-sensitive K^+ uptake [17–19], which eventually leads to RVI. In contrast to other cell types, $Na/K–2Cl$ co-transport appears not to participate appreciably in hepatic RVI, even though the carrier probably exists in the hepatic cell membrane and its activation by insulin is followed by an increase of cell volume [29,30].

Neither RVI nor RVD, however, completely restores the initial liver cell volume, i.e. the hepatocytes are left in either a slightly shrunken or a slightly swollen state. The extent of this volume deviation apparently acts as a signal which modifies cellular function.

Osmolytes

In addition to the ionic mechanisms of cell volume regulation, some cell types specifically accumulate or release organic compounds, so-called organic osmolytes, in response to cell shrinkage or cell swelling. Osmolytes need to be non-perturbing solutes that do not interfere with protein function even when occurring

Abbreviations used: RVD, regulatory volume decrease; RVI, regulatory volume increase; MDCK cells, Madin–Darby canine kidney cells; MAP kinase, mitogen-activated protein kinase; cAMP, cyclic AMP; PEPCK, phosphoenolpyruvate carboxykinase; SAP kinase, stress-activated kinase.
* Dedicated to Professor Dr. Dr. h.c. Wolfgang Gerok on the occasion of his 70th birthday.

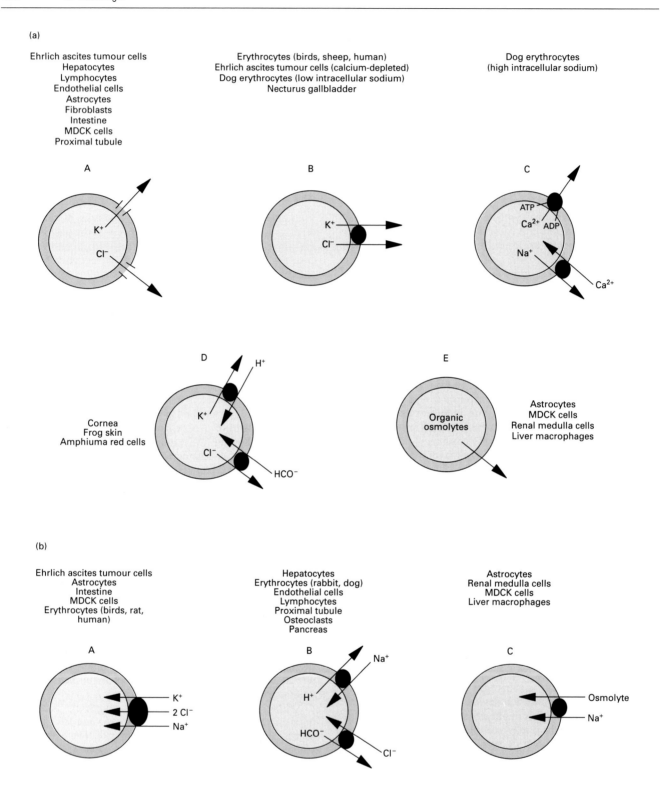

Figure 1   Major mechanisms of (a) RVD and (b) RVI following hypo-osmotic and hyperosmotic exposure respectively

The different ionic mechanisms of cell volume regulation are cell type- and species-dependent; in addition, some cell types use the accumulation or release of osmolytes. (**a**) Mechanisms of RVD: parallel activation of K^+ and Cl^- channels (A); K^+–Cl^- symport (B); parallel activation of Ca^{2+}-ATPase and Na^+/Ca^{2+} exchange (C); parallel activation of K^+/H^+ antiport and Cl^-/HCO_3^- exchange (D); release of organic osmolytes (E). (**b**) Mechanisms of RVI: activation of K^+/Na^+–$2Cl^-$ co-transport (A); parallel activation of Na^+/H^+ exchange and Cl^-/HCO_3^- antiport (B); cumulative Na^+-dependent uptake of organic osmolytes (C).

at high intracellular concentrations [4,7,31–33]. Such a prerequisite may explain why only a few classes of organic compounds, i.e. polyols such as inositol and sorbitol (e.g. in astro-

cytes, renal medulla and lens epithelial cells), methylamines such as betaine and α-glycerophosphocholine (e.g. in renal medulla) and certain amino acids such as taurine (e.g. in Ehrlich ascites

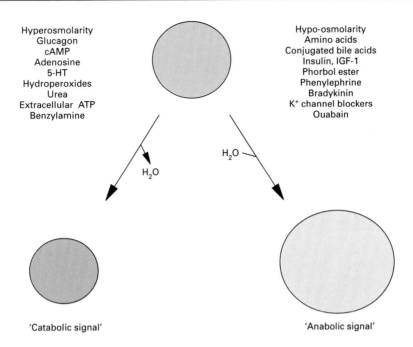

Hyperosmolarity
Glucagon
cAMP
Adenosine
5-HT
Hydroperoxides
Urea
Extracellular ATP
Benzylamine

Hypo-osmolarity
Amino acids
Conjugated bile acids
Insulin, IGF-1
Phorbol ester
Phenylephrine
Bradykinin
K$^+$ channel blockers
Ouabain

H_2O

H_2O

'Catabolic signal'

'Anabolic signal'

Figure 2   Modulators of liver cell hydration

Abbreviations: 5-HT, 5-hydroxytryptamine; IGF-1, insulin-like growth factor-1.

tumour cells) have evolved as osmolytes in living cells. Different mechanisms contribute to the intracellular accumulation of osmolytes during hyperosmotic stress: (i) decreased degradation (α-glycerophosphocholine), (ii) increased synthesis (induction of aldose reductase) and (iii) increased uptake following induction of specific Na$^+$-coupled transporters (e.g. for *myo*-inositol, betaine and taurine). Osmolytes in the renal medulla are especially important, since medullary fluid osmolarity can increase up to 3800 mosmol/l during antidiuresis and decrease to 170 mosmol/l during diuresis [34]. In the antidiuretic state (high extracellular osmolarity in the renal medulla), intracellular osmolarity increases in renal medullary cells as a result of the accumulation of inositol and betaine, which are taken up via Na$^+$-dependent transporters, and also as a result of increased synthesis of sorbitol and α-glycerophosphocholine. This process of intracellular osmolyte accumulation can produce intracellular organic osmolyte concentrations of several hundred mmol/l and intra/extracellular osmolyte concentration gradients of up to 1000 [32]. The enhanced synthesis of sorbitol from glucose by aldose reductase under these conditions involves an increased expression of the enzyme due to hyperosmotic activation of the encoding gene [35,36]. Likewise, the Na$^+$-dependent transporters for inositol and betaine are induced upon hyperosmotic exposure [33,36–39].

Organic osmolytes have not been identified with certainty in hepatocytes; however, betaine is an osmolyte in liver macrophages (Kupffer cells) and the expression of the betaine transporter is regulated by osmolarity in these cells (F. Zhang, U. Warskulat and D. Häussinger, unpublished work). In addition, a volume-activated taurine channel is present in skate hepatocytes [40], perfused rat liver releases small amounts of taurine in response to hypo-osmotic exposure [41], and rat hepatocytes possess a Na$^+$-coupled taurine transporter in the plasma membrane [42]. It is, however, unclear whether this taurine transporter in hepatocytes is osmoregulated (as is the taurine-transporting system β in mouse preimplantation conceptuses [43]) and whether

taurine accumulation inside the hepatocytes is at all quantitatively relevant for cell volume regulation.

PHYSIOLOGICAL MODULATORS OF LIVER CELL VOLUME

Aniso-osmotic exposure may primarily be seen as an experimental tool to modify liver cell volume, although during intestinal absorption of water portal venous blood may become slightly hypotonic [44], and clinically relevant severe aniso-osmolarities (220–400 mosmol/l) due to hypo- and hyper-natraemia in plasma have been documented [45,46]. Physiologically more important, however, are cell volume changes due to cumulative substrate uptake [47–51], hormones and oxidative stress (Figure 2). Na$^+$-dependent amino acid transporters in the plasma membrane (for review see [52]) can build up intra/extracellular amino acid concentration gradients of up to 20-fold. Na$^+$ entering the hepatocyte together with the amino acid is extruded in exchange for K$^+$ by the electrogenic Na$^+$/K$^+$-ATPase. The accumulation of amino acids and K$^+$ in the cells leads to hepatocyte swelling, which in turn triggers volume-regulatory K$^+$ efflux [47,48,50,51] (Figure 3). This RVD, however, only prevents cell swelling from becoming excessive, and the hepatocytes remain in a swollen state as long the amino acid load continues. Cessation of amino acid infusion is followed by rapid cell shrinkage, and an RVI finally brings cell volume back to the starting level (Figure 3). Importantly, amino acid-induced cell swelling and volume-regulatory responses occur upon exposure to amino acids in the physiological concentration range [51], and physiological fluctuations in the portal amino acid concentration are accompanied by parallel alterations of liver cell volume. The degree of amino acid-induced cell swelling seems to be related largely to the steady-state intra/extracellular amino acid concentration gradient. This gradient, and accordingly the degree of cell swelling, is modified by hormones and the nutritional state in a complex way due to effects on the expression of plasma membrane transport systems, the electrochemical Na$^+$ gradient as a driving

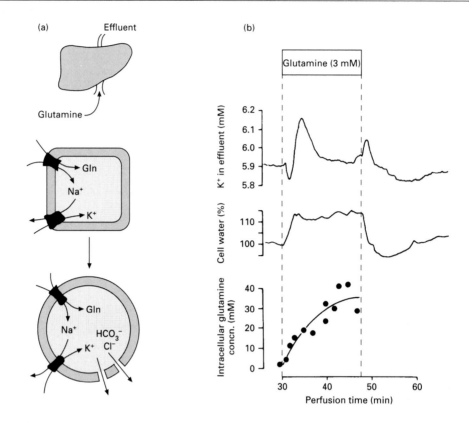

Figure 3   Effect of glutamine (3 mM) addition to the influent perfusate of isolated, single-pass perfused rat liver on intracellular glutamine accumulation, cell volume and volume-regulatory K+ fluxes

(a) Glutamine leads to hepatocyte cell swelling due to cumulative, Na+-dependent, uptake into liver cells and activates RVD due to K+, Cl− and HCO₃⁻ efflux. (b) Addition of glutamine to isolated perfused rat liver creates within about 12 min an intra/extracellular glutamine concentration gradient of about 12-fold. During the first 2 min of glutamine accumulation into the hepatocytes the liver cell volume increases rapidly, and hepatic net K+ uptake during this phase is probably due to extrusion of Na+ by the Na+/K+-ATPase. Thereafter, no further increase in cell volume is observed, despite the continuing accumulation of glutamine inside the cell. This is achieved by a volume-regulatory K+ efflux from the liver which occurs until the steady-state intracellular glutamine concentration of about 35 mM is reached. The liver cell remains in a swollen state as long as glutamine is infused. The extent of cell swelling modifies cellular function. Reproduced from [48], with permission.

force for Na+-coupled transport and alterations of intracellular amino acid metabolism. For example, in livers from fed rats, insulin, which is known to stimulate amino acid transport via system A, enhances glycine-induced cell swelling [29]. The concentrative uptake of conjugated bile acids also induces cell swelling [53].

Hormones are potent modulators of liver cell volume by affecting the activity of volume-regulatory ion transport systems [29,30,54–56]. In liver, insulin stimulates amiloride-sensitive Na+/H+ exchange [57], loop-diuretic-sensitive Na/K–2Cl co-transport [29,30] and the Na+/K+-ATPase [57,58], i.e. transport systems which are also turned on for RVI in liver and many other tissues. The concerted activation of these transporters leads to cellular accumulation of potassium, sodium and chloride, and consequently cell swelling. Insulin-induced cell swelling and cellular K+ accumulation are abolished in the presence of bumetanide plus amiloride. Glucagon activates Na+/K+-ATPase [59], but simultaneously depletes cellular K+ [30,54,55], probably due to a simultaneous opening of Ba²⁺- and quinidine-sensitive K+ channels [30]. As a result of the cellular Na+, K+ and probably Cl− depletion, hepatocytes shrink (Figure 4). The physiological relevance is underlined by the finding that half-maximal effects of insulin and glucagon on liver cell hydration are found at hormone concentrations normally present in portal venous blood *in vivo*, i.e. 1.0 and 0.1 nM respectively [55]. Other hormones also modify

hepatocellular hydration (Figure 2). Insulin-induced cell swelling is counteracted by glucagon. Insulin plus phenylephrine induces a 20 % increase in cell volume, whereas vasopressin plus glucagon leads to an approx. 20 % decrease. In order to achieve comparable changes of cell volume by aniso-osmotic exposure, extracellular osmolarity changes of about 100 mosmol/l are required [55].

Glucagon and Ca²⁺-mobilizing hormones were also identified as modulators of mitochondrial matrix volume [5,6,60], but may affect the cytosolic and mitochondrial water spaces in opposite directions. For example, glucagon induces cell shrinkage and simultaneously swells the mitochondria. On the other hand, both cell and mitochondrial swelling occur under the influence of phenylephrine. For the mechanisms underlying the hormone-induced increase of mitochondrial matrix volume and the arising functional consequences, the reader is referred to refs. [5,6,60].

Oxidative stress exerted by hydroperoxides induces hepatocellular shrinkage due to the opening of Ba²⁺-sensitive K+ channels [61,62]. Cell shrinkage and K+ channel opening also occur when hydrogen peroxide is generated intracellularly during the oxidation of monoamines. Evidence has been presented that the balance between intracellular metabolic H_2O_2 generation and its removal by detoxification systems such as catalase and glutathione peroxidase is one determinant for hepatocellular K+ balance, and accordingly cell volume [62]. Oxidative stress may also lead to cell shrinkage in other cell types, as H_2O_2 stimulates

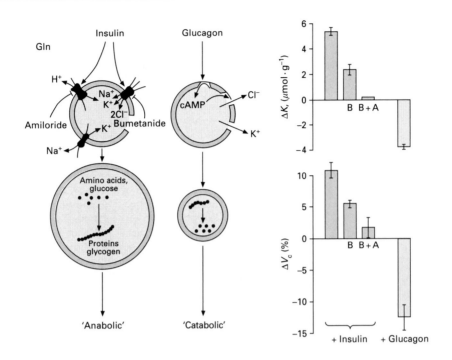

Figure 4   Modulation of liver cell volume and cellular K⁺ balance by insulin and glucagon

Insulin activates amiloride-sensitive Na⁺/H⁺ exchange, bumetanide-sensitive Na/K–2Cl co-transport and the Na⁺/K⁺-ATPase, resulting in the cellular accumulation of K⁺ (ΔK_i), Na⁺ and Cl⁻ and cell swelling (ΔV_c). Both cellular K⁺ accumulation and cell swelling induced by insulin are abolished in the presence of amiloride (A) plus bumetanide (B). Conversely, glucagon and cAMP deplete cellular K⁺ by activating Cl⁻ channels and quinidine/Ba²⁺-sensitive K⁺ channels, thereby inducing cell shrinkage. The hormone-induced cell volume alterations act like another signal which participates in mediating the hormone effects on hepatic metabolism: cell swelling triggers an anabolic pattern and cell shrinkage a catabolic pattern of cellular function. Reproduced from [59a], with permission.

K⁺ conductance in pancreatic B-cells [63] and oxidative stress inhibits Na/K–2Cl co-transport in vascular endothelial cells [64]. K⁺ channel opening under the influence of urea at concentrations found in uraemia also induces cell shrinkage [65].

CELL VOLUME CHANGES AS A SIGNAL MODULATING CELL FUNCTION

Recent evidence suggests that the cellular hydration state is an important determinant of cell function and that hormones, oxidative stress and nutrients exert their effects on metabolism and gene expression in part by a modification of cell volume. The concept that cellular hydration acts as an independent signal on liver cell function is based on the following observations. (i) Persistent alterations of metabolism occur within minutes in response to aniso-osmotic cell volume changes (Table 1); these are fully reversible upon restoration of the resting cell volume. There is a dose–response relationship between the extent of the cell hydration change that remains after completion of volume-regulatory ion fluxes and the metabolic response. (ii) Intracellular signal transduction pathways are activated in response to changes in cell hydration. Their interruption by suitable inhibitors abolishes some anisotonicity-induced metabolic responses [66]. (iii) Several long-known, but mechanistically poorly understood, effects of amino acids which could not be related to their metabolism, such as stimulation of glycogen synthesis [67,68] or inhibition of proteolysis (for reviews see [69,70]), can be quantitatively mimicked by swelling the cells in hypo-osmotic media to the same extent as that induced by the amino acids [49,71–74]. This indicates that the metabolic effects exerted by these amino acids are due to their swelling potency. (iv) Several metabolic

hormone effects can be mimicked by equipotent aniso-osmotic cell swelling or shrinkage, and some hormone effects disappear when the hormone-induced cell volume changes are prevented.

Thus cell hydration changes in response to physiological stimuli are an important and until recently unrecognized signal which helps to adapt cellular metabolism to alterations of the environment (substrate, tonicity, hormones) [1,73]. Consequently, Na⁺-dependent amino acid transport systems in the plasma membrane should be viewed not merely as amino acid translocators; these transporters also act as transmembrane signalling systems by altering cellular hydration in response to substrate supply. Such a signalling role may shed new light on the long-known heterogeneity of transport systems among different cell types and their differential expression during development (for review see [52]), rendering specific amino acids as more or less potent signals for cell function. Likewise, transmembrane ion movements under the influence of hormones are an integral part of hormonal signal transduction mechanisms, with alterations of cellular hydration acting as another 'second messenger' of hormone action [1]. However, the exact place of hormone-induced cell volume changes in the network of known hormone-activated intracellular messenger systems remains to be established.

For example, insulin is capable of stimulating the mitogen-activated protein (MAP) kinase cascade through its own receptor tyrosine kinase. This initiates a series of other protein phosphorylations which bring about the effects of insulin on cell function. One of these may be activation of membrane ion transporters such as Na⁺/H⁺ exchange and Na/K–2Cl co-transport, resulting in cell swelling. Thus cell swelling in response to insulin may be a consequence rather than a cause of insulin

Table 1   Effects of hypo-osmotic cell swelling on liver function

Opposite functional patterns are triggered by hyperosmotic cell shrinkage. In general, liver cell swelling acts as an anabolic signal, whereas cell shrinkage is catabolic.

Liver cell swelling increases:
  Protein synthesis [84]
  Glycogen synthesis [56,71,90,91,95]
  Lactate uptake [19]
  Pentose phosphate shunt [74,92]
  Amino acid uptake [48,104]
  Glutamine breakdown [48]
  Glycine oxidation [74,85]
  Ketoisocaproate oxidation [85]
  Acetyl-CoA carboxylase [93,97]
  Lipogenesis [95]
  Urea synthesis from amino acids [88]
  MAP kinase activity [66]
  Glutathione (GSH) efflux [88]
  Taurocholate excretion into bile [24,53,74,107]
  Actin polymerization [124]
  Microtubule stability [76]
  Exocytosis [24,74,87,108,109]
  pH in vesicular compartments [25,82,112–114]
  mRNA levels of c-*jun* [128], ornithine decarboxylase [125], β-actin [124] and tubulin [76]

Liver cell swelling decreases:
  Proteolysis [29,49,54,72,73,76,77]
  Glycogenolysis [9,19,88]
  Glucose-6-phosphatase activity [170]
  Carnitine palmitoyltransferase I activity [96]
  Glutamine synthesis [48]
  Urea synthesis from NH_4^+ [88]
  Biliary GSSG release [92]
  Cytosolic pH [14,16,82,113]
  mRNA levels for PEPCK [127] and tyrosine aminotransferase[*]
  Viral replication [131]

[*] U. Warskulat and D. Häussinger, unpublished work.

action. On the other hand, hypo-osmotic cell swelling was shown to activate MAP kinases (see below) by itself, which may potentially result in a very complex interaction between cell volume and hormone action.

CELL-VOLUME-SENSITIVE METABOLIC PATHWAYS

Protein turnover

Hepatocellular hydration is a major point of proteolysis control in liver [73]: cell swelling inhibits, and conversely cell shrinkage stimulates, protein breakdown under conditions when the proteolytic pathway is not already fully activated [49,54,72,73]. There is a close relationship between proteolytic activity and hepatocellular hydration, regardless of whether the latter is modified by hormones, glutamine, glycine, alanine, bile acids, the K^+ channel blocker Ba^{2+} or anisotonic exposure (Figure 5), indicating that hydration changes are the common mechanism underlying proteolysis control by these heterogeneous effectors. The effects of glutamine, alanine, glycine, insulin, insulin-like growth factor-1 and glucagon on proteolysis can be mimicked quantitatively when the cell volume changes that occur in response to these effectors are induced to the same degree by aniso-osmotic exposure. Apparently the known anti-proteolytic effect of insulin and several (but not all) amino acids is transmitted in large part by agonist-induced cell swelling, whereas stimulation of proteolysis by glucagon is apparently mediated by cell shrinkage [29,49,54,73]. In line with this, the anti-proteolytic action of insulin disappears when insulin-induced cell swelling is

prevented in the presence of inhibitors of the Na^+/H^+ antiporter and the $Na^+/K^+–2Cl^-$ co-transporter [73]. Furthermore, inhibition of proteolysis by glutamine and glycine is additive to the same extent as is observed with respect to cell swelling induced by both amino acids [49]. The nutritional state exerts its control on proteolysis by determining the swelling potencies of hormones and amino acids. For example, the anti-proteolytic effect of glycine in the fed state is only about one-third of that found after 24 h starvation, due to an approx. 3-fold higher swelling potency of glycine during starvation, which is explained by an up-regulation of the glycine-transporting amino acid transport system A [75]. Likewise, both the swelling potency and the anti-proteolytic effect of insulin are diminished in parallel by 60–70 % during starvation. It should be noted that not all amino acids exert their anti-proteolytic effect via changes of cell hydration; for example, leucine and phenylalanine are potent inhibitors of proteolysis, yet they exert little effect on cell volume. Here other mechanisms of proteolysis control apparently come into play.

The mechanisms by which cellular hydration exerts control on proteolysis are not clear; however, intact microtubular structures are required [76–78]. Disruption of microtubules by colchicine abolishes the anti-proteolytic action of hypo-osmotic, amino acid- or insulin-induced cell swelling [76,77], although colchicine itself has little effect on proteolysis. A requirement for intact microtubules may also explain why hypo-osmotic cell swelling was found to be without effect on proteolysis in freshly isolated rat hepatocytes [79]. However, the cell volume sensitivity of proteolysis reappears when cytoskeletal structures are reconstituted in culture [77]. The formation of autophagic vacuoles was shown to be controlled by protein phosphorylation/dephosphorylation [80,81], and current evidence suggests a role for protein phosphorylation in mediating the volume sensitivity of proteolysis. This could reside at the levels of ribosomal protein S6 phosphorylation [81], acidification of pre-lysosomal endocytotic/autophagic vesicular compartments [25,82] and possibly, but not yet proven, at the level of phosphorylation of microtubule-associated proteins.

Liver cell hydration also affects protein synthesis in the opposite direction to proteolysis: cell shrinkage inhibits, whereas cell swelling stimulates, protein synthesis [79,83,84]. The cyclic AMP (cAMP)-induced inhibition of protein synthesis in liver may, at least in part, be ascribed to cell shrinkage, because equipotent hyperosmotic cell shrinkage decreases protein synthesis in isolated rat hepatocytes to a comparable extent [84]. A close relationship exists between cell shrinkage and inhibition of protein synthesis under the influence of hyperosmotic exposure, cAMP and vasopressin [84]. On the other hand, insulin and phenylephrine have no significant effect on protein synthesis. However, when phenylephrine is added together with cAMP, the cAMP-induced inhibition of protein synthesis is potentiated although phenylephrine counteracts cAMP-induced cell shrinkage. As expected, cell volume changes may be only one factor among several mediating the hormonal control of protein synthesis.

Amino acid and ammonia metabolism

In rat liver, hypo-osmotic cell swelling switches the hepatic glutamine balance from net release to net uptake. This is due to a stimulation of flux through glutaminase in periportal hepatocytes and a simultaneous inhibition of glutamine synthesis in perivenous hepatocytes [48]. Opposite effects are found upon hyperosmotic exposure. Swelling-induced activation of glutaminase is most likely due to simultaneous mitochondrial swelling, which probably alters the attachment of the enzyme to

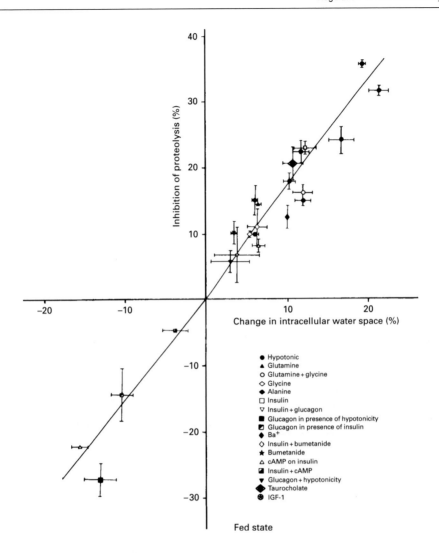

Figure 5   Relationship between cell volume and proteolysis in liver

Cell volume in perfused liver was determined as intracellular water space [112] and proteolysis was assessed as [³H]leucine release in the effluent perfusate from perfused rat livers, which were prelabelled *in vivo* by intraperitoneal injection of [³H]leucine 16 h prior to the perfusion experiment. Cell shrinkage stimulates proteolysis, whereas cell swelling inhibits it. It should be noted that proteolysis is already maximally activated in the absence of hormones and amino acids and cannot be stimulated further by hyperosmotic or glucagon-induced cell shrinkage. The proteolysis-stimulating effect of these cell-shrinking manoeuvres, however, becomes apparent when proteolysis is pre-inhibited by either amino acids or insulin. Cell volume changes were induced by insulin, cAMP, glucagon, amino acids, Ba²⁺, taurocholate or aniso-osmotic exposure. Abbreviation: IGF-1, insulin-like growth factor-1. Adapted from [73], with permission.

the mitochondrial inner membrane (for reviews see [5,6]). Likewise, hypo-osmotic swelling stimulates glycine oxidation in the perfused rat liver [85] and in isolated mitochondria [86]. These findings suggest that alterations of amino acid metabolism following aniso-osmotic cell volume changes can, at least in part, be explained by parallel alterations of mitochondrial matrix volume. Indeed, morphometric studies revealed transient mitochondrial swelling in isolated hepatocytes following hypo-osmotic exposure [87]. The effects of glucagon, cAMP and Ca²⁺-mobilizing hormones on glutamine and glycine oxidation are also due to hormone-induced increases in mitochondrial matrix volume (for reviews see [5,6]), although these agents simultaneously induce liver cell shrinkage. Thus mitochondrial pathways, such as glutamine and glycine oxidation and respiration, are stimulated not only by hypo-osmolarity but also by cAMP, glucagon and Ca²⁺-mobilizing hormones, despite opposing effects of these agents on cell volume.

In livers from fed rats, hypo-osmotic cell swelling stimulates urea synthesis and ammonia formation from amino acids [88],

but inhibits these pathways when ammonia is used as the sole substrate for urea synthesis, although swelling of isolated mitochondria was shown to stimulate citrulline synthesis (for reviews see [5,6]). Inhibition of urea synthesis from ammonia by cell swelling is probably due to a block of the urea cycle at the step of argininosuccinate synthesis, as it is accompanied by an increase in citrulline but a decrease in aspartate, glutamate and malate tissue levels [88]. Simultaneously under these conditions the lactate/pyruvate ratio is increased whereas the β-hydroxy-butyrate/acetoacetate ratio is decreased. These findings were interpreted to reflect a cell-swelling-induced disturbance of the transfer of reducing equivalents via the malate/aspartate shuttle and impaired aspartate regeneration during cell swelling. In line with this, the swelling-induced inhibition of urea synthesis from NH₄Cl was overcome by the addition of lactate and pyruvate. These substrates allowed intramitochondrial regeneration of oxaloacetate via pyruvate carboxylase, a reaction which was also shown to be stimulated by swelling of isolated mitochondria (for reviews see [5,6]).

Carbohydrate and fatty acid metabolism

Like protein turnover, carbohydrate metabolism in the liver is critically dependent upon the hepatocellular hydration state (Table 1). Hepatocyte swelling inhibits glycogenolysis, glycolysis [9,19,89] and glucose-6-phosphatase activity [90], but simultaneously stimulates glycogen synthesis [56,71,90,91], flux through the pentose phosphate pathway [92] and lipogenesis [93]. The opposite effects occur in response to cell shrinkage. It has become clear that the stimulatory effect of glutamine and other amino acids on glycogen synthesis and lipogenesis is due to amino acid-induced cell swelling [71]. Glycogen synthesis and lipogenesis are controlled by the activity of glycogen synthase and acetyl-CoA carboxylase; both of these enzymes are subject to regulation by phosphorylation/dephosphorylation. Swelling of isolated rat hepatocytes activates glycogen synthase in parallel with acetyl-CoA carboxylase [24,71,93,94] and decreases glycogen phosphorylase a activity, suggestive of interference by cellular hydration with protein phosphorylation. It was proposed that activation of glycogen synthase in response to hypo-osmotic cell swelling is, at least in part, due to a lowering of the intracellular chloride concentration, which relieves the inhibition of glycogen synthase phosphatase [90]. Activation of glycogen synthase by glutamate may contribute during glutamine-induced, but not during hypo-osmolarity-induced, cell swelling [90,93]. The swelling-induced stimulation of flux through the pentose phosphate shunt, and accordingly an increased NADPH provision for glutathione reductase, may explain why, during oxidative stress, the cellular losses of oxidized glutathione are smaller when the hepatocellular hydration state increases [92].

In the perfused rat liver, hypo-osmotic cell swelling stimulates, whereas hyperosmotic cell shrinkage inhibits, ketogenesis from ketoisocaproate, whereas ketogenesis from tyrosine, octanoate and glucose is largely uninfluenced [85]. However, with these substrates, hypo-osmotic (hyperosmotic) cell swelling (shrinkage) led to a decrease (increase) in the β-hydroxybutyrate/acetoacetate ratio [85]. The cell swelling-induced decrease in the β-hydroxybutyrate/acetoacetate ratio may be explained by a stimulation of the respiratory chain, as it occurs in response to mitochondrial swelling [5]. The inhibition of hepatic ketogenesis by glutamine, proline, alanine and asparagine is probably not due to amino acid-induced cell swelling, since no effect on ketogenesis was observed with the non-metabolizable amino acid analogue 2-aminoisobutyrate, despite cell swelling [95].

Hypo-osmotic incubation of hepatocytes slightly stimulates lipogenesis from glucose [95] and inhibits carnitine palmitoyltransferase [96]. Acetyl-CoA carboxylase, a key enzyme in fatty acid synthesis, is activated in response to hypo-osmotic- and amino acid-induced cell swelling [24]. As with glycogen synthase, activation of acetyl-CoA carboxylase is due to de-inhibition of a protein phosphatase, which occurs in response to a lowering of intracellular chloride and/or an increased intracellular concentration of glutamate and aspartate [97]. Lipogenesis is stimulated by glutamine, proline and alanine, but not by aminoisobutyrate, histidine or asparagine [95], and no correlation was detectable between the potency of these amino acids to swell the hepatocytes on the one hand and their potency to stimulate lipogenesis on the other [95]. This may suggest that volume changes *per se* may not play a major role in mediating the lipogenic effect of amino acids.

Plasma membrane transport

Not only are amino acids potent modulators of liver cell volume but, conversely, cell volume can exert control on amino acid transport. One obvious example is the induction of Na^+-dependent transport systems for neutral amino acids in a variety of cell types in response to hyperosmolarity ([98–101]; for review see [102]). Here, neutral amino acids apparently function as osmolytes in order to counteract cell shrinkage. However, amino acid transport may also be stimulated in response to cell swelling. In liver, the glutamine-transporting system N was shown to be activated by amino acids in a cycloheximide-insensitive way [103]. This stimulation was not explained by *trans*-stimulation and, in addition to substrates of system N, alanine, serine and the non-metabolizable analogue aminoisobutyrate, i.e. substrates of the Na^+-dependent system A, were also effective. This amino acid-dependent short-term stimulation of amino acid transport is due to cell swelling [104]. Hypo-osmotic cell swelling increases hepatic alanine and glutamine uptake [48], whose intracellular degradation rate was shown to be controlled by transport rather than by metabolism.

A swelling-induced hyperpolarization of the membrane [9] may not only augment Na^+-coupled substrate transport but may also explain why sinusoidal glutathione efflux from the liver is increased following hypo-osmotic liver cell swelling [88], since glutathione release was shown to be under the control of the membrane potential [105].

Bile excretion

In rat liver, conjugated bile acids are taken up across the canalicular membrane by a Na^+-dependent transporter, whereas canalicular excretion is thought to be predominantly accomplished by an ATP-dependent transport system (for reviews see [24,106]) and to represent the rate-controlling step for overall transcellular bile acid transport. In the perfused rat liver, the rate of transcellular taurocholate transport from the sinusoidal space into the biliary lumen is critically dependent upon the hydration state of the hepatocyte [53,107]. Cell shrinkage inhibits, whereas cell swelling stimulates, taurocholate excretion into bile, regardless of whether the cell volume is modified by anisotonic exposure or by amino acids [53]. The swelling-induced stimulation of taurocholate excretion into bile is due to an increase in transport capacity. An increase of hepatocellular hydration of about 10% leads within minutes to a doubling of the V_{max} of taurocholate excretion into bile [53]. This increase in V_{max} is not explained by changes in the cellular ATP content or the membrane potential, but is abolished in the presence of colchicine, indicating a requirement for intact microtubules for the interaction between cellular hydration and taurocholate excretion into bile [74]. From this, and from the substantial evidence for a transient stimulation of exocytosis as an early response to hypo-osmotic cell swelling [24,74,87,108,109], it was postulated that alterations in hepatocellular hydration induce rapid changes in the taurocholate secretion capacity due to a microtubule-dependent insertion/retrieval of canalicular bile acid transporter molecules into/from the canalicular membrane [24,53,74,108]. These transporters may be stored in an intracellular vesicular compartment underneath the canalicular membrane and could correspond to the known bile acid-containing vesicles, which were seen in the past to reflect 'vesicular transcellular bile acid transport'. However, the role of these vesicles may reside in the transport of the bile acid transporter molecules rather than in the transport of bile acids. Interestingly, the swelling-induced stimulation of taurocholate excretion is abolished in the presence of tyrosine kinase inhibitors or after pretreatment with cholera or pertussis toxin [110]. These inhibitors were also shown to block the swelling-induced activation of MAP kinases in rat hepatocytes [66,110], and this suggests a causal relationship between swelling-induced MAP kinase activation and the stimulation of bile acid transport. It is not yet clear whether cell volume also affects other canalicular transport ATPases, such as the multi-specific organic anion transporter. However, recent studies on cysteinyl leukotriene excretion into bile in endotoxinaemia suggest this to be the case [111].

Acidification of endocytotic vesicles

In the liver, cell swelling (shrinkage) leads to a rapid alkalinization (acidification) of intracellular vesicular compartments, as revealed in studies on Acridine Orange fluorescence [112,113] and the fluorescence of endocytosed fluorescein isothiocyanate-labelled dextran [25,78,82]. The effects on vesicular pH occurred regardless of whether cell volume was modified by anisotonicity, insulin, amino acids, hydroperoxides or the K^+ channel blocker Ba^{2+} [82]. Subsequent studies revealed that the cell volume sensitivity of the vesicular pH reflects the response of an early endocytotic compartment (pH around 6), but not of late, more acidic, compartments (pH around 5), which are probably lysosomal [25]. Given the important role of vesicular acidification for receptor–ligand sorting, exocytosis and protein targeting [114,115], one is tempted to speculate that cellular hydration may also interfere with these processes. In addition, cell volume may interfere with receptor-mediated endocytosis: hyperosmotic exposure inhibits galactose-receptor-mediated endocytosis, but not fluid-phase endocytosis, in isolated hepatocytes [116]. Interestingly, pH control by the hydration state of endocytotic vesicles accessible to fluorescein isothiocyanate-labelled dextran is abolished after disruption of microtubules with colchicine [78,82], indicating again the involvement of the cytoskeleton.

The mechanism by which cell volume affects pH in endocytotic vesicles is not fully understood; however, it is mediated by a G-protein- and tyrosine-kinase-dependent, but Ca^{2+}- and cAMP-independent, mechanism [25]. This inhibitor sensitivity of pH regulation by cell volume in endocytotic vesicles again resembles that of the swelling-induced activation of MAP kinases [66]. Vesicular acidification requires the presence of a chloride conductance in the vesicular membrane in order to dissipate the membrane potential generated by the H^+ pump and to augment the acidification process [117]. Evidence suggests that chloride channel activity and acidification in endosomes prepared from calf brain or rabbit proximal tubule is regulated by protein kinase A-dependent phosphorylation [118,119], and phospho-proteins have been observed in endosomal membranes [120]. Although protein kinase A is not involved in the regulation of vesicular pH in liver [25], it is conceivable that other swelling-activated protein kinases (e.g. MAP kinases [25,66]) mediate the volume sensitivity of vesicular acidification by modulation of chloride channel activity [25].

CELLULAR HYDRATION AND GENE EXPRESSION

Cellular hydration affects cellular metabolism also on a long-term time scale by modifying gene expression. This involves not only osmoregulatory genes (whose mRNA levels increase in response to hypertonic stress), such as genes for aldose reductase or for osmolyte transporters such as the Na^+-coupled *myo*-inositol (SMIT) and betaine (BGT1) transporters in renal cells and astrocytes [32,37,121–123], but also the expression of genes coding for proteins which are not necessarily linked to osmo-regulation. Examples of the latter are the hypo-osmolarity-induced increases in mRNA levels for β-actin [124], tubulin [76] and ornithine decarboxylase [125], the hyperosmolarity-induced stimulation of cyclo-oxygenase-2 expression in activated liver macrophages [126] and the cell volume-dependent expression in

liver of phosphoenolpyruvate carboxykinase (PEPCK) [127], a key gluconeogenic enzyme. PEPCK mRNA levels markedly increase in response to hyperosmotic cell shrinkage but decrease in response to cell swelling both in the intact perfused rat liver and in cultured rat hepatoma H4IIE cells. Stimulation of proteolysis and induction of a key gluconeogenic enzyme in response to hypertonic cell shrinkage may be seen as an example of the joint regulation of functionally linked processes by cell volume. Aniso-osmotic exposure affects the expression of early immediate genes; examples are the increase in c-*jun* (but not c-*fos*) mRNA levels in response to liver cell swelling [128] and the induction of Egr-1 and c-fos mRNA following hyperosmotic treatment of MDCK cells [129]. In fibroblasts, amino acid deprivation leads to a cycloheximide-sensitive adaptive increase in the activity of amino acid transport system A [98]. This adaptive increase is potentiated by hyperosmotic cell shrinkage and counteracted by hypo-osmolar cell swelling, indicating again that cell volume modifies the expression of amino acid transport systems. Recent studies have drawn attention to amino acid-regulated gene expression in eukaryotic cells (for review see [130]). Although many signalling pathways may exist which link amino acid supply to gene expression, it is likely that amino acid-induced changes in cellular hydration contribute to this regulation.

Viral replication depends upon host cell hydration [131–134]. For example, hypo-osmotic swelling of duck hepatocytes inhibited replication of duck hepatitis B virus by about 50%, whereas hyperosmotic shrinkage stimulated its replication 4–5-fold [131]. Surprisingly, cell shrinkage increased viral protein synthesis, whereas the synthesis of host cell proteins was decreased. In addition, evidence has been presented that viruses may interfere with ion transport across the plasma membrane [135–137], which could influence host cell hydration.

The mechanisms by which cell volume changes affect gene expression are largely unknown, but changes in ionic composition [35,138–140], the cytoskeleton [141] and protein phosphorylation [25,142] are likely candidates. Recent studies with MDCK cells have identified a hypertonic stress-responsive element in the 5′-flanking region of the mammalian BGT1 (betaine transporter) gene; however, the *trans*-acting factor(s) remain to be characterized [143]. A protein kinase C-dependent activation of MAP kinases in response to hypertonic stress has been described in MDCK cells [144]. However, the role of MAP kinases in inducing the betaine transporter is doubtful, as osmolyte transporter mRNA accumulation is still stimulated by hypertonicity in protein kinase C-depleted MDCK cells [145]. On the other hand, the swelling-induced induction of c-*jun* mRNA in hepatoma cells [128] may be due to MAP kinase activation [66]. Regulation of PEPCK mRNA levels by cellular hydration does not involve protein kinase C activation or changes in cAMP levels [127], but is sensitive to the protein kinase inhibitor H7 (U. Warskulat and D. Häussinger, unpublished work). The role of MAP kinases in the cell volume-dependent regulation of the PEPCK gene remains to be established. In yeast, evidence has been presented for a role of MAP kinase pathways in the regulation of the transcriptional activation of the glycerol synthetic pathway in response to high-salt conditions [146].

Alterations in gene expression, however, may also in turn affect cellular volume. This was suggested by an approx. 30% increase in the resting-state cell volume following expression of the *ras* oncogene in NIH fibroblasts [147]. The growth factor-independent proliferation of cells expressing the *ras* oncogene is sensitive to amiloride and furosemide, i.e. to blockers of Na^+/H^+ antiport and Na/K–2Cl co-transport, suggesting a role for cell swelling induced by activation of these transporters in cell proliferation. Furthermore, in lymphocytes, mitogenic signals activate these transporters and may shift the set-point of cell volume regulation to higher resting values [148]; this cell volume increase may be an important prerequisite for cellular proliferation.

Gene expression is affected not only by the cellular hydration state but also by the ambient colloid osmotic pressure. Evidence has been presented that the activity of a dominant liver transcription factor, hepatic nuclear factor-1α (HNF-1α), which controls the transcription of several liver-specific genes, is modulated by fluctuations in the level of oncotically active macromolecules [149].

SENSING OF THE CELLULAR HYDRATION STATE

In view of the multiple effects of cell hydration on cell function, the question arises as to how cell volume changes are sensed and how the signal is transduced to the level of cell function. Little is known yet about the structures that sense the changes in cell hydration. Because cell volume/hydration is a physical property of the cell, sensing should occur physically and/or mechanically. One hypothesis on physical volume sensing is that hydration changes will affect the concentrations of one or more intracellular constituents, which may act to influence volume-regulatory transport systems and/or intracellular signalling pathways. One intriguing model postulates that the extent of 'macromolecular crowding', i.e. the cytosolic protein concentration, will determine the tendency of intracellular macromolecules to associate with the plasma membrane and consequently their enzymic activity [150–152]. It is conceivable that cellular hydration may in such a way interfere with the activity of protein kinases and phosphatases, and changes in protein phosphorylation may trigger not only volume-regulatory responses [148,153] but also the volume-dependent alterations in cellular metabolism and gene expression.

Candidates for mechanical cell volume sensing are the cytoskeleton, recently identified ion-conductance regulator proteins [28] and so-called stretch-activated cation and anion channels, which have been identified in a variety of cell types and whose open-probability increases with membrane tension (for review see [154]). The molecular mechanisms of the stretch-activation of these channels are still unclear, although there is little doubt that such channels participate in cell volume regulation. Liberation of fatty acids from the membrane and interactions with the cytoskeleton have been discussed as initial events in the stretch-activation of ion channels. Recently, histidine kinases have been identified in yeast which are putative integral membrane proteins and may act as osmosensors [155], with the signal being transduced by autophosphorylation and subsequent phosphate transfer to an aspartate residue in the receiver domain of a cognate response regulator molecule in order to regulate a MAP-kinase-like protein kinase cascade.

INTRACELLULAR SIGNALLING EVENTS

A little more is known about the intracellular signalling events that couple changes in cell hydration to cell function, although the picture is incomplete and complicated by the fact that cell volume signalling may depend on the cell type under study. In addition, it is conceivable that the mechanism of how cell swelling is achieved (hypo-osmotic- versus amino acid-induced swelling) will influence cell volume signalling; for example, in jejunal enterocytes the RVD in response to cumulative substrate uptake was sensitive to inhibitors of protein kinase C, whereas the RVD following hypo-osmotic exposure was not [156]. Besides

protein phosphorylation, the cytoskeleton and intracellular ions may be important for the link between cell hydration and cell function.

An important role for protein phosphorylation in cell volume regulation is suggested by the fact that volume-regulatory ion transporters such as the Na/K–Cl co-transporter and the Na^+/H^+ antiporter are regulated by phosphorylation. Okadaic acid, an inhibitor of protein phosphatases, increases cell volume in lymphocytes and hepatocytes, probably via phosphorylation-mediated activation of these transporters. Depending on the cell type under study, various protein kinases and phosphatases have been implicated in the mechanisms of cell volume regulation (for reviews see [2,3,148,153,155,157]); all of these could also participate in the regulation of cell function by cell volume. Current interest focuses on the regulation of MAP kinases and related protein kinases, such as Jnk [157], by osmotic stress. Hyperosmotic stress activates MAP kinases in yeast [147,157] and MDCK cells [144], whereas in rat hepatoma cells [66], rat hepatocytes [110,158], the human intestine 407 cell line [20] and primary astrocytes (F. Schliess, R. Sinning and D. Häussinger, unpublished work) MAP kinases are activated in response to hypo-osmotic cell swelling. A signal transduction sequence, which is initiated by the osmotic water shift across the plasma membrane and ultimately leads to changes in cell function, was recently identified in rat hepatoma and liver cells [66,110]. Here, hypo-osmotic cell swelling results within 1 min in a pertussis toxin-, cholera toxin- and genistein-sensitive, but protein kinase C- and Ca^{2+}-independent, phosphorylation of the MAP kinases Erk-1 and Erk-2 [66,110]. This suggests that liver cell swelling leads to a G-protein-mediated activation of an as-yet-unidentified tyrosine kinase, which acts to activate a pathway towards MAP kinases (Figure 6). The functional significance of this volume signalling pathway in liver is suggested by the findings that not only hypo-osmotic MAP kinase activation but also swelling-induced alkalinization and stimulation of bile acid excretion can be inhibited at upstream events, i.e. at the G-protein and tyrosine kinase level [25,65,110]. The finding that some metabolic responses to hypo-osmotic liver swelling are completely abolished by G-protein and tyrosine kinase inhibitors indicates not only that simple dilution of intracellular substrates following the osmotic water shifts cannot explain the cell volume-dependence of metabolism, but also that cell volume signalling may start at the plasma membrane. The intracellular signalling cascade which is initiated in response to liver cell swelling (Figure 6) resembles that triggered by growth factor receptor activation [159]. This similarity may explain why cell swelling acts like an anabolic signal in liver with respect to protein and carbohydrate metabolism (Table 1). MAP kinases have multiple protein substrates [159], such as the microtubule-associated proteins MAP-2 and Tau, other protein kinases, such as S6 kinase, and transcription factors such as c-Jun. In fact, the swelling-induced activation of MAP kinases is followed by increased phosphorylation of c-Jun, which may explain (due to autoregulation of the c-*jun* gene) the increase in c-*jun* mRNA levels 30 min after the onset of cell swelling [66,128]. However, besides Erk-1 and Erk-2, other Jun kinases may also be activated by cell swelling. A swelling-induced phosphorylation of transcription factors such as c-Jun may explain the influence of cell hydration on gene expression. A new subfamily of protein kinases, the stress-activated protein (SAP) kinases, has recently been described [160]. SAP kinases are activated by different forms of intra- and extra-cellular stress and act as c-Jun kinases; however, it remains to be established whether SAP kinases are also activated by osmotic stress. Protein phosphatases also participate in the regulation of cell function by cell hydration: hypo-osmotic hepatocyte swelling lowers the

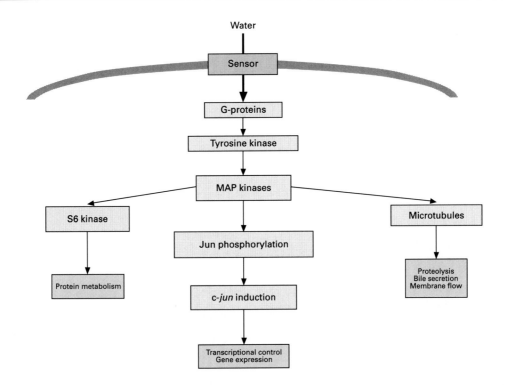

Figure 6   Cell volume signalling in the hepatocyte

The hypothetical scheme focuses on the role of MAP kinases, but it should be kept in mind that the signalling sequence depicted here is incomplete.

intracellular chloride concentration, thereby leading to a de-inhibition of glycogen synthase phosphatase [90].

Microtubules apparently play an important role in transducing some metabolic alterations in response to changes in cellular hydration. For example, disruption of microtubules by colchicine abolishes the swelling-induced alkalinization of endocytotic vesicles [82,78], inhibition of proteolysis [76–78] and stimulation of transcellular bile acid transport [74] in liver. It remains to be established to what extent changes in the phosphorylation of microtubule-associated proteins are involved in microtubule- and MAP-kinase-dependent cell volume signalling. On the other hand, other pathways which are activated in response to cell swelling, such as stimulation of glycine oxidation or of the pentose phosphate shunt, are not affected following microtubule disruption. Intact microtubules are not required for the swelling-induced activation of MAP kinases (F. Schliess and D. Häussinger, unpublished work). This suggests that microtubular structures are required during cell volume signalling at a step downstream of MAP kinases. Cell swelling leads within 1 min to an increased polymerization state of β-actin [124] and increases the stability of microtubules [76]; however, the relevance of actin filaments for cell volume signalling in liver is unknown. Microtubule stabilization found in *Ha-ras-1*-transformed Buffalo rat liver cells [162] may be explained by the increase in cellular hydration following *ras* oncogene expression [147].

Whereas in many cell types hypo-osmotic exposure is followed by an increase in intracellular Ca^{2+} (for review see [163]), this has not been observed consistently in hepatocytes [15,21,23,25,66]. Liver cell swelling decreases the intracellular pH [16,82,113], transiently hyperpolarizes the cell membrane potential [9] and stimulates inositol 1,4,5-trisphosphate formation [23,161]; how-ever, translocation of protein kinase C to the membrane could not be detected (G. Finkenzeller and D. Häussinger, unpublished

work). The significance of these alterations for the observed effects on hepatic metabolism remains to be established.

CLINICAL IMPLICATIONS

Many pathophysiological implications may arise from the in-teraction between cell hydration and cell function (for review see [164]); only a few can be addressed here. Contemporary clinical medicine pays careful attention to the hydration state of the extracellular space but not enough to cellular hydration, probably because of the lack of routinely applicable techniques for the assessment of cell volume in patients. The role of cellular hydration as an important determinant of protein turnover sheds new light on the understanding of protein-catabolic states in disease, and it has been hypothesized that cell shrinkage in skeletal muscle and liver may be the common end-path triggering protein catabolism in a variety of diseases [165]. Indeed, a close relationship between the cellular hydration state in skeletal muscle and the negativity of nitrogen balance was shown in severely ill patients, irrespective of the underlying disease [165].

The proposed dynamic system of bile acid carrier insertion/retrieval into/from the canalicular membrane in response to small changes in hepatocellular hydration may explain the beneficial effect of tauroursodeoxycholate in cholestasis [166]. In presence of taurocholate, tauroursodeoxycholate at low concen-trations induces cell swelling, which leads to an increase in the V_{max} of taurocholate excretion into bile [53,74]. Hypo-osmotically swollen cells are apparently more resistant to hydroperoxide-induced cell damage [92,167], and in liver considerably higher concentrations of bile acids are required to induce the known cholestatic effect of taurocholate in swollen than in shrunken liver cells [53]. The role of cell hydration in organ preservation has not yet been elucidated.

Ammonia induces glial swelling as a result of glutamine accumulation in astrocytes. Both glial swelling and an increase in intracranial pressure were recognized as major events leading to brain dysfunction or death following acute ammonia intoxication or acute fulminant liver failure. In animal models of acute liver failure ammonia toxicity is reduced by inhibitors of glutamine synthetase [168], which also prevent glial swelling. Recent proton-NMR studies on the human brain *in vivo* indicate that astrocyte swelling is an early event in chronic hepatic encephalopathy, although an increase in intracranial pressure is clinically not detectable. Here a decrease in *myo*-inositol, a glial osmolyte, is found, indicative of glial swelling [46,169]. It was hypothesized that swelling-induced alterations in glia function may trigger disturbances of glial–neuronal communication, thereby provoking the clinical picture of chronic hepatic encephalopathy.

It is hoped that this review will stimulate further research in the area of cellular hydration, not only in basic but also in clinical science.

Our own work reported herein was supported by Deutsche Forschungsgemeinschaft, the Gottfried-Wilhelm-Leibniz-Prize, the Schilling-Stiftung and the Fonds der Chemischen Industrie.

REFERENCES

1   Häussinger, D. and Lang, F. (1992) Trends Pharmacol. Sci. **13**, 371–373

2   Lang, F. and Häussinger, D. (eds.) (1993) Interaction of Cell Volume and Cell Function, Springer Verlag, Heidelberg

3   Strange, K. (ed.) (1994) Cellular and Molecular Physiology of Cell Volume Regulation, CRC Press, Boca Raton

4   Häussinger, D., Gerok, W. and Lang, F. (1994) Am. J. Physiol. **267**, E343–E355

5   Halestrap, A. P. (1989) Biochim. Biophys. Acta **973**, 355–382

6   Halestrap, A. P. (1993) in Interactions of Cell Volume and Cell Function (Lang, F. and Häussinger, D., eds.), pp. 279–307, Springer Verlag, Heidelberg

7   Chamberlin, M. E. and Strange, K. (1989) Am. J. Physiol. **257**, C159–C173

8   Lang, F., Völkl, H. and Häussinger, D. (1990) Comp. Physiol. **4**, 1–25

9   Graf, J., Haddad, P., Häussinger, D. and Lang, F. (1988) Renal Physiol. Biochem. **11**, 202–220

10 Hoffmann, E. K. and Simonsen, L. K. (1989) Physiol. Rev. **69**, 315–382

11 McCarty, N. A. and O'Neil, R. G. (1992) Physiol. Rev. **72**, 1037–1061

12 Sarkadi, B. and Parker, J. C. (1991) Biochim. Biophys. Acta **1071**, 407–427

13 Fitz, G. (1994) Gastroenterology **107**, 1906–1907

14 Corasanti, J. G., Gleeson, D., Gautam, A. and Boyer, J. L. (1990) Renal. Physiol. Biochem. **13**, 164

15 Corasanti, J. G., Gleeson, D. and Boyer, J. L. (1990) Am. J. Physiol. **258**, G290–G298

16 Gleeson, D., Corasanti, J. G. and Boyer, J. L. (1990) Am. J. Physiol. **258**, G299–G307

17 Haddad, P., Thalhammer, T. and Graf, J. (1989) Am. J. Physiol. **256**, G563–G569

18 Häussinger, D., Stehle, T. and Lang, F. (1990) Hepatology **11**, 243–254

19 Lang, F., Stehle, T. and Häussinger, D. (1989) Pflügers Arch. **413**, 209–216

20 Tilly, B. C., van den Berghe, N., Tertoolen, L. G. J., Edixhoven, M. J. and de Jonge, H. R. (1993) J. Biol. Chem. **268**, 19919–19922

21 Bear, C. E. (1990) Am. J. Physiol. **258**, C421–C428

22 Bear, C. E. and Petersen, O. H. (1987) Pflügers Arch. **410**, 342–344

23 Baquet, A., Meijer, A. J. and Hue, L. (1991) FEBS Lett. **278**, 103–106

24 Boyer, J. L., Graf, J. and Meier, P. J. (1992) Annu. Rev. Physiol. **54**, 415–438

25 Schreiber, R. and Häussinger, D. (1995) Biochem. J. **309**, 19–24

26 Henderson, R. M., Graf, J. and Boyer, J. L. (1989) Am. J. Physiol. **256**, G1028–G1035

27 Sackin, H. (1987) Am. J. Physiol. **253**, F1253–F1262

28 Krapivinsky, G. B., Ackerman, M. J., Gordon, E. A., Krapivinsky, L. D. and Clapham, D. E. (1994) Cell **76**, 439–448

29 Hallbrucker, C., vom Dahl, S., Lang, F., Gerok, W. and Häussinger, D. (1991) Eur. J. Biochem. **199**, 467–474

30 Hallbrucker, C., vom Dahl, S., Lang, F., Gerok, W. and Häussinger, D. (1991) Pflügers Arch. **418**, 519–521

31 Yancey, P. H., Clark, M. E., Hand, S. C., Bowlus, R. D. and Somero, G. N. (1982) Science **217**, 1214–1222

32 Garcia-Perez, A. and Burg, M. B. (1991) Physiol. Rev. **71**, 1081–1115

33 Burg, M. B. (1994) J. Exp. Zool. **268**, 171–175

34 Graf, J. (1993) in Interaction of Cell Volume and Cell Function (Lang, F. and Häussinger, D., eds.), pp. 67–117, Springer Verlag, Heidelberg

35 Uchida, S., Garcia-Perez, A., Murphy, H. and Burg, M. (1989) Am. J. Physiol. **256**, C614–C620

36 Ferraris, J. D., Williams, C. K., Martin, B. M., Burg, M. B. and Garcia-Perez, A. (1994) Proc. Natl. Acad. Sci. U.S.A. **91**, 10742–10746

37 Kwon, H. M., Yamauchi, A., Uchida, S., Preston, A. S., Garcia-Perez, A., Burg, M. B. and Handler, J. S. (1992) J. Biol. Chem. **267**, 6297–6301

38 Takenaka, M., Preston, A. S., Kwon, H. M. and Handler, J. S. (1994) J. Biol. Chem. **269**, 29379–29381

39 Petronini, P. G., De Angelis, E., Borghetti, A. F. and Wheeler, K. (1994) Biochem. J. **300**, 45–50

40 Ballatori, N., Simmons, T. W. and Boyer, J. L. (1994) Am. J. Physiol. **267**, G285–G291

41 Brand, H. S., Meijer, A. J., Gustafson, L. A., Jorning, G. G., Leegwater, A. C., Maas, M. A. and Chamuleau, R. A. (1994) Biochem. Cell Biol. **72**, 8–11

42 Berkowitz, D., Hug, P., Sleight, R. G. and Bucuvalas, J. C. (1994) Am. J. Physiol. **267**, G932–G937

43 Van Winkle, L. J., Patel, M., Wasserlauf, H. G., Dickinson, H. R. and Campione, A. L. (1994) Biochim. Biophys. Acta **1191**, 244–255

44 Haberich, F. J., Aziz, O. and Nowacki, P. E. (1966) Pflügers Arch. **285**, 73–89

45 Lee, J. H., Arcinue, E. and Ross, B. D. (1994) N. Engl. J. Med. **331**, 439–442

46 Häussinger, D., Laubenberger, J., vom Dahl, S., Ernst, T., Bayer, S., Langer, M., Gerok, W. and Hennig, J. (1994) Gastroenterology **107**, 1475–1480

47 Bakker-Grunwald, T. (1983) Biochim. Biophys. Acta **731**, 239–242

48 Häussinger, D., Lang, F., Bauers, K. and Gerok, W. (1990) Eur. J. Biochem. **188**, 689–695

49 Hallbrucker, C., vom Dahl, S., Lang, F., Gerok, W. and Häussinger, D. (1991) Eur. J. Biochem. **197**, 717–724

50 Kristensen, L. O. and Folke, M. (1984) Biochem. J. **221**, 265–268

51 Wettstein, M., vom Dahl, S., Lang, F., Gerok, W. and Häussinger, D. (1990) Biol. Chem. Hoppe-Seyler **371**, 493–501

52 Kilberg, M. S. and Häussinger, D. (eds.) (1992) Mammalian Amino Acid Transport, Plenum Press, New York

53 Häussinger, D., Hallbrucker, C., Saha, N., Lang, F. and Gerok, W. (1992) Biochem. J. **288**, 681–689

54 vom Dahl, S., Hallbrucker, C., Lang, F., Gerok, W. and Häussinger, D. (1991) Biochem. J. **278**, 771–777

55 vom Dahl, S., Hallbrucker, C., Lang, F., Gerok, W. and Häussinger, D. (1991) Biochem. J. **280**, 105–109

56 Al-Habori, M., Peak, M., Thomas, T. H. and Agius, L. (1992) Biochem. J. **282**, 789–796

57 Jakubowski, J. and Jakob, A. (1990) Eur. J. Biochem. **193**, 541–549

58 Fehlmann, M. and Freychat, P. (1981) J. Biol. Chem. **256**, 7449–7453

59 Moule, S. K. and McGivan, J. D. (1990) Biochim. Biophys. Acta **1031**, 383–397

59a Häussinger, D. and Lang, F. (1991) Biochim. Biophys. Acta **1071**, 331–350

60 Halestrap, A. P., Davidson, A. M. and Potter, W. D. (1990) Biochim. Biophys. Acta **1018**, 278–281

61 Hallbrucker, C., Ritter, M., Lang, F., Gerok, W. and Häussinger, D. (1993) Eur. J. Biochem. **211**, 449–458

62 Saha, N., Schreiber, R., vom Dahl, S., Lang, F., Gerok, W. and Häussinger, D. (1993) Biochem. J. **296**, 701–707

63 Krippeit-Drews, P., Lang, F., Häussinger, D. and Drews, G. (1994) Pflügers Arch. **426**, 552–554

64 Elliott, S. J. and Schilling, W. P. (1992) Am. J. Physiol. **263**, H96–H102

65 Hallbrucker, C., vom Dahl, S., Ritter, M., Lang, F. and Häussinger, D. (1994) Pflügers Arch. **428**, 552–560

66 Schliess, F., Schreiber, R. and Häussinger, D. (1995) Biochem. J. **309**, 13–17

67 Katz, J., Golden, S. and Wals, P. A. (1976) Proc. Natl. Acad. Sci. U.S.A. **73**, 3433–3437

68 Lavoinne, A., Baquet, A. and Hue, L. (1987) Biochem. J. **248**, 429–437

69 Mortimore, G. E. and Pösö, A. R. (1987) Annu. Rev. Nutr. **7**, 539–564

70 Seglen, P. O. and Gordon, P. B. (1984) J. Cell Biol. **99**, 435–444

71 Baquet, A., Hue, L., Meijer, A. J., van Woerkom, G. M. and Plomp, P. J. A. M. (1990) J. Biol. Chem. **265**, 955–959

72 Häussinger, D., Hallbrucker, C., vom Dahl, S., Lang, F. and Gerok, W. (1990) Biochem. J. **272**, 239–242

73 Häussinger, D., Hallbrucker, C., vom Dahl, S., Decker, S., Schweizer, U., Lang, F. and Gerok, W. (1991) FEBS Lett. **283**, 70–72

74 Häussinger, D., Saha, N., Hallbrucker, C., Lang, F. and Gerok, W. (1993) Biochem. J. **291**, 355–360

75 Hayes, M. R. and McGivan, J. D. (1982) Biochem. J. **204**, 365–368

76 Häussinger, D., Stoll, B., vom Dahl, S., Theodoropoulos, P. A., Markogiannakis, E., Gravanis, A., Lang, F. and Stournaras, C. (1994) Biochem. Cell Biol. **72**, 12–19

77   vom Dahl, S., Stoll, B., Gerok, W. and Häussinger, D. (1995) Biochem. J. **308**, 529–536

78   Busch, G. L., Schreiber, R., Dartsch, P. C., Völkl, H., vom Dahl, S., Häussinger, D. and Lang, F. (1994) Proc. Natl. Acad. Sci. U.S.A. **91**, 9165–9169

79   Meijer, A. J., Gustafson, L. A., Luiker, J. J. F. P., et al. (1993) Eur. J. Biochem. **215**, 449–454

80   Holen, I., Gordon, P. B. and Seglen, P. O. (1993) Eur. J. Biochem. **215**, 113–122

81   Luiken, J. F. P., Blommaart, E. F. C., Boon, L., van Woerkom, G. M. and Meijer, A. J. (1994) Biochem. Soc. Trans. **22**, 458–461

82   Schreiber, R., Stoll, B., Lang, F. and Häussinger, D. (1994) Biochem. J. **303**, 113–120

83   Kruppa, J. and Clemens, M. J. (1984) EMBO J. **3**, 95–100

84   Stoll, B., Gerok, W., Lang, F. and Häussinger, D. (1992) Biochem. J. **287**, 217–222

85   Häussinger, D., Stoll, B., Morimoto, Y., Lang, F. and Gerok, W. (1992) Biol. Chem. Hoppe-Seyler **373**, 723–734

86   Brosnan, J. T., Jois, M. and Hall, B. (1990) in Amino Acids: Biochemistry, Biology and Medicine (Lubec, G. and Rosenthal, G. A., eds.), pp. 896–902, ESCOM Science Publishers BV, Leiden

87   Pfaller, W., Willinger, C., Stoll, B., Hallbrucker, C., Lang, F. and Häussinger, D. (1993) J. Cell. Physiol. **154**, 248–253

88   Häussinger, D., Lang, F., Bauers, K. and Gerok, W. (1990) Eur. J. Biochem. **193**, 891–898

89   Meijer, A. J., Baquet, A., Gustafson, L., van Woerkom, G. M. and Hue, L. (1992) J. Biol. Chem. **267**, 5823–5828

90   Grant, A., Tosh, D. and Burchell, A. (1992) Biochem. Soc. Trans. **21**, 39S

91   Peak, M., Al-Habori, M. and Agius, L. (1992) Biochem. J. **282**, 797–805

92   Saha, N., Stoll, B., Lang, F. and Häussinger, D. (1992) Eur. J. Biochem. **209**, 437–444

93   Baquet, A., Maisin, L. and Hue, L. (1991) Biochem. J. **278**, 887–890

94   Hue, L. (1994) Biochem. Soc. Trans. **22**, 505–508

95   Baquet, A., Lavoinne, A. and Hue, L. (1991) Biochem. J. **273**, 57–62

96   Guzman, M., Velasco, G., Castro, J. and Zammit, V. A. (1994) FEBS Lett. **344**, 239–241

97   Baquet, A., Gaussin, V., Bollen, M., Stalmans, W. and Hue, L. (1993) Eur. J. Biochem. **217**, 1083–1089

98   Gazzola, G. C., Dall'Asta, V., Nucci, F. A., Rossi, P. A., Bussolati, O., Hoffmann, E. K. and Guidotti, G. G. (1991) Cell. Physiol. Biochem. **1**, 131–142

99   Yamauchi, A., Miyai, A., Yokoyama, K., Itoh, T., Kamada, T., Ueda, N. and Fujiwara, Y. (1994) Am. J. Physiol. **267**, C1493–C1500

100   Soler, C., Felipe, A., Casado, J., McGivan, J. D. and Pastor-Anglada, M. (1993) Biochem. J. **289**, 653–658

101   Chen, J. G., Klus, L. R., Steenbergen, D. K. and Kempson, S. A. (1994) Am. J. Physiol. **267**, C529–C536

102   McGivan, J. D. and Pastor-Anglada, M. (1994) Biochem. J. **299**, 321–334

103   Weissbach, L. and Kilberg, M. S. (1984) J. Cell. Physiol. **121**, 133–138

104   Bode, B. and Kilberg, M. S. (1991) J. Biol. Chem. **266**, 7376–7381

105   Fernandez-Checa, J. C., Ren, C., Awe, T. Y., Ookhtens, M. and Kaplowitz, N. (1988) Am. J. Physiol. **255**, G403–G408

106   Nathanson, M. H. and Boyer, J. L. (1991) Hepatology **14**, 551–566

107   Hallbrucker, C., Lang, F., Gerok, W. and Häussinger, D. (1992) Biochem. J. **281**, 593–595

108   Bruck, R., Haddad, P., Graf, J. and Boyer, J. L. (1992) Am. J. Physiol. **262**, G806–G812

109   van Rossum, G. D. V., Russo, M. A. and Schisselbauer, J. C. (1987) Curr. Top. Membr. Transp. **30**, 45–74

110   Häussinger, D., Noe, B., Wettstein, M. and Schliess, F. (1995) in Cell Biology and Molecular Basis of Liver transport (Wehner, F. and Petringer, E., eds.), pp. 85–91, Project Verlag, Dortmund

111   Wettstein, M., Noe, B. and Häussinger, D. (1995) Hepatology, **22**, 235–240

112   Völkl, H., Friedrich, F., Häussinger, D. and Lang, F. (1993) Biochem. J. **295**, 11–14

113   Völkl, H., Busch, G. L., Häussinger, D. and Lang, F. (1994) FEBS Lett. **338**, 27–30

114   Dautry-Versat, A., Ciechanover, A. and Lodish, H. F. (1983) Proc. Natl. Acad. Sci. U.S.A. **80**, 2258–2262

115   Tager, J. M., Aerts, J. M. F. G., Oude-Elferink, R. J. A., Groen, A. K., Hollemans, M. and Schram, A. W. (1988) in pH Homeostasis (Häussinger, D., ed.), pp. 123–162, Academic Press, London

116   Oka, J. A., Christensen, M. D. and Weigel, P. H. (1989) J. Biol. Chem. **264**, 12016–12024

117   van Dyke, R. (1993) Am. J. Physiol. **265**, C901–C917

118   Bae, H. R. and Verkman, A. S. (1990) Nature (London) **348**, 637–639

119   Rasenick, M. M., Wang, N. and Yan, K. (1990) in The Biology and Medicine of Signal Transduction (Nishizuka, Y., ed.), pp. 381–386, Raven Press, New York

120   Mulberg, A. E., Tulk, B. M. and Forgac, M. (1991) J. Biol. Chem. **266**, 20590–20593

121   Bedford, J. J., Bagnasco, S. M., Kador, P. F., Harris, W. F. and Burg, M. B. (1987) J. Biol. Chem. **262**, 14255–14259

122   Paredes, A., McManus, M., Kwon, H. M. and Strange, K. (1993) Am. J. Physiol. **263**, C1282–C1288

123   Yamauchi, A., Uchida, S., Preston, A. S., Kwon, H. M. and Handler, J. S. (1993) Am. J. Physiol. **264**, F20–F23

124   Theodoropoulos, T., Stournaras, C., Stoll, B., Markogiannakis, E., Lang, F., Gravani, A. and Häussinger, D. (1992) FEBS Lett. **311**, 241–245

125   Tohyama, Y., Kameji, T. and Hayashi, S. (1991) Eur. J. Biochem. **202**, 1327–1331

126   Zhang, F., Wettstein, M., Warskulat, U., Schreiber, R., Henninger, P., Decker, K. and Häussinger, D. (1995) Biochem. J. **312**, 135–143

127   Newsome, W. P., Warskulat, U., Noe, B., Wettstein, M., Stoll, B., Gerok, W. and Häussinger, D. (1994) Biochem. J. **304**, 555–560

128   Finkenzeller, G., Newsome, W. P., Lang, F. and Häussinger, D. (1994) FEBS Lett. **340**, 163–166

129   Cohen, D. M., Wasserman, J. C. and Gullans, S. R. (1991) Am. J. Physiol. **261**, C594–C601

130   Kilberg, M. S., Hutson, R. G. and Laine, R. O. (1994) FASEB J. **8**, 13–19

131   Offensperger, W. B., Offensperger, S., Stoll, B., Gerok, W. and Häussinger, D. (1994) Hepatology **20**, 1–7

132   Agol, V. I., Lipskaya, G. Y., Tolskaya, E. A., Voroshilova, M. K. and Romanova, L. I. (1970) Virology **41**, 533–540

133   Waite, M. R. F. and Pfefferkorn, E. R. (1968) J. Virol. **2**, 759–760

134   Bishop, J. M., Maldonado, R. L., Garry, R. F., Allen, P. T., Bosem H. R. and Waite, M. R. F. (1976) J. Virol. **17**, 446–452

135   Norrie, D. H., Wolstenholme, J., Howcroft, H. and Stephen, J. (1982) J. Gen. Virol. **62**, 127–136

136   Rey, O., Rossi, J. P. F. C., Lopez, R., Iapalucci-Espinoza, S. J. and Franze-Fernandez, M. T. (1988) J. Gen. Virol. **69**, 951–954

137   Pinto, L. H., Holsinger, L. J. and Lamb, R. A. (1992) Cell **69**, 517–528

138   Broeck, J. V., De Loof, A. and Callaerts, P. (1992) Int. J. Biochem. **24**, 1907–1916

139   Leake, R. E., Trench, M. E. and Barry, J. M. (1972) Exp. Cell Res. **71**, 17–26

140   Higgins, C. F., Cairney, J., Stirling, D. A., Sutherland, L. and Booth, I. R. (1987) Trends Biochem. Sci. **12**, 339–344

141   Hesketh, J. E. and Pryme, J. F. (1991) Biochem. J. **277**, 1–10

142   Santell, L., Rubin, R. L. and Levin, E. G. (1993) J. Biol. Chem. **268**, 21443–21447

143   Takenaka, M., Preston, A. S., Kwon, H. M. and Handler, J. S. (1994) J. Biol. Chem. **269**, 29379–29381

144   Itoh, T., Yamauchi, A., Miyai, A., Yokoyama, K., Kamada, T., Ueha,. N. and Fujiwara, Y. (1994) J. Clin. Invest. **93**, 2387–2392

145   Nakanishi, T., Balaban, R. S. and Burg, M. B. (1988) Am. J. Physiol. **255**, C181–C191

146   Brewster, J. L., de Valoir, T., Dwyer, D., Winter, E. and Gustin, M. C. (1993) Science **259**, 1760–1763

147   Lang, F., Ritter, M., Wöll, E., Weiss, E., Häussinger, D., Maly, K. and Grunicke, H. (1992) Pflügers Arch. Physiol. **420**, 424–427

148   Bianchini, L. and Grinstein, S. (1993) in Interaction of Cell Volume and Cell Function (Lang, F. and Häussinger, D., eds.), pp. 249–277, Springer Verlag, Heidelberg

149   Pietrangelo, A. and Shafritz, D. A. (1994) Proc. Natl. Acad. Sci. U.S.A. **91**, 182–186

150   Minton, A. P., Colclasure, G. C. and Parker, J. C. (1992) Proc. Nat. Acad. Sci. U.S.A. **89**, 10504–10506

151   Parker, J. C. (1993) Am. J. Physiol. **265**, C1191–C1200

152   Colclasure, G. and Parker, J. C. (1991) J. Gen. Physiol. **98**, 881–887

153   Grinstein, S., Furuya, W. and Bianchini, L. (1992) News Physiol. Sci. **7**, 232–237

154   Sackin, H. (1994) in Cellular and Molecular Physiology of Cell Volume Regulation (Strange, K., ed.), pp. 215–240, CRC Press, Boca Raton

155   Maeda, T., Wurgler-Murphy, S. M. and Saito, H. (1994) Nature (London) **369**, 242–245

156   MacLeod, R. J., Lembessis, P. and Hamilton, J. R. (1992) Am. J. Physiol. **262**, C950–C957

157   Galcheva-Gargova, Z., Derijard, B., Wu, I. H. and Davis, R. J. (1994) Science **265**, 806–811

158   Agius, L., Peak, M., Beresford, G., Al Habori, M. and Thomas, T. (1994) Biochem. Soc. Trans. **22**, 516–522

159   Davis, R. J. (1993) J. Biol. Chem. **268**, 14553–14556

160   Kyriakis, J. M., Banerjee, P., Nikolakaki, E. et al. (1994) Nature (London) **369**, 156–160

161   vom Dahl, S., Hallbrucker, C., Lang, F. and Häussinger, D. (1991) Eur. J. Biochem. **198**, 73–83

162   Theodoropoulos, T., Gravanis, A., Saridakis, I. and Stournaras, C. (1992) Cell Biochem. Funct. **10**, 281–288

163 Foskett, K. J. (1994) in Cellular and Molecular Physiology of Cell Volume Regulation (Strange, K., ed.), pp. 259–277, CRC Press, Boca Raton

164 McManus, M. L. and Churchwell, K. B. (1994) in Cellular and Molecular Physiology of Cell Volume Regulation (Strange, K., ed.), pp. 63–77, CRC Press, Boca Raton

165 Häussinger, D., Roth, E., Lang, F. and Gerok, W. (1993) Lancet **341**, 1330–1332

166 Kitani, K. and Kanai, S. (1982) Life Sci. **30**, 515–523

167 Martins, E. A. and Meneghini, R. (1994) Biochem. J. **299**, 137–140

168 Hawkins, R. A., Jessy, J., Mans, A. M. and De Joseph, M. R. (1993) J. Neurochem. **60**, 1000–1006

169 Kreis, R., Ross, B. D., Farrow, N. A. and Ackerman, Z. (1992) Radiology **182**, 19–27

170 Grant, A., Tosh, D. and Burchell, A. (1992) Biochem. Soc. Trans. **21**, 39S

Biochem. J. (1996) **317**, 633–641 (Printed in Great Britain)

REVIEW ARTICLE
Cell cycle regulation in *Aspergillus* by two protein kinases

Stephen A. OSMANI and Xiang S. YE
Weis Center For Research, Geisinger Clinic, Danville, PA 17822-2617, U.S.A.

Great progress has recently been made in our understanding of the regulation of the eukaryotic cell cycle, and the central role of cyclin-dependent kinases is now clear. In *Aspergillus nidulans* it has been established that a second class of cell-cycle-regulated protein kinases, typified by NIMA (encoded by the *nimA* gene), is also required for cell cycle progression into mitosis. Indeed, both $p34^{cdc2}$/cyclin B and NIMA have to be correctly activated before mitosis can be initiated in this species, and $p34^{cdc2}$/cyclin B plays a role in the mitosis-specific activation of NIMA. In addition, both kinases have to be proteolytically destroyed before mitosis can be completed. NIMA-related kinases may also regulate the cell cycle in other eukaryotes, as expression of NIMA can promote mitotic events in yeast, frog or human cells.

Moreover, dominant-negative versions of NIMA can adversely affect the progression of human cells into mitosis, as they do in *A. nidulans*. The ability of NIMA to influence mitotic regulation in human and frog cells strongly suggests the existence of a NIMA pathway of mitotic regulation in higher eukaryotes. A growing number of NIMA-related kinases have been isolated from organisms ranging from fungi to humans, and some of these kinases are also cell-cycle-regulated. How NIMA-related kinases and cyclin-dependent kinases act in concert to promote cell cycle transitions is just beginning to be understood. This understanding is the key to a full knowledge of cell cycle regulation.

INTRODUCTION

Dividing cells traverse the cell cycle in order to duplicate their constituents and then undergo division [1]. During the cell cycle, nuclear DNA is replicated during S phase and segregated equally into two daughter nuclei during mitosis. Cytokinesis then segregates the divided nuclei, along with cytoplasmic components, into two separate cells. Cells therefore have to make a continuum of decisions about when to start and when to stop cell-cycle-specific functions. They also have to accurately detect that nuclear DNA has been replicated exactly once in each turn of the cell cycle, and that DNA has passed quality controls, before division, to avoid transmission of defective genetic information. Finally, multicellular organisms need to restrict the number and pattern of divisions that their cells are allowed to complete, to ensure normal growth and morphogenesis. How cells orchestrate these different levels of regulation over the cell cycle is just beginning to be unravelled, and will continue to come under intensive study due to the implications for human diseases such as cancer.

Great insight into the regulation of the cell cycle has been obtained from the study of conditional cell-cycle-specific mutants isolated in model genetic systems [2–5]. By utilizing molecular genetics it has been possible to clone such genes and then analyse their products biochemically. The combined approaches of classical genetics, molecular genetics and biochemistry have proven extremely powerful for unravelling how cells regulate their cell cycles biochemically. The realization has also been made that cell cycle regulatory mechanisms are highly conserved from lower genetic systems, such as fungi, through evolution to humans. In this review we will cover recent contributions made utilizing the experimentally amenable fungus *Aspergillus nidulans* and try to integrate this information with that derived from other systems. For an earlier review, see Doonan [6].

Analysis of cell cycle regulation utilizing *A. nidulans* was pioneered by N. Ronald Morris, who isolated a collection of temperature-sensitive mutations affecting a range of biological processes including cell cycle progression, septation and nuclear movement [2,7] (Figure 1). Of the 1000 temperature-sensitive strains analysed by Morris (Figure 1), 23 were characterized as being required for interphase progression (*nim*, for **n**ever **i**n **m**itosis mutants) and six for progression through mitosis (*bim*, for **b**locked **i**n **m**itosis mutants). Notably, multiple alleles of only one *nim* gene were isolated (four alleles of *nimA*), indicating that the screen for interphase-specific functions was not saturating. In addition to mutations specifically affecting cell cycle progression, four mutations affecting the deposition of septa (*sep* mutants) and two affecting nuclear migration (*nud* mutants) were identified. Several of these genes have been isolated by complementation, and their analysis has provided unique insights into the regulation of the eukaryotic cell cycle [8–19].

THE NIMA KINASE

In the field of cell cycle research much attention has focused on cyclin-dependent kinases, as these kinases play key roles in cell cycle regulation, during initiation of both DNA replication and mitosis, in organisms ranging from fungi to humans [20,21]. However, there is accumulating evidence that other levels of regulation do exist [22–26]. In *A. nidulans* much attention has focused on the role of *nimA* in the initiation and completion of mitosis. Early studies demonstrated that *nimA* function is required specifically at the G_2–M transition [8,27,28]. This was most dramatically demonstrated by temperature shift experiments. At the restrictive temperature for *nimA5* the function of NIMA (the protein product of *nimA*; see below) is destroyed and cells accumulate, within one cell cycle, in late G_2 but continue to grow to a considerable size (Figure 1; see *nim* mutants at 42 °C).

Abbreviations used: APC, anaphase-promoting complex; Clb, B-type cyclin; Cln, G_1 cyclin.

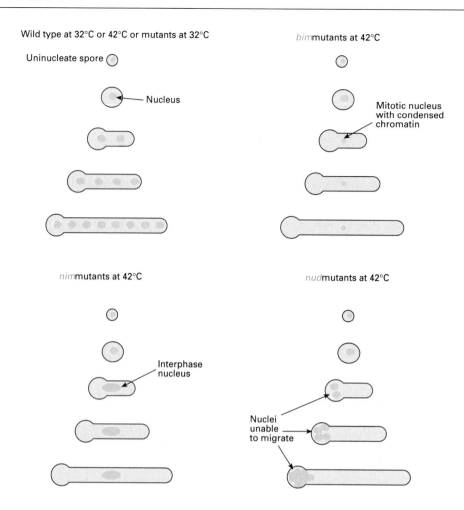

Figure 1   Terminal phenotypes of *nim*, *bim* and *nud* mutants of *Aspergillus nidulans*

The growth characteristics and nuclear divisions are shown for conidial spores inoculated into media at either a restrictive temperature (42 °C) or a permissive temperature (32 °C).

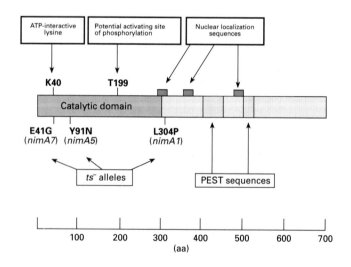

Figure 2   Potential functional domains and important residues in NIMA

Such cells show no cytological signs of mitosis, as microtubules are present in stable interphase arrays and nuclear DNA is not condensed. Upon return to a permissive temperature, NIMA function is restored, cytoplasmic microtubules disassemble, the mitotic spindle is formed and chromosomes condense as nuclei rapidly and synchronously enter mitosis. The function of *nimA* is therefore specifically required to initiate all cytological aspects of mitosis [8,27,28].

Molecular cloning of *nimA* by complementation of the *nimA5* temperature-sensitive phenotype [6] and sequence analysis indicated that it encodes a protein (699 amino acids) containing the hallmarks of serine/threonine protein kinases [9,29], which was designated NIMA (Figure 2). NIMA is the founding member of a growing family of protein kinases with members isolated from *Saccharomyces cerevisiae*, *Neurospora crassa*, *Trypanosoma*, mice and humans [30–34]. All contain their catalytic domains in the N-terminus and all have a very high isoelectric point (> 10) (Figure 2).

Biochemical analysis of NIMA, either isolated by immunoprecipitation from *A. nidulans* or expressed in *Escherichia coli*, has shown it to contain serine/threonine protein kinase activity which is independent of second messengers [35,36]. Of the commonly used artificial protein kinase substrates tested, NIMA has a clear preference for β-casein, and it has a pH optimum ranging from 7.5 to 9.5 [35,36]. Screening of a peptide library for potential artificial substrates identified a good peptide substrate for NIMA [PLM-(58–72): GTFRSSIRRLSTRRR] which is phosphorylated at Ser-6 [37]. Mutational analysis of residues around the phosphorylation site of the peptide indicated that a key determinant of NIMA specificity is the Phe residue at position -3 N-terminal to the site of phosphorylation.

CELL-CYCLE-SPECIFIC REGULATION OF NIMA

The level of NIMA protein kinase activity is regulated throughout the cell cycle, being low during G_1 and S phase and then increasing during G_2 to reach a maximum during late G_2 and early mitosis [35]. If cells are arrested in mitosis, e.g. by depolymerizing microtubules with nocodazole, then NIMA kinase activity remains at an elevated level. When cells are allowed to progress through mitosis, however, NIMA kinase activity is abolished [38]. Therefore progression through interphase and into early mitosis correlates with accumulation of active NIMA, but progression through mitosis inactivates NIMA.

If normal regulation of NIMA is overcome by induction from an inducible promoter then mitotic events are promoted, even if cells are first arrested in S phase, leading to death by untimely mitosis [9]. This appears to be a universal property of NIMA, as expression of *A. nidulans* NIMA in the fission yeast *Schizosaccharomyces pombe* [39], *Xenopus* oocytes or human cells [40] promotes mitotic events in these heterologous systems. This suggests that the NIMA specific substrates that mediate mitotic events are conserved from fungi through to humans, and there is mounting evidence that NIMA-like protein kinases are involved in cell cycle regulation in many, if not all, eukaryotes (see below).

ROLE OF NIMA ACCUMULATION DURING THE CELL CYCLE

Like the NIMA kinase activity, the level of NIMA protein also increases during interphase [38], peaking during mitosis, as does the level of *nimA* mRNA [8]. However, a clear understanding of the cell-cycle-specific expression of *nimA* is not available. For instance, changes in the level of *nimA* mRNA could be due to increased rates of transcription or decreased rates of mRNA degradation during cell cycle progression. Similarly, the accumulation of NIMA protein may be influenced by increased rates of translation or by a decreased rate of NIMA proteolysis. However, some information regarding the role of other cell cycle regulators in the expression of NIMA protein is available. In particular, the relationship between *nimA* and p34^{cdc2} has been under study. First isolated from *S. cerevisiae* [41] and subsequently from *S. pombe* [42], the p34^{cdc2} protein kinase is the founding member of the cyclin-dependent family of protein kinases [43], and is a key regulator of mitosis [20,44].

It was initially concluded that p34^{cdc2} plays no role in the expression of *nimA*, as inactivation of p34^{cdc2} to arrest cells in G_2 did not prevent accumulation of apparently fully active NIMA [22]. This analysis was performed using the *nimT23*^{cdc25} mutation, as a mutation in the p34^{cdc2} homologue of *A. nidulans* was not isolated in the original screen by Morris. *nimT*^{cdc25} encodes the Cdc25 type tyrosine phosphatase of *A. nidulans* and is required to dephosphorylate and activate p34^{cdc2} [10]. Consequently, mutations in *nimT*^{cdc25} arrest at G_2 because they impair the mitosis-promoting function of p34^{cdc2} [22,45–48]. At the G_2 arrest point of *nimT23*^{cdc25}, active NIMA protein accumulates, thus demonstrating that activation of p34^{cdc2} by tyrosine dephosphorylation is not required for expression of active NIMA kinase [22].

The subsequent isolation of a functional homologue of *cdc2*, called *nimX*^{cdc2}, from *A. nidulans* [49] enabled the consequences of direct inactivation of p34^{cdc2} on *nimA* expression to be addressed. Direct inactivation of p34^{cdc2} caused arrest of the cell

cycle in both G_1 and G_2 and also prevented the accumulation of NIMA protein [38,49]. Therefore, although inactivation of the mitotic form of p34^{cdc2} does not prevent accumulation of NIMA protein, direct inactivation of p34^{cdc2} to arrest cells in G_1 or G_2 does prevent the accumulation of NIMA.

This situation arises because p34^{cdc2} is the catalytic subunit of more than one protein kinase complex [50–52], and different p34^{cdc2}-containing complexes are thought to promote the various stages of the cell cycle in fungi [53–56]. For example, in *S. cerevisiae*, Cdc28 (the p34^{cdc2} homologue of budding yeast) binds to G_1 cyclins (Clns) to promote G_1 progression and then binds to B-type cyclins (Clbs) to promote S phase and mitosis [52,53,57–60]. In *A. nidulans*, p34^{cdc2} function is also required at least twice during the cell cycle, to promote both the G_1–S and the G_2–M transitions [49]. At present only one cyclin homologue, *nimE*^{cyclinB} [10], has been isolated from *A. nidulans*, but it is likely that NIMX^{cdc2} associates with different partners to promote the G_1–S and G_2–M transitions in this species also, or at least exists in two distinct forms during G_1 and G_2. By inactivating *nimT*^{cdc25} only one particular function of p34^{cdc2} is therefore impaired. However, direct inactivation of p34^{cdc2} is likely to impair all of its functions, both in G_1–S and in G_2–M. It can therefore be concluded that the mitotic form of p34^{cdc2}/cyclin B is not required for expression of *nimA* but that some other, non-mitotic, form of p34^{cdc2} is essential for *nimA* expression, presumably a form that is required for progression through G_1–S–G_2. The level at which p34^{cdc2} is required for expression of active NIMA is currently not known.

ROLE OF PHOSPHORYLATION IN NIMA REGULATION

The kinase activity of NIMA is regulated not only through the cell cycle by accumulation and degradation of NIMA protein, but also by phosphorylation [38]. Both recombinant NIMA and that isolated from *A. nidulans* are inactivated by enzymic dephosphorylation. This suggests that NIMA may need to be activated by a NIMA-activating kinase. A potential activating phosphorylation site, analogous to the autophosphorylation site of cAMP-dependent protein kinase, has been defined by mutation in NIMA [61]. This site is required for normal activation of the cAMP-dependent kinases [62], and mutation of the analogous site (Thr-199) in NIMA to a non-phosphoralatable residue also renders NIMA non-functional and inactive as a casein kinase [61]. It is noteworthy that the proposed activating phosphorylation site in NIMA is conserved in all NIMA-related kinases and that this site conforms to a minimal NIMA phosphorylation site [37] (FXXT/S) in NIMA and some NIMA-related kinases. This suggests that this residue may be a site for autophosphorylation. Indeed, when expressed in *E. coli*, NIMA undergoes autophosphorylation and so activates itself. However, if dephosphorylated *in vitro* and then subjected to autophosphorylation, NIMA is unable to activate itself as a kinase [36]. This suggests that NIMA has a limited capacity to activate itself by autophosphorylation, leaving open the possibility that there is a NIMA-activating kinase required for its efficient activation in *A. nidulans*.

Biochemical analyses of NIMA thus indicate that it could be regulated by phosphorylation, and recent evidence suggest that this is indeed the case. The level of NIMA phosphorylation has been followed during the cell cycle of *A. nidulans* [38] by observing mobility shifts on SDS/PAGE and by probing with the MPM-2 monoclonal antibody, which detects certain mitosis-specific phosphoproteins [63]. Two major changes in the phosphorylation state of NIMA occur during the cell cycle (Figure 3). NIMA is synthesized early in interphase in an apparently

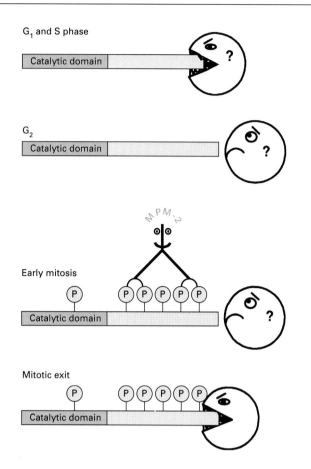

Figure 3   Degradation and phosphorylation of NIMA through the cell cycle

During G$_1$ and S phase, NIMA is in low abundance and may not be phosphorylated. During G$_2$ NIMA becomes phosphorylated and its level increases. During the G$_2$–M transition NIMA is hyperphosphorylated and becomes reactive to the MPM-2 monoclonal antibody that detects mitotic phosphoproteins. As cells exit from mitosis, NIMA is proteolytically destroyed.

unphosphorylated form during S phase. It then becomes phosphorylated during G$_2$, which can be visualized as a small retardation of NIMA on SDS/PAGE, and this phosphorylation allows partial activation of NIMA as a protein kinase. This initial site of phosphorylation is likely to be Thr-199. Upon the G$_2$–M transition, and only after activation of p34^{cdc2}/cyclin B, NIMA becomes hyperphosphorylated and shows marked retardation on SDS/PAGE. The hyperphosphorylation of NIMA further activates its level of activity in G$_2$ and also generates the MPM-2 epitope on NIMA. Soon after its hyperphosphorylation during mitosis, NIMA is proteolytically destroyed (Figure 3).

The phosphorylation of NIMA during mitosis, which generates the MPM-2 antigen, is not only dependent upon activation of p34^{cdc2}/cyclin B but may be carried out directly by the mitotic form of the p34^{cdc2} kinase. This is suggested because, *in vitro*, phosphorylation of a kinase-negative version of NIMA by p34^{cdc2}/cyclin B causes the molecular mass shift of NIMA detected on SDS/PAGE and also generates the MPM-2 antigen [38]. Thus, using purified p34^{cdc2}, it is possible to reconstruct *in vitro* the characteristics of mitotic NIMA phosphorylation promoted after activation of p34^{cdc2} H1 kinase *in vivo*. This role for p34^{cdc2} in the function of NIMA was initially not observed [22] because NIMA is very sensitive to inactivation by dephosphorylation during isolation procedures. However, incorporation of okadaic acid into isolation buffers was found to

markedly stabilize the activity of NIMA during its isolation [38]. This suggests the existence of an okadaic acid-sensitive phosphatase that has high activity towards NIMA, perhaps at the proposed Thr-199 phosphorylation site.

NIMA DEGRADATION DURING MITOTIC EXIT

The level of NIMA protein falls dramatically during progression through mitosis and is stabilized if cells are arrested in mitosis [38]. Mitotic destruction of NIMA may therefore be important for mitotic exit, and data have been presented to support this notion. A C-terminal truncated form of NIMA has a considerably longer half-life than full-length NIMA when expressed in *A. nidulans*. The truncated protein retains the ability to promote mitosis but is resistant to degradation during mitotic progression and is therefore highly toxic [39,64]. The importance of the mitotic degradation of NIMA has been investigated by expression of the truncated NIMA in cells arrested at G$_2$ [64]. The level of expression was kept low so that the truncated NIMA did not promote mitotic events. Then, by releasing the G$_2$ arrest, the ability of cells to traverse mitosis in the presence of nondegradable NIMA was determined. The cells were able to enter mitosis but were unable to complete it in the normal fashion. This indicates that mitotic proteolysis of NIMA is required for mitotic exit [64].

NIMA is not the first protein to be identified that is specifically destroyed during mitosis and whose destruction is required for mitotic exit. The first such protein identified was cyclin B [65–70]. The role of cyclin B in mitotic regulation has been well documented in many systems, including *A. nidulans*, both biochemically and genetically [10,53,59,60,71–87]. It functions as an essential subunit of the p34^{cdc2}/cyclin B H1 kinase, acting to first activate the p34^{cdc2} H1 kinase during interphase and then to inactivate this kinase during mitosis by being proteolytically destroyed. The N-terminus of cyclin B contains a conserved motif, called the cyclin destruction box, which acts to target cyclin B for ubiquitin-mediated proteolysis during mitosis. The cyclin destruction box acts to target lysine residues for polyubiquitination specifically during mitosis. The poly-ubiquitinated cyclin B is then thought to be proteolysed by a 26 S proteosome complex [88–90]. Thus, as with NIMA, truncation of cyclin B to remove the cyclin destruction box renders the protein mitotically stable (due to the lack of ubiquitination during mitosis) which then prevents normal mitotic exit.

The relationship between the mitotic destruction of NIMA and cyclin B remains to be elucidated. However, these two proteins, although both degraded during mitosis, are proteolysed with different kinetics, with cyclin B being degraded earlier than NIMA, perhaps suggesting different mechanisms of degradation [64]. The C-terminal domain of NIMA that contains determinants of instability does not contain a cyclin B-type destruction box, so NIMA is unlikely to be degraded by a mechanism involving this type of motif. However, the C-terminal portion of NIMA that is involved in specifying rapid degradation is rich in PEST sequences. Such motifs are rich in proline (P), glutamate (E), serine (S) and threonine (T) residues, and are involved in targeting proteins for destruction [91].

Recent work investigating the stability of the *S. cerevisiae* G$_1$ cyclin Cln3, and the transcription factor Gcn4, suggests that PEST sequences may be another type of signal that targets proteins for ubiquitin-mediated proteolysis [92]. Cln3 is very unstable and is essential for progression through Start in the G$_1$ phase of *S. cerevisiae*. It is also rich in PEST sequences in its C-terminus, and mutations that remove the PEST sequences stabilize Cln3 and also shorten G$_1$ [93,94].

It has been shown that the PEST-containing region of Cln3 can confer instability to a reporter protein, such as β-galactosidase, and that degradation occurs through the ubiquitin-mediated pathway of degradation [92]. Similarly, the transcription factor Gcn4 is degraded via the ubiquitin-mediated pathway and also contains PEST sequences important to its instability [95]. NIMA may therefore be degraded by the ubiquitin pathway of proteolysis during mitosis, or potentially G_1, by a mechanism directed by PEST sequences. What makes this possibility even more interesting is that it has been proposed that PEST sequences may act to couple phosphorylation with ubiquitination, as Cln3 protein is stable if the Cdc28 kinase is inactivated. Furthermore, if a Cdc28 consensus phosphorylation site is mutated within the PEST domain of the Cln3–β-galactosidase fusion protein, this protein is stabilized. Therefore Cdc28-mediated phosphorylation may target Cln3 for ubiquitination and subsequent degradation [92]. This scenario has obvious similarities to the situation with NIMA degradation. NIMA also contains PEST sequences which have been implicated in its mitotic degradation, and NIMA is also phosphorylated by a Cdc2(28)-dependent mechanism just prior to its degradation (Figure 3). NIMA also contains numerous Cdc2 phosphorylation sites in the region rich in PEST sequences. Thus phosphorylation of NIMA during mitosis may mark it for ubiquitin-dependent proteolysis in a manner analogous to the PEST-directed degradation of Cln3 during G_1.

Two lines of evidence suggest that proteins other than cyclin B need to be degraded in order for cells to complete anaphase. Using mitotic extracts derived from *Xenopus* oocytes it has been shown that the addition of non-degradable cyclin B does not prevent initiation of anaphase [67]; nor does stable cyclin B prevent anaphase progression in *S. cerevisiae* [96]. However, if ubiquitin-mediated proteolysis is inhibited then anaphase cannot be initiated in *Xenopus* extracts [67]. This indicates that proteins other than cyclin B need to be degraded during mitosis by the ubiquitin pathway of degradation in order for anaphase to be initiated.

Secondly, mutation of either *CDC16* or *CDC23* stabilizes the Clb2 B-type cyclin in *S. cerevisiae* during anaphase and G_1, suggesting that they are involved in the degradation pathway leading to destruction of Clb2 [97]. Both Cdc16 and Cdc23 are found in a complex with Cdc27 in *S. cerevisiae* [98], and each is a member of a family of proteins termed the TPR [99,100] proteins. The TPR (tetratricopeptide repeat) motif is thought to be involved in protein–protein interactions. Inactivation of any one of these three TPR proteins prevents the onset of anaphase in organisms ranging from fungi to humans [11,97,101,102]. In addition, both Cdc27 and Cdc16 have been found by biochemical analysis to be components of a large multi-protein complex which acts to ligate ubiquitin to cyclin B during mitosis and has been called the anaphase-promoting complex (APC) [103]. However, although the APC can function as a mitosis-specific cyclin B ubiquitin ligase, this complex may target other proteins for mitotic degradation and so promote anaphase, as cyclin B degradation is not required to initiate anaphase [67,96]. In addition, a mutation of *CDC27* in *S. cerevisiae* does not stabilize cyclin B but does arrest cells prior to anaphase [97], further implicating the proteolysis of non-cyclin proteins in the initiation of anaphase.

It has been proposed that some proteins active in binding sister chromatids together during early mitosis are degraded during the metaphase-to-anaphase transition [67]. Alternatively a regulatory protein, such as NIMA, may function to promote sister chromatid attachment, and destruction of NIMA may then consequently allow sister chromatid separation during anaphase. Further study will reveal the relationship between NIMA degra-

dation and the ubiquitin pathway of proteolysis and the onset of anaphase.

There is another relationship evident between cyclin B and NIMA, as activation of each is apparently required for the degradation of the other. For instance, if cells are arrested in G_2 by inactivation of NIMA, then cyclin B is stable, even though the p34^{cdc2}/cyclin B H1 kinase activity is at an elevated mitotic level. Only when NIMA is activated do cells enter mitosis and trigger cyclin B degradation. This suggests that activation of the degradation pathway leading to the destruction of cyclin B requires mitotic NIMA kinase activity. Conversely, if cyclin B is inactivated then NIMA protein accumulates and is stable, and only becomes unstable after activation of p34^{cdc2} H1 kinase activity. Using the same logic, it is clear that p34^{cdc2}/cyclin B activation helps to trigger cyclin B degradation and that NIMA activation helps to trigger NIMA degradation. It appears that some common triggering mechanism, which becomes activated only after mitotic activation of both NIMA and p34^{cdc2}/cyclin B protein kinases, may be responsible for the degradation of both cyclin B and NIMA. The ubiquitin ligase for cyclin B (APC [103] or cyclosome [104]; see above) is regulated by p34^{cdc2}/cyclin B kinase [78,103,105] and, given that NIMA activation is required for cyclin B degradation, it is possible that NIMA could also play a role in regulating the APC.

Further insight into the mode of degradation of NIMA may be obtained utilizing mutations of *bimA*, a functional homologue of *CDC27*. Temperature-sensitive mutations of *bimA* cause arrest in metaphase at the restrictive temperature [2,11]. Affinity-purified antibodies directed against BIMA^{CDC27} (the protein product of *bimA*) detect both BIMA^{CDC27} and a post-translationally modified form of BIMA^{CDC27} on Western blotting [106]. A similar situation has been reported for the *Xenopus* counterpart of BIMA^{CDC27} (*Xenopus* CDC27), with the higher-molecular-mass form corresponding to phosphorylated *Xenopus* CDC27 [103]. The phosphorylation of *Xenopus* CDC27 occurs during mitosis and may function to activate its ubiquitin ligase activity. It will be interesting to determine if BIMA^{CDC27} is also phosphorylated during mitosis and to ascertain the potential role of NIMA in this phosphorylation.

BIMA^{CDC27} was found to be localized to the spindle pole body in *A. nidulans* [106], and human CDC27 has a similar subcellular localization, being localized to the centrosome and also the mitotic spindle [102]. These proteins are localized to the spindle pole body or centrosome throughout the cell cycle, further suggesting that their mitotic function may be regulated by post-translational modification after activation of both p34^{cdc2} and NIMA.

The kinase activity of NIMA is thus subjected to several levels of regulation which ensure that it is activated at the appropriate time during transition through interphase and that it is then irreversibly inactivated during mitosis, presumably to guarantee that cells do not attempt a second mitosis before DNA has been replicated (Figure 3). There are obvious parallels that can be drawn between the regulation of mitotic p34^{cdc2} and NIMA activities. The levels of both kinase activities fluctuate through the cell cycle in very similar ways, peaking during G_2 and early mitosis and then being inactivated during mitotic exit. The kinase activity of mitotic p34^{cdc2} depends upon binding to cyclin B, but no essential partner protein is known for NIMA; however, it may need to form a homo-oligomer for activity. Both kinases are regulated by cell-cycle-specific phosphorylation and both are also irreversibly inactivated during mitosis by proteolysis. For the mitotic p34^{cdc2}/cyclin B complex, inactivation by proteolysis functions through destruction of the activating cyclin subunit, but for NIMA the whole kinase is destroyed. These two kinases are therefore regulated by very similar mechanisms to achieve similar patterns of activity through the cell cycle.

ROLE OF NIMA DURING MITOSIS

By studying the conditional loss of function alleles of *nimA*, it has been shown that inactivation of NIMA still allows p34[cdc2] to become fully activated as an H1 kinase, but this activated form of p34[cdc2] cannot induce any cytological aspects of mitosis, such as chromosome condensation or spindle formation [22]. This demonstrates that NIMA plays no role in the activation of p34[cdc2] as an H1 protein kinase. It does not, however, preclude the possibility that NIMA function is required to stimulate the mitosis-promoting functions of activated p34[cdc2] H1 kinase. For instance, active p34[cdc2] may not be able to get to its mitotic substrates in the nucleus in the absence of NIMA function. Alternatively, key mitotic substrates of p34[cdc2] may have to be phosphorylated not only by p34[cdc2] but also by NIMA in order to manifest mitotic events. If NIMA has no role in stimulating the mitosis-promoting activity of p34[cdc2] (which seems unlikely), this would place NIMA function downstream of p34[cdc2], making NIMA the kinase responsible for the phosphorylation of all proteins required to promote mitosis, a role currently thought to be played directly by p34[cdc2].

As NIMA expression is dependent on p34[cdc2], and NIMA is phosphorylated during mitosis after activation of mitotic p34[cdc2], it is likely that p34[cdc2] promotes mitosis, at least in part, by activating NIMA, which may then promote some specific aspects of mitosis (Figure 4). This favours the view that p34[cdc2] promotes mitosis by regulating at least one other mitotic regulator rather than by directly causing mitotic events such as chromosome condensation and spindle formation. Another kinase shown to act downstream of p34[cdc2] during mitosis is the Plo1 kinase of *S. pombe* [107]. This kinase, which is related to the Polo kinase of *Drosophila* [108] and Cdc5 kinase of *S. cerevisiae* [109], plays a role in spindle formation and septum formation. When over-expressed in the absence of p34[cdc2] function Plo1 is able to promote septum formation, suggesting that, like NIMA, the Plo1 kinase is responsible for a particular part of the transition through G$_2$–M–G$_1$ which is normally only executed after activation of mitotic p34[cdc2]. However, lack of Plo1 (or Polo or Cdc5) does not prevent initiation of mitosis; it leads to an abnormal mitosis. In contrast, lack of NIMA actually prevents the initiation of mitosis.

The initiation of S phase in *S. cerevisiae*, as in *S. pombe* [54] and *A. nidulans* [49], is also controlled by a p34[cdc2] homologue, Cdc28[cdc2] [57]. Cdc28[cdc2] promotes S phase by regulating the expression of a range of genes involved in the enzymology of DNA replication [109,110] and also by directly or indirectly activating a second protein kinase, Cdc7 [111–116]. Lack of Cdc7 prevents initiation of S phase after activation of Cdc28[cdc2] [117]. This suggests that a second commitment point exists, after activation of Cdc28[cdc2] during G$_1$, through which yeast cells have to pass in order to initiate DNA replication. This situation is analogous to the requirement for the NIMA kinase to promote mitosis after activation of p34[cdc2]/cyclin B in *A. nidulans*. This suggests that cyclin-dependent protein kinases can promote different cell-cycle-specific events not only by promoting the expression of different cell-cycle-specific genes but also by activating other cell-cycle-specific protein kinases.

One mitosis-specific role of NIMA may be to promote chromatin condensation, which it is able to do in the absence of p34[cdc2] when overexpressed [38]. However, it has long been accepted that the way in which p34[cdc2] mediates chromosome condensation is by direct mitosis-specific phosphorylation of histone H1. This conclusion is based largely upon the close correlation observed between p34[cdc2] H1 kinase activation and chromosome condensation, and on the observation that the sites of phosphorylation of histone H1 *in vivo* are the same as those phosphorylated by p34[cdc2] *in vitro* (see Guo et al. [118] for references). The likelihood that p34[cdc2] directly causes chromosome condensation by phosphorylation of histone H1 has, however, been recently questioned. Removal of histone H1 by immunoprecipitation does not prevent chromosome condensation in *Xenopus* mitotic extracts, suggesting that histone H1 does not play an essential role in chromosome condensation [119]. Secondly, it has been shown that complete inactivation of mitotic p34[cdc2] does not affect chromosome condensation induced by phosphatase inhibitors [118]. These studies were carried out using the FT210 mouse cell line, which contains temperature-sensitive p34[cdc2] and arrests in G$_2$ at the restrictive temperature. Treatment of the G$_2$-arrested cells with the phosphatase inhibitors fostricin or okadaic acid causes full chromosome condensation in the absence of p34[cdc2] activation and in the absence of any detectable histone H1 phosphorylation. Interestingly, chromosome condensation in this system was sensitive to inhibition of protein kinases using the protein kinase inhibitor staurosporine [118]. This indicates that a staurosporine-sensitive kinase, whose activity can be stabilized by okadaic acid, may be responsible for mediating chromosome condensation. One excellent candidate kinase with appropriate characteristics is NIMA. Use of dominant-negative versions of NIMA, as described below, should help to investigate this possibility.

IS THERE A NIMA-LIKE PATHWAY OF MITOTIC REGULATION IN OTHER EUKARYOTES?

Although several NIMA-related protein kinases have been identified in other eukaryotic cells, including humans, only one functional homologue has so far been isolated, that from another filamentous ascomycete, *N. crassa* [61], termed *nim-1*. *N. crassa nim-1* was isolated by hybridization using *nimA* cDNA as a probe, and a single copy of *nim-1* is able to fully complement a temperature-sensitive mutation in *nimA*. Although highly conserved over their catalytic domains, NIMA and NIM-1 are not as highly conserved in their C-terminal domains. This therefore suggests that homologues in higher eukaryotes are

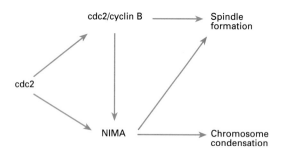

Figure 4   Relationship between p34[cdc2] and NIMA

A form of p34[cdc2] that functions during interphase is required for the expression of NIMA. For initiation of mitosis p34[cdc2] binds to cyclin B to form pre-mitosis promoting factor (pre-MPF). Activation of pre-MPF by tyrosine dephosphorylation to form MPF then leads to the hyperphosphorylation of NIMA. Expression of NIMA from an inducible promoter to a high level can promote chromatin condensation in the absence of active p34[cdc2] in both *A. nidulans* and human cells, suggesting that this protein normally plays a direct role in chromosome condensation. Induction of NIMA has also been shown to promote spindle formation.

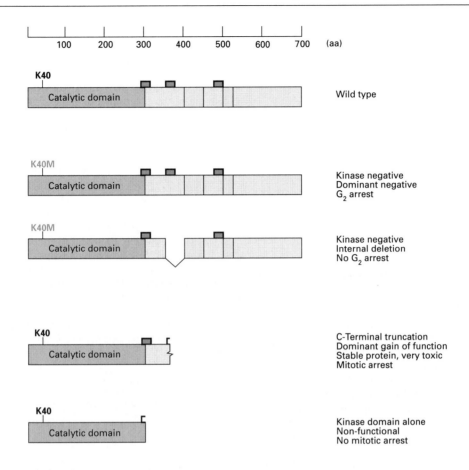

Figure 5   Mutational analysis defining residues ∼ 300–400 of NIMA as functionally important

Potential functional domains are depicted as in Figure 2. Mutation of Lys-40 to Met inactivates the kinase activity of NIMA and generates a dominant-negative version of NIMA that causes G$_2$ arrest. Deletion of residues 345–395 renders the kinase-negative NIMA unable to affect cell cycle progression. Truncation of NIMA C-terminal to residue 381 generates a dominant-positive version of NIMA that is toxic and causes mitotic arrest. However, further removal of residues 295–381 eliminates the toxicity of the truncated NIMA which is then unable to promote mitosis.

unlikely to be significantly conserved outside their catalytic domains at the primary amino acid sequence level.

The isolation of the functional homologue *nim-1* from *N. crassa* indicates that a *nimA* pathway of regulation exists in other multicellular filamentous fungi. The situation in unicellular fungi is less clear, as the *nimA*-related KIN3 protein [34,120,121] of budding yeast is non-essential. This perhaps indicates a redundant function for KIN3 which would suggest the existence of other *nimA*-related kinases in yeast. The isolation of synthetically lethal mutants in the KIN3-deleted strain may therefore identify other *nimA*-related kinases in yeast, or, perhaps, as with the cyclin genes of *S. cerevisiae*, multiple NIMA-related genes will have to be deleted to observe the phenotype caused by loss of NIMA function.

No complementation of *nimA* function in *A. nidulans* has as yet been reported using either higher eukaryote or yeast *nimA* homologues. These kinases have been termed *nek* (for ***nimA*-related kinase) or *nrk* (also for ***nimA*-related kinase), due to their protein sequence similarity with *A. nidulans* NIMA [30,31,33,40,122,123]. Two such genes have also been isolated from *Trypanosoma* [32]. As mentioned above, the yeast KIN3 kinase is not essential, but there is little functional information regarding the role of other NIMA-related kinases. Mouse *nek1* is highly expressed in germline cells, suggesting a role in meiosis [30]. The level of human Nek2 protein and activity is regulated through the cell cycle, reaching a maximum during S/G$_2$,

suggesting that it may play a role earlier in the cell cycle than does NIMA. These data are suggestive of a cell-cycle-specific function for NIMA-related kinases, but none of those so far isolated and studied have the characteristics of a mitosis regulator such as NIMA.

Several lines of indirect evidence indicate that a NIMA-related pathway of mitotic regulation does exist in other cell types. Expression of NIMA in *A. nidulans* from a strong inducible promoter induces several mitosis-specific responses. Cytoplasmic microtubules are depolymerized, abnormal mitotic spindles are transiently formed [8] and chromatin becomes condensed [38]. A more recent study, involving the expression of a truncated form of NIMA, observed similar effects, although spindles were not noted upon induction of truncated NIMA [39]. This may indicate that the C-terminus of NIMA has an essential function in spindle formation. It is clear, however, that induction of NIMA can promote mitotic events in the absence of p34^{cdc2} and can therefore promote chromatin condensation in cells arrested in S phase or in cells lacking p34^{cdc2} function [8,38,39].

Expression of NIMA in *S. pombe*, *Xenopus* oocytes or HeLa cells can also induce mitotic events, most notably chromatin condensation, from any point in the cell cycle [39,40]. The ability of induced NIMA to promote chromatin condensation across such wide species boundaries strongly indicates conservation of NIMA substrates in all eukaryotes. Furthermore, phosphorylation of these substrates by NIMA has the same

Coiled-coil regions

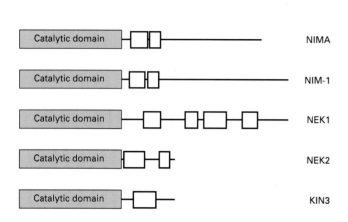

Figure 6   Coiled-coil domains found in NIMA and related protein kinases

effect of condensing the chromatin of fungi and human cells, strongly indicating the presence of a NIMA pathway of mitotic regulation in higher eukaryotes [40].

Further support for the existence of a NIMA pathway of mitotic regulation has been obtained by the use of dominant alleles of NIMA (Figure 5). It is possible to generate a dominant-negative form of NIMA by inactivating its protein kinase activity. This has been done by mutating the catalytic site of NIMA and then expressing the kinase-negative form of NIMA from an inducible promoter in *A. nidulans*, which leads to a specific arrest of the cell cycle in G_2 in a manner identical to that induced by the temperature-sensitive alleles of *nimA* [124]. The dominant-negative allele is therefore able to interfere with wild-type NIMA function, leading to G_2 arrest. It has been suggested that the effect of kinase-negative NIMA is not caused by its ability to complex with wild-type NIMA to generate a kinase-inactive complex. Rather, the dominant-negative effects are thought to occur by the binding of the mutant NIMA to a docking site in the cell normally occupied by NIMA, which thus prevents wild-type NIMA from working correctly. The dominant-negative phenotype can also be generated by expression of the C-terminal portion of NIMA which lacks the kinase domain, suggesting that the proposed docking site resides in the C-terminus [124].

The dominant-negative alleles of *nimA* have been shown to affect cell cycle progression in a human cell line [40]. Transient expression of kinase-negative NIMA, or the C-terminal portion of NIMA, led to an increase in the percentage of cells in G_2. By deletion analysis a small region of NIMA just C-terminal to the catalytic domain (residues 345–395) has been implicated in the ability both of dominant-negative alleles of NIMA to cause G_2 delay and of wild-type NIMA to promote mitotic events in human cells (Figure 5). This region also contains two potential bipartite nuclear localization signals which may be functional, as NIMA expressed in human cells is localized to the nucleus. Removal of the nuclear localization motifs both decreases the nuclear localization of active NIMA and eliminates the chromatin condensation phenotype. Their removal also impairs the dominant-negative effects of inactive NIMA. Nuclear localization of NIMA is therefore potentially important to its function. However, artificially targeting truncated NIMA to the nucleus, by tagging it with a Simian Virus 40 nuclear localization sequence, did not restore the capacity of truncated NIMA to promote mitosis. Some other structural feature of NIMA, in addition to the nuclear localization signal, is therefore important for its ability to promote mitotic events [40].

Truncation analysis of NIMA in *A. nidulans* also identified a region in the C-terminus of the kinase domain (residues 295–381) as being important to NIMA function [64]. As noted above, deletion of the C-terminal domain of NIMA generates a stable protein kinase which promotes mitotic events and, due to its increased stability, is very toxic. However, if this construct undergoes further deletion to leave just the kinase domain then the protein is no longer toxic and it cannot functionally replace *nimA5*. Clearly the non-catalytic domain of NIMA in the region of residues 295–400 contains important functional features. One such motif, found not only in NIMA but also in several NIMA-related kinases (NIM-1, NEK1 NEK2 and KIN3), is coiled-coil domains (Figure 6) [125]. Among other functions, such motifs have been shown to be involved in both hetero- and homo-dimerization [126,127]. When one considers the similarities of the phenotypes caused by overexpression of active NIMA, or of dominant-negative NIMA, in fungi and human cells, the potential binding partner(s) for these coiled-coil domains are of great interest as these may be conserved components of cell cycle regulation. It is likely that such binding partners will soon be isolated, either in genetic screens or by utilizing the two-hybrid system in yeast.

We thank Henk-Jan Bussink, Liping Wu, Russell Fincher, Aysha Osmani and Matthew Puncher for their comments on this Review, and Gang Xu for the coiled-coil analysis. This work was supported by NIH grant GM 42564 and by funds from the Geisinger Clinic Foundation.

REFERENCES

1   Murray, A. and Hunt, T. (1993) The Cell Cycle: An Introduction, W. H. Freeman and Co., New York
2   Morris, N. R. (1976) Genet. Res. **26**, 237–254
3   Hartwell, L. H. (1991) Genetics **129**, 975–980
4   Hartwell, L. H., Culotti, J. and Reid, B. (1970) Proc. Natl. Acad. Sci. U.S.A. **66**, 352–359
5   Nurse, P., Thuriaux, P. and Nasmyth, K. (1976) Mol. Gen. Genet. **146**, 167–178
6   Doonan, J. H. (1992) J. Cell Sci. **103**, 599–611
7   Morris, N. R. (1976) Exp. Cell Res. **98**, 204–210
8   Osmani, S. A., May, G. S. and Morris, N. R. (1987) J. Cell Biol. **104**, 1495–1504
9   Osmani, S. A., Pu, R. T. and Morris, N. R. (1988) Cell **53**, 237–244
10 O'Connell, M. J., Osmani, A. H., Morris, N. R. and Osmani, S. A. (1992) EMBO J. **11**, 2139–2149
11 O'Donnell, K. L., Osmani, A. H., Osmani, S. A. and Morris, N. R. (1991) J. Cell Sci. **99**, 711–719
12 Osmani, A. H., Osmani, S. A. and Morris, N. R. (1990) J. Cell Biol. **111**, 543–551
13 Doonan, J. H. and Morris, N. R. (1989) Cell **57**, 987–996
14 Osmani, S. A., Engle, D. B., Doonan, J. H. and Morris, N. R. (1988) Cell **52**, 241–251
15 Engle, D. B., Osmani, S. A., Osmani, A. H., Rosborough, S., Xiang, X. and Morris, N. R. (1990) J. Biol. Chem. **265**, 16132–16137
16 Enos, A. P. and Morris, N. R. (1990) Cell **60**, 1019–1027
17 Xin, X., Osmani, A., Osmani, S. A., Xin, M. and Morris, N. R. (1990) Mol. Biol. Cell **6**, 297–310
18 Xiang, X., Beckwith, S. and Morris, N. R. (1994) Proc. Natl. Acad. Sci. U.S.A. **91**, 2100–2104
19 May, G. S., McGoldrick, C. A., Holt, C. L. and Denison, S. H. (1992) J. Biol. Chem. **267**, 15737–15743
20 Nurse, P. (1990) Nature (London) **344**, 503–508
21 Norbury, C. and Nurse, P. (1992) Annu. Rev. Biochem. **61**, 441–470
22 Osmani, A. H., McGuire, S. L. and Osmani, S. A. (1991) Cell **67**, 283–291
23 Amon, A., Surana, U., Muroff, I. and Nasmyth, K. (1992) Nature (London) **355**, 368–371
24 Sorger, P. K. and Murray, A. W. (1992) Nature (London) **355**, 365–367
25 Stueland, C. S., Lew, D. J., Cismowski, M. J. and Reed, S. I. (1993) Mol. Cell. Biol. **13**, 3744–3755
26 Geneviere-Garrigues, A., Barakat, A., Doree, M., Moreau, J. and Picard, A. (1995) J. Cell Sci. **108**, 2693–2703
27 Oakley, B. R. and Morris, N. R. (1983) J. Cell Biol. **96**, 1155–1158

28 Bergen, L. G., Upshall, A. and Morris, N. R. (1984) J. Bacteriol. **159**, 114–119

29 Hanks, S. K., Quinn, A. M. and Hunter, T. (1988) Science **241**, 42–52

30 Letwin, K., Mizzen, L., Motro, B., Ben-David, Y., Bernstein, A. and Pawson, T. (1992) EMBO J. **11**, 3521–3531

31 Schultz, S. J. and Nigg, E. A. (1993) Cell Growth Differ. **4**, 821–830

32 Gale, J. M. and Parsons, M. (1993) Mol. Biochem. Parasitol. **59**, 111–122

33 Levedakou, E. N., He, M., Baptist, E. W., Craven, R. J., Cance, W. G., Welcsh, P. L., Simmons, A., Naylor, S. L., Leach, R. J., Lewis, T. B., Bowcock, A. and Liu, E. T. (1994) Oncogene **9**, 1977–1988

34 Jones, D. G. L. and Rosamond, J. (1990) Gene **90**, 87–92

35 Osmani, A. H., O'Donnell, K., Pu, R. T. and Osmani, S. A. (1991) EMBO J. **10**, 2669–2679

36 Lu, K. P., Osmani, S. A. and Means, A. R. (1993) J. Biol. Chem. **268**, 8769–8776

37 Lu, K. P., Kemp, B. E. and Means, A. R. (1994) J. Biol. Chem. **269**, 6603–6607

38 Ye, X. S., Xu, G., Pu, P. T., Fincher, R. R., McGuire, S. L., Osmani, A. H. and Osmani, S. A. (1995) EMBO J. **14**, 986–994

39 O'Connell, M. J., Norbury, C. and Nurse, P. (1994) EMBO J. **13**, 4926–4937

40 Lu, K. P. and Hunter, T. (1995) Cell **81**, 413–424

41 Nasmyth, K. A. and Reed, S. I. (1980) Proc. Natl. Acad. Sci. U.S.A. **77**, 2119–2123

42 Beach, D., Durkacz, B. and Nurse, P. (1982) Nature (London) **300**, 705–709

43 Meyerson, M., Enders, G. H., Wu, C. L., Su, L. K., Gorka, C., Nelson, C., Harlow, E. and Tsai, L.-H. (1992) EMBO J. **11**, 2909–2917

44 Lewin, B. (1990) Cell **61**, 743–752

45 Millar, J. B. A., McGowan, C. H., Lenaers, G., Jones, R. and Russell, P. (1991) EMBO J. **10**, 4301–4309

46 Russell, P. and Nurse, P. (1986) Cell **45**, 145–153

47 Moreno, S., Nurse, P. and Russell, P. (1990) Nature (London) **344**, 549–552

48 Lee, M. S., Ogg, S., Xu, M., Parker, L. L., Donoghue, D. J., Maller, J. L. and Piwnica-Worms, H. (1992) Mol. Biol. Cell **3**, 73–84

49 Osmani, A. H., van Peij, N., Mischke, M., O'Connell, M. J. and Osmani, S. A. (1994) J. Cell Sci. **107**, 1519–1528

50 Krek, W. and Nigg, E. A. (1991) EMBO J. **10**, 305–316

51 Draetta, G. (1990) Trends Biochem. Sci. **15**, 378–383

52 Nasmyth, K. (1993) Curr. Opin. Cell Biol. **5**, 166–179

53 Surana, U., Robitsch, H., Price, C., Schuster, T., Fitch, I., Futcher, A. B. and Nasmyth, K. (1991) Cell **65**, 145–161

54 Nurse, P. and Bisset, Y. (1981) Nature (London) **292**, 558–560

55 Piggott, J. R., Rai, R. and Carter, B. L. A. (1982) Nature (London) **298**, 391–393

56 Reed, S. I. and Wittenberg, C. (1990) Proc. Natl. Acad. Sci. U.S.A. **87**, 5697–5701

57 Reed, S. (1992) Annu. Rev. Cell Biol. **8**, 529–561

58 Hadwiger, J. A., Wittenberg, C., Richardson, H. E., Lopes, M. D.-B. and Reed, S. I. (1989) Proc. Natl. Acad. Sci. U.S.A. **86**, 6255–6259

59 Richardson, H., Lew, D. J., Henze, M., Sugimoto, K. and Reed, S. I. (1992) Genes Dev. **6**, 2021–2034

60 Ghiara, J. B., Richardson, H. E., Sugimoto, K., Henze, M., Lew, D. J., Wittenberg, C. and Reed, S. I. (1991) Cell **65**, 163–174

61 Pu, R. T., Gang, X., Wu, L., Vierula, J., O'Donnell, K., Ye, X. and Osmani, S. A. (1995) J. Biol. Chem. **271**, 18110–18116

62 Steinberg, R. A., Cauthron, R. D., Symcox, M. M. and Shuntoh, H. (1993) Mol. Cell. Biol. **13**, 2332–2341

63 Davis, F. M., Tsao, T. Y., Fowler, S. K. and Rao, P. N. (1983) Proc. Natl. Acad. Sci. U.S.A. **80**, 2926–2930

64 Pu, R. T. and Osmani, S. A. (1995) EMBO J. **14**, 995–1003

65 Evans, T. E., Rosenthal, J., Youngbloom, K., Distel, K. and Hunt, T. (1983) Cell **33**, 389–396

66 Glotzer, M., Murray, A. W. and Kirschner, M. W. (1991) Nature (London) **349**, 132–138

67 Holloway, S. L., Glotzer, M., King, R. W. and Murray, A. W. (1993) Cell **73**, 1393–1402

68 Gallant, P. and Nigg, E. A. (1992) J. Cell Biol. **117**, 213–224

69 van der Velden, H. M. W. and Lohka, M. J. (1993) Mol. Cell. Biol. **13**, 1480–1488

70 Draetta, G., Luca, F., Westendorf, J., Brizuela, L., Ruderman, J. and Beach, D. (1990) Cell **56**, 829–838

71 Booher, R. N., Alfa, C. E., Hyams, J. S. and Beach, D. H. (1989) Cell **58**, 485–497

72 Alfa, C. E., Ducommun, B., Beach, D. and Hyams, J. S. (1990) Nature (London) **347**, 680–682

73 Gallagher, I. M., Alfa, C. E. and Hyams, J. S. (1993) Mol. Biol. Cell **4**, 1087–1096

74 Knoblich, J. A. and Lehner, C. F. (1993) EMBO J. **12**, 65–74

75 Epstein, C. B. and Cross, F. R. (1992) Genes Dev. **6**, 1695–1706

76 Grandin, N. and Reed, S. I. (1993) Mol. Cell. Biol. **13**, 2113–2125

77 Pines, J. and Hunter, T. (1991) J. Cell Biol. **115**, 1–17

78 Luca, F. C., Shibuya, E. K., Dohrmann, C. E. and Ruderman, J. V. (1991) EMBO J. **10**, 4311–4320

79 Peeper, D. S., Parker, L. L., Ewen, M. E., Toebes, M., Hall, F. L., Xu, M., Zantema, A., van der Eb, A. J. and Piwnica-Worms, H. (1993) EMBO J. **12**, 1947–1954

80 Lehner, C. H. and O'Farrell, P. H. (1990) Cell **61**, 535–547

81 Meijer, L., Arion, D., Golsteyn, R., Pines, J., Brizuela, L., Hunt, T. and Beach, D. (1989) EMBO J. **8**, 2275–2282

82 Minshull, J., Blow, J. J. and Hunt, T. (1989) Cell **56**, 947–956

83 Murray, A. and Kirschner, M. W. (1989) Nature (London) **339**, 275–280

84 Pines, J. and Hunter, T. (1989) Cell **58**, 833–846

85 Westendorf, J. M., Swenson, K. I. and Ruderman, J. V. (1989) J. Cell Biol. **108**, 1431–1444

86 Amon, A., Irniger, S. and Nasmyth, K. (1994) Cell **77**, 1037–1050

87 Hunt, T., Luca, F. C. and Ruderman, J. V. (1992) J. Cell Biol. **116**, 707–724

88 Ciechanover, A. (1994) Cell **79**, 13–21

89 Hochstrasser, M. (1995) Curr. Opin. Cell Biol. **7**, 215–223

90 Peters, J. M. (1994) Trends Biochem. Sci. **19**, 377–382

91 Rogers, S., Wells, R. and Rechsteiner, M. (1986) Science **234**, 364–368

92 Yaglom, J., Linskens, M. H. K., Sadis, S., Rubin, D. M. and Futcher, B. (1995) Mol. Cell. Biol. **15**, 731–741

93 Nash, R., Tokiwa, G., Anaud, S., Erickson, K. and Futcher, B. (1988) EMBO J. **7**, 4335–4346

94 Tyers, M., Tokiwa, G., Nash, R. and Futcher, B. (1992) EMBO J. **11**, 1773–1784

95 Kornitzer, D., Raboy, B., Kulka, R. G. and Fink, G. R. (1994) EMBO J. **13**, 6021–6030

96 Surana, U., Amon, A., Dowzer, C., McGrew, J., Byers, B. and Nasmyth, K. (1993) EMBO J. **12**, 1969–1978

97 Irniger, S., Simonetta, P., Michaelis, C. and Nasmyth, K. (1995) Cell **81**, 269–277

98 Lamb, J. R., Michaud, W. A., Sikorski, R. S. and Hieter, P. A. (1994) EMBO J. **13**, 4321–4328

99 Hirano, T., Kinoshita, N., Morikawa, K. and Yanagida, M. (1994) Cell **60**, 319–328

100 Sikorski, R. S., Boguski, M. S., Goebl, M. and Hieter, P. (1990) Cell **60**, 307–317

101 Hirano, T., Hiraoka, Y. and Yanagida, M. (1990) J. Cell Biol. **106**, 307–317

102 Tugendreich, S., Tomkiel, J., Earnshaw, W. and Heiter, P. (1995) Cell **81**, 261–268

103 King, R. W., Peters, J. M., Tugendreich, S., Rolfe, M., Hieter, P. and Kirschner, M. W. (1995) Cell **81**, 279–288

104 Sudakin, V., Ganoth, D., Dahan, A., Heller, H., Hershko, J., Luca, F. C., Ruderman, J. V. and Hershko, A. (1995) Mol. Biol. Cell **6**, 185–198

105 Felix, M.-A., Labbe, J.-C., Doree, M., Hunt, T. and Karsenti, E. (1990) Nature (London) **346**, 379–382

106 Mirabito, P. M. and Morris, N. R. (1993) J. Cell Biol. **120**, 959–968

107 Schwob, E., Bohm, T., Mendenhall, M. and Nasmyth, K. (1994) Cell **79**, 233–244

108 Llamazares, A., Moreira, A., Tavares, A., Girdam, C., Spruce, B. A., Gonzalez, C., Karess, R. E., Glover, D. M. and Sunkel, C. E. (1991) Genes Dev. **5**, 2153–2165

109 Kitada, K., Johnson, A. L., Johnston, L. H. and Sugino, A. (1993) Mol. Cell. Biol. **13**, 4445–4457

110 Toyn, J. H., Toone, W. M., Morgan, B. A. and Johnston, L. H. (1995) Trends Biochem. Sci. **20**, 70–73

111 Hollingsworth, Jr., R. E. and Sclafani, R. A. (1990) Proc. Natl. Acad. Sci. U.S.A. **87**, 6272–6276

112 Jackson, A. L., Pahl, P. M., Harrison, K., Rosamond, J. and Sclafani, R. A. (1993) Mol. Cell. Biol. **13**, 2899–2908

113 Yoon, H.-J., Loo, S. and Campbell, J. L. (1993) Mol. Biol. Cell **4**, 195–208

114 Yoon, J. and Campbell, J. L. (1991) Proc. Natl. Acad. Sci. U.S.A. **88**, 3574–3578

115 Buck, V., White, A. and Rosamond, J. (1991) Mol. Gen. Genet. **227**, 452–457

116 Dowell, S. J., Romanowski, P. and Diffley, J. F. X. (1994) Science **265**, 1243–1246

117 Hereford, L. M. and Hartwell, L. H. (1974) J. Mol. Biol. **84**, 445–461

118 Guo, X. W., Th'ng, J. P. H., Swank, R. A., Anderson, H. J., Tudan, C., Bradbury, E. M. and Roberge, M. (1995) EMBO J. **14**, 976–985

119 Oshumi, K., Katagiri, C. and Kishimoto, T. (1993) Science **262**, 2033–2035

120 Barton, A., Davies, C., Hutchison, III, C. and Kaback, D. (1992) Gene **117**, 137–140

121 Schweitzer, B. and Philippsen, P. (1992) Mol. Gen. Genet. **234**, 164–167

122 Schultz, S. J., Fry, A. M., Sutterlin, C., Ried, T. and Nigg, E. A. (1994) Cell Growth Differ. **5**, 1–11

123 Fry, A. M., Schultz, S. J., Bartek, J. and Nigg, E. A. (1995) J. Biol. Chem. **270**, 12899–12905

124 Lu, K. P. and Means, A. R. (1994) EMBO J. **13**, 2103–2113

125 Cohen, C. and Parry, D. A. (1986) Trends Biochem. Sci. **11**, 245–248

126 Ho, C., Gordon Adamson, J., Hodges, R. S. and Smith, M. (1994) EMBO J. **13**, 1403–1413

127 O'Shea, E. K., Rutkowski, R. and Kim, P. S. (1989) Science **243**, 538–542

Biochem. J. (1996) **320**, 697–711 (Printed in Great Britain)

REVIEW ARTICLE
Quinoprotein-catalysed reactions

Christopher ANTHONY

Biochemistry Department, University of Southampton, Southampton SO16 7PX, U.K.

This review is concerned with the structure and function of the quinoprotein enzymes, sometimes called quinoenzymes. These have prosthetic groups containing quinones, the name thus being analogous to the flavoproteins containing flavin prosthetic groups. Pyrrolo-quinoline quinone (PQQ) is non-covalently attached, whereas tryptophan tryptophylquinone (TTQ), topa-quinone (TPQ) and lysine tyrosylquinone (LTQ) are derived from amino acid residues in the backbone of the enzymes. The mechanisms of the quinoproteins are reviewed and related to their recently determined three-dimensional structures. As expected, the quinone structures in the prosthetic groups play important roles in the mechanisms. A second common feature is the presence of a catalytic base (aspartate) at the active site which initiates the reactions by abstracting a proton from the substrate, and it is likely to be involved in multiple reactions in the mechanism. A third common feature of these enzymes is that the first part of the reaction produces a reduced prosthetic group; this part of the mechanism is fairly well understood. This is followed by an oxidative phase involving electron transfer reactions which remain poorly understood. In both types of dehydrogenase (containing PQQ and TTQ), electrons must pass from the reduced prosthetic group to redox centres in a second recipient protein (or protein domain), whereas in amine oxidases (containing TPQ or LTQ), electrons must be transferred to molecular oxygen by way of a redox-active copper ion in the protein.

INTRODUCTION TO QUINOPROTEINS

Quinoproteins are enzymes whose catalytic mechanisms involve quinone-containing prosthetic groups in their active sites; these may be bound non-covalently or derived from amino acids in the protein backbone of the enzyme (Figure 1). The only non-covalently bound example is pyrrolo-quinoline quinone (PQQ), which is the prosthetic group in a number of bacterial dehydrogenases. Tryptophan tryptophylquinone (TTQ) is derived from two tryptophan residues and occurs in bacterial amine dehydrogenases. Topa-quinone (TPQ) is derived from tyrosine and is the prosthetic group of the copper-containing amine oxidases found in bacteria, yeasts and plants. Lysine tyrosylquinone (LTQ) is also derived from tyrosine together with a lysyl residue, and has been very recently described in a special type of copper-containing amine oxidase, lysyl oxidase. The last few years has seen the identification of these prosthetic groups and determination by X-ray crystallography of examples of each type of enzyme. This review aims to summarize these achievements, to consider the mechanisms proposed for the enzymes and to relate these to each other and to significant features seen in the X-ray structures.

The history of the quinoproteins began in the 1960s with the characterization of the novel prosthetic groups of glucose dehydrogenase by Hauge [1] and methanol dehydrogenase by Anthony and Zatman [2]. More than 10 years later Duine, Frank and co-workers demonstrated that the prosthetic group of methanol dehydrogenase is a quinone structure containing two nitrogen atoms [3,4], and Kennard's group showed it to be PQQ by X-ray diffraction analysis [5]. A number of other bacterial dehydrogenases were subsequently shown, by the groups of

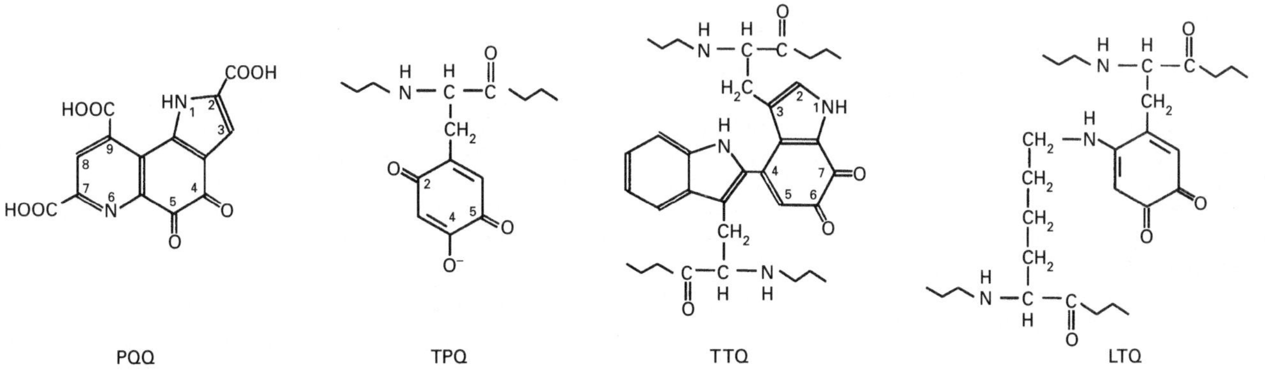

Figure 1   Prosthetic groups of quinoproteins

PQQ is the prosthetic group of some bacterial dehydrogenases. TPQ is the prosthetic group of the copper-containing amine oxidases of bacteria, plants and animals. TTQ is the prosthetic group of bacterial amine dehydrogenases. LTQ is the prosthetic group of lysyl oxidase, a specific copper-containing amine oxidase occurring in animals.

Abbreviations used: LTQ, lysine tyrosylquinone; PQQ, pyrrolo-quinoline quinone; TPQ, topa quinone (6-hydroxyphenylalanine); TTQ, tryptophan tryptophylquinone.

Duine and Frank in Delft and of Ameyama and Adachi in Yamaguchi, to contain PQQ, and the term quinoprotein was coined to include all these proteins [6]. About this time, incorrect identification of PQQ as the prosthetic group of many other enzymes led to considerable confusion, which has now been resolved with the determination of the structure of TTQ by McIntire and his colleagues [7], and of TPQ [8] and LTQ [8a] by Klinman, Dooley and their colleagues.

THE PQQ-CONTAINING DEHYDROGENASES

The only enzymes containing PQQ are bacterial enzymes in which this prosthetic group is tightly, but not covalently, bound; the best known of these quinoproteins catalyse the oxidation of alcohols and glucose in the periplasm of bacteria. They are usually assayed with artificial electron acceptors (A) such as phenazine ethosulphate, when the following reaction is catalysed:

$$R_1R_2CHOH + A \rightarrow R_1R_2C{=}O + AH_2$$

The physiological electron acceptor is a soluble cytochrome c in the case of methanol dehydrogenase and some ethanol dehydrogenases; it is protein-bound haem c in the quinohaemoprotein alcohol dehydrogenases, and ubiquinone in the membrane-bound glucose dehydrogenase. These enzymes all have a bivalent cation at the active site, and their catalytic subunits are very similar, as indicated by modelling the alcohol dehydrogenase and glucose dehydrogenase sequences on to the co-ordinates of the methanol dehydrogenase of *Methylobacterium extorquens* [9,10].

Methanol dehydrogenase and its prosthetic group (PQQ)

This enzyme catalyses the oxidation of methanol to formaldehyde in the periplasm of methylotrophic bacteria (for general reviews see [11–15], for reviews of relevant electron transport systems see [16,17], and for a review of the biosynthesis of the dehydrogenase and associated electron transport components see [18]). It has an $\alpha_2\beta_2$ tetrameric structure; the α-subunit containing the PQQ has a molecular mass of 66 kDa, and the β-subunit is very small (8.5 kDa) and was not discovered until 25 years after the first description of the enzyme [19]. The subunits cannot be reversibly dissociated and no function has been ascribed to the small subunit. Each α-subunit contains a calcium ion, which is essential for maintaining the PQQ in its active configuration in the active site [20–22].

The prosthetic group (Figure 1) was first isolated, purified and characterized by Anthony and Zatman [2], the X-ray structure of an acetone adduct was determined by Kennard and her colleagues [5], and an extensive chemical characterization was achieved by Frank and Duine and their co-workers (summarized in [12–14]). The pH-dependence of the midpoint redox potential of the PQQ/PQQH$_2$ couple (+90 mV at pH 7.0) indicates that it acts as a 2e$^-$/2H$^+$ redox carrier [23]. Resonance Raman spectroscopy of the isolated PQQ and other quinones, their derivatives and quinoproteins has been extensively reviewed by Dooley and Brown [24,25], and methods for the foolproof measurement of PQQ were reviewed by Klinman and Mu [26].

X-ray structure of methanol dehydrogenase

This is the only PQQ-containing enzyme for which a structure is available [21,22,27–30], the highest resolution structure (1.94 Å) being that of the enzyme from *Methylobacterium extorquens* (the numbering system for amino acids is for this enzyme) [22]. It is an $\alpha_2\beta_2$ tetramer structure, with the small β-subunits folding around the surface of the α-subunits. The α-subunit is a super-barrel made up of eight β-sheets arranged with radial symmetry

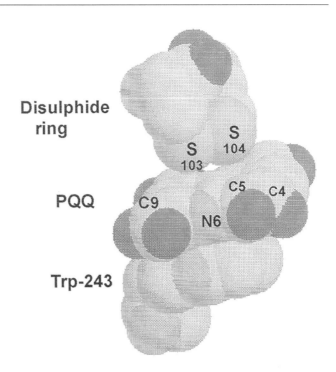

Figure 2   Novel disulphide ring in the active site of methanol dehydrogenase

The ring is formed by disulphide bond formation between adjacent cysteine residues. The PQQ is 'sandwiched' between this ring and the tryptophan that forms the floor of the active-site chamber. The calcium ion (not shown) is co-ordinated between the C-9 carboxylate, the N-6 of the PQQ ring and the carbonyl oxygen at C-5. The oxygen of the C-4 carbonyl is clearly out of the plane of the ring. The full structure is given in [22].

(the 'propeller fold') and held together by novel tryptophan docking motifs [22,30]. The PQQ is buried in the interior of the superbarrel within a chamber that communicates with the exterior through a shallow hydrophobic funnel-shaped depression in the surface. The floor of the chamber is formed by the plane of a tryptophan residue, the ceiling being formed by a novel ring structure arising from a disulphide bridge between adjacent cysteine residues joined by a novel non-planar peptide bond (Figures 2 and 3). In addition, the PQQ is bonded by many equatorial reactions as shown in Figure 4, which also indicates the co-ordination of the calcium ion to the PQQ and the protein. The calcium is co-ordinated to Glu-177 (both carboxylate oxygen atoms) and Asn-261, as well as N-6, the C-5 quinone oxygen and the C-7 carboxyl group of PQQ. Figures 3 and 4 also show a likely active-site base (Asp-303) and indicate that the C-4 carbonyl oxygen atom is clearly out of the plane of the ring, perhaps consistent with the semiquinone state of the prosthetic group in this enzyme [20,22,31].

Reductive half-reaction of methanol dehydrogenase

The reaction mechanism of methanol dehydrogenase has been more difficult to elucidate by kinetic and spectroscopic investigations than those of the other types of quinoprotein. This is because the isolated enzyme contains PQQ in the fully reduced form or as the semiquinone, and so addition of substrate (a two-electron donor) does not lead to its reduction; furthermore, 'endogenous substrate' immediately reduces any oxidized enzyme produced during experiments. Finally, the enzyme becomes inactivated when oxidized with artificial electron acceptors [11,13,14,32]. It catalyses a Ping-Pong reaction, consistent with

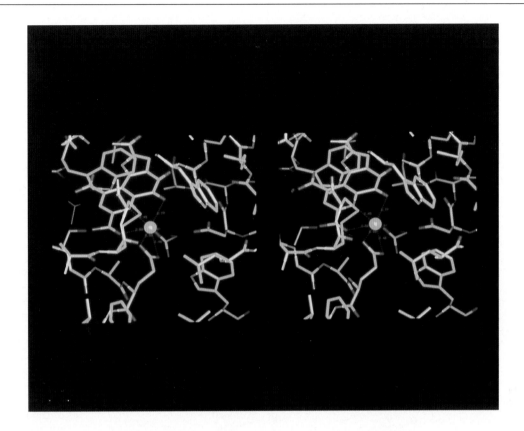

Figure 3   Stereo view of the active site of methanol dehydrogenase centred on the likely substrate-binding site

The PQQ, sandwiched between Trp-243 and the novel ring structure formed by the Cys-103–Cys-104 disulphide bridge, can be seen at the top left. The Ca^{2+} is shown as a yellow sphere ligated by O-5, N-6 and O-7 of PQQ and the side chains of Glu-177 and Asn-261. It was suggested that the proposed active-site base is Asp-303, which is shown at centre right. Reproduced with permission from [22].

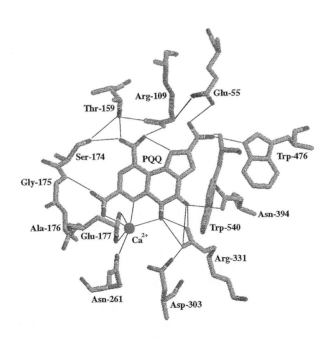

Figure 4   Co-ordination of the calcium ion and PQQ at the active site of methanol dehydrogenase

In addition to the equatorial interactions shown, PQQ is sandwiched between the indole ring of Trp-243 and the disulphide ring structure shown in Figures 2 and 3. The full structure is given in [22].

reduction of PQQ by substrate and release of product, followed by two sequential single-electron transfers to the cytochrome c_L, during which the $PQQH_2$ is oxidized back to the quinone by way of the free radical semiquinone [33–35]. The rate-limiting step is the conversion of the oxidized complex containing the substrate into the reduced enzyme plus product, and is the only step requiring the activator ammonia.

The C-5 carbonyl of isolated PQQ is very reactive towards nucleophilic reagents, adducts being formed with methanol, aldehydes, ketones, urea, cyanide, ammonia and amines [36,37], and this has encouraged the assumption that a covalent PQQ–substrate complex may be important in the reaction mechanism. The reaction of MDH with cyclopropanol gave a C-5 propanal adduct, indicating that the mechanism consists of proton abstraction by a base at the active site, followed by rearrangement of the cyclopropoxy anion into a ring-opened carbanion, and attack of this on the electrophilic C-5 of PQQ [32,38]. It was concluded that during oxidation of methanol a similar proton abstraction must occur, followed by formation of a carbon–oxygen bond to give a hemiketal intermediate. The only direct evidence for this point is the slight change in the spectrum of a possible intermediate seen during reaction with deuterated methanol [34,35]. No evidence was obtained for a covalent intermediate in studies with the barium-containing enzyme, which has some advantages for study of the mechanism of methanol dehydrogenase [39].

In the structure of the enzyme from *Methylophilus* W3A1, a solvent molecule is seen near the O-5 of PQQ. It is suggested that the hydroxy group of methanol could form hydrogen bonds to

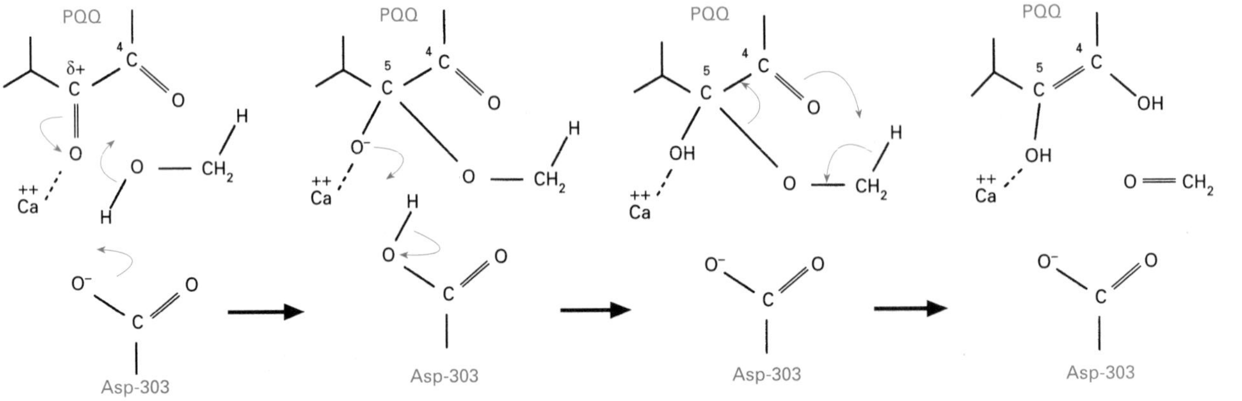

Figure 5   Reaction mechanism for methanol dehydrogenase involving a hemiketal intermediate

Proton abstraction by the base leads to an oxyanion form of the substrate which attacks the electrophilic C-5, giving the hemiketal intermediate from which the methyl proton is abstracted; this in turn gives the quinone and product formaldehyde.

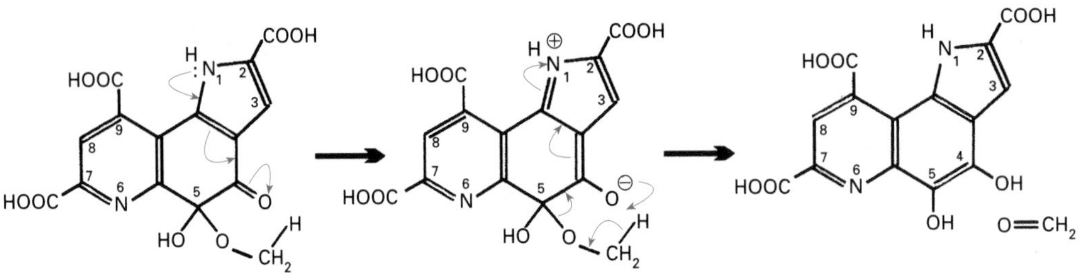

Figure 6   Possible involvement of the pyrrole nitrogen in the reaction mechanism of methanol dehydrogenase

This is a modification of the mechanism shown in Figure 5. In this variation the difficult proton abstraction from the methyl group is facilitated by the ionization of the C-4 carbonyl oxygen, which is made possible by the presence of the pyrrole nitrogen atom.

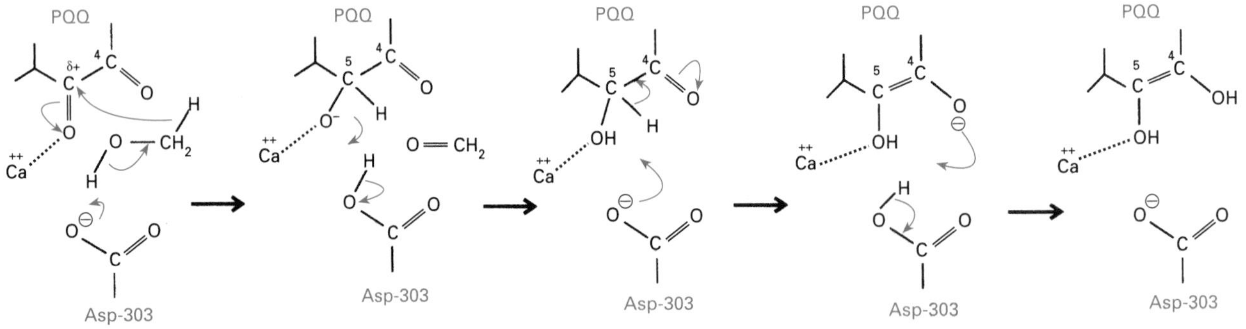

Figure 7   Reaction mechanism for methanol dehydrogenase involving hydride transfer

This mechanism is suggested as an alternative to that shown in Figure 5. The key difference is that there is no covalent bonding of substrate. In this case the initial proton abstraction is the same, but the electrophilic C-5 is involved directly in removal of the methyl hydrogen as a hydride. The active-site base (Asp-303) acts twice in the mechanism.

both O-5 and Asp-303 in the same way as this solvent molecule, and that the substrate methyl group might be accommodated by a hydrophobic cavity, bounded by two tryptophans, a leucine and the disulphide ring [30]. It is probable that Asp-303 (Figures 3 and 4) provides the catalytic base which initiates the reaction by abstraction of a proton from the alcohol substrate (Figures 5–7). In these mechanisms the Ca^{2+} ion is given a role in addition to a structural role in maintaining PQQ in an active configuration; it is proposed that the Ca^{2+} acts as a Lewis acid by way of its co-ordination to the C-5 carbonyl oxygen of PQQ, thus providing the electrophilic C-5 for attack by an oxyanion or hydride [15].

It is also possible that the Ca^{2+} ion co-ordinates to the substrate oxygen atom. The role of Ca^{2+} in the mechanism has been given support by a study of a Sr^{2+}-containing methanol dehydrogenase produced by growing bacteria in a high concentration of Sr^{2+} [40], and by investigations using an active enzyme containing Ba^{2+} instead of Ca^{2+} [39]. This is the first example of an enzyme in which barium plays an active catalytic role; the modified enzyme has a relatively low affinity for methanol (K_m 3.4 mM instead of 10 μM) and for its activator ammonia, but its activation energy is half (and its V_{max} twice) that of the normal Ca^{2+}-containing enzyme. We have suggested that this may be due to a

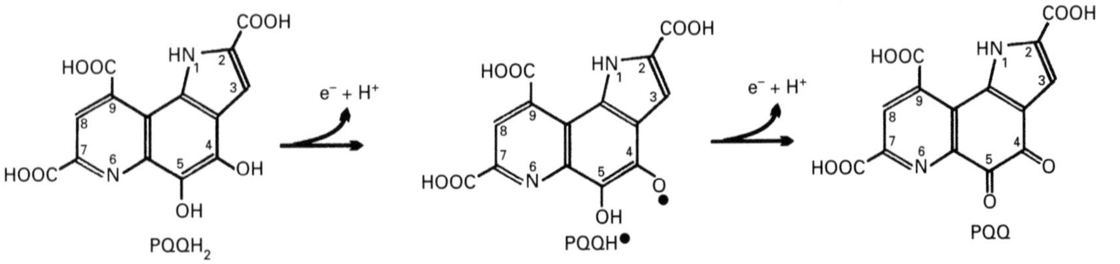

Figure 8   Oxidative half-reaction of methanol dehydrogenase

The free-radical semiquinone (PQQH·) is indicated with the unpaired electron on the C-4 carbonyl oxygen, which is perhaps consistent with the fact that this oxygen is seen to be out of the plane of the rest of the PQQ (see Figure 2). The electron acceptor is either a dye such as phenazine ethosulphate or the natural electron acceptor cytochrome c_L.

change in conformation at the active site, leading to a decrease in the free energy of binding and hence to a decrease in activation energy [39].

In the mechanism shown in Figure 5 the oxyanion produced by proton abstraction attacks the electrophilic C-5, leading to formation of the proposed hemiketal intermediate. The subsequent reduction of the PQQ with release of product aldehyde is likely to be facilitated by prior ionization of the hemiketal complex, which might involve the pyrrole N atom (Figure 6). An alternative mechanism (Figure 7) is a simple acid/base-catalysed hydride transfer in which Asp-303 again provides the base and Ca^{2+} again acts as a Lewis acid.

The large deuterium isotope effect (approx. 6) observed during the reductive phase of the reaction is consistent with either mechanism; in both cases the step affected will be the breaking of the C–H bond, and it is this step that is affected by the activator ammonia [34,39]. Ammonia (the free base) is required as activator when the enzyme is assayed with artificial electron acceptors, but not usually when it is assayed with cytochrome [11,14,34,35,41]. This activation is confined to the step involving hyrogen transfer from the substrate to PQQ, but its mechanism of action is not known [34,35,39]. An obvious possibility would involve formation of a covalent iminoquinone adduct, and one of the first mechanisms proposed for this enzyme suggested that methanol binds as a methoxy group to the same carbon atom (C-4) as the ammonia and is then released as the aldehyde during reduction of the PQQ [13,42]. Although ammonia is able to form an adduct at the C-5 position of isolated PQQ [36,37,43], there is no convincing evidence that this occurs during the enzyme reaction. No nitrogen-containing adduct of PQQ has been isolated from the enzyme; added ammonia is easily removed by gel filtration to give inactive enzyme, which then requires added ammonia for activity. In the reaction with cytochrome this activator is not always required, and some alcohol dehydrogenases which are likely to have essentially similar mechanisms (see below) do not require ammonia as activator.

Oxidative half-reaction of methanol dehydrogenase

This involves two separate single-electron transfer steps to cytochrome c_L, which is a novel class of large acidic c-type cytochromes [16,44]. Cytochrome c_L contains five α-helices, the central three of which envelop the haem group and correspond to analogous helices in most other c-type cytochromes [45].

There is considerable evidence that the interaction between the dehydrogenase and cytochrome c_L is electrostatic in nature [16,35], and chemical modification studies have indicated that the interaction is by way of a small number of lysyl residues on the dehydrogenase and carboxylates on the cytochrome [46,47].

However, hydrophobic interactions might also be important, as indicated by the X-ray structure of the dehydrogenase. This shows that the PQQ is buried within an internal chamber which communicates with the exterior of the protein by way of a hydrophobic funnel in the surface, which is perhaps the most likely place for interaction with the cytochrome [15,22,41]. A study of the initial 'docking' interaction confirmed the role of electrostatic interactions but, surprisingly, showed that this interaction is not inhibited by 50 μM EDTA, which is sufficient to inhibit the overall electron transfer process [48]. It was therefore suggested that EDTA inhibits by binding to nearby lysyl residues, thus preventing movement of the 'docked' cytochrome to its optimal position for electron transfer, which probably involves interaction with the hydrophobic funnel in the surface of the dehydrogenase [48]. An extensive kinetic investigation of the interaction of the two proteins has led to a similar conclusion [49].

It is reasonable to assume that electron transfer from the quinol form of PQQ to the cytochrome electron acceptor occurs in two single-electron transfer steps, with the semiquinone form of PQQ being produced after the first of these transfers (Figure 8). The protons are released from the reduced PQQ into the periplasmic space, thus contributing to the protonmotive force [17,50]. It was once thought that an intermediary in this process was the novel disulphide bridge between adjacent cysteines in the active site; this novel structure is very readily reduced with Cleland's reagent (dithiothreitol), yielding enzyme that is inactive with cytochrome but active with phenazine ethosulphate [28]. However, the activity with phenazine ethosulphate occurs because the two thiols are rapidly oxidized by this reagent back to the original disulphide; furthermore, no free thiols were ever detected during the reaction cycle, and reaction with iodoacetate led to active enzyme containing carboxymethylated cysteine residues that could no longer take part in oxidation/reduction reactions [31]. This type of disulphide ring structure has not been observed previously in an active enzyme, and its rarity would suggest some special biological function. It is not present in the quinoprotein glucose dehydrogenase, in which electrons are transferred to membrane ubiquinone from the quinol $PQQH_2$ and in which the semiquinone free radical is unlikely to be involved as a stable intermediate. It has been suggested, therefore, that this novel structure might function in stabilization or protection from solvent at the entrance to the active site of the free-radical PQQ semiquinone in methanol dehydrogenase [31].

PQQ-containing dehydrogenases for alcohols and glucose

There are three types of PQQ-containing alcohol dehydrogenases that are distinct from methanol dehydrogenase [51], but they all contain a calcium ion and the reductive parts of their mechanisms are likely to be similar. The first type, such as that in *Pseudomonas aeruginosa*, is almost identical with methanol dehydrogenase, except for its substrate specificity [52]. The other types are quinohaemoproteins, having an in-built electron acceptor in the form of haem *c*.

The quinohaemoprotein alcohol dehydrogenase from acetic acid bacteria is membrane-bound and contains three types of subunit, but no subunit equivalent to the small *β*-subunit of methanol dehydrogenase [51,53]. The primary sequence of the catalytic subunit shows an N-terminal region (600 residues) with an additional C-terminal extension containing a haem-binding site [54,55]. In the N-terminal region there is 31% identity with the sequence of methanol dehydrogenase, and it was possible to model the structure using the co-ordinates of methanol dehydrogenase [9]. Although the model predicts considerable differences in the external loops, particularly those involved in the formation of the shallow funnel leading to the active site, the active-site region was highly conserved, including the tryptophan and the disulphide ring on opposite sides of the plane of the PQQ, and also most of the equatorial co-ordinations to the PQQ (Figures 2–4). Especially important with respect to the mechanism is the conservation of the active-site base (Asp-303 in methanol dehydrogenase) and all the co-ordinations to the calcium ion. This suggests that the mechanism of this alcohol dehydrogenase is essentially similar to that of methanol dehydrogenase. Comparison of the protein sequence of the soluble quinohaemoprotein ethanol dehydrogenase from *Comomonas testosteroni* with that of methanol dehydrogenase leads to a similar conclusion for that enzyme [56].

There are two completely different types of PQQ-containing glucose dehydrogenases in bacteria [13,51,53]. Little is known about the structure or mechanism of the soluble (periplasmic) enzyme, which will not be considered further. The membrane-bound glucose dehydrogenase catalyses the oxidation of the pyranose form of D-glucose and other monosaccharides to the lactone. The reaction occurs in the periplasm, and the electron acceptor is ubiquinone in the membrane [57,58]. The enzyme is an intrinsic monomeric membrane protein (molecular mass ~ 87 kDa) for which a bivalent cation is probably necessary for activity. The N-terminal region (residues 1–154) forms a membrane anchor with five transmembrane segments, and this region is likely to contain the ubiquinone-binding site [59], which is not very similar to that in NADH dehydrogenase in mitochondria [60]. The remaining periplasmic region (residues 155–796) shows 26% sequence identity with the *α*-subunit of methanol dehydrogenase, and it has been possible to model its structure using the co-ordinates of methanol dehydrogenase [10]. In the model structure, the novel disulphide ring is replaced by a histidine residue which maintains the position of PQQ in the active site, consistent with the previous demonstration that a histidine residue is essential for binding of PQQ [61]. There are fewer equatorial interactions between the protein and PQQ, perhaps explaining why it is possible to effect the reversible dissociation of PQQ from the glucose dehydrogenase, but not from methanol dehydrogenase [51,62]. The ligation of calcium is similar, suggesting that the calcium plays a similar role in the two enzymes, i.e. that of a Lewis acid through co-ordination to the C-5 carbonyl oxygen, thus providing the electrophilic C-5 of PQQ. The proposed active-site base is conserved, suggesting that the reaction is initiated by abstraction of a proton from the anomeric hydroxy group of the pyranose ring. This would be followed by attack by the resulting oxyanion to form a hemiketal intermediate, or attack by a hydride from the glucose oxyanion leading directly to formation of the lactone and the quinol form

of PQQ. That the mechanism might be different, however, is indicated by the fact that Mg^{2+} can replace Ca^{2+} [62,63], which is not possible in methanol dehydrogenase [20]. A previous suggestion [61] that the reaction with glucose is initiated by proton abstraction by a cysteine residue is unlikely to be correct, because there is no cysteine residue within the appropriate region of the active site and all cysteines are involved in disulphide bond formation.

The oxidative half-reaction is likely to be completely different from that in the dehydrogenases for methanol and ethanol. In those enzymes there must be two single-electron transfers to haem *c*, whereas in glucose dehydrogenase two hydrogen atoms must be transferred to the acceptor ubiquinone. It is not necessary, therefore, for a stable semiquinone to be formed, and indeed no semiquinone has ever been observed. The active-site funnel is not hydrophobic, and there is no suggestion from the model structure or from the primary sequence that there is a hydrophobic region of the protein that could interact with the membrane except for the N-terminal transmembrane segments.

TTQ-CONTAINING AMINE DEHYDROGENASES

These enzymes catalyse the oxidative deamination of primary amines to the aldehyde, ammonia and water in the presence of an electron acceptor, which is usually a specific blue copper protein. The dye phenazine methosulphate or ethosulphate (A) is usually used as a convenient artificial electron acceptor in the reaction:

$$RCH_2NH_2 + H_2O + A \rightarrow RCHO + AH_2 + NH_3$$

The first of these enzymes to be described was the methylamine dehydrogenase responsible for methylamine oxidation in the same species of methylotrophic bacteria in which the PQQ-containing methanol dehydrogenase was first discovered; it was first described, and its prosthetic group (TTQ) characterized, by Eady and Large around 1970 [64,65]. It is a periplasmic enzyme and usually uses a blue copper protein called amicyanin as electron acceptor, the reduction of which leads to release of protons into the periplasm, thus contributing to the protonmotive force and hence to ATP synthesis [17,50,66]. Methylamine dehydrogenase has been characterized in detail, and shown to be similar in a number of different methylotrophic bacteria [64,65,67–71]. Davidson has published an outstanding review of the mechanistic and historical aspects of this enzyme [72].

Prosthetic group of methylamine dehydrogenase

When methylamine dehydrogenase was first described, its absorption spectra in the presence and absence of substrate, its sensitivity to carbonyl reagents and its probable Schiff-base formation suggested that a pyridoxal derivative might be acting in a novel fashion as the prosthetic group [65]. This group was subsequently claimed to be a covalently attached form of PQQ, but this possibility was excluded by MS of a derivative of the isolated prosthetic-group peptide [70], and by analysis of X-ray data (at 2.25 Å) obtained for the whole enzyme which suggested a precursor form of PQQ (pro-PQQ) [73]. The structure of the prosthetic group (TTQ; Figure 1) was finally solved by McIntire and his colleagues by an extensive rigorous analysis of the semicarbazide derivative by ^{13}C-NMR and MS [7,74]. Their proposed structure was consistent with the sequence of the gene for the light subunit of methylamine dehydrogenase, indicating that the prosthetic group arose by post-translational modification

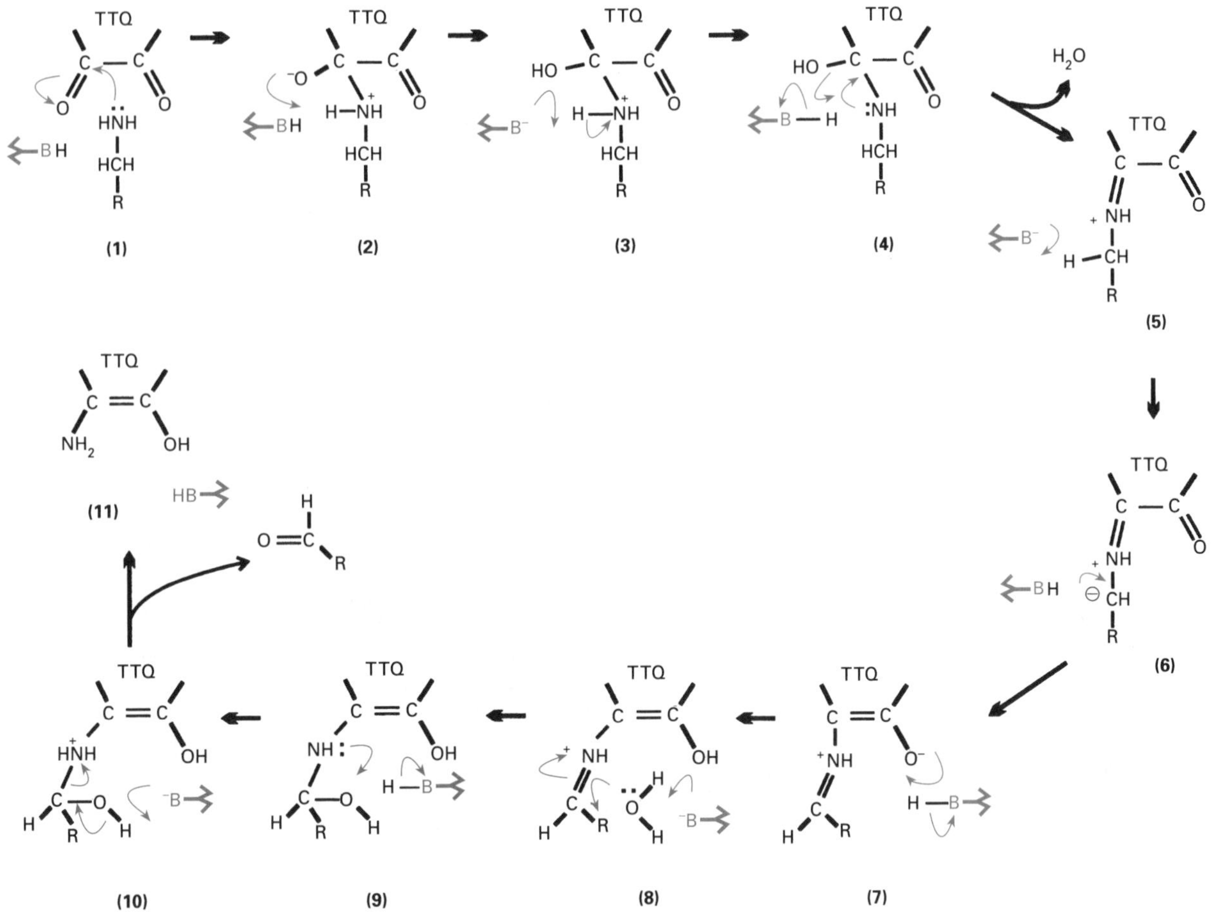

Figure 9   Reductive half-reaction catalysed by amine dehydrogenase

The initial nucleophilic attack by amine substrate gives the carbinolamine (**3**), which loses water to give the iminoquinone Schiff base (**5**). A key feature of this mechanism is the subsequent abstraction by the active-site base (−B⁻) of a proton from the α-carbon atom of the amine substrate to give the carbanion intermediate (**6**). This leads to reduction of the TTQ ring system with production of the product Schiff base (**7**), which is hydrolysed to the aldehyde product and the aminoquinol form of the enzyme. This Figure shows a base involved in a total of five separate steps in the overall sequence; this may be the same base, but a second base (or acid) may be involved.

of two tryptophans [75]. The structure proposed by McIntire was subsequently shown to fit the X-ray data for methylamine dehydrogenase [76], and resonance Raman spectroscopy has confirmed the same structure in a number of other methylamine dehydrogenases [24,25,77,78].

Reductive half-reaction of methylamine dehydrogenase

Steady-state kinetic studies have demonstrated a Ping-Pong mechanism in which the aldehyde product is released before reaction with phenazine methosulphate (PMS), which oxidizes the enzyme with the concomitant release of ammonia. This evidence, together with measurements of absorption spectra, led to the following postulated reaction sequence [65]:

The proposal that the prosthetic group is an orthoquinone similar to PQQ led to a rationalization of the kinetic and spectral data, leading to the suggestion that the reaction proceeds by initial Schiff-base formation and subsequent production of an aminoquinol [79]. These intermediates have now been confirmed in a wide range of studies involving stopped-flow kinetics and modification with inhibitors and substrates, which have led to the mechanistic proposals described in Figure 8 (see reviews by Davidson and co-workers [72,80] for a full description and analysis of this work, together with methods for determination of the intermediates). The deuterium kinetic isotope effect of 3, measured in steady-state kinetic studies, implies a mechanism involving rate-limiting abstraction of a methyl proton [81]. Subsequent stopped-flow kinetics confirmed this suggestion and demonstrated the presence of two kinetically significant intermediates, a relatively fast transition due to reduction of TTQ by substrate and a slower transition due to release of the aldehyde product [82]. In these experiments, and in similar experiments using the aromatic amine dehydrogenase, an exceptionally large primary deuterium isotope effect was observed [83–85]. The origin of this is unknown, but it is similar to that observed in the proton abstraction step in the TPQ-containing amine oxidase, which was attributed to the phenomenon of quantum-mechanistic tunnelling [86].

Benzylamines are not full substrates of methylamine dehydrogenase, but are competitive inhibitors able to reduce TTQ; analysis of Hammett plots for these reactions led to the proposal

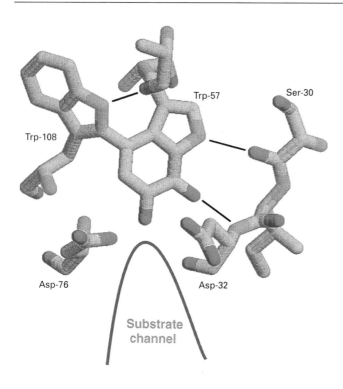

Figure 10   TTQ in the active site of methylamine dehydrogenase from *Paracoccus denitrificans*

The hydrogen bonds to the TTQ are shown together with Asp-76, which is likely to be the active-site base. The full structure of the enzyme is described in [76,93,101].

that a key intermediate in the reductive half-reaction is the carbanion (intermediate **6** in Figure 9) [87]. The involvement of a carbanion in the mechanism is supported by the results of a detailed investigation of a second TTQ-containing enzyme, an aromatic amine dehydrogenase [84,85,88]. It is proposed by Davidson [87] that the reaction is initiated by a nucleophilic attack by the substrate amine nitrogen on the quinone carbon to give the carbinolamine intermediate (**3**); further evidence for this intermediate comes from resonance Raman spectroscopy [24,25,77]. The carbinolamine loses water to form the imino-quinone (**5**), which undergoes nucleophilic attack by an active-site base to give the carbanionic intermediate (**6**) concomitant with the reduction of TTQ. The X-ray structure of the enzyme indicates that the base is probably Asp-76 (Figure 10), and also shows that the C-6 carbonyl is the reactive carbonyl likely to

undergo nucleophilic attack by the substrate [89]. This is close to a cation-binding site, and resonance Raman spectroscopy has indicated that the ammonium group of the substrate binds in this location adjacent to the C-6 carbonyl [90]. Hydrolysis of the carbanionic intermediate leads to release of the aldehyde product and formation of the aminoquinol (**11**). It is this final step that is very slow with benzylamines compared with the normal aliphatic substrates. My proposed reaction sequence in Figure 9 (adapted from Davidson's mechanism) involves an active-site base, not only in the key proton abstraction step (**5–6**) but also in the formation of the carbinolamine (**3**), the removal and subsequent addition of water, and the formation of the aldehyde product. Similarly, the oxidative half-reaction (see below) is also likely to involve an acid- or base-facilitated reaction during removal of the ammonia from the aminoquinol. It is likely, given the limitation of space in the active site, that a single residue is involved in most of these reactions.

Investigations using a number of inhibitors all support the mechanism described in Figure 9. Ammonia is a reversible competitive inhibitor, reacting to produce either an imino-quinone adduct [69,91] or a carbinolamine adduct [77,87]. The product depends on which enzyme is used, probably reflecting slight differences in the active site near the TTQ [77]. Various nucleophilic amines, such as phenylhydrazine, semicarbazide and hydroxylamine, are irreversible inhibitors, reacting with TTQ to form covalent adducts [65,69,92]. A likely mechanism for these reactions is the same as that with normal substrates, but there is no α-carbon for subsequent proton abstraction and so the reaction stops at the semiquinone stage (**5**) [92]. Cyclo-propylamine acts as an irreversible mechanism-based inhibitor, leading to covalent modification of TTQ and cross-linking of the α- and β-subunits [92]. This cross-linking indicates that the α-subunit must also play a role in providing the environment of the active site [92], and this is supported by the X-ray structure, which shows the TTQ located in a narrow channel at an interface between the subunits [93].

Oxidative half-reaction of methylamine dehydrogenase

The physiological electron acceptor (usually amicyanin) is a single-electron acceptor, leading to a semiquinone intermediate during oxidation of the aminoquinol that was produced in the reductive half-reaction (Figure 11). That a stable semiquinone intermediate is produced during the reaction has been known for some time [69,94], and this has been confirmed by resonance Raman spectroscopy, which also provided evidence for the formation of the hydroxy and carbinolamine adducts [77,78]. It was initially suggested that the first step in the oxidation of the

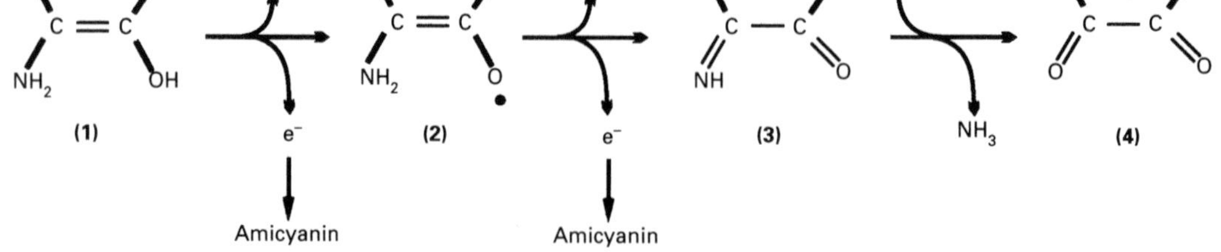

Figure 11   Oxidative half-reaction of methylamine dehydrogenase

The unpaired electron in (**2**) is delocalized in the indole ring of the TTQ. The hydrolysis of (**3**) to (**4**) is likely to involve an acid or base at the active site, which may be the same base(s) involved in the reductive half-reaction (Figure 9).

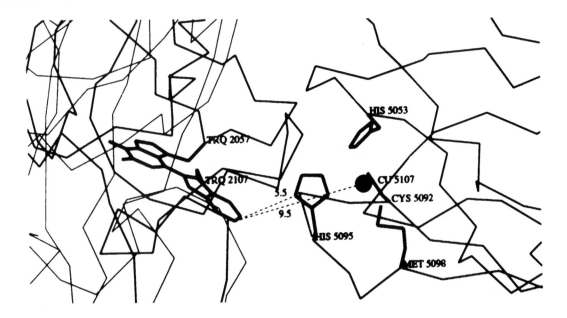

Figure 12   Interaction between methylamine dehydrogenase (left) and amicyanin (right)

The quinone part of the TTQ is on Trp-57 (TRQ 2057 in this Figure). The second tryptophan ring, Trp-108 (TRQ 2157 in this Figure) is nearer the electron acceptor (Cu 5107). The shortest distances between the TTQ and the copper atom and to the histidine (His-5095) are given in Å. Reproduced with permission from [138].

aminoquinol would involve the release of ammonia with production of the semiquinone, with the second electron transfer to amicyanin producing the oxidized quinone [72,95]. However, subsequent studies have shown that the nitrogen atom from the amine substrate is attached to the semiquinone, indicating that release of ammonia takes place after, or in concert with, the oxidation of the aminosemiquinone (Figure 11) [96,97,97a]. Spectroscopy of the aminosemiquinone revealed that it is deprotonated, and that the conformation of the tryptophans in relation to each other (dihedral angle of about 40°) is the same as in the crystal structure (Figure 10) [97]. These studies also showed that the delocalization of electrons into the ring system of the semiquinone is consistent with the route for departure of electrons from the indole ring that was proposed on the basis of the X-ray structure [98] and kinetic studies [99].

Three-dimensional structure of methylamine dehydrogenase and its complex with amicyanin

Methylamine dehydrogenase has an $\alpha_2 \beta_2$ structure; the heavy α-subunits are approx. 40 kDa and the light β-subunits are approx. 13 kDa. Its three-dimensional structure has been determined using the enzymes from *Thiobacillus versutus* and *Paracoccus denitrificans* [76,93,100,101]. There is little interaction between the α-subunits of methylamine dehydrogenase, but extensive interactions exist between the α-and β-subunits, with the active site lying in a hydrophobic channel between them. The two indole rings of the TTQ are not co-planar, but lie at a dihedral angle of about 42° [101] (Figures 10 and 12). The larger α-subunit has a propeller-fold structure (seven-bladed), similar to that of the larger α-subunit of methanol dehydrogenase (eight-bladed) [22,30]. The fact that both of these quinoproteins have a propeller-fold structure is intriguing, because the related structures have no related catalytic function; in methanol dehydrogenase the prosthetic group (PQQ) is in the large subunit that has this propeller fold, but in methylamine dehydrogenase the TTQ is in the small subunit.

The complex formed between the dehydrogenase and amicyanin, solved by Mathews and his colleagues [98], has two molecules of amicyanin per $\alpha_2 \beta_2$ tetramer of dehydrogenase; the amicyanin molecules have no contact with each other but are in contact with both subunits of the dehydrogenase. The greatest area of contact is between the amicyanin and the smaller β-subunit which contains the TTQ. Inhibition by high ionic strength had previously emphasized the importance of electrostatic interactions between the two proteins [102], but in the complex hydrophobic interactions predominate, perhaps reflecting the fact that the complex was crystallized in high-ionic-strength solutions. It is possible, however, that docking of the proteins is by electrostatic interactions and that formation of the specific electron transfer complex involves hydrophobic interactions, as has been suggested for the interaction of methanol dehydrogenase with its specific cytochrome *c* [48,49].

In the complex, the tryptophan that does not contain the *o*-quinone (Trp-108) lies near the surface, only 9.5 Å away from the amicyanin copper atom. His-95, one of the four copper ligands, is on the surface of the protein and about half-way between the copper and the TTQ, with the shortest distance between them being approx. 5.5 Å (Figure 12). It was therefore suggested that this histidine might mediate electron transfer between the redox centres, thus forming an electron transfer triad [98]. On the basis of stopped-flow kinetics, an electron transfer pathway has been suggested that involves a 3.6 Å jump through space from TTQ to the carbonyl of Pro-94 and passage through six covalent bonds to the copper (a total distance of 14 Å) [99]. An alternative pathway was suggested for electron transfer in a ternary complex with cytochrome *c*, which is moderately more efficient but depends critically on the presence of an intracomplex water molecule [45]. Measurements by polarized absorption spectroscopy of single crystals of these binary and ternary complexes have demonstrated that electron transfer between the redox centres does occur within the complexes [103], but some of the relatively low rates perhaps indicate that the orientation of the proteins is not necessarily optimal.

AMINE OXIDASES CONTAINING COPPER AND TPQ

Copper-containing amine oxidases catalyse the oxidative de-amination of primary amines, the actual substrate (mono-, di- or poly-amine) depending on the source of the enzyme:

$$RCH_2NH_2 + O_2 + H_2O \rightarrow RCHO + H_2O_2 + NH_3$$

Although many amine oxidases with a great diversity of functions have been described, they are likely to be similar in overall structure and mechanism. They are important, for example, in processes as different as bacterial growth on amines, secondary metabolism in plants and the oxidation of histamine and neurotransmitters in animals. This diversity and its physiological and pharmacological significance has been described, together with extensive discussion of the history and mechanism of these enzymes, in comprehensive and authoritative reviews by Mondovi [104], McIntire and Hartmann [105], Knowles and Dooley [106] and Hartmann and Dooley [107].

Prosthetic group and copper site in amine oxidase

For nearly 60 years it has been known that amine oxidases have an organic, covalently bound, prosthetic group, and from about 1950 this has been assumed to be an unusual form of pyridoxal phosphate, as was also proposed for amine dehydrogenase. It was subsequently claimed to be a covalent form of PQQ, as also proposed for the amine dehydrogenases; this was a premature suggestion that proved to be incorrect, but which had the merit of directing attention to the problem of its true nature. This subsequently led Klinman and her colleagues [8,108,109] to identify the prosthetic group of bovine serum amine oxidase as the quinone of 2,4,5-trihydroxyphenylalanine (topa or 6-hydroxydopa); this is now known as TPQ (Figure 1). It should be noted that resonance Raman spectroscopy has demonstrated that there is substantial electron delocalization between the C-2 and C-4 oxygens and that only the C-5 oxygen has predominantly carbonyl character [110].

The review by Klinman and Mu [26] should be consulted for the authorized version of their exciting discovery of the structure of TPQ. The initial problem of obtaining sufficient active-site chromophoric peptide after proteolysis was overcome by first reacting the enzyme with [^{14}C]phenylhydrazine and then releasing the phenylhydrazone derivative by proteolysis using thermolysin in 2 M urea; the structure was then determined by MS and NMR of the pure peptide and synthetic compounds [8,111]. Resonance Raman spectra of the derivatized protein and the labelled peptide isolated from it were consistent with this structure [8,24,25,112], and this technique has been used to demonstrate TPQ in all the copper-containing amine oxidases examined [24,25,106]. A simple spectrophotometric assay has now been developed for identification of TPQ in amine oxidases which is based on the pH-dependence of the absorption spectra of TPQ p-nitrophenylhydrazone in the intact enzyme [108,113].

An extensive range of experiments and physical techniques has contributed to our knowledge of the structure of the copper site in amine oxidases, which has been summarized in an excellent recent review by Knowles and Dooley [106]. These techniques include EXAFS [3,114,115], ESR [115], electron nuclear double resonance [116], electron spin echo envelope modulation [117] and nuclear magnetic resonance dispersion [118,119] spectroscopy. From these investigations it was concluded that three histidine residues form ligands, together with two water ligands in a distorted square pyramidal configuration. Knowles and Dooley [106] pointed out that there was also the possibility that TPQ forms a weak axial co-ordination to the copper through one

of its carbonyl oxygen atoms. During the turnover of the enzyme it is probable that an intermediate oxidation state is formed, with TPQ in the semiquinone form bonded to reduced Cu(I); there is, however, very little information available on the co-ordination of the copper centre in this oxidation state.

Reductive phase of the catalytic cycle of amine oxidase

Steady-state and stopped-flow kinetic studies have demonstrated an aminotransferase type of Ping-Pong mechanism [120]. The reaction with substrate leads to reduced enzyme and production of aldehyde; the enzyme is then oxidized by molecular oxygen with release of ammonia:

(i) $Enz\text{-}CHO + RCH_2NH_2 \rightarrow Enz\text{-}CH_2NH_2 + RCHO$

(ii) $Enz\text{-}CH_2NH_2 + O_2 + H_2O \rightarrow Enz\text{-}CHO + NH_3 + H_2O_2$

Perhaps not surprisingly, many features of the mechanism are similar to those described for the TTQ-containing amine dehydrogenases. Figure 13 gives a scheme for the reductive phase of the catalytic cycle, as deduced for the oxidation of benzylamines catalysed by bovine plasma amine oxidase. The reaction is initiated by a nucleophilic attack by the amine substrate, giving a carbinolamine intermediate (3) from which water is removed, leading to covalent attachment of substrate to TPQ at the C-5 carbonyl by way of a Schiff base (5), which can be trapped by reduction with NaCNBH₃ [121,122]. It was suggested that the low pK_a of the C-4 hydroxy group (~ 3) provides electrostatic stabilization of the initially formed Schiff base. Base-catalysed proton abstraction then gives a carbanion transition intermediate (6) [123–126]. The large primary isotope effect observed during benzylamine oxidation has led to the conclusion that significant quantum tunnelling occurs during this process. This is the first example of quantum mechanical tunnelling in an enzyme-catalysed proton abstraction mechanism [86,127], and it has subsequently been proposed for the TTQ-containing amine dehydrogenases. After formation of the carbanion (6) and transfer of electrons into the ring of TPQ, the pK_a of the C-4 hydroxy group undergoes a large increase, leading to proton transfer from the active-site base to the oxyanion at C-4 (6–7). This eliminates the electrostatic stabilization of the product Schiff-base imine complex (7), leading to its rapid hydrolysis and formation of the product aldehyde (7–11). It has been suggested that the presence of the C-4 hydroxy group increases the redox potential by 300 mV compared with a simple quinone such as dopa-quinone, thus increasing the driving force for the reaction [26,128]. The transient intermediate product Schiff base (7) was the most difficult intermediate to demonstrate, because of the rapid rate of its hydrolysis. It was eventually confirmed as its quinonoid tautomer by using benzylamines with electron-releasing substituents in rapid-scanning stopped-flow studies [129]. After release of the aldehyde product, the nitrogen from the substrate remains on the TPQ, which is thus now in the aminoquinol form [125,126,130]. Excellent comprehensive descriptions of the methods used for the synthesis and detection of the intermediates shown in Figure 13 have recently been published by Mure and Klinman [131] and Hartmann and Dooley [107].

In mechanisms of this sort, attention is usually directed towards the key catalytic step, which is proton abstraction from the α-carbon of the substrate. In addition to this step, I suggest that there are five other reactions that also require the involvement of an active-site base (or acid) (Figure 13). These include the transfer of the proton from the nitrogen to the oxygen (2–3), the removal of water from the resulting carbinolamine (3–5) and the

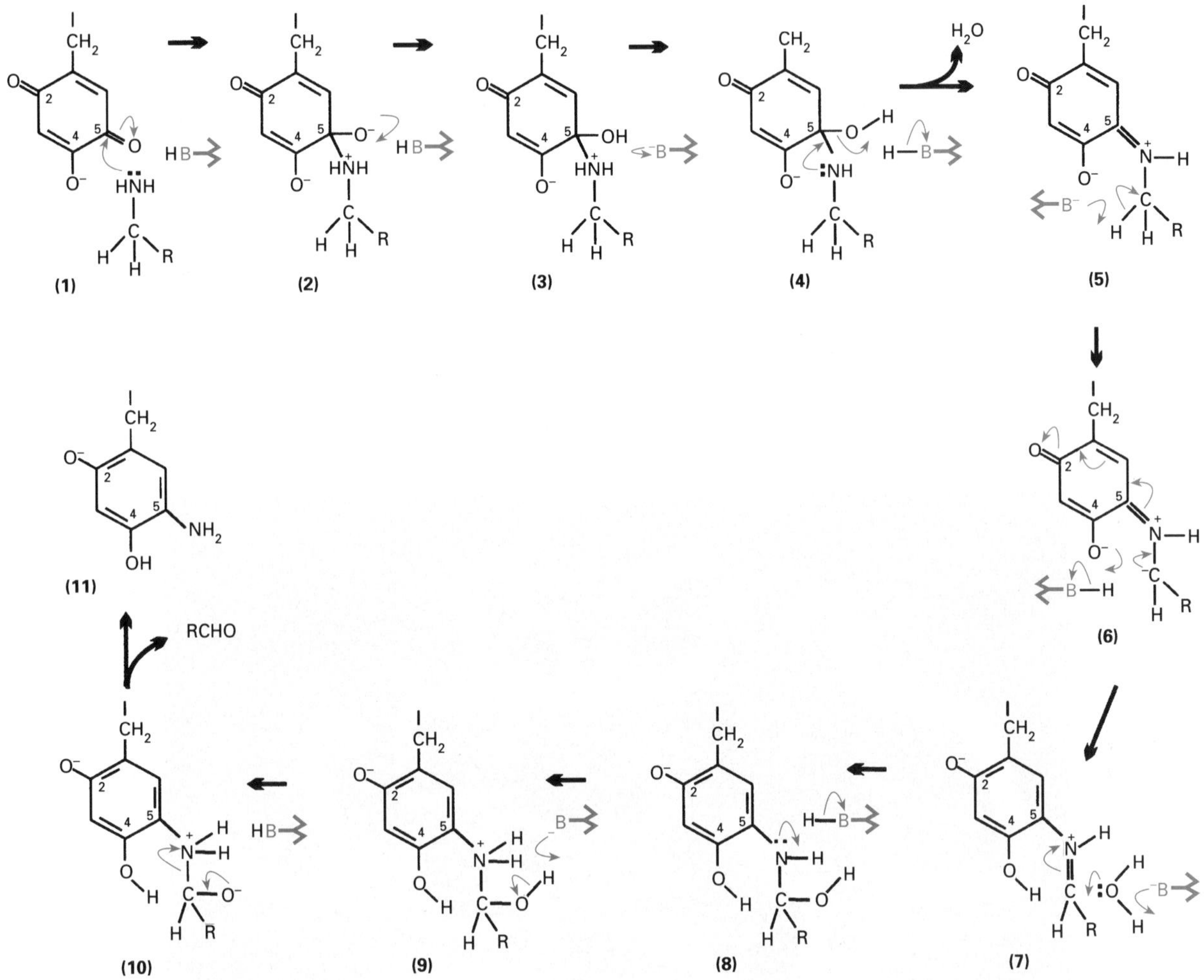

Figure 13   Reductive phase of the catalytic cycle of amine oxidase

Many of the intermediates are analogous to those in the mechanism for methylamine dehydrogenase shown in Figure 9. The key reaction is the proton abstraction by the active-site base (-B) from the α-carbon of the amine substrate. An active-site base (or possibly acid) is also likely to be involved in initial formation of the carbinolamine (**3**), removal of water from it to give the substrate Schiff base (**5**), addition of water to the product Schiff base (**7**) and production of aldehyde and the aminoquinol form of the prosthetic group (**11**).

later hydrolysis of the product Schiff base (**7–11**). It is possible that the same base (Asp-383) may be involved in all of these reactions. The same active-site base may also be involved in the removal of water from the aminoquinol product (**11**) during the oxidative half-reaction shown in Figure 14 (see below).

Oxidative phase of the catalytic cycle of amine oxidase

Oxidation of the aminoquinol form of TPQ, back to the quinone, involves the two-electron reduction of oxygen to hydrogen peroxide and the release of ammonia [120,132] (Figure 14). The inhibitory effects of azide or cyanide and their effects on the modification of the EPR spectrum provided evidence for the involvement of copper in this process [132,133], and established the occurrence of a free-radical intermediate form of enzyme containing the semiquinone form of TPQ and the reduced form of copper, Cu(I) [118]. Direct measurements of this intermediate led to the proposal that the reduced TPQ is oxidized by way of copper in two single-electron steps, the Cu(I)–semiquinone intermediate reacting with oxygen to form a transient superoxide intermediate [118,134–136]. It was initially proposed that the

semiquinone form of TPQ is the iminosemiquinone [106,134], and this has been confirmed by resonance Raman spectroscopy [136]. Using the enzyme from *Escherichia coli*, and methylamine as a slow substrate, it was possible to show that the substrate amine group remains bound to the TPQ in the semiquinone state and that addition of cyanide leads to formation of a Cu(I)–cyanide complex with the nitrogen-containing semiquinone. Little is known about the actual site of oxygen reduction, and almost no information is available about the proton transfer steps that must be involved in oxygen reduction and peroxide release [106].

Active site of amine oxidase as seen in the X-ray structures of the *E. coli* and pea seedling enzymes

The first X-ray structure of a copper- and TPQ-containing amine oxidase has recently been obtained (at 2.0 Å) for the enzyme from *E. coli* by the group of Knowles and Phillips at Leeds [137]. Each subunit of the mushroom-shaped dimer consists of an extensive β-sandwich domain which contains the active site and provides the dimer interface, and three smaller α/β domains

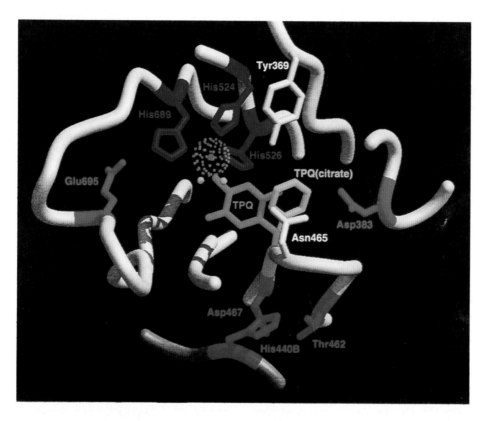

Figure 14   Oxidative phase of the catalytic cycle of amine oxidase

The copper is shown to be directly co-ordinated to the oxygen at position 2 of TPQ. If this distance is too great, a water molecule may come between the TPQ oxygen and the copper atom. A similar mechanism could be drawn, if structural observations suggest it, in which the copper is co-ordinated to the oxygen at position 4 of TPQ. The hydrolysis of intermediate (**4**) is likely to involve an active-site acid or base as discussed in the legends to Figures 9 and 13. This may or may not be the same base(s) involved in the reactions in Figure 13.

Figure 15   Active site of *E. coli* amine oxidase

Shown are the co-ordination of the copper, and the two conformations of the TPQ in the active sites of the two crystal types. In the enzyme crystallized from citrate, the TPQ (green) is co-ordinated to the copper. In the enzyme crystallized from ammonium salts, the precise location of the TPQ ring could not be completely determined and its general location is indicated by a yellow phenyl ring, close to the putative catalytic base (Asp-383, in red). In this form the TPQ is not a copper ligand and the copper co-ordination is completed by two water molecules, shown in yellow. Reproduced with permission from [137].

around the outside of the molecule. Two structures were solved, an active lower-resolution form and an inactive higher-resolution structure; these were essentially similar but with important differences in the active site. All of the identified residues of the active site are buried in the interior of the molecule between the sheets of a β-sandwich. The TPQ ring orientation is not clear in the active form, but it is orientated away from the copper and towards the putative active-site base Asp-383; it cannot be seen, however, which is the nearer out of the C-2 carbonyl and the C-5 carbonyl of TPQ, which is the carbonyl most likely to be involved in catalysis (Figure 15). In the inactive form, Asp-383 is

within about 5.7 Å of O-2 of TPQ. Comparison with the pea seedling enzyme structure (see below) indicates that the possibility of rotation of the TPQ in the active site may be an important aspect of the catalytic function of these enzymes.

In the *E. coli* enzyme the copper atom lies about 12 Å below the molecular surface, co-ordinated to three histidine ligands (His-524, His-526 and His-689), their positions being very similar in the two crystal types. In the active enzyme, co-ordination of the copper is a distorted square pyramid, and is close to that proposed from spectroscopic studies described above. The apical ligand is water, and the three histidines plus a second water molecule form the distorted base of the pyramid (Figure 15); TPQ was not within co-ordination distance in this active form of the enzyme. In the inactive form there was no water in the co-ordination sphere of the copper. The co-ordination geometry was described as distorted tetrahedral, with His-524 in an 'axial-like' position and the other three ligands 'equatorial-like'; these ligands are His-526 and His-689 plus TPQ co-ordinated via the oxygen at the 4-position. There was no electron density to suggest the presence of a second axial-like ligand, a small empty pocket being present at this location.

The first structure of a eukaryotic (pea seedling) amine oxidase has now been solved at 2.2 Å [139] and shown to be very similar to that of the *E. coli* enzyme. The TPQ aromatic group is located approx. 6 Å away from the Cu atom, its location being different from that in both the active and inactive forms of the *E. coli* enzyme. It has been suggested [139] that the combination of evidence from the two structures indicates that the TPQ side chain is sufficiently flexible to permit the aromatic group to rotate about the Cβ–Cγ bond, and to move between bonding and non-bonding positions with respect to the Cu atom. Some conformational flexibility is also required at the surface of the molecule to allow the substrates access to the active site, which is inaccessible to solvent, as expected for an enzyme that uses radical chemistry. The potential for rotation of the TPQ is also likely to be relevant to the mechanism for production of the modified tyrosine which involves the copper atom at the active site [139].

LYSYL OXIDASE: A SPECIAL AMINE OXIDASE CONTAINING COPPER AND LTQ

Lysyl oxidase plays a major role in the development and repair of connective tissues. It catalyses the oxidation of lysyl residues in collagen and elastin to form the peptidyl α-aminoadipic-δ-semialdehyde:

Protein-(CH$_2$)$_3$-CH$_2$-NH$_2$ + O$_2$ → Protein-(CH$_2$)$_3$-CHO

+ H$_2$O$_2$ + NH$_3$

Aldehyde residues produced by this reaction spontaneously condense with nearby aldehydes or ε-amino groups, giving inter- or intra-chain covalent cross-linkages, thus leading to the insoluble fibres in the extracellular matrix [140]. Earlier work indicated that the enzyme contains copper and that catalysis involves a carbonyl group; most subsequent work on its mechanism has indicated that it is very similar to other copper-containing amine oxidases [140]. The discovery of TTQ as the prosthetic group of amine oxidases drew attention to the possibility that lysyl oxidase also contained TTQ, although those working in the field were sensibly wary of warmly embracing this idea (see [141]). The recent publication of the structure of the prosthetic group of lysyl oxidase by Klinman, Dooley, Kagan and their colleagues [8a] has justified this caution. The structure is derived from the cross-linking of the ε-amino group of a peptidyl lysine with the modified side chain of a tyrosyl residue, and has been designated LTQ (Figure 1). It appears to be the only example of a mammalian cofactor formed from the cross-linking of two amino acid side chains. It has been pointed out [8a] that alternative TPQ analogues might occur in which other amino acids substitute for the lysyl residue in LTQ. There is some evidence that this does, indeed, occur in the prosthetic group of one of the amine oxidases of the fungus *Aspergillus niger*; in the isolated prosthetic group of this enzyme the modified tyrosine residue is esterified to the γ-carboxyl of glutamate [142].

CONCLUDING REMARKS

This review summarizes the results of a great range of different approaches that have been applied to the study of the mechanisms of the four types of quinoprotein, and the conclusions from this work have been satisfyingly supported by the three-dimensional structures elucidated by X-ray crystallography. As expected, the quinone structures in the prosthetic groups play important roles in all of the mechanisms. A second common feature is the presence of a catalytic base (aspartate) in the active sites which initiates the reactions by abstracting a proton from the substrate; it is also apparent that this same base is likely to be involved in multiple reactions in all of the mechanisms. Another common feature of these enzymes is that the first part of the reaction produces a reduced prosthetic group, this part of the mechanism being fairly well understood; this is followed by an oxidative phase involving electron transfer reactions which are far less well understood. In the case of the dehydrogenases, electrons must pass one at a time from the prosthetic group to redox centres in a second recipient protein (or protein domain), whereas in the amine oxidases electrons must be transferred to molecular oxygen by way of a redox-active copper ion in the enzyme itself. It is this area that is most likely to occupy future investigators, together with attempts to understand further the details of the enzyme mechanisms; this must include determining the location of the substrates and activators, and understanding their interactions in the active sites. This review has been written at the start of a new phase in the study of these fascinating proteins in which a wide range of physical techniques will continue to be focused upon the problem, coupled with extensive studies using the techniques of site-directed mutagenesis and X-ray crystallography.

I thank the BBSRC (U.K.), The Wellcome Trust and The Royal Society for supporting my work that is included in this review. I also thank M. Surya Prakash and the members of the Biochemistry Department, S.V. University, Tirupati, India, for their support during preparation of the article.

REFERENCES

1 Hauge, J. G. (1964) J. Biol. Chem. **239**, 3630–3639
2 Anthony, C. and Zatman, L. J. (1967) Biochem. J. **104**, 960–969
3 Duine, J. A., Frank, J. and Westerling, J. (1978) Biochim. Biophys. Acta **524**, 277–287
4 Duine, J. A. and Frank, J. (1980) Biochem. J. **187**, 221–226
5 Salisbury, S. A., Forrest, H. S., Cruse, W. B. T. and Kennard, O. (1979) Nature (London) **280**, 843–844
6 Duine, J. A., Frank, J. and Verwiel, P. E. J. (1980) Eur. J. Biochem. **108**, 187–192
7 McIntire, W. S., Wemmer, D. E., Chistoserdov, A. and Lidstrom, M. E. (1991) Science **252**, 817–824
8 Janes, S. M., Mu, D., Wemmer, D., Smith, A. J., Kaur, S., Maltby, D., Burlingame, A. L. and Klinman, J. P. (1990) Science **248**, 981–987
8a Wang, S. X., Mure, M., Medzihradszky, K. F., Burlingame, A. L., Brown, D. E., Dooley, D. M., Smith, A. J., Kagan, H. M. and Klinman, J. P. (1996) Science **273**, 1078–1084
9 Cozier, G. E., Giles, I. G. and Anthony, C. (1995) Biochem. J. **307**, 375–379
10 Cozier, G. E. and Anthony, C. (1995) Biochem. J. **312**, 679–685
11 Anthony, C. (1986) Adv. Microb. Physiol. **27**, 113–210
12 Duine, J. A., Frank, J. and Jongejan, J. A. (1987) Adv. Enzymol. **59**, 169–212
13 Duine, J. A. (1991) Eur. J. Biochem. **200**, 271–284
14 Anthony, C. (1993) in Principles and Applications of Quinoproteins (Davidson, V. L., ed.), pp. 17–45, Marcel Dekker, New York
15 Anthony, C., Ghosh, M. and Blake, C. C. F. (1994) Biochem. J. **304**, 665–674
16 Anthony, C. (1992) Biochim. Biophys. Acta **1099**, 1–15
17 Anthony, C. (1993) in Principles and Applications of Quinoproteins (Davidson, V. L., ed.), pp. 223–244, Marcel Dekker, New York
18 Goodwin, P. M. and Anthony, C. (1995) Microbiology **141**, 1051–1064
19 Nunn, D. N., Day, D. J. and Anthony, C. (1989) Biochem. J. **260**, 857–862
20 Richardson, I. W. and Anthony, C. (1992) Biochem. J. **287**, 709–715

21 White, S., Boyd, G., Mathews, F. S., Xia, Z. X., Dai, W. W., Zhang, Y. F. and Davidson, V. L. (1993) Biochemistry **32**, 12955–12958

22 Ghosh, M., Anthony, C., Harlos, K., Goodwin, M. G. and Blake, C. C. F. (1995) Structure **3**, 177–187

23 Duine, J. A., Frank, J. and Verwiel, P. E. J. (1981) Eur. J. Biochem. **118**, 395–399

24 Dooley, D. M. and Brown, D. E. (1993) in Principles and Applications of Quinoproteins (Davidson, V. L., ed.), pp. 132–140, Marcel Dekker, New York

25 Dooley, D. M. and Brown, D. E. (1995) Methods Enzymol. **258**, 132–140

26 Klinman, J. P. and Mu, D. (1994) Annu. Rev. Biochem. **63**, 299–344

27 Xia, Z. X., Dai, W. W., Xiong, J. P., Hao, Z. P., Davidson, V. L., White, S. and Mathews, F. S. (1992) J. Biol. Chem. **267**, 22289–22297

28 Blake, C. C. F., Ghosh, M., Harlos, K., Avezoux, A. and Anthony, C. (1994) Nature Struct. Biol. **1**, 102–105

29 Anthony, C., Ghosh, M. and Blake, C. C. F. (1994) Biochem. J. **304**, 665–674

30 Xia, Z., Dai, W., Zhang, Y., White, S. A., Boyd, G. D. and Mathews, F. S. (1996) J. Mol. Biol. **259**, 480–501

31 Avezoux, A., Goodwin, M. G. and Anthony, C. (1995) Biochem. J. **307**, 735–741

32 Frank, J., Dijkstra, M., Balny, C., Verwiel, P. E. J. and Duine, J. A. (1989) in PQQ and Quinoproteins (Jongejan, J. A. and Duine, J. A., eds.), pp. 13–22, Kluwer Academic Publishers, Dordrecht

33 Duine, J. A. and Frank, J. (1980) Biochem. J. **187**, 213–219

34 Frank, J., Dijkstra, M., Duine, A. J. and Balny, C. (1988) Eur. J. Biochem. **174**, 331–338

35 Dijkstra, M., Frank, J. and Duine, J. A. (1989) Biochem. J. **257**, 87–94

36 Dekker, R. H., Duine, J. A., Frank, J., Verwiel, P. E. J. and Westerling, J. (1982) Eur. J. Biochem. **125**, 69–73

37 Ohshiro, Y. and Itoh, S. (1993) in Principles and Applications of Quinoproteins (Davidson, V. L., ed.), pp. 309–329, Marcel Dekker, New York

38 Frank, J., van Krimpen, S. H., Verwiel, P. E. J., Jongejan, J. A., Mulder, A. C. and Duine, J. A. (1989) Eur. J. Biochem. **184**, 187–195

39 Goodwin, M. G. and Anthony, C. (1996) Biochem. J. **318**, 673–679

40 Harris, T. K. and Davidson, V. L. (1994) Biochem. J. **300**, 175–182

41 Harris, T. K. and Davidson, V. L. (1993) Biochemistry **32**, 14145–14150

42 Forrest, H. S., Salisbury, S. A. and Kilty, C. G. (1980) Biochem. Biophys. Res. Commun. **97**, 248–251

43 Itoh, S., Ogino, M., Fukui, Y., Murao, H., Komatsu, M., Ohshiro, Y., Inoue, T., Kai, Y. and Kasai, N. (1993) J. Am. Chem. Soc. **115**, 9960–9967

44 Nunn, D. N. and Anthony, C. (1988) Biochem. J. **256**, 673–676

45 Chen, L. Y., Durley, R. C. E., Mathews, F. S. and Davidson, V. L. (1994) Science **264**, 86–90

46 Chan, H. T. C. and Anthony, C. (1991) Biochem. J. **280**, 139–146

47 Cox, J. M., Day, D. J. and Anthony, C. (1992) Biochim. Biophys. Acta **1119**, 97–106

48 Dales, S. L. and Anthony, C. (1995) Biochem. J. **312**, 261–265

49 Harris, T. K., Davidson, V. L., Chen, L. Y., Mathews, F. S. and Xia, Z. X. (1994) Biochemistry **33**, 12600–12608

50 Anthony, C. (1988) in Bacterial Energy Transduction (Anthony, C., ed.), pp. 293–316, Academic Press, London

51 Matsushita, K. and Adachi, O. (1993) in Principles and Applications of Quinoproteins (Davidson, V. L., ed.), pp. 47–63, Marcel Dekker, New York

52 Schrover, J. M. J., Frank, J., van Wielink, J. E. and Duine, J. A. (1993) Biochem. J. **290**, 123–127

53 Matsushita, K., Toyama, H. and Adachi, O. (1994) Adv. Microb. Physiol. **36**, 247–301

54 Inoue, T., Sunagawa, M., Mori, A., Imai, C., Fukuda, M., Takagi, M. and Yano, K. (1989) J. Bacteriol. **171**, 3115–3122

55 Inoue, T., Sunagawa, M., Mori, A., Imai, C., Fukuda, M., Takagi, M. and Yano, K. (1990) J. Ferment. Bioeng. **70**, 58–60

56 Stoorvogel, J., Kraayveld, D. E., van Sluis, C. A., Jongejan, J. A., Devries, S. and Duine, J. A. (1996) Eur. J. Biochem. **235**, 690–698

57 Matsushita, K., Shinagawa, E., Inoue, T., Adachi, O. and Ameyama, M. (1986) FEMS Microbiol. Lett. **37**, 141–144

58 Matsushita, K., Shinagawa, E., Adachi, O. and Ameyama, M. (1989) J. Biochem. (Tokyo) **105**, 633–637

59 Yamada, M., Sumi, K., Matsushita, K., Adachi, O. and Yamada, Y. (1993) J. Biol. Chem. **268**, 12812–12817

60 Sakamoto, K., Miyoshi, H., Matsushita, K., Nakagawa, M., Ikeda, J., Ohshima, M., Adachi, O., Akagi, T. and Iwamura, H. (1996) Eur. J. Biochem. **237**, 128–135

61 Imanaga, Y. (1989) in PQQ and Quinoproteins (Jongejan, J. A. and Duine, J. A., eds.), pp. 87–96, Kluwer Academic Publishers, Dordrecht

62 Ameyama, M., Nonobe, M., Hayashi, M., Shinagawa, E., Matsushita, K. and Adachi, O. (1985) Agric. Biol. Chem. **49**, 1227–1231

63 Buurman, E. T., Tenvoorde, G. J. and Demattos, M. J. T. (1994) Microbiology **140**, 2451–2458

64 Eady, R. R. and Large, P. J. (1968) Biochem. J. **106**, 245–255

65 Eady, R. R. and Large, P. J. (1971) Biochem. J. **123**, 757–771

66 Tobari, J. and Harada, Y. (1981) Biochem. Biophys. Res. Commun. **101**, 502–508

67 Shirai, S., Matsumoto, T. and Tobari, J. (1978) J. Biochem. (Tokyo) **83**, 1599–1607

68 Matsumoto, T., Hiraoka, B. Y. and Tobari, J. (1978) Biochim. Biophys. Acta **522**, 303–310

69 Kenney, W. C. and McIntire, W. S. (1983) Biochemistry **22**, 3858–3868

70 McIntire, W. S. and Stults, J. T. (1986) Biochem. Biophys. Res. Commun. **141**, 562–568

71 Husain, M. and Davidson, V. L. (1987) J. Bacteriol. **169**, 1712–1717

72 Davidson, V. L. (1993) in Principles and Applications of Quinoproteins (Davidson, V. L., ed.), pp. 73–95, Marcel Dekker, New York

73 Vellieux, F. M. D. and Hol, W. G. J. (1989) FEBS Lett. **255**, 460–464

74 McIntire, W. S. (1995) Methods Enzymol. **258**, 149–164

75 Chistoserdov, A. Y., Tsygankov, Y. D. and Lidstrom, M. E. (1990) Biochem. Biophys. Res. Commun. **172**, 211–216

76 Chen, L. Y., Mathews, F. S., Davidson, V. L., Huizinga, E. G., Vellieux, F. M. D., Duine, J. A. and Hol, W. G. J. (1991) FEBS Lett. **287**, 163–166

77 Backes, G., Davidson, V. L., Huitema, F., Duine, J. A. and Sanders-Loehr, J. (1991) Biochemistry **30**, 9201–9210

78 McIntire, W. S., Bates, J. L., Brown, D. E. and Dooley, D. M. (1991) Biochemistry **30**, 125–133

79 Anthony, C. (1982) The Biochemistry of Methylotrophs, p. 207, Academic Press, London

80 Davidson, V. L., Brooks, H. B., Graichen, M. E., Jones, L. H. and Hyun, Y. L. (1995) Methods Enzymol. **258**, 176–190

81 Davidson, V. L. (1989) Biochem. J. **261**, 107–111

82 McWhirter, R. B. and Klapper, M. H. (1989) in PQQ and Quinoproteins (Jongejan, J. A. and Duine, J. A., eds.), pp. 259–268, Kluwer Academic Publishers, Dordrecht

83 Brooks, H. B., Jones, L. H. and Davidson, V. L. (1993) Biochemistry **32**, 2725–2729

84 Hyun, Y. L. and Davidson, V. L. (1995) Biochim. Biophys. Acta Protein Struct. Mol. Enzymol. **1251**, 198–200

85 Hyun, Y. L. and Davidson, V. L. (1995) Biochemistry **34**, 816–823

86 Grant, K. L. and Klinman, J. P. (1989) Biochemistry **28**, 6597–6605

87 Davidson, V. L., Jones, L. H. and Graichen, M. E. (1992) Biochemistry **31**, 3385–3390

88 Govindaraj, S., Eisenstein, E., Jones, L. H., Sanders-Loehr, J., Chistoserdov, A. Y., Davidson, V. L. and Edwards, S. L. (1994) J. Bacteriol. **176**, 2922–2929

89 Huizinga, E. G., Vanzanten, B. A. M., Duine, J. A., Jongejan, J. A., Huitema, F., Wilson, K. S. and Hol, W. G. J. (1992) Biochemistry **31**, 9789–9795

90 Moënne-Loccoz, P., Nakamura, N., Itoh, S., Fukuzumi, S., Gorren, A. C. F., Duine, J. A. and Sanders-Loehr, J. (1996) Biochemistry **35**, 4713–4720

91 McIntire, W. S. (1987) J. Biol. Chem. **262**, 11012–11019

92 Davidson, V. L. and Jones, L. H. (1992) Biochim. Biophys. Acta **1121**, 104–110

93 Mathews, F. S. (1995) Methods Enzymol. **258**, 191–216

94 de Beer, R., Duine, J. A., Frank, J. and Large, P. J. (1980) Biochim. Biophys. Acta **622**, 370–374

95 van Wielink, J. E., Frank, J. and Duine, J. A. (1989) in PQQ and Quinoproteins (Jongejan, J. A. and Duine, J. A., eds.), pp. 269–278, Kluwer Academic Publishers, Dordrecht

96 Warncke, K., Brooks, H. B., Babcock, G. T., Davidson, V. L. and McCracken, J. (1993) J. Am. Chem. Soc. **115**, 6464–6465

97 Warncke, K., Brooks, H. B., Lee, H. I., McCracken, J., Davidson, V. L. and Babcock, G. T. (1995) J. Am. Chem. Soc. **117**, 10063–10075

97a Bishop, G. R., Brooks, H. B. and Davidson, V. L. (1996) Biochemistry **35**, 8948–8954

98 Chen, L. Y., Durley, R., Poliks, B. J., Hamada, K., Chen, Z. W., Mathews, F. S., Davidson, V. L., Satow, Y., Huizinga, E., Vellieux, F. M. D. and Hol, W. G. J. (1992) Biochemistry **31**, 4959–4964

99 Brooks, H. B. and Davidson, V. L. (1994) Biochemistry **33**, 5696–5701

100 Vellieux, F. M. D., Huitema, F., Groendijk, H., Kalk, K. H., Frank, J., Jongejan, J. A., Duine, J. A., Petratos, K., Drenth, J. and Hol, W. G. J. (1989) EMBO J. **8**, 2171–2178

101 Chen, L. Y., Mathews, F. S., Davidson, V. L., Huizinga, E. G., Vellieux, F. M. D. and Hol, W. G. J. (1992) Protein Struct. Funct. Genet. **14**, 288–299

102 Gray, K. A., Davidson, V. L. and Knaff, D. B. (1988) J. Biol. Chem. **263**, 13987–13990

103 Merli, A., Brodersen, D. E., Morini, B., Chen, Z. W., Durley, R. C. E., Mathews, F. S., Davidson, V. L. and Rossi, G. L. (1996) J. Biol. Chem. **271**, 9177–9180

104 Mondovi, B. (1986) Structure and Function of Amine Oxidases, CRC Press, Baton Rouge

105 McIntire, W. S. and Hartmann, C. (1993) in Principles and Applications of Quinoproteins (Davidson, V. L., ed.), pp. 97–171, Marcel Dekker, New York

106 Knowles, P. F. and Dooley, D. M. (1994) in Metal Ions in Biological Systems (Seigel, H., ed.), pp. 361–403, Marcel Dekker, New York

107 Hartmann, C. and Dooley, D. M. (1995) Methods Enzymol. **258**, 69–90

108 Janes, S. M., Palcic, M. M., Scaman, C. H., Smith, A. J., Brown, D. E., Dooley, D. M., Mure, M. and Klinman, J. P. (1992) Biochemistry **31**, 12147–12154

109 Janes, S. M. and Klinman, J. P. (1995) Methods Enzymol. **258**, 20–34

110 Nakamura, N., Matsuzaki, R., Choi, Y. H., Tanizawa, K. and Sanders-Loehr, J. (1996) J. Biol. Chem. **271**, 4718–4724

111 Adams, G. W., Mayer, P., Medzihradszky, K. F. and Burlingame, A. L. (1995) Methods Enzymol. **258**, 90–114

112 Brown, D. E., McGuirl, M. A., Dooley, D. M., Janes, S. M., Mu, D. and Klinman, J. P. (1991) J. Biol. Chem. **266**, 4049–4051

113 Palcic, M. M. and Janes, S. M. (1995) Methods Enzymol. **258**, 34–38

114 Knowles, P. F., Strange, R. W., Blackburn, N. J. and Hasnain, S. S. (1989) J. Am. Chem. Soc. **111**, 102–107

115 Barker, R., Boden, N., Cayley, G., Charlton, S. C., Henson, R., Holmes, M. C., Kelly, I. D. and Knowles, P. F. (1979) Biochem. J. **177**, 289–301

116 Barker, G. J., Knowles, P. F., Pandeya, K. B. and Rayner, J. B. (1986) Biochem. J. **237**, 609–612

117 McCracken, J., Peisach, J. and Dooley, D. M. (1987) J. Am. Chem. Soc. **109**, 4064–4072

118 Dooley, D. M., McGuirl, M. A., Brown, D. E., Turowski, P. N., McIntire, W. S. and Knowles, P. F. (1991) Nature (London) **349**, 262–264

119 Dooley, D. M., McGuirl, M. A., Cote, C. E., Knowles, P. F., Singh, I., Spiller, M., Brown, R. D. and Koenig, S. H. (1991) J. Am. Chem. Soc. **113**, 754–761

120 Ruis, F. X., Knowles, P. F. and Pettersson, G. (1984) Biochem. J. **220**, 767–772

121 Hartmann, C. and Klinman, J. P. (1987) J. Biol. Chem. **262**, 962–965

122 Bushnell, G. W., Louie, G. V. and Brayer, G. D. (1990) J. Mol. Biol. **214**, 585–595

123 Lovenberg, W. and Beaven, M. A. (1971) Biochim. Biophys. Acta **251**, 452–455

124 Neumann, R., Hevey, R. C. and Abeles, R. H. (1975) J. Biol. Chem. **250**, 6362–6367

125 Farnum, M., Palcic, M. and Klinman, J. P. (1986) Biochemistry **25**, 1898–1904

126 Hartmann, C. and Klinman, J. P. (1991) Biochemistry **30**, 4605–4611

127 Palcic, M. and Klinman, J. P. (1983) Biochemistry **22**, 5957–5966

128 Mure, M. and Klinman, J. P. (1993) J. Am. Chem. Soc. **115**, 7117–7127

129 Hartmann, C., Brzovic, P. and Klinman, J. P. (1993) Biochemistry **32**, 2234–2241

130 Janes, S. M. and Klinman, J. P. (1991) Biochemistry **30**, 4599–4605

131 Mure, M. and Klinman, J. P. (1995) Methods Enzymol. **258**, 39–52

132 Lindstrom, A., Olsson, B. and Pettersson, G. (1974) Eur. J. Biochem. **48**, 237–243

133 Dooley, D. M., McGuirl, M. A., Peisach, J. and McCracken, J. (1987) FEBS Lett. **214**, 274–278

134 McCracken, J., Peisach, J., Cote, C. E., McGuirl, M. A. and Dooley, D. M. (1992) J. Am. Chem. Soc. **114**, 3715–3722

135 Turowski, P. N., McGuirl, M. A. and Dooley, D. M. (1993) J. Biol. Chem. **268**, 17680–17682

136 Moënne-Loccoz, P., Nakamura, N., Steinebach, V., Duine, J. A., Mure, M., Klinman, J. P. and Sanders-Loehr, J. (1995) Biochemistry **34**, 7020–7026

137 Parsons, M. R., Convery, M. A., Wilmot, C. M., Yadav, K. D. S., Blakeley, V., Corner, A. S., Phillips, S. E. V., McPherson, M. J. and Knowles, P. F. (1995) Structure **3**, 1171–1184

138 Mathews, F. S. and Hol, W. G. J. (1993) in Principles and Applications of Quinoproteins (Davidson, V. L., ed.), pp. 245–273, Marcel Dekker, New York

139 Kumar, V., Dooley, D. M., Freeman, H. C., Guss, J. M., Harvey, I., McGuirl, M. A., Wilce, M. C. J. and Zubak, V. M. (1996) Structure **4**, 945–955

140 Kagan, H. M. and Trackman, P. C. (1991) Am. J. Respir. Cell Mol. Biol. **5**, 206–210

141 Dooley, D. M. and Brown, D. E. (1993) in Principles and Applications of Quinoproteins (Davidson, V. L., ed.), pp. 132–140, Marcel Dekker, New York

142 Frebort, I., Pec, P., Luhova, L., Toyama, H., Matsushita, K., Hirota, S., Kitagawa, T., Ueno, T., Asano, Y., Kato, Y. and Adachi, O. (1996) Biochim. Biophys Acta Protein Struct. Mol. Enzymol. **1295**, 59–72

Biochem. J. (1996) **315**, 1–9 (Printed in Great Britain)

REVIEW ARTICLE
The inter-α-inhibitor family: from structure to regulation

Jean-Philippe SALIER*, Philippe ROUET, Gilda RAGUENEZ and Maryvonne DAVEAU
INSERM Unit-78 and Institut Fédératif de Recherches Multidisciplinaires sur les Peptides, B.P. 73, 76233 Boisguillaume, France

Inter-α-inhibitor (IαI) and related molecules, collectively referred to as the IαI family, are a group of plasma protease inhibitors. They display attractive features such as precursor polypeptides that give rise to mature chains with quite distinct fates and functions, and inter-chain glycosaminoglycan bonds within the various molecules. The discovery of an ever growing number of such molecules has raised pertinent questions about their patho-physiological functions. The knowledge of this family has long been structure-oriented, whereas the structure/function and structure/regulation relationships of the family members and their genes have been largely ignored. These relationships are now being elucidated in events such as gene transcription, precursor processing, changes in plasma protein levels in health and disease and binding capacities that involve hyaluronan as well as other plasma proteins as ligands. This review presents some recent progress made in these fields that paves the way for an understanding of the functions of IαI family members *in vivo*. Finally, given the wealth of heterogeneous, complicated and sometimes contradictory nomenclatures and acronyms currently in use for this family, a new, uniform, nomenclature is proposed for IαI family genes, precursor polypeptides and assembled proteins.

INTRODUCTION

In higher organisms, a delicate balance between proteases and their natural inhibitors participates in the control of activation and catabolism of many intra- and extra-cellular proteins. In mammals, the bloodstream is a major carrier for many glyco-proteins that act as protease inhibitors. Inter-α-inhibitor (IαI) and related molecules, collectively referred to as the IαI family, are a fascinating group of such plasma protease inhibitors. As detailed below, they display unique and interesting features such as precursor polypeptides that give rise to mature chains with quite distinct fates and functions, and inter-chain glycosamino-glycan bonds within the various molecules. The discovery of an increasing number of such molecules has raised numerous questions about their pathophysiological functions. Until re-cently, the knowledge of the IαI family was mostly structure-oriented. Indeed, IαI was for many years described as a single entity, and a major breakthrough in the late 1980s was the emergence of a set of related molecules. Six years have now elapsed since a review on IαI summarized structural data about these molecules and outlined the emergence of what is now an acknowledged family of proteins [1]. However, detailed in-formation about the relevant gene structures and mRNA/protein sequences was not available at that time. Moreover, the regu-latory steps involved in the synthesis and processing of the chains and molecules had not been studied. Hence the structure/function and structure/regulation relationships of the members of this family, and their genes, were largely ignored. Recently these relationships have been elucidated in events such as gene tran-scription, precursor processing, changes in plasma protein levels in health and disease, and protein–ligand binding. Therefore this review presents some recent progress made in these fields, as the current research tendency to shift from structure to regulation paves the way for a final understanding of the functions of IαI family members *in vivo*.

As detailed below, an ever-growing number of proteins in the IαI family are being described by various research teams. Concomitantly, an ever-growing difficulty in finding further informative names that decipher the chain assemblies within these proteins has been encountered. At the moment, research into the IαI family is plagued by a wealth of heterogeneous, complicated and sometimes contradictory nomenclatures and acronyms. Clearly, the need for an improved nomenclature comes of age. Accordingly, we wish to take advantage of the present review to propose a uniform nomenclature. For the purpose of clarity some former protein names will be used at the beginning of this review, and the proposed new nomen-clature will be introduced later on.

GENES, PRECURSORS AND GLYCOPROTEINS: A STRUCTURAL OVERVIEW

At least four genes, designated *H1*, *H2*, *H3* and *H4*, are involved in the synthesis of IαI family members (Figure 1). The *H1*, *H2* and *H3* genes and cDNAs have been known for some time [1], whereas the H4 cDNA was cloned quite recently [2]. The cDNA sequences of these genes are clearly homologous [1–3], which implies the occurrence of several gene duplication events from a shared ancestral *H* gene [3,4]. As detailed in Figure 1, each of the *H* loci has been mapped to two homologous areas on human and mouse chromosomes, which indicates that all of the *H* gene duplication events took place prior to the human–rodent di-vergence [5]. On the basis of protein sequence homologies, the *H2* gene diverged from an *H* ancestor approx. 300 million years ago, while the *H1* and *H3* genes separated later, some 230 million years ago [3]. Indeed, *H1* and *H3* are still located in close physical proximity [4–6]: the human *H1* gene comprises 22 exons, the last one being only 2.7 kb away from exon 1 of *H3* [7,8]. The approximate timing of the separation of the *H4* gene is still

---

Abbreviations used: IαI, inter-α-inhibitor; α1m, α1-microglobulin; AMBP, α1m/bikunin precursor; PαI, pre-α-inhibitor; IαLI, inter-α-like-inhibitor; PGP, protein–glycosaminoglycan–protein; HNF, hepatocyte nuclear factor; vWA, von Willebrand factor type-A; HA, hyaluronic acid.
* To whom correspondence should be addressed.

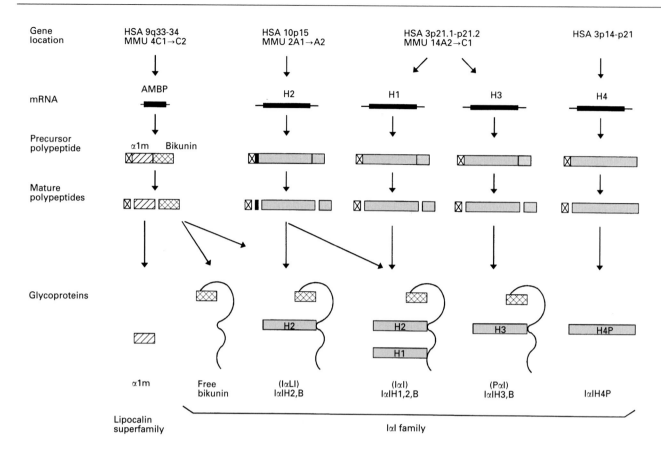

Figure 1   A current view of the IαI family, from genes to glycoproteins

This current view of the biosynthetic pathways in the IαI family is modified from a former proposal [1]. The homologous areas of the human and mouse chromosomes to which a given gene maps are indicated by the species (HSA, *Homo sapiens*; MMU, *Mus musculus*) and the chromosome band numbers (*H4* gene mapping: L. Jean, M. G. Mattei, F. Banine and J. P. Salier, unpublished work). Symbols read as follows. mRNAs: thin line, non-coding sequence; black box, open reading frame. Precursor polypeptides: crossed box, signal peptide; hatched box, α1m; double-hatched box, bikunin; black box, N-terminal cleavage product of H precursor; pink box, mature H chain; grey box, C-terminal cleavage product of H precursor. Glycoproteins: wavy line, covalent PGP cross-link. Only the major members of the IαI family are depicted; for the sake of clarity, their former, usual, acronyms as well as the newly proposed nomenclature (see Table 2) are given.

Table 1   Sizes of the H1 to H4 precursor polypeptides and mature chains

HxP, precursor polypeptide; Hx, mature chain found in some IαI-related plasma proteins. The sizes are expressed as the number of amino acid residues, as inferred from the cDNAs and protein sequences published and/or discussed in [2,3,9,10,11,63,64]. The signal peptides as defined in [2,3] are not included in the precursor sizes indicated. Note that the order H2P, H1P, H3P, H4P, as shown, reflects the most closely related polypeptide pairs: H1P and H3P; H3P and H4P [2–4]. A possible trimming of a 36-amino-acid propeptide at the N-terminus of H2P [3,9] is included in the calculations. The H4P polypeptide can undergo some cleavage steps after its secretion into plasma, eventually resulting in a major 633-amino-acid peptide (see text and [2]). n.a., not applicable; n.d., not determined.

	No. of amino acids							
	H2P	H2	H1P	H1	H3P	H3	H4P	H4
Human	928	648	882	643	868	630	902	n.a.
Mouse	928	648	881	642	868	628	n.d.	n.d.

unknown. The mRNAs of these genes are 2.9–3.3 kb in size [1,2,6].

The cDNA-deduced H1, H2 and H3 polypeptide sequences, but not the H4 sequence, are considered to represent precursors [2,3,9,10] (Table 1). Indeed, these precursors undergo extensive trimming of their C-terminal ends, with removal of 240–280 amino acid residues (Figure 1; Table 1), as judged from the sequences of the mature H polypeptides found in the purified plasma proteins. Furthermore, the H2 precursor displays a stretch of 36 amino acids located between its signal peptide and the N-terminus of the mature H2 chain, which suggests a further trimming of a propeptide at the H2 N-terminus [3,9,11] (Figure 1; Table 1). Based upon the N-terminal amino acid sequences of the purified H chains [11] and alignment of the cDNA-deduced sequences, the presence of a similar propeptide in the H1, H3 and H4 precursors is unlikely, as discussed elsewhere [3].

In addition to the four *H* genes, another gene, designated the α1-microglobulin (α1m)/bikunin precursor (*AMBP*) gene, is involved in the synthesis of most IαI family members (Figure 1; Table 2). The AMBP mRNA encodes a precursor for two polypeptides, α1m and bikunin. The *AMBP* gene is comprised of six α1m-encoding exons and four bikunin-encoding exons separated by a large (7 kb) intron, suggesting that the modern *AMBP* gene resulted from the fusion of two ancestral *α1m* and *bikunin* genes [12]. As homologous AMBP mRNAs have been cloned in many vertebrates, including human, pig, rodents and fishes [13–20], this gene fusion event must have taken place prior to the teleost–tetrapod divergence that occurred some 400 million years ago. The *AMBP*-gene-encoded polypeptide undergoes a cleavage step (see below) that releases α1m and bikunin; these

Table 2   Acknowledged or suggested nomenclatures for the genes, polypeptides and proteins of the Iαl family

1. Genes

Accepted symbol[1]			Former symbols		
Human	Mouse	Encoded polypeptide	Human	Mouse	References
AMBP	*Ambp*	α1m/bikunin precursor	*ITIL*	*Intin-4*	[6,65]
ITIH1	*Itih-1*	Heavy chain 1 precursor	*ITIH1*	*Intin-1*	[6,66a]
ITIH2	*Itih-2*	Heavy chain 2 precursor	*ITIH2*	*Intin-2*	[6,66a]
ITIH3	*Itih-3*	Heavy chain 3 precursor	*ITIH3*	*Intin-3*	[6,66a]
ITIH4	*Itih-4*	Heavy chain 4	*ITIHL1*	None	[66b]

2. Precursor polypeptides

Proposed acronym	Polypeptide	Former designations	References
AMBP	α1m/bikunin precursor	α1m/bikunin precursor	[5]
H1P	Heavy chain 1 precursor	Heavy chain 1; H1	[4]
H2P	Heavy chain 2 precursor	Heavy chain 2; H2	[4]
H3P	Heavy chain 3 precursor	Heavy chain 3; H3	[4]
H4P	Heavy chain 4[2]		

3. Mature polypeptides

Proposed name/acronym	Polypeptide	Former designations	References
α1m	α1m[3]	HC protein ('heterogenous in charge'); α1m	[21]
Bikunin	Double-headed, Kunitz-type, protease inhibitor	HI-30; ITI light (L) chain; ITI subunit 1 or α; bikunin	[4,27,67]
H1	Mature heavy chain 1	Heavy chain 1; H1; HC1; ITI subunit 2 or β	[4,27,67]
H2	Mature heavy chain 2	Heavy chain 2; H2; HC2; ITI subunit 3 or γ	[4,27,67]
H3	Mature heavy chain 3	Heavy chain 3; H3; HC3; Pαl subunit δ	[4,27,67]
H4P	Heavy chain 4[2]	PK-120[4]; IHRP[4]	[3,10]

4. Proteins[5]

Proposed name/acronym	Combination of mature polypeptides	Former designations	References
Bikunin	Free bikunin	HI-30; light chain (L); bikunin; trypstatin; urinary trypsin inhibitor	[1,67,68]
Iαl[6] or IαlH1,2,B	H1 + H2 + bikunin	Inter-α-trypsin inhibitor (Iαl; ITI; IαTI); extracellular matrix stabilizing factor	[1,49,67]
Pαl[6] or IαlH3,B	H3 + bikunin	Pre-α-trypsin inhibitor (Pαl); HC3/bikunin	[27,30]
IαlH2,B	H2 + bikunin	IαLI; Iαl-related HC2/bikunin; extracellular matrix stabilizing factor; p126	[27,31,46]
IαlH2,3,B[7]	H2 + H3 + bikunin	p236	[46]
IαlH1,B	H1 + bikunin	Iαl-related HC1/bikunin	[27]
IαlH1,2	H1 + H2 (human)	SHAP[4]	[47a]
IαlH2,3	H2 + H3 (bovine)[8]	SHAP[4]	[47a]
IαlH4P	Free H4P[2]	PK-120[4]; IHRP[4]	[3,10]

[1] These gene symbols are currently recommended by the international nomenclature committees for human and mouse genes.

[2] Apart from its signal peptide cleavage, no maturation step of the H4P polypeptide prior to its secretion into plasma has been observed. The single-chain H4P protein can undergo kallikrein-induced cleavages in plasma [2].

[3] Originates from AMBP but is not to be included in the Iαl family.

[4] PK-120, plasma kallikrein-sensitive glycoprotein 120; IHRP, inter-α-trypsin inhibitor family heavy-chain-related protein; SHAP, serum-derived, hyaluronan-associated, protein.

[5] Including some proteins that might be natural or *in vitro* by-products of other Iαl family members.

[6] Given the current lack of information about which enzyme(s) are physiological target(s) for this molecule, the former indication 'trypsin' is currently omitted [67].

[7] Described in bovine only.

[8] Mistakenly designated H1 + H2 in [47a].

two proteins are structurally and functionally unrelated. $\alpha 1m$ belongs to a large set of hydrophobic ligand carriers called lipocalins [21–23], and is found in a free state as well as complexed with plasma proteins such as IgAs in the human or $\alpha 1$-inhibitor-3 and fibronectin in the rat [24]. Bikunin is a Kunitz-type protease inhibitor found in most IαI family molecules (Figure 1). Kunitz-type inhibitors, whose chief member is the well known pancreatic trypsin inhibitor, usually have a low relative molecular mass, a basic isoelectric point and one or several inhibitory domain(s) with a typical disulphide bond arrangement and a broad spectrum of activity towards serine proteases. Two such tandemly arranged, Kunitz-type, inhibitory domains are present in bikunin, the target enzymes of which include trypsin, chymotrypsin, cathepsin G, leucocyte elastase, acrosin and plasmin. However, these enzymes are more efficiently inhibited by other, quantitatively major, plasma inhibitors, which makes the genuine target(s) of bikunin, and hence the biological significance of IαI family members as protease inhibitors, still elusive (reviewed in [1]).

The IαI family of plasma proteins is made up of various structurally related molecules that were initially designated IαI, pre-α-inhibitor (PαI), inter-α-like-inhibitor (IαLI) etc., based on their behaviour in non-denaturing electrophoresis at pH values greater than 8. The major members of this family are depicted in Figure 1, and a full list of the molecules described so far is provided in the final section of Table 2. All evolutionarily related members in the family contain one or more H chains that identify them. For instance, IαI and PαI result from the processing and assembly of H1 + H2 + bikunin and H3 + bikunin respectively. Notably, most proteins in the IαI family contain the bikunin moiety and are protease inhibitors. Bikunin is also found in a free state in plasma, urine and hepatocyte cultures [1,25]. For practical reasons the free bikunin molecule is quoted as a member of the IαI family, although it is not evolutionarily related to the H chains.

The inter-chain link in the molecules is quite unusual. It is covalent and made up of a chondroitin 4-sulphate glycan bond that has been termed a protein–glycosaminoglycan–protein (PGP) cross-link [26]. The involvement of this PGP in the cross-linking of bikunin to mature H chains in IαI, PαI and IαLI is proven [26–28]. The anchor sites on the chains for this PGP bond will be described further below. The PGP cross-link is resistant to reduction but it can be cleaved by chondroitinase, hyaluronidase or mild treatment with NaOH [26,27,29–31]. The complete structure of this chondroitin 4-sulphate glycan bond has been recently reported to be ΔHexA-α1-3GalNAc(4-sulphate)β1-4GlcAβ1-3Gal(4-sulphate)β1-3Galβ1-4Xyl-ol, where ΔHexA is 4-deoxy-α-L-*threo*-hex-4-enepyranosyluronic acid and Xyl-ol is xylitol [32]. Elucidation of the number and sequence of other, more usual, N- or O-linked glycan moieties on the chains of IαI family members still awaits detailed analysis.

FROM STRUCTURE TO NOMENCLATURE REGULATIONS

In Table 2, we propose a uniform nomenclature for IαI-related genes, gene products and proteins. It has been elaborated with advice from others (see the Acknowledgements section) and with the following considerations in mind. (1) Electrophoretic migration as part of a protein name is of limited usefulness, and this parameter has not been evaluated for all IαI family members. Protein designations should instead be informative in terms of structure/assembly, and yet they should remain simple. (2) A new nomenclature should preferably take into account all IαI-related polypeptides and molecules whether or not they are physiologically relevant, as it may be useful to quote a non-

physiological by-product. (3) Ambiguous acronyms such as 'HC', which can currently stand for $\alpha 1m$ (synonym: HC protein) as well as a mature H chain, must be avoided; however, a polypeptide precursor and its mature by-products should be distinguished. (4) As exemplified by H4, not all proteins in the family contain bikunin, and therefore a nomenclature of 'bikunin proteins', as recently proposed [33], cannot encompass all IαI family members. (5) Addition of further genes, polypeptides and proteins in the future should not compromise the internal consistency of this new nomenclature.

The proposed nomenclature regulations rely in particular on structure. The proposed protein acronyms clearly identify the H chain(s) involved by their number(s) and indicate whether precursors or mature H chains are involved. Indeed, the H chain precursors are designated 'HxP' whereas the mature H chains are designated 'Hx' (a C-terminal by-product from an HxP could be quoted 'cHx' if required in the future). The letter H applies to all subsequently listed H numbers (i.e. x), and these H numbers are separated by commas (in case more than ten H chains are discovered in future investigations). In the protein acronyms the presence of bikunin is marked 'B' whenever appropriate. The IαI symbol is retained for the family name and is further included in all acronyms, which allows the reader to immediately identify any protein as an IαI family member. For instance IαLI, comprising H2 + bikunin, is now IαIH2,B. Only in two instances, namely the IαIH1,2,B and IαIH3,B proteins that are historical leaders in the family, do we propose to also use their original, complete, names, i.e. inter-α-inhibitor and pre-α-inhibitor respectively, when appropriate.

FROM GENE REGULATION TO STRUCTURE

Various physiological or pathological phenomena such as the tissue-restricted expression of some proteins, and their quantitative changes during development and organogenesis or an acute inflammatory response, allow one to observe concomitant changes in gene expression and to infer some structural hints about how the transcription of a given gene is controlled. These aspects have been recently investigated within the IαI family.

Gene expression in liver and brain

The pattern of expression of the AMBP and H mRNAs in various tissues has been studied in detail in primates, pigs and rodents [1,3,7,10,15,16,34], and the overall conclusion is clearcut. The *AMBP* and *H* genes are primarily transcribed in the liver, which is the source of all plasma proteins of the IαI family. Transcription of the *AMBP* gene is tightly regulated, as no other tissue has been found to express the AMBP mRNA, apart from its possible weak expression in pig stomach [15]. The pattern of expression of the *H* genes is also tightly regulated, but it involves two different organs, namely the liver and, to a lower extent, the brain (and cerebellum). The quantitative expression of the H1, H2 and H3 mRNAs is roughly similar in the liver [3,34,35], whereas an obvious H3 > H2 hierarchy is seen in the brain, where the H1 mRNA is undetectable [3,34]. The transcription of the *H4* gene in brain has yet not been studied. The implications of these observations in terms of gene structure are severalfold.

First, the liver-restricted expression of the *AMBP* and *H* genes strongly suggests the presence of similar regulatory elements in these genes. Addressing this issue requires a functional analysis of the 5′ flanking regions in all IαI family genes. A first, major, advance in this field has been reported recently [36,37]. Indeed, the tissue-specific pattern of expression of the *AMBP* gene is accounted for by a potent and liver-specific enhancer that drives

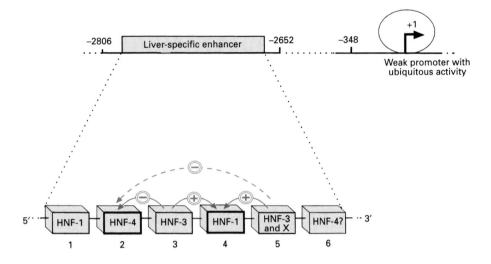

Figure 2   Regulation of *AMBP* gene transcription by a liver-specific enhancer

The 5′ flanking region of the *AMBP* gene contains an enhancer located 2.7 kb from the promoter. This promoter displays a ubiquitous activity and therefore does not account for the liver-specific transcription of the gene. The enhancer comprises six 'boxes', numbered 1–6, namely two HNF-1-, two HNF-3- and two HNF-4-binding sites, which together increase by about 50-fold the basal activity of the promoter when the gene is in an hepatic context. The various HNF sites act to a variable extent: the quantitative participation of a given box/factor unit in the overall enhancer activity is roughly proportional to the thickness of the line bounding the corresponding rectangle. Within the enhancer, a positive (+) or negative (−) influence of a box/factor unit on another one is proven (solid red arrow) or suspected (broken red arrow). X is an unidentified nuclear protein that binds box 5. The cognate factor of box 6 may be a member of the HNF-4 family but has not been completely identified. The black arrow indicates the start site of gene transcription. After [36,37].

a weak promoter of ubiquitous activity. The *AMBP* enhancer is made up of six functional binding sites for transcription factors, the so-called hepatocyte nuclear factors (HNFs) (Figure 2). Three major HNF proteins, designated HNF-1, -3 and -4, are currently known and a family of related proteins exists for each of them [38]. The HNF-1, -3 and -4 proteins are enriched in the liver and are known to be activators of numerous genes in this organ [38]. A tight cluster of six binding sites for three distinct HNF proteins, as seen in the *AMBP* enhancer (Figure 2), has not been reported for other genes. This enhancer is functional exclusively in a liver cell context, which accounts for the liver-restricted expression of the *AMBP* gene [36,37].

Secondly, the transcription of the *H3* and *H2* genes in the brain suggests that one or more homologous sites for the binding of brain-specific transcription factor(s) to the *H3* and *H2* regulatory regions are likely to mediate the activation of these genes in this tissue. Alternatively, one or more binding sites for a brain-specific factor might be located in both the *AMBP* and *H1* sequences and involved in a brain-specific 'turn-off' of these two genes. Given the shared ancestry of the *H* genes (see above), the first situation is more likely; nevertheless, it implies a late mutation in the *H1* gene with a concomitant loss of a binding site for an activator brain protein.

Gene expression and development

The developmental pattern of expression of the AMBP mRNA was first investigated in pig liver [15] but the results should be regarded with caution, as the changes in AMBP mRNA levels were normalized against the albumin mRNA level, which itself is particularly prone to development-associated changes. More recently, the developmental pattern of expression of the *AMBP*, *H1*, *H2* and *H3* genes has been investigated in the mouse [34,39]. In the mouse embryo, all four mRNAs are clearly seen in the liver in the late gestation period. mRNA levels peak 4–5 days after birth and eventually return to a significantly lower value in adulthood. The transient and co-ordinated increase in these

mRNAs seen after birth is transcriptionally regulated, which suggests that all four genes are perinatally triggered at the level of similar control elements by development-associated nuclear factors [34]. This conclusion is supported by the strongly impaired perinatal expression of these four genes in a mutant mouse strain that is deficient in transcription factors such as HNF-1 and HNF-4 [34]. Therefore our current view is that similar HNF-binding boxes in the *AMBP* and *H* genes might participate, whether directly or indirectly via interactions with other box/factor units, in a co-ordinated expression of the IαI family genes during development.

Gene expression and acute inflammation

The acute inflammatory condition is known to up- and down-regulate transcription of the so-called positive and negative acute-phase genes respectively in liver. The behaviour of IαI family members in inflammation has been debated for years and a prevailing but false view has long regarded free bikunin as a positive acute-phase protein. Indeed, many contradictory results were published when (i) molecules such as IαIH3,B were still unknown and likely to be confused with free bikunin due to their similar electrophoretic migration and a lack of H-chain-specific antisera, and (ii) cDNAs for the various H mRNAs remained to be cloned. Therefore only recently have we been able to undertake a detailed analysis of the behaviour of the whole set of IαI family mRNAs and proteins (apart from H4) in acute inflammation using appropriate probes [31,35]. Using human serum samples and liver biopsies, as well as hepatoma cell cultures, we demonstrated that the *AMBP* and *H2* genes are down-regulated and the corresponding molecules (bikunin, IαIH1,2,B and IαIH2,B) are negative acute-phase proteins, whereas the *H3* gene is up-regulated and the corresponding IαIH3,B molecule is a positive acute-phase protein. The *H1* gene does not seem to be affected by acute inflammation. Finally, recent data obtained in the rat and pig suggest that the *H4* gene, too, is up-regulated in acute inflammation [40]. The major pro-inflammatory cytokines interleukin-1 and interleukin-6 mediate the concomitant changes in gene transcription [35]. Our results provided the first example in humans of positive and negative acute-phase plasma proteins that are encoded by evolutionarily related genes [35]. The opposite regulation of the evolutionarily related *H2* and *H3* genes in acute inflammation does not contradict our proposal of homologous elements in the transcriptional control regions of these genes. Indeed, in different genes similar box/factor units can undergo opposite regulation by virtue of further neighbouring (in the present case, interleukin-1- or interleukin-6-driven) box(es) (e.g. [41]).

Overall, our current view is that homologous regulatory elements, e.g. some HNF-binding boxes, allow for a spatially (liver) and temporally (development) co-ordinated expression of the set of genes that synthesize the various chains that are ultimately assembled as IαI family molecules. In unusual circumstances (e.g. acute inflammation), where the amount of a particular precursor is increased compared with the others (e.g. H3P, which drives IαIH3,B synthesis), further, opposite and therefore gene-specific regulations takes place and modulate the basal HNF-mediated transcriptions by virtue of neighbouring box/factor units that differ from gene to gene.

FROM PROTEIN STRUCTURE TO REGULATION

In contrast to the situation for the IαI family genes, a great deal of structural data is available for the corresponding polypeptides and proteins; however, comparatively little is known about how the primary structures participate in precursor processing or protein activities. Indeed, the unusual cleavages of the H precursors, the resulting chain assemblies via a PGP cross-link and the numerous possibilities of such assemblies that are a basis for

diversity in the IαI family raise provocative questions: What are the sequence/function relationships within a chain? How are the precursor cleavages controlled? How are some preferred chain assemblies controlled? What is the purpose of diversity in sequences and molecules within the family? Some of these issues are currently being investigated by various groups.

From chain sequences towards functional domains

The well-known pair of Kunitz-type inhibitory domains are still the only biologically active sites that have been identified in the bikunin sequence. More recently, some progress has been made in the search for putative functional domains within the H1P to H4P sequences (Figure 3). First, all four H precursors and the resulting mature H chains harbour a so-called von Willebrand type-A (vWA) domain that is conserved in the human and mouse [3,42]. A heterophilic binding capacity for proteins with one or more vWA domain(s) seems to be the rule, and the targets for proteins with a vWA domain include such varied molecules as integrins, collagen, proteoglycans and heparin [43]. Secondly, H1P and H3P harbour a multicopper oxidase domain located within their C-terminal segment that is trimmed off during chain assembly [3]. The blue copper-containing oxidases and related proteins form an extremely ancient and diverse group of quite distantly related molecules that have retained or lost the ability to bind copper. Of the plasma proteins, caeruloplasmin and coagulation factors V and VIII are examples of such molecules. Whether the multicopper oxidase domain in H1P and H3P actually binds copper is currently unknown. This domain is absent from H2P and H4P, which is in keeping with the evolutionary distance between H sequences as mentioned above.

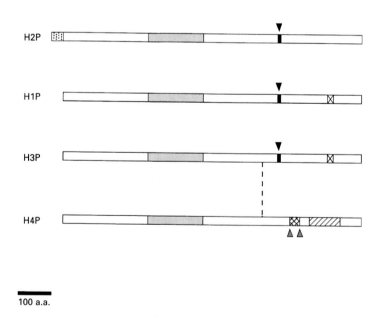

Figure 3   Structural comparison of the four H chain precursors and their putative functional domains

The four polypeptides are listed in the order H2P, H1P, H3P and H4P as a function of sequence similarities (see Table 1). A vertical broken line separates the N-terminal two-thirds of H4P which is homologous to the H3P, H1P and H2P counterparts from the C-terminal one-third of H4P that is not homologous to the other three polypeptides. Arrowheads indicate cleavage sites: black arrowhead, processing of H1P, H2P and H3P; red arrowhead, plasma kallikrein-sensitive site. Other symbols are as follows: stippled box, N-terminal propeptide [3,9]; pink box, vWA domain [3,42]; black box, Asp-(Pro-His-Phe-Ile-Ile) anchor site for a PGP cross-link (see text); crossed box, multicopper oxidase domain [3]; double-hatched box, plasma kallikrein-released, bradykinin-like, peptide (sequence: Pro-Gly-Val-Leu-Ser-Ser-Arg-Glu-Leu-Gly-Leu-Pro-Gly-Pro-Pro-Asp-Val-Pro-Asp-His-Ala-Ala-Tyr-His-Pro-Phe-Arg) [2]; hatched box, ATP-dependent-protease-like domain [10]. The relative sizes of all polypeptides and domains are drawn to scale, but minor differences in the numbers of amino acids (a.a.) between polypeptides or between species are not taken into account. The signal peptides are omitted.

Thirdly, the H4P C-terminal sequence harbours (i) a proline-rich, bradykinin-like domain (see the legend to Figure 3) bracketed by two Arg residues that allow it to be released by plasma kallikrein [2], and (ii) a domain with some similarity to an ATP-dependent protease [10]. The significance of this last observation remains to be confirmed, as the putative active site of the ATP-dependent protease appears to lie outside the region of similarity (as inferred from a similarity search made in our laboratory). More generally, it must be stressed that all the domains delineated so far in the H precursors are inferred from sequence similarity searches made against protein data banks or published sets of consensus amino acid sequences for domains with a given biological function. Therefore, even though the domains present in the H polypeptides are significant from an evolutionary viewpoint, they are only suggestive of functional activities that remain to be demonstrated experimentally.

From precursor structure to regulation of processing and assembly

The series of events that leads to the maturation, glycosylation and assembly of AMBP, H precursors and mature chains thereof takes place intracellularly in the liver [25,33,44,45]. Given that α1m and bikunin have no known functional connection, their co-synthesis as a single precursor, i.e. AMBP, is unique compared with other pro-proteins such as pro-complement components or pro-hormones. Hence whether or not the co-expression of α1m and bikunin is of functional significance is an important issue. The cleavage of AMBP into α1m and bikunin takes place at the C-terminal end of a connecting tetrapeptide comprising a conserved motif including three basic amino acids [13–20]. As this motif is known to be a typical target for pro-protein processing proteases such as furin, this protease was suspected to be responsible for AMBP processing [13,44]. However, although the cleavage of AMBP by furin is possible *in vitro*, the regulatory protease involved *in vivo* is unlikely to be furin; indeed, a recombinant AMBP expressed in a furin-deficient mammalian cell line is still cleaved into α1m and bikunin [45]. The cleavage step takes place during AMBP transport in the Golgi apparatus of liver cells [44]. Most importantly, recombinant bikunin molecules that are cleaved from recombinant AMBP or are synthesized alone display a difference in size that is accounted for by a difference in glycosylation and chondroitin 4-sulphate content [45]. Therefore the glycosylation and sulphation of bikunin, which are known to take place prior to AMBP cleavage in the Golgi [25], might be modulated by the adjacent α1m sequence, as recently proposed [45]. Furthermore, the H–PGP–bikunin assemblies that follow bikunin sulphation and provide the IαI family members seem to be formed prior to AMBP cleavage [25]. Therefore the regulation of assemblies containing bikunin by the adjacent α1m sequence, in as much as it controls the chondroitin 4-sulphate attachment, is an attractive possibility.

The processing of H precursors and the formation of their assemblies with bikunin is quite unusual as it involves the shedding of a large C-terminal peptide (240–280 amino acid residues) from the H precursors and the attachment of an inter-chain covalent PGP cross-link, as mentioned above. While the fate and biological function of the C-terminal peptide released from the H precursors remain unknown, other requirements and events involved in these chain assemblies have been investigated in detail. In the HepG2 hepatoma cell line, which is defective in the synthesis and assembly of most IαI family members, the abnormal presence of a mature but PGP-free H2 chain, in addition to the expected IαIH2,B protein, indicates that the mechanism responsible for the C-terminal cleavage of H pre-

cursors and the formation of the PGP cross-link is divided into two discrete steps, involving proteolytic cleavage followed by carbohydrate attachment [33]. This PGP cross-link is mediated by a chondroitin 4-sulphate chain attached to Ser[10] of bikunin by an O-glycosidic link and esterified to the C-terminal Asp of H1, H2 or H3 by an N-acetylgalactosamine in the chondroitin 4-sulphate chain [26,27]. The C-terminal Asp residue on the mature H chains becomes so located by shedding of the C-terminal peptide from the H precursors. It has been observed that this Asp residue is part of a homologous Asp-Pro-His-Phe-Ile-Ile sequence that is found in H1P, H2P and H3P from humans and rodents [2,26,27]. Therefore this conserved sequence in H precursors appears to be a good candidate as a recognition unit and/or a target for one or more processing enzymes involved in C-terminal cleavage at the Asp-Pro junction and addition of PGP/bikunin to the Asp residue. This view has been supported by the recent discovery of a natural mutant of an H precursor, namely H4P. Indeed, the Asp-Pro-His-Phe-Ile-Ile sequence is absent from H4P [3,10], which correlates with the physiological lack of H4P C-terminal cleavage and H4/bikunin assembly in liver cells (Figure 1; Table 2).

If the chain assemblies were not controlled further, the IαI family would comprise a large variety of molecules. A theoretical possibility exists that some or all of the plasma molecules listed in Table 2 are by-products induced *in vitro* from a (fragile) super-molecule such as IαIH1,2,3,B that could give rise to all possible chain combinations. This proposal is quite unlikely, however, given (i) the unequal amounts of IαIH1,B, IαIH2,B and IαIH3,B in plasma [27,30], (ii) some differences between species, as exemplified by the IαIH2,3,B molecule that has been described in bovine plasma [46] but is unknown in the human (Table 2), and (iii) the simultaneous decrease in plasma IαIH1,2,B and increase in IαIH3,B in inflammation (see above) that cannot be accounted for by degradation of a super-molecule as a shared precursor. Therefore it is reasonable to consider that the limited variety of molecules in the IαI family is because certain chain assemblies (e.g. IαIH1,2,B and IαIH3,B) are preferred to others (e.g. IαIH1,3,B, which is as yet unknown) in the biosynthetic pathways. In this respect, an unequal synthesis of H1P, H2P and H3P in the liver is a simple way to favour some assemblies and, hence, the relative amounts of IαI family members. Indeed, a limited number of studies suggest that H2 mRNA and H2 may predominate slightly in liver and plasma respectively [31,34,35], which is in keeping with IαIH1,2,B and IαIH2,B being the quantitatively predominant plasma molecules, as seen in both human [30,31,35] and mouse (J. P. Salier and L. Jean, unpublished work). Apart from this relatively simple, quantitative, control of H precursors that in turn regulates the relative amounts of plasma molecules, it remains to be determined whether some sequence-borne information that differs between the various H precursors participates in the fine tuning of chondroitin 4-sulphate chain attachment and further regulates the occurrence of preferred assemblies.

From protein structure to regulatory functions: the hyaluronic acid (HA) connection

Significant progress has recently been made in the search for the biological functions of some IαI family members. Notably, a HA-binding capacity has been ascribed to some of them. Although serum IαIH1,2,B is not bound to HA under normal circumstances, adding HA to human or bovine serum promotes the occurrence of a bikunin-free, covalent, HA–protein complex that involves H1 + H2 in human [47a,48] and H2 + H3 in bovine [47a] serum (in [47a] an H3 sequence was incorrectly described as

H1). Likewise a natural, bikunin-free, complex comprising HA and H chains, including H1, can be observed *in vivo* in HA-rich media, e.g. normal human follicular fluid or synovial fluid from rheumatoid arthritis patients [48]. It has been proposed that these complexes originate from human IαIH1,2,B or bovine IαIH2,3,B from which bikunin has been be shed and replaced by HA [47,47b].

The physiological relevance of a HA-binding capacity for IαI family members has been established with various cell types that display a HA-containing pericellular coat. Notably, the involvement of IαI family members in the process of oocyte maturation has been studied. During the pre-ovulatory period the cumulus cells of the so-called cumulus–oocyte complex secrete large amounts of a HA-enriched extracellular matrix and dissociate from each other as a consequence of the endogenous gonadotropin surge (reviewed in [49,50]). One or more IαI family members (e.g. IαIH1,2,B as well as IαIH3,B, IαIH2,B or bovine IαIH2,3,B) are absolutely required to limit this cumulus–oocyte complex expansion, which otherwise results in a deleterious loss of HA and cumulus cells [46,49–51]. A charge-mediated, non-covalent, binding of HA to IαI family members is likely to participate in this control of cumulus–oocyte complex expansion [50,52], whereas a major role for the enzyme inhibitory activity of bikunin has been ruled out [50]. The IαI family members involved in this physiological control of cumulus–oocyte complex expansion most likely originate from the blood. Indeed, IαIH1,2,B and IαIH3,B are present in the follicular fluid and ovarian follicles in both human and mouse after chorionic-gonadotropin-stimulated ovulation [49], and IαIH1,2,B molecules injected intravenously into the mouse are exclusively localized to the vasculature until an experimental surge of luteinizing hormone allows them to pass the blood–follicle barrier within seconds [51]. These results outline the importance of blood-borne IαI family molecules in the regulation of extracellular matrices that may lie outside the blood vessels. This regulatory function of IαI as a stabilizer of HA-containing pericellular coats has also been seen *in vitro* with other cell types, namely fibroblasts and mesothelial cells in culture [53].

The HA-binding capacity of IαI family members has also been demonstrated to be significant in inflammatory conditions. In patients with rheumatoid arthritis and some other arthritic diseases, the so-called TSG-6 glycoprotein, an anti-inflammatory molecule and a member of the hyaladherin family of HA-binding proteins, is found at high levels in the synovial fluid and, at a lower concentration, in the plasma [54–56]. A unique feature of TSG-6 is its propensity to form a reduction-resistant complex with IαIH1,2,B; this complex involves the glycosaminoglycan bond harboured by IαIH1,2,B [53]. When TSG-6 and purified IαIH1,2,B are mixed *in vitro*, the resulting complex has been shown to comprise TSG-6, H2 and bikunin, which suggests shedding of an H1 chain [55]. Likewise, within complete plasma the involvement of IαIH1,2,B in a complex with TSG-6 is clear, even though the number and identity of other IαI family member(s) possibly participating in further complex(es) with TSG-6 *in vivo* should be borne in mind [55]. Indeed, in acute inflammation, plasma IαIH3,B is replaced by a heavier molecule that is resistant to reducing conditions and may carry a ligand [35]. If so, the ligand could well be TSG-6. Therefore whether IαIH3,B and TSG-6, two up-regulated acute-phase proteins [35,54], are indeed a source for a covalent TSG-6 + H3 + bikunin complex *in vivo* remains an important issue. Finally, the biological significance of the complex between TSG-6 and IαIH1,2,B is being deciphered, as this complex enhances the plasmin-inhibitory activity of bikunin, a very weak plasmin inhibitor [56]. Also, in a model of experimentally induced inflammation in the mouse, TSG-6 mutants that have lost their anti-inflammatory activity are unable to potentiate plasmin inhibition, although they still form a complex with IαIH1,2,B [56]. These data suggest the involvement of several acute-phase plasma proteins, namely TSG-6 and IαI family members, in a negative feedback loop whereby plasmin, itself a central molecule in the activation of inflammation-associated enzyme cascades, is inhibited [56]. Such a regulatory loop may prove to be important in a number of tissues, given for instance the high contents of HA, TSG-6 and IαI family members in the synovial fluid of inflamed joints [54,57].

Finally, finding a link between one or more of the potential domains in the H precursors and/or mature chains (Figure 3) and a proven biological function of some IαI family members is a major aim of current studies, yet for the most part such links still remain to be established. However, in the *in vitro* induced H1 + H2/HA complex detailed above [47a], the HA-bound region has been tentatively located within the C-terminal half of H1 [47a], i.e. in the middle of the vWA domain (Figure 3). As targets for proteins with a vWA domain include proteoglycans and heparin [43], we have proposed that the vWA domain in H chains is involved in the HA-liganding activity [3]. If this is true, all H chains are likely to bind HA by virtue of their vWA domain (Figure 3), which is consistent with: (i) the diversity of complexes induced *in vitro*, i.e. H1 + H2/HA and H2 + H3/HA [47a], and (ii) the diversity of IαI family members that are able to stabilize HA-containing cell coats, as detailed above. This vWA domain-mediated binding of HA to H chains, if it occurred, would be transient, charge-mediated and non-covalent (see above) and followed *in vivo* by a final, covalent linkage. Indeed, such a covalent linkage between HA and the H chains, as recently seen in arthritic joints, has been tentatively localized to the C-terminal Asp residue of H chains [47b]. This suggests that a complex of HA with members of the IαI family may be formed by transferring some H chains from their PGP cross-link to HA by a substitution reaction [47b]. This last event is likely to be enzymically controlled (L. Chen, H. Zhang, R. W. Powers, P. T. Russell and W. J. Larsen, unpublished work).

CONCLUSIONS AND FUTURE TRENDS

In the last 6 years of research into IαI family members, shifting from a mostly structural description of a new family to an analysis of the fine-tuned regulation associated with genes and molecules thereof has proved rewarding. We now know, at least partly, where, when and why the *AMBP* and *H* genes are co-expressed. We are gaining some preliminary hints as to how the precursor polypeptides are processed and assembled. The precursors and molecules of the IαI family harbour some previously unexpected regulatory functions that have shed light on the importance of the H chains, in addition to the well-known protease-inhibitory capacity of bikunin. Finally, some of the above mechanisms have now been evaluated under pathological conditions.

Not surprisingly, as a spin-off of these recent studies further complexity has been brought to the field. While previous questions are not yet fully resolved, we are now facing new ones. Among these the need, if any, for combined HA-binding activity and protease-inhibitory activity within a molecule is an interesting one. Other regulation-oriented issues that warrant investigation include: (i) possible control of the activity of genes, mRNAs or proteins by sequence polymorphisms, as the latter have been described in detail by several groups (e.g. [6,7,58]); (ii) a role for bikunin-free H chains in the blood [2,10], brain [3] and possibly other tissues; (iii) the fate and functions of the H chain C-terminal peptides released during formation of H+bikunin assemblies; (iv) the fate and functions of blood-borne bikunin after it enters tissues such as normal skin [59] or brains of Alzheimer patients [60]; and (v) the capacity of bikunin to inhibit a plasmin-driven proliferation of tumour cells that harbour surface receptors for bikunin (reviewed in [61,62]). Some of these forthcoming research topics will undoubtedly prove beneficial to the elucidation of major pending issues that include the co-regulation of genes, the likely differences in biological activity between chains and between members of the IαI family and, ultimately, the precise biological function(s) of all IαI family members *in vivo*.

We acknowledge the kind participation of our colleagues J. J. Enghild (Durham, NC, U.S.A.), E. Fries (Uppsala, Sweden), W. J. Larsen (Cincinnati, OH, U.S.A.), J. Vilcek and H. G. Wisniewski (New York, NY, U.S.A.) and an anonymous referee in discussions and improvements of the new nomenclature. We thank W. J. Larsen, J. Vilcek and H. G. Wisniewski for making some unpublished results and manuscripts available to us. We are indebted to Dr. S. Claeyssens for critical reading of the manuscript. The work carried out in the authors' laboratory was supported in part by grants from N.A.T.O. and from Groupement de Recherches et d'Etudes sur les Génomes to J.P.S., and by the University of Rouen.

REFERENCES

1 Salier, J. P. (1990) Trends Biochem. Sci. **15**, 435–439
2 Nishimura, H., Kakizaki, I., Muta, T., Sasaki, N., Pu, P. X., Yamashita, T. and Nagasawa, S. (1995) FEBS Lett. **357**, 207–211
3 Chan, P., Risler, J. L., Raguenez, G. and Salier, J. P. (1995) Biochem. J. **306**, 505–512
4 Diarra-Mehrpour, M., Bourguignon, J., Sesboué, R., Mattei, M. G., Passage, E., Salier, J. P. and Martin, J. P. (1989) Eur. J. Biochem. **191**, 131–139
5 Salier, J. P., Simon, D., Rouet, P., Raguenez, G., Muscatelli, F., Gebhard, W., Guenet, J. L. and Mattei, M. G. (1992) Genomics **14**, 83–88
6 Salier, J. P., Verga, V., Doly, J., Diarra-Mehrpour, M. and Erickson, R. P. (1992) Mamm. Genome **2**, 233–239
7 Bost, F., Bourguignon, J., Martin, J. P., Sesboue, R., Thiberville, L. and Diarra-Mehrpour, M. (1993) Eur. J. Biochem. **218**, 283–291
8 Diarra-Mehrpour, M., Bourguignon, J., Sarafan, N., Bost, F., Sesboué, R., Muschio-Bonnet, F. and Martin, J. P. (1994) Biochim. Biophys. Acta **1219**, 551–554
9 Gebhard, W., Schreitmuller, T., Hochstrasser, K. and Wachter, E. (1988) FEBS Lett. **229**, 63–67
10 Saguchi, K. I., Tobe, T., Hashimoto, K., Sano, Y., Nakano, Y., Miura, N. H. and Tomita, M. (1995) J. Biochem. (Tokyo) **117**, 14–18
11 Malki, N., Balduyck, M., Maes, P., Capon, C., Mizon, C., Han, K. K., Tartar, A., Fournet, B. and Mizon, J. (1992) Biol. Chem. Hoppe-Seyler **373**, 1009–1018
12 Diarra-Mehrpour, M., Bourguignon, J., Sesboué, R., Salier, J. P., Léveillard, T. and Martin, J. P. (1990) Eur. J. Biochem. **191**, 131–139
13 Kaumeyer, J. F., Polazzi, J. O. and Kotik, M. P. (1986) Nucleic Acids Res. **14**, 7839–7850
14 Gebhard, W., Schreitmuller, T., Vetr, H., Wachter, E. and Hochstrasser, K. (1990) FEBS Lett. **269**, 32–36
15 Tavakkol, A. (1991) Biochim. Biophys. Acta **1088**, 47–56
16 Lindqvist, A., Bratt, T., Altieri, M., Kastern, W. and Akerstrom, B. (1992) Biochim. Biophys. Acta **1130**, 63–67
17 Chan, P. and Salier, J. P. (1993) Biochim. Biophys. Acta **1174**, 195–200
18 Ide, H., Itoh, H. and Nawa, Y. (1994) Biochim. Biophys. Acta **1209**, 286–292
19 Hanley, S. and Powell, R. (1994) Gene **147**, 297–298
20 Leaver, M. J., Wright, J. and George, S. G. (1994) Comp. Biochem. Physiol. **108**, 275–281
21 Akerstrom, B. and Logdberg, L. (1990) Trends Biochem. Sci. **15**, 240–243
22 Flower, D. R. (1994) FEBS Lett. **354**, 7–11
23 Chan, P., Simon-Chazottes, D., Mattei, M. G., Guenet, J. L. and Salier, J. P. (1994) Genomics **23**, 145–150
24 Falkenberg, C., Enghild, J. J., Thogersen, I. B., Salvesen, G. and Akerstrom, B. (1994) Biochem. J. **301**, 745–751
25 Sjoberg, E. M. and Fries, E. (1992) Arch. Biochem. Biophys. **295**, 217–222
26 Enghild, J. J., Salvesen, G., Hefta, S., Thogersen, I. B., Rutherfurd, S. and Pizzo, S. V. (1991) J. Biol. Chem. **266**, 747–751
27 Enghild, J. J., Salvesen, G., Thogersen, I. B., Valnickova, Z., Pizzo, S. V. and Hefta, S. A. (1993) J. Biol. Chem. **268**, 8711–8716
28 Morelle, W., Capon, C., Balduyck, M., Sautiere, P., Kouach, M., Michalski, C., Fournet, B. and Mizon, J. (1994) Eur. J. Biochem. **221**, 881–888
29 Jessen, T. E., Faarvang, K. L. and Ploug, M. (1988) FEBS Lett. **230**, 195–200
30 Enghild, J. J., Thogersen, I. B., Pizzo, S. V. and Salvesen, G. (1989) J. Biol. Chem. **264**, 15975–15981
31 Rouet, P., Daveau, M. and Salier, J. P. (1992) Biol. Chem. Hoppe-Seyler **373**, 1019–1024
32 Yamada, S., Oyama, M., Kinugasa, H., Nakagawa, T., Kawasaki, T., Nagasawa, S., Khoo, K. H., Morris, H. R., Dell, A. and Sugahara, K. (1995) Glycobiology **5**, 335–341
33 Thogersen, I. B. and Enghild, J. J. (1995) J. Biol. Chem. **270**, 18700–18709
34 Salier, J. P., Chan, P., Raguenez, G., Zwingman, T. and Erickson, R. P. (1993) Biochem. J. **296**, 85–91
35 Daveau, M., Rouet, P., Scotte, M., Faye, L., Hiron, M., Lebreton, J. P. and Salier, J. P. (1993) Biochem. J. **292**, 485–492
36 Rouet, P., Raguenez, G., Tronche, F., Yaniv, M., N'guyen, C. and Salier, J. P. (1992) J. Biol. Chem. **267**, 20765–20773
37 Rouet, P., Raguenez, G., Tronche, F., Mfou'ou, V. and Salier, J. P. (1995) Nucleic Acids Res. **23**, 395–404
38 Tronche, F. and Yaniv, Y. (1994) Liver Gene Expression, R. G. Landes, Austin, TX
39 Jean, L., Lyoumi, S. and Salier, J. P. (1996) Anal. Biochem., in the press
40 Gonzalez-Ramon, N., Alava, M. A., Sarsa, J. A., Pineiro, M., Escartin, A., Garcia-Gil, A., Lampreave, F. and Pineiro, A. (1995) FEBS Lett. **371**, 227–230
41 Fey, G. H., Hocke, G. M., Wilson, D. R., Ripperger, J. A., Juan, T. S. C., Cui, M. Z. and Darlington, G. J. (1994) in The Liver: Biology and Pathobiology, 3rd edn. (Arias, I. M., Boyer, J. L., Fausto, N., Jakoby, W. B., Schachter, D. A. and Shafritz, D. A., eds.), pp. 113–143, Raven Press, New York
42 Bork, P. and Rohde, K. (1991) Biochem. J. **279**, 908–910
43 Colombatti, A. and Bonaldo, P. (1991) Blood **77**, 2305–2315
44 Bratt, T., Olsson, H., Sjoberg, E. M., Jergil, B. and Akerstrom, B. (1993) Biochim. Biophys. Acta **1157**, 147–154
45 Bratt, T., Cedervall, T. and Akerstrom, B. (1994) FEBS Lett. **354**, 57–61
46 Castillo, G. M. and Templeton, D. M. (1993) FEBS Lett. **318**, 292–296
47a Huang, L., Yoneda, M. and Kimata, K. (1993) J. Biol. Chem. **268**, 26725–26730
47b Zhao, M., Yoneda, M., Ohashi, Y., Kurono, S., Iwata, H., Ohnuki, Y. and Kimata, K. (1995) J. Biol. Chem. **270**, 26657–26663
48 Jessen, T. E., Odum, L. and Johnsen, A. H. (1994) Biol. Chem. Hoppe-Seyler **375**, 521–526
49 Chen, L., Mao, S. J. T. and Larsen, W. J. (1992) J. Biol. Chem. **267**, 12380–12386
50 Chen, L., Mao, S. J. T., McLean, L. R., Powers, R. W. and Larsen, W. J. (1994) J. Biol. Chem. **269**, 28282–28287
51 Powers, R. W., Chen, L., Russell, P. T. and Larsen, W. J. (1995) Am. J. Physiol. **269** (Endocrinol. Metab. **32**), 290–298
52 Camaioni, A., Hascall, V. C., Yanagishita, M. and Salustri, A. (1993) J. Biol. Chem. **268**, 20473–20481
53 Blom, A., Pertoft, H. and Fries, E. (1995) J. Biol. Chem. **270**, 9698–9701
54 Wisniewski, H. G., Maier, R., Lotz, M., Lee, S., Klampfer, L., Lee, T. H. and Vilcek, J. (1993) J. Immunol. **151**, 6593–6601
55 Wisniewski, H. G., Burgess, W. H., Oppenheim, J. D. and Vilcek, J. (1994) Biochemistry **33**, 7423–7429
56 Wisniewski, H. G., Hua, J. C., Poppers, D. M., Naime, D., Vilcek, J. and Cronstein B. N. (1996) J. Immunol., in the press
57 Hutadilok, N., Ghosh, P., and Brooks, P. M. (1988) Ann. Rheum. Dis. **47**, 377–385
58 Luckenbach, C., Kompf, J. and Ritter, H. (1991) Hum. Genet. **87**, 89–90
59 Sjoberg, E. M., Blom, A., Larsson, B. S., Alston-Smith, J., Sjöquist, M. and Fries, E. (1995) Biochem. J. **308**, 881–887
60 Yoshida, E., Yoshimura, M., Ito, Y. and Mihara, H. (1991) Biochim. Biophys. Res. Commun. **174**, 1015–1021
61 Kobayashi, H., Gotoh, J., Fujie, M. and Terao, T. (1994) J. Biol. Chem. **269**, 20642–20647
62 Kobayashi, H., Gotoh, J., Hirashima, Y., Fujie, M., Sugino, D. and Terao, T. (1995) J. Biol. Chem. **270**, 8361–8366
63 Diarra-Mehrpour, M., Bourguignon, J., Bost, F., Sesboué, R., Muschio, F., Sarafan, N. and Martin, J. P. (1992) Biochim. Biophys. Acta **1132**, 114–118
64 Bourguignon, J., Diarra-Mehrpour, M., Thiberville, L., Bost, F., Sesboue, R. and Martin, J. P. (1993) Eur. J. Biochem. **212**, 771–776
65 Leveillard, T., Bourguignon, J., Sesboue, R., Hanauer, A., Salier, J. P. and Martin, J. P. (1988) Nucleic Acids Res. **16**, 2744
66a Leveillard, T., Salier, J. P., Sesboue, R., Bourguignon, J., Diarra-Mehrpour, M. and Martin, J. P. (1988) Nucleic Acids Res. **16**, 11852
66b Tobe, T., Saguchi, K., Hashimoto, K., Miura, N. H., Tomita, M., Li, F., Wang, Y., Minoshima, S. and Shimizu, N. (1995) Cytogenet. Cell Genet. **71**, 296–298
67 Gebhard, W., Hochstrasser, K., Fritz, H., Enghild, J. J., Pizzo, S. V. and Salvesen, G. (1990) Biol. Chem. Hoppe-Seyler **371**, 13–22
68 Itoh, H., Ide, H., Ishikawa, N. and Nawa, Y. (1994) J. Biol. Chem. **269**, 3818–3822

Biochem. J. (1996) **317**, 1–11 (Printed in Great Britain)

REVIEW ARTICLE
The denaturation and degradation of stable enzymes at high temperatures

Roy M. DANIEL*, Mark DINES and Helen H. PETACH†
Department of Biological Sciences, The University of Waikato, Hamilton, New Zealand

Now that enzymes are available that are stable above 100 °C it is possible to investigate conformational stability at this temperature, and also the effect of high-temperature degradative reactions in functioning enzymes and the inter-relationship between degradation and denaturation. The conformational stability of proteins depends upon stabilizing forces arising from a large number of weak interactions, which are opposed by an almost equally large destabilizing force due mostly to conformational entropy. The difference between these, the net free energy of stabilization, is relatively small, equivalent to a few interactions. The enhanced stability of very stable proteins can be achieved by an additional stabilizing force which is again equivalent to only a few stabilizing interactions. There is currently no strong evidence that any particular interaction (e.g. hydrogen bonds, hydrophobic interactions) plays a more important role in

proteins that are stable at 100 °C than in those stable at 50 °C, or that the structures of very stable proteins are systematically different from those of less stable proteins. The major degradative mechanisms are deamidation of asparagine and glutamine, and succinamide formation at aspartate and glutamate leading to peptide bond hydrolysis. In addition to being temperature-dependent, these reactions are strongly dependent upon the conformational freedom of the susceptible amino acid residues. Evidence is accumulating which suggests that even at 100 °C deamidation and succinamide formation proceed slowly or not at all in conformationally intact (native) enzymes. Whether this is the case at higher temperatures is not yet clear, so it is not known whether denaturation or degradation will set the upper limit of stability for enzymes.

INTRODUCTION

During the 1980s, organisms were discovered that are capable of survival and growth at temperatures up to 110 °C [1]. Most biochemists accept that an integral feature of such extremely thermophilic organisms are enzymes that are stable *in vivo* at these temperatures. With the discovery of such enzymes that are stable above 100 °C [2,3] (Table 1), it is no longer feasible to confine the study of temperature effects on enzymes to denaturation and the effects of heat on activity. Degradation is likely to play a major role in the loss of enzyme activity above 80 °C, and an increasing number of studies are now describing such effects. The present review will deal with both denaturation and degradation of enzymes at high temperatures, and the inter-relation-

ship of these with each other and with activity. We define denaturation here as loss of tertiary (and often secondary) protein structure not involving covalent bond cleavage, and as being, in principle at least, reversible. Degradation is the loss of primary structure with associated covalent bond cleavage and/or formation, and is irreversible. Because we wish to look at the upper limits of enzyme stability, we will confine ourselves to events taking place readily above 80 °C in aqueous conditions. Much of this work has dealt with the stability of enzymes from extreme thermophiles, since other enzymes stable above 80 °C for useful periods are uncommon. Throughout the review we have attempted to deal with enzyme stability rather than protein stability. While in general we expect the review to apply to all proteins, it is possible that those proteins that have structural, binding or transport functions have a range of structural and dynamic properties which only partially overlap the properties of catalytic proteins, broad though these are.

Enzyme denaturation has been the subject of much work over an extended period (e.g. [4,5]), although relatively few detailed studies have been carried out above 80 °C. What is new in this field is the study of enzyme stability at temperatures significantly higher than 100 °C [6,7]. As Figure 1 shows, enzymes are now available which, with or without stabilizing treatments or conditions, have half-lives in excess of 10 min at 130 °C. Furthermore, enzyme activity has been measured at these same high temperatures [8–11]. This is important because many observations of high-temperature enzyme stability have been performed simply by heating the enzyme, rapidly cooling a sample, and assaying for residual activity at a lower temperature (e.g. 95 °C). This procedure is open to criticism because of the possibility that the enzyme may be denatured at high temperature

Table 1   Some enzymes stable above 100 °C

Data from [2,3].

Enzyme	Source	Half-life (h)	Temperature (°C)
Glyceraldehyde-3-phosphate dehydrogenase	*Thermotoga maritima*	> 2	100
Hydrogenase	*Pyrococcus furiosus*	2	100
Amylase	*P. woesei*	6	100
DNA–RNA polymerase	*Thermoproteus tenax*	2	100
Glutamate dehydrogenase	*P. furiosus*	10	100
Cellobiohydrolase	*Thermotoga* sp.	1.1	108
Amylase	*P. furiosus*	2	120

Abbreviation used: ΔG, the difference in free energy between the folded and unfolded states of a protein.
* To whom correspondence should be addressed.
† Present address: Department of Chemistry, University of Colorado at Denver, Denver, CO 80217-3364, U.S.A.

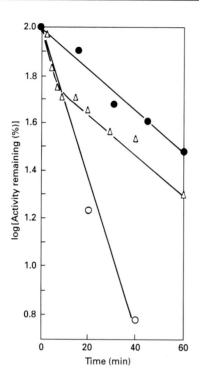

Figure 1   Stability of three enzymes at 130 °C

The plot shows the loss of activity with time of the unpurified amylase from *Pyrococcus furiosus* (○), the α-glucosidase from *Thermococcus* strain ANI in 90% sorbitol (●), and the immobilized xylanase from *Thermotoga maritima* sp. FjSS3BI in molten sorbitol (△). (Data from [8], [11] and [9] respectively.)

and then renatured (thus regaining activity) rapidly enough to display high activity in the cooled assayed sample. While we know of no cases where rapid renaturation of a very thermophilic enzyme has occurred in this way, high-temperature measurements of activity greatly reduce the likelihood that an apparent high-temperature stability is actually caused by reversible denaturation. Data on enzyme stability at high temperatures have been presented by some authors as 'T_{opt}', i.e. the peak temperature on a graph of product produced over a fixed time against temperature. These data are not very helpful in ascertaining enzyme stability since, in addition to the complicating effect of substrate, the T_{opt} will be highly dependent on assay duration (not always given), on stability and on the effect of temperature on activity. The latter may vary from Q_{10} values of 0.4 [i.e. an activity increase of 40% for a 10 °C rise in temperature (for a thermophilic xylanase; C. Aebischer and R. M. Daniel, unpublished work)] to 4.5 (for a thermophilic proteinase [12]). Other temperature-dependent factors such as buffer pK_a, water activity and dielectric constant may also have an effect. Furthermore, such T_{opt} graphs are often derived from measurements of product after a single fixed reaction time, at temperatures where the rate of product formation is almost certainly not linear with respect to time.

Temperature-driven degradation is a newer field of study. Over the time spans we are considering here (minutes to hours), and at neutral pH, such degradation is only significant above 80 °C.

A variety of 'physiological' agents can act to stabilize enzymes against heat, although none will stabilize all enzymes. Calcium, for example, is a well known stabilizer of many enzymes at both high and low temperatures. A number of agents (including

protein) which act as stabilizers at high concentration may do so by altering the water activity/dielectric content of the medium. In particular, at very low hydration levels (below about 0.4 g of water/g of protein [13]), proteins are greatly stabilized, both as solids [14] and in non-aqueous solution [15]. At high temperatures the most notable stabilizing agents are probably cyclic 2,3-diphosphoglycerate and di-*myo*-inositol 1,1′-phosphate, found in *Methanothermus fervidus* and *Pyrococcus woesei* respectively. In the presence of potassium, these agents increase the half-lives of some enzymes by up to 130-fold at 90 °C [16].

CONFORMATIONAL STABILITY OF ENZYMES

When stable enzymes from thermophiles became available, they were seized upon as tools for the investigation of protein conformational stability. Much of this work compared stable enzymes from extreme thermophiles with less stable enzymes of similar structure and function from mesophiles. Amino acid composition data were often available, sequences less frequently so, and structural information was relatively rare. On the basis of these compositional comparisons a number of fairly specific proposals were made for amino acid changes associated with increased thermal stability, for example an increased arginine/lysine ratio [17,18]. Now enzymes that are stable over an even greater temperature range are available, and more composition (and sequence) data have been obtained, it has become apparent that these 'traffic rules' lack predictive value and statistical significance. The most unequivocal evidence of this comes from a study of the amino acid sequences and compositions of 26 glyceraldehyde phosphate dehydrogenases from organisms growing optimally over the temperature range 20–102 °C, carried out by Bohm and Jaenicke [19] (Figure 2). The data may suggest low serine and glycine and high isoleucine content in the more thermophilic enzymes. However, although the serine content of the *Thermus aquaticus* enzyme is the lowest found (3.5%), the even more thermophilic enzymes have a serine content identical to that of the *Escherichia coli* enzyme; and while the isoleucine and glycine values for the archaeal hyperthermophiles fall outside the mesophilic range, values for the thermophilic bacteria do not [19]. The archaeal origin of the enzymes thus seems to be a more likely cause of amino acid compositional variation than does thermophily.

It is clear that, if there are amino acid changes associated with thermophile enzymes, they are small and probably not universal, and with the number of sequences currently available are not statistically detectable above the background of amino acid changes which are evident for taxonomic or functional reasons. Even in the cases where there is specific evidence bearing on the way a particular amino acid may influence stability, as in the cases of proline, asparagine and cysteine for example (see below), there are enough significant exceptions to cast serious doubt on the universality of any proposed amino acid substitution 'rules' which take no account of the position of the amino acid in the structure.

Structural data give much more specific information. Using a mesophilic ferridoxin structure, Perutz and Raidt [20] compared ferridoxins from mesophiles and thermophiles, and attributed half-life differences of around three orders of magnitude to a quite small number of specific interactions. More recent and detailed structural studies of another low-molecular-mass non-haem iron protein, rubredoxin from the archaeal hyperthermophile *Pyrococcus furiosus* [21,22], indicated that the structure is very similar to that of mesophilic rubredoxins. The authors suggest that additional hydrogen bonds within the β-sheet and a

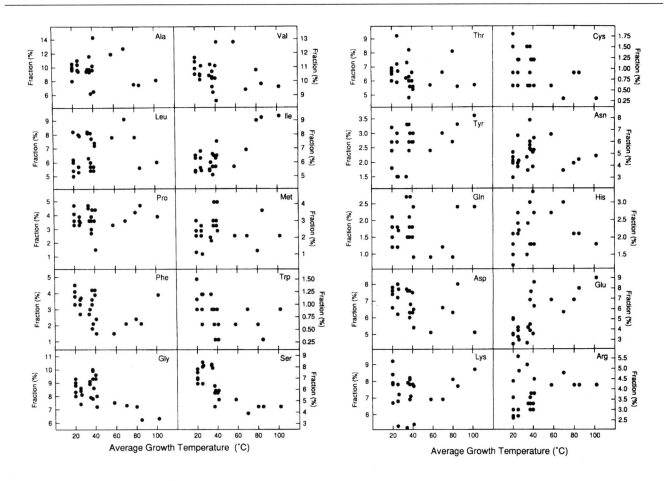

Figure 2   Amino acid occurrences in the database of D-glyceraldehyde-3-phosphate dehydrogenases related to the growth temperature of the enzyme source

Reprinted with permission from Bohm, G. and Jaenicke, R. Int. J. Peptide Protein Res. **43**, 97–106 [19], ©1994 Munksgaard International Publishers Ltd., Copenhagen, Denmark.

few extra electrostatic interactions at the protein surface may account for the greater stability of the *Pyrococcus* rubredoxin.

However, accounting for the difference in stability between two homologous enzymes in this way has not enabled the formulation of systematic mechanisms by which thermostability can be, or even has been, achieved. For this to be the case the structural mechanisms proposed (e.g. loss of surface loops, restriction of N-terminus mobility, stabilization of α-helix dipoles) should be more frequent in the more stable variants of enzymes of similar structure and function, and less frequent in less stable enzymes, and this should be observable in a variety of different enzyme groups. There is not good evidence for this at present. It is possible that more data will resolve this issue, but there currently seems no good reason why a single mechanism or small subgroup of mechanisms should be widely favoured out of the variety of potentially stabilizing structural mechanisms, except in enzymes which have a high degree of sequence and structural identity.

The most structurally similar proteins from mesophiles and extreme thermophiles show multiple amino acid differences, and even with structural information it is difficult to identify those changes arising from changed stability rather than from taxonomic or minor functional differences. This problem can be overcome by using point mutations as a probe of stability. In an early study [23], single mutations spread along the β-galactosidase gene of *E. coli* were found to be mostly destabilizing (18 out of 56 decreasing the half-life by more than 10-fold) or had little

effect (37 of the 56). Only one mutation increased the half-life by more than 2-fold. The destabilizing mutations tended to be grouped into particular parts of the sequence. Much more detailed research, especially by Matthews and colleagues, who have coupled the use of point mutation with detailed structural studies, has confirmed and greatly extended these earlier results (for reviews, see [24–26]).

A variety of studies [27–31] have shown that the majority of point mutations have little effect on stability. For example, of over 2000 mutations made to T4 lysozyme, 91% had no significant effect on thermal stability [24,25]. Most of these mutations appeared to be at the surface of the protein. However, for those mutations that alter stability, destabilization is a more common effect. For a number of enzymes where enough data have been gathered [28–31], including lysozyme [28] and subtilisin [31], the occurrence of random point mutations which cause stabilization is of the order of 1 in 10000. The destabilizing mutations in lysozyme (about 9% of the total) all occurred at amino acids which have restricted mobility [24,25,32]. These are likely to be in the interior of the protein, and to be stabilized by hydrophobic interactions. This tends to support the view that hydrophobic interactions are the most important interaction in protein conformation [33]. On the other hand, there is a correlation between a large change in the partial heat capacity of unfolding of a protein (which is proportional to the interior hydrophobicity) and low net stability [34]. This may indicate that hydrogen bonding and electrostatic interactions are more important than van der Waals interactions and the hydrophobic effect.

Conformational stability can be affected by altering the stability of the unfolded as well as of the native form of the protein. Thus a mutation which decreases the conformational freedom of the unfolded state will raise its free energy and increase ΔG (the difference in free energy between the folded and unfolded states), stabilizing the native protein. The substitution of proline for other amino acid residues has been proposed to have a general stabilizing effect [35–37], in some cases due to restriction of the conformational freedom of the unfolded state [36]. However, this is clearly not a universal stabilization rule [19], and the position of proline in the structure is probably a dominating factor [37].

Although the work of Matthews [25] and others has led to great increases in our understanding of the way amino acid substitutions affect structure and stability, variation in protein structure is so great, and the intramolecular interactions so complex, that there has been little success in developing simple and widely applicable amino acid substitution 'rules' by which protein stability may be increased. However, for more tightly defined and specific elements which occur widely in protein structure, e.g. the α-helix, success may be closer. Introduction of a negatively charged residue near the N-terminus of an α-helix has a stabilizing effect due to enhanced electrostatic and hydrogen bonding interactions within the helix [25,33,38–41]. Given the fundamental role played by α-helices, this may be a useful stabilizing strategy, although substitution at the ends of the α-helices may require a very good knowledge of the local stereochemistry to allow such a stabilization strategy to be used reliably.

CONFORMATIONAL STABILITY AT HIGH TEMPERATURES: THEORETICAL CONSIDERATIONS

It has been established for some time that the tertiary structure of proteins is only marginally stable (e.g. [4]). The conformational stability of a protein (defined as the difference in free energy between the folded and unfolded states, ΔG) is the sum of a large number of weak, non-covalent, interactions, including hydrogen bonds, van der Waal interactions, salt bridges and the hydrophobic effect, and the destabilizing forces arising largely from conformational entropy. All of these forces are affected by environmental conditions, including, for example, solvent and temperature, in a complex way.

In an 'average' protein the sum of the stabilizing interactions is large, about 1 MJ·mol⁻¹. Destabilizing forces are also large, and ΔG, the difference between the two, is only of the order of 40 kJ·mol⁻¹. A single weak interaction, for example, may contribute up to 25 kJ·mol⁻¹ [25]. Small changes in the number or strength of the stabilizing interactions will thus have a major effect on stability.

From these considerations (and with hindsight), the very high stability of enzymes from extreme thermophiles does not seem so remarkable. Perutz and Raidt [20] have suggested that, assuming the same ground-state free energy for two similar proteins, the corresponding difference in the free energy of activation, $\Delta G\ddagger$, is given by $\Delta G\ddagger = RT\ln.\Delta t_{\frac{1}{2}}$, where $\Delta t_{\frac{1}{2}}$ is the difference in half-lives of the two proteins. This can be used to approximate the stabilizing energy difference between two proteins. On this basis, comparison of the denaturation rates at 70 °C of two highly similar enzymes, one from a mesophile and one from an extreme thermophile, might well give us a ratio of the order of $1:10^6$, and thus a difference in stabilizing energies of the two proteins of

about 40 kJ·mol⁻¹ at 70 °C. This could correspond to a rather small number of well placed stabilizing interactions.

Relatively good experimental evidence is available for this contention in terms of destabilization. A number of examples are known [25] where a point mutation has lowered ΔG by more than 20 kJ·mol⁻¹. However, such evidence is less readily available in terms of stabilization. We can calculate approximately the sum of the stabilizing interactions responsible for the integrity of the tertiary structure of a particular protein, and the magnitude of some individual interactions in a particular protein, but it is difficult to estimate the extent to which all the amino acids in a protein participate in stabilizing interactions, so we do not know the possibilities for developing additional interactions and thus achieving higher stability by point mutation. On the one hand we have the fact that the majority of point mutations of proteins have little or no effect on stability, suggesting that only a minority of amino acids are involved in significantly stabilizing interactions. On the other hand, quite extensive efforts to use single point mutations to enhance protein stability have been relatively unsuccessful in increasing half-lives by amounts corresponding to stabilizing energy differences (see above and [20]) of as much as 5 kJ·mol⁻¹. Nevertheless, accumulated point mutations have led to stability increases of more than 15 kJ·mol⁻¹ [42], and there does not seem to be any theoretical objection to proteins of very much greater conformational stability than those currently known.

From the above considerations it is evident that structural changes are not a requirement for changed stability. There is also no strong evidence that evolutionary changes in enzyme stability have proceeded by any particular strategy. There is a trend towards an increase in the average hydrophobicity of amino acids in some thermophilic enzymes [17] and, given that the hydrophobic effect is strongest at around 70 °C (and is weaker at temperatures above and below this), this might seem to indicate a broadly applicable stabilization strategy. However, this increase in hydrophobicity is also seen in hyperthermophilic enzymes which are stable above 100 °C [43]. The hydrophobic effect plays a major role in enzyme stability, and it would be surprising if in many very stable enzymes an increase in hydrophobicity could not be detected. Examples of increased enzyme stability caused by other weak interactions are also known [30,41,44].

A variety of strategies have been successful in achieving modest increases in protein stability, in addition to point mutations. These have included the incorporation of metal binding sites [45], elimination of a buried solvent molecule [46] and addition of a peptide β-hairpin [47]. Detailed comparison of mesophilic and extremely thermophilic enzymes has led to suggestions that the elimination of cavities within the enzyme structure and a decrease in the number of surface loops [48] may be ways of achieving greater stability. However, in addition to the difficulty of distinguishing which structural changes are associated with stability rather than, say, activity, comparisons are often further complicated by the likelihood that the extremely thermophilic enzyme will be from an Archaea, which is taxonomically separated from bacteria and possibly more primitive [7,48]. As is the case for amino acid substitutions, there is not yet evidence that any of these mechanisms are of very broad applicability. In structural terms the differences between stable and less stable enzymes are no greater than those found between enzymes of similar stability.

INTER-RELATIONSHIP OF ENZYME STABILITY AND ACTIVITY

Relatively few studies have been carried out on the dynamics of very stable enzymes, but evidence from hydrogen–deuterium exchange shows that, at a given temperature, thermostable enzymes are less flexible than thermolabile ones [49], and more specifically that enzymes from extreme thermophiles are, at room temperature, less flexible than those from mesophiles [50,51]. Theoretical studies [52] confirm this. However, at the temperatures at which they have evolved to act, the flexibility of enzymes from mesophiles and extreme thermophiles is similar.

There is considerable evidence that enzyme activity is dependent upon enzyme flexibility [53–56]. It is implicit in the concept of induced fit [57,58] that an enzyme must flex over a time scale in keeping with its catalytic-centre activity, and that changes between conformational substates allow catalytic activity. Enzyme activity has been demonstrated at temperatures below the glass transition [59], where mobility does not occur [60–63], but catalytic-centre activities are very low, and the significance of this activity is not yet clear.

Mesophilic enzymes from different sources carrying out the same reaction display a very wide range of specific activities. Once enough thermophilic enzymes have been studied, we may well find their range of specific activities to be large also. It is therefore difficult to validly compare enzymes of similar structure and function from mesophiles and thermophiles in terms of their specific activities. Such a comparison is further complicated because the 'natural' (*in vivo*) growth temperatures of many thermophiles may be different from those found using relatively rich growth media. It is fairly clear that an *E. coli* strain isolated from the human gut and growing optimally *in vitro* at 37 °C has evolved to function at that temperature. However, we do not know with any degree of certainty the temperatures at which most thermophiles (and indeed most mesophiles) have evolved. For example, *Thermus* strains isolated from hot pools at 85 °C may grow optimally *in vitro* at 70 °C. With these reservations, there is not currently good evidence to suggest that enzymes from thermophiles and mesophiles have different specific activities at their respective growth temperatures, apart from those special cases where changes in the susceptibility of the substrate occur at different temperatures. A typical example of this is the high specific activity of thermophilic proteinases when (denatured) mesophilic protein is the substrate, but not when small peptides are the substrate [64].

The activities of very stable enzymes thus correlate well with what is known of their dynamics: at temperatures where their flexibility is similar to that of less stable enzymes, there is no evidence for differences in specific activity. This correlation extends to conformational stability. Although enzymes from organisms growing at high temperatures are more stable than their counterparts from mesophiles, they are still denatured fairly quickly 20 °C or so above the optimum growth temperature for the organism [2]. The stabilities *in vivo* of the enzymes from both mesophiles and thermophiles are generally similar; or, to put it another way, the stabilities are similar at similar levels of molecular flexibility.

We can therefore correlate in general terms conformational stability, flexibility and specific activity, and together with the general structural and functional identity of stable and less stable enzymes, this leads to the view that the instability of enzymes from mesophiles is a functional requirement, rather than because of any restraint on achieving higher stability. It is required so that enzymes have sufficient flexibility to perform their catalytic functions. An additional requirement for instability can be inferred from the finding that, irrespective of whether or not they are denatured, stable proteins are more resistant to proteolysis [65]. Excessively stable enzymes may therefore hinder the normal cellular turnover of enzymes.

While enzyme activity is dependent upon flexibility, a less flexible enzyme will be more stable. Techniques that restrain free movement of the enzyme at the molecular level, such as immobilization and intramolecular cross-linking, tend to stabilize enzymes [66], including enzymes from extreme thermophiles [9,11]. However, excessive flexibility is the first step towards denaturation, so although an enzyme must be sufficiently rigid to have a reserve of stability, it must also have the flexibility needed for effective catalysis. These are of course generalizations, since stability is a global property of an enzyme [25] whereas the flexibility required for activity may well be localized [54–56]. They do, however, explain why enzymes tend in general to be denatured at temperatures not very far above their 'design' temperature. Too much stability would mean not enough flexibility for effective catalysis, whereas too little stability means too short a useful lifetime [7]. Point mutations which increase enzyme stability often lower specific activity [44,67–69], and vice versa. There is, therefore, strong evidence for the contention that enzyme stability, flexibility and activity are closely interrelated, and that a balance between stabilizing and destabilizing interactions is required to meet the conflicting demands of stability on the one hand, and catalytic function and cellular turnover on the other. As a consequence, if we consider only conformational stability, we may postulate that the reason we have not found proteins that are stable much above 130 °C is because we have not found organisms growing above 110 °C (rather than vice versa).

PROTEIN DEGRADATION ABOVE 80 °C

Although conformational stability can be maintained for thermophilic proteins at high temperatures [2,3], irreversible degradative processes can lead to enzyme inactivation [70,71]. In contrast to denaturation, the irreversible processes of protein inactivation arise from changes in covalent bonding. Deamidation of the amide side chain of Asn and Gln residues, succinimide formation at Glu and Asp, and oxidation of His, Met, Cys, Trp and Tyr are the most facile and common amino acid degradations. These mechanisms of protein degradation are greatly accelerated at high temperatures, and can thus play an important role in the thermo-inactivation of enzymes.

Asparagine residues

Asn residues are labile due to the deamidation of the side-chain amine, which yields Asp (Figure 3). This Asn deamidation in proteins and peptides in aqueous solution can proceed at a much higher rate than is observed for hydrolysis of an amide linkage of a peptide bond [70,72,73], and the increased rate suggests an intramolecular reaction [74,75]. Some deamidation in peptides and proteins is known to occur through intramolecular attack of the carboxy-peptide-nitrogen on the side-chain γ-carbonyl carbon, resulting in the formation of a succinimide ring [72,76]. The succinimide is unstable in aqueous solution and is hydrolysed at either carboxy group, producing a mixture of aspartyl and isoaspartyl peptides (Figure 3). The succinimide can further react by hydrolysis or racemization and yield L- and D-aspartyl and L- and D-isoaspartyl peptides [72].

The mechanism of deamidation in peptides is pH-dependent. Slow deamidation in peptides at pH 1–2 appears to largely bypass the succinimide intermediate [77]. Maximum stability of Asn residues within peptides was observed between pH 2 and 5. Between pH 5 and 12, the reaction proceeds entirely through a succinimide intermediate and is dependent on the concentration of OH^- (nucleophiles appearing to catalyse the reaction).

Furthermore, the identity of the amino acids in the vicinity of the Asn residue can affect rates of deamidation [72], because of

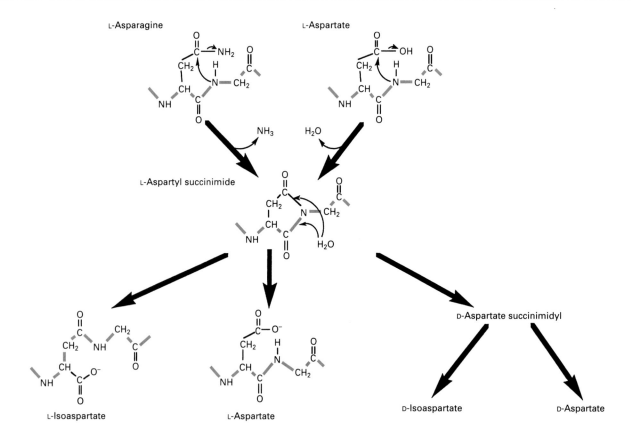

Figure 3   Mechanism for succinimide formation in asparagine and aspartate residues

The coloured bonds indicate the peptide backbone. See [81].

the conformational constraints imposed by the tertiary structure. Trypsin deamidates at three residues, Asn-48, -95 and -115, all of which have similar main-chain and side-chain conformations, characterized by the Oδ1 oxygen of the side chain forming a hydrogen bond with the $(n+2)$ peptide nitrogen [78]. This type of interaction is observed in peptide turns where Asn is found in the first position [79]. Modelling of the Asn environments suggests that with minimal main-chain movement ($+20$ ° phi, psi) the side chain can be rotated by 80 ° around the Cα–Cβ bond, which places the amide-carbonyl carbon in an orientation where it is readily accessible to attack by the peptide nitrogen of the succeeding $(n+1)$ residue. Additionally, the $n+1$ peptide nitrogen must be accessible and deprotonated. Each susceptible Asn is followed by a Ser with the side-chain hydroxy group in a position to hydrogen-bond to the Asn Oδ1 and Asn Nδ2 hydrogens, or the $n+1$ peptide nitrogen (which may aid in deprotonation of the nitrogen). These hydrogen bonds are thought to facilitate the formation of the cyclic imide intermediate [78].

Conformation alone, however, does not predict the propensity for deamidation. In trypsin, Asn-34 has a similar conformation to Asn-48, -95 and -115 and is followed by a Ser, yet no deamidation is observed. The succeeding peptide nitrogen of Asn-34 is hydrogen-bonded to a carbonyl oxygen. Further, the N34 to O64 and O34 to N64 hydrogen-bond network inhibits conformational change. Thus a combination of steric and hydrogen-bonding interactions prevent productive succinimide geometry at Asn-34. Other Asn residues are stable, since succinimide formation would require rotation of 180 ° around the

peptide bond. Structurally constraining hydrogen bonds protect other Asn residues in the highly structured Ca^{2+} binding loop [78]; Asn-34, -79, -97, -143 and -233 all hydrogen-bond to the $n+1$ peptide chain nitrogen (Figure 4), hindering the side-chain geometry required for succinimide formation. Hydrogen bonding of the main chain appears to inhibit deamidation more than hydrogen bonding of the amide side chain. Deamidation may, however, be limited when hydrogen bonding requires the side-chain amide moiety as a donor or where multiple hydrogen bonds are made with side chains [78]. Based on site proximity, the size of the $n+1$ amino acid seems to be more influential in deamidation than the identity of the $n-1$ residue or β-carboxy linkages [78].

The differing rates of succinimide formation between similar sequences in peptides and proteins may be explained by the precise positioning of relevant functional groups in rigid regions of the protein [76]. By constraining reactive groups to a limited range of orientations, the rate of deamidation can increase by as much as 10000-fold [80]. The optimal alignment of the peptide nitrogen and side-chain carbonyl oxygen in a three-dimensional protein structure could allow succinimide formation in sequences that are rarely deamidated in peptides [76]. Conversely, rigid suboptimal alignment might be expected to decrease the rate of succinimide formation. The orientation of both Asn and Asp residues in proteins is generally such that succinimide formation does not occur. However, residue combinations known from peptide studies to be susceptible to deamidation are found in proteins as frequently as less reactive combinations [81]. The effect of protein conformation is apparent when considering that

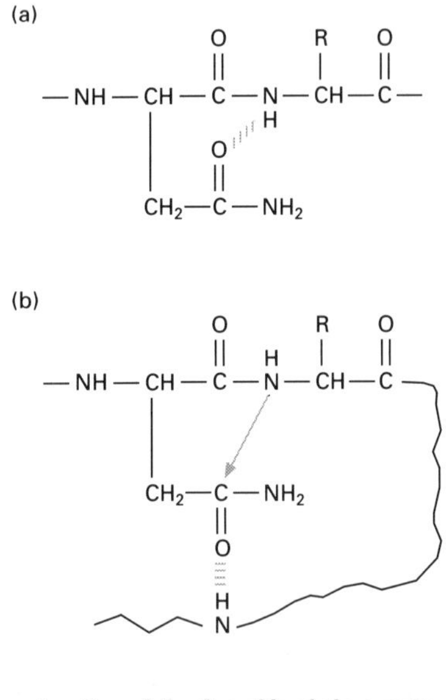

(a)

(b)

Figure 4   Hydrogen bonding of the Asn side-chain oxygen

Hydrogen bonding is shown to (**a**) the $n+1$ peptide-chain nitrogen (which hinders succinimide formation), and (**b**) a more distant peptide-chain nitrogen, which leaves the $n+1$ peptide chain nitrogen available for succinimide formation (coloured arrow).

the deamidation of Asn-67 in native ribonuclease A at 37 °C and pH 8 was found to be more than 30 times slower than that in the reduced and unfolded protein [82]. Furthermore, the presence of trifluoroethanol, known to stabilize secondary structures in proteins, suppressed deamidation of lysozyme at pH 6 and 100 °C [83].

The closely related structure of Gln is also able to deamidate, but at a much lower rate than for Asn. Gln deamidation is thought to proceed via analogous intramolecular reactions as for Asn, except that a six-membered ring leads to a glutarimide. The less favourable six-membered ring is expected to be formed approx. 10 times more slowly than the analogous five-membered succinimide [84]. However, as discussed above, deamidation in proteins is dependent on conformation; thus Gln deamidation could be a significant process producing glutamyl, isoglutamyl or γ-glutamyl products [85].

Although the most common mechanism for degradation at Asn is deamidation, Asn succinimide formation may cause cleavage of polypeptides under physiological conditions, and is known to occur in peptides at physiological pH [72,86]. The side-chain amide nitrogen attacks the peptide bond, releasing the carboxy-flanking peptide and forming a C-terminal succinimide (Figure 5). Asn cleavage has been noted in α-crystallin, where an Asn has the correct orientation to be both cleaved and de-amidated [87].

The reason for the greatly increased rate of Asn deamidation at high temperatures may be the activation energy, which was found to be 88.7 kJ·mol^{-1} for deamidation of peptides at pH 7.4 [72], indicating a marked temperature-dependence. The high energy of activation suggests that deamidation will occur at higher temperatures and explains the observation that, above 90 °C, deamidation is a major form of irreversible thermo-inactivation in both lysozyme (pH 4, 6 and 8) [70] and ribo-nuclease (pH 6 and 8) [71].

Aspartate residues

Asp is known to form succinimide intermediates which may be

hydrolysed to yield a break in the peptide chain and thus lead to irreversible degradation of the protein structure. Although succini-mide formation at Asp is thought to occur by a similar mechanism to that at Asn (Figure 3), the rate of Asp succinimide formation at 37 °C and pH 7.4 was measured in peptides and found to be 13–35 times less than that for the corresponding Asn peptide [76]. However, in proteins a significant fraction of succinimide forma-tion may originate from Asp residues. Indeed, this is the case in lysozyme [70] and ribonuclease [71], where significant irreversible thermo-inactivation was noted at pH 4 above 90 °C due to Asp hydrolysis. In calmodulin, succinimides are formed preferentially from Asp residues rather than Asn residues. At pH 7.4 the Asp side chain has only 1 in every 3000 carboxy groups protonated and thus available for succinimide formation [88]. The rate of succinimide formation for a protonated Asp would, therefore, approach that of an ester and would be much more rapid than the amide [76]. Thus not only conformation, but also the relative degree of protonation, will affect Asp succinimide formation [76].

Investigations of Asp and Asn degradation in calmodulin showed that, when Ca^{2+} was present, only Asp-2 and Asp-78 formed succinimides despite the presence of six Asn residues [89]. Two of the stable Asn residues form Gly-Asn sequences [89], which are labile in peptides [72,75] but not in protein structure. Succinimide formation at Asp and Asn is thus highly regulated by three-dimensional structure, due to the requirement that the peptide nitrogen atom be in a position to attack the side-chain carbonyl of the Asx residue [72]. The Ca^{2+}–calmodulin crystal structure indicates that the peptide-bond nitrogen of Thr-79 is not in a position to attack the side-chain carbonyl carbon of Asp-78 and form a succinimide [90]; however, for calmodulin, the crystal structure [90] and solution studies [91] indicate that Asp-78 and Thr-79 are in flexible regions, allowing succinimide formation. The location of the labile Asp-2 is the flexible N-terminal region of calmodulin [89]. In the absence of Ca^{2+}, the Ca^{2+} binding sites are less constrained and the Asx residue can form succinimides. The removal of Ca^{2+} may allow new contact between Asn and neighbouring residues that may catalyse succinimide formation [89].

Studies at 37 °C of human growth hormone indicated that isoaspartate formation at Asp-130 and Asn-149 occurs when followed by Ser. However, another Asn-Ser sequence does not form isoaspartate, suggesting that the structure of the protein influences isoaspartate formation [92]. In the X-ray crystal structure of the pig analogue of this protein, which is known to have extensive sequence identity to the human hormone, amino acids 128–151 are known to lack well defined structure [93], and thus there may be considerable freedom of rotation around that portion of the peptide backbone. Furthermore, when peptides were constructed to investigate rates of isoaspartate formation, Asp-130 had a $t_\frac{1}{2}$ value similar to that of the peptide, indicating similar conformational flexibility. The $t_\frac{1}{2}$ of Asn-149 was twice that of the peptide, suggesting that the conformation of the protein prevents succinimide formation in some cases [92].

The well known rapid hydrolysis at Asp is thought to proceed by a similar mechanism to hydrolysis at Asn, with rapid cleavage after formation of a five-membered ring [94] (Figure 5). The observation that Asp is the only amino acid released when proteins are heated in weak acids has long been known [95], and results of hydrolysis in dilute HCl at pH 4 of cytochrome c, wool proteins and egg yolk apovitellenins suggest that the majority of Asp bonds are cleaved within 2 h. This cleavage is dependent on the pH, as this affects the proton donor ability of the Asp side-chain carboxy group. Asp hydrolysis accounts for approx. 80 % of the irreversible thermoinactivation of ribonuclease at pH 4

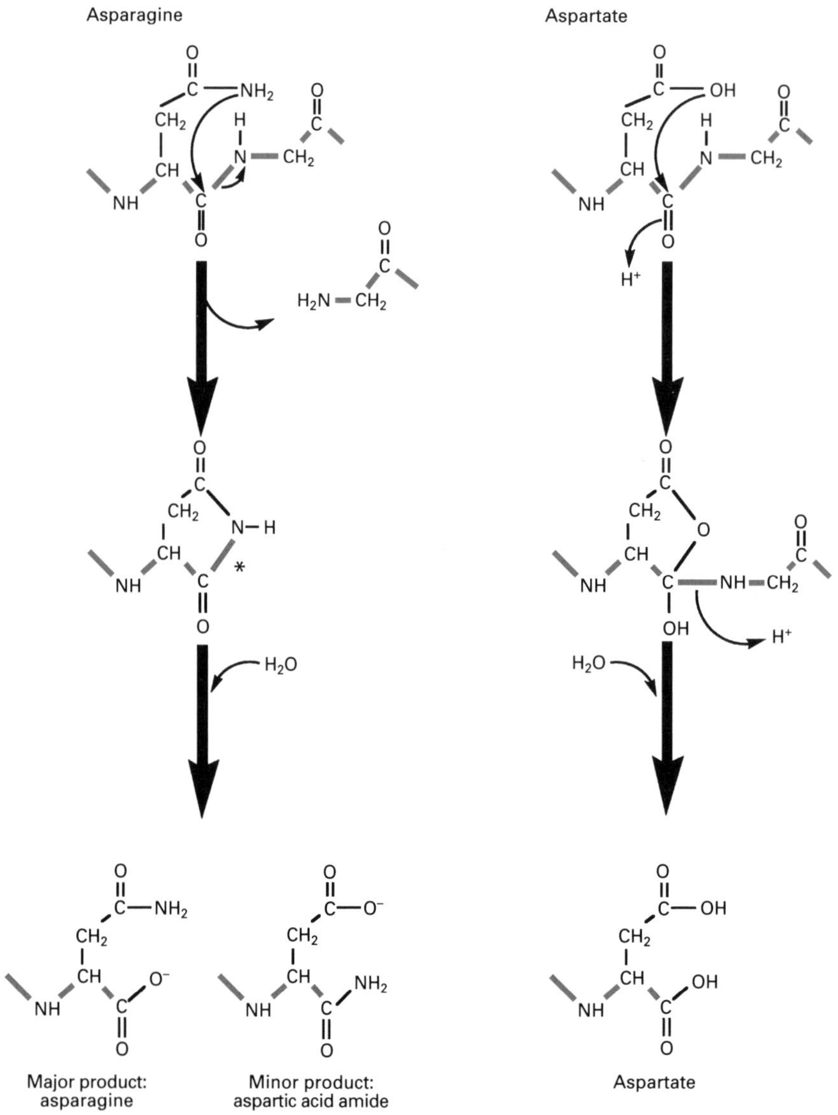

Figure 5   Ring formation and subsequent cleavage of asparagine [87] and aspartate [94] residues in proteins

The coloured bonds indicate the peptide backbone. The asterisk indicates the site of preferential hydrolysis.

and 90 °C [71], and has been noted in lysozyme at pH 4 and 100 °C [70].

Other residues

Oxidation of amino acid side chains also leads to irreversible thermo-inactivation in proteins. Potential sites of oxidation in proteins are the side chains of His, Met, Cys, Trp and Tyr residues [96]. The reactivity of a given Met residue towards oxidation seems to be dependent on its position. For example, human growth hormone Met-170 was found to be completely resistant to oxidation by hydrogen peroxide [97]. The thiol group of cysteine can also be oxidized. Even molecular oxygen will oxidize thiols in the presence of trace amounts of metal ions [98]. The nature of neighbouring groups greatly influences the oxidation of thiols [99]. The side chains of His, Tyr, Met, Cys and Trp residues can be oxidized via photo-oxidation in the presence of dyes [96].

Thermal destruction processes involving disulphides and thiol groups are also important in both structure and catalysis. β-Elimination of cystine under alkaline conditions forms dehydroalanine, which may go on to react with a nearby Lys residue [100]. β-Elimination is not limited to Cys; Ser, Thr, Phe and Lys can also be degraded in alkaline conditions [96]. Cys residues are also susceptible to disulphide exchange, in which disulphide bonds are reduced and later re-form between different Cys residues.

Aggregation of denatured proteins remains an important mechanism of irreversible thermo-inactivation, but often requires substantial disruption of the tertiary structure. The other mechanisms discussed above may occur at discrete sites within the protein while the conformation remains intact or nearly so.

In summary, numerous reactions which irreversibly change protein structure and thus inactivate enzymes are facilitated by increased temperatures (for specific examples, see Table 2). Clearly, the mechanisms of degradation vary dramatically in

Table 2 Major degradative processes determined by experiment in selected enzymes

Enzyme	Conditions	Prevalent degradative processes	Ref.
Cellobiohydrolase I (*Trichoderma reesei*)	70 °C, pH 4.8	Aggregation, deamidation	101
Lysozyme	100 °C, pH 6	Deamidation	83
Ribonuclease A	90 °C, pH 4	Deamidation, hydrolysis at Asp	71
α-Amylase (*Bacillus amyloliquefaciens*)	90 °C, pH 8	Deamidation	102
α-Amylase (*B. stearothermophilus*)	90 °C, pH 8	Oxidation of Cys	102

different enzymes since they are mediated by protein structure and can be highly dependent upon local flexibility and neighbouring amino acid residues.

INTERACTIONS BETWEEN DEGRADATION AND DENATURATION

The susceptibility of a protein to high-temperature degradative reactions seems to be dependent upon the conformational integrity of the protein at that temperature. The chemical mechanisms for irreversible degradation in proteins require a certain local molecular flexibility. For example, at 37 °C the rate of deamidation has been shown to be higher for small peptides with high flexibility than for proteins when comparing the same amino acid sequence [103], and higher in denatured proteins [82]. A survey of environments around Asp and Asn residues in known three-dimensional protein structures suggests that the rigidity of the folded protein greatly decreases the intramolecular imide formation necessary for further degradation. In the numerous X-ray crystal structures studied, the peptide-bond nitrogen could not approach the side-chain carbonyl carbon closely enough to form the succinimide ring [81].

Experimental evidence has linked the resistance to degradation of a protein with its conformational integrity. Thermally stable proteins have been used for these studies, so that the conformation is known to be retained for significant periods at the high temperatures at which degradative reactions occur. Hensel and colleagues have shown that the rate of deamidation for a thermostable glyceraldehyde phosphate dehydrogenase from

Pyrococcus woesei at 100 °C is increased after denaturation of the enzyme [104] (and after dialysing away the denaturing guanidinum hydrochloride). Furthermore, for the same enzyme from *Methanothermus fervidus* deamidation occurs more readily at 85 °C in a less stable form of the enzyme in which the protein is formed from a recombinant DNA comprised of both thermophilic and mesophilic genes. Similarly, the addition of phosphate, known to stabilize the conformation of the dehydrogenase, decreases the rate of peptide bond hydrolysis at temperatures ranging from 85 to 100 °C [104].

Support for the view that the loss of conformation precedes irreversible degradative reactions comes from studies of deamidation (ammonia release) and loss of activity of the very stable xylanase from *Thermotoga* strain FjSS 3B1 in the range 95–110 °C (C. Aebischer and R. M. Daniel, unpublished work). Both the onset and progress of deamidation occurred later than those of activity loss, consistent with a dependence of deamidation upon loss of conformation (Figure 6).

Studies on peptide bond hydrolysis in native and denatured myokinase at 95 °C (M. Dines, H. H. Petach and R. M. Daniel, unpublished work) give similar results. The rate of peptide bond hydrolysis is always lower than the rate of activity loss, although agents which affect the rate of activity loss such as SDS plus mercaptoethanol (faster) and substrate (slower) have a similar effect on peptide bond hydrolysis. These correlations are consistent with a dependence of degradation on loss of conformation.

The irreversible loss of activity of glucose isomerase at high temperatures may also be dependent upon protein unfolding. The temperature-dependence and large activation energy observed for the activity loss of the isomerase are uncharacteristic of covalent reactions. Thus Volkin and Klibanov [105] suggest that the deamidating residues may be located in the centre of the protein, and thus unfolding precedes deamidation, giving rise to large activation energies.

CONCLUSION

Until comparatively recently, an assumption that loss of enzyme activity at 100 °C was simply a more rapid version of that taking place at 60 °C would have seemed very reasonable. However, during the 1980s it became clear that at 100 °C and above a number of irreversible reactions, including deamidation and peptide bond cleavage, occur in addition to denaturation [70,71]. Although it is difficult to place a theoretical upper temperature limit on the stability of proteins to denaturation, because of their universal applicability these irreversible degradative reactions would seem to limit protein stability to temperatures below, say, 120 °C. This prediction would be in keeping with the failure of extensive efforts to find living organisms growing optimally at temperatures above 110 °C, since our concept of life rests heavily on the functioning of enzymes. However, enzymes have been found that have significant half-lives at 130 °C [8,10], and there is growing evidence that the degradative reactions to which proteins are subject are slower or do not occur in conformationally intact proteins. In other words, the upper temperature limit for protein stability may after all be determined by the conformational integrity of the protein, although we must bear in mind that little work on conformational or degradative stability has taken place above 100 °C.

If this is so, and since there is no obvious upper temperature limit on conformational stability, why have more stable proteins not been found? The answer presumably lies in the requirement for sufficient flexibility (instability) for activity and turnover, coupled with the lack of living organisms found growing above 110 °C.

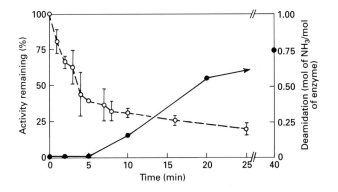

Figure 6 Denaturation and degradation of xylanase from *Thermotoga maritima* sp. FjSS3B1 at 100 °C

The plot shows activity loss (○) (means ± S.D.) and deamidation (●) (C. Aebischer and R. M. Daniel, unpublished work).

So what is the upper temperature limit for enzyme stability? Experiments carried out at 250 °C indicate that the half-life of some peptide bonds is only a few minutes at this temperature, giving rise to a calculated protein half-life of about 1 s [106]. Even though this work was carried out on small peptides, it is difficult to envisage proteins that are stable enough to withstand this temperature. The most stable proteins found so far have half-lives in excess of 10 min at 130 °C [8], and greater stabilities have been shown for very stable enzymes immobilized and in the presence of stabilizing agents [9,11], so that prospects exist for characterizing denaturation and degradation up to 140 °C or so. Such stable proteins could also be starting points for attempts to engineer higher stability. A variety of strategies are available [107]. Conformational stability can be increased by rigidifying the enzyme structure by such means as cross-linking and immobilization [66], but such a loss of flexibility is also likely to reduce enzyme activity. Thus the most useful approach might be to rigidify the enzyme outside the active-site region, and thus leave the active site flexible enough for high activity. Unfortunately, this may leave the active region of the enzyme as the most susceptible to degradation. Strategies are also available for reducing degradation, for example by mutating susceptible residues such as Asp to the less labile, but structurally similar, amino acid Glu [108,109].

However, the suggestion that enzyme activity is possible below the glass transition [59] (i.e. in a rigid protein structure), if confirmed, may indicate that enzyme activity does not absolutely require a flexible structure and thus one of limited stability. If so, this could hold out prospects of stabilizing the whole enzyme without too large a loss of catalytic activity. Few attempts have been made to further stabilize very stable enzymes. The stabilizations so far achieved in less stable proteins by point mutation and other techniques, while individually relatively small, should, when applied in concert to enzymes stable at 120 °C, be quite sufficient to confer stability at temperatures raised by 30 °C or so. Overall, it does not seem at all unreasonable to expect to see within the next 10 years enzymes that are active and stable at around 150 °C.

REFERENCES

1 Stetter, K. A., Fiala, G., Huber, G., Huber, R. and Segerer, A. (1990) FEMS Microbiol. Rev. **75**, 117–124
2 Coolbear, T., Daniel, R. M. and Morgan, H. W. (1992) Adv. Biochem. Eng. Biotech. **45**, 57–98
3 Adams, M. W. W. (1993) Annu. Rev. Microbiol. **47**, 627–658
4 Brandts, J. F. (1967) in Thermobiology (Rose, A. H., ed.), pp. 25–72, Academic Press, New York and London
5 Privalov, P. L. (1979) Adv. Protein Chem. **33**, 167–241
6 Jaenicke, R. (1991) Eur. J. Biochem. **202**, 715–728
7 Daniel, R. M. (1996) Enzyme Microb. Technol., in the press
8 Koch, R., Zoblowski, P., Spreinat, A. and Antranikian, G. (1990) FEMS Microbiol. Lett. **71**, 21–26
9 Simpson, H. D., Haufler, U. R. and Daniel, R. M. (1991) Biochem. J. **277**, 413–417
10 Brown, S. H. and Kelly, R. M. (1993) Appl. Environ. Microbiol. **59**, 2614–2621
11 Piller, K., Daniel, R. M. and Petach, H. H. (1996) Biochim. Biophys. Acta **1292**, 197–205
12 Toogood, H.S, Prescott, M. and Daniel, R. M. (1995) Biochem. J. **307**, 783–789
13 Rupley, J. A. and Careri, G. (1991) Adv. Protein Chem. **41**, 37–172
14 Mullaney, P. F. (1966) Nature (London) **210**, 953
15 Mozhaev, V. V., Poltevsky, K. G., Slepner, V. I., Badun, G. A. and Levashorv, A. V. (1991) FEBS Lett. **292**, 159–161
16 Hensel, R. and Jakob, L. (1994) Syst. Appl. Microbiol. **16**, 742–745
17 Argos, P., Rossman, M. G., Gnau, U. M., Zuber, H., Frank, G. and Tratschim, J. D. (1979) Biochemistry **18**, 5698–5703
18 Menendez-Arias, L. and Argos, P. (1989) J. Mol. Biol. **206**, 397–406
19 Bohm, G. and Jaenicke, R. (1994) Int. J. Peptide Protein Res. **43**, 97–106
20 Perutz, M. F. and Raidt, H. (1975) Nature (London) **255**, 256–259
21 Day, M. W., Hsu, B. T., Joshua-Tor, L., Park, J.-B., Zhou, Z. H., Adams, M. W. W. and Rees, D. C. (1992) Protein Sci. **1**, 1494–1507
22 Blake, P. R., Park, J.-B., Zhou, Z. H., Hare, D. R., Adams, M. W. W. and Summers, M. F. (1992) Protein Sci. **1**, 1508–1521
23 Langridge, J. (1968) J. Bacteriol. **96**, 1711–1717
24 Alber, T. (1989) Annu. Rev. Biochem. **58**, 765–798
25 Matthews, B. W. (1993) Annu. Rev. Biochem. **62**, 139–160
26 Matthews, B. W. (1993) Curr. Opin. Struct. Biol. **3**, 589–593
27 Pakula, A. A., Young, V. B. and Sauer, R. T. (1986) Proc. Natl. Acad. Sci. U.S.A. **83**, 8829–8833
28 Bryan, P. N., Rollance, M. L., Pantoliano, M. W., Wood, J., Finzel, B. C., Gilliland, G. L., Howard, A. J. and Poulos, T. L. (1986) Proteins **1**, 326–334
29 Alber, T. and Wozniak, J. A. (1985) Proc. Natl. Acad. Sci. U.S.A. **82**, 747–750
30 Matsumura, M. and Aiba, S. (1985) J. Biol. Chem. **260**, 15298–15303
31 Liao, H., Mackenzie, T. and Hageman, R. (1986) Proc. Natl. Acad. Sci. U.S.A. **83**, 576–580
32 Rennel, D., Bouvier, S. E., Hardy, L. W. and Poteete, A. R. (1991) J. Mol. Biol. **222**, 67–87
33 Baldwin, E. P. and Matthews, B. W. (1994) Curr. Opin. Biotechnol. **5**, 396–402
34 Creighton, T. E. (1993) Proteins, 2nd edn., Freeman, New York
35 Suzuki, Y., Oishi, K., Nakano, H. and Nagayama, T. (1987) Appl. Microbiol. Biotechnol. **26**, 546–551
36 Matthews, B. W., Nicholson, H. and Becktel, W. J. (1987) Proc. Natl. Acad. Sci. U.S.A. **84**, 6663–6667
37 Watanabe, K., Masuda, T., Ohashi, H., Mihara, H. and Suzuki, Y. (1994) Eur. J. Biochem. **226**, 277–293
38 Serrano, L. and Fersht, A. R. (1989) Nature (London) **342**, 296–299
39 Bell, J. A., Becktel, W. J., Sayer, U., Baase, W. H. and Matthews, B. W. (1992) Biochemistry **31**, 499–511
40 Dasgupta, S. and Bell, J. A. (1993) Int. J. Peptide Protein Res. **41**, 499–511
41 Tidor, B. (1994) Proteins **19**, 310–323
42 Pantoliano, M. W., Whitlow, M., Wood, J. F., Dodd, S. W., Handnum, K. D., Rollence, M. L. and Bryan, P. N. (1989) Biochemistry **28**, 7205–7219
43 Hensel, R. (1993) in The Biochemistry of the Archaea (Kates, M., Kushner, D. J. and Matheson, A. T., eds.), pp. 209–221, Elsevier, Amsterdam and New York
44 Matsumura, M., Yasumura, S. and Aiba, S. (1986) Nature (London) **323**, 356–358
45 Kuroki, R., Kawakita, H., Nakamura, H. and Yutani, K. (1992) Proc. Natl. Acad. Sci. U.S.A. **89**, 6803–6807
46 Vriend, G., Bevendsen, H. J. C., Van Der Zee, J. R., Van Den Berg, B., Venema, G. and Eijsink, V. G. H. (1991) Protein Eng. **4**, 941–945
47 Eijsink, V. G. H., Vriend, G., Van Den Berg, B., Van Der Zee, J. R., Veltman, O. R., Stulp, B. K. and Venema, G. (1992) Protein Eng. **5**, 157–163
48 Russell, R. J. M., Hough, D. W., Danson, M. J. and Taylor, G. L. (1994) Structure **2**, 1157–1167
49 Wagner, G. and Wuthrich, K. (1979) J. Mol. Biol. **130**, 31–37
50 Wrba, A., Schweiger, A., Schultes, V., Jaenicke, R. and Zavodszky, P. (1990) Biochemistry **29**, 7584–7592
51 Varley, P. G. and Pain, R. H. (1991) J. Mol. Biol. **220**, 531–538
52 Vihinen, M. (1987) Protein Eng. **1**, 477–480
53 Lipscombe, W. N. (1970) Acc. Chem. Res. **3**, 81–89
54 Frauenfelder, H., Petsko, G. A. and Tsernoglou, D. (1979) Nature (London) **280**, 558–563
55 Artymuik, P. J., Blake, C. C. F., Grace, D. E. P., Oatley, S. J., Phillips, D. C. and Sternberg, M. J. E. (1979) Nature (London) **280**, 563–568
56 Huber, R. and Bennett, W. S. (1983) Biopolymers **22**, 261–279
57 Neet, K. E. and Koshland, D. E. (1966) Proc. Natl. Acad. Sci. U.S.A. **56**, 1606–1611
58 Koshland, D. E. (1970) Enzymes 3rd Ed. **1**, 342–396
59 More, N., Daniel, R. M. and Petach, H. H. (1995) Biochem. J. **305**, 17–20
60 Parak, F., Frolov, E. N., Kononenko, A. A., Mössbauer, R. L., Goldanskii, V. I. and Rubin, A. B. (1980) FEBS Lett. **117**, 368–372
61 Parak, F., Knapp, E. W. and Kucheida, D. (1982) J. Mol. Biol. **161**, 177–194
62 Hartmann, H., Parak, F., Steigemann, W., Petsko, G. A., Ringe Ronzi, D. and Frauenfelder, H. (1982) Proc. Natl. Sci. U.S.A. **79**, 4967–4971
63 Bauminger, E. R., Cohen, S. G., Nowik, I., Ofer, S. and Yariv, J. (1983) Proc. Natl. Acad. Sci. U.S.A. **80**, 736–740
64 Cowan, D. A., Daniel, R. M. and Morgan, H. W. (1987) Int. J. Biochem. **19**, 741–743
65 Daniel, R. M., Cowan, D. A., Morgan, H. W. and Curran, M. P. (1982) Biochem. J. **207**, 641–644
66 Germain, P., Slagmolen, T. and Crighton, R. R. (1982) Biotechnol. Bioeng. **33**, 563–569
67 Wagas, T., Makino, Y., Yamamoto, K., Urabe, I. and Ohada, H. (1989) FEBS Lett. **253**, 113–116

68 Makino, Y., Negoro, I., Urabe, I. and Ohaka, H. (1989) J. Biol. Chem. **264**, 6381–6385

69 Shoichet, B. K., Baase, W. A., Kurohi, R. and Matthews, B. W. (1995) Proc. Natl. Acad. Sci U.S.A. **92**, 452–456

70 Ahern, T. J. and Klibanov, A. M. (1985) Science **228**, 1280–1283

71 Zale, S. E. and Klibanov, A. M. (1986) Biochemistry **25**, 5432–5444

72 Geiger, T. and Clarke, S. (1987) J. Biol. Chem. **262**, 785–794

73 Kahne, D. and Still, W. C. (1988) J. Am. Chem. Soc. **110**, 7529–7534

74 Bernhard, S. A., Berger, A., Carter, J. H., Katchalski, E., Sela, M. and Shalitin, Y. (1962) J. Am. Chem. Soc. **84**, 2421–2434

75 Bornstein, P. and Balian, G. (1977) Methods Enzymol. **47**, 132–145

76 Stephenson, R. C. and Clarke, S. (1989) J. Biol. Chem. **264**, 6164–6170

77 Patel, K. and Borchardt, R. T. (1990) Pharm. Res. **7**, 787–793

78 Kossiakoff, A. A. (1988) Science **240**, 191–194

79 Richardson, J. S. (1981) Adv. Protein Chem. **34**, 168–330

80 Storm, D. R. and Koshland, P. E. (1972) J. Am. Chem. Soc. **94**, 5805–5814

81 Clarke, S. (1987) Peptide Protein Res. **30**, 808–821

82 Wearne, S. J. and Creighton, T. E. (1989) Proteins Struct. Funct. Genet. **5**, 8–12

83 Tomizawa, H., Yamada, H., Wada, K. and Imoto, J. (1995) J. Biochem. (Tokyo) **117**, 635–640

84 Terwilliger, T. C. and Clarke, S. (1981) Biochemistry **256**, 3067–3076

85 Clarke, S., Stephenson, R. C. and Lowenson, J. D. (1992) in Stability of Protein Pharmaceuticals (Ahern, T. and Manning, M. C., eds.), Part A, pp. 1–29, Plenum Press, New York

86 Blodgett, J. K., London, G. M. and Collins, K. D. (1985) J. Am. Chem. Soc. **107**, 4305–4313

87 Voorter, C. E., de Haard-Hoekman, W. A., van der Oetelaar, P. J. M., Bloemendal, H. and de Jong, W. W. (1988) J. Biol. Chem. **263**, 19020–19023

88 Bernhard, S. A. and Shalitin Y. (1966) J. Am. Chem. Soc. **88**, 4711–4721

89 Ota, I. M. and Clarke, S. (1989) Biochemistry **28**, 4020–4027

90 Babu, Y. S., Bugg, C. E. and Cook, W. (1988) J. Mol. Biol. **204**, 191–204

91 Persechini, A. and Kretsinger, R. H. (1988) J. Biol. Chem. **263**, 12175–12178

92 Johnson, B. A., Shriokawa, J. M., Hancock, W. S., Spellman, M. W., Basd, L. J. and Aswad, D. W. (1989) J. Biol. Chem. **264**, 14262–14271

93 Abdel-Meguid, S.S, Shieh, H, Smith, W. W, Dayringer, H. E., Violand, B. N. and Bentle, L. A. (1987) Proc. Natl. Acad. Sci. U.S.A. **84**, 6434–6437

94 Inglis, A. S. (1983) Methods Enzymol. **91**, 322–324

95 Partridge, S. M. and Davis, H. F. (1950) Nature (London) **165**, 62–63

96 Manning, M. C., Patel, K. and Borchardt, T. T. (1989) Pharm. Res. **6**, 903–918

97 The, L. C., Murphy, L. S., Hua, N. L., Surus, A. S., Friesen, H. G., Lazarus, L. and Chapman, G. E. (1987) J. Biol. Chem. **262**, 6472–6477

98 Lamfrom, H. and Nielson, S. O. (1957) J. Am. Chem. Soc. **79**, 1966–1970

99 Overberger, C. G. and Ferraro, J. J. (1962) J. Org. Chem. **27**, 3539–3545

100 Ahern, T.J and Klibanov, A. M. (1987) Methods Biochem. Anal. **33**, 91–127

101 Jimenez, J., Dominguez, J. M., Castillon, M. P. and Acebal, C. (1995) Carbohydr. Res. **268**, 257–266

102 Tomazic, S. J. and Klibanov, A. M. (1988) J. Biol. Chem. **263**, 3092–3096

103 Brennan, T. V., Anderson, J. W., Jua, Z., Waygood, E. B. and Clarke, S. (1994) J. Biol. Chem. **269**, 24586–24595

104 Hensel, R., Jakob, I., Scheer, H. and Lottspeich, R. (1992) Biochem. Soc. Symp. **58**, 127–133

105 Volkin, D. B. and Klibanov, A. M. (1989) Biotechnol. Bioeng. **33**, 1104–1111

106 White, R. H. (1984) Nature (London) **310**, 430–432

107 Mozhaev, V. V. (1993) Trends Biotechnol. **11**, 88–95

108 George-Nascimento, C., Lowenson, J., Borissenko, M., Calderon, M., Medina-Selby, A., Kuo, J., Clarke, S. and Randolph, A. (1990) Biochemistry **29**, 9584–9591

109 Ahern, T. J., Casal, J. I., Petsko, G. A. and Klibanov, A. M. (1987) Proc. Natl. Acad. Sci. U.S.A. **84**, 675–679

Biochem. J. (1996) **320**, 345–357 (Printed in Great Britain)

REVIEW ARTICLE
Mammalian mitochondrial β-oxidation

Simon EATON, Kim BARTLETT* and Morteza POURFARZAM
Sir James Spence Institute of Child Health, Royal Victoria Infirmary, Newcastle-upon-Tyne NE1 4LP, U.K.

The enzymic stages of mammalian mitochondrial β-oxidation were elucidated some 30–40 years ago. However, the discovery of a membrane-associated multifunctional enzyme of β-oxidation, a membrane-associated acyl-CoA dehydrogenase and characterization of the carnitine palmitoyl transferase system at the protein and at the genetic level has demonstrated that the enzymes of the system itself are incompletely understood. Deficiencies of many of the enzymes have been recognized as important causes of disease. In addition, the study of these disorders has led to a greater understanding of the molecular mechanism of β-oxidation and the import, processing and assembly of the β-oxidation enzymes within the mitochondrion. The tissue-specific regulation, intramitochondrial control and supramolecular organization of the pathway is becoming better understood as sensitive analytical and molecular techniques are applied. This review aims to cover enzymological and organizational aspects of mitochondrial β-oxidation together with the biochemical aspects of inherited disorders of β-oxidation and the intrinsic control of β-oxidation.

1 INTRODUCTION

β-Oxidation is the major process by which fatty acids are oxidized, thus providing a major source of energy for the heart and for skeletal muscle [1,2]. Hepatic β-oxidation serves a different role by providing ketone bodies (acetoacetate and β-hydroxybutyrate) to the peripheral circulation. Ketone bodies are an important fuel for extra-hepatic organs, especially the brain, when blood glucose levels are low. For this reason, β-oxidation is stimulated when glucose levels are low, for instance during starvation or endurance exercise, essentially as postulated in the Randle cycle [2a].

Investigation of the pathway of β-oxidation of long-chain fatty acids commenced with the celebrated early studies of Knoop [3], who fed dogs ω-phenyl odd or even carbon-number long-chain fatty acids. He inferred from the excretion of phenylacetylglycine and benzoylglycine respectively, that the metabolism of fatty acids proceeded by a process of successive removal of two-carbon fragments. These observations were subsequently confirmed by Dakin [4]; however, detailed elucidation of the pathway was not achieved until the discovery of CoA and the realization that acetyl-CoA is the product of β-oxidation [5,6]. The chemical synthesis of acyl-CoA substrates and the elucidation of the cofactor requirements of soluble extracts of mitochondria carrying out β-oxidation [7] was rapidly followed by series of papers from the laboratories of Green, Lynen and Ochoa respectively, and the discovery of the basic sequence of steps: FAD-linked dehydrogenation, hydration, NAD$^+$-linked dehydrogenation and thiolytic cleavage, to yield acetyl-CoA, e.g. [7a,7b]. The role of carnitine in the transport of fatty acids across the inner mitochondrial membrane (see [8]), and the function of malonyl-CoA as a regulator of transport, together with the discovery of auxiliary systems for the metabolism of polyunsaturated fatty acids (PUFA) have been more recent developments. The basic enzymology of β-oxidation is shown in Figure 1. In the present review we discuss the recent developments in the

enzymology of the pathway, the discovery of a class of inherited metabolic disorders of mitochondrial β-oxidation and the organization and control of the pathway.

2 ENZYMES OF MITOCHONDRIAL β-OXIDATION

2.1 Acyl-CoA synthases

The enzymes of mitochondrial β-oxidation are summarized in Table 1. The enzymes of β-oxidation all act on CoA esters, so a preliminary to β-oxidation is the ATP-dependent formation of fatty acyl-CoA esters, catalysed by acyl-CoA synthase. Several acyl-CoA synthases are associated with mammalian mitochondria. Of these, the short-chain acyl-CoA synthases are found within the matrix and are important in ruminants [9]. Two medium-chain acyl-CoA synthases are also found in the mitochondrial matrix [10,11]. Long-chain acyl-CoA synthase activity is found in the mitochondrial outer membrane [12] and it appears to be a transmembrane protein with at least the CoA-binding domain on the cytosolic face [13].

2.2 Carnitine palmitoyltransferases (CPTs) and the acylcarnitine-carnitine translocase

Although the CPT system for the entry of acyl moieties into the mitochondrion has been known since the 1960s, until recent years the enzymology of the system has been controversial. Essentially, the controversy was over: (a) whether the outer CPT (henceforth CPT I) and the inner CPT (henceforth CPT II) were different polypeptides; and (b) whether the malonyl-CoA binding of CPT I was due to the catalytic polypeptide or a separate subunit. Now, hepatic CPT I has been purified and immunologically characterized as distinct from CPT II [14] and has been cloned and sequenced, allowing expression in yeast and demonstration of malonyl-CoA binding by the catalytic polypeptide [15]. An isoform of CPT I immunologically distinct from that in liver is present in skeletal muscle, heart and adipose tissue

---

Abbreviations used: CPT I, outer carnitine palmitoyl transferase; CPT II, inner carnitine palmitoyl transferase; PUFA, polyunsaturated fatty acids; SCAD, short-chain acyl-CoA dehydrogenase; MCAD, medium-chain acyl-CoA dehydrogenase; LCAD, long-chain acyl-CoA dehydrogenase; VLCAD, very-long-chain acyl-CoA dehydrogenase; ETF, electron transfer flavoprotein; ETF:QO, electron transfer flavoprotein:ubiquinone oxidoreductase; hsp, heat-shock protein; ACD, acyl-CoA dehydrogenase; LCHAD, long-chain 3-hydroxyacyl-CoA dehydrogenase; SCHOAD, short-chain 3-hydroxyacyl-CoA dehydrogenase; NEFA, non-esterified fatty acids.
* To whom correspondence should be addressed.

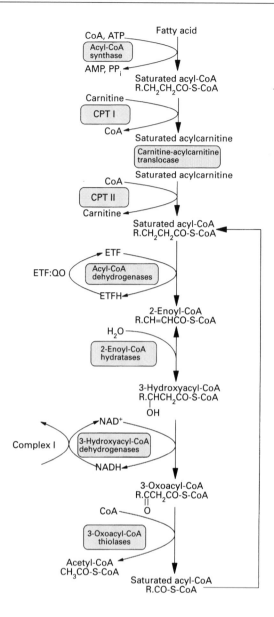

Figure 1   Enzymes of mitochondrial β-oxidation

Abbreviations: CPT, carnitine palmitoyl transferase; ETF, electron transfer flavoprotein; ETF:QO, ETF:ubiquinone oxidoreductase; ETFH, reduced ETF.

Table 1   Enzymes of mitochondrial β-oxidation

Enzyme	Structure	Molecular mass (kDa)	Reference
Acyl-CoA synthase	Unknown	78	[223]
Carnitine palmitoyl transferase I (liver)	Unknown	88	[14]
Carnitine palmitoyl transferase I (muscle)	Unknown	82	[17,18]
Carnitine acylcarnitine translocase	Unknown	32.5	[20]
Carnitine palmitoyl transferase II	Unknown	68	[224]
Very-long-chain acyl-CoA dehydrogenase	Homodimer	150	[22]
Long-chain acyl-CoA dehydrogenase	Homotetramer	180	[225]
Medium-chain acyl-CoA dehydrogenase	Homotetramer	180	[226]
Short-chain acyl-CoA dehydrogenase	Homotetramer	168	[227]
Long-chain 3-hydroxyl-CoA dehydrogenase	Heterooctomer	460	[49]
Long-chain 2-enoyl-CoA hydratase			
Long-chain 3-oxoacyl-CoA thiolase			
(Trifunctional protein)			
Short-chain 2-enoyl-CoA hydratase	Homohexamer	164	[32]
Short-chain 3-oxoacyl-CoA thiolase	Homotetramer	169	[228]
General 3-oxoacyl-CoA thiolase	Homotetramer	200	[48]
Electron transfer flavoprotein (ETF)	Heterodimer	57	[227]
ETF:ubiquinone oxidoreductase	Monomer	68	[30]
2,4-Dienoyl-CoA reductase	Homotetramer	124	[229]
Short-chain Δ^3,Δ^2-enoyl-CoA isomerase	Homodimer	70	[230]
Long-chain Δ^3,Δ^2-enoyl-CoA isomerase	Unknown	200	[231]
$\Delta^{3,5},\Delta^{2,4}$-dienoyl-CoA isomerase	Homotetramer	126	[232]

2.3 Acyl-CoA dehydrogenase (ACD), electron-transfer flavoprotein (ETF) and ETF:ubiquinone oxidoreductase

There are multiple enzymes for each of the constituent steps of the pathway, which vary in their chain-length specificity. In the case of acyl-CoA dehydrogenation there are four enzymes: short-chain acyl-CoA dehydrogenase (SCAD, active with C_4 and C_6), medium-chain acyl-CoA dehydrogenase (MCAD, active with C_4 to C_{12}), long-chain acyl-CoA dehydrogenase (LCAD, active with C_8 to C_{20}) and very-long-chain acyl-CoA dehydrogenase (VLCAD, active with C_{12} to C_{24}). Each of these enzymes catalyses the formation of 2-enoyl-CoA from the corresponding saturated ester. SCAD, MCAD and LCAD are homotetramers located in the matrix. VLCAD, however, is a homodimer and is located in the inner mitochondrial membrane. Until recently it had been assumed that there are only three ACDs involved in mitochondrial β-oxidation; SCAD, MCAD and LCAD. However, the isolation and purification of VLCAD [22], and the demonstration that patients previously thought to have inherited deficiencies of LCAD were in fact suffering from VLCAD deficiency [23], has shown that there are in fact four enzymes. Recently a novel member of this family has been identified by expression of a cDNA with an open reading frame of 1.3 kb encoding a precursor polypeptide of 431 amino acids, which was processed to a mature protein of 399 amino acids [24]. This enzyme had 38% homology with SCAD and lesser homology with other members of the ACD family. The enzyme is active towards branched and straight short-chain substrates. The role of this enzyme in β-oxidation remains unclear. MCAD is the best characterized of the ACD family, and the structure of the protein at 3 Å resolution, a dimer of dimers, was reported some years ago [25]. The mechanism of action of this group of flavoproteins appears to be very similar, with the concerted removal of the pro-R-α-hydrogen from the acyl-CoA as a proton and elimination of the corresponding pro-R-β-hydrogen to the N-5 position of the flavin as a hydride equivalent [26].

Reoxidation of the FAD prosthetic group of the ACDs requires a matrix FAD-linked protein, the ETF ([27]), which in turn

[14,16], and Weis et al. have demonstrated, on the basis of inhibitor-binding studies, that two isoforms of CPT I, the liver isoform and the skeletal muscle isoform, are simultaneously expressed in heart tissue [17]. They have subsequently shown, using the differential sensitivity of the two isoforms to DNP-etomoxir [18], that the contribution of the liver form to total heart CPT I activity decreases from 25% in the neonatal period to 2–3% in adult rats [19], and they hypothesize that the reason for this is that the markedly different kinetic characteristics of the two isoforms with respect to carnitine and to malonyl-CoA inhibition overcome the low perinatal carnitine levels in the heart.

The carnitine acylcarnitine translocase, which had previously only been partially purified, has been purified and found to catalyse a slow unidirectional transport of carnitine in addition to its translocase activity [20,21].

passes reducing equivalents to ETF:ubiquinone oxidoreductase (ETF:QO [28]) and thence to the mitochondrial respiratory chain at the level of ubiquinone. ETF contains 1 mol of bound FAD per mol of dimer [29], and EFT:QO is a 68 kDa iron–sulphur flavoprotein of the inner membrane [28,30].

2.4 2-Enoyl-CoA hydratase

Similarly it appears that there are three 2-enoyl-CoA hydratases. One, a soluble matrix enzyme, is most active towards short-chain substrates, although it will act on substrates up to C_{16} at a much slower rate and with higher K_m values [31–33]. This was the first enzyme of mammalian mitochondrial β-oxidation to be purified (crotonase; short-chain-enoyl-CoA hydratase, EC 4.2.1.17 [34]). The long-chain enzyme (EC 4.2.1.74) is most active with C_6 to C_{10} substrates and virtually inactive with crotonyl-CoA (C_4), the preferred substrate of crotonase [35,36]. It is now apparent that the long-chain enzyme is in fact a constituent of the trifunctional enzyme described below. Studies of patients with inherited disorders suggested the existence of a third, medium-chain enzyme [37], and this activity has now been partially characterized from pig and human liver [38].

2.5 3-Hydroxyacyl-CoA dehydrogenase

The third step of the pathway, L-3-hydroxyacyl-CoA dehydrogenation, is catalysed by two enzymes with overlapping chain-length specificities. The short-chain enzyme is a soluble matrix enzyme which will act on substrates of chain-length C_4 to C_{16} although, as with crotonase, the shorter chain-length substrates are preferred [39–42]. A long-chain 3-hydroxyacyl-CoA dehydrogenase was first demonstrated by El-Fakhri and Middleton [43]. This enzyme is firmly associated with the inner mitochondrial membrane and is active with medium- and long-chain substrates, C_{16} being the preferred substrate. As with the long-chain-enoyl-CoA hydratase described above, the long-chain 3-hydroxyacyl-CoA dehydrogenase is a constituent of the trifunctional protein.

2.6 3-Oxoacyl-CoA thiolase

The final step of the pathway, thiolytic cleavage of 3-oxoacyl-CoA to yield acetyl-CoA and a chain-shortened intermediate, is catalysed by three enzymes. Two soluble activities have been identified. One is specific for acetoacetyl-CoA and 2-methylacetoacetyl-CoA (EC 2.3.1.9 [44–46]). The second thiolase, the 'general' thiolase, is active with all substrates from C_6 to C_{16} to an approximately equal extent [47,48]. The third activity is part of a newly described trifunctional enzyme which also comprises the long-chain 2-enoyl-CoA hydratase and long-chain 3-hydroxyacyl-CoA dehydrogenase activities described above.

2.7 Trifunctional protein

The trifunctional protein complex, a hetero-octomer made up of four α-units with long-chain-enoyl-CoA hydratase and 3-hydroxyacyl-CoA dehydrogenase activities and 4 β-units with long-chain 3-oxothiolase activity, is closely associated with the inner mitochondrial membrane, and was described by Uchida et al. [49] and rapidly confirmed by others [50,51]. Evidence for the existence of such a complex was first suggested from our studies of a child with an inherited disorder of β-oxidation. Analysis of acyl-CoA and acyl-carnitine esters derived from incubations of mitochondrial fractions with [U-^{14}C]hexadecanoate revealed the presence of 3-oxo-, 3-hydroxy- and 2-enoyl- derivatives. Subsequent enzyme measurement demonstrated a total absence of

long-chain 3-oxoacyl-CoA thiolase activity and markedly diminished long-chain enoyl-CoA hydratase and 3-hydroxyacyl-CoA dehydrogenase activities [37]. The α-unit is required for membrane binding and a mutant trifunctional protein with an absent α-unit resulted in mislocation of β-unit 3-oxoacyl-CoA thiolase activity in the matrix [52].

A complex of activities associated with CPT II has also been reported by Kerner and Bieber [53]. However, in that study the constituent activities were not characterized with any precision, and it is difficult to know if Bieber's complex also involved the trifunctional enzyme. In any event CPT II, VLCAD and the trifunctional protein are associated with the inner mitochondrial membrane together with Complex I, ETF:QO and the remainder of the respiratory chain. The implications of this arrangement are discussed below in relation to intra-mitochondrial control of β-oxidation.

2.8 Auxiliary enzymes of PUFA oxidation

PUFA are probably β-oxidized at a low rate due to the low activity of CPT towards PUFA-CoA [54]. Complete mitochondrial oxidation is important, as PUFA-CoA can act as inhibitors of β-oxidation [55]. PUFA present two main problems to the classical enzymes of β-oxidation. The first of these is that many PUFA contain *cis* double bonds at even-numbered carbon atoms. After chain shortening, a *cis*-2 double-bonded fatty acid is formed. This can be hydrated, but yields a D-3-hydroxyacyl-CoA ester which is not a substrate for the L-specific 3-hydroxyacyl-CoA dehydrogenases, and would require epimerization to the L-form before further oxidation. However, it is now generally accepted that mitochondrial oxidation of PUFA proceeds via Δ^2,Δ^4-dienoyl-CoA reduction to 3-enoyl-CoA, followed by the action of Δ^3,Δ^2-enoyl-CoA isomerase, rather than the epimerase-dependent route [56]. The second problem is the presence of *cis*-5 double bonds in PUFA, which will eventually yield 2-*trans*,5-*cis*-dienoyl-CoA. These could be oxidized by chain shortening (to 3-enoyl-CoA), followed by the action of Δ^3,Δ^2-enoyl-CoA isomerase and complete oxidation. However, a further enzyme, $\Delta^{3,5},\Delta^{2,4}$-dienoyl-CoA isomerase, has been purified. This alternative route involves isomerization of 2-*trans*,5-*cis*-dienoyl-CoA to Δ^3,Δ^5–dienoyl-CoA (by Δ^3,Δ^2-enoyl-CoA isomerase) and a further isomerization to $\Delta^{2,4}$-dienoyl-CoA by the novel enzyme, followed be the Δ^2,Δ^4-dienoyl-CoA reductase-dependent route [57,58]. This appears to be the major route operative in intact mitochondria [59].

3 INHERITED DISORDERS OF MITOCHONDRIAL β-OXIDATION

No attempt is made to review all known cases of β-oxidation disorders in the literature. The interested reader is referred to specialized reviews for a comprehensive survey of the literature [60–65].

3.1 CPT and carnitine deficiency

CPT deficiency with predominantly muscle involvement usually presents in adolescence or adulthood, and usually in males, although the inheritance is autosomal [66]. Recurrent myoglobinuria and rhabdomyolysis (muscle breakdown) induced by exercise and fasting are common presenting features. The defect is specifically of CPT II [67], and very little immunoreactive CPT II was found in the patient concerned or in the series of patients reported on by Demaugre et al. [68].

However, CPT II deficiency presenting in infancy with hypoketonaemia, hypoglycaemia and cardiomyopathy has also been described [69]. The authors postulate that in the previous cases of

CPT II deficiency, the residual activity (25%) is sufficient for significant activity of hepatic, but not muscle, β-oxidation, whereas the profound deficiency in their patient (activity 10% of controls) resulted in hepatic involvement. However, in other CPT II patients with profound deficiency, there was no liver involvement (e.g. [67]). In addition there are problems of interpretation, due in part to differences in assay methods [67] and in part to the occurrence of kinetic mutants [70], which makes the comparison of data from assays carried out at different substrate concentrations problematic. The human CPT II gene has been assigned to chromosome 1p32 and spans approx. 20 kb comprising five exons of 81–1305 bp. A series of studies has elucidated the molecular characteristics of this disorder, and a common S113L mutation appears to be the most frequent, but by no means the only, disease-causing mutation [71,72].

The hepatic/infantile phenotype, first described by Bougnères et al. [73], presents with fasting hypoglycaemia and hypoketonaemia in the presence of normal insulin levels. CPT I activity is absent in liver and fibroblasts, although not in muscle [74].

Only a few cases of deficiency of the carnitine/acylcarnitine translocase have been reported. The first, a boy, presented with fasting hypoketotic hypoglycaemia, muscle weakness and cardiomyopathy. Carnitine was highly esterified in both fed and fasted states [75]. A further four cases have now been described [76–78].

Some patients were originally characterized as having lipid storage myopathies associated with carnitine deficiency; however, it is now apparent that the majority of these cases are primary β-oxidation disorders with secondary carnitine deficiency. Several inherited and acquired disorders are known to cause carnitine deficiency: inborn errors of β-oxidation, of branched-chain amino acid catabolism and of the respiratory chain, alcoholic cirrhosis, valproate therapy and extended haemodialysis have all been reported to cause carnitine deficiency [79].

Primary carnitine deficiency is rare, the only well-characterized defect is of cellular carnitine uptake [80,81]. These patients present in infancy or childhood with muscle weakness and cardiomyopathy, and diagnosis is made by measurement of carnitine transport into fibroblasts. The usual lack of liver involvement could suggest a different transport mechanism in liver, or reflect increased endogenous synthesis.

3.2 VLCAD/LCAD deficiency

VLCAD/LCAD deficiency was first described by Hale et al. [82]. The defect often has a severe clinical presentation with nonketotic hypoglycaemia, muscle weakness and hepato- and cardio-megaly; six patients died early in infancy although others had a milder clinical course [83], which taken with the biochemical evidence [84] suggests heterogeneity of the molecular defect.

Studies by Indo et al. [85] show that normal immunoreactive protein is present in most cases studied, suggesting a point mutation(s) to be the cause of the defect. However, further studies revealed normal LCAD cDNA in several patients with apparent LCAD deficiency, i.e. patients with greatly reduced activity with respect to hexadecanoyl-CoA. Immunoblot analysis of VLCAD in fibroblasts from these patients revealed absent protein, and it is now apparent that most, if not all, of the cases previously characterized as having a deficiency of LCAD, were in fact deficient in VLCAD [23,86].

3.3 MCAD deficiency

MCAD deficiency is probably the most common β-oxidation defect, with over 100 cases reported since the original patient of

Kølvraa et al. [87]. Illness is often precipitated by prolonged fasting or illness, resulting in non-ketotic hypoglycaemia, elevated plasma non-esterified fatty acids (NEFA) and a spectrum of abnormal urinary and plasma metabolites, including dicarboxylic acids (DC_6–DC_{10}), suberylglycine, hexanoylglycine and mediumchain acylcarnitines (see [88,89] and the literature cited therein; [90]), with secondarily low tissue acylcarnitines. cis-4-Decenoic acid (a linoleic acid metabolite) has been reported to be a specific plasma metabolite present during remission [91], as have plasma octanoate [92] and plasma acyl-carnitines [93]. A number of other diagnostic procedures have been described, for example analysis of [^{14}C]acyl-CoA esters and [^{14}C]acylcarnitines generated by incubation of tissue preparations with [U-^{14}C]hexadecanoate (Figure 2), allowing the identification of several disorders of mitochondrial β-oxidation. The use of tandem MS to detect the presence of characteristic pathognomic acylcarnitines in body fluids has the advantage of being applicable to whole population screening. This is an important consideration, since MCAD deficiency is readily treated by the avoidance of fasting.

Much effort has been devoted to the study of the molecular basis of the deficiency, aided by the recognition of a common point-mutation at position 985 of the MCAD cDNA [94], which is responsible for up to 90% of patients with MCAD deficiency. This mutation only appears in Caucasians [95], and this finding, together with the strong association of this point mutation with a particular intron haplotype [96], suggests that a mutational hot-spot at 985 is unlikely and a founder effect is more probable. Many other mutations have been identified e.g. [97,98].

The common mutation causing MCAD deficiency results in the substitution of glutamate for lysine at position 304 (K304E), and the mature protein is not detectable in tissues from affected patients. It has been suggested that, since the Lys-304 residue is located in the domain involved in dimer–dimer interaction to form the native homotetramer, the substitution by a glutamate residue in the mutant protein causes a failure of assembly and consequent instability [99–101]. However, more recent studies have suggested that the chaperonin-mediated folding pathway, which involves mitochondrial import of a 421 amino acid precursor, cleavage of a 25 amino acid leader sequence, formation of a transient complex with mitochondrial heat-shock protein 70 ($hsp70_{mit}$) followed by transfer of the polypeptide to $hsp60_{mit}$ and final assembly of the catalytically active homotetramer, is impaired [102]. Specifically, the presence of the lysine to glutamate substitution results in the formation of a $K304E$–$hsp60_{mit}$ complex, which is more stable than the corresponding wild-type complex, with consequent attenuation of tetramer formation. Similarly Bross et al. [103], using an Escherichia coli expression system, have demonstrated that chaperonin-mediated folding is affected, but also suggest that oligomer assembly and stability are involved.

The prevalence of MCAD has been estimated at 1 in 18500 by Blakemore et al. [104] for G985 homozygotes in England, and Matsubara et al. [95] have reported an even higher figure (1 in 6400). This is of the same order as phenylketonuria and suggests that neonatal screening should be considered from Guthrie spots, either using the PCR technique (which would only detect 85–90% of affected individuals) or by detection of specific remission metabolites, for instance by tandem MS of carnitine esters.

3.4 SCAD deficiency

SCAD deficiency is a rare defect, only having been described in a few patients since the first report by Turnbull et al. [105]; some

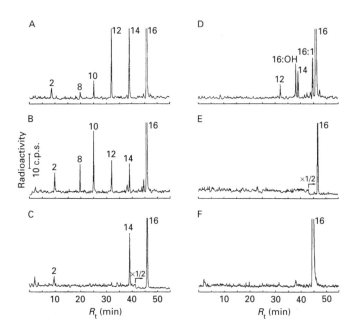

Figure 2   Radio-HPLC chromatogram of acyl-CoA esters generated from incubation of [U-¹⁴C]hexadecanoate with fibroblasts from patients with inherited disorders of mitochondrial β-oxidation

(**A**) Control; (**B**) MCAD deficiency; (**C**) VLCAD deficiency; (**D**) trifunctional enzyme deficiency; (**E**) CPT II deficiency; (**F**) CPT I deficiency. Peak identification: carbon numbers are as indicated; 2,3-enoyl-CoA esters are indicated by the suffix 1; and 3-hydroxyacyl-CoA esters are indicated by the suffix OH. Taken from [221] with permission.

of the cases described may be riboflavin-responsive multiple defects. There seem to be two forms: first, a severe infantile systemic form [106,107]; and secondly, a mild, late-onset phenotype with predominantly muscle involvement [105,108]. The disorder is not expressed in the fibroblasts of patients with the muscle phenotype but is in the infantile variant [109].

The heterogeneity of the described cases has been confirmed by immunochemical studies: the first myopathic case had no immunoreactive protein in muscle [109], whereas in the three other cases, normal sized mRNA was synthesized and immunoreactive protein was present in two of the three cases [110]. In the patient with no immunodetectable protein, a heterogeneous molecular lesion was found: one allele had a point mutation at 319 bp (corresponding to a strongly conserved residue in the ACD family) and the other had a point mutation at 136 bp [111]. Recently Bhala et al. [112] have characterized the defect in fibroblasts from six patients and demonstrated marked heterogeneity in the degree of enzyme attenuation. In one patient (case 2) there was 40% residual activity with little abnormal metabolite excretion; indeed the criteria for deciding what is a deficiency and what is not can sometimes be difficult to delineate. A mouse (BALB/cByJ) model for SCAD deficiency, due to a 278 bp deletion which resulted in a mis-spliced mRNA, has been described [113].

3.5 ETF and ETF:QO deficiencies

Deficiency in either ETF or ETF:QO is termed glutaric aciduria type II [114] and was also known as multiple acyl-CoA dehydrogenation deficiency until recognition of the primary defect(s) by Christensen et al. [115] and Frerman and Goodman [116]. The

activities of all dehydrogenases served by ETF and ETF:QO are impaired [117].

Defects have been recognized in three groups of patients: a fatal neonatal form associated with congenital abnormalities and cardiomyopathy [118]; an infantile or childhood form associated with episodic hypoglycaemia, metabolic acidosis and hepatomegaly [119]; and a late onset form with marked muscle involvement [120]. The molecular lesion is heterogeneous. In ETF-defective patients Ikeda et al. [121] demonstrated α-ETF to be deficient, Loehr et al. [119] have shown β-ETF to be deficient, and Yamaguchi et al. [122] have demonstrated a defect in β-ETF biosynthesis and another defect in which both α- and β-ETF were very labile. ETF:QO patients may have normal, low or absent immunoreactive protein [119].

3.6 Long-chain 3-hydroxyacyl-CoA dehydrogenase (LCHAD) deficiency (trifunctional protein deficiency)

This disorder is probably the second most common β-oxidation disorder and many cases have been described ([37,123–125] and citations therein). The age at onset of symptoms ranges from 3 days to 3 years, and clinical presentations include recurrent episodes of non-ketotic hypoglycaemia, sudden infant death and cardiomyopathy. In most cases hepatic dysfunction is a prominent feature and muscle weakness is also a common finding. The disorder was fatal in several cases although successful treatment with medium-chain triacylglycerols has been reported. The cases described by Jackson et al. [37,123] provide evidence of autosomally recessive inheritance, and in addition, long-chain enoyl-CoA hydratase and thiolase activities were diminished. It now seems certain that cases previously identified as LCHAD deficiency are in fact disorders of the trifunctional protein [37]. This only becomes apparent if enzymes are assayed with the physiologically relevant substrates or if the functional integrity of the pathway is assessed by direct measurement of the intermediates of the pathway. However, detailed studies of the molecular basis of this disorder have demonstrated the presence of a common 1528 G → C mutation, which results in a glutamate to glutamine substitution at amino acid 510 [126,127] and loss of 3-hydroxyacyl-CoA dehydrogenase activity [128]. A rather unusual finding is the association of acute fatty liver of pregnancy in mothers who are obligate heterozygotes for LCHAD deficiency [126].

3.7 Short-chain 3-hydroxyacyl-CoA dehydrogenase (SCHOAD) deficiency

SCHOAD deficiency has been described in two cases, one of them only in abstract form. The patient of Hale et al. [129] presented at the age of 9 months with a recurrent Reye-like illness and a dicarboxylic aciduria. The defect was expressed in fibroblasts. The patient reported by Tein et al. [130] presented at 16 years with hypoketotic hypoglycaemic encephalopathy, myoglobinuria and cardiomyopathy and excreted small amounts of dicarboxylic acids and a trace of hydroxydodecanedioic acid. The defect was expressed in muscle and presumably liver, but not in fibroblasts.

4 REGULATION, CONTROL AND ORGANIZATION OF MITOCHONDRIAL β-OXIDATION

4.1 Physiological regulation of mitochondrial β-oxidation

The physiological (extrinsic) regulation of mitochondrial β-oxidation depends, at least in part, on the organ in question. The liver is capable of high rates of β-oxidation and ketogenesis, or of lipogenesis and esterification of fatty acids. The regulation of carbon flux is therefore very important so that substrate cycling does not take place. Under fed conditions, circulating glucose levels are high, NEFA levels are low, and CPT I is inhibited by high levels of malonyl-CoA. Hence, carbon flux in the liver is from glucose to *de novo* lipogenesis via citrate and malonyl-CoA. This is shown in Figure 3. During starvation, circulating levels of NEFA rise due to the action of adipose tissue triacylglycerol lipase in response to a raised [glucagon]/[insulin] ratio. Hepatic malonyl-CoA levels are lowered, due both to slower efflux of citrate from the mitochondrion and to the phosphorylation and consequent inactivation of acetyl-CoA carboxylase by AMP-dependent protein kinase, again in response to a raised [glucagon]/[insulin] ratio [131]. Hence β-oxidation and keto-genesis are activated and a rise in ketone body levels is observed. The rise in plasma NEFA levels and ketone body levels as glucose availability falls during starvation is diagnostically very useful for β-oxidation disorders; an inappropriate ketone body/ NEFA ratio is suggestive of an inborn error of hepatic β-oxidation or of ketogenesis [132]. Further detailed consideration of the regulation of hepatic fatty acid metabolism is outside the scope of this review and the reader is directed to other recent reviews [133,134]. We will consider below, however, the effects that take place at the level of CPT I or intramitochondrially.

In extra-hepatic tissues in which there is no active lipogenesis, such as heart and skeletal muscle, β-oxidation serves to provide contractile energy. In these tissues, the rate of β-oxidation has been described as demand led, in that an increased work rate and ATP demand leads to faster oxidative phosphorylation and tricarboxylic acid cycle activity. NADH and acetyl-CoA levels diminish, thus increasing β-oxidation flux [135,136]. However, a role for malonyl-CoA inhibition of muscle and heart CPT I in control of β-oxidation flux has been postulated, and this is discussed below. Again the interested reader is referred to recent reviews for a more thorough discussion of the regulation of fatty-acid metabolism in the heart and skeletal muscle [137,138].

4.2 Hepatic regulation of mitochondrial β-oxidation at the level of CPT I

Much of the control and regulation of the rate of hepatic mitochondrial β-oxidation appears to reside at the level of the entry of acyl groups into the mitochondrion. Inhibition of CPT I by malonyl-CoA in the fed state was first demonstrated by McGarry and Foster [139], and has been shown to have significant control over β-oxidation flux in isolated mitochondria [140] and in intact hepatocytes under different metabolic conditions [141].

Several factors in addition to the cytosolic concentration of malonyl-CoA affect CPT I activity in the hepatocyte.

(i) The sensitivity of CPT I inhibition to malonyl-CoA itself alters [142,143] over a relatively long time-scale [144]. This may be due to a change in membrane fluidity rather than an attenuation in numbers of malonyl-CoA binding sites [145]. Phosphatidylglycerol or cardiolipin are similar in their effect to cholate extracts of mitochondria with respect to their ability to

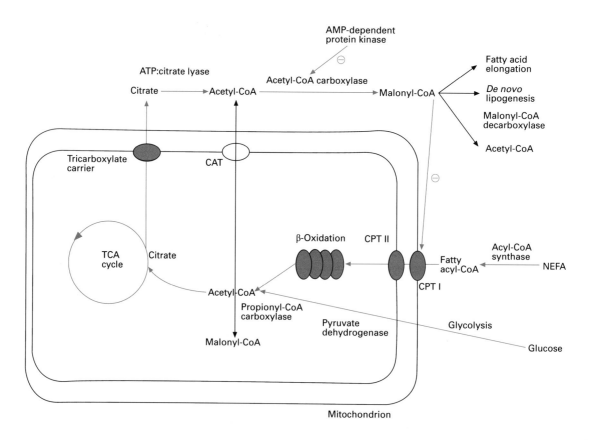

Figure 3   Modulation of CPT I activity by carbohydrates

The pathway shown in red is the established pathway for the inhibition of hepatic CPT activity by carbohydrates. The other pathways shown are alternative routes that may be operative in heart (see text). Abbreviations: CPT, carnitine palmitoyl transferase; CAT, carnitine acetyl transferase.

increase malonyl-CoA sensitivity of CPT I [146] and may account for the previous observations implying a separate malonyl-CoA binding subunit of CPT I [147].

(ii) The affinity of CPT I for its acyl-CoA substrate can alter under ketogenic conditions [148,149], although not directly in response to changes in the insulin/glucagon ratio [144].

(iii) Phosphorylation has been suggested to modulate CPT I activity. Although Harano et al. [150] showed that phosphorylation of CPT in hepatocytes in response to glucagon and forskolin increased its activity, anti-CPT II antibodies were used in this study and total CPT activity was measured. Okadaic acid, the protein phosphatase 1 and 2A inhibitor, was found to increase CPT I activity and palmitate oxidation in hepatocytes, implying phosphorylation of CPT I [151,152]. However, it has been subsequently shown that the okadaic acid-induced rise in CPT I activity is unlikely to be due to phosphorylation of CPT I, as it could not be antagonized by exogenous phosphatases 1 and 2A, was not modulated by fluoride ions and did not lead to labelling of CPT I with [^{32}P]ATP [153].

(iv) Changes in hepatocyte cell volume [154] or cytosolic pH [155] may be responsible for modulation of CPT I activity, as the effect of okadaic acid on CPT I activity is dependent on cell volume, and the sensitivity of CPT I to malonyl-CoA is strongly pH dependent [156]. All these effects may be mediated through the cytoskeleton [157], which would explain the sensitivity to the method of hepatocyte permeablization used [153].

(v) Longer term changes in CPT I activity can be brought about by changes in the expression of CPT I: insulin has been shown to inhibit transcription of CPT I mRNA [158].

4.3 Regulation of β-oxidation by modulation of CPT I activity in extrahepatic tissues

The role of malonyl-CoA in the modulation of CPT I activity in extra-mitochondrial tissues is still controversial. In the heart and skeletal muscle, β-oxidation flux is often thought of as being demand led [135,136]. However, in extrahepatic tissues, CPT I is very much more sensitive to malonyl-CoA than in the liver [159–161], and the measured levels of malonyl-CoA [161,162] would suggest that CPT I activity is almost completely inhibited in the intact tissue. However, much of the malonyl-CoA measured in heart may be intramitochondrial, due to the action of propionyl-CoA carboxylase on acetyl-CoA [163,164]. It is also possible that binding of extra-mitochondrial malonyl-CoA to mitochondrial low-affinity sites [165] or to cytosolic binding proteins, such as that described in liver [166], would also prevent malonyl-CoA from inhibiting CPT I. In order to inhibit CPT I under physiological conditions, malonyl-CoA must be formed in response to carbohydrate feeding. In liver, citrate efflux from mitochondria oxidizing carbohydrate is followed by the actions of ATP:citrate lyase and acetyl-CoA carboxylase to form malonyl-CoA, which inhibits CPT I and thus mitochondrial β-oxidation, and is itself the substrate for fatty acid synthase [139]. Muscle and heart obtain 75 % of their energy requirements from fatty acids [1,2], but β-oxidation can still be suppressed to a certain extent by carbohydrate [167]. Involvement of inhibition of CPT I activity during oxidation of lactate has been inferred from the measurement of long-chain acylcarnitine levels in the perfused heart [168], and a role for regulation of β-oxidation flux by modulation of CPT I activity is suggested by the finding that AMP-dependent protein kinase phosphorylates and inhibits acetyl-CoA carboxylase during increased workload in skeletal muscle [169] and the heart [170], as it does in the liver in response to changes in glucagon/insulin ratios. However, in muscle and heart, there is only a slow efflux of citrate from mitochondria

[171], although there is ATP:citrate lyase activity [172] and acetyl-CoA carboxylase activity [173,174] from a different isoform (280 kDa) than that predominant in liver (265 kDa). There is no obvious removal route for malonyl-CoA, as fatty acid synthase is absent from heart; however, Awan and Saggerson have demonstrated a fatty-acid elongation system which utilizes malonyl-CoA [172], and it has been suggested that an extra-mitochondrial malonyl-CoA decarboxylase may be responsible for disposal of malonyl-CoA [175]. Lopaschuk and co-workers suggest that a high mitochondrial efflux of acetylcarnitine during pyruvate oxidation [176] and carnitine acetyl transferase activity provide the source of extra-mitochondrial acetyl-CoA [177]. An extra-mitochondrial carnitine acetyl transferase could not, however, be detected in a variety of tissues [178], although a transient association of carnitine acetyl transferase with mitochondrial contact sites has recently been demonstrated histochemically [179].

Strong correlations between acetyl-CoA and malonyl-CoA concentrations and between malonyl-CoA concentrations and the β-oxidation rate have been found in the intact heart perfused with varying concentrations of dichloroacetate to provide a range of pyruvate dehydrogenase activities [180]. However, since the subcellular localizations of the measured acetyl-CoA and malonyl-CoA are unknown, these results could equally well fit with the model of the transfer of citrate across the membrane and the subsequent action of ATP:citrate lyase to form acetyl-CoA; a slow transfer of citrate across the mitochondrial membrane [171] could well provide enough malonyl-CoA for inhibition of CPT I activity during pyruvate oxidation. These alternative possible routes for inhibition of CPT I activity are shown in Figure 3.

4.4 Intramitochondrial control over β-oxidation

Although it appears that much of the control of mitochondrial β-oxidation flux resides in CPT I, intramitochondrial controls may be important physiologically. As mitochondrial β-oxidation consists of several enzymes of overlapping chain-length-specificities, some of which are membrane bound (the tri-functional protein and the VLCAD) and transfer reducing equivalents to the respiratory chain, the possibility of supra-molecular organization of the β-oxidation enzymes and auxiliary systems should be considered. The ACDs appear to have by far the lowest activity of the enzymes of β-oxidation in rat and human tissues [37,123,181,182]. However, comparison of isolated enzyme activities may be misleading, since assays of the enzymes of β-oxidation are carried out under non-physiological conditions: for example, 3-hydroxacyl-CoA dehydrogenase is measured in the reverse direction, and ACDs are measured with artificial electron acceptors with which they may have very much lower turnover numbers than with the physiological acceptor ETF. In the absence of specific inhibitors, it is difficult to determine the control strengths of the enzymes of β-oxidation, although a preliminary analysis has been performed [183]. Another method is selective overexpression of the enzymes of β-oxidation and determination of changes in flux. This has been carried out for VLCAD, in which VLCAD cDNA was introduced into three rat hepatoma cell lines. In two of the three cell lines, β-oxidation flux was increased 3-fold, although the amount of enzyme activity in the transfected cell lines was not measured [184].

4.4.1 Control by feedback inhibition and the acylation state

The ACDs have a high affinity both for their acyl-CoA substrates and for their enoyl-CoA products, resulting in product inhibition

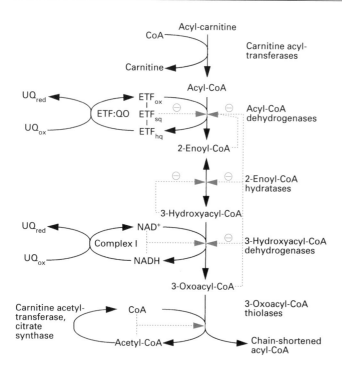

Figure 4   Mitochondrial β-oxidation showing the sites of intra-mitochondrial control in red

Abbreviations: UQ_{red}, reduced ubiquinone; Uq_{ox}, oxidized ubiquinone; ETF_{ox}, oxidized ETF; ETF_{sq}, ETF semiquinone; ETF_{hq}, reduced ETF; Complex I, NADH:ubiquinone oxidoreductase.

[185,186]. In addition, short-, medium- and long-chain ACDs are inhibited by 3-oxoacyl-CoA esters [187]. Similarly, crotonase is strongly inhibited by acetoacetyl-CoA [33], and although long-chain enoyl-CoA hydratase activity is not inhibited by acetoacetyl-CoA [36] it is not known whether it is inhibited by long-chain 3-oxoacyl-CoA esters. 3-Hydroxyacyl-CoA dehydrogenases are subject to product inhibition by 3-oxoacyl-CoA esters [42,187], and the enoyl-CoA hydratase reaction is similarly inhibited by its 3-hydroxyacyl-CoA products [188]. Hence inhibition of 3-hydroxyacyl-CoA dehydrogenase activity would lead to inhibition of 2-enoyl-CoA hydratase activity, so that accumulation of 3-oxoacyl-CoA esters within the mitochondrion would inhibit each of the individual reactions both directly and by a feedback mechanism (see Figure 4) and so be highly inhibitory to β-oxidation. The general 3-oxoacyl-CoA thiolase is inhibited by acetyl-CoA [189], so that were disposal of acetyl-CoA to ketogenesis, to the tricarboxylic acid cycle or to acetyl-carnitine inhibited, feedback inhibition of β-oxidation would result. This has been suggested to be of possible regulatory importance [190], and the observation that whereas 3-oxoacyl-CoA esters are readily observed as intermediates of peroxisomal β-oxidation [191], we have never observed accumulation of 3-oxoacyl-CoA esters in mitochondrial incubations [192,193], would suggest that the accumulation of 3-oxoacyl-CoA esters is strongly prevented and that an excess of thiolase activity 'pulls' β-oxidation, as the 3-oxoacyl-CoA thiolases are not inhibited by their acyl-CoA products. It does appear that β-oxidation can be partly controlled by disposal of acetyl-CoA to ketogenesis [194], and this may be important in regulation in the immediate post-natal period in allowing ketogenesis to take place in response to raised glucagon levels, which de-succinylate and stimulate 3-hydroxymethylglutaryl-CoA synthase [195]. Connected with

the inhibition of 3-oxoacyl-CoA thiolase by acetyl-CoA is the potential control of β-oxidation by the acylation state of mitochondrial CoA. As the mitochondrial CoA pool is limited, depletion of free CoA will inhibit both CPT II and the 3-oxoacyl-CoA thiolase. Other intramitochondrial enzymes dependent on unesterified CoA will also be inhibited. As these include pyruvate dehydrogenase, the branched-chain oxo-acid dehydrogenase and 2-oxoglutarate dehydrogenase, the complete acylation of the mitochondrial CoA pool would result in the breakdown of mitochondrial oxidative metabolism. Garland et al. [196] found that 95% of intramitochondrial CoA was acylated during maximal β-oxidation flux, so that only a small amount of free CoA can sustain β-oxidation. The effect, observed by Wang et al. [190], of the acetyl-CoA/CoA ratio on β-oxidation flux could be due to complete acylation. Some inhibitors of β-oxidation have been postulated to act by sequestration of intramitochondrial CoA [197,198].

4.4.2 Control by the respiratory chain

β-Oxidation is linked to the respiratory chain at two stages, that of the 3-hydroxyacyl-CoA dehydrogenase to complex I via NAD^+/NADH, and the ACDs to ubiquinone via ETF and its oxidoreductase. Inhibition of either of these stages leads to inhibition of β-oxidation [199,200]. ETF-semiquinone, which is the partially reduced form of ETF, can accumulate when the ubiquinone pool is reduced [201] and is a potent inhibitor of ACD [202]. However, ETF-semiquinone disproportionates to the fully oxidized and fully reduced forms in a reaction catalysed by ETF:QO [30], so that the levels of the various ETF species occurring in intact mitochondria are unknown. Hence the activity of the ACDs could be responsive to the redox state of the ubiquinone pool either via ETF and ETF-semiquinone or by complex I and accumulation of 3-hydroxyacyl-CoA esters (which would lead to the accumulation of 2-enoyl-CoA esters and inhibition of the ACDs). Work by Kunz [183,203] has suggested control of β-oxidation at the ETF level, and it has been suggested that the ETF reduction state is responsible for changes in β-oxidation flux with osmolality [193,204,205].

4.4.3 Organization of the enzymes of β-oxidation

The supramolecular organization of the enzymes of β-oxidation has been postulated, for which however, there is little direct evidence. Sumegi and Srere demonstrated that purified SCHOAD, crotonase and acetoacetyl-CoA thiolase bound to the inner mitochondrial membrane, although with differing dependencies on ionic strength and pH and with different binding stochimetries [206], and Kispal et al. [207] subsequently isolated a short-chain 3-hydroxyacyl-CoA binding protein of subunits 69 and 71 kDa from the inner mitochondrial membrane. The protein bound SCHOAD with a stoichiometry of 1:1 and increased SCHOAD activity. However, the authors did not examine crotonase or thiolase binding activity. Recently, Furuta and Hashimoto [208] demonstrated that rat liver mitochondrial membranes bound SCHOAD, and they subsequently isolated an SCHOAD binding protein, a homodimer of subunit mass 60 kDa. The protein bound SCHOAD and 3-oxoacyl-CoA thiolase, although it did not bind crotonase or short-, medium- or long-chain ACDs, with a stoichiometry of 1 mol of binding protein:2 mol of enzyme. However, SCHOAD activity was inhibited on binding. The relationship between these two binding proteins and their function in the intact mitochondrion remain to be established. The β-oxidation enzyme complex isolated from rat heart mitochondria [53] included 3-hydroxyacyl-CoA dehydrogenase, crotonase and acetoacetyl-CoA thiolase activities associated with a 68 kDa protein with CPT activity. This corresponds to the mass of CPT II, but was described as having malonyl-CoA sensitivity. The chain-length dependence of the enzyme activities was not investigated, and it is possible that the complex may have been the subsequently described trifunctional protein [49,50]. Further characterization of this complex is required.

4.4.4 Intermediates of β-oxidation and the question of channelling

A method for deducing the intramitochondrial control of β-oxidation is analysis of the CoA and carnitine ester intermediates. Early studies used radio-GLC of saponified fatty acids and demonstrated that saturated esters were predominant, suggesting the ACDs to have a high control strength over β-oxidation in the intact mitochondrion [209–211]. 3-Hydroxyacyl- and 2-enoyl-moieties were only found during conditions in which the respiratory chain was inhibited, and 3-oxoacyl-CoA esters were not found. However, when radio-HPLC methods for the direct measurement of intact CoA and carnitine esters were established [212,213], 2-enoyl- and 3-hydroxyacyl-CoA esters were observed in incubations of rat skeletal muscle and liver mitochondria with [U-^{14}C]hexadecanoate under State 3 conditions [192,193], but not in human skeletal muscle or fibroblast mitochondria [37,200]. 2-Enoyl- and 3-hydroxyacylcarnitine esters also accumulated, although in lower proportions than in the CoA ester fractions, probably reflecting the lower activity of the CPT system towards 2-enoyl- and 3-hydroxyacyl-CoA esters than towards saturated CoA esters [214,215]. When the amounts of NAD$^+$ and NADH accumulating during oxidation of hexadecanoate were measured directly, it was found that in rat skeletal muscle mitochondria there was very little reduction of NAD(H), whereas a steady 30 % reduction level was reached in rat liver mitochondria. The presence of 3-hydroxyacyl- and 2-enoyl-CoA esters in rat skeletal muscle mitochondria was implied to be either due to a small pool of rapidly turning over NAD$^+$/NADH [192,193] channelled between the trifunctional protein and complex I, as has been described for other dehydrogenases in contact with complex I [216], or due to high sensitivity of 3-hydroxyacyl-CoA dehydrogenase activity of the trifunctional protein to NADH [217]. The presence of saturated acyl-CoA, 2-enoyl- and 3-hydroxyacyl-CoA esters under State 3 conditions suggests that control over β-oxidation flux is shared between different steps.

The presence or absence of channelling of CoA esters in mitochondrial β-oxidation has not yet been established. It has often been assumed that because of the low concentrations of CoA esters observed by early workers, channelling of CoA esters occurred. However, the role of channelling in reducing pool size is controversial [218,219], and our more recent studies have demonstrated measurable amounts of the CoA ester intermediates of β-oxidation [192,193]. An important observation, originally made by measurement of NEFA [209,210] and subsequently by measurement of intact CoA esters [192,193], is that the amounts of the intermediates of β-oxidation are not in steady state during a pulse of β-oxidation, despite steady-state production of acetyl- units or consumption of oxygen, and do not behave as true 'intermediates'. This led to the 'leaky hosepipe' model, in which there is a small pool of rapidly turning over intermediates and the CoA esters observed are 'leaks' from the main flux of β-oxidation, which although they do not represent the true 'intermediates' must represent points at which some control over β-oxidation flux is exerted.

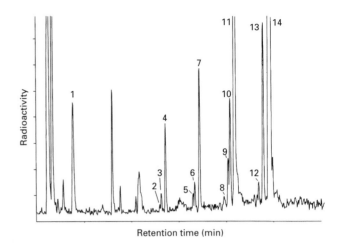

Figure 5   Radio-HPLC chromatograms showing the accumulation of CoA esters from [U-^{14}C]hexadecanoyl-CoA by rat heart mitochondria

Rat heart mitochondria were incubated with 90 μM [U-^{14}C]hexadecanoyl-CoA, and CoA esters were extracted and analysed by radio-HPLC [192]. Peak identification: 1, acetyl-; 2, dec-2-enoyl-; 3, 3-hydroxydodecanoyl-; 4, decanoyl-; 5, dodec-2-enoyl-; 6, 3-hydroxytetradecanoyl-; 7, dodecanoyl-; 8, tetradec-3-enoyl-; 9, tetradec-2-enoyl-; 10, 3-hydroxyhexadecanoyl-; 11, tetradecanoyl-; 12, hexadec-3-enoyl-; 13, hexadec-2-enoyl-; 14, hexadecanoyl-.

It has been suggested that the failure to detect long-chain intermediates of the degradation of [17,17,18,18-^{2}H$_4$]linoleic acid by cultured skin fibroblasts provides evidence for intermediate channelling in mitochondrial β-oxidation [220]. This hypothesis is appealing since the acyl-carnitine/carnitine translocase, carnitine palmitoyltransferase, VLCAD and trifunctional protein are bound to the inner mitochondrial membrane and would constitute a β-oxidation metabolon. Futhermore, the long-chain acyl-CoA intermediates are amphipathic in nature and it would appear advantageous to avoid having these highly surface-active agents in the mitochondrial-matrix bulk phase. There are other theoretical advantages of substrate channelling within the mitochondrial compartment, such as solvation, archipelago effects and avoidance of non-productive substrate binding, particularly in the context of β-oxidation. Nada et al. [220] carried out incubations of intact cells over a period of 96 h and, in the case of control cell lines, detected labelled butyryl-carnitine, hexanoyl-carnitine, octanoyl-carnitine and decenoyl-carnitine but no acyl-carnitines of longer chain-length. These findings are most probably due to the very prolonged incubation times employed by these workers with consequent substrate depletion or incorporation into lipid, and in our view their findings are of little relevance to the presence or otherwise of substrate channelling. We have shown in a series of papers dealing with the characteristics of β-oxidation in isolated liver, skeletal muscle, fibroblast and cardiac mitochondrial fractions from rat and man, that irrespective of whether acyl-CoA esters or acyl-carnitines are measured, long-chain intermediates are readily detected [192,193,200,221], see Figure 5. Although the detailed distribution of intermediates with respect to chain-length and type (acyl, 2,3-enoyl, 3-hydroxyacyl) is dependent upon tissue, duration of incubation, redox state and respiratory state, in no case have we ever failed to detect long-chain intermediates under conditions where there remained substrate to oxidize. These findings were true of the β-oxidation of dicarboxylates as well as monocarboxylates. Thus a limited form of channelling may exist, as exemplified by the 'leaky hosepipe' model. In the case of acyl-CoA esters it is not possible to distinguish intermediates in

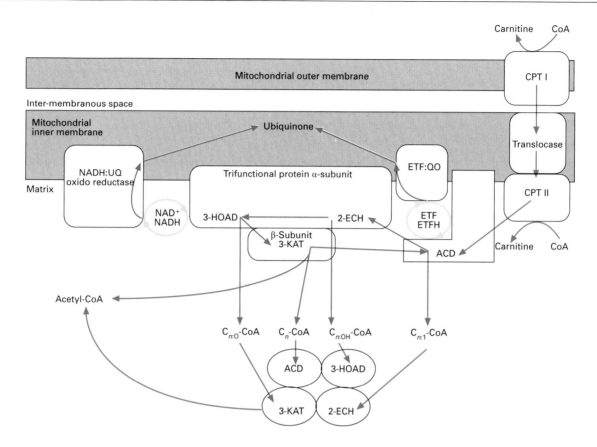

Figure 6 A model for mammalian mitochondrial β-oxidation

C_n refers to number of carbon atoms; 2,3-enoyl-CoA esters are indicated by the subscript 1; 3-hydroxyacyl-CoA esters are indicated by the subscript OH and 3-oxoacyl-CoA esters by the subscript O. Abbreviations: 2-ECH, 2-enoyl-CoA hydratase; 3-HOAD, 3-hydroxyacyl-CoA dehydrogenase; 3-KAT, 3-oxoacyl-CoA thiolase.

the matrix bulk phase from those which might be tightly bound (channelled). In the case of the acylcarnitine fraction, the intermediates seen reflect movement of acyl groups laterally from the pathway and possibly out of the mitochondrial compartment, although this will depend on the specificities of the carnitine acyltransferases and the translocase. The question of channelling in β-oxidation has recently been discussed in more detail [222].

4.5 Model for mitochondrial β-oxidation

Given the recent observations concerning the presence of membrane-bound enzymes of β-oxidation and the close relationship between β-oxidation and the respiratory chain, together with the 'leaky-hosepipe model', we propose the model of mitochondrial β-oxidation illustrated in Figure 6. In this model, long-chain acyl- moieties enter the mitochondrion and are acted on by membrane-bound enzymes of β-oxidation. Reducing equivalents NAD$^+$ and ETFH$_2$ are channelled to complex I and ETF:QO respectively, and the concentrations of the CoA ester intermediates are kept low either by substrate channelling between the active sites of the enzymes or because of the lipophilic nature of long-chain acyl-CoA esters. The long-chain CoA esters we observe, together with chain-shortened CoA esters, are 'leaked' from this pool of rapidly turning over intermediates and are acted on by medium- and short-chain enzymes in the matrix, or are loosely associated with the inner mitochondrial membrane as suggested by Sumegi and Srere [206].

5 CONCLUSIONS

There have been significant recent advances in our knowledge of the enzymology of mitochondrial β-oxidation with the description of the VLCAD, the trifunctional protein and the medium-chain enoyl-CoA hydratase. These studies, and also the identification of inherited disorders in man, have led to the emergence of a model of the pathway that is in two parts: a complex of activities bound to the inner mitochondrial membrane and a matrix system. Paradoxically it is the longer-chain intermediates which are experimentally detectable.

The British Heart Foundation is gratefully thanked for support.

REFERENCES

1   Neely, J. R. and Morgan, H. E. (1974) Annu. Rev. Physiol. **36**, 413–459
2   Felig, P. and Wahren, J. (1975) N. Engl. J. Med. **293**, 1078–1084
2a Randle, P. J., Garland, P. B., Hales, C. N. and Newsholme, E. A. (1963) Lancet **i**, 785–789
3   Knoop, F. (1905) Beitr. Chem. Physiol. Pathol. **6**, 150–162
4   Dakin, H. D. (1909) J. Biol. Chem. **6**, 203–220
5   Lynen, F. and Reichert, E. (1951) Z. Angew. Chem. **63**, 47–48
6   Lipmann, F. (1953) Bacteriol. Rev. **17**, 1–16
7   Drysdale, G. R. and Lardy, H. A. (1953) J. Biol. Chem. **202**, 119–136
7a Wakil, S. J., Green, D. E., Mii, S. and Mahter, H. R. (1954) J. Biol. Chem. **207**, 631–638
7b Lynen, F. and Ochoa, S. (1953) Biochim. Biophys. Acta **12**, 299–314
8   Bremer, J. (1983) Physiol. Rev. **63**, 1420–1480
9   Bergman, E., Reid, R., Murray, M., Brockway, J. and Whitelaw, F. (1965) Biochem. J. **97**, 53–58

10 Mahler, H., Wakil, S. and Bock, R. (1953) J. Biol. Chem. **204**, 453–467
11 Killenberg, P. G., Davidson, E. D. and Webster, Jr., L. (1971) Mol. Pharmacol. **7**, 260–268
12 Norum, K. R., Farstad, M. and Bremer, J. (1966) Biochem. Biophys. Res. Commun. **22**, 797–804
13 Hesler, C., Olymbios, C. and Haldar, D. (1990) J. Biol. Chem. **265**, 6600–6605
14 Kolodziej, M. P., Crilly, P. J., Corstorphine, C. G. and Zammit, V. A. (1992) Biochem. J. **282**, 415–421
15 Brown, N. F., Esser, V., Foster D. W. and McGarry, J. D. (1994) J. Biol. Chem. **269**, 26438–26442
16 Esser, V., Brown, N. F., Cowan, A. T., Foster, D. W. and McGarry, J. D. (1996) J. Biol. Chem. **271**, 6872–6977
17 Weis, B. C., Esser, V., Foster, D. W. and McGarry, J. D. (1994) J. Biol. Chem. **269**, 18712–18715
18 Weis, B. C., Cowan, A. T., Brown, N., Foster, D. W. and McGarry, J. D. (1994) J. Biol. Chem. **269**, 26443–26448
19 Brown, N. F., Weis, B. C., Husti, J. E., Foster, D. W. and McGarry, J. D. (1995) J. Biol. Chem. **270**, 8952–8957
20 Indiveri, C., Tonazzi, A. and Palmieri, F. (1990) Biochim. Biophys. Acta **1020**, 81–86
21 Indiveri, C., Tonazzi, A. and Palmieri, F. (1991) Biochim. Biophys. Acta **1069**, 110–116
22 Izai, K., Uchida, Y., Orii, T., Yamamoto, S. and Hashimoto, T. (1992) J. Biol. Chem. **267**, 1027–1033
23 Yamaguchi, S., Indo, Y., Coates, P. M., Hashimoto, T. and Tanaka, K. (1993) Pediatr. Res. **34**, 111–113
24 Rozen, R., Vockley, J., Zhou, L., Milos, R., Willard, J., Fu, K. and Vicanek, C. (1994) Genomics **24**, 280–287
25 Kim, J. J. and Wu, J. (1988) Proc. Natl. Acad. Sci. U.S.A. **85**, 6677–6681
26 Thorpe, C. and Kim, J. (1995) FASEB J. **9**, 718–725
27 Crane, F. and Beinert, H. (1956) J. Biol. Chem. **218**, 717–731
28 Ruzicka, F. J. and Beinert, H. (1977) J. Biol. Chem. **252**, 8440–5844
29 Furuta, S., Miyazawa, S. and Hashimoto, T. (1981) J. Biochem. (Tokyo) **90**, 1739–50
30 Beckmann, J. D. and Frerman, F. E. (1985) Biochemistry **24**, 3913–3921
31 Stern, J. R. and del Campillo, A. (1956) J. Biol. Chem. **218**, 985–1002
32 Hass, G. M. and Hill, R. L. (1969), J. Biol. Chem. **244**, 6080–6086
33 Waterson, R. M. and Hill, R. L. (1972) J. Biol. Chem. **247**, 5258–5265
34 Stern J. R., del Campillo, A. and Raw, I. (1956) J. Biol. Chem. **218**, 971–983
35 Wit-Peeters, E. M., Scholte, H. T., Van den Akker, F. and De Nie, I. (1971) Biochim. Biophys. Acta **231**, 23–31
36 Fong, J. C. and Schulz, H. (1977) J. Biol. Chem. **252**, 542–547
37 Jackson, S., Kler, R. S., Bartlett, K., Briggs, H., Bindoff, L. A., Pourfarzam, M., Gardner-Medwin, D. and Turnbull, D. M. (1992) J. Clin. Invest. **90**, 1219–1225
38 Jackson, S., Schaefer, J., Middleton, B. and Turnbull, D. (1995) Biochem. Biophys. Res. Commun. **214**, 247–253
39 Stern, J. R. (1957) Methods Enzymol. **1**, 559–567
40 Bradshaw, R. A. and Noyes, B. E. (1975) Methods Enzymol. **35**, 122–128
41 Osumi, T. and Hashimoto, T. (1980) Arch. Biochim. Biophys. **203**, 372–383
42 He, X.-Y., Yang, S. Y. and Schulz, H. (1989) Anal. Biochem. **180**, 105–109
43 El-Fakhri, M. and Middleton, B. (1982) Biochim. Biophys. Acta **713**, 270–279
44 Middleton, B. (1972) Biochem. Biophys. Res. Commun. **46**, 508–515
45 Middleton, B. (1973) Biochem. J. **132**, 717–730
46 Middleton, B. and Bartlett, K. (1983) Clin. Chim. Acta **128**, 291–305
47 Seubert, W., Lamberts, I., Kramer, R. and Ohly, B. (1968) Biochim. Biophys. Acta, **164**, 498–517
48 Staack, H., Binstock, J. F. and Schulz, H. (1978) J. Biol. Chem., **253**, 1827–1831
49 Uchida, Y., Izai, K., Orii, T. and Hashimoto, T. (1992) J. Biol. Chem. **267**, 1034–1041
50 Carpenter, K., Pollitt, R. J. and Middleton, B. (1992) Biochem. Biophys. Res. Commun. **183**, 443–448
51 Luo, M. J., He, X. Y., Sprecher, H. and Schulz, H. (1993) Arch. Biochem. Biophys. **304**, 266–271
52 Weinberger, M., Rinaldo, P., Strauss, A. and Bennett, M. (1995) Biochem. Biophys. Res. Commun. **209**, 47–52
53 Kerner, J. and Bieber, L. (1990) Biochemistry. **29**, 4326–4334
54 Gavino, G. R. and Gavino, V. C. (1991) Lipids **26**, 266–270
55 Osmundsen, H. and Bjornstad, K. (1985) Biochem. J. **230**, 329–337
56 Schulz, H. and Kunau, W.-H. (1987) Trends Biochem. Sci. **12**, 403–406
57 Smeland, T. E., Nada, M., Cuebas, D. and Schulz, H (1992) Proc. Natl. Acad. Sci. U.S.A. **89**, 6673–6677
58 Chen, L.-S., Jin, S.-J. and Tserng, K.-Y. (1994) Biochemistry **33**, 10527–10534
59 Tserng, K., Jin, S. and Chen, L. (1996) Biochem. J. **313**, 581–588
60 Vianey-Liaud, C., Divry, P., Gregersen, N. and Matthieu, M. (1987) J. Inherited Metab. Dis. **10** (suppl. 1), 159–198
61 Bartlett, K. (1993) Baillieres Clin. Endocrinol. Metab. **7**, 643–667
62 Hale, D. E. and Bennett, M. J. (1992) J. Pediatr. **121**, 1–11

63 Vockley, J. (1994) Mayo Clin. Proc. **69**, 249–257
64 Gregersen, N., Andresen, B. S., Bross, P., Bolund, L. and Kølvraa, S. (1994) in New Horizons in Neonatal Screening (Farriaux, J.-P. and Dhondt, J.-L., eds.), pp. 247–255, Elsevier, Amsterdam
65 Roe, C. R. and Coates, P. M. (1995) in The Metabolic Basis of Inherited Disease (Scriver, C. R., Beaudet, L., Sly, W. S. and Valle, D., eds.), pp. 889–914, McGraw-Hill Inc., New York
66 Angelini, C., Freddo, L., Battistella, P., Bresolin, N, Pierobon-Bormioli, S., Armani, M. and Vergani, L. (1981) Neurology **31**, 883–886
67 Singh, R., Shepherd, I. M., Derrick, J. P., Ramsay, R. R., Sherratt, H. S. A. and Turnbull, D. M. (1988) FEBS Lett. **241**, 126–130
68 Demaugre, F., Bonnefont, J. P., Cepanec, C., Scholte, J., Saudubray, J.-M. and Leroux, J. P. (1990) Pediatr. Res. **27**, 497–500
69 Demaugre, F., Bonnefont, J. P., Colonna, M., Cepanec, C., Leroux, J. P. and Saudubray, J.-M. (1991) J. Clin. Invest. **87**, 859–864
70 Zierz, S. and Engel, A. G. (1985) Eur. J. Biochem. **149**, 207–214
71 Taroni, F., Verderio, E., Dworzak, F., Willems, P., Cavadini, P. and DiDonato, S. (1993) Nature (Genetics) **4**, 314–319
72 Taroni, F., Verderio, E., Fiorucci, S., Cavadini, P., Finocchiaro, G. and Uziel, G. (1992) Proc. Natl. Acad. Sci. U.S.A. **89**, 8429–8433
73 Bougnères, P. F., Saudubray, J. M., Marsac, C., Bernard, O., Odievre, M. and Girard, J. (1981) J. Pediatr. **98**, 742–746
74 Tein, I., Demaugre, F., Bonnefont, J. P. and Saudubray, J. M. (1989) J. Neurol. Sci. **92**, 229–245
75 Stanley, C. A., Hale, D. E., Berry, G. T., Deleeuw, S., Boxer, J. and Bonnefont, J. P. (1992) N. Engl. J. Med. **327**, 19–23
76 Pande, S., Brivet, M., Slama, A., Demaugre, F., Aufrant, C. and Saudubray, J. (1993) J. Clin. Invest. **91**, 1247–1252
77 Niezen-Koning, K., Vanspronsen, F., Ijlst, L., Wanders, R. and Brivet, M. (1995) J. Inherited Metab. Dis. **18**, 230–232
78 Brivet, M., Slama, A., Ogier, H., Boutron, A., Demaugre, F. and Saudubray, J. (1994) J. Inherited Metab. Dis. **17**, 271–274
79 Stanley, C. A. (1987) Adv. Pediatr. **34**, 59–84
80 Treem, W. R., Stanley, C. A., Finegold, D. N., Hale, D. E. and Coates, P. M. (1988) N. Engl. J. Med. **319**, 1331–1336
81 Stanley, C. A., Treem, W. R., Hale, D. E. and Coates, P. M. (1990) Progr. Clin. Biol. Res. **321**, 457–464
82 Hale, D. E., Batshaw, M..L, Coates, P. M., Frerman, F. E., Goodman, S. I., Singh, I. and Stanley, C. A. (1985) Pediatr. Res. **19**, 666–671
83 Hale, D. E., Stanley, C. A. and Coates, P. M. (1990) Progr. Clin. Biol. Res. **321**, 303–311
84 Amendt, B. A., Moon, A., Teel, L. and Rhead, W. J. (1988) Pediatr. Res. **23**, 603–605
85 Indo, Y., Coates, P. M., Hale, D. E. and Tanaka, K. (1991) Pediatr. Res. **30**, 211–215
86 Strauss, A., Powell, C., Hale, D., Anderson, M., Ahuja, A. and Brackett, J. (1995) Proc. Natl. Acad. Sci. U.S.A. **92**, 10496–10500
87 Kølvraa, S., Gregersen, N., Christensen, E. and Hobolth, N. (1982) Clin. Chim. Acta **126**, 53–67
88 Millington, D. S., Norwood, D. L., Kodo, N., Roe, C. R. and Inoue, F. (1989) Anal. Biochem. **180**, 331–339
89 Millington, D. S., Kodo, N., Terada, N., Roe, D. and Chace, D. H. (1991) Int. J. Mass Spectrom. Ion Proc. **111**, 211–228
90 Bhuiyan, A. K. M. J., Watmough, N. J., Turnbull, D. M., Aynsley-Green, A. and Bartlett, K., (1987) Clin. Chim. Acta **165**, 39–44
91 Duran, M., Bruinvis, L., Ketting, D., de Klerk, J. B. C. and Wadman, S. K. (1988) Clin. Chem. **34**, 548–551
92 Pourfarzam, M., Naughten, E., Cahalane, S., Bhuiyan, A. K. M. J. and Bartlett, K. (1990) in Stable Isotopes in Paediatric Nutrition and Metabolic Research (Chapman, T. E., Berger, R., Reijngoud, D. J. and Okken, A., eds.), pp. 249–255, Intercept Press, Andover, Hampshire
93 Rinaldo, P., O'Shea, J. J., Coates, P. M., Hale, D. E., Stanley, C. A. and Tanaka, K. (1988) N. Engl. J. Med. **319**, 1308–1313
94 Matsubara, Y., Narisawa, K., Miyabayashi, S., Tada, K. and Coates, P. M. (1990) Lancet **335**, 1589
95 Matsubara, Y., Narisawa, K., Tada, K., Ikeda, H., Yao, Y. Q., Danks, D. M., Green, A. and McCabe, E. R. (1991) Lancet **338**, 552–553
96 Kolvraa, S., Gregersen, N., Blakemore, A. I., Schneidermann, A. K., Winter, V., Andresen, B. S., Curtis, D., Engel, P. C., Pricille, D. and Rhead, W. (1991) Hum. Genet. **87**, 425–428
97 Yokota, I., Tanaka, K. Coates, P. M. and Ugarte, M. (1990) Lancet **336**, 748
98 Ding, J. H., Yang, B. Z., Bao, Y., Roe, C. R., Chen, Y. T. (1992) Am. J. Hum. Genet. **50**, 229–233

99   Kelly, P. D., Whelan, A. J., Ogden, M. L., Alpers, R., Zang, Z., Bellus, G., Gregersen, N., Dorland, L. and Strauss, A. W. (1990) Proc. Natl. Acad. Sci. U.S.A. **87**, 9236–9240

100 Yokota, I., Indo, Y., Coates, P. M. and Tanaka, K. (1990) J. Clin. Invest. **86**, 1000–1003

101 Gregersen, N., Andresen, B., Brosse, P., Rüdiger, N., Engst, S., Christensen, E., Kelly, D., Strauss, A. W., Kølvraa, S., Bolund, L. and Ghisla, S. (1991) Hum. Genet. **86**, 545–551

102 Saijo, T., Welch, W. J. and Tanaka, K. (1994) J. Biol. Chem. **269**, 4401–4408

103 Bross, P., Jespersen, C., Jensen T. G., Andresen, B. S., Kristensen, M. J., Winter, V., Nandy, A., Kräutle, F., Ghisla, S., Bolund, L. et al. (1995) J. Biol. Chem. **270**, 10284–10290

104 Blakemore, A. I. F., Singleton, H., Pollitt, R. J., Engel, P. C., Kølvraa, S., Gregersen, N. and Curtis, D. (1991) Lancet **337**, 298–299

105 Turnbull, D. M., Bartlett, K., Stevens, D. L., Alberti, K. G. M. M., Gibson, G. J., Johnson, M. A., McCulloch, A. J. and Sherratt, H. S. A. (1984) N. Engl. J. Med. **311**, 1232–1236

106 Amendt, B. A., Greene, C., Sweetman, L., Cloherty, J., Shih, V., Moon, A., Teel, L. and Rhead, W. J. (1987) J. Clin. Invest. **79**, 1303–1309

107 Coates, P. M., Hale, D. E., Finocchiaro, G., Tanaka, K. and Winter, S. C. (1988) J. Clin. Invest. **81**, 171–175

108 DiDonato, S., Gellera, C., Peluchetti, D., Uzieli, G., Antonelli, A., Lus, G. and Rimoldi, M. (1989) Ann. Neurol. **25**, 479–484

109 Farnsworth, L., Shepherd, I. M., Johnson, M. A., Bindoff, L. A. and Turnbull, D. M. (1990) Ann. Neurol. **28**, 717–720

110 Naito, E., Indo, Y. and Tanaka, K. (1989) J. Clin. Invest. **84**, 1671–1674

111 Naito, E., Indo, Y. and Tanaka, K. (1990) J. Clin. Invest. **85**, 1575–1582

112 Bhala, A., Willi, S., Rinaldo, P., Bennett, M., Schmidt-Sommerfeld, E. and Hale, D. (1995) J. Pediatr. **126**, 910–915

113 Hinsdale, M., Hamm, D. and Wood, P. (1996) Biochem. Mol. Med. **57**, 106–115

114 Przyrembel, H., Wendel, U., Becker, K., Bremer, H. J., Bruinvis, L., Ketting, D. and Wadman, S. (1976) Clin. Chim. Acta **66**, 227–239

115 Christensen, E., Kolvraa, S. and Gregersen, N. (1984) Pediatr. Res. **18**, 663–667

116 Frerman, F. E. and Goodman, S. I. (1985) Proc. Natl. Acad. Sci. U.S.A. **82**, 4517–4520

117 Rinaldo, P., Welch, R. D., Previs, S. F., Schmidt-Sommerfeld, E., Gargus, J. J., O'Shea, J. J. and Zinn, A. B. (1991) Pediatr. Res. **30**, 216–221

118 Lehnert, W., Wendel, U., Lindenmaier, S. and Bohm, N. (1982) Eur. J. Pediatr. **139**, 56–59

119 Loehr, J. P., Goodman, S. I. and Frerman, F. E. (1990) Pediatr. Res. **27**, 311–315

120 Dusheiko, G., Kew, M., Joffe, B., Lewin, J., Mantagos, S. and Tanaka, K. (1979) N. Engl. J. Med. **301**, 1405–1409

121 Ikeda, Y., Keese, S. M. and Tanaka, K. (1986) J. Clin. Invest. **78**, 997–1002

122 Yamaguchi, S., Orii, T., Suzuki, Y., Maeda, K., Oshima, M. and Hashimoto, T. (1991) Pediatr. Res. **29**, 60–63

123 Jackson, S., Bartlett, K., Land, J., Moxon, E. R., Pollitt, R. J., Leonard, J. V. and Turnbull, D. M. (1991) Pediatr. Res. **29**, 406–411

124 Duran, M., Wanders, R. J. A., de Jeger, J. P., Dorland, L., Bruinis, L., Ketting, D., Ijlst, L. and VanSprang, F. J. (1991) Eur. J. Paediatr. **150**, 190–195

125 Rocchiccioli, F., Wanders, R. J. A. Aubourg, P., Vianey-Liaud, C., Ijlst, L., Fabre, M., Cartier, N. and Bougneres. P. F. (1990) Pediatr. Res. **28**, 657–662

126 Sims, H., Brackett, J., Powell, C., Treem, W., Hale, D. and Bennett, M. (1995) Proc. Natl. Acad. Sci. U.S.A. **92**, 841–845

127 Ijlst, L., Wanders, R., Ushikubo, S., Kamijo, T. and Hashimoto, T. (1994) Biochim. Biophys. Acta **1215**, 347–350

128 Ijlst, L., Ruiter, J., Hoovers, J., Jacobs, M. and Wanders, R. J. A. (1996) J Clin Invest **98**, 1028–1033

129 Hale, D. E., Thorpe, C. and Braat, K. (1989) Pediatr. Res. **25**, 199A (abstract)

130 Tein, I., De Vivo, D. C., Hale, D. E., Clarke, J. T. R., Zinman, H., Laxer, R., Shore, A. and DiMauro, S. (1991) Ann. Neurol. **30**, 415–419

131 Hardie, D. (1992) Biochim. Biophys. Acta **1123**, 231–238

132 Bartlett, K., Aynsley-Green, A., Leonard, J. V. and Turnbull, D. M. (1991) in Inborn Errors of Metabolism (Schob, J., Van Hoof, F. and Vis, H. L., eds.), pp. 19–41, Raven Press, New York

133 Zammit, V. (1996) Biochem. J. **314**, 1–14

134 Guzman, M. and Geelen, M. (1993) Biochim. Biophys. Acta **1167**, 227–241

135 Neely, J. R., Bowman, R. H. and Morgan, H. E. (1969) Am. J. Physiol. **216**, 804–811

136 Oram, J. F., Bennetch, S. L. and Neely, J. R. (1973) J. Biol. Chem. **248**, 5299–5309

137 Vandervusse, G., Glatz, J., Stam, H. and Reneman, R. (1992) Physiol. Rev. **72**, 881–940

138 Lopaschuk, G., Belke, D., Gamble, J., Itoi, T. and Schonekess, B. (1994) Biochim. Biophys. Acta **1213**, 263–276

139 McGarry, J. D. and Foster, D. W. (1980) Annu. Rev. Biochem. **49**, 395–420

140 Quant, P. and Makins, R. (1994) Biochem. Soc. Trans. **22**, 441–446

141 Drynan, L., Quant, P. and Zammit, V. (1996) Biochem. J. **317**, 791–795

142 Cook, G. A., Otto, D. A. and Cornell, N. W. (1980) Biochem. J. **192**, 955–958

143 Cook, G. A., Stephens, T. W., and Harris, R. A. (1984) Biochem. J. **219**, 337–339

144 Grantham, B. D., and Zammit, V. A. (1988) Biochem. J. **249**, 409–414

145 Kolodziej, M. P. and Zammit, V. A. (1990) Biochem. J. **272**, 421–5

146 Mynatt, R. L., Greenhaw, J. J. and Cook, G. A. (1994) Biochem. J. **299**, 761–767

147 Ghadiminejad, I. and Saggerson, E. D. (1990) Biochem. J. **270**, 787–94

148 Brady, L. J., Silverstein, L. J., Hoppel, C. L. and Brady, P. S. (1985) Biochem. J. **232**, 445–450

149 Grantham, B. D. and Zammit, V. A. (1986) Biochem. J. **239**, 485–488

150 Harano, Y., Kashiwagi, A., Kojima, H., Suzuki, M., Hashimoto, T. and Shigeta, Y. (1985) FEBS Lett. **188**, 267–72

151 Guzman, M. and Castro, J. (1991) FEBS–Lett. **291**, 105–108

152 Guzman, M. and Geelen, M. J. H. (1992) Biochem. J. **287**, 487–492

153 Guzman, M., Kolodziej, M. P., Caldwell, A., Corstorphine, C. G. and Zammit, V. A. (1994) Biochem. J. **300**, 693–699

154 Guzman, M., Velasco, G., Castro, J. and Zammit, V. A. (1994) FEBS Lett. **344**, 239–241

155 Moir, A. M. B. and Zammit, V. A. (1995) Biochem. J. **305**, 953–958

156 Mills, S. E., Foster, D. W. and McGarry, J. D. (1984) Biochem. J. **219**, 601–608

157 Velasco, G., Sanchez, C., Geelen, M. and Guzman, M. (1996) Biochem. Biophys. Res. Commun. **224**, 754–759

158 Park, E. A., Mynatt, R. L., Cook, G. A. and Kashfi, K. (1995) Biochem. J. **310**, 853–858

159 Saggerson, E. D. and Carpenter, C. A. (1981) FEBS Lett. **129**, 229–232

160 Cook, G. A. (1984) J. Biol. Chem. **259**, 2030–2033

161 McGarry, J. D., Mills, S. E., Long, C. S. and Foster, D. W. (1983) Biochem. J. **214**, 21–28

162 Singh, B., Stakkestad, J. A., Bremer, J. and Borrebaek, B. (1984) Anal. Biochem. **138**, 107–111

163 Hulsmann, W. C. (1966) Biochim. Biophys. Acta **178**, 137–144

164 Scholte, H. R., Luyt Houwen, I. E., Dubelaar, M. L. and Hulsmann, W. C. (1986) FEBS Lett. **198**, 47–50

165 Bird, M. I. and Saggerson, E. D. (1984) Biochem. J. **222**, 639–647

166 Dugan, R. E., Osterlund, B. R., Drong, R. F. and Swenson, T. L. (1987) Biochem. Biophys. Res. Commun. **147**, 234–241

167 Taegtmeyer, H., Hems, R. and Krebs, H. (1980) Biochem. J. **186**, 701–11

168 Bielefeld, D. R., Vary, T. C. and Neely, J. R. (1985) J. Mol. Cell. Cardiol. **17**, 619–625

169 Winder, W. and Hardie, D. (1996) Am. J. Physiol. **33**, E299–E304

170 Kudo, N., Barr, A., Barr, R., Desai, S. and Lopaschuk, G. (1995) J. Biol. Chem. **270**, 17513–17520

171 England, P. J. and Robinson, B. H. (1969) Biochem. J. **112**, 8P

172 Awan, M. M. and Saggerson, E. D. (1993) Biochem. J. **295**, 61–66

173 Thampy, K. G. (1989) J. Biol. Chem. **264**, 17631–17634

174 Bianchi, A., Evans, J. L., Iverson, A. J., Nordlund, A. C., Watts, T. D. and Witters, L. A. (1990) J. Biol. Chem. **265**, 1502–1509

175 Kudo, N., Barr, A. J., Barr, R. L., Desai, S. and Lopaschuk, G. D. (1995) J. Biol. Chem. **270**, 17513–17520

176 Lysiak, W., Toth, P. P., Suelter, C. H. and Bieber, L. L. (1986) J. Biol. Chem. **261**, 3698–3703

177 Lopaschuk, G. D., Belke, D. D., Gamble, J., Itoi, T. and Schonekess, B. O. (1994) Biochim. Biophys. Acta **1213**, 263–276

178 Edwards, Y. H., Chase, J. F., Edwards, M. R. and Tubbs, P. K. (1974) Eur. J. Biochem. **46**, 209–215

179 Bakker, A., Biermans, W., Vanbelle, H., Debie, M., Bernaert, I. and Jacob, W. (1994) Biochim. Biophys. Acta **1185**, 97–102

180 Saddik, M., Gamble, J., Witters, L. A. and Lopaschuk, G. D. (1993) J. Biol. Chem. **268**, 25836–25845

181 Reichmann, H. and DeVivo, D. C. (1991) Comp. Biochem. Physiol. **98B**, 327–331

182 Melde, K., Jackson, S., Bartlett, K., Sherratt, H. S. A. and Ghisla, S. (1991) Biochem. J. **274**, 395–400

183 Kunz, W. S. (1991) Biomed. Biochim. Acta **50**, 1143–1157

184 Aoyama, T., Ueno, I., Kamijo, T. and Hashimoto, T. (1994) J. Biol. Chem. **269**, 19088–19094

185 Davidson, B. and Schulz, H. (1982) Arch. Biochem. Biophys. **213**, 155–162

186 Powell, P. J., Lau, S. M., Killian, D. and Thorpe, C. (1987) Biochemistry **26**, 3704–3710

187 Schifferdecker, J. and Schulz, H. (1974) Life Sci. **14**, 1487–1492

188 He, X.-Y., Yang, S. Y. and Schulz, H. (1992) Arch. Biochem. Biophys. **298**, 527–531

189 Olowe, Y. and Schulz, H. (1980) Eur. J. Biochem. **109**, 425–429

190 Wang, H. Y., Baxter, Jr., C. F. and Schulz, H. (1991) Arch. Biochem. Biophys. **289**, 274–280

191 Sleboda, J., Pourfarzam, M., Bartlett, K. and Osmundsen, H. (1995) Biochim. Biophys. Acta **1258**, 309–318

192 Eaton, S., Bhuiyan, A. K. M. J., Kler, R. S., Turnbull, D. M. and Bartlett, K. (1993) Biochem. J. **289**, 161–168

193 Eaton, S., Turnbull, D. M. and Bartlett, K. (1994) Eur. J. Biochem. **220**, 671–681

194 Quant, P. A., Robin, D., Robin, P., Girard, J. and Brand, M. D. (1993) Biochim. Biophys. Acta **1156**, 135–143

195 Quant, P. A., Robin, D., Robin, P., Ferre, P., Brand, M. D. and Girard, J. (1991) Eur. J. Biochem. **195**, 449–454

196 Garland, P. B., Shepherd, D. and Yates, D. W. (1965) Biochem. J. **97**, 587–594.

197 Pacanis, A., Strzelecki, T. and Rogulski, J. (1981) J. Biol. Chem. **256**, 3035–3038

198 Turnbull, D. M., Bone, A. J., Bartlett, K., Koundakjian, P. P. and Sherratt, H. S. A. (1983) Biochem. Pharmacol. **32**, 1887–1892

199 Bremer, J. and Wojtczak, A. B. (1972) Biochim. Biophys. Acta. **280**, 515–530

200 Singh Kler, R., Jackson, S., Bartlett, K., Bindoff, L. A., Eaton, S., Pourfarzam, M., Frerman, F. E., Watmough, N. J. and Turnbull, D.M (1991) J. Biol. Chem. **266**, 22932–22938

201 Frerman, F. E. (1987) Biochim. Biophys. Acta. **893**, 161–169

202 Beckmann, J. D., Frerman, F. E. and McKean, M. C. (1981) Biochem. Biophys. Res. Commun. **102**, 1290–1294

203 Kunz, W. S. (1988) Biochim. Biophys. Acta. **932**, 8–16

204 Halestrap, A. P. and Dunlop, J. L. (1986) Biochem. J. **239**, 559–565

205 Halestrap, A. P. (1987) Biochem. J. **244**, 159–164

206 Sumegi, B. and Srere, P. A. (1984) J. Biol. Chem. **259**, 8748–8752

207 Kispal, G., Sumegi, B. and Alkonyi, I. (1986) J. Biol. Chem. **261**, 14209–14213

208 Furuta, S. and Hashimoto, T. (1995) J. Biochem. (Tokyo) **118**, 810–818

209 Stanley, K. K. and Tubbs, P. K. (1974) FEBS Lett. **39**, 325–328

210 Stanley, K. K. and Tubbs, P. K. (1975) Biochem. J. **150**, 77–88

211 Lopes-Cardozo, M., Klazinga, W. and Bergh, S. G. (1978) Eur. J. Biochem. **83**, 629–634

212 Watmough, N. J., Turnbull, D. M., Sherratt, H. S. A. and Bartlett, K. (1989) Biochem. J. **262**, 261–269

213 Bhuiyan, A. K. M. J., Jackson, S., Turnbull, D. M., Aynsley-Green, A., Leonard, J. V. and Bartlett, K. (1992) Clin. Chim. Acta. **207**, 185–204

214 Al-Arif, A. and Blecher, M. (1971) Biochim. Biophys. Acta. **248**, 406–415

215 Mahadevan, S., Malaiyandi, M., Erfle, J. D. and Sauer, F. (1970) J. Biol. Chem. **245**, 4585–4595

216 Fukushima, T., Decker, R. V., Anderson, W. M. and Spivey, H. O. (1989) J. Biol. Chem. **264**, 16483–16488

217 Middleton, B. (1994) Biochem. Soc. Trans. **22**, 427–431

218 Cornish-Bowden, A. and Cardenas, M. (1993) Eur. J. Biochem. **213**, 87–92

219 Mendes, P., Kell, D. and Westerhoff, H. (1996) Biochim. Biophys. Acta **1289**, 175–186

220 Nada, M. A., Rhead, W. J., Sprecher, H., Schulz, H. and Roe, C. R. (1995) J. Biol. Chem. **270**, 530–535

221 Pourfarzam, M., Schaefer, J., Turnbull, D. M. and Bartlett, K. (1994) Clin. Chem. **40**, 2267–2275

222 Osmundsen, H., Bartlett, K., Pourfarzam, M., Eaton, S. and Sleboda, J. (1996) in Channelling in Intermediary Metabolism (Agius L. and Sherratt H. S. A., eds.) Portland Press, London, in the press

223 Suzuki, H., Kawarabayasi, Y., Kondo, J., Abe, T., Nishikawa, K. and Kimura, S. (1990) J. Biol. Chem. **265**, 8681–8685

224 Clarke, P. R. and Bieber, L. L. (1981) J. Biol. Chem. **256**, 9861–9868

225 Ikeda, Y., Dabrowski, C. and Tanaka, K. (1985) J. Biol. Chem. **258**, 1066–1076

226 Finocchiario, G., Ito, K. and Tanaka, K. (1987) J. Biol. Chem. **262**, 7982–7989

227 Furuta, S., Miyazawa, S. and Hashimoto, T (1981) J. Biochem. (Tokyo) **90**, 1739–1750

228 Gehring, U. and Repertinger, C. (1968) Eur. J. Biochem., **6**, 281–292

229 Dommes, V. and Kunau, W. H. (1984) J. Biol. Chem. **259**, 1789–1798

230 Palossari, P. M., Kilponnen, J. M., Sormunen, R. T., Hassinen, I. E. and Hiltunen, J. K. (1990) J. Biol. Chem. **265**, 3347–3353

231 Kilponen, J. M., Palosaari, P. M. and Hiltunen, J. K. (1990) Biochem. J. **269**, 223–226

232 Luo, M. J., Smeland, T. E., Shoukry, K. and Shulz, H. (1994) J. Biol. Chem. **269**, 2384–2388

Biochem. J. (1996) **314**, 1–14 (Printed in Great Britain)

REVIEW ARTICLE
Role of insulin in hepatic fatty acid partitioning: emerging concepts

Victor A. ZAMMIT
Hannah Research Institute, Ayr KA6 5HL, Scotland, U.K.

INTRODUCTION

The effects of insulin action on hepatic lipid metabolism are increasingly emerging as central factors in the integration of the metabolic status of the whole animal. In humans, insulin resistance results in derangements of liver metabolism which are associated with such clinically important conditions as obesity, diabetes and atherosclerosis. The perceived role of the liver in the aetiology of these conditions has attracted extensive research in the area, and several reviews have appeared in recent years that deal with aspects of the role of the hormone in the control of liver lipid metabolism (see, e.g., [1–3]). The aim of the present one is not to duplicate these, but rather to suggest possible ways in which newly acquired information can be combined with other well-established phenomena in the formulation of an integrated theory of control of the hepatic partitioning of fatty acid metabolism by insulin. The approach adopted is purposely speculative, in the anticipation that this is most likely to stimulate novel ideas and work.

The liver plays a central role in orchestrating the delivery of substrates to peripheral tissues. In the process, it partly determines the circulating concentrations of metabolites and lipoproteins. The traffic is bidirectional and, under any particular set of physiological conditions, a balance is established as a result of the rates of secretion by the liver, uptake by peripheral tissues and re-uptake by the liver. This is perhaps best exemplified by lipid substrates, because the liver plays such a central role in the metabolism of fatty acids and cholesterol. Both can be synthesized by the liver, and are also delivered to it in the circulation. Fatty acids are esterified to produce triacylglycerols (TAGs) and phospholipids which, together with unesterified cholesterol and cholesteryl esters, make up the lipid components of very-low-density lipoproteins (VLDL) secreted by the liver. Their secretion, combined with the ability of the peripheral tissues to metabolize them and of the liver itself to clear the remnants or products of the metabolism of VLDL (and chylomicrons secreted from the gut), determines the levels of such important parameters as the plasma concentrations of triacylglycerol, esterified and unesterified cholesterol and of the remnants themselves. Hepatic fatty acid and cholesterol metabolism is therefore intimately linked, and insulin plays an important role in determining the outcome of their metabolism within the liver.

Insulin affects hepatic lipid metabolism directly through its actions to modulate the expression of enzymes or secretory proteins [e.g. apolipoprotein B (apoB); see below], and indirectly by determining the rate of delivery of esterified or non-esterified fatty acids to the liver, through its antilipolytic effect on adipose tissue and its influence on the expression of lipoprotein lipase activity. Much of the interest in the effects of insulin has centred on the role of the liver in the aetiology of non-insulin-dependent diabetes mellitus and its associated insulin resistance. In this respect, it has become increasingly appreciated that a deficient response of lipid metabolism to diurnal changes in insulin secretion accompanying food intake may directly affect postprandial triglyceridaemia and influence the development of other aspects of insulin resistance, such as the impaired response of glycogen synthesis to insulin in muscle [2,3].

Under normal conditions, the liver is regularly exposed, via the portal circulation, to major increases in insulin concentration in response to food intake. Consequently, the most physiologically relevant studies of the response of hepatic lipid metabolism to the effects of insulin are those that involve meal-induced changes in the secretion of the hormone. However, the effects of insulin during such physiological changes are difficult to study because of the concomitant changes in counter-regulatory hormone concentrations. Consequently, frequently the effects of insulin have had to be studied by using the isolated perfused liver or cultured hepatocyte preparations, with the limitations that these impose. As a result, considerable controversy has arisen in the literature over the significance of certain effects of insulin *in vitro* when different preparations are used in different laboratories. Attempts to overcome these difficulties have been made through the development of methods to study hepatic fatty acid metabolism *in vivo* (through the specific labelling of hepatic fatty acids [4]), and particularly the effects thereon of changes in insulin status of the animals, either when meal-induced or after pharmacological manipulation.

ENZYMES INVOLVED IN THE PARTITIONING OF FATTY ACIDS BETWEEN OXIDATION AND GLYCEROLIPID FORMATION

Insulin favours the esterification of fatty acids to form glycerolipids. In all liver preparations used *in vitro*, insulin has always been found to stimulate TAG synthesis [5–9]. The enzymes that catalyse the first reactions that commit long-chain acyl-CoA either to glycerolipid formation (the glycerol-3-phosphate acyltransferases, GPATs) or to intramitochondrial β-oxidation (overt carnitine palmitoyltransferase, CPT I) necessarily compete for their common substrate, and the outcome of this competition is directly influenced by insulin [10].

The hepatic mitochondrial and microsomal forms of GPAT respond differently to insulin-deficient states in the rat, suggesting that their expression may be differentially affected by insulin *in vivo*. Thus it is specifically the activity of mitochondrial GPAT that is decreased in the diabetic- or starved-rat liver [11,12]. Moreover, insulin acutely increases mitochondrial GPAT activity in the perfused rat liver [13]. This effect is analogous to the effect of insulin on cardiomyocyte and adipocyte GPAT [14,15]. These may be considered to be the direct actions of insulin to favour esterification. However, insulin also exerts indirect effects on β-oxidation of fatty acids (see Figure 1).

The esterification of the long-chain acyl moiety to carnitine is catalysed by a family of carnitine acyltransferases [known commonly as carnitine palmitoyl- or octanoyl-transferases (CPTs

Abbreviations used: ACC, acetyl-CoA carboxylase; apoB, apolipoprotein B; CPT, carnitine palmitoyltransferase; CT, CTP:phosphocholine cytidylyltransferase; DAG, diacylglycerol; DGAT, DAG acyltransferase; ER, endoplasmic reticulum (rER, rough ER; sER, smooth ER); GPAT, glycerol-3-phosphate acytransferase; mHMG-CoA, intramitochondrial 3-hydroxy-3-methylglutaryl-CoA; NEFA, non-esterified fatty acids; PDH, pyruvate dehydrogenase; TAG, triacylglycerol; VLDL, very-low-density lipoproteins.

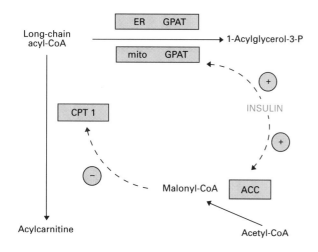

Figure 1   Direct and indirect effects of insulin on the partitioning of cytosolic acyl-CoA between oxidation and esterification

Insulin activates mitochondrial GPAT and also ACC (see the text). The latter effect results in the elevation of intrahepatic malonyl-CoA concentration, which indirectly inhibits β-oxidation of fatty acids through inhibition of CPT I. Abbreviation: mito, mitochondrial.

or COTs respectively)], which are found in mitochondria, the microsomal fraction and peroxisomes. They are all immunologically distinct [16–19], and sequencing of the cDNA of several of them has so far indicated that they are separate gene products [20–23]. Interestingly, the pattern that appears to be emerging from ongoing studies is that each of these membrane systems possess two isoforms of CPT, the activity of only one of which is sensitive to malonyl-CoA inhibition [23–29]. The functions of the microsomal and peroxisomal enzymes can only be speculated on, at present. In peroxisomes, the carnitine acyltransferases may be involved in the transfer, out of organelles, of the chain-shortened products of the oxidation of very-long-chain fatty acids [30,31]. The presence of two CPTs in the microsomal fraction, a membrane-bound cytosol-facing one which is malonyl-CoA-sensitive, and a second one that appears to be lumenal [24,29], suggests that carnitine-dependent transfer of acyl moieties across the microsomal membrane may occur. Alternatively, the lumenal enzyme may be involved in the buffering of intralumenal CoA concentrations. If net carnitine-dependent transfer does occur *in vivo* across the endoplasmic-reticular membrane, it would be important to determine in which direction net transfer is mediated. Intralumenal acyl-CoA esters may be required for processes such as protein acylation [32]. However, it is likely that significant fluxes would only be involved if the transported acyl chains were to be used as intermediates in the synthesis of a major secretory product, e.g. TAG. The possible involvement of such a mechanism in the synthesis of secreted TAG is discussed below. Moreover, it should be emphasized that alternative functions (phospholipase, thiol protease) have been proposed for at least one of these proteins that display CPT activity (see discussion in [24, 33]).

Mitochondrial β-oxidation is quantitatively by far the most important route for the oxidation of the most abundant fatty acids and, consequently, plays a dominant role in determining the partitioning of hepatic fatty acid metabolism. The CPT system of mitochondria (overt CPT I and latent CPT II) is much better understood than the other membrane systems of the cell. The activity of CPT I to convert acyl-CoA into acylcarnitine, and the reversal of this reaction by CPT II in the inner-membrane-

matrix compartment (connected by the transport of the acyl-carnitine by a specific carrier located in the inner membrane), results in the effective transfer of long-chain fatty-acyl molecules across the inner membrane while leaving the acyl-CoA pools in the cytosolic and matrix compartments separate. CPT I resides in the outer membrane of the mitochondria [34] and is an integral membrane protein [35,36], whereas CPT II is a peripheral protein of the inner membrane, with its active site facing the matrix space. CPT I is the one that is sensitive to malonyl-CoA. This property not only makes the reaction it catalyses a potentially important control site, but also provides a mechanism through which insulin can effect acute control over acylcarnitine synthesis, and thus β-oxidation of fatty acids.

Malonyl-CoA is the product of the reaction catalysed by acetyl-CoA carboxylase (ACC), which commits cytosolic acetyl-CoA to the synthesis of fatty acids. Because of the high fatty acid synthase/ACC activity ratio, and the relatively low affinity of fatty acid synthase for malonyl-CoA, the concentration of the inhibitor of CPT I is determined largely by the activity of ACC [37]. Insulin can potentially activate ACC through a variety of mechanisms (see [38]), including dephosphorylation of serine residues in the protein [39], the phosphorylation of which results in enzyme inhibition [40], direct (activatory) phosphorylation at an 'insulin site' [41], or possibly through the action of a low-molecular-mass effector [42]. However, it is not certain that all these mechanisms are necessarily applicable to the liver enzyme, as most of the studies on post-translational modification of the enzyme have been conducted on adipose-tissue ACC. The net result, however, is that conditions characterized by high insulin/glucagon concentration ratios in the portal circulation are accompanied by high hepatic activity of ACC and elevated concentrations of malonyl-CoA, whereas absolute or relative insulin deficiency results in low malonyl-CoA concentrations [43–47]. Consequently, the rate of fatty acid oxidation would be expected to be inhibited at the CPT I reaction simply through the effects of insulin on ACC activity. However, acylcarnitine synthesis has the potential to be much more sensitive to the effects of insulin, because hypoinsulinaemic or insulin-resistant states result in a desensitization of CPT I to malonyl-CoA [48–53]. Consequently, the effects of changes in the absolute concentration of malonyl-CoA in the liver are amplified by the adaptation of CPT I kinetic properties during conditions characterized by absolute or relative insulin deficiency [38,54,54a].

The ability of CPT I to alter its properties under different physiological conditions is thought to result directly from its strong interaction between the hydrophobic regions present in its primary structure [22] (including two potential membrane-spanning domains [55]) and the lipid components of the mitochondrial outer membrane. The changes in its kinetic characteristics observed in mitochondria isolated from rats in different conditions probably occur as a result of changes in the composition of the annular lipids that interact directly with these and other hydrophobic stretches in the primary sequence of the protein. Thus the densitization to malonyl-CoA observed *in vivo* can be mimicked *in vitro* by decreasing the lipid order (i.e. increasing the fluidity of the lipid bilayer) in purified outer-membrane preparations, whether this is induced chemically [56] or thermally [56–58]. In addition, studies on the overall lipid composition of outer membrane isolated from normal fed, starved and streptozotocin-diabetic rats indicate that there is a correlation between the mobility of a hydrophobic molecular probe (diphenylhexatriene) within the membrane (a measure of 'fluidity'), the cholesterol/phospholipid ratio in the membrane and the degree of desensitization of CPT I to malonyl-CoA (C. G. Corstorphine and V. A. Zammit, unpublished work). Addition of phospholipids to mitochondria [59], or treatment with phospholipases [60], has also shown that the sensitivity of CPT I to malonyl-CoA can be altered *in vitro*. Insulin may be involved in modulating membrane lipid composition, e.g. through its effects on cholesterol metabolism [61,62] and/or the remodelling of membrane phospholipids. It is noteworthy that the malonyl-CoA-sensitive CPTs in the other membrane systems of hepatocytes (microsomes, peroxisomes) also undergo changes in sensitivity similar to those of the CPT I of mitochondria [24,25].

The levels of expression of mitochondrial CPT I and II proteins have been shown to increase several-fold in ketogenic conditions, the effect being much more marked for CPT I [17]. Experiments using cultured hepatocytes have shown that fatty acids and cyclic AMP-elevating agents increase the rate of transcription of the CPT I gene, but not of the CPT II gene. Insulin antagonizes the effect of cyclic AMP and inhibits the expression of CPT I [63]. Incubation of cultured foetal rabbit hepatocytes with long-chain fatty acids also induces the expression of CPT I [64]. Similarly, studies on neonatal-rat liver have shown that the increase in insulin/glucagon concentration ratio that accompanies weaning of the pups on to a high-carbohydrate diet is accompanied by a rapid decline in the CPT I mRNA content, but not in that of CPT II [65]. Interestingly, the expression of CPT I protein showed a much more gradual decline than that of its mRNA, indicating that the protein has a long half-life. The same inference was drawn from measurement of CPT I activity in mitochondria isolated from livers of starved–refed [66] or insulin-treated streptozotocin-diabetic rats [67].

The slowness of the changes in the expression of CPT I activity during physiological transitions is matched by that shown by changes in the sensitivity to malonyl-CoA. Thus, it takes 12–14 h of starvation of rats before any decrease in sensitivity to malonyl-CoA occurs (L. Drynan and V. A. Zammit, unpublished work). This is considerably later than the large decrease in plasma insulin concentrations that occurs after 6 h of starvation [68]. Similarly, reversal of the effects of starvation or streptozotocin-induced diabetes in rats only starts 6 h after refeeding of the animals and is not complete even after 12 h after the start of refeeding [66] or insulin treatment [67] respectively. The requirement for several hours to achieve the observed modification of CPT I kinetics has been confirmed in experiments using cultured hepatocytes exposed to insulin or glucagon [63,69] and in rats subjected to a hyperinsulinaemic/euglycaemic clamp [70]. The time scale over which these changes occur is commensurate with the involvement of changes in lipid composition of the mitochondrial outer membrane (see above) in determining the properties of CPT I. It also has important implications for the distribution of metabolic control over the fatty acid oxidative pathway, especially during the reversal of oxidative (ketogenic) conditions. When starved rats are refed [66] or when diabetic animals are treated with insulin [67], there is a rapid decline in circulating ketone-body concentrations, reflecting a decreased rate of ketone-body production in the liver. The questions arise, therefore, as to the involvement of insulin in this response, as well as to the site at which the effect is exerted.

If, as the enzymological data appear to indicate, the expression of CPT I remains high in spite of the rapid decline in CPT I mRNA (because of the long half-life of the protein) and its desensitized state with respect to malonyl-CoA inhibition persists during the reversal of ketogenic states [66,67], then it would be anticipated that this reaction no longer exerts major control over the rate of fatty acid oxidation during these metabolic transitions. It has been suggested that intramitochondrial 3-hydroxy-3-methylglutaryl-CoA (mHMG-CoA) synthase becomes an im-

portant locus of control under these conditions [71,72]. Insulin treatment of starved or diabetic rats results in the rapid decline in the hepatic concentration of mHMG-CoA synthase mRNA and protein [72,73]. In addition, the enzyme reverts to the succinylated (inhibited) state during refeeding of starved rats [74]. Although these changes suggest that a lower proportion of intramitochondrial acetyl-CoA is diverted towards ketone-body formation under these conditions, it is unlikely that the overall partitioning of fatty acids between oxidation and glycerolipid formation is greatly influenced by this. Indeed, when metabolic-control-analysis experiments are performed by using the CPT-specific inhibitor tetradecylglycidic acid (TDGA) to titrate CPT I activity in hepatocytes isolated from fed, starved and starved–refed or insulin-treated starved rats, the control strength of CPT I over the ketogenic rate from exogenously provided palmitate is uniformly high (L. Drynan and V. A. Zammit, unpublished work). These data therefore suggest that if control is lost from the reaction catalysed by CPT I, it is transferred to a step that is proximal to the reaction catalysed by CPT I (e.g. fatty acid supply to the liver) rather than one distal to it (such as mHMG-CoA synthase).

The inference to be drawn from the above is that the potential of the oxidative pathway to compete for cytosolic acyl-CoA remains high during the initial phases of reversal of episodes characterized by high rates of fatty acid oxidation. Such an inference is difficult to test *in vivo*, although evidence for it has existed for some time. Thus insulin treatment of diabetic rats does not result in an immediate reversal of the enhanced ability of isolated hepatocytes or perfused liver to oxidize fatty acids; reversal of the high oxidative capacity requires about 6 h of insulin treatment [75]. Similarly, the ability of hepatocytes isolated from diabetic rats to oxidize fatty acids remains high even after several hours of insulin treatment of the rats from which the hepatocytes are prepared [76]. It is difficult to demonstrate that this lag in the response of the liver actually occurs *in vivo*, because such sustained partitioning of fatty acids in favour of oxidation has to be discerned over and above the rapidly declining rates of absolute rates of fatty acid oxidation, due to the curtailment of non-esterified fatty acid (NEFA) delivery to the liver through the antilipolytic action of insulin on adipose tissue.

The way this difficulty can be overcome is to monitor the partitioning of long-chain fatty acid metabolism, rather than the actual flux. This can be achieved *in vivo* through the specific labelling of the hepatic fatty acid pool, by using cholesteryl [1-14C]oleate-labelled remnants of previously TAG-rich lipoproteins (VLDL and/or chylomicrons) to deliver the ester to the hepatocyte population [4,77]. The labelled ester is rapidly and selectively taken up by these cells and hydrolysed with a half-life of about 15 min [78] to yield an effective pulse of intrahepatic radioactively labelled oleate [4,77]. The rate of the label can then be monitored by measuring the accumulation of label in the oxidation and esterification products in the liver, plasma and exhaled gases of the animals (see [79,80] for further details). This approach has been used to try to elucidate the role of insulin in the partitioning of hepatic fatty acid metabolism *in vivo*, not only between oxidation and esterification but also at the main branch-points of glycerolipid metabolism (see below).

Two experimental models in which acute and well-defined increases in insulin concentration occur are the starved–refed rat and the meal-fed rat. The latter animals are trained to consume their daily food intake within 3 h every day for 15 days before the experiment. Consequently, they are effectively starved for 21 h diurnally before being allowed to consume their diet. The insulin-secretory response to feeding is more pronounced, although other physiological adaptations (e.g. increased absorption from the gut) are also apparent [81]. In both models, the ingestion of a meal results in a rapid decline in circulating NEFA concentrations, thus curtailing the availability of substrate (acyl-CoA) for both oxidation and esterification. But when the partitioning of hepatic acyl-CoA between oxidation and esterification is monitored *in vivo*, it is apparent that the response of the liver is very different between the two conditions. Whereas in starved–refed rats acylcarnitine synthesis remains able to compete successfully for cytosolic acyl-CoA for several hours after the start of refeeding, in meal-fed rats the ingestion of food brings about a much more rapid switch of utilization of acyl-CoA from oxidation to esterification [79,80]. The difference is due to a combination of the adaptation in the properties of CPT I (it is desensitized and over-expressed in starved–refed, but not in meal-fed, animals) and to the response of hepatic malonyl-CoA concentrations to food intake. The latter rises only moderately upon refeeding of starved rats, but overshoots the 'fed' value within 2 h of the start of feeding in meal-fed animals [80]. Evidently, the greater expression of ACC and its more extensive kinetic changes, as well as the accelerated dephosphorylation of pyruvate dehydrogenase (PDH), in meal-fed rat livers result in a much higher flux through the lipogenic pathway [82]; hence the rapid rise in malonyl-CoA concentrations. The effect is amplified by the relatively greater sensitivity of CPT I to the inhibitor in meal-fed animals [80]. Consequently, whereas in starved–refed animals, as expected, the inhibition of acylcarnitine synthesis (i.e. CPT I activity) plays little or no role in mediating the very rapid decline in the absolute rate of fatty acid oxidation (i.e. CPT I no longer exhibits high control strength), in meal-fed animals control is shared, much earlier during the prandial/early-absorptive period, between extra- and intra-hepatic mechanisms, namely antilipolysis in adipose tissue and inhibition of CPT I. The possible physiological rationale behind this difference in terms of the different requirements, in the two models, for continued utilization of pyruvate for the synthesis of glucosyl units during the prandial period has been discussed elsewhere [77]. Of interest in the present context is the role of insulin. The rapid activation of PDH in meal-fed rats is unlikely to be due to the direct action of the hormone, as insulin is known to activate PDH only to a small extent, if at all, in the liver [83,84,84a]. So the PDH effect is likely to be substrate-driven, e.g. through a rise in intramitochondrial pyruvate concentration, which will inhibit PDH kinase [85]. The rapid decline in oxidation of fatty acids will also result in a decrease in the intramitochondrial acetyl-CoA/CoA ratio, which would favour dephosphorylation of PDH [85]. However, the extrahepatic control exerted on this parameter (i.e. NEFA supply) is not likely to be very different between the two models, such that the role of insulin in mediating extrahepatic control is likely to be very similar.

It was of interest to determine whether, if the insulin-secretory response to a meal is highly attenuated, hepatic metabolism can still respond by altering the partitioning of fatty acids between oxidation and esterification. When starvation–refeeding experiments were performed on severely diabetic rats and the fate of [1-^{14}C]oleate was monitored, it was found that the liver was able to switch off acylcarnitine synthesis very rapidly, and before any significant rise in hepatic malonyl-CoA concentration occurred [86]. One possibility to explain this finding is that substrate-driven effects are sufficient to bring about the rapid diversion of fatty acyl-CoA flux from oxidation to esterification entirely through intrahepatic mechanisms [86]. The net result *in vivo* was that, whereas NEFA concentrations rose in these adipose-tissue-depleted starved diabetic rats upon refeeding, the circulating ketone-body concentrations did not increase, just as expected

from the partitioning data [86]. In these experiments, a late, modest and transient rise in malonyl-CoA concentration was observed, which suggested that the desensitization of CPT I to malonyl-CoA inhibition, characteristic of the diabetic state [54a], was somehow overcome. A possible mechanism may be a putative acute decrease in the cytosolic pH during the prandial/absorptive period, such as might be associated with the observed rapid increase in hepatic lactate/pyruvate ratio. This would result in the effective sensitization of CPT I to malonyl-CoA, as the enzyme is much more sensitive to malonyl-CoA inhibition at the lower end of the physiological range of pH [53]. Another mechanism may involve the inhibition of CPT I activity by increased hepatocyte cell volume [87], which may accompany any increase in the portal concentrations of osmotically active metabolites and/or ions (see below).

ACUTE CHANGES IN GLYCEROLIPID PARTITIONING DURING THE PRANDIAL/EARLY-ABSORPTIVE PHASE

The secretion of hepatic VLDL-TAG is an important source of fatty acids for peripheral tissues. Even though the expression of lipoprotein lipase in peripheral tissues may be diminished in insulin-deficient and/or -resistant states, they still utilize VLDL-TAG for an important part of their energy requirements [88,89]. Consequently, under these conditions hepatic VLDL secretion is substantial in absolute terms, even if the fraction of fatty acids metabolized by the liver that is esterified is lower than in insulin-replete states. The increased rates of delivery of fatty acids to the liver makes this possible. (Although it has been found repeatedly that rates of VLDL-TAG secretion by isolated cultured hepatocytes and the perfused livers obtained from starved and diabetic rats have a much lower rate of secretion of TAG [90–92], these experiments may have underestimated the rates attainable, due to the inadequacy of the mixture of substrates provided to these preparations *in vitro*, as demonstrated in [93].)

The same considerations may apply to the fasted/post-absorptive state, i.e. a substantial rate of VLDL-TAG secretion would be expected to be maintained by the liver in the post-absorptive state [94]. Upon resumption of food intake, the absorption of fat from the gut results in the secretion of TAG-rich chylomicrons that compete with VLDL for lipolysis by lipoprotein lipase [95,96]. Consequently, if the liver were to continue to secrete VLDL at pre-prandial rates, the hypertriglyceridaemia experienced during the absorptive phase would be exacerbated [96]. It would be expected that, when the plasma NEFA concentrations are acutely decreased upon refeeding, due to the anti-lipolytic action of insulin on adipose tissue, the magnitude of the flux of acyl chains entering the TAG-synthetic route would be automatically curtailed and that this would be sufficient to inhibit hepatic TAG synthesis and secretion. However, the liver does not appear to depend solely on diminished substrate supply to achieve a decline in TAG secretion during the prandial period.

Evidence for this comes from several lines of study. Thus studies in humans have shown that intraportal infusion of insulin in humans results in a decrease in the secretion of VLDL-TAG [97]. Similarly, when humans are treated with insulin, the secretion of apoB$_{100}$-associated TAG is decreased [98]. Although a large proportion of this effect can be attributed to the decrease in circulating NEFA concentrations due to the anti-lipolytic action of insulin, even when the concentration of NEFA was kept high, insulin still gave a decrease in hepatic VLDL secretion rate of about 30 % [98]. In humans, the concentrations of both apoB$_{48}$ (from the gut) and of apoB$_{100}$ (from the liver) increase with very similar time courses in the post-prandial state [99]. Although this had previously been considered to demonstrate that hepatic VLDL-TAG secretion increases during the absorptive phase because [³H]leucine incorporation into circulating apoB$_{100}$ was increased [100], it is now appreciated that the reason for the increase in apoB$_{100}$ during the post-prandial period is that chylomicron TAG competes with VLDL-TAG for hydrolysis by lipoprotein lipase in peripheral tissues [101]. The fall in apoB$_{100}$ content of the intermediate-density lipoprotein fraction during peak post-prandial triglyceridaemia probably reflects both this competition between chylomicrons and VLDL, and an inhibition of VLDL-TAG secretion rate by the liver [96], but direct evidence for the latter is difficult to obtain in humans. In experiments in which the fatty acid pool of the liver in awake rats is specifically labelled *in vivo* [4,77], changes are observed in the partitioning of label between TAG and phospholipid synthesis, as well as in the fractional rate of secretion of newly labelled TAG, that suggest that acute inhibition in the proportion of synthesized TAG that is secreted occurs during the ingestion of a meal [79,80]. Finally, studies *in vitro* also suggest that a prandial increase in insulin secretion should inhibit TAG secretion by the liver, e.g. culture of rat hepatocytes or HepG2 cells in the presence of insulin stimulates TAG synthesis, but inhibits its secretion.

Insulin may act at several different loci to exert such acute effects. In order to be able to discuss these possibilities, a brief summary of the metabolic models that have recently emerged to explain the route taken by fatty acyl moieties in reaching the secretory TAG pool is given.

ASSEMBLY OF VLDL PARTICLES IN THE LIVER

VLDL particles are made of a core of hydrophobic lipids (TAGs, cholesteryl esters) associated with a molecular of apoB, a large protein with many, relatively short, hydrophobic stretches of amino acid sequence that interact with this core lipid. (In humans apoB$_{100}$ is the hepatic form secreted; in species like the rat both apoB$_{48}$ and B$_{100}$ are synthesised and secreted by the liver.) In addition, a layer of polar, surface, lipids, including cholesterol and phospholipids, covers the surface area left unoccupied by apoB and provides the interaction required with the aqueous phase in the plasma [102]. The requirements for cholesterol [103,104] and for phosphatidylcholine synthesis [105] for the efficient production and secretion of VLDL by isolated perfused liver preparation or cultured hepatocytes have been described. Some general principles about apoB assembly with the hydrophobic core have been well established, but others remain to be determined.

ApoB is co-translationally inserted into the rough endoplasmic reticular (rER) membrane [106], but only a fraction of the nascent polypeptide is translocated into the lumen of the rER. This translocation is accompanied by the acquisition of a TAG core, and thus the formation of a nascent lipoprotein particle. A model that has been suggested to account for this postulates that, during insertion of apoB across the membrane and into the lumen of the rER, the TAG passes through the disrupted membrane and is accepted into a hydrophobic pocket formed by the apoB molecule itself [102,107,108] (see inset in Figure 2). Assembly of TAG-rich apoB lipoproteins has an obligatory requirement for a lumenal microsomal triacylglycerol transfer protein [109]. It catalyses the transfer of TAG and other non-polar lipids between membranes and liposomes *in vitro* [110] and its function *in vivo* may be to facilitate the transfer of TAG across the ER membrane and into the nascent apoB polypeptide. It is noteworthy that this lumenal protein occurs as a heterodimer in association with protein disulphide-isomerase (which is essential for the correct folding of secretory proteins involving disulphide bridges) in the lumen of the ER [110]. Its apparently obligatory

involvement in VLDL assembly and secretion may reflect a putative dual function in the correct folding of apoB and the concomitant transfer of TAG to the nascent VLDL particle.

The subsequent events involved in the acquisition of the full complement of core lipids by apoB have proved controversial to determine. Several accounts of the detailed arguments in favour of the different models have appeared (see, e.g., [108,111–113]). It will suffice to summarize the three models that have emerged as being the most plausible, each apparently based on good experimental evidence. In one model, nascent VLDL obtains its full complement of core lipids in the rER. This is supported by lipid-composition studies on apoB-containing particles isolated from the ER lumen [112]. A two-step model (see Figure 2) has also been suggested in which part of the TAG content of the mature particles is incorporated into the apoB-containing particles in the rER during the co-translational insertion of the apoprotein into the ER lumen, whereas the rest is derived through enlargement with intralumenal TAG that occurs in globules which reside within the smooth ER (sER) [102]. This is supported by electron-micrographic evidence showing that TAG globules, which are not associated with apoB, occur within the lumen of the sER [114]. A third model is an elaboration of the two-step hypothesis and envisages that the lipoprotein particles originating in the rER are relatively lipid-poor, and that core lipids are added as the particles traverse the secretory system from ER through to the Golgi [115]. Interestingly, the work of Boren et al. [116], in which McA-RH7777 cells were used, indicated that apoB$_{48}$-VLDL is almost exclusively assembled through a two-step process, whereas a substantial proportion of apoB$_{100}$-containing particles formed during the co-translational insertion of apoB$_{100}$ into the sER lumen occur in the VLDL density range. More recent experimental data obtained from the work on rat liver [116a] showed that, even in the less dense nascent particles (d 1.006), the amount of TAG associated with each particle was only approximately half that present in the mature VLDL present in the Golgi, and presumably secreted as VLDL. A denser type of particle within the rER (d 1.006–1.020), which contained almost exclusively apoB$_{48}$, was almost devoid of TAG. These data suggest that, if rER particles are precursors to Golgi VLDL, additional core lipids would need to be added to them to bring their composition up to that present in secreted VLDL. This would imply, in agreement with [102], that co-translational insertion of apoB into the ER lumen is accompanied by substantial acquisition of an amount of core lipid dictated by the size of the nascent polypeptide, with additional core lipid being added in a second step. The work in [116a] suggests that the two-step core-lipidation process could be extended to encompass not only the rER and sER, but also the Golgi, as previously suggested [117].

In all three models, it is suggested that the amount of apoB that is translocated into the lumen is only a fraction of the total synthesized, and that the protein destined for early degradation remains in the outer leaflet of the ER membrane [115]. In common with [102], the work in [115] suggests that a sizeable proportion of nascent particles that progress through the secretory pathway are not secreted, but are targeted for degradation. Both this suggestion, and the proposal that a separate pool of TAG exists within the sER within non-apoB-associated globules, raise the prospect that TAG metabolism (synthesis and/or lipolysis) within the lumen of the microsomal fraction can occur independently of that of apoB. For example, if the TAG within the nascent particles targeted for degradation is hydrolysed, the existence of an intralumenal lipase activity would need to be envisaged, and the fate of the constituent acyl chains (e.g. whether they are transported out of the lumen back into the

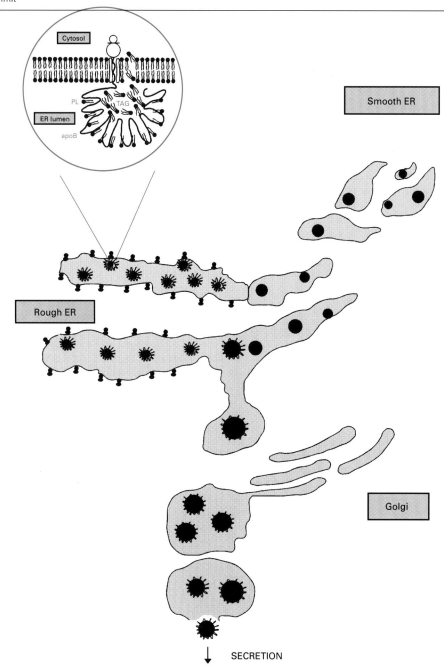

Figure 2   Diagrammatic representation of the two-step model for the assembly of VLDL in the liver

The inset shows the way in which co-translational insertion of apoB$_{100}$ into the lumen of the rER may be accompanied by the stabilization of the protein through the simultaneous acquisition of a hydrophobic core of TAG. The addition of TAG to nascent particles after fusion of sER and rER could be particularly important for the formation of apoB$_{48}$-VLDL. The diagram summarizes the suggestions made in [102,108,114,116,117].

cytosol; see discussion on the possible roles of microsomal carnitine acyltransferases, above) would need to be considered.

POOLS OF FATTY ACIDS, DIACYLGLYCEROL (DAG) AND TAG IN THE LIVER

There is heterogeneity with respect to the use of different intrahepatic pools of fatty acids for phospholipid or TAG synthesis. Fatty acids synthesized *de novo* by the liver are

preferentially utilized for phospholipid synthesis, and these phospholipids are also preferentially secreted within VLDL [118]. Heterogeneity also exists in the route through which fatty acyl moieties reach the secretory TAG pool(s). The existence of multiple pools of TAG in the cytosol, ER membrane and ER lumen, each with distinct rates of turnover, has been established for some time [119–122]. The relationship between the cytoplasmic 'storage' pool of TAG and secreted TAG has been difficult to elucidate. It is evident from experiments conducted on

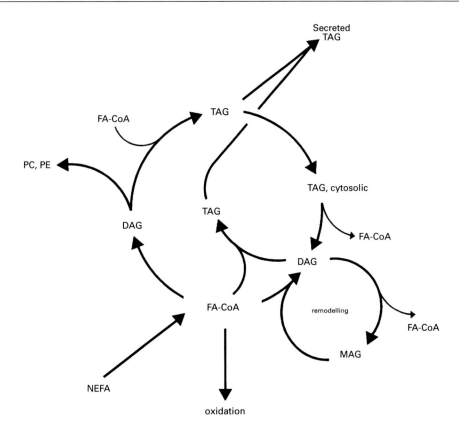

Figure 3   Schematic representation of the possible inter-relationships between different pools of intracellular TAG and DAG with respect to the secretion of VLDL-TAG by the liver

The scheme is based on proposals by several authors (see [120–125]). Further abbreviations: MAG, monoacylglycerol; PC, phosphatidylcholine; PE, phosphatidylethanolamine.

isolated hepatocytes [123] and *in vivo* [124] that only a minor proportion of the flux leading from cytosolic TAG to secreted TAG is due to transfer *en bloc* of intact molecules of TAG. Rather, the cytosolic TAG has to undergo a degree of hydrolysis and re-esterification before being utilized for secretion. The work of Yang et al. [125], in which the acyl-chain composition of secreted TAG was compared with that of intrahepatic glycerolipids, suggested that the hydrolysis of cytosolic TAG only proceeds down to the DAG level, rather than completely to the component fatty-acyl moieties. It further suggests that, after remodelling of the DAG acyl composition, it is re-esterified to TAG, which is then secreted. Accordingly, there should exist at least two pools of DAG within microsomal membranes (in agreement with [121]), as well as two distinctly localized activities of diacylglycerol acyltransferase (DGAT). Evidence for the latter has emerged from fractionation studies performed on HepG2 cells [115]. It is of interest that the proportion of secreted TAG that has been calculated to be derived from cytosolic TAG in [125] is very similar to those calculated from experiments performed on isolated rat liver perfused with exogenous fatty acids [126, 127]. In this preparation, labelled fatty acids abstracted from the perfusate do not equilibrate with the cytosolic pool before they appear in the secreted TAG [126,127]. These same experiments all suggest, however, that a cycle of synthesis (with consequent dilution of label) and mobilization of cytosolic TAG accounts for about half of the overall TAG secretion rate, even when the liver is perfused with relatively high concentrations of fatty acid [126]. Moreover, this proportion does not appear to depend on the initial level of cytosolic TAG in the livers [126]. A

model that would accommodate the recent and the previous observations is given in Figure 3, in which it is envisaged that TAG is synthesized *de novo* from component fatty acyl chains and glycerol phosphate, via phosphatidate, on the cytoplasmic face of the ER. This pool of TAG is only partly used directly for assembly into nascent lipoprotein particles by association with apoB during the protein's co-translational insertion through the membrane into the lumen of the ER. The TAG that is not immediately utilized for this purpose is directed into the cytosolic pool, which is constantly used to generate a second, distinct, pool of DAG. This pool of DAG, after remodelling of its acyl chain composition, is used to form TAG, presumably through the involvement of a second DGAT activity. It is not known whether this second pool of DAG is more likely to give rise to the non-apoB-associated TAG in the sER, where, in density gradients of homogenates of HepG2 cells, a distinct high-specific-activity peak of DGAT occurs [115]. In this context, it is noteworthy that the inability of HepG2 cells to secrete large, TAG-rich, VLDL [128] coincides with their inability to form apoB-free TAG particles (i.e. the absence of a 'second' step) [102], and that this is associated with the existence of a much lower rate of turnover (hydrolysis) of their cytosolic pool of TAG [129].

The scheme depicted in Figure 3 suggests that the phosphatidate-derived DAG pool that gives rise to the initial synthesis of TAG would also be available for phospholipid (phosphatidylcholine, phosphatidylethanolamine) synthesis, and therefore would represent a branch-point at which the control of partitioning between TAG and phospholipid synthesis can be exerted. In addition, as indicated in Figure 4, it would

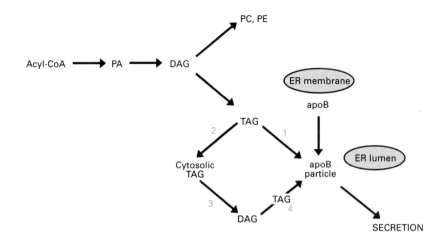

Figure 4   Dual routes may be involved in the incorporation of TAG into VLDL in the liver to determine the number and size of the particles secreted

TAG synthesized on the ER is considered to be partitioned between association with apoB to form nascent particles (step 1) and the formation of cytosolic TAG (step 2). The latter is hydrolysed (step 3) to form partial glycerides and fatty acyl-CoA. TAG is synthesized from the remodelled DAG (step 4) to give a second source of secretory TAG. This scheme is based on proposals presented in [123,161]. Further abbreviations: PA, phosphatidic acid; PC, phosphatidylcholine; PE, phosphatidylethanolamine.

imply that the partitioning of TAG between retention in the liver and secretion into the plasma can be achieved through two mechanisms: (i) by altering the fraction of apoB available for assembly of TAG into nascent particles in the rough ER (i.e. between steps 1 and 2 in Figure 4), and (ii) by altering the rate of cytosolic TAG hydrolysis (step 3). Such dual loci of regulation would be expected to result in the control of both the size of the cytosolic TAG pool and the size of the secreted VLDL particles. There is no evidence at present as to whether distinct TAG pools are preferentially utilized for the 'first' and 'second' step phases of incorporation of TAG into the hydrophobic core of VLDL. If they were to be so used, then the existence of separate routes for TAG secretion would enable control to be exerted on the number and size of the secreted particles, and thus allow these parameters to be differentially affected by the supply of fatty acid and carbohydrate substrates to the liver [130,131]. Insulin is known to decrease the availability of apoB for secretion by increasing the rate of early degradation of the newly synthesized protein (see [132]). In addition, insulin appears to be able to inhibit utilization of the cytosolic pool of TAG for secretion [9,133], although, paradoxically, the hormone does not affect the rate of turnover of cytosolic TAG [123] (but see below).

As mentioned above, the existence of a separate pool of DAG that is utilized to form non-apoB-associated TAG within the lumen of the ER would imply the existence of a second population of DGAT enzyme molecules. Evidence for two peaks of DGAT activity, one associated with the rER and one (the larger) with the sER, already exists (see above). As pointed out in [125], although the formation of DAG through hydrolysis of cytosolic TAG would have the advantage of producing a membrane-permeant intermediate, the current consensus is that DGAT activity is exclusively located on the cytosolic face of the ER membrane. In fact, this is not entirely supported by the data in [134], which show that a substantial proportion of DGAT activity may be latent in intact rat liver microsomes. This is in contrast with the case for the activities of all the other enzymes involved in glycerolipid synthesis, with the possible exception of DGAT cholinephosphotransferase [134]. It is possible, therefore, that DAG generated from cytosolic TAG permeates the ER membrane and is used for TAG resynthesis on the lumenal side of the membrane, i.e. acyl moieties would have to be transported across

the ER membrane for esterification of DAG to occur. The possibility that a substantial flux of acyl chains may occur across the microsomal membrane has emerged from the above-mentioned observation that microsomal membranes have separate cytosol-facing and lumenal carnitine acyltransferase activities [24,29]. If a carnitine–acylcarnitine carrier analogous to that present in the mitochondrial inner membrane exists in the microsomal membrane, then it would be possible for carnitine-dependent transfer of acyl chains to occur across the membrane and regenerate acyl-CoA within the lumen of the ER. However, there is no suggestion that these activities are enriched in either rER or sER [29], and the chain-length specificity of the cytosol-facing enzyme suggests that it is not particularly active with long-chain acyl-CoA esters as substrates [29]. However, teleologically, carnitine-dependent transfer of acyl chains across the ER membrane would provide an additional locus of control by insulin, as the cytosol-facing microsomal carnitine acyltransferase is sensitive to malonyl-CoA inhibition [24,29]. Consequently, insulin, by raising cytosolic malonyl-CoA concentrations, would simultaneously increase the overall rate of fatty acid esterification (by inhibiting fatty acid oxidation at the reaction catalysed by mitochondrial CPT I; see above) and inhibit the transfer of fatty acids into the lumen for re-synthesis of TAG from DAG derived from the hydrolysis of cytosolic TAG. Such a role would resolve the paradox as to how insulin is able to inhibit the utilization of cytosolic TAG for secretion [9,133] without diminishing the rate of turnover of this pool [123]. If insulin were to be able not only to enhance apoB degradation, but also to inhibit secretory TAG resynthesis from DAG (step 4 in Figure 4), it would be able simultaneously to decrease the number and size of the VLDL particles secreted. Evidence that insulin achieves the inhibition of TAG secretion in cultured hepatocytes through both mechanisms has been presented [135]. However, these suggestions are necessarily highly speculative at present.

EXPERIMENTAL OBSERVATIONS FOR A ROLE OF INSULIN IN TAG SECRETION

The effect of insulin on hepatic TAG secretion has been controversial to determine. Thus, whereas experiments conducted with isolated perfused rat liver showed that insulin stimulates both synthesis and secretion of TAG [136,137], many subsequent studies conducted on cultured hepatocytes have found that insulin, although promoting synthesis of TAG, inhibits its secretion [6,8,133,138,139]. The reasons for this discrepancy between the two sets of data may reside in the use of different experimental systems (perfused liver, cultured cells) and in the choice of the precise perfusion or culture conditions, by different laboratories. Thus Topping et al. [137] have pointed out that the perfusion rate, haematocrit and level of oxygenation of the medium used to perfuse the liver *in vitro* all affect the action of insulin. Laker and Mayes [136] suggest that cultured cells are not very secretion-competent compared with the perfused liver and may not be the best system in which to study the effects of insulin. Conversely, Sparks and Sparks [113] suggest that the stimulatory effect of insulin on TAG secretion observed in the perfused liver may have been related to glucocorticoid effects. In addition, it is apparent that TAG synthesis and secretion respond to different concentrations of insulin in cell-culture experiments, higher ones being required to inhibit secretion [8]. There also seems to be disagreement as to whether insulin only affects apoB degradation [139], or whether it also affects apoB synthesis [140]. Moreover, prolonged exposure (5 days) of cultured rat hepatocytes to insulin enhanced apoB$_{48}$ secretion by promoting apoB mRNA editing [141].

Experiments using cultured rat hepatocytes and HepG2 cells have indicated that insulin could have a biphasic effect on TAG secretion. Culture of hepatocytes with insulin for 6–48 h results in the inhibition of secretion, whereas more prolonged culture (up to 72 h) results in stimulation of secretion (over a diminishing control value) [142–144]. Although this phenomenon has not always been observed [6,145], it has been interpreted as indicating that insulin acts acutely on the liver to inhibit TAG secretion, whereas, when the liver is chronically exposed to high levels of insulin, it becomes resistant to the hormone and the effect is lost. Consequently, it has been suggested that acute exposure of the liver to surges in portal insulin concentration (such as would occur during the prandial period) would result in the inhibition of TAG secretion, whereas more prolonged exposure of the liver to hyperinsulinaemic conditions (e.g. in insulin-resistant states) would make it refractory or resistant to this effect (see, e.g., [8,132,146]). If these observations can be extrapolated to situations *in vivo* characterized by chronic hyperinsulinaemia and insulin resistance (e.g. non-insulin-dependent diabetes mellitus, obesity), they would offer an explanation for the apparently paradoxical observations that hyperinsulinaemic conditions are normally associated with hypertriglyceridaemia, partly due to increased hepatic TAG secretion [147,148]. They would imply that, as the rate of delivery of NEFA to the liver is increased, because of the diminished anti-lipolytic action of insulin on adipose tissue, the increased rate of TAG synthesis would not be counteracted by insulin-mediated inhibition of secretion. Consequently, insulin resistance would be accompanied by an increased rate of hepatic TAG secretion, which would contribute towards the dyslipidaemia associated with these conditions because of the failure (through insulin resistance) of both extra- and intra-hepatic mechanisms [2,3].

In order to start addressing these questions *in vivo*, experiments in my laboratory have been performed on acute increases in insulin concentrations and prolonged insulin deficiency. Acute exposure of the liver to insulin occurs during the prandial period. Consequently, experiments using the specific labelling of hepatic fatty acids *in vivo* [4,77] have been aimed at addressing the question as to whether the partitioning of hepatic glycerolipid flux is altered during the prandial period upon refeeding of starved rats or in animals subjected to a meal-feeding regime.

Detailed time courses were obtained for the partitioning of acyl-CoA (see above), DAG (between the labelling of TAG and of the major phospholipids) and of TAG between retention within the liver and secretion into the circulation. These experiments led to the description of concomitant acute decreases in the proportion of DAG that is used for the synthesis of TAG, and of the latter that is secreted [4,77,79,80]. The effects were apparent within 1 h of the refeeding of starved rats, reached a peak at 2–3 h and were relatively rapidly reversed thereafter, although they persisted longer in meal-fed animals. The time courses for the two effects are very similar, suggesting that they are mediated by the same mechanism or hormonal action. The synchronization of the two effects is noteworthy in that the two metabolic branch points concerned occur sequentially. Consequently, the overall effect of the two acute changes in metabolic partitioning is multiplicative; the proportion of the acyl chains esterified to the glyceryl moiety that is secreted by the liver as VLDL-TAG would be expected to be acutely and markedly diminished. In absolute terms, the rate of secretion of TAG would additionally be affected by the acute decrease in the rate of delivery of NEFA to the liver during the prandial period [149]. The role for intrahepatic mechanisms that specifically curtail TAG synthesis and fractional rate of secretion may arise from the inability to decrease NEFA supply to the liver rapidly enough and to a sufficiently low level (see [4]) to achieve the required degree of inhibition.

Two questions arise about these acute changes in DAG and TAG partitioning within the liver. Firstly, which enzymes/processes are likely to be involved? Secondly, are the changes in partitioning mediated (solely) by insulin?

PARTITIONING OF DAG BETWEEN PHOSPHOLIPID AND TAG SYNTHESIS

Two mechanisms could be involved in mediating the change in partitioning of DAG observed *in vivo*. It is possible that the affinity of DGAT for DAG in the microsomal membrane is lower than that of CDP-choline and CDP-ethanolamine acyltransferases, as suggested in [150,151], such that a decreased rate of DAG synthesis (owing to a decreased rate of delivery of NEFA to the liver) would automatically favour its partitioning into phospholipids [152–154]. As fatty acids are known to activate DGAT [154], the sharp decline in NEFA supply to the liver upon refeeding of starved rats could result in the attenuation of DGAT activity (see below). Evidence from experiments on permeabilized hepatocytes indicates that phosphatidylcholine and TAG syntheses utilize the same pool of DAG [155], although, as discussed above, at least one additional pool of microsomal membrane DAG may exist [121,122,125]. Alternatively, altered DAG partitioning could involve acute changes in the activities of the enzymes that exert major control over the respective pathways through post-translational mechanisms. Little is known about the possible mechanisms involved in the acute control of phosphatidylethanolamine biosynthesis. However, in the case of phosphatidylcholine biosynthesis, the properties of the enzyme that catalyses the main rate-controlling step (CTP:phosphocholine cytidylyltransferase, CT) are well characterized (see [156] for recent review). The activity of CT is affected by the availability of fatty acids, which enhance the association of this ambiquitous enzyme with a membranous fraction of the cell, where its substrate resides, thus effectively recruiting inactive, soluble, enzyme to become active at the site of phosphatidylcholine synthesis [157]. This translocation on to the membrane could involve an increase in the DAG content of the membrane [158,159], although, as intimated above, in normal rats during the prandial/early-absorptive period, this is unlikely to be due to an increase in the supply of substrate for DAG synthesis. However, other mechanisms for increasing membrane DAG may exist (see below). Interactions of CT with membrane lipids (through a hydrophobic stretch in its primary amino acid sequence) result in its activation [160]. The particulate fraction with which the enzyme becomes associated has been conventionally assumed to be the microsomal membrane, but relatively recent immunocytochemical evidence suggests that translocation occurs to the nuclear membrane [161,162]. The significance of such a localization is not clear at present. The translocation process has also been suggested to be affected by the reversible phosphorylation of CT by cyclic AMP-dependent protein kinase [163]. Although addition of cyclic AMP analogues to intact hepatocytes or the elevation of endogenous cyclic AMP concentrations does not appear to affect either the phosphorylation or the membrane association of CT with the membrane [164], it has been suggested that the effects of increased cellular cyclic AMP could be mediated through a decrease in membrane content of DAG [158]. Therefore, it may be relevant that in starved-rat liver the activity of cyclic AMP-dependent protein kinase and the concentration of cyclic AMP are increased, and that during refeeding they are both rapidly decreased [68,165]. Evidently experiments need to be conducted, on liver samples obtained *in situ*, to find out directly whether changes in membrane DAG content and translocation of CT are involved in the acute increase in the partitioning of TAG towards phospholipids.

The activity of DGAT is thought to be modulated by reversible phosphorylation; incubation of rat liver microsomal fractions with ATP and Mg^{2+} inhibits the activity of DGAT [166,167]. Conversely, the activity of the enzyme is increased by fatty acids [153,154]. Consequently, it is possible that the decreased availability of NEFA to the liver during the prandial period could be involved in lowering the activity of DGAT, and hence the relative flux of DAG directed towards TAG formation. No information is available as to whether DGAT activity is acutely affected by exposure of the liver to insulin.

PARTITIONING OF TAG BETWEEN SECRETION AND RETENTION IN THE LIVER

The possibility that TAG may follow two separate routes to join the secretory pathway raises the prospect that partitioning of TAG between secretion and retention in the liver (i.e. the fractional rate of secretion) is determined through control at two loci. As discussed above, an amount of TAG is required for the formation of nascent apoB-containing particles during the co-translational insertion of the apoprotein through the ER membrane. Consequently, the proportion of apoB available for secretion would be expected to be involved in determining the fractional rate of secretion of TAG at this site, with the 'excess' TAG being diverted towards the cytosolic pool. This concept is supported by early work on the isolated perfused liver which showed that TAG secretion is proportional to the rate of fatty acid uptake (i.e. perfusate fatty acid concentration) until a maximal rate is reached, beyond which TAG starts to accumulate in the liver [168]. Insulin may exert a major effect at this step, as it enhances the rate of degradation of apoB, thus making less of the protein available to the secretory pathway. ApoB is synthesized at a rate in excess of that required to meet the requirements of its rate of secretion in VLDL. Its rate of secretion does not appear to be determined at the transcriptional level (see [169] for review), although there is one report that insulin inhibits secretion of apoB partly through the inhibition of apoB synthesis and partly through the stimulation of its degradation [140]. In general, however, the main determinant of the

Table 1   Effects of starvation and refeeding or insulin treatment of normal or diabetic rats on the partitioning of [^{14}C]oleate label between phospholipid and TAG labelling and on the fractional rate of [^{14}C]TAG secretion *in vivo*

Rats were starved for 24 h before being refed (for 2 h) or given an intraperitoneal injection of insulin (15 units/kg) 2 h before being used. [^{14}C]Oleate label was delivered specifically to the liver as described in [77]. Abbreviations: PL, phospholipids; GL, glycerolipids. Data are from [77,79,80,86,172,173].

	^{14}C in PL as % of ^{14}C in total GL	[^{14}C]TAG secreted as % of total labelled
Fed (6)	14.2 ± 2.0	57.1 ± 3.3
Starved (6)	16.7 ± 1.8	60.5 ± 2.4
Starved, refed (4)	33.3 ± 2.1	39.2 ± 3.4
Diabetic (4)	15.9 ± 1.1	68.9 ± 4.6
Diabetic, starved (4)	24.0 ± 3.0	75.2 ± 3.0
Diabetic, starved–refed (3)	78.3 ± 6.3	27.4 ± 12.9
Diabetic, insulin treated (3)	23.8 ± 2.5	57.9 ± 1.7

rate of apoB secretion appears to be the rate of intracellular degradation of the protein. Newly synthesized polypeptide that remains associated with the ER membrane, rather than being translocated into the lumen, is degraded. In addition, a proportion of apoB-containing nascent particles within the lumen are also targeted for degradation [108]. This possibility is supported by the results of experiments on HepG2 cells, in which a 70 kDa product of apoB degradation was found associated partly with the lumenal contents of the ER [170]. Insulin-enhanced degradation only affects a proportion of total apoB and appears to have different characteristics from constitutive degradation [113,140]. However, it is not known whether it is the availability of apoB that determines the fraction of TAG that is channelled towards secretion, or vice versa; evidently increases in TAG secretion can occur in the absence of any increase in apoB secretion [130]. From experiments conducted on cultured hepatocytes obtained from fed or starved rats, Davis et al. [171] concluded that the availability of apoB$_{48}$, which is the major B apoprotein in rat liver, determines the rate of TAG secretion. The question has also been addressed by Boren et al. [108], working on HepG2 cells. They concluded that it is the availability of TAG that determines the proportion of apoB that is committed to secretion and spared degradation within the lumen of the ER; increased availability of TAG did not determine the fraction of apoB translocated into the ER lumen, but increased the number of mature VLDL particles formed [108]. More recent work from this group [116] has further supported this conclusion, and the two-step hypothesis in general.

When the meal-induced effects on the partitioning of DAG and on the fractional rate of secretion of TAG were first described for rat liver *in vivo* [77,79,80], it was assumed that they were necessarily mediated by first-phase insulin release [113,132]. This conclusion appeared to be supported by the fact that the effects were slightly more pronounced and longer-lasting in meal-fed rats, in which the prandial release of insulin is more pronounced. In an attempt to determine more directly the role of insulin in the diversion of acyl-CoA away from TAG synthesis and secretion during the prandial period *in vivo*, refeeding experiments were performed on starved–refed streptozotocin-diabetic rats [172]. In spite of a much attenuated insulin-secretory response to food intake [86], the liver of these animals showed the same pattern of acute changes in the partitioning of DAG and in the fractional rate of secretion of TAG (see Table 1 and Figure 5). Indeed, the effects occurred more rapidly and were longer-lasting than in

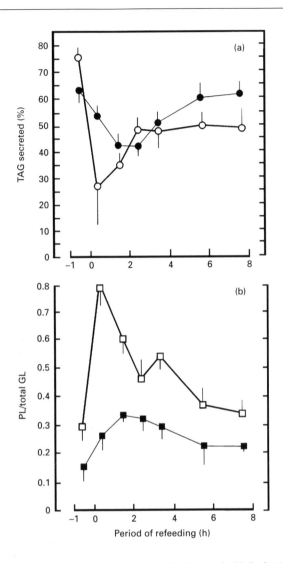

Figure 5   Comparison of the time-courses for changes in (a) the fractional rate of secretion of newly synthesized [^{14}C]TAG, and (b) the partitioning of [^{14}C]oleate between phospholipids (PL) and total glycerolipids (GL) after specific labelling of hepatic oleate *in vivo* in 24 h-starved diabetic (○, □) and normal (●, ■) rats refed for the periods indicated

Data are from [77] and [172].

acid oxidation and the induction of gene transcription (see [175] for recent review). Experiments have therefore been conducted to determine whether the presumed hepatocyte swelling that would be induced *in vivo* during the absorptive phase could provide the signal whereby DAG is diverted towards phospholipid synthesis and the fractional rate of TAG secretion is lowered. Isolated rat hepatocytes were incubated *in vitro* under conditions that increase their water content and the intracellular volume (i.e. either by incubation with the glutamine or in hypo-osmotic media) to check whether cell swelling can mimic the effects on glycerolipid metabolism observed *in vivo* during the prandial/early-absorptive phase. Several key observations have emerged from these studies [176]. Hepatocyte swelling results in (i) an increase in the proportion of exogenously supplied [^{14}C]palmitate that is incorporated into phospholipids, at the expense of TAG labelling (this effect is counteracted by concentrations of fatty acids above 0.5 mM), and (ii) an inhibition of the overall rate of [^{14}C]TAG secretion, as well as of the fraction of newly synthesized TAG that is secreted [176]. Thus, both effects observed *in vivo* on the partitioning of DAG and TAG can be mimicked *in vitro* by exposure of isolated hepatocytes to conditions that result in a 10–20 % increase in cell volume. These observations may explain why the changes in glycerolipid metabolism that accompany food intake are unimpaired in severely diabetic rats [172], especially since the rate of Na$^+$ co-transport of amino acids such as glutamine and alanine is induced several-fold in diabetic-rat hepatocytes [177]. They do not exclude the possibility, however, that in the normal animal the effects of increased portal substrate/ion concentrations on cell volume could act synergistically with the effects of insulin. Indeed, insulin itself could exert part of its action on TAG secretion through the effects it exerts on cell volume in its own right, as it increases cell volume in the isolated perfused rat liver [178] and cultured rat hepatocytes [179]. The hormone may therefore act to decrease apoB and TAG secretion through an effect on hepatocyte volume. It may be relevant that both insulin treatment and hypotonic swelling of cultured rat hepatocytes result in the activation of several, and not entirely overlapping, peaks of mitogen-activated protein (MAP) kinase activity in hepatocyte extracts [180]. Thus insulin could act synergistically with portally delivered substrates to induce hepatocyte volume increases to feed into overlapping signalling mechanisms. In insulin-deficient or -resistant states, the substrate-mediated insulin-independent effects would still be operative, provided that the peak concentrations of volume-active metabolites and/or ions in the portal vein are sufficiently high. As mentioned above, if activation of osmolyte amino acid transport across the plasma membrane of hepatocytes is associated with relative or absolute insulin deficiency (e.g. starvation, diabetes [177,181,182]), the effects of osmolyte amino acids in the portal circulation could be amplified to compensate for the attenuation or absence of insulin action. It would therefore be expected that the composition of the diet is likely to be important in determining whether the liver is able to respond by decreasing its rate of TAG secretion during the prandial period, especially in insulin-deficient or -resistant states. The physiological requirement for the curtailment of hepatic TAG secretion during the prandial/early-absorptive phase is possibly more acute in individuals suffering from these conditions, owing to the exacerbation of hypertriglyceridaemia due to a diminished rate of clearance of chylomicron-TAG by peripheral tissues [183,184]. It would therefore be of interest to study the dependence of the extent of apoB$_{100}$-associated post-prandial hypertriglyceridaemia, in response to a given fat load in humans, on the ability of the other components of the meal to generate osmotically active substrates for hepatocyte plasma-membrane

normal rats. By contrast, when diabetic rats were treated with insulin, there was no inhibition of the fractional rate of secretion of TAG, and only a minor decrease in the proportion of total glycerolipid labelling that was accounted for by TAG [173]. Consequently, it is unlikely that insulin has an obligatory role in mediating the effects of food intake on hepatic glycerolipid partitioning. Other mechanism(s) must exist to act in synergy with the hormone in normal animals, and/or to compensate for the attenuation of its putative effect in diabetic animals.

As discussed above for the case of acyl-CoA partitioning between oxidation and esterification, an additional mechanism through which liver metabolism could be affected during the prandial phase involves acute changes in hepatocyte hydration state (i.e. cell volume). Increased concentrations of osmolyte amino acids or metal ions (e.g. K$^+$) would increase hepatocyte water content (see [174]). Increased cell volume is known to affect hepatocyte metabolism in many respects, including the stimulation of glycogen and fatty acid synthesis, inhibition of fatty transporters. Such considerations are important, in view of the association between the degree of post-prandial lipaemia and the development of coronary heart disease [101,185]. The effects of increased hepatocyte volume on TAG secretion suggest that detrimental effects of absolute or relative insulin deficiency on the severity of post-prandial lipaemia can be counteracted by appropriate dietary intake.

CONCLUDING REMARKS

As the complexities of the regulation of hepatic fatty acid metabolism are unravelled, it becomes increasingly evident that the control of the pathways involved is distributed over several loci. This is true both for the partitioning of acyl-CoA between oxidation and esterification, and for control within these two major pathways themselves. An example of this has emerged recently from studies on the distribution of control exerted *in vivo* by dietary fatty acids on the different partitioning steps discussed in this review [186]. Insulin can affect hepatic fatty acid metabolism at several different points in the two pathways. Moreover, it can act on the liver indirectly through its effects on NEFA supply to the tissue. The observation that NEFA concentration can interact with the effects of insulin on glycerolipid metabolism in the perfused liver [7], and with the effects of cell swelling in isolated rat hepatocytes [176], as well as on apoB degradation and secretion [7,113] in cultured cells, emphasizes the degree of integration that is possible between the intra- and extra-hepatic effects of the hormone. The emergence of the modulation of cell volume as a regulatory mechanism through which portal substrates, as well as insulin itself (and other hormones), may partly exert their metabolic effects has raised the possibility that substrate-driven effects may act synergistically with the hormone to achieve acute changes in hepatic metabolism in response to food intake. Both fatty acid oxidation and TAG synthesis and secretion appear to be candidates for this mode of synergy between the effects of insulin and of the hydration state of the hepatocyte.

The work performed in my laboratory was supported by the British Diabetic Association, British Heart Foundation, The Leverhume Trust and the Scottish Office Agricultural and Fisheries Division.

REFERENCES

1 Dixon, J. L. and Ginsberg, H. N. (1993) J. Lipid Res. **34**, 167–179
2 Reaven, G. M. (1995) Diabetologia **38**, 3–13
3 McGarry, J. D. (1994) J. Cell. Biochem. **55**, 29–38
4 Zammit, V. A. and Moir, A. M. B. (1994) Trends Biochem. Sci. **19**, 313–317
5 Duerden, J. M., Bartlett, S. M. and Gibbons, G. F. (1989) Biochem. J. **263**, 937–943
6 Byrne, C. D., Brindle, N. P. J., Wang, T. W. M. and Hales, C. N. (1991) Biochem. J. **280**, 99–104
7 Laker, M. E. and Mayes, P. A. (1984) Biochim. Biophys. Acta **795**, 427–430
8 Sparks, C. E., Sparks, J. D., Bolognino, M., Salhamick, A., Strumph, P. S. and Amatruda, J. M. (1986) Metab. Clin. Exp. **35**, 1128–1136
9 Emmison, N., Zammit, V. A. and Agius, L. (1992) Biochem. J. **285**, 655–660
10 Topping, D. L. and Mayes, P. A. (1982) Biochem. J. **204**, 433–439
11 Bates, E. J. and Saggerson, E. D. (1977) FEBS Lett. **84**, 229–232
12 Zammit, V. A. (1981) Biochem. J. **198**, 75–83
13 Bates, E. J., Topping, D. L., Sooranna, S. P., Saggerson, E. D. and Mayes, P. A. (1977) FEBS Lett. **84**, 225–228
14 Farese, R. V., Standaert, M. L., Yamada, K., Huang, L. C., Zhang, C., Cooper, D. R., Wang, Z., Yang, Y., Suzuki, S., Toyota, T. and Larner, J. (1994) Proc. Natl. Acad. Sci. U.S.A. **91**, 11040–11044
15 Vila, M. C., Milligan, G., Standaert, M. L. and Farese, R. V. (1990) Biochemistry **29**, 8735–8740
16 Derrick, J. P. and Ramsay, R. R. (1989). Biochem. J. **262**, 801–806
17 Kolodziej, M. P., Crilly, P. J., Corstorphine, C. G. and Zammit, V. A. (1992) Biochem. J. **282**, 415–421.
18 Murthy, M. S. R. and Bieber, L. L. (1992) Protein Expression Purif. **3**, 75–79
19 Nic A'Bhaird, N. and Ramsay, R. R. (1992) Biochem. J. **286**, 637–640
20 Chatterjee, B., Song, C. S., Kim, J. M. and Roy, A. K. (1988) Biochemistry **27**, 9000–9006
21 Woeltje, K. F., Esser, V., Weis, B. C., Sen, A., Cox, W. F., McPhaul, M. J., Slaughter, C. A., Foster, D. W. and McGarry, J. D. (1990) J. Biol. Chem. **265**, 10720–10725
22 Esser, V., Britton, C. H., Weis, B. C., Foster, D. W. and McGarry, J. D. (1993) J. Biol. Chem. **268**, 5817–5822
23 Chung, C.-D. and Bieber, L. L. (1993) J. Biol. Chem. **268**, 4519–4524
24 Murthy, M. S. R. and Pande, S. V. (1994) J. Biol. Chem. **269**, 18283–18286
25 Pande, S. V., Bhuiyan, A. K. M. and Murthy, M. S. R. (1992) in Current Concepts in Carnitine Research (Carter, A. L., ed.), pp. 165–178, CRC Press, Boca Raton, FL
26 Murthy, M. S. R. and Pande, S. V. (1994) Biochem. J. **304**, 31–34
27 Bhuiyan, A. K. M., Murthy, M. S. R. and Pande, S. V. (1994) Biochem. Mol. Biol. Int. **34**, 493–503
28 Lilly, K., Sugaisky, G. E., Umeda, P. K. and Bieber, L. L. (1990) Arch. Biochem. Biophys. **280**, 167–174
29 Broadway, N. M. and Saggerson, E. D. (1995) Biochem. J. **310**, 989–995
30 Ramsay, R. R. (1994) Essays Biochem. **28**, 47–61
31 Farrell, S. O. and Bieber, L. L. (1983) Arch. Biochem. Biophys. **222**, 123–132
32 Huang, G., Lee, D. M. and Singh, S. (1988) Biochemistry **27**, 1395–1400
33 Murthy, M. S. R. and Pande, S. V. (1993) Mol. Cell. Biochem. **122**, 133–138
34 Murthy, M. S. R. and Pande, S. V. (1987) Proc. Natl. Acad. Sci. U.S.A. **84**, 378–382
35 Zammit, V. A. and Corstorphine, C. G. (1985) Biochem. J. **230**, 389–394
36 Woeltje, K. F., Kuwajima, M., Foster, D. W. and McGarry, J. D. (1987) J. Biol. Chem. **262**, 9824–9827
37 Brindle, N. P. J., Zammit, V. A. & Pogson, C. I. (1985) Biochem. J. **232**, 177–182
38 Zammit, V. A. (1994) Diabetes Rev. **2**, 132–155
39 Witters, L. A., Watts, T. D., Daniels, D. L. and Evans, J. L. (1988) Proc. Natl. Acad. Sci. U.S.A. **85**, 5473–5477
40 Hardie, D. G. (1989) Prog. Lipid Res. **20**, 117–146
41 Borthwick, A. C., Edgell, N. J. and Denton, R. M. (1990) Biochem. J. **270**, 795–801
42 Haystead, T. A. J. and Hardie, D. G. (1986) Biochem. J. **240**, 99–106
43 Moir, A. M. B. and Zammit, V. A. (1990) Biochem. J. **272**, 511–517
44 Cook, G. A., King, M. T. and Veech, R. L. (1978) J. Biol. Chem. **253**, 2529–2531
45 McGarry, J. D., Mannaerts, G. P. and Foster, D. W. (1978) Biochim. Biophys. Acta **530**, 305–313
46 Singh, B., Stakkestad, J. A., Bremer, J. and Borrebaek, B. (1984) Anal. Biochem. **138**, 107–111
47 Elayan, I. M., Cartmill, D. C., Eckersell, C. B., Wilkin, J. and Winder, W. W. (1991) Proc. Soc. Exp. Biol. Med. **198**, 569–578
48 Cook, G. A., Otto, D. A. and Cornell, N. W. (1980) Biochem. J. **192**, 955–958
49 Ontko, J. A. and Johns, M. L. (1980) Biochem. J. **192**, 959–962
50 Bremer, J. (1981) Biochim. Biophys. Acta **665**, 628–631
51 McGarry, J. D. and Foster, D. W. (1981) Biochem. J. **200**, 217–223
52 Saggerson, E. D., Carpenter, C. A. and Tselentis, B. S. (1982) Biochem. J. **208**, 667–672
53 Stephens, T. W., Cook, G. A. and Harris, R. A. (1983) Biochem. J. **212**, 521–524
54 Robinson, L. N. and Zammit, V. A. (1982) Biochem. J. **206**, 177–179
54a Zammit, V. A. (1984) Prog. Lipid Res. **23**, 39–67
55 Kolodziej, M. P. and Zammit, V. A. (1993) FEBS Lett. **327**, 294–296
56 Kolodziej, M. P. and Zammit, V. A. (1990) Biochem. J. **272**, 421–425
57 Zammit, V. A. (1984) Biochem. J. **218**, 379–386
58 Zammit, V. A., Corstorphine, C. G. and Gray, S. R. (1984) Biochem. J. **222**, 335–342
59 Mynatt, R. L., Greenshaw, J. J. and Cook, G. A. (1994) Biochem. J. **299**, 761–767
60 Murthy, M. S. R. and Pande, S. V. (1987) Biochem. J. **248**, 727–733
61 Easom, R. A. & Zammit, V. A. (1985) Biochem. J. **230**, 747–752
62 Zammit, V. A. & Easom, R. A. (1987) Biochim. Biophys. Acta **927**, 223–228
63 Park, E. A., Mynatt, R. L., Cook, G. A. and Kashfi, K. (1995) Biochem. J. **310**, 853–898
64 Prip-Buus, C., Thumelin, S., Chatelain, F., Pegorier, J.-P. and Girard, J. (1995) Biochem. Soc. Trans. **23**, 500–506
65 Thumelin, S., Esser, V., Charvy, D., Kolodziej, M. P., Zammit, V. A., McGarry, J. D., Girard, J. and Pegorier, J.-P. (1994) Biochem. J. **300**, 583–587
66 Grantham, B. D. and Zammit, V. A. (1986) Biochem. J. **239**, 485–488
67 Grantham, B. D. and Zammit, V. A. (1988) Biochem. J. **249**, 409–414
68 Munday, M. R., Milic, M. R., Takhan, S., Holness, M. J. and Sugden, M. C. (1991) Biochem. J. **280**, 733–737
69 Prip-Buus, C., Pegorier, J.-P., Duee, P. H., Kohl, C. and Girard, J. (1990) Biochem. J. **269**, 309–415
70 Penicaud, L., Robin, D., Robin, P., Kande, J., Picon, L., Girard, J. and Ferré, P. (1991) Metab. Clin. Exp. **40**, 873–876
71 Quant, P. A. (1994) Essays Biochemistry **28**, 13–25
72 Serra, D., Casals, N., Asions, G., Royo, T., Ciudad, C. J. and Hegardt, F. G. (1993) Arch. Biochem. Biophys. **307**, 40–45
73 Casals, N., Roca, N., Guerrero, M., Gil-Gomez, A., Ayte, J., Ciudad, C. J. and Hegardt, F. G. (1992) Biochem. J. **283**, 261–264

74 Quant, P. A. (1990) Biochem. Soc. Trans. **18**, 994–995

75 Woodside, W. F. and Heimberg, M. (1976) J. Biol. Chem. **251**, 13–23

76 Koloyianni, M. and Freedland, R. A. (1990) Int. J. Biochem. **22**, 159–164

77 Moir, A. M. B. and Zammit, V. A. (1992) Biochem. J. **283**, 145–149

78 Holder, J. C., Zammit, V. A. and Robinson, D. S. (1990) Biochem. J. **272**, 735–741

79 Moir, A. M. B. and Zammit, V. A. (1993) Biochem. J. **289**, 49–55

80 Moir, A. M. B. and Zammit, V. A. (1993) Biochem. J. **291**, 241–246

81 Ip, M. I., Ip, C., Tepperman, H. M. and Tepperman, J. (1977) J. Nutr. **107**, 49–55

82 Holness, M. J. and Sugden, M. C. (1989) Biochem. J. **262**, 321–325

83 Stansbie, D., Brownsey, R. W., Crettaz, M. and Denton, R. M. (1976) Biochem. J. **160**, 413–416

84 Mukkerjee, C. and Jungas, R. L. (1975) Biochem. J. **148**, 229–235

84a Assimacopoulos-Jeannet, M., McCormack, J. G., Prentki, M., Jeanrenaud, B. and Denton, R. M. (1982) Biochim. Biophys. Acta **717**, 86–90

85 Randle, P. J., Kerberg, A. L. and Espinal, J. (1988) Diabetes Metab. Rev. **4**, 623–638

86 Moir, A. M. B. and Zammit, V. A. (1995) Biochem. J. **305**, 953–958

87 Guzman, M., Velasco, G., Castro, J. and Zammit, V. A. (1994) FEBS Lett. **344**, 239–241

88 Howard, B. V. (1989) in Complications of Diabetes Mellitus (Drazin, B., Melmed, S. and Le Roith, D., eds.), pp. 59–67, A. Liss, New York

89 Woelfe, R. R. and Durkot, M. J. (1985) J. Lipid Res. **26**, 210–217

90 Reaven, G. M. and Mondon, C. E. (1984) Horm. Metab. Res. **16**, 230–232

91 Amatruda, J. M. and Chang, C. L. (1993) Metab. Clin. Exp. **32**, 224–229

92 Sparks, J. D., Sparks, C. E., Bolognino, M., Roncone, A. M., Jackson, T. K. and Amatruda, J. M. (1988) J. Clin. Invest. **82**, 37–43

93 Duerden, J. M. and Gibbons, G. R. (1993) Biochem. J. **294**, 167–171

94 Gibbons, G. F. (1986) Clin. Sci. **71**, 477–486

95 Potts, J. L., Fisher, R. M., Humphreys, S. M., Coppack, S. W., Gibbons, G. F. and Frayn, K. N. (1991) Clin. Sci. **81**, 621–626

96 Zilversmit, D. B. (1995) Clin. Chem. **41**, 153–158

97 Voegelberg, K. H., Gries, F. A. and Moschinski, D. (1980) Horm. Metab. Res. **12**, 688–694

98 Lewis, G. F., Uffelman, K. D., Szeto, L. W., Weller, B. and Steiner, G. (1995) J. Clin. Invest. **95**, 158–166

99 Schneeman, B. O., Kotite, L., Todd, K. M. and Havel, R. J. (1993) Proc. Natl. Acad. Sci. U.S.A. **90**, 2069–2073

100 Cohn, J. S., Wagner, D. A., Cohn, S. D., Miller, J. S. and Schaefer, E. J. (1990) J. Clin. Invest. **85**, 804–811

101 Havel, R. J. (1994) Curr. Opin. Lipidol. **5**, 102–109

102 Spring, D. J., Chen-Liu, L. W., Chatterton, J. E., Elorson, J. and Schumaker, V. N. (1992) J. Biol. Chem. **267**, 14839–14845

103 Khan, B. K., Wilcox, H. G. and Heimberg, M. (1989) Biochem. J. **259**, 807–816

104 Kosykh, V. S., Prerbrazhensky, S. N., Fuki, I. V., Zaikin, O. E., Tsibulsky, V. P., Repin, V. S. and Smirhov, V. N. (1985) Biochim. Biophys. Acta **836**, 385–389

105 Yao, Z. and Vance, D. E. (1988) J. Biol. Chem. **263**, 2998–3004

106 Pease, R. J., Harrison, G. B. and Scott, J. (1991) Nature (London) **353**, 448–450

107 Boren, J., Graham, L., Wettesten, M., Scott, J., White, A. and Olofsson, S.-O. (1992) J. Biol. Chem. **267**, 9858–9867

108 Boren, J., Wettesten, M., Rustaeus, S., Andersson, M. and Olofsson, S.-O. (1993) Biochem. Soc. Trans. **21**, 487–493

109 Gregg, R. E. and Wetterau, J. R. (1994) Curr. Opin. Lipidol. **5**, 81–86

110 Jamil, H., Dickson, J. K., Chu, C.-H., Lago, M. W., Rinehart, J. K., Biller, S. A., Gregg, R. E. and Wetterau, J. R. (1995) J. Biol. Chem. **270**, 6549–6554

111 Hamilton, R. L. and Havel, R. J. (1993) Hepatology **18**, 460–463

112 Rusinol, A., Verkade, H. and Vance, J. E. (1993) J. Biol. Chem. **268**, 3555–3562

113 Sparks, J. E. and Sparks, C. D. (1994) Biochim. Biophys. Acta **1215**, 9–23

114 Alexander, C. A., Hamilton, R. L. and Havel, R. J. (1976) J. Cell Biol. **69**, 241–263

115 Boren, J., Wettesten, M., Sjoberg, A., Thorlin, T., Bondjers, G., Wiklund, O. and Olofsson, S.-O. (1990) J. Biol. Chem. **265**, 10556–10564

116 Boren, J., Rustaeus, S. and Olofsson, S.-O. (1994) J. Biol. Chem. **269**, 25879–25888

116a Swift, L. L. (1995) J. Lipid Res. **36**, 395–406

117 Higgins, J. A. (1988) FEBS Lett. **232**, 405–408

118 Gibbons, G. F., Bartlett, S. M., Sparks, C. E. and Sparks, J. D. (1992) Biochem. J. **287**, 749–753

119 Fukuda, N., Azain, M. J. and Ontko, J. A. (1982) J. Biol. Chem. **257**, 14066–14072

120 Glaumann, H., Bergstrand, A. and Ericsson, J. L. E. (1975) J. Cell Biol. **64**, 356–377

121 Hande, W., Wagner, H., Theil, S., Haase, H. and Humicke, G. (1972) Acta Biol. Med. Ger. **28**, 963–975

122 Kondrup, J., Damgaard, S. E. and Fleron, P. (1979) Biochem. J. **184**, 73–81

123 Wiggins, D. and Gibbons, G. F. (1992) Biochem. J. **284**, 457–462

124 Francone, O. L., Kalopissis, A.-D. and Giffaton, G. (1989) Biochim. Biophys. Acta **1002**, 28–36

125 Yang, L.-Y., Kiksis, A., Myker, J. J. and Steiner, G. (1995) J. Lipid Res. **36**, 125–136

126 Azain, M. J., Fukuda, N., Chao, F.-F., Yamamoto, M. and Ontko, J. A. (1985) J. Biol. Chem. **260**, 174–181

127 Yamamoto, M., Yamamoto, I., Tanaka, Y. and Ontko, J. A. (1987) J. Lipid Res. **28**, 1156–1165

128 Thrift, R. N., Forte, T. M., Cahoon, B. E. and Shore, V. G. (1986) J. Lipid Res. **27**, 236–250

129 Gibbons, G. F., Khurana, R., Odwell, A. and Seelaender, M. C. L. (1994) J. Lipid Res. **35**, 1801–1808

130 Sniderman, A. D. and Cianflone, K. (1993) Arterioscler Thromb. **13**, 629–636

131 Dashti, N. (1992) J. Biol. Chem. **267**, 7160–7169

132 Sparks, J. D. and Sparks, C. E. (1993) Curr. Opin. Lipidol. **4**, 177–186

133 Duerden, J. M. and Gibbons, G. F. (1990) Biochem. J. **272**, 583–587

134 Coleman, R. and Bell, R. M. (1978) J. Cell Biol. **76**, 245–253

135 Patsch, W., Franz, S. and Schonfeld, G. (1983) J. Clin. Invest. **71**, 1161–1174

136 Laker, M. and Mayes, P. A. (1984) Biochim. Biophys. Acta **795**, 427–430

137 Topping, D. L., Storer, G. B. and Trimble, R. P. (1988) Am. J. Physiol. **255**, E306–E314

138 Durrington, P. N., Newton, R. S., Weistein, D. B. and Steinberg, D. (1982) J. Clin. Invest. **70**, 63–73

139 Jackson, T. K., Salhanick, A. I., Lorson, J., Deichman, M. L. and Amatruda, J. M. (1990) J. Clin. Invest. **86**, 1746–1751

140 Sparks, J. D. and Sparks, C. E. (1990) J. Biol. Chem. **265**, 8854–8862

141 Thorngate, F. E., Raghow, R., Wilcox, H. G., Werner, C. S., Heimberg, M. and Elam, M. B. (1994) Proc. Natl. Acad. Sci. U.S.A. **91**, 5392–5396

142 Bartlett, S. M. and Gibbons, G. F. (1988) Biochem. J. **249**, 37–43

143 Dashti, N., Williams, D. L. and Alauporic, P. (1989). J. Lipid Res. **30**, 1365–1373

144 Bjornsson, O. G., Duerden, J. M., Bartlett, S. M., Sparks, J. D., Sparks, C. E. and Gibbons, G. F. (1992) Biochem. J. **281**, 381–386

145 Arrol, S., Mackness, M. I., Laing, I. and Durrington, P. N. (1994) Diabetes Nutr. Metab. **7**, 263–271

146 Durrington, P. N. (1990) Curr. Opin. Lipidol. **1**, 463–464

147 Kazumi, T., Vranic, M. and Steiner, G. (1985) Endocrinology **117**, 1145–1150

148 Reaven, G. M. and Chen, Y.-D. I. (1988) Diabetes Metab. Rev. **4**, 639–652

149 McGarry, J. D., Meier, J. M. and Foster, D. W. (1973) J. Biol. Chem. **248**, 270–278

150 Cornell, R. (1989) in Phosphatidylcholine Metabolism (Vance, D. E., ed.), pp. 47–64, CRC Press, Boca Raton, FL

151 Sundler, R. and Akkenson, B. (1977) Biochem. Soc. Trans. **5**, 43–45

152 Groener, J. E. M. and van Golde, L. M. G. (1978) Biochim. Biophys. Acta **529**, 88–95

153 Azain, M. J., Fukuda, N., Chao, F.-F., Yamamoto, M. and Ontko, J. A. (1985) J. Biol. Chem. **260**, 174–181

154 Haagsman, H. P. and van Golde, L. M. G. (1981) Arch. Biochem. Biophys. **208**, 395–402

155 Stals, H. K., Top, W. and Declercq, P. E. (1994) FEBS Lett. **343**, 99–102

156 Tronchére, H., Record, M. F. and Chap, H. (1994) Biochim. Biophys. Acta **1212**, 137–151

157 Weinhold, P. A., Charles, L., Rounsifer, M. E. and Feldman, D. A. (1991) J. Biol. Chem. **266**, 6093–6100

158 Jamil, H., Utal, A. K. and Vance, D. E. (1992) J. Biol. Chem. **267**, 1752–1760

159 Utal, A. K., Jamil, H. and Vance, D. E. (1991) J. Biol. Chem. **266**, 24084–24091

160 Kalmar, G. B., Kay, R. J., Lachance, A., Aebersald, R. and Cornell, R. B. (1990) Proc. Natl. Acad. Sci. U.S.A. **87**, 6029–6033

161 Yang, Y., Sweitzer, T. D., Weinhold, P. A. and Kent, C. (1993) J. Biol. Chem. **268**, 5899–5904

162 Wang, Y., MacDonald, J. L. S. and Kent, C. (1995) J. Biol. Chem. **270**, 354–360

163 Saughera, J. S. and Vance, D. E. (1989) J. Biol. Chem. **264**, 1215–1223

164 Watkins, J. D., Wang, Y. and Kent, C. (1992) Arch. Biochem. Biophys. **292**, 360–367

165 Selawry, H., Gutman, R., Fink, G. and Recant, L. (1973) Biochem. Biophys. Res. Commun. **51**, 198–204

166 Ide, H. and Weinhold, P. A. (1982) J. Biol. Chem. **257**, 14926–14931

167 Haagsman, H. P., de Haas, C. G. M., Geelen, M. J. H. and van Golde, L. M. G. (1982) J. Biol. Chem. **257**, 10593–10598

168 Woodside, W. F. and Heimberg, M. (1978) Metab. Clin. Exp. **27**, 1763–1777

169 Yao, Z. and McLeod, R. S. (1994) Biochim. Biophys. Acta **1212**, 152–166

170 Sallach, S. M. and Adeli, K. (1995) Biochim. Biophys. Acta **1265**, 29–32

171 Davis, R. A., Roogaerts, J. R., Bochardt, R. A., Malone-McNeal, M. and Archambault-Schexnayder, J. (1985) J. Biol. Chem. **260**, 14137–14144

172 Moir, A. M. B. and Zammit, V. A. (1995) FEBS Lett. **370**, 255–258

173 Moir, A. M. B. and Zammit, V. A. (1994) Biochem. J. **304**, 177–182

174 Agius, L., Peak, M. and Al-Habori, M. (1991) Biochem. J. **276**, 843–845

175 Haussinger, D. and Schleiss, F. (1995) J. Hepatol. **22**, 94–100

176 Zammit, V. A. (1995) Biochem. J. **312**, 57–62

177 Barber, E. F., Handlogten, M. F., Vida, T. A. and Kilberg, M. S. (1982) J. Biol. Chem. **257**, 14960–14967

178 Vom Dahl, S., Hallbrucker, C., Lang, F. and Haussinger, D. (1991) Biochem. J. **280**, 105–109

179 Peak, M., Al-Habori, M. and Agius, L. (1992) Biochem. J. **282**, 797–805

180 Agius, L., Peak, M., Beresford, G., Al-Habori, M. and Thomas, T. H. (1994) Biochem. Soc. Trans. **22**, 516–521

181 Samson, M., Fehlman, M., Dolais-Kitabgi, J. and Freychet, P. (1980) Diabetes **29**, 996–1000

182 McGivan, J. D. and Pastor-Anglada, M. (1994) Biochem. J. **299**, 321–334

183 Howard, B. V., Reitman, J. S., Vasquez, B. and Zech, L. (1983) Diabetes **32**, 271–276

184 Tomkin, G. H. and Owens, D. (1994) Diabetes Metab. Rev. **10**, 225–252

185 Zilversmit, D. B. (1979) Circulation **60**, 473–485

186 Moir, A. M. B., Park, S.-B. and Zammit, V. A. (1995) Biochem. J. **208**, 537–542

Biochemical Society Symposia

Mammary Development and Cancer
NEW

Edited by **P S Rudland**, **D G Fernig** and **S J Leinster**, *University of Liverpool.*

Biochemical Society Symposium No. 63

This book is concerned with how the mammary gland grows and differentiates at the molecular level under the influence of circulating hormones and locally produced growth factors and how this molecular machinery can be either utilised by man for product production or subverted by nature in the formation of cancer.

This book has been designed to appeal to a wide audience making use of the fact that the mammary gland is probably the most intensively studied developmental system, for a number of reasons:

(1) The mammary gland develops after birth so making many experiments more tractable; (2) The hormones that control the process are well known and are important in a major industrial product, milk; (3) Genetic engineering of small mammals and of suitable farm animals for industrial use is the most advanced in the mammary gland; (4) The disease of breast cancer is one of the most common cancers; (5) Breast cancer is virtually incurable when it is metastasized in the body.

These five reasons for intensive research have led to discoveries and models which are not only applicable to the mammary gland itself, but also to many other organs as well. Thus the book will have a general educational role in biology as a whole, in addition to being of interest to mammary gland/breast cancer researchers (scientific medical, and veterinary workers), as well as to biochemists/molecular biologists who are interested in developmental processes.

Contents: The individual sections of the book reflect the five reasons for intensive research in the mammary gland and are as follows: Control of mammary development; Regulation of milk production; Mammary transgenics; Genetic Changes in mammary cancer; Mechanisms of mammary metastasis.
1 85578 087 9 Hard May 1997 250 pages £65.00/US$110.50

Extracellular Regulators of Differentiation and Development

Edited by **K E Chapman**, *University of Edinburgh,* **S P Jackson**, *Wellcome/CRC Institute, Cambridge* & **D G Wilkinson**, *Laboratory of Developmental Neurobiology, NIMR, London.*

Biochemical Society Symposium No. 62

Extracellular Regulators of Differentiation and Development pulls together works from a number of world experts in the general field of controlling cell function by extracellular regulators. It covers all aspects of this important and complicated field ranging from receptors for regulators at the cell surface to effectors of the response in the cell nucleus.
This book will be of interest to lecturers, PhD students, postdoctoral fellows and final-year undergraduates studying the subjects of biochemistry, pharmacology, gene regulation, molecular genetics and developmental biology.
1 85578 070 4 Hard July 1996 192 pages £65.00/US$110.50

✂

Please supply me with [ ] copies of **Mammary Development and Cancer** @ £65.00/US$110.50 each

Please supply me with [ ] copies of **Extracellular Regulation of Differentiation and Development** @ £65.00/US$110.50 each

Name:

Address:

[ ] Cheque payable to Portland Press Ltd

[ ] Proforma Invoice

[ ] Visa/Mastercard/Access/AmEx (delete as appropriate)

Card No.

Exp. Date

Signature

Date

I enclose a payment of £

Postage: UK customers add £2.50 per book to a maximum of £7.50

Send orders to: Portland Press, Commerce Way, Colchester, CO2 8HP
Tel: 01206-796351 Fax: 01206-799331 email: sales@portlandpress.co.uk

ATS/0497/A

Student Textbooks

Life Chemistry & Molecular Biology

By **E J Wood**, **C A Smith** and **W R Pickering**

"crafted by experienced teachers who know the elements of conveying core information to students, [this book] has the potential to redefine the use of textbooks in teaching biochemistry."
                                        ASBMB Inc Newsletter.

This is a new biology textbook which uses a unique and innovative format, consisting of a series of annotated diagrams with linking text, to make it an ideal study-guide for students as well as a valuable tool for teachers. Each chapter includes further reading suggestions and also examination questions. Biological principles and their application in commercial, medical, ecological and physiological contexts are explained in the book. The text covers information for the newly proposed A-level syllabuses, and is equally useful for undergraduates and students of vocational life-science courses.

Contents:

Life Chemistry and the Ecosystem; Biological Molecules; Enzymes; Obtaining Energy; Using Metabolic Energy; DNA: Dealing with Information; Molecular Biology and Applied Biochemistry

ISBN: 1 85578 064 X Paper November 1996 240 pages £16.00/US$27.00

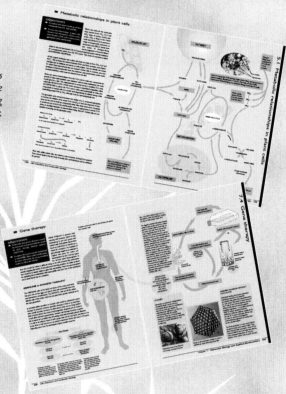

Glossary of Biochemistry and Molecular Biology

By **D M Glick**

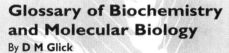

"it will make a welcome addition to the reference shelf."
                                        SFEP Newsletter

Glossary of Biochemistry and Molecular Biology compiles nearly 3000 terms and gives succinct definitions to assist those who wish to converse with biochemists or molecular biologists in their native tongue! Some of the terms appear only in earlier literature, some are very current, some are common terms invested with new meanings, some are lab. jargon.

References are included with many of the entries, to assist the reader in searching for additional information. These citations, frequently review articles, are offered as a convenient introduction to the literature on the subject. The glossary should be of interest to undergraduate and post-graduate students, to new researchers and to other professions having to interface with modern biochemistry or molecular biology.

ISBN: 1 85578 088 7 Paper October 1996 220 pages £16.50/US$28.00

Basic Chemistry for the Biological Sciences:

A Self-Directed Study Aid
Edited by **C Wynn**

This study aid has been designed, after undergoing extensive trials, to help students studying biology with fundamental chemistry. The study aid is arranged in the form of graded questions, with the answers at the end of each section. Questions based directly on biological examples are included, so that the connections and relevance to biology can be appreciated. Students studying A-level biology and those taking vocational science courses, as well as, undergraduate students in the biological sciences and the allied health sciences will find this guide invaluable.

Contents:

Atomic Structure and Bonding; Chemical Equilibrium; Acids and Bases -pH-Buffers; Rate of Reaction; Oxidation and Reduction; Basic Organic Chemistry; Ring Compounds; Supplementary reading; Index

ISBN: 1 85578 105 0 Loose leafs within a folder £12.50/copy
ISBN: 1 85578 115 8 Loose leafs only £10/copy
ISBN: 1 85578 116 6 Teacher sets (one folder + 15 sets loose leafs) £125/set
(postage free)

Send orders to: Portland Press, Commerce Way, Colchester, CO2 8HP
Tel: 01206-796351 Fax: 01206-799331 email: sales@portlandpress.co.uk

ATW/0497/A

Frontiers in Neurobiology

Amino Acid Neurotransmission

Edited by **F A Stephenson**, *University of London* and
A J Turner, *University of Leeds*.

Frontiers in Neurobiology No. 3

This book concerns all aspects of amino acid neurotransmission in the brain. It covers an integrated approach to inhibitory and excitatory neurotransmission including release of the transmitter, receptor subtypes – molecular pharmacology and molecular biology, inactivation via uptake systems and their involvement in disease processes. The book is written by international authorities in each field giving up-to-date information in a fast moving area.

Amino Acid Neurotransmission includes both pre and post synaptic mechanisms.

Contents: Perspective, A J Turner and F A Stephenson; Neurotransmitter release mechanisms, D G Nicholls and J Sánchez-Prieto; Structure, function and regulation of sodium coupled neurotransmitter transporters, B I Kanner; Electrophysiology of GABA$_A$ receptors, T G Smart; Molecular structure of GABA$_A$ receptors, F A Stephenson; The inhibitory glycine receptor, C-M Becker and D Langosch; Metabotropic glutamate receptors, J M Henley, R Burton and S A Richmond; Non-*N-methyl-D*-aspartate (NMDA) glutamate receptors: molecular properties, R J Wenthold and R S Petralia; Molecular biology of *N-methyl-D*-aspartate(NMDA)-type glutamate receptors, P J Whiting and T Priestley; Receptor regulation by phosphorylation, L Raymond; Excitatory amino acids and neurodegeneration, A M Palmer; Long term potentiation in the hippocampus, Z I Bashir.

1 85578 080 1 Hard May 1997 250 pages £65.00/US$110.50

Nerve Growth and Guidance

Edited by **C D McCaig**, *University of Aberdeen*.

Frontiers in Neurobiology No. 2

This book considers in detail mechanisms underlying nerve growth and guidance. There is considerable coverage of intracellular second messenger involvement, both in guiding growth and collapse of growth cones. All those interested in developmental neurobiology in general, and nerve guidance in particular, from senior undergraduate level through to research levels will find this book extremely useful.

1 85578 085 2 Hard May 1996 180 pages £65.00/US$110.50

Neuropeptide Gene Expression

Edited by **A J Turner**, *University of Leeds*.

Frontiers in Neurobiology No. 1

Neuropeptides are a rapidly growing class of signal molecules. In this book, leading scientists survey the recent progress made in the identification of neuropeptides and the factors regulating their expression including transcription factors, enzymes involved in post-translational processing as well as agents modulating peptide hormone release at synaptic terminals. It will be of interest to both neuroscientists and those studying gene expression in general.

1 85578 044 5 Hard July 1994 260 pages £65.00/US$110.50

Landmarks in Intracellular Signalling

by R D Burgoyne and O H Peterson

The intracellular signalling pathways that control cell function have been, and still are, one of the most intensively studied aspects of biology. In recent years the detailed characterization of the multiple cell-signalling pathways by many laboratories has resulted in a bewildering increase in knowledge in this field. For this reason, students and others learning about this topic for the first time are increasingly overwhelmed by the mass of information and frequently are unable to find time to read and digest the key original papers.

The idea behind **Landmarks in Intracellular Signalling** is to provide full reproductions of a set of key papers which have been chosen as landmark papers in the various aspects of intracellular signalling. The selected papers have all resulted in significant advances in one or other aspect of intracellular signalling. Readers of **Landmarks in Intracellular Signalling** will now have easy, ready available access to the original literature from one source. The papers are accompanied by commentaries that describe why the papers are significant, how the work came about and summarize the advances that have been made up to the present time as a consequence of the original paper. The commentaries will also serve as mini-reviews of many aspects of cell regulation and can be read on their own.

The area of intracellular signalling is relevant to many areas of biology and the basic principles need to be understood by undergraduates in many disciplines. Background knowledge of this area is also important for postgraduate students in many fields as well as more senior research workers and academics.

Contents:

1 85578 101 8   June 1997   Paperback 250 pages   £20.00/US$34.00

Please supply me with ☐ copies of Landmarks in Intracellular Signalling @ £20.00/US$34.00 each

Name:

Address:

I enclose a payment of £

☐ Cheque payable to Portland Press ☐ Proforma Invoice ☐ Visa/Mastercard/ Access/AmEx

Card No. Exp. Date

Signature Date

Send orders to: Portland Press, Commerce Way, Colchester, CO2 8HP
Tel: 01206 796351 Fax: 01206 799331 email: sales@portlandpress.co.uk.
Postage: UK: please add £2.50 per book to a maximum of £7.50

ATU/0497/A

Landmarks in Gene Regulation

Edited by D S Latchman

Twenty years ago there was no known example in eukaryotes of a defined regulatory protein which bound to a specific DNA sequence in a target gene and regulated its expression. Since that time a bewildering number of such regulatory proteins, or transcription factors, have been defined and their roles in regulating the expression of specific genes analysed. When faced with such a vast array of information on individual factors, it is all to easy to forget that the foundations of the study of such factors — and indeed of eukaryote gene regulation in general — have been laid by a relatively small number of seminal papers which establish key points in this area.

The aim of this book is to reproduce a series of key papers illustrating the increase in our understanding of gene regulation. Each paper is followed by a commentary section which places the paper in context of what was known at the time and what has been established subsequently. This will allow both the student and the more experienced research worker to see how the field has developed over time and how the principles were elucidated.

As well as providing access to an important set of classic papers, **Landmarks in Gene Regulation** will be a useful source of information to those working in various life science fields. By emphasizing the process of 'scientific discovery' (i.e. problem-solving, how one piece of research is built upon another, etc.), the volume will also be a useful student resource.

Contents:

1 85578 109 3   March 1997   Paperback   250 pages   £20.00/US$34.00

Please supply me with ☐ copies of Landmarks in Gene Regulation @ £20.00/US$34.00 each

Name:

Address:

I enclose a payment of £

☐ Cheque payable to Portland Press   ☐ Proforma Invoice   ☐ Visa/Mastercard/Access/AmEx

Card No.   Exp. Date

Signature   Date

Send orders to: Portland Press, Commerce Way, Colchester, CO2 8HP
Tel: 01206 796351 Fax: 01206 799331 email: sales@portlandpress.co.uk.
Postage: UK: please add £2.50 per book to a maxiumum of £7.50

ATV/0497/A